AF685177

20012671077

TK 145 .E488 2007

Electromechanical devices &
components illustrated
sourcebook

rmain

Dallas Co. Community College District
C20012671077
RICHLAND COLLEGE
of the
DALLAS COUNTY
COMMUNITY COLLEGE DISTRICT

JUN 10 2013

ELECTROMECHANICAL DEVICES & COMPONENTS ILLUSTRATED SOURCEBOOK

ELECTROMECHANICAL DEVICES & COMPONENTS ILLUSTRATED SOURCEBOOK

BRIAN S. ELLIOTT

McGraw Hill

New York • Chicago • San Francisco • Lisbon • London • Madrid
Mexico City • Milan • New Delhi • San Juan • Seoul
Singapore • Sydney • Toronto

8
0248909
100.95

The McGraw-Hill Companies

Cataloging-in-Publication Data is on file with the Library of Congress

McGraw-Hill books are available at special quantity discounts to use as premiums and sales promotions, or for use in corporate training programs. For more information, please write to the Director of Special Sales, Professional Publishing, McGraw-Hill, Two Penn Plaza, New York, NY 10121-2298. Or contact your local bookstore.

Electromechanical Devices & Components Illustrated Sourcebook

Copyright © 2007 by The McGraw-Hill Companies. All rights reserved. Printed in the United States of America. Except as permitted under the Copyright Act of 1976, no part of this publication may be reproduced or distributed in any form or by any means, or stored in a database or retrieval system, without the prior written permission of publisher.

1 2 3 4 5 6 7 8 9 0 QPD/QPD 0 1 9 8 7

ISBN-13: 978-0-07-147752-9
ISBN-10: 0-07-147752-7

Sponsoring Editor
Kenneth P. McCombs

Editorial Supervisor
Janet Walden

Project Manager
Arushi Chawla

Copy Editor
Megha Ghai

Proofreader
Nigel O'Brien

Indexer
Brian S. Elliott

Production Supervisor
George Anderson

Composition
International Typesetting and Composition

Illustration
Brian S. Elliott

Cover Designer
The Gazillion Group

Dallas Co. Community College District
C20012671077

Information has been obtained by McGraw-Hill from sources believed to be reliable. However, because of the possibility of human or mechanical error by our sources, McGraw-Hill, or others, McGraw-Hill does not guarantee the accuracy, adequacy, or completeness of any information and is not responsible for any errors or omissions or the results obtained from the use of such information.

ABOUT THE AUTHOR

Brian S. Elliott has over 30 years of experience designing scientific, commercial. and industrial products. During his long career, he has held positions as technician through managing engineer. He is currently serving as vice president of engineering and manufacturing for Air Options, Inc. Brian is responsible for managing all engineering, manufacturing, and field operations within the company. He also oversees all product engineering specifications, R&D, prototyping, testing, and production design activities on all new products. In addition to this book, he is the author of the *Compressed Air Operations Manual*, also published by McGraw-Hill.

CONTENTS

LIST OF ILLUSTRATIONS

CHAPTER 1

CHAPTER 2

CHAPTER 3

CHAPTER 4

CHAPTER 5

CHAPTER 6

CHAPTER 7

CHAPTER 8

CHAPTER 9

CHAPTER 10

CHAPTER 11

CHAPTER 12

CHAPTER 13

CHAPTER 14

CHAPTER 15

CHAPTER 16

CHAPTER 17

CHAPTER 18

PREFACE

Most of us take our comfortable life styles for granted, unable to fully comprehend or appreciate the incredibly complex technologies that surrounds us. The very technologies that allow the society we are immersed in to exist. We go through our lives without really noticing the enormous infrastructure that supports our comfort, expectations, and even our demands.

Technology plays a critical role in our societal system. Without it, our world would be a very different place. It is difficult to find anyone who gives the slightest thought to what happens when they flip a light switch on in the morning. We click the switch and expect the light to come on. If it doesn't, we're annoyed that we have to replace the bulb. The subtle complexities of a modern light bulb, let alone that of the power generation and distribution system that allows it to do its job, are out of the realm of comprehension for most people, and even most engineers.

Those of us that occasionally think about the light bulb, generally considered it to be an electrical device. After all, the most common ratings on a light bulb are volts and watts, electrical terms. The fact of the matter is that a light bulb, like most other electrical appliances, is actually a mechanical device that is designed to use electricity as its energy source. The globe is designed to deal with extreme heat, rough handling, rapid cooling, light diffusion, protection from the atmosphere, dirt buildup, and convenience. The filament must be designed to withstand various shocks, repeated cycling and longevity, while still producing the light that we require. All of these parameters are mechanical in nature. The modern incandescent light bulb is truly a marvel of electromechanical technology.

There are very few manufactured things in the world that do not require the marriage of electrical and mechanical disciplines. You may consider an ordinary garden rake a purely mechanical device and you would be right. However, what you may not have considered is the complex electromechanical system that was required to manufacture the rake, the transportation system that delivered the rake to your local garden store, the cash register that was used to ring up the transaction, and your car that transported you and your new rake right up to your yard.

The internal combustion engine would not exist if it weren't for an electrical ignition system. Higher horsepower engines couldn't be started without an electric starter. The starter's battery would go dead in very short order if the engine didn't include an electrical generator for recharging. Modern pollution standards and fuel economy could not be met without applying sophisticated electrical controls to the mechanical systems of an automobile engine.

In any appliance there are varying degrees of electrical and mechanical requirements. A lawn mower engine for instance has a very small electric component in the form of a simple magneto ignition system. On the other hand, a desktop computer is heavily electrical in nature and the mechanical systems are there only to support the electrical functions. Aircraft of the early 1900s had a very small electrical component, oftentimes limited to the engine's ignition system. Modern aircraft simply could not exist without the complex electrical and electronic control systems that help manage the aircraft's flight envelope. The startling difference between a military fighter plane of the first world war and a modern jet fighter is a clear indication of how electromechanical technologies have impacted our civilization.

ACKNOWLEDGMENTS

Special thanks goes to Rozanne Hill for her support in getting me through the effort of writing this book. As I worked on the project her input was invaluable. I don't suppose that I could have finished without her help.

I would also like to thank my brother, Timothy Elliott, for lending his expertise in the area of electrical and electronics.

I'll know that I have achieved literary greatness on the day that I am able to purchase a copy of one of my books, in a garage sale, for a dollar.

INTRODUCTION

This book is intended to introduce engineers and technicians to equipment and components that use both mechanical and electrical elements. In the world of technical book publishing, this is a particularly neglected field, even though the marriage of these two disciplines is what give us the technologies that surround us every day.

This book is arranged in a logical progression of the topic. Starting with the basics of electrical and mechanical, and ending with standard and preferred engineering methods.

Chapter 1 reviews the basics of electricity, covering current, voltage, resistance, and the application of Ohms law. Simple types of circuits are discussed using easy to understand terms. Finally the difference between direct current (DC) and alternating current (AC) are discussed.

Chapter 2 reviews the basics of mechanics. All of the simple machine are discussed and the two principal energies. Simple examples of how the basic machines are applied are shown through out.

Chapter 3 introduces the reader to electrical power sources, ranging from batteries to uninterruptible power supplies. A solid review of these power sources are provided giving the reader a clear grounding in this topic.

Chapter 4 provides information on electrical controls, both devices intended to control electricity and devices intended to use electricity as their control element. As this is a very important subtopic, this chapter is the most extensive of the book. It covers switches, relays, contactors, timers, resistors, capacitors, and the likes.

Chapter 5 reviews magnetic components, which is principally transformers and inductors. This is an extremely important subtopic and should be studied carefully by the reader. Other topics covered in this chapter are magnetic recording, eddy currents, and solenoids.

Chapter 6 deals with rotating equipment. The different types of electric motors are presented in a format that will leave you with a solid knowledge of DC, AC, and three-phase motors. In addition to motors, generators are reviewed.

Chapter 7 reviews the common application of heating elements. A description of the different types of heaters is provided and how to apply them. A brief discussion of induction and microwave heating is also included.

Chapter 8 takes a look at circuit protection. It covers both fuses and circuit breakers and provides the reader with a clear understanding of this topic. Motor protection and the associated methods are discussed. Lighting protection is also discussed.

Chapter 9 reviews connectors. Although this is considered a mundane subtopic, it is, however, a very important one. The information starts with basic wire splicing and finishes with sophisticated multi pin connectors. Standard uses of connectors are also discussed.

Chapter 10 covers the most fundamental electromechnical device, the wire. Mock like the connector this is a neglected topic which is responsible for bringing every electronic, electrical, and electromechnical device together.

Chapter 11 reviews acoustic devices. Any electromechnical component device that is intended to make a noise or vibration falls into this category. Starting with a simple buzzer and following through with sophisticated loudspeakers, this chapter provides the reader with a base knowledge of the area.

Chapter 12 examines lighting technologies, from Edison's first electric light bulb to modern sodium vapor lamps. A review is provided of the various techniques used to generate light and the packages that these technologies are delivered in.

Chapter 13 reviews the very important topic of meters. Meters are the tool that we rely on to measure, troubleshoot, monitor, and understand the electronic, electrical, and electromechnical world. This chapter deals primarily with voltmeters and how to configure them. In addition a brief review of higher level test equipment is provided.

Chapter 14 provides a cursory introduction to vacuum tubes. This is a well documented topic so only a brief description of these devices is provided.

Chapter 15 discusses sensors and there application. Sensors are the way we monitor the technology significant world that surrounds us. This chapter reviews various sensors and how their detections can be converted into electrical signals.

Chapter 16 briefly discusses the world of electrostatics. Electrostatics is the study of electrical energy at rest. For the most part, individuals that are involved with electrostatics are dealing with high voltage and the particulars involved with that pursuit.

Chapter 17 reviews a few common electromechnical movements and assemblies that aren't covered in the previous chapters.

Chapter 18 is a review of electrical schematics and how to use them. A complete list of standard symbols and a discussion of standard methodologies are provided. In addition, a review of representational schematics is provided.

CHAPTER 1
BASIC ELECTRICITY

Most of us have some basic knowledge of electricity because of the many electrical gadgets and appliances that we use every day. We know that when we flip on a light switch, electricity is supplied to a lamp and it produces light. We turn on the blender and the motor starts to turn. Flashlights use batteries as their power source and when they run down, they don't produce enough electricity to light the bulb.

In fact most of us have a slightly more advanced knowledge of electricity than we may realize. You have probably noticed that most electrical devices have two wires feeding power to them, or that light bulbs have a brilliant glow and they are very hot during operation. If you leave the car's lights on overnight you know that the battery will be dead the next morning. You may also be aware that a car operates on 12 volts DC and the outlets in the house supply 120 volts AC.

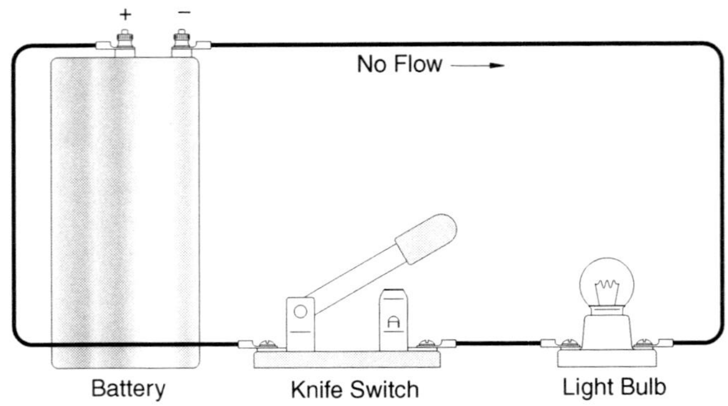

Figure 1-2 Open Circuit

The Electric Circuit

In order to better utilize the information presented in this book, it is necessary to have a basic understanding of electricity and how it can be applied to work with, or provide control for, mechanical devices.

To better understand electricity, let's first consider the basic electric circuit. All electrical circuits have two things in common, they have a power source and something that is using that power, generally referred to as the load. In Figure 1-1 the power source is a battery and the load is a light bulb. The battery supplies energy to the light bulb and the bulb's filament glows. In this arrangement, the stored energy of the battery is transformed into something useful, light. Electrons flow from the negative terminal (–) of the battery through the light bulb filament and back to the positive terminal (+).

To make the basic electrical a little more friendly, a switch is applied to the circuit. The switch is much like a valve for electrons. When the switch is open, as shown in Figure 1-2, the flow of electrons is interrupted and the lamp is off. This mode is usually referred to as an open circuit.

To turn the lamp on, the switch is closed and the electrons can flow again, as shown in Figure 1-3. This mode is generally referred to as a closed circuit. The switch provides ready convenience in operating the circuit.

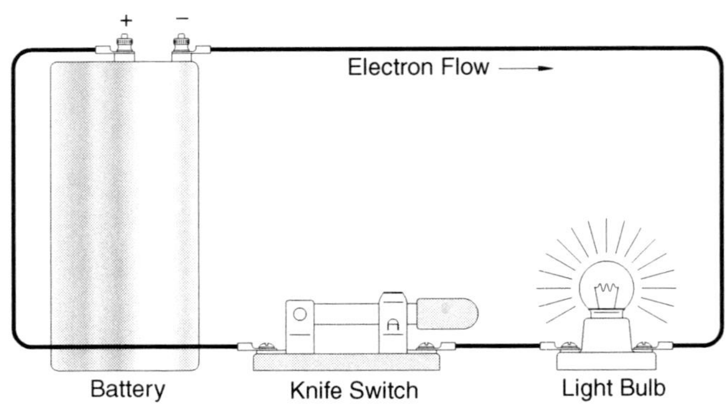

Figure 1-3 Closed Circuit

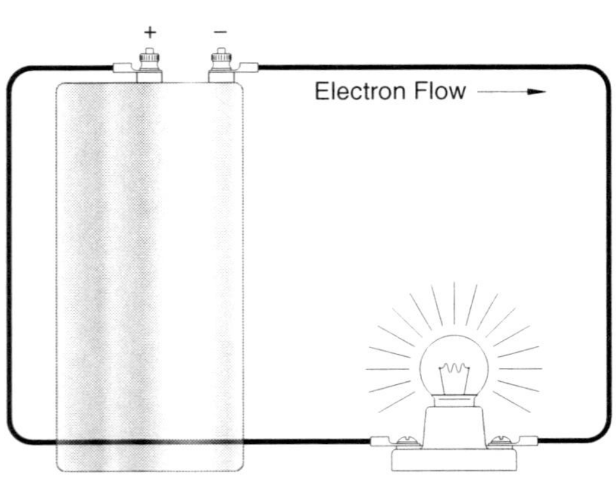

Figure 1-1 Basic Electric Circuit

Voltage, Current, and Resistance

Now that we have an understanding of the basic electric circuit, there are a few items that will provide us with a better working knowledge of how electricity does what it does. When considering electric circuits there are three different parameters that are important. Each of these parameters works in conjunction with the other two, therefore all three must be considered in every electrical circuit. These parameters are voltage, current, and resistance. Voltage is the parameter that is most commonly known. Cars generally operate on 12 volts, the outlets in your house supply 120 volts and major appliances like stoves and dryers operate on 220 volts. The second parameter is current. The unit of measure for current is amperes or amps. The current rating found on the circuit breakers in the electrical panel at your house are rated in amperes. A circuit breaker which is labeled 20 is designed to carry a maximum current of 20 amps. The small fuses in your car are also rated in amps. If a fuse is printed with 15 it is designed to carry a maximum current of 15 amps. The third, and least known, parameter is resistance. Resistance is just what the word implies, resistance to the flow of electricity. The unit of measure for resistance is ohms. Since resistance is what allows us to extract work from an electrical circuit, it is an exceptionally important parameter to understand.

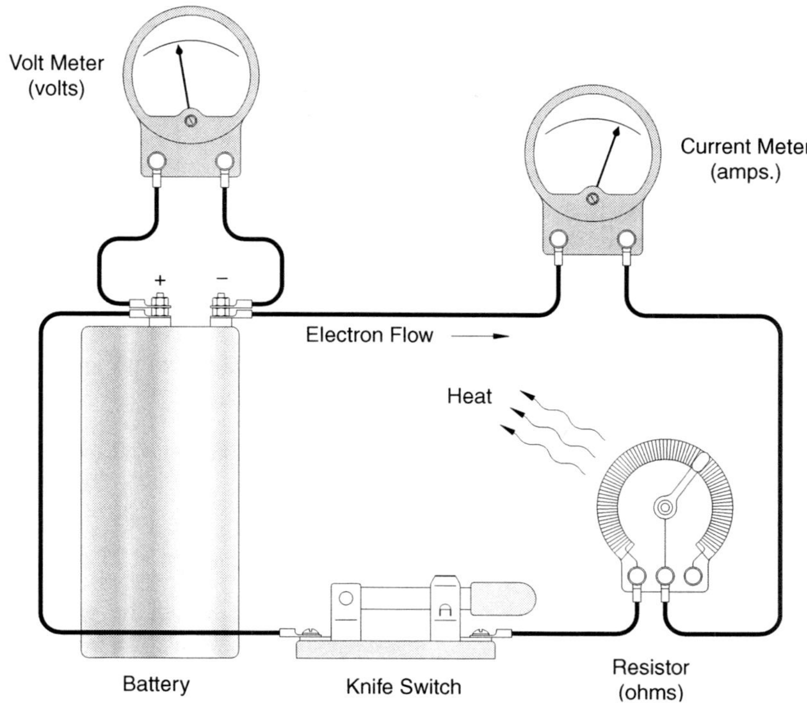

Figure 1-4 Voltage, Current, and Resistance

Figure 1-4 shows how these different parameters apply to our basic electric circuit. Volts are measured at the battery terminals as shown. The voltage will not principally change unless the battery is depleted or a load is placed on it that is greater than it is capable of supplying. Current is measured in amps by placing a meter in the electrical loop. All current used flows through the meter and, therefore, it provides a visual indicator of the current required for the circuit. Resistance is the load. In the illustration the light bulb is replaced with a variable resistor. The work generated is in the form of heat.

To get a clearer understanding of the circuit, we can compare it to the basic water system shown in Figure 1-5. An elevated bucket filled with water represents a battery. It has a stored mass, which can be released when necessary. The water that flows out of the bottom of the bucket has some pressure at this point. This pressure can be equated to voltage. The higher the pressure, the higher the voltage. The flow rate through the pipes can be compared to current. Greater flow equates to higher amperage. The paddle wheel represents resistance. The more load placed on the paddle wheel, the more resistance to flow it will have. The valve can be viewed as an on/off switch. When it is closed there is no flow.

By increasing the amount of water in the bucket, you get greater capacity from your battery (charge). When you increase the bucket's elevation, you increase the pressure (voltage). By using larger pipes you will get higher flow rates

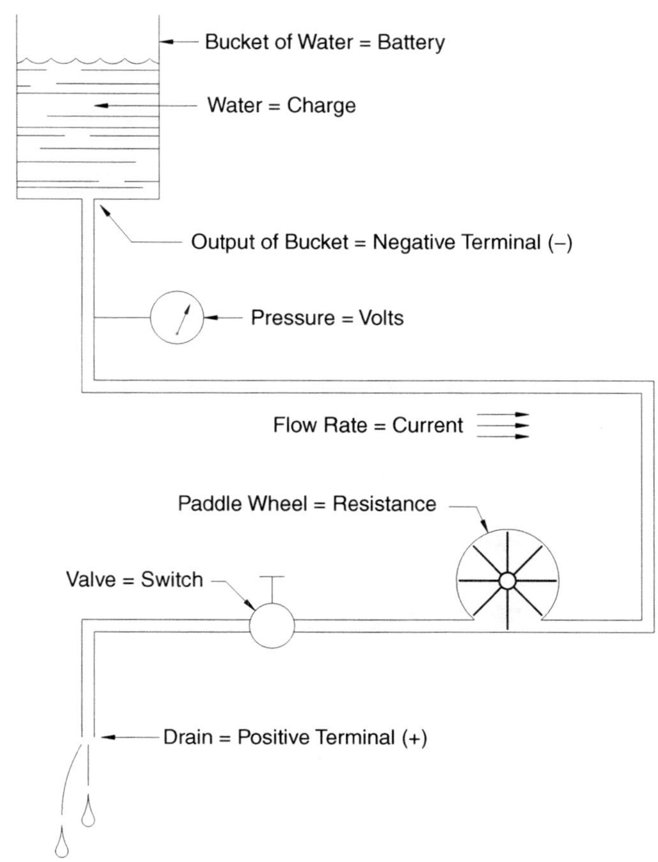

Figure 1-5 Water System Comparison

(current), and by reducing the load on the paddle wheel you reduce resistance to flow (ohms).

It can be seen that by increasing the pressure on the system (voltage), you increase flow through the pipe and the paddle wheel goes faster. By increasing the pipe size (current), the flow can be higher and the paddle wheel will go faster. By decreasing the load on the paddle wheel (ohms), its speed will increase. In this manner you can see how the three parameters are interrelated.

Ohm's Law

A complete understanding of how voltage, current, and resistance interact with one another can be gained through the use of Ohm's law. Ohm's law states that in any circuit the current is equal to the voltage divided by the resistance, as shown in Figure 1-6.

$$\text{Current} = \frac{\text{Volts}}{\text{Resistance}}$$

Figure 1-6 Ohm's Law

When working with Ohm's law, the standard letter symbols are: I (amperes), E (volts), and R (ohms). It may seem odd to use letter symbols that are not the same as the parameter that they represent. In fact, the letters relate to a more basic description of the parameter. I represents the word *current* (amperes), E represents *electromotive force* (volts) and R represents the word *resistance* (ohms). Therefore, the formula may be written as shown in Figure 1-7.

$$I = \frac{E}{R}$$

Figure 1-7 Ohm's Law

If the basic circuit shown in Figure 1-4 uses a 1.5 volt battery and the resistor is 100 ohms, then the current will be 0.015 amperes.

$$1.5 \text{ (E, volts)} \div 100 \text{ (R, ohms)} = 0.015 \text{ (I, amperes)}$$

If we change the resistance to 50 ohms, then the current will be 0.03 amperes.

$$1.5 \text{ (E, volts)} \div 50 \text{ (R, ohms)} = 0.03 \text{ (I, amperes)}$$

If we change the resistance to 200 ohms, then the current will be 0.0075 amperes.

$$1.5 \text{ (E, volts)} \div 200 \text{ (R, ohms)} = 0.0075 \text{ (I, amperes)}$$

The basic formula of Ohm's law can be manipulated to calculate any of the three parameters if the other two are known. Figure 1-8 shows a handy quick reference graphic that electrical engineers and electricians use as a guide to Ohm's law. By covering the symbol of the parameter that you do not know, the graphic tells you which two symbols to use and how to execute the math. As an example, say we know that a circuit incorporates a resistance of 25 ohms, has a current of 3 amps, and we do not know the voltage. Place your finger over the E symbol (volts) and the remaining two symbols show the calculation:

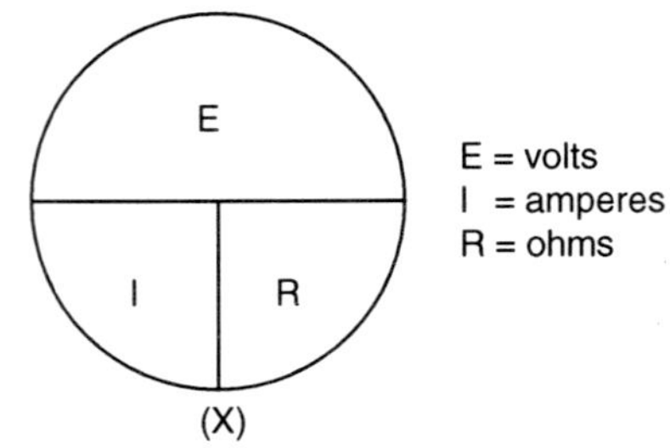

Figure 1-8 Ohm's Law

$$\text{I (amps)} \times \text{R (ohms)} = \text{E (volts)}$$

or

$$3 \text{ amps} \times 25 \text{ ohms} = 75 \text{ volts}$$

The vertical line in the graphic should be considered a multiplication symbol. The horizontal line should be considered a division symbol.

As second example, say we know that the circuit is using 50 volts, has a current of 5 amps, and we do not know the resistance. Place your finger over the R symbol (ohms) and the remaining two symbols show the calculation:

$$\text{E (volts)} \div \text{I (amps)} = \text{R (ohms)}$$

$$50 \text{ (volts)} \div 5 \text{ (amps)} = 10 \text{ (ohms)}$$

Figures 1-9 through 1-11 show three examples of how Ohm's law is applied and how the Ohm's law graphic is used.

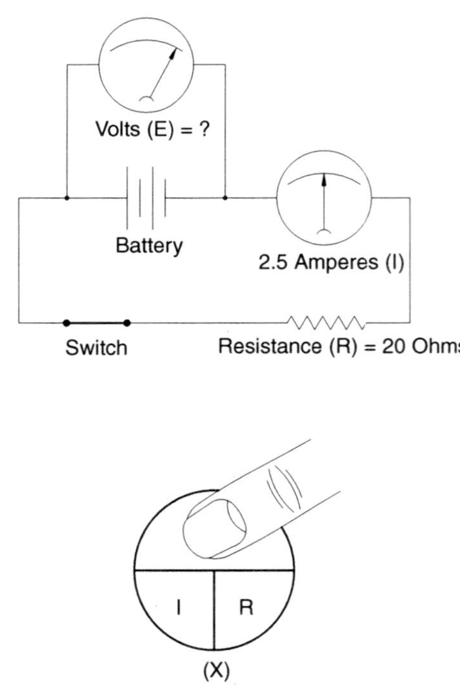

2.5 Amps x 20 Ohms = 50 Volts

Figure 1-9 Finding Voltage

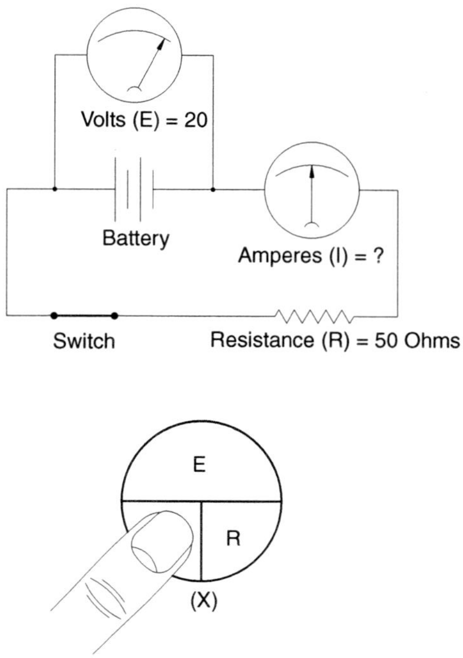

20 Volts ÷ 50 Ohms = 0.4 Amps

Figure 1-10 Finding Current

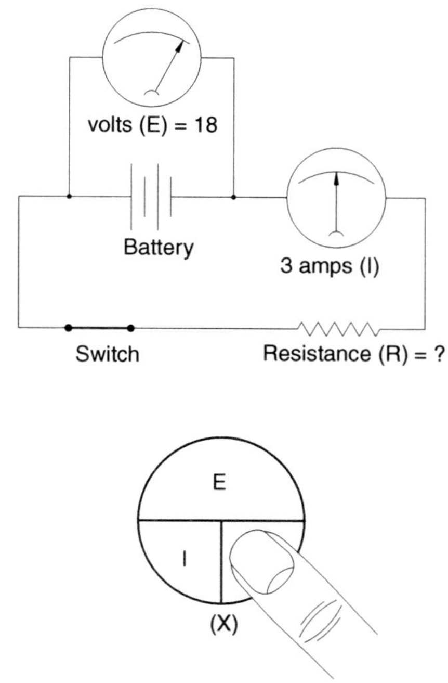

18 volts ÷ 3 amps = 6 ohms

Figure 1-11 Finding Resistance

To become more comfortable with using Ohm's law, try selecting different known values and calculating the unknown value.

Circuit Types

There are two basic types of electrical circuits, parallel and series. Both types of circuits are commonly used and it is important that you have a clear understanding of the difference between the two.

A parallel circuit is simply two different circuits that use some common component(s). Figure 1-12 shows a parallel circuit with two light/switch loops that are supplied with power from a common battery. By far, the most common use for parallel circuits is power distribution. A perfect example of parallel circuitry is house wiring. A primary power source is connected to the house and routed into the breaker box. The breakers are connected to the primary power source in parallel branch circuits. The house's outlets are then connected to the output of each breaker in further parallel branch circuits.

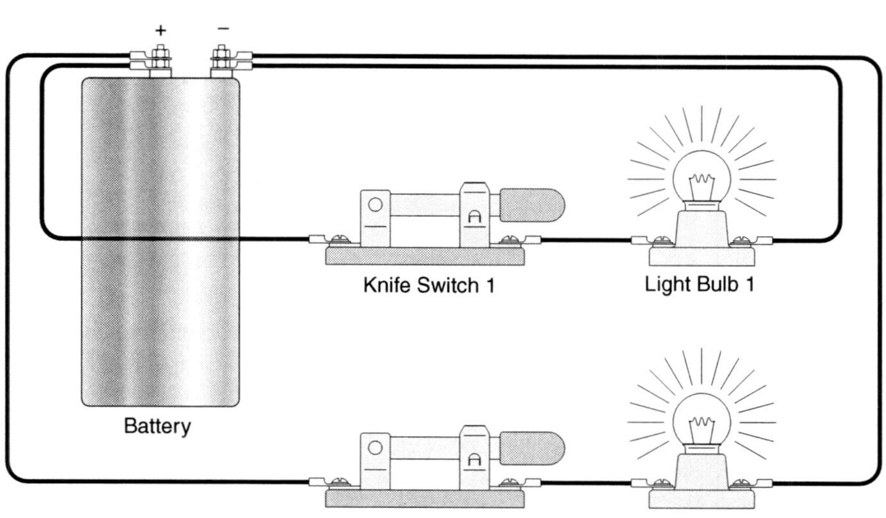

Figure 1-12 Parallel Circuit

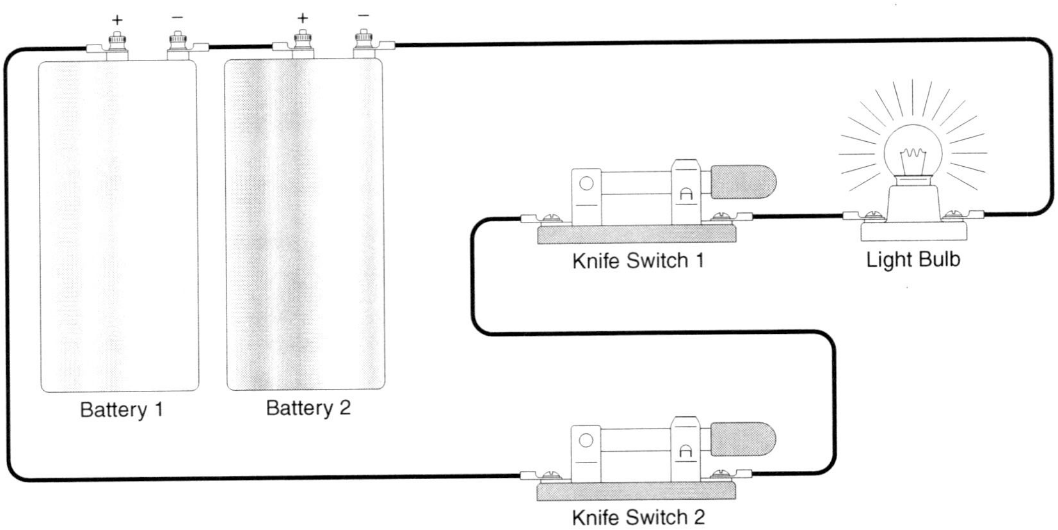

Figure 1-13 Series Circuit

Power distribution systems in automobiles are also parallel circuits. In the case of the automobile there are two power sources, the battery and the alternator or generator. When the car is not running, the battery provides necessary power to start the engine, turn on the lights, play the radio, and so forth. Once the car has been started, the alternator takes over as the primary power source.

Most everyone has heard of a parallel printer port. A parallel printer port uses several parallel circuits to transfer data simultaneously. In this way, the data transfer rate is considerably faster.

Figure 1-13 shows an example of a series circuit. Notice that there is only one loop in the circuit. It can be said that the switches and the batteries are wired in a series arrangement.

Series circuits are most commonly found in the devices that produce work from electricity. As an example, a common lamp has a light bulb and a switch wired in series. A heater has a heating element, thermostat, and on/off switch wired in series. Most of us have experienced those annoying strings of holiday lights that fail completely when one bulb burns out. These lights are wired in series and fail completely because when one bulb burns out the loop is broken and current can't flow.

It should also be noted that in almost any complex electrical system there are examples of both parallel and series circuits. By referring back to Figure 1-12, it can be seen that the switches and the light bulbs are placed into a series arrangement within two parallel circuits. In most applications of electricity, some combination of parallel and series circuits are necessary to accomplish the required outcome.

Reversing Circuits

A very common circuit in industry is the reversing circuit. The purpose of the reversing circuit is to switch the polarity of the power source. This type of circuit has many applications. Figure 1-14 shows a reversing circuit setup to change the rotation of a permanent magnet DC motor. The battery is connected to the reversing switch through a knife switch, which can be used to turn on and off the motor regardless of its rotation. The reversing switch is set up to route the positive and negative sides of the battery to the matching terminals on the motor when the lever is in the down position. When the lever is in the up position, the positive and negative sides of the battery are connected to the opposite terminals of the motor. By reversing the polarity of the battery the motor will run in the reverse direction. Study the illustration carefully, the concept of the reversing circuit is very important to understand.

Alternating Current (AC)

Up until this point we have reviewed circuits that use direct current (DC). Direct current is electricity that flows in one direction only, from the negative terminal to the positive terminal.

Alternating current is the antithesis of DC. In AC, the electricity flows in both directions. The flow is in one direction for a short period of time and then reverses and flows in the opposite direction for a short period of time. The polarity changes, or alternates, many times a second, hence is termed alternating current. You can consider that AC is like continuously switching the reversing switch, shown in Figure 1-14, back and forth.

Figure 1-15 shows how a reversing switch might be actuated from a crank shaft to produce AC. The battery provides DC to the input of the switch. The plunger rod is pulled up and down by the crank shaft, which switches the output polarity of the circuit.

Alternating current is generally represented graphically as shown in Figure 1-16. The horizontal line represents

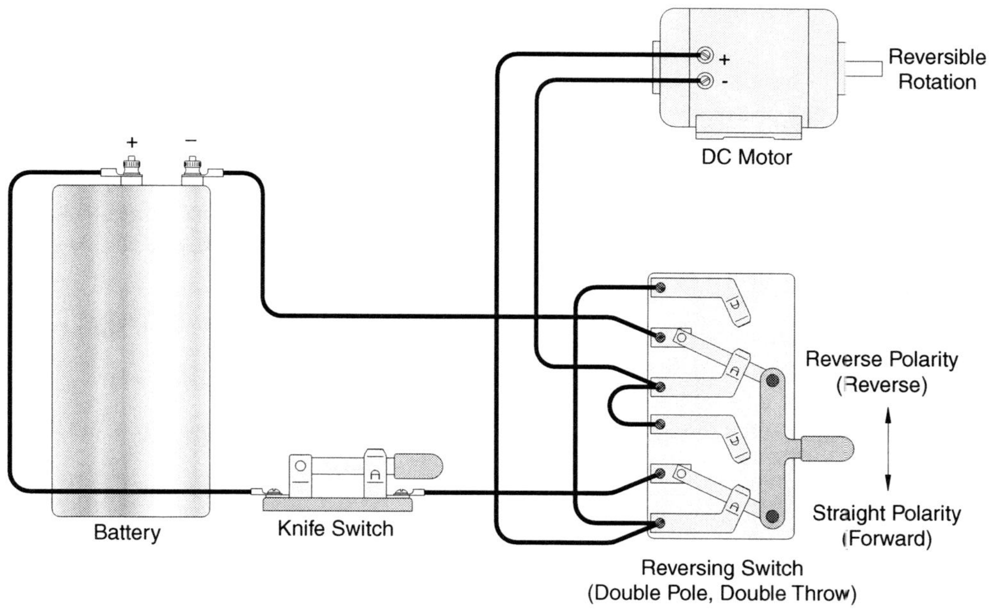

Figure 1-14 Reversing Circuit

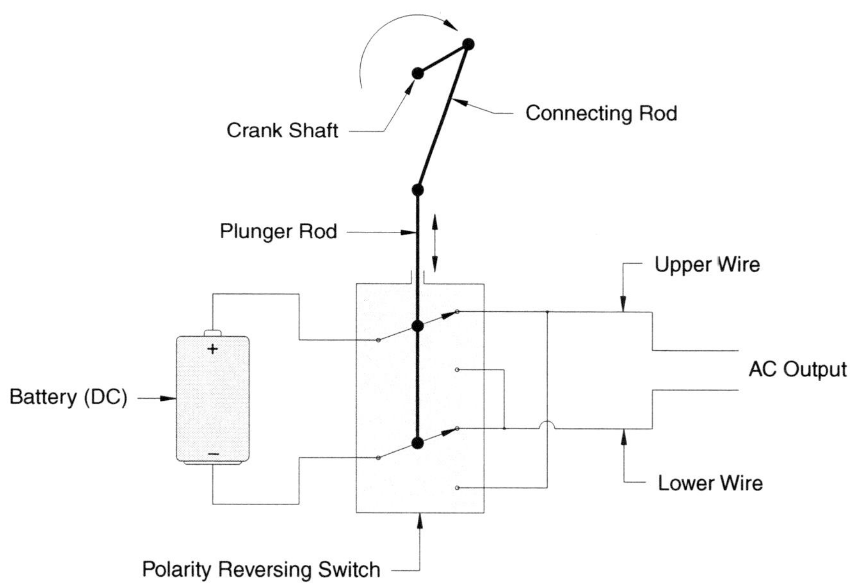

Figure 1-15 Synthesizing Alternating Current

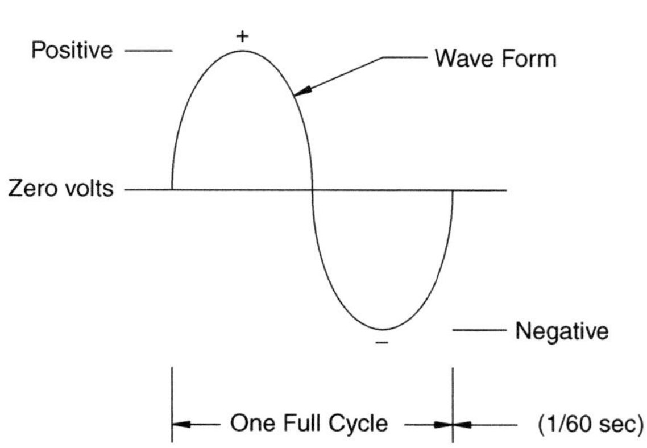

Figure 1-16 Alternating Current Cycle

zero volts. The data points located above or below the line represent volts, with points above the line—positive volts, and points below the line—negative volts. In the United States, AC power is delivered at 60 cycles per second, which means that a full AC cycle is $^1/_{60}$ of a second. The illustration shows that for the first half of the cycle the polarity is positive and for the second half of the cycle the polarity is negative.

Alternating current power has many useful attributes that will be discussed further in the following chapters of the book.

Watts

Watt is an important unit of measure to have knowledge of when discussing basic electricity. It is one of the most common terms used in electricity. We have all considered what wattage the light bulb should be when replacing a burned out one. Most of us have heard of stereo equipment rated in watts. Hair dryers and heaters are usually rated in watts. We all know that a 2000 watt heater produces more heat than a 1000 watt unit. But what is a watt?

The watt is a unit of measure that describes work. When considering electrical circuits, the total watts are derived by multiplying the voltage by the current. As an example, if a light bulb operates on 120 volts and requires 0.83 amps then it will produce 100 watts of work. In the case of a light bulb, the work is in the form of light emissions.

$$120 \text{ volts} \times 0.83 \text{ amps} = 100 \text{ watts}$$

It is also important to understand that if you know the wattage of a device and either the voltage or current, then you can use Ohm's law to determine the resistance.

To derive current from wattage divide the watts by the voltage.

$$100 \text{ watts} \times 120 \text{ volts} = 0.83 \text{ amps}$$

To find the resistance with Ohm's law.

$$120 \text{ volts (E)} \times 0.83 \text{ amps (I)} = 144.6 \text{ ohms (R)}$$

CHAPTER 2
BASIC MECHANICS

If we think back, most of us can remember learning about mechanics in our high school physics classes. Basic mechanics and its application is imperative to our day-to-day lifestyle. There are numerous examples of mechanical devices surrounding us in our homes, cars, and at work.

Much like electricity, most of us give little or no thought to the machines that surround us. When our desktop printer runs out of ink, we mutter under our breaths "stupid piece of junk," failing to appreciate the truly spectacular technology at our disposal. We go out and mow our lawns every week with a piece of equipment that we have almost no understanding of. Everyday we drive to work in a piece of equipment that is so complex, it is beyond the comprehension of the vast majority of the populace. We sit at our desks, lounge in our living rooms, shop at our favorite stores, and float in our swimming pools, never realizing it is all because of these simple machines.

Energy

Before discussing basic machines let's take a moment to review energy. There are two types of mechanical energy, potential and kinetic. Potential energy is associated with objects at rest, while kinetic energy is associated with objects in motion.

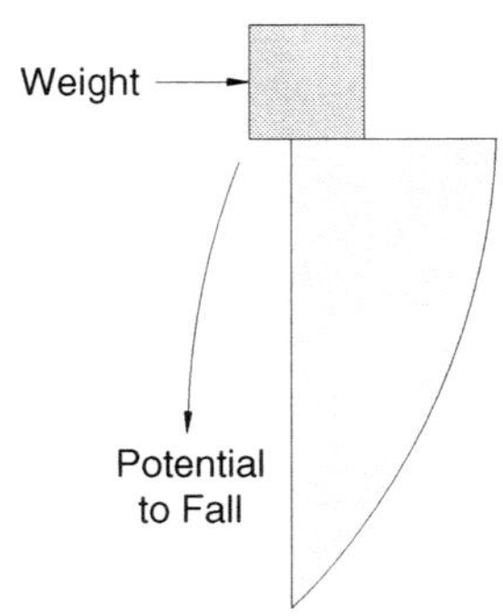

Figure 2-1 Potential Energy

Figure 2-1 shows an example of potential energy. The weight is at rest on top of the platform and energy is stored in reference to gravity. If the weight is pushed off the edge of the platform, the stored energy propels the weight to the ground.

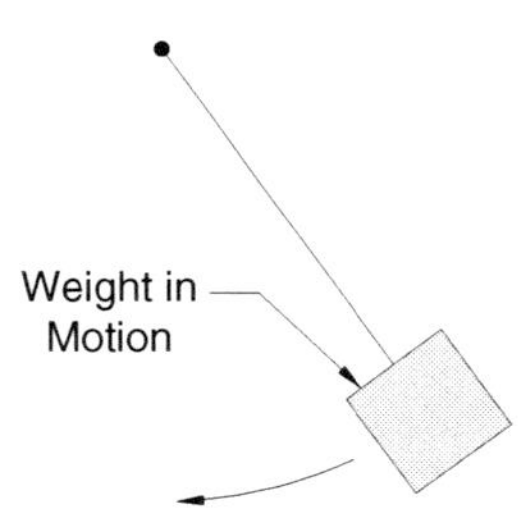

Figure 2-2 Kinetic Energy

Figure 2-2 shows an example of kinetic energy. The energy is stored in the weight in reference to its motion. If the weight is suddenly stopped, the stored energy is released in the form of an impact shock.

Simple Machines

Without knowing it, most people have an intuitive understanding of simple machines. We've all pried the lid off a paint can with a screw driver, a simple lever. Most of us have had a ride in a car, the wheel. We've casually watched our neighbor pull start his lawn mower, the pulley. How about pushing a bicycle up a hill, the inclined plane. Removing the lid from a jar of peanut butter, the screw. And who among us hasn't sat at their desk and played with a rubber band, the spring. If you just take a minute and look around, you'll see hundreds of examples of simple machines.

The Lever

The most basic machine is the lever. The lever is simply a beam that is toggled on a pivot point, or fulcrum. Figure 2-3 shows a simple balanced lever. If 1 pound of force is applied to the right end of the beam, then 1 pound of force is generated on the left end of the beam. A seesaw, or teeter-totter, is a good example of the basic lever.

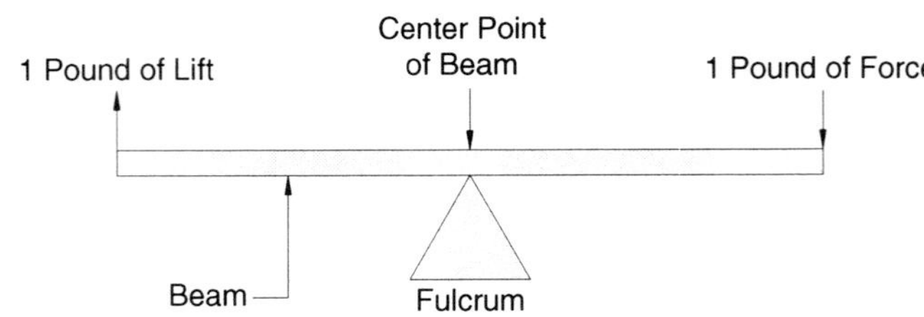

Figure 2-3 Basic Lever

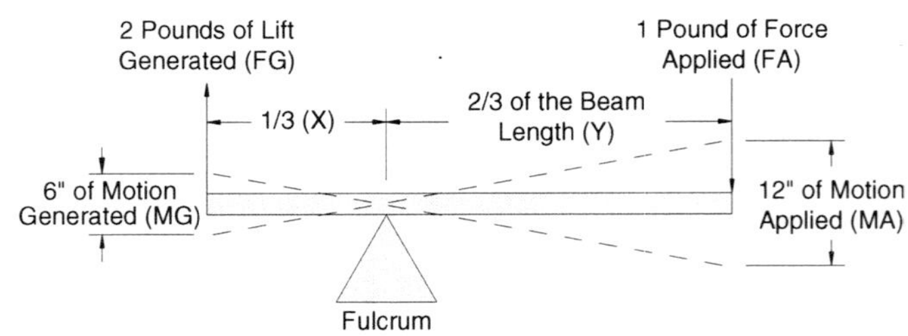

Figure 2-4 First Class Lever

Figure 2-4 shows a first class lever. In the illustration, the fulcrum is placed two-thirds from the right end of the beam. In placing the fulcrum off-center, a certain amount of mechanical advantage can be achieved. By placing 1 pound of force (FA) on the right end of the beam, 2 pounds of force (FG) is generated on the left end of the beam. The amount of travel is also transformed. In the illustration, 12 inches of motion (MA) on the right end of the beam translates to 6 inches of motion (MG) on the left end of the beam. To calculate both force and motion the formula is:

$$\text{For Force: } (Y \div X) \times FA = FG$$
$$\text{For Motion: } (Y \div X) \times MA = MG$$

Figure 2-5 shows a second class lever. A second class lever has the fulcrum placed at the end of the beam and the force is applied at the opposite end. The force generated is at a point between the force applied and the fulcrum. By manipulating the point along the beam, a certain amount of mechanical advantage can be achieved. Placing 1 pound of force (FA) on the left end of the beam, generates 3 pounds of force (FG) at the point within the beam. The amount of travel is also transformed. In the illustration, 12 inches of motion (MA) on the left end of the beam translates to 4 inches of motion (MG) at

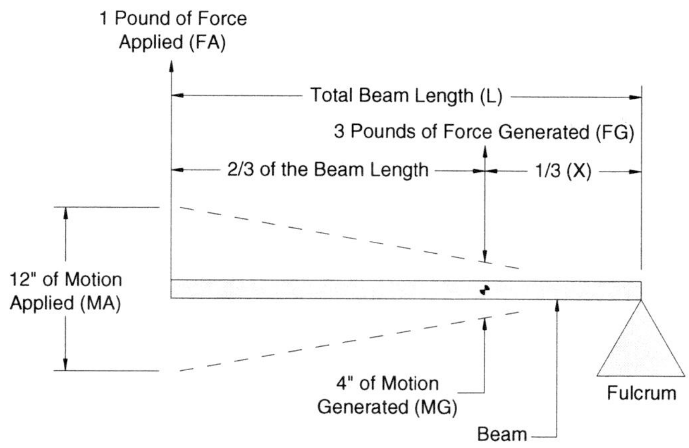

Figure 2-5 Second Class Lever

the point within the beam. To calculate both force and motion the formula is:

For Force: (L ÷ X) × FA = FG

For Motion: (L ÷ X) × MA = MG

Figure 2-6 shows a third class lever. A third class lever is similar to a second class lever, except the points at which the force is applied and the force is generated are reversed. The force is applied at a point between the fulcrum and the force generation point. By manipulating the point of force along the beam a certain amount of mechanical advantage can be achieved. Placing 3 pounds of force (FA) at the point within the beam, 1 pound of force (FG) is generated at the left end of the beam. The amount of travel is also transformed. In the illustration, 12 inches of motion (MG) on the left end of the beam is generated when 4 inches of motion (MA) is applied at the point within the beam. To calculate both force and motion the formula is:

For Force: (L ÷ X) × FA = FG

For Motion: (L ÷ X) × MA = MG

The usefulness of levers can be greatly enhanced with the application of connecting rods. Figure 2-7 shows two first

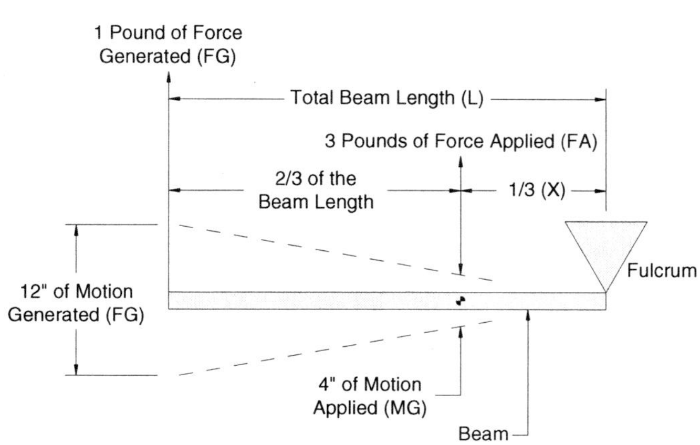

Figure 2-6 Third Class Lever

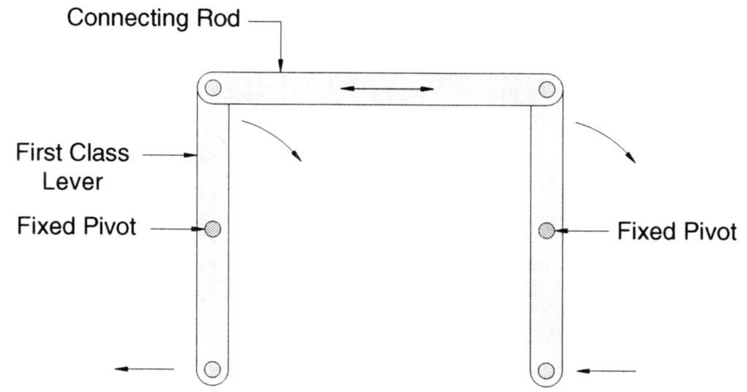

Figure 2-7 Levers and Connecting Rods

class levers, which are interconnected with a connecting rod. In doing so, the motion of the left lever is duplicated in the right lever.

Bell cranks are 90°, first class levers. They are typically used to change direction of motion in a linkage system. Most bell cranks have a ratio of 1 to 1; however, they are often applied with other ratios that will increase or decrease the mechanical advantage. Figure 2-8 shows a simple bell crank.

Figure 2-9 shows an arrangement of levers and connecting rods. The motion is supplied from the solenoid at the upper

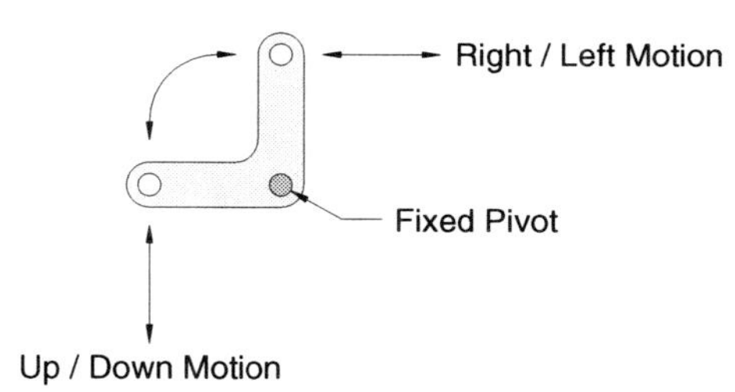

Figure 2-8 90° Bell Crank

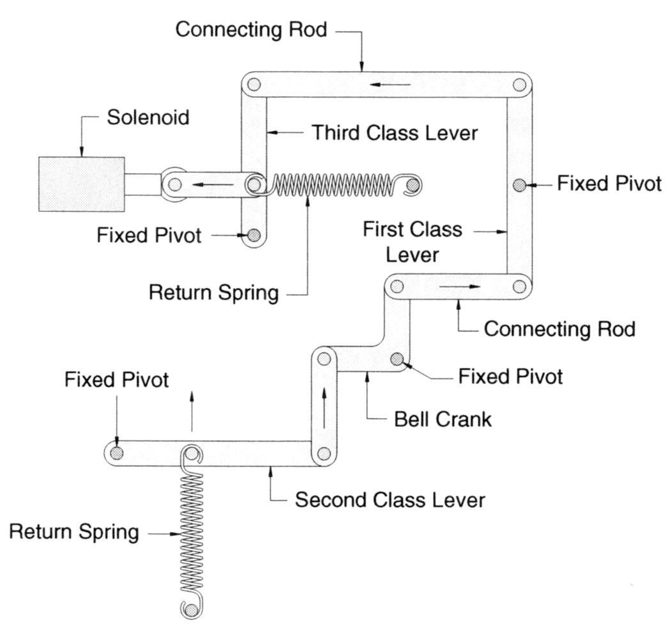

Figure 2-9 Levers and Connecting Rods

left. Study the illustration carefully and follow the motion of the linkage. Also notice that return springs are applied at both the solenoid and at the far end of the linkage.

The Wheel

The wheel is a variation of a first class lever. Consider the lever in Figure 2-3 and instead of a beam, imagine a circular disk with the fulcrum at the center. If the disk is allowed to rotate freely, then you have a wheel. The basic wheel is encountered in two versions, those with a rotating axle, as shown in Figure 2-10, and those with a fixed axle, as shown in Figure 2-11. Both variations have broad applications.

Figure 2-12 shows an example of a gear train. The gear on the motor carries a rotating axle, which allows the rotation of the motor to be transferred to the gear. The rotation of the drive gear is transferred to the idler gear, which has a fixed axle. The rotation of the idler gear is transferred to the driven gears, which have rotating axles and are used to conduct some sort of work.

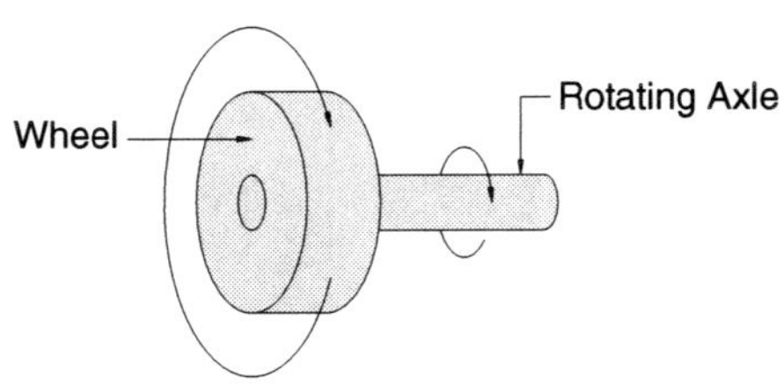

Figure 2-10 Wheel and Rotating Axle

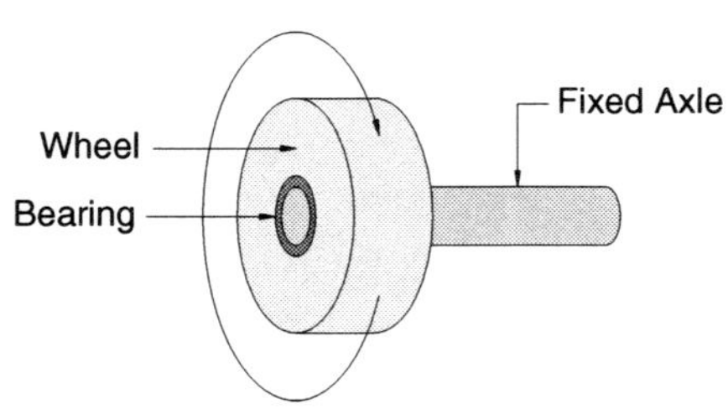

Figure 2-11 Wheel and Fixed Axle

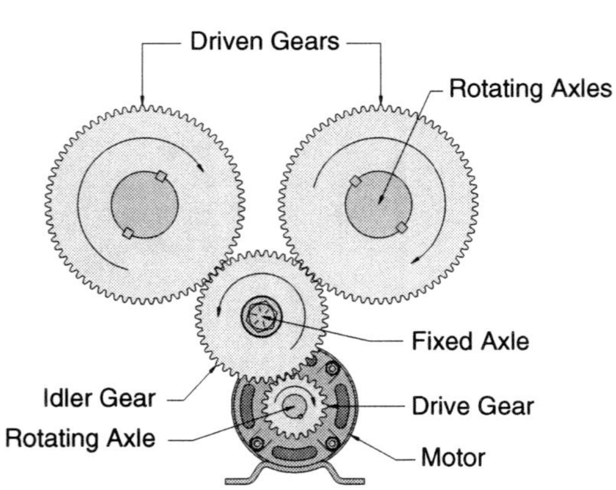

Figure 2-12 Gear Train

Pulleys

The pulley is a variation of the wheel. The pulley is a wheel with a circular groove, which provides a guide for a rope draped over the outside diameter. If one end of the rope is pulled with 1 pound of force, then the other end generates 1 pound of lift. Figure 2-13 illustrates a simple pulley.

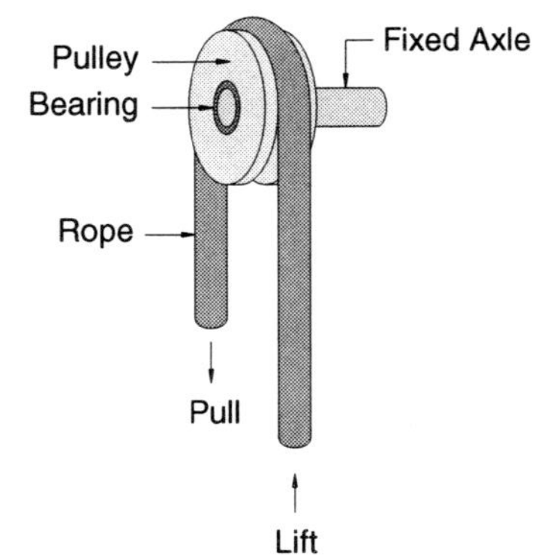

Figure 2-13 Pulley

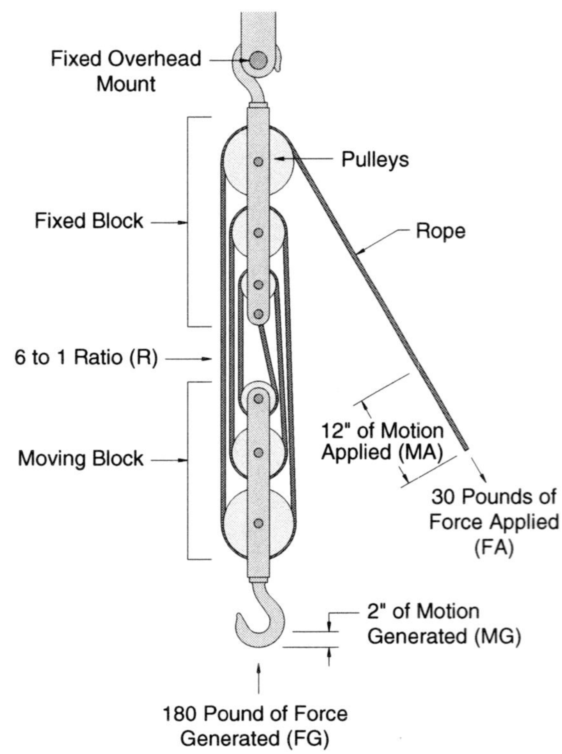

Figure 2-14 Block and Tackle

By arranging pulleys in a progressive manner, significant mechanical advantage can be realized. Figure 2-14 shows a typical block and tackle intended to provide higher lifting force. To determine the lifting force for a block and tackle, divide one by the number of vertical ropes between the pulleys. The illustration shows a block and tackle arrangement with a 6-to-1 ratio. To calculate the force generated (FG), multiply the applied force (AF) by the ratio.

30 pounds (AF) × 6 = 180 pounds of force generated (FG)

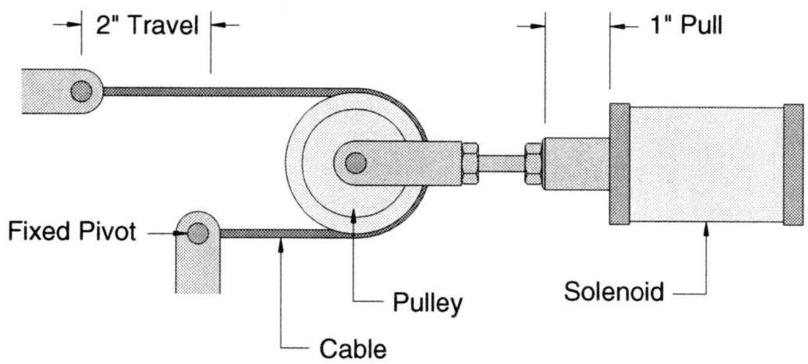

Figure 2-15 Doubling Solenoid Throw

To calculate the motion generated (MG), divide the motion applied (MA) by the ratio.

12" (MA) ÷ 6 = 2" (MG)

Figure 2-15 shows how a single moving pulley can be used as a motion multiplier in conjunction with a solenoid. The pulley is mounted to the plunger and when the solenoid is activated, the motion is multiplied by 2.

One of the most common uses for pulleys is vee belt drives, used in power transmission. Figure 2-16 shows a typical vee belt drive. The ratio of the drive can be determined by dividing the diameter of the driven pulley by the diameter of the drive pulley. The revolutions per minute (RPM) of the driven pulley can be determined by dividing the motor RPM by the drive ratio.

1725 (motor RPM) ÷ (6 ÷ 3) = 862.5 (driven pulley RPM)

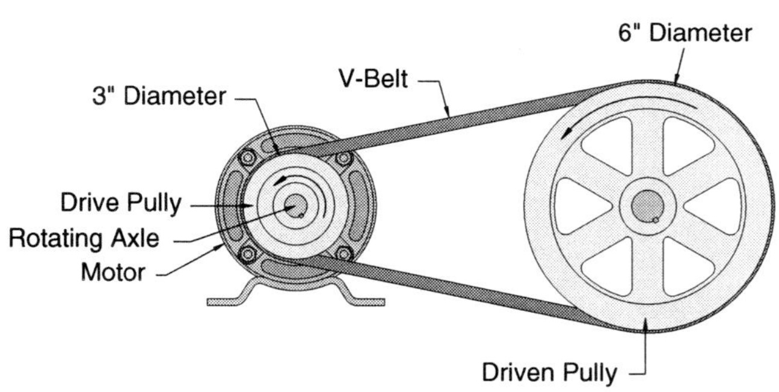

Figure 2-16 Vee Belt Drive

The traction drive is another method of using the pulley. These drives are most commonly found as the lift mechanism for elevators. In a traction drive (Figure 2-17), a simple pulley is set up. The load is placed on one end of the cable and a counter-weight is placed on the opposite end. To move the load, the pulley, or traction spool, is driven by an electric motor and reduction drive. A traction drive has the advantage of applying a uniform load on the drive motor, regardless of the length of the cable.

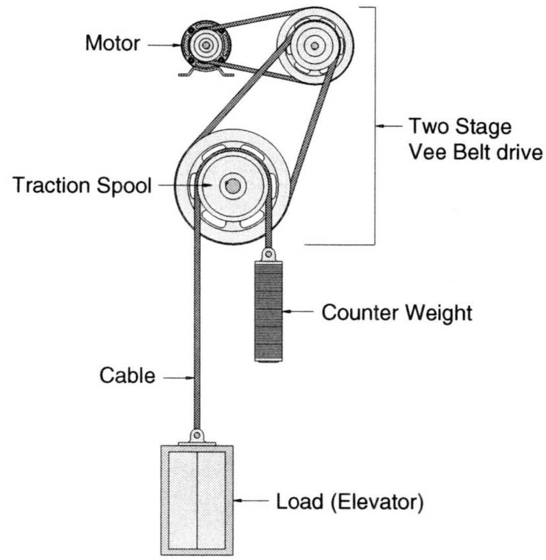

Figure 2-17 Traction Drive

Another variation of the pulley is the spool. The spool is most commonly used on winching equipment. The spool is a pulley that continuously wraps the cable around the drum, as shown in Figure 2-18. Normally, the spool is driven through the axle via a reversible electric motor and reduction drive. When the motor is operated in forward, the cable extends and when the motor is reversed, the cable retracts.

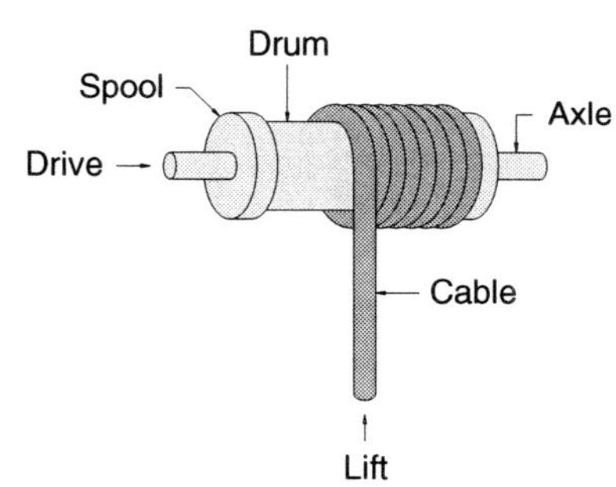

Figure 2-18 Cable Spool

Cable actuation systems are most commonly found on aircraft. Cable actuators can offer an exceptionally lightweight method to transmit power and motion. Figure 2-19 shows the basic elements of a cable actuation system. Notice that the return spring is located at the opposite end of the cable from the solenoid. This is necessary to maintain a tension on the cable at all times.

Figure 2-20 shows how a cable and pulley actuation system may be approximated by a cable within a sheath. A common example of cable and sheath actuators can be seen on most bicycles.

The Inclined Plane

The inclined plane is a simple machine that can provide us with a significant mechanical advantage. Inclined planes are most commonly found on and around loading docks where

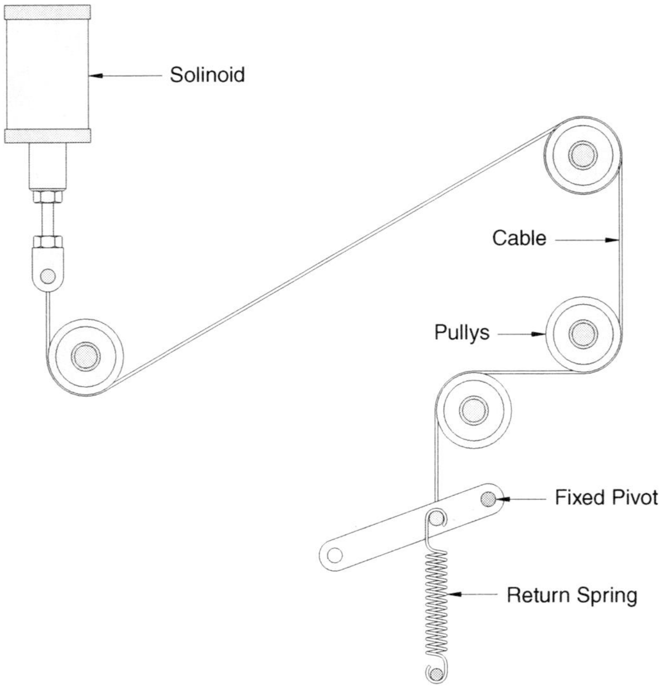

Figure 2-19 Cable and Pulley Actuator System

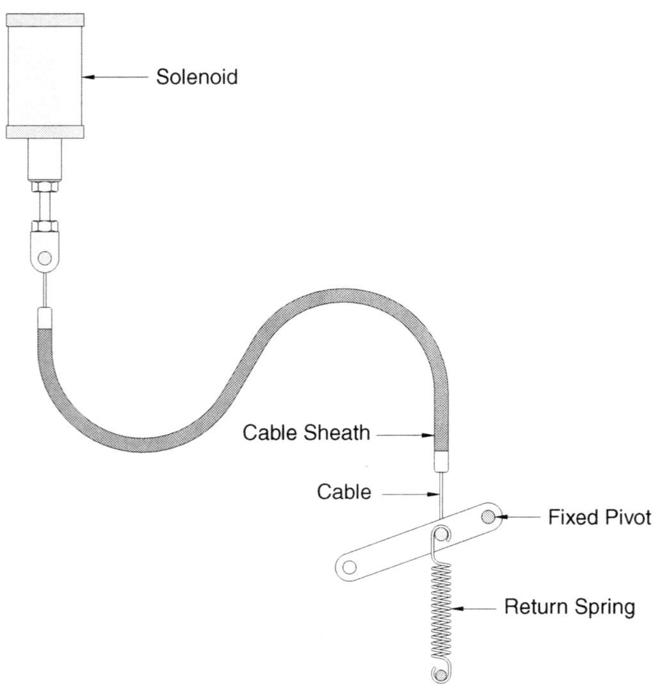

Figure 2-20 Cable and Sheath Actuator System

workers shuttle their two wheelers, pallet jacks, and fork lifts up and down the ramps that lead into the truck and the dock.

Figure 2-21 shows a schematic representation of an inclined plane and a rolling load. The force necessary to elevate the load is reduced in direct proportion to the angle of the plane. In the illustration, the 100 pound load (W) can be raised with only 25 pounds of applied force (FA) because the plane has an angle of 22.5° (A). The amount of force required

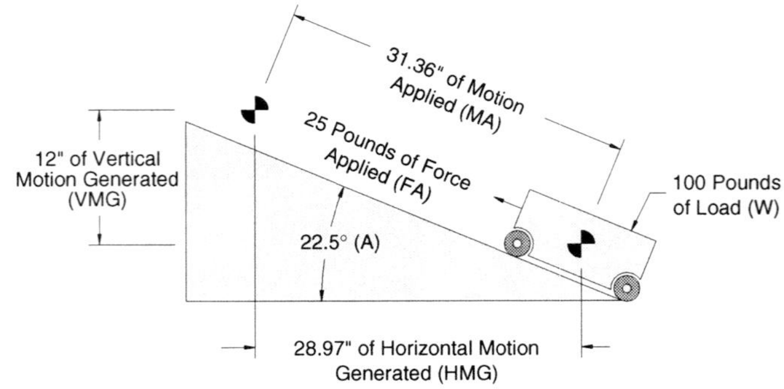

Figure 2-21 Inclined Plane

to pull the load up the plane can be calculated by the following formula:

$$100 \text{ pounds (W)} \div (90 \div 22.5 \text{ (A)}) = 25 \text{ pounds (FA)}$$

The vertical motion generated can be calculated by the following right angle triangle formula:

$$12'' \text{ (VMG)} = 31.36 \text{ (MA)} \times \sin 22.5° \text{ (A)}$$

The horizontal motion generated can be calculated by the following right angle triangle formula:

$$29'' \text{ (HMG)} = 31.36 \text{ (MA)} \times \text{cosine } 22.5° \text{ (A)}$$

The Screw

The screw is one of the most important mechanical devices ever devised. The basic screw can be considered to be an inclined plane that has been wrapped around a round shaft. In this way a spiral inclined plane is created, or a screw. Figure 2-22 shows how a simple screw is generated with an inclined plane. The progression of the spirals is referred to as the pitch, and the pitch is generally referred to in threads per inch (TPI).

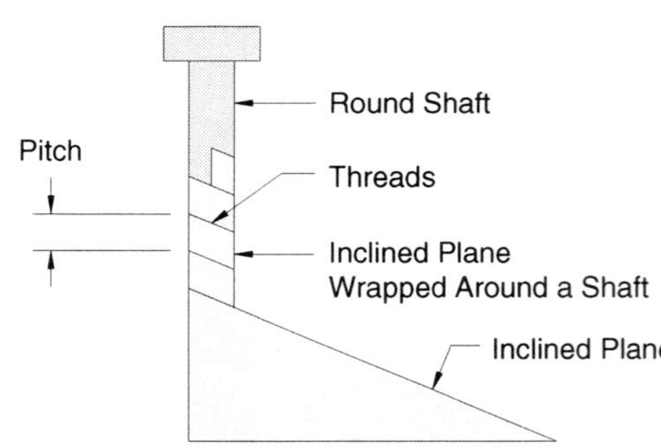

Figure 2-22 Basic Screw Thread

In addition to their fastening value, screws are used extensively for motion control. Figure 2-23 shows a sectional view of a motorized screw thread actuator. In this case a piston, with a threaded nut at its base, is inserted into a guide tube. When the threaded shaft is turned, the piston moves in and out, depending on the rotation. The threaded shaft is driven by a toothed belt connected to a direct current (DC) motor. By reversing the polarity of the motor, the actuator can be

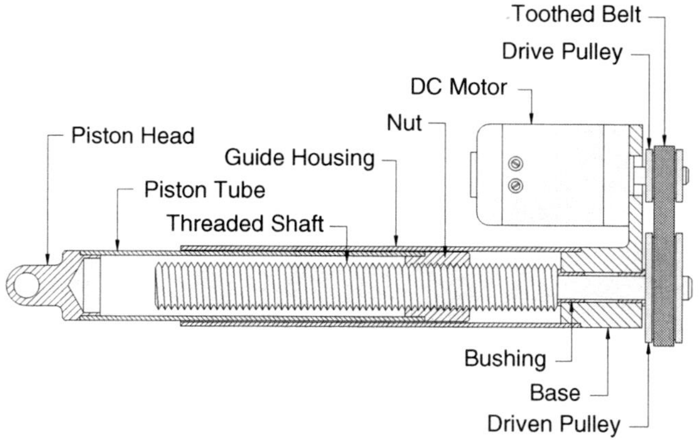

Figure 2-23 Motorized Screw Thread Actuator

extended or retracted. The speed of the piston can be controlled by the speed of the motor and the ratio of the belt drive.

Springs

Although springs are generally not considered simple machines, they are a critical primary element in machine design. A spring is a mechanical device that can store energy through elastic deflection. There are many different springs designs that are intended to fulfill a variety of applications. We are all familiar with the loud report of a screen door slamming and with the long spring that is responsible for that action. The ordinary screen door spring is an excellent example of an extension spring.

Figure 2-24 shows several of the more common spring designs that may be found in electromechanical devices.

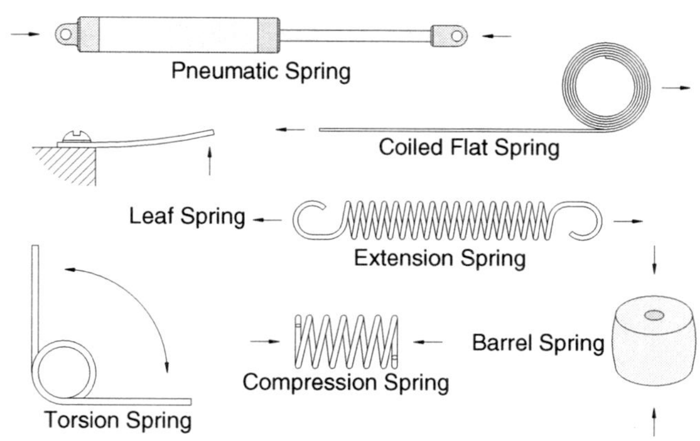

Figure 2-24 Common Springs

CHAPTER 3
POWER SOURCES

When dealing with electromechanical equipment, it is necessary to have a clear understanding of the various types of electrical power that are at our disposal. There are two basic types of electrical power, direct current (DC) and alternating current (AC).

Direct Current (DC)

Direct current is electrical power which maintains a flow of electrons in one direction only. The basic circuit shown in Figure 3-1 illustrates DC. Electrons flow from the negative terminal on the battery, through the light bulb, and back to the positive terminal. Most devices that use batteries operate using DC. Your automobile, calculator, flash light, transistor radio, camera, and wall clock all operate on DC.

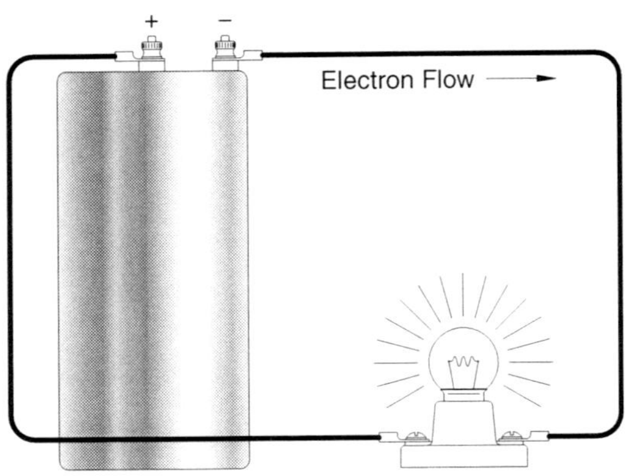

Figure 3-1 Direct Current Electron Flow

Alternating Current (AC)

The second type of electrical power is AC. The way this type of power works is not so intuitive as DC power and can be a little difficult to understand when discussing the various ways it is applied.

Alternating current is the type of power that is delivered to homes and business. It is referred to as AC because its polarity, and thus the direction of electron flow, is constantly reversing or "alternating." Alternating current is generally represented graphically as shown in Figure 3-2. The horizontal line represents zero volts. The data points located above or below the line represent volts, with points above the line being positive volts, with electrons flowing in one direction, and points below the line being negative volts, with electrons flow in the opposite direction. In the United States, AC power is delivered at 60 cycles per second, which means that a full AC cycle is $\frac{1}{60}$ of a second long. The illustration shows that for the first half of the cycle ($\frac{1}{120}$ sec) the polarity is positive and for the second half of the cycle ($\frac{1}{120}$ sec) the polarity reverses and is negative.

Alternating current has a number of significant attributes that will be discussed further in the following chapters of the book.

For most applications there are four standard AC voltages. These are 24, 120, 240, and 480 volts. AC voltage of 24 volts doesn't represent a significant shock hazard, so it is preferred as a control voltage in most commercial and industrial equipments. It is also used as the primary voltage for model trains and slot cars. These toys have exposed terminals and safety considerations dictate that nonlethal voltages are used.

AC voltage of 120 volts is commonly found at the receptacles inside our houses. Most small appliances in the United States operate on 120 volts. Additionally, 120 volts is a common control voltage in industrial equipment where it is used primarily to limit the current and, consequently, the wire size that is used for the circuit of equipment.

Voltage of 240 volts is typically supplied to homes. This voltage can be reduced to 120 volts for use in the home's outlets. Major appliances and larger motors will generally operate on 240 volts. The higher voltage provides the same power with lower current and, therefore, smaller wire. It should be noted that 240 volts represents a significant shock hazard. Severe injury or death can occur if this voltage is handled improperly. Before working with any 240-volt circuit, be certain that the power is turned off and locked out.

Voltage of 480 volts is generally reserved for industrial applications. It carries the voltage/current advantage a step further. Generally speaking, motors over 25 horsepower, induction furnaces, arc welders, and overhead cranes will be operating on 480 volts. Even more than 240 volts, 480 volts, can be very dangerous to work with. Severe injury or death can occur if this voltage is handled improperly. Only properly trained electricians should work with 480-volt circuits. As with other circuits the power should be disconnected and locked out before working with 480 volt equipment.

Figure 3-3 shows the four different AC voltages that are generally encountered. Notice that the frequency is the same

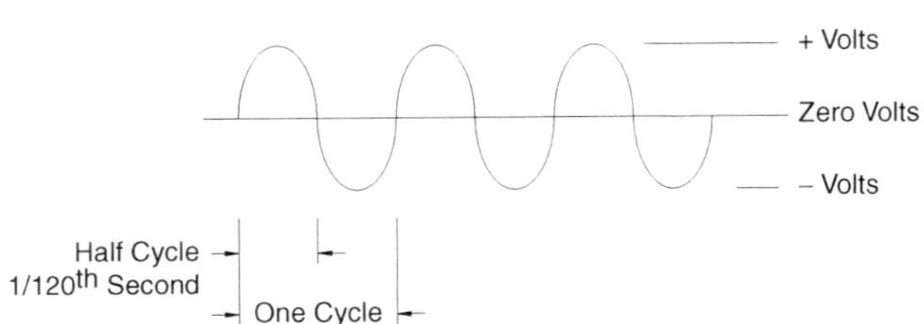

Figure 3-2 60-Cycle Wave Form

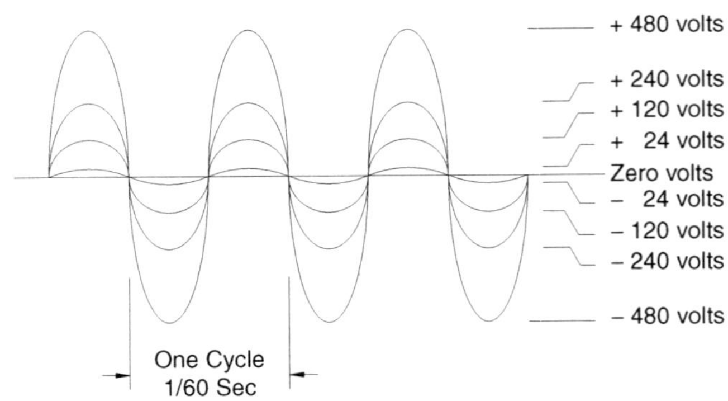

Figure 3-3 24, 120, 240, and 480-VAC 60 Hz

for each voltage, only the amplitude changes. A visual reference for the level of power that each voltage can supply can be gauged by comparing the area above the zero-volt line and within the curve, with a larger area representing higher power.

Some countries around the world use 50 Hz AC power. For all practical purposes there is very little difference between 50 and 60 Hz Power; the most significant difference is that induction motors operate at a slower speed when using 50 Hz power. Most heating equipment provides the same output with 60 or 50 Hz. Other industrial equipment, such as arc welding machines and power supplies, show very little difference in their output.

To limit the weight of electromagnetic equipment, 400 Hz power is utilized. Power of 400 Hz is used primarily on aircraft where low weight is critical.

Figure 3-4 shows 50, 60, and 400 Hz wave forms. Note that the voltage, or amplitude, is the same for all three frequencies. It should also be noted that the overall power is the same for each frequency. The difference in frequency is only critical when considering the amount of iron that is used in the component to be driven. This consideration will be discussed in greater detail in Chapter 5.

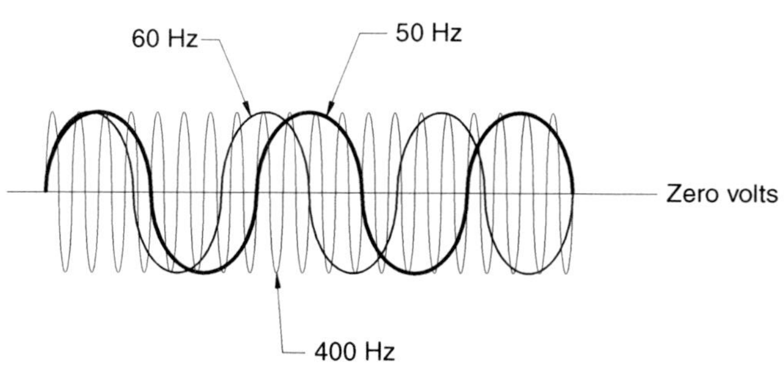

Figure 3-4 50, 60, and 400 Hz Waveform

Three-Phase

Alternating current power is generally delivered in two different forms, single-phase and three-phase. All of the previous examples of AC shown in this chapter of the book are single-phase. Generally, three-phase is used for power generation and distribution. Applications for three-phase power are normally in the commercial and industrial communities.

Three-phase power is effectively three single-phase circuits combined so that they may use three wires. If three single-phase circuits were used, then a total of six wires would be required. This can be accomplished if the phase angle of each phase is controlled to 120°. Figure 3-5 shows a graphical representation of three-phase power. Note that there are thee different wave forms. The second phase starts $^1/_{180}$ of a second after the first phase, or 120°. The third phase starts $^1/_{90}$ of a second after the first phase or 240°. By controlling the phase angles, the voltage potential between each phase will equal zero and, therefore, the phases can share common wires.

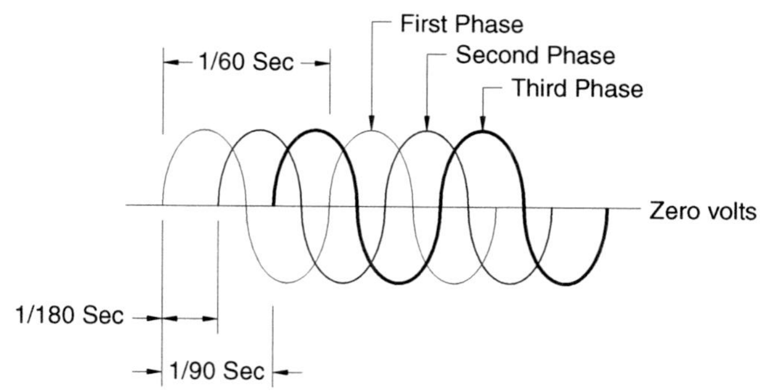

Figure 3-5 Three-Phase Wave Form

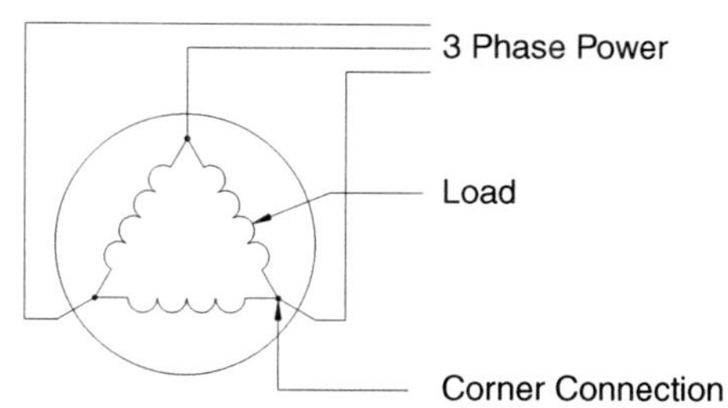

Figure 3-6 Delta Configuration

There are two different types of connections that can be used with three-phase. These configurations are referred to as Delta and Wye. Figure 3-6 shows a Delta configuration. In this setup the three loads are electrically arranged in a triangle and the power connections are made to the three corners. Figure 3-7 shows a Wye configuration. With this arrangement the three loads are electrically arranged in a Y and the power connections are made at the end of the loads. Wye configurations generally have a fourth "return" wire as shown.

Three-phase power also has significant advantages when dealing with induction motors. This attribute will be discussed further in Chapter 6.

In the same way that three-phase power is utilized, higher numbers of phases can also be configured. Figure 3-8 shows a six-phase graphic. The frequency is 60 Hz and the delay time is $^1/_{360}$ of a second or a phase angle of 60°.

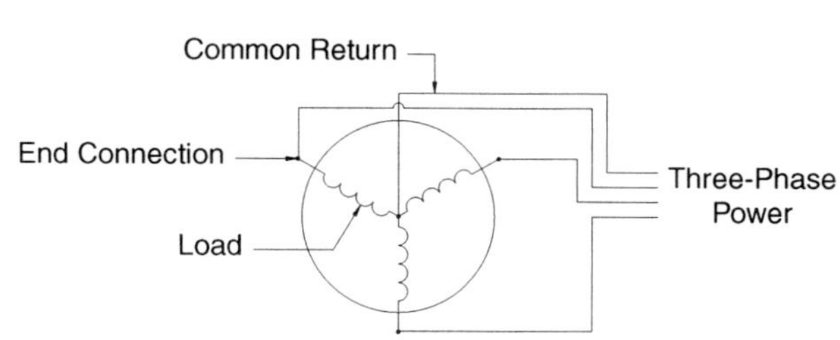

Figure 3-7 Wye Configuration

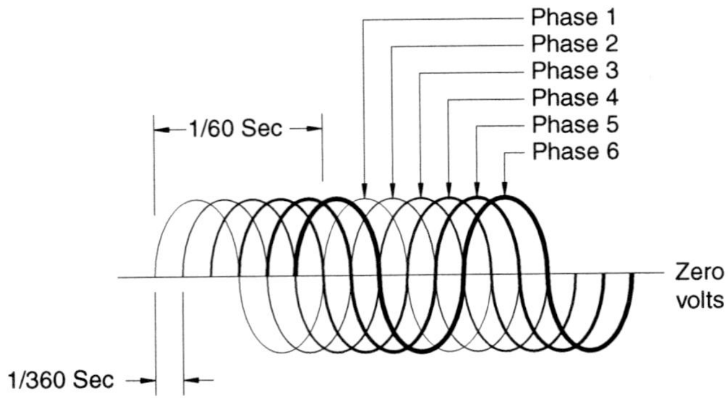

Figure 3-8 Six-Phase Wave Form

Batteries

The most common and the most intuitively understandable electrical power source is the battery. Batteries are an excellent source of DC electricity. They are easy to understand and make sense to the casual observer. They are inexpensive, reasonably light weight, and are available in a variety of different configurations that are appropriate for all manner of applications.

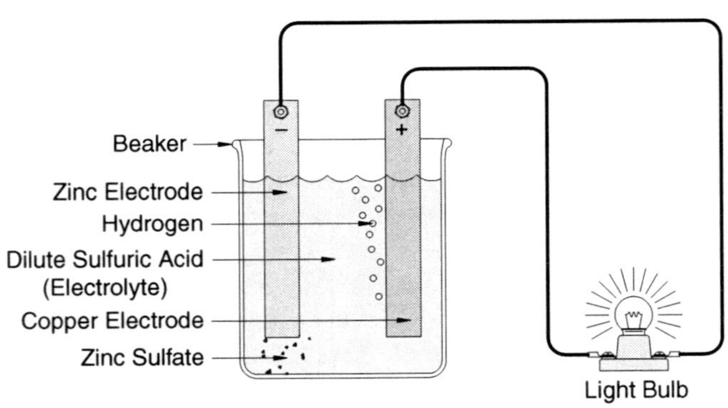

Figure 3-9 Simple Cell

Figure 3-9 shows how a simple storage battery operates. Two electrodes are immersed into an electrolyte bath. The negative terminal is zinc, the positive terminal is copper, and the electrolyte is dilute sulfuric acid. When the electrodes are placed into the electrolyte they both have the same potential. When a power source is connected to the electrodes for charging, a chemical process takes place. As the acid acts on the zinc, zinc sulfate forms and drops off the electrode, leaving behind an excess number of electrons. As a surplus of electrons builds up on the zinc, or negative electrode, a difference in potential is established. When the battery is fully charged the stored energy is available for applications. It should also be noted that during the charging operation, hydrogen forms on the copper, or positive electrode, and bubbles up to the top of the electrolyte.

Batteries are generally rated using two parameters. The first, and most obvious, is voltage. The second, and less intuitive, is amp-hours. Amp-hours indicate the maximum current that a battery can continuously deliver for a period of 1 hour. When a battery is discharged at this rate, usually it's full charge will be expended. A battery that has a 250 amp-hour rating is capable of delivering 250 amps at the batteries full voltage for 1 hour. A battery that has a 500 mA-hour rating is capable of delivering 1/2 amp for 1 hour. It should also be noted that the amp-hour rating is an indication of the capacity of battery. If our 250 amp-hour battery is discharged at a rate of 2 amps, its charge life will be 125 hours. Similarly, if we discharge the battery at 375 amps, the charge life will be 0.66 hours or 39.6 minutes.

$$\text{Amp-Hours} \div \text{Discharge Rate} = \text{Charge Life}$$

Another figure we see on automotive batteries is "cold cranking amps." This figure is generally higher than the amp-hour rating. This rating refers to the maximum current that the battery can deliver at full charge for a short period of time. This is a loose standard and shouldn't be relied on when selecting a battery for peak demand applications. It should also be noted that when a battery is pressed into this type of service, it can get fairly hot and a long cool down period is required.

Lead/Acid Batteries

Figure 3-10 shows a cut-away view of a typical 6-volt lead/acid battery. Because a battery of this type will only produce 2 volts, commercial batteries are actually several batteries connected in series. A 6-volt battery will have 3 cells, a 12-volt battery will have 6 cells, and a 48-volt battery will have 24 cells. Note the bridge conductors on the top of the battery case. These conductors connect the negative terminal of one cell to the positive terminal of the adjacent cell. Although it is hard to imagine an application that would require the construction rather than the purchase of a battery, it is, however, an excellent exercise to construct a battery to gain a better

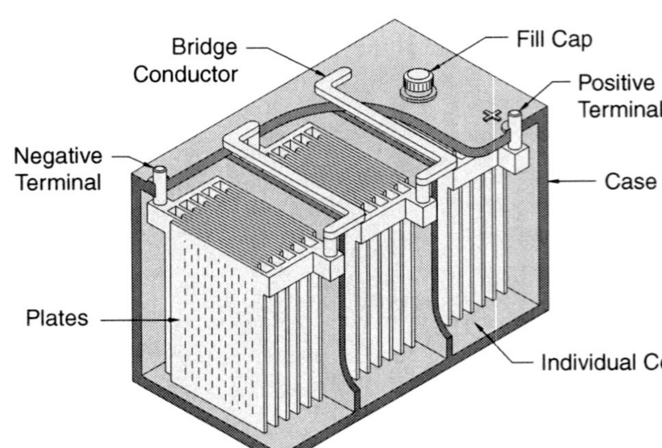

Figure 3-10 6-Volt Lead/Acid Battery

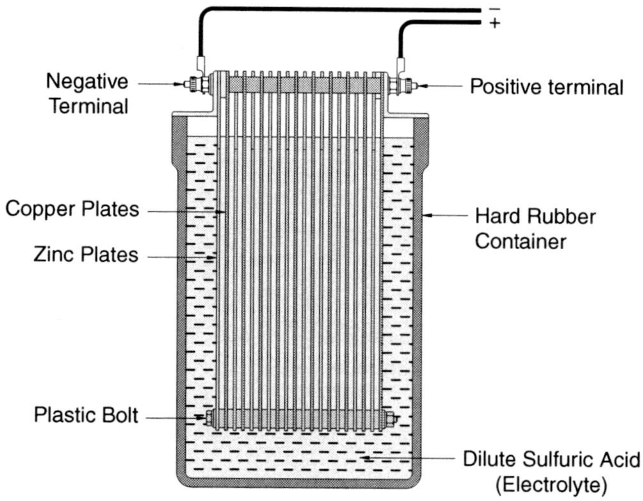

Figure 3-11 Bench Built Lead/Acid Storage Battery

understanding of their interworkings. Figure 3-11 shows a sectional view of a bench built lead/acid storage battery. The container should be a nonbreakable unit that is impervious to sulfuric acid, such as hard rubber. The copper and zinc plates are alternated in an array and immersed into the acid solution. The electrolyte is made by blending 18% sulfuric acid with 82% distilled water.

Figure 3-12 shows an exploded view of the battery assembly. The plastic-threaded rods at the bottom of the plates are there to control spacing. The brass rods at the top of the plates act as battery terminals as well as clamping bolts.

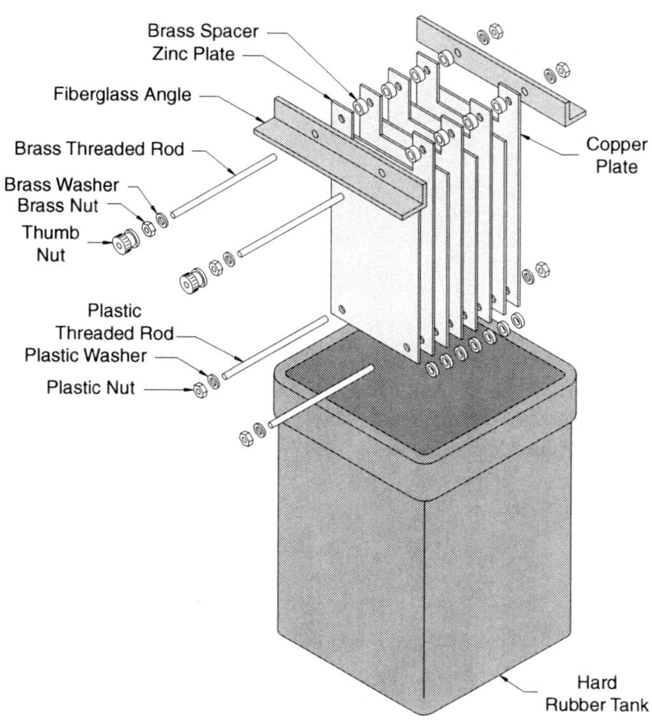

Figure 3-12 Bench Built Storage Battery

It should be noted that the higher the surface area of the plates, the greater the battery's current capacity will be. Surface area can be gained by using either larger plates or a higher plate count. The proximity of the plates to one another will also affect current capacity. The closer the plates are to one another, the higher the battery's current capacity will be.

Figure 3-13 shows a few common automotive batteries. Obviously, the larger the equipment, the greater the electrical load and, therefore, the larger the battery. As an example, motorcycles generally have very small batteries that can fit in the palm of your hand. On the other hand, heavy construction equipment may use batteries that weigh several hundred pounds.

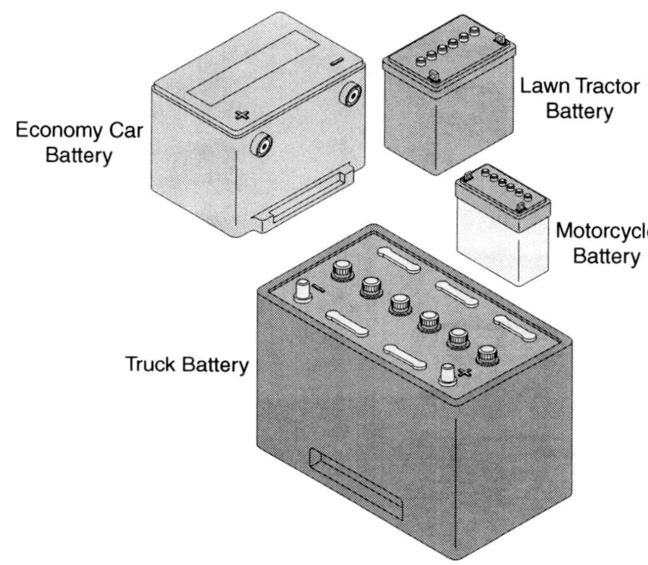

Figure 3-13 12-Volt Automotive Batteries

There are quite a number of different lead/acid formulations commonly used in commercial batteries. The type of battery to be selected should be based entirely on the application. As an example, a standard automotive battery is formulated to deliver very high current for short periods of time. Typically, these batteries should not be significantly discharged. If they are subjected to repeated deep discharging and recharging, their life will be significantly shortened. On the other hand, a deep-cycle battery is formulated specifically for repeated deep discharging and recharging. Furthermore, if deep-cycle batteries are subjected to repeated high current loads, their life will be significantly shortened. To make matters worse, the outward appearance of these two types of batteries is almost identical, so a clear understanding of the intended application of the battery is critical.

When designing a piece of equipment that requires a battery, recommendations should be solicited from the battery manufacturer. The recommendations passed down by the manufacturer should be rigidly adhered to. Taking the advice of the manufacture will greatly improve the service life of the selected battery.

Dry Cells

Dry cells are the batteries that we are most familiar with. Dry cells power most of our personal appliances like flashlights, calculators, cameras, and cell phones. The term "dry cell" can be a little deceiving. The electrolyte of these batteries are not really dry, rather it is a paste. Figure 3-14 shows an ordinary 1.5-volt dry cell battery. The positive terminal is a carbon rod. The negative terminal is a zinc container. The container has a liner made of blotting paper. The electrolyte is a paste of sal-ammoniac and manganese dioxide. The top of the battery has a plastic sealing cap and a cardboard cover protects the zinc container.

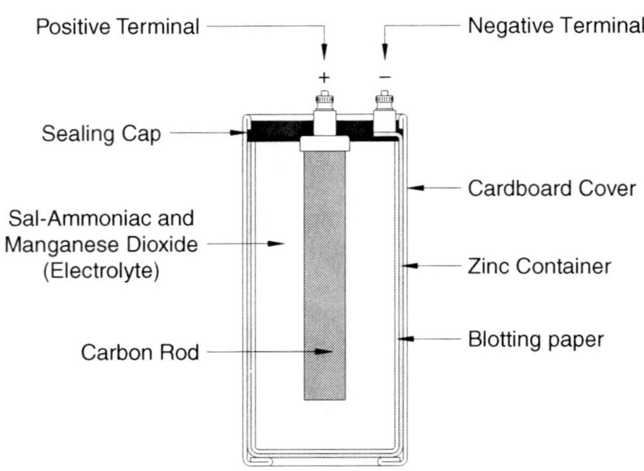

Figure 3-14 Dry Cell

There are a variety of dry cell formulations that are commonly in the market. These include carbon/zinc, alkaline, silver oxide, mercury, lithium, nickel/cadmium, and nickel/metal hydride. All of these different formulations have different applications. As with the lead/acid batteries, recommendations should be solicited from the battery manufacturer before designing a battery into a piece of equipment. The recommendations passed down by the manufacturer should be rigidly adhered to. The advice of the manufacturer is generally intended to improve the service life of the selected battery.

Figure 3-15 shows various dry cell battery types. AA, A, C, D, and PP3 are very common battery sizes and may be purchased in nearly every grocery and convenience store in the United States. Coin cells can be purchased in most electronic and drug stores. Standard and lantern cells can be purchased in most hardware stores. Sizes F, G, and J are less common and usually must be purchased from an industrial supply house.

Battery Packs

Since most dry cell batteries produce 1.5 volts, it is necessary to arrange them in series to produce higher voltages. Figure 3-16 shows how to connect four 1.5-volt cells to produce a 6-volt output. If a 12-volt output is required, the same arrangement would be used, except with eight batteries. 18 volts would use 12 batteries, 24 volts would use 16 batteries, and so on.

Building a high-voltage battery is rather simple. Figure 3-17 shows how eight G cells are shrink wrapped into a single

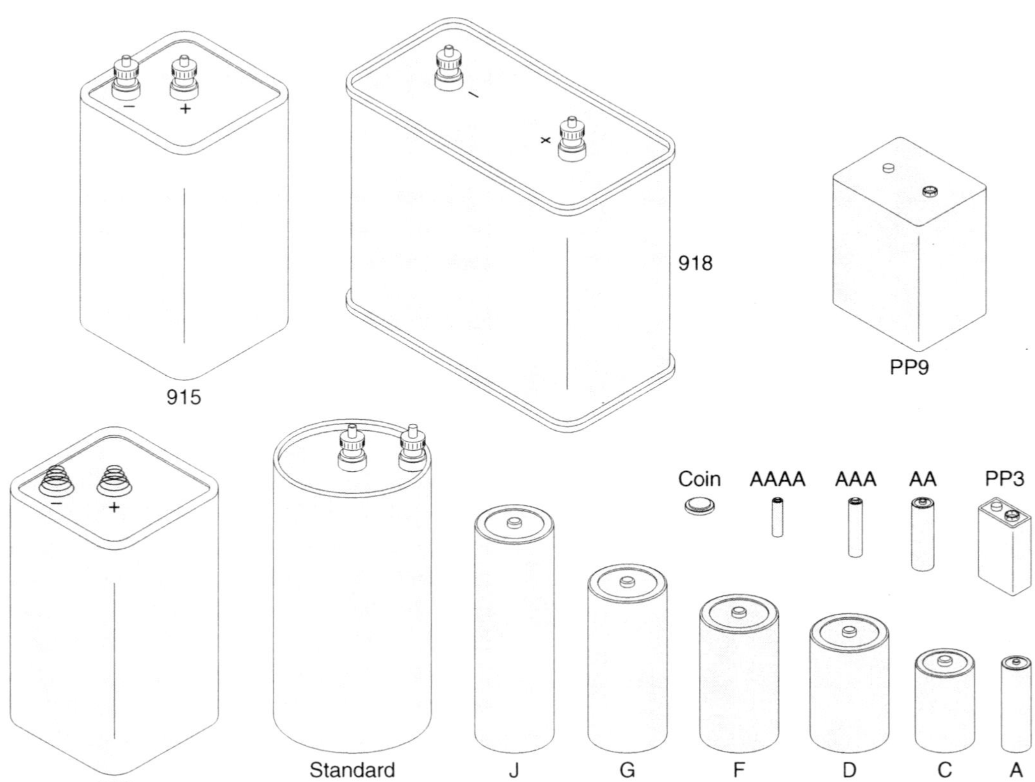

Figure 3-15 Various Dry Cell Batteries

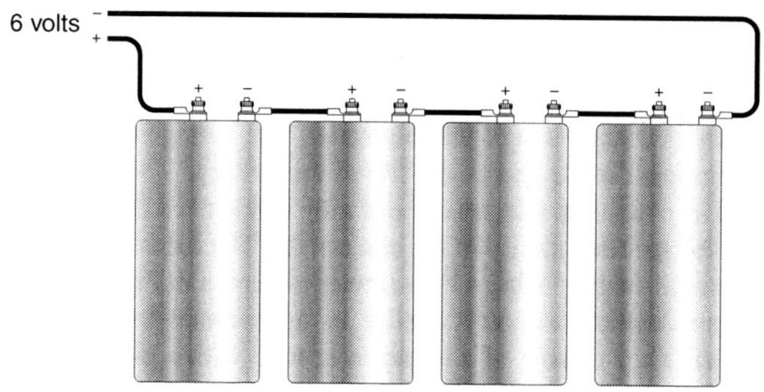

Figure 3-16 Series Connected Batteries

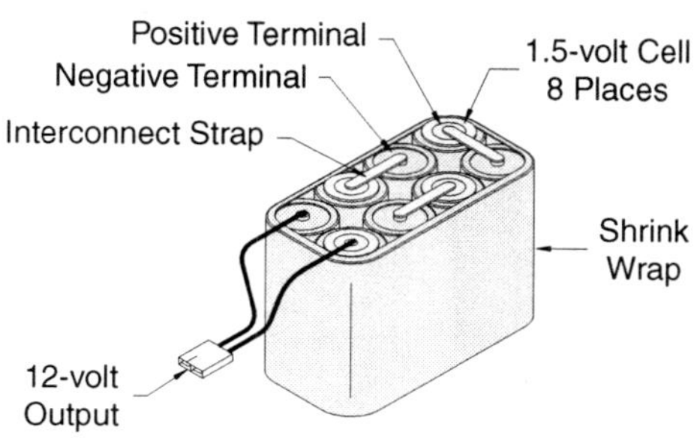

Figure 3-17 Shrink Wrap Battery Pack

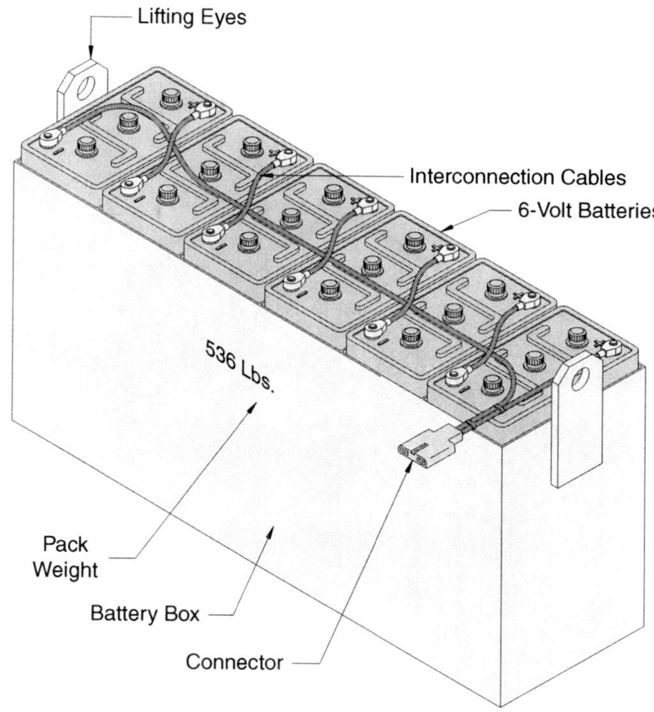

Figure 3-19 36-Volt Industrial Fork Truck Battery Pack

bundle. Interconnections are accomplished by soldering copper strips to the cell's terminals. The output is a standard modular connector that is soldered to the battery terminals as shown. Figure 3-18 shows an exploded view of the pack. Battery bundles like this are commonly found in larger motorized toys, uninterruptible power supplies, and test equipment.

Industrial automotive applications require very large battery packs that are capable of delivering high currents for extended periods of time. Fork trucks and floor cleaning equipment are the most common use for these battery packs.

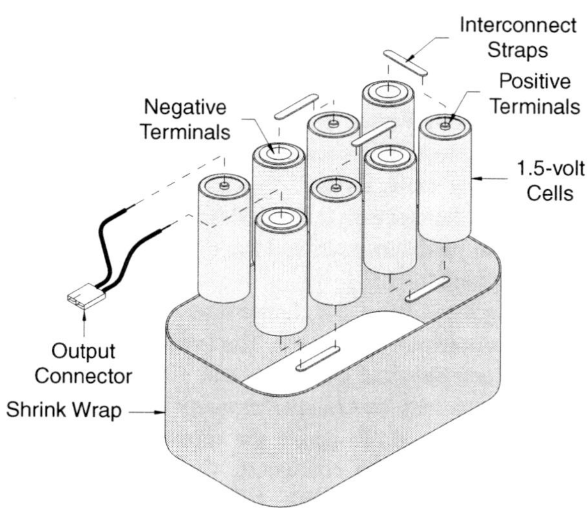

Figure 3-18 Shrink Wrap Battery Pack Exploded View

Figure 3-19 shows a 36-volt industrial fork truck battery pack. It is made from six 6-volt lead/acid batteries that are placed into a steel battery box. The batteries are interconnected with short cables carrying a terminal clamp on each end. The output of the pack is a high current industrial connector. Battery packs like the one shown can be very heavy and usually incorporate two lifting eyes on either end of the box. During operation, the pack is plugged into the electrical system of the truck. At the end of the shift the truck is parked in close proximity to a battery charger. The battery pack is unplugged from the fork truck and plugged into the charger for the night. In the morning, the battery is fully charged and the truck is placed back into service for the next shift.

For fixed applications, arrays of batteries are generally installed into some sort of framework, as shown in Figure 3-20. An array like this might be used as a back-up system for telephone or radio communications equipment. Take note of the continuous interconnect buss bar. These bars are generally not insulated and great care should be exercised when working around these arrangements. Even though most battery arrays are set up in a parallel arrangement and do not produce lethal voltages, their current capabilities can be extremely high. If an ordinary wrench is inadvertently dropped onto the top of the buss bars, these arrays can deliver enough current to literally vaporize the wrench and, probably, a sizable chunk of the buss bar. For arrays that are set up in series, great care should be exercised when working with these systems. Most people see a battery and don't consider it to be particularly dangerous. However, if two hundred 1.5-volt dry cells are set up in series, the array will produce 300 volts! This is more than enough voltage to be lethal.

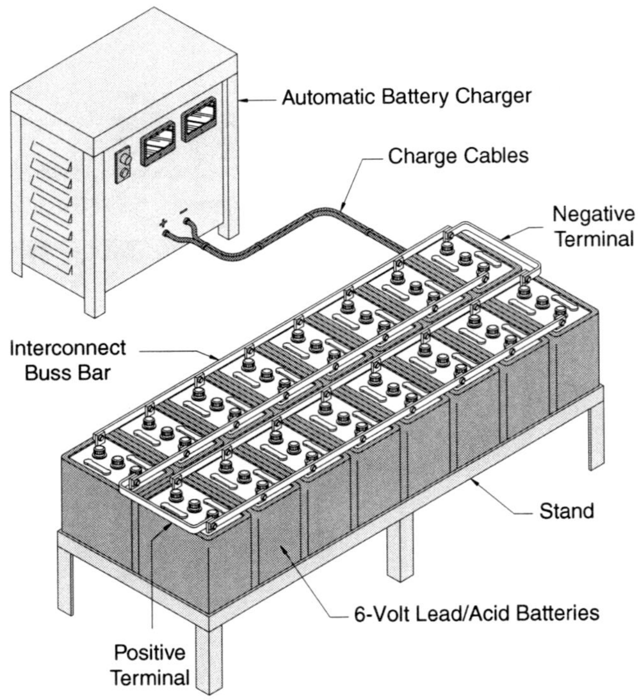

Figure 3-20 6-Volt Industrial Battery Bank with Automatic Charger

Testing Batteries

Being able to assess the charge and condition is important to the effective use of batteries. Most of us gauge the charge and condition of our batteries by noticing that the appliance that they power isn't working anymore. Time to change the batteries! This really isn't a bad method for small personal appliances such as flashlights, calculators, cameras, and cell phones. Unfortunately, most of us use this same method for our more important batteries. Discovering that your car battery has failed is a great way to top off an evening at the symphony.

For standard dry cells, the general condition can be assessed by simply measuring the voltage. Figure 3-21 shows a typical voltage measurement. Measuring the voltage will only give a general assessment of the condition of the battery. This is because the battery will produce its full voltage even when its charge is substantially spent. The voltage only starts to diminish at the end of the life of the battery.

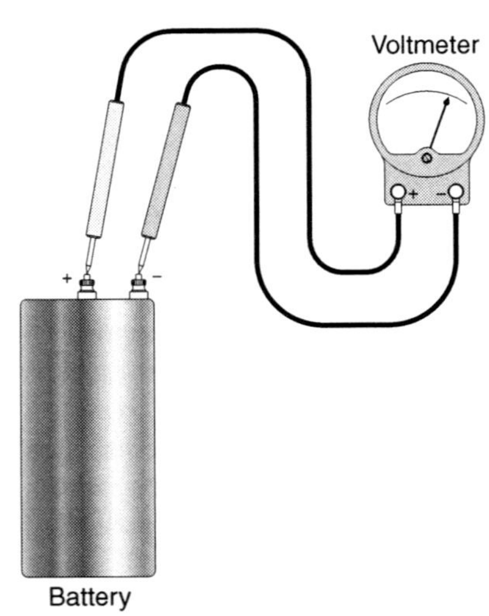

Figure 3-21 Testing Voltage

The charge on a lead/acid cell may be checked by measuring the specific gravity of the electrolyte. The measurement is accomplished by using a hydrometer as shown in Figure 3-22. These devices are very similar in appearance to a turkey baster. They consist of a transparent tube with a pickup tube on one end and a squeeze bulb is attached to the other end. Inside the transparent tube is a calibrated float. When electrolyte is drawn up into the transparent tube, the float adopts a level that corresponds to the specific gravity.

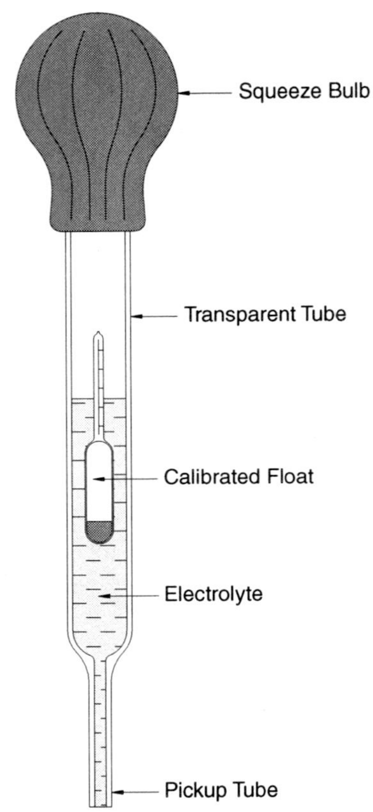

Figure 3-22 Hydrometer

The float itself, Figure 3-23, is a glass ampoule with a weight at the bottom. The neck of the ampoule is marked with a specific gravity scale. By reading the point on the scale that corresponds to the electrolyte level, the specific gravity of the electrolyte can be determined and the charge state of the battery can be gauged.

Measuring the voltage and charge state of a battery will not tell you the whole story, however. The internals of the battery may be in poor physical condition and although the charge state and voltage may be OK, the capacity of the battery may be greatly diminished. To gauge the capacity of a battery, a current load test must be conducted. Current load testing is only practical on rechargeable batteries because the test requires a substantial amount of current flow. This creates an

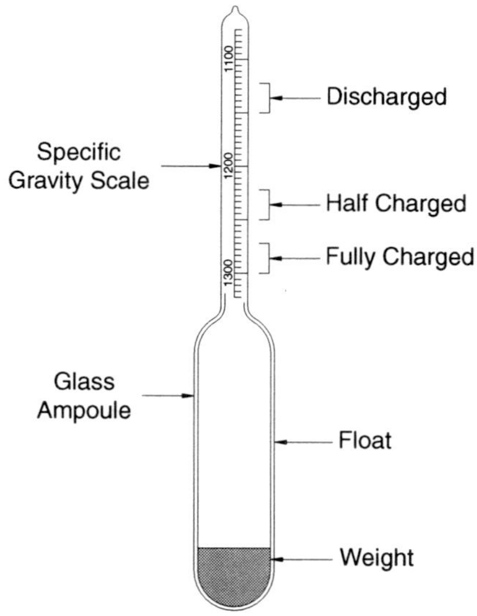

Figure 3-23 Hydrometer Float

unacceptable loss for non-rechargeable batteries such as standard dry cells. The test is most commonly carried out on lead/acid batteries and is a very common test for car batteries. Almost any auto parts store in the United States can conduct a current load test on a battery.

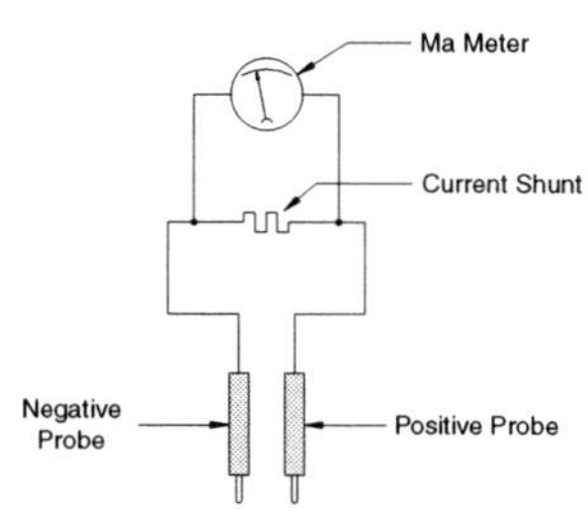

Figure 3-24 Current Load Battery Tester Schematic

Figure 3-24 shows a schematic representation of a current load tester. The unit consists of a current shunt and a meter. The shunt is actually a heating element that may glow red hot during the test. The meter measures the current flow across the heater. During the test, any given battery must maintain a specific current during a measured amount of time. If the current level drops below an acceptable lower limit during the test time the battery is deemed unusable and must be replaced.

Figure 3-25 shows a typical commercial current load tester. The units are typically portable so that they may be

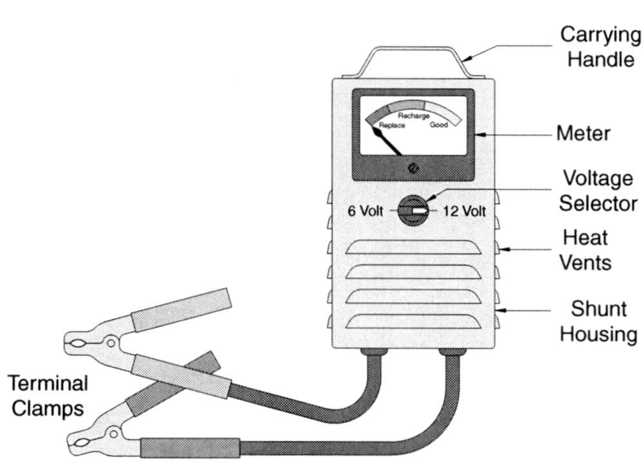

Figure 3-25 Current Load Battery Tester

used under the hood of a car or truck. They have two short cables with high current terminal clamps. The case of the tester is ventilated to release the heat that is generated during testing.

Battery Charging

Standard dry cells are not rechargeable and must be discarded after the charge is spent. However, there are a variety of rechargeable batteries. To recharge a battery, a charger must be used. For lead/acid batteries the charger is little more than a basic DC power supply. Figure 3-26 shows a schematic representation of a typical automobile battery charger. The charger has a step-down transformer which has a 14-volt output. The output is routed through a current limiting resistor (to prevent the transformer from overloading), an amp meter (to gauge the charge on the battery), and a diode (to change the AC output to DC). The input of the transformer is equipped with a voltage selector switch and a timer (to prevent overcharging).

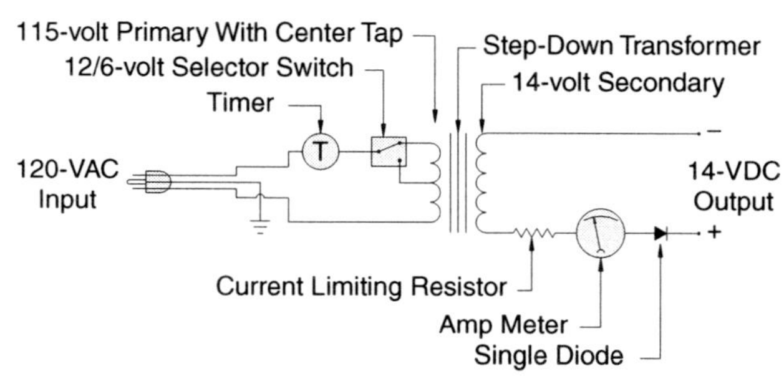

Figure 3-26 Automobile Battery Charger Schematic

Lead/acid batteries have very little internal resistance when they are discharged. Their internal resistance increases as the charge state increases. Therefore, the charge state of the battery can be gauged by monitoring the current drawn from the battery charger. When the battery stops drawing current, it is fully charged.

Figure 3-27 shows a commercial automobile battery charger. The voltage is selected (6 or 12 volts), terminal clamps are connected to the battery, the charger is plugged in, and the timer is set. The charge meter shows the relative charge on the battery. When the battery is fully charged the needle will be pointing to the green zone on the left hand side of the meter.

Electric fork trucks are dependent on recharging their batteries everyday of operation. Normally, the company operating the fork truck will have a service area which includes a charging station. At the end of the shift, the fork truck is parked in the service area and the battery pack is disconnected from the truck and connected to an automatic charger. Figure 3-28 shows a typical fork truck battery charger. The charger shown can recharge two trucks simultaneously.

A particularly common type of battery charger is the trickle charger. This type of charger is designed to produce a

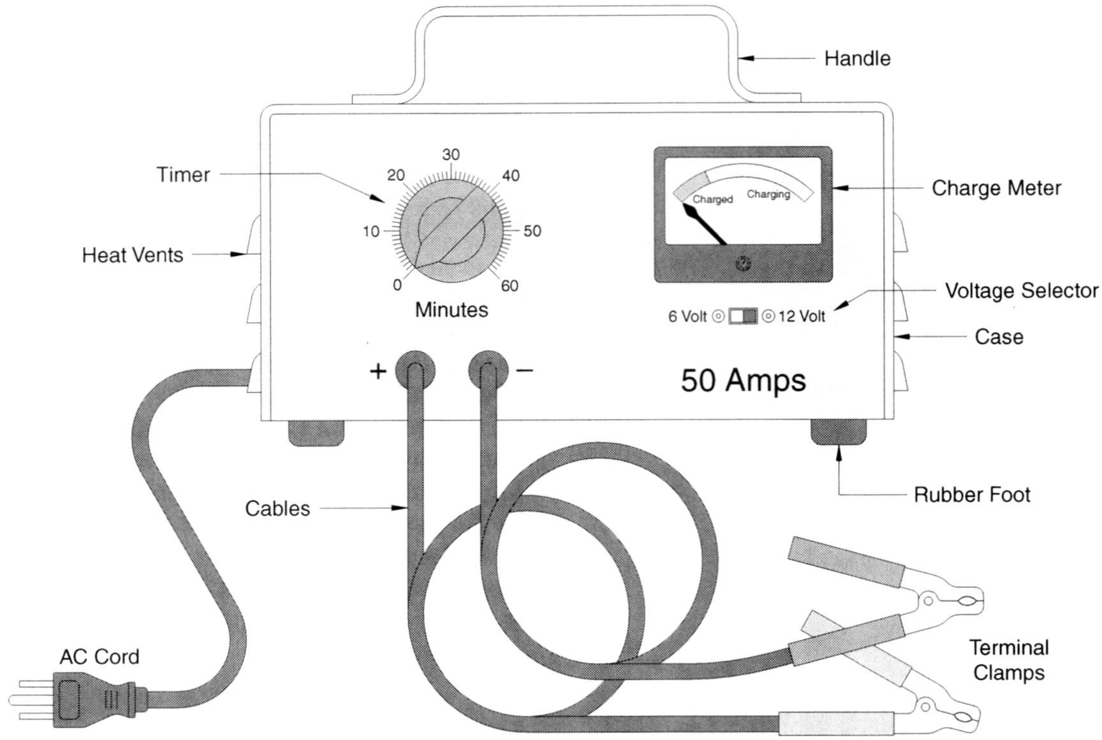

Figure 3-27 Commercial Automobile Battery Charger

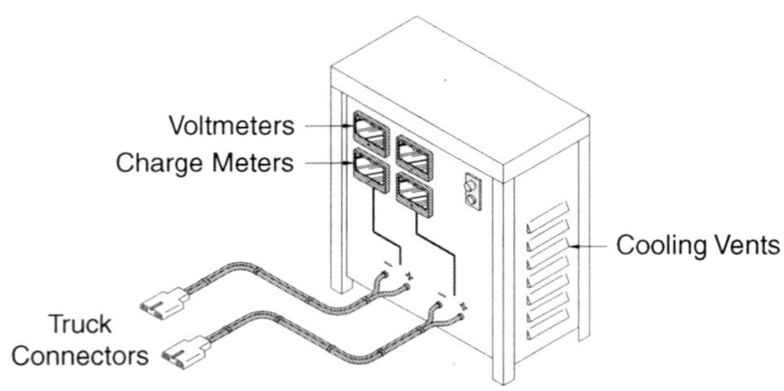

Figure 3-28 36-Volt, Dual-Output Fork Truck Battery Charger

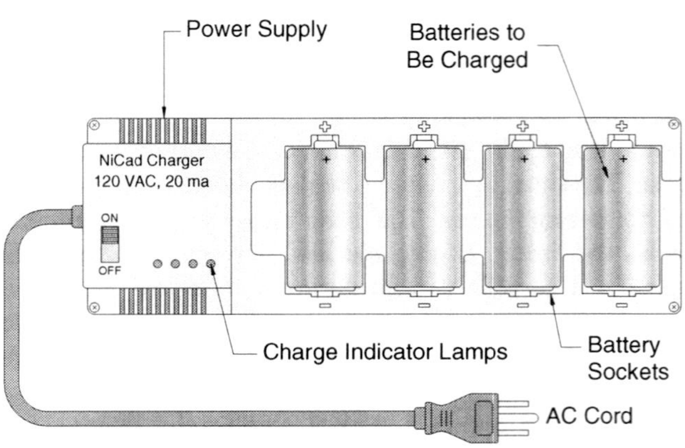

Figure 3-30 Commercial Trickle Charger

continuous charge rate low enough that it cannot damage the battery, regardless of how long the battery is charged. This charger consists of a small transformer with a current limiting resistor and diode, as shown in Figure 3-29.

Figure 3-30 shows a commercial trickle charger, which is usually used for recharging nickel/cadmium and nickel/metal

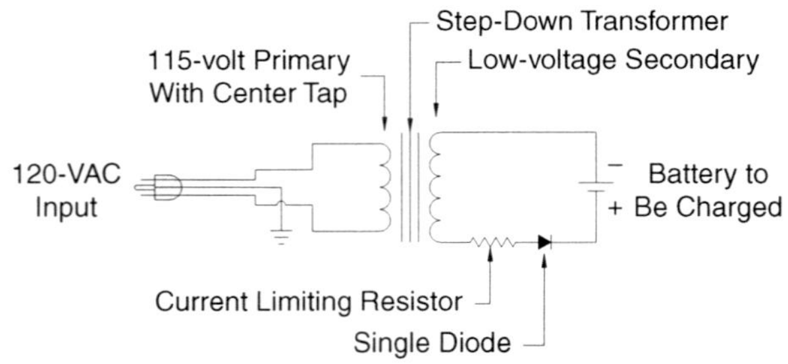

Figure 3-29 Trickle Charger Schematic

hydride batteries. The batteries to be recharged are placed into the sockets, the charger is plugged in to an AC receptacle and 6 to 14 hours later the batteries are at full charge. These types of chargers are very inexpensive and work rather well if the application allows the batteries to charge overnight.

For faster charging a slightly more sophisticated charger is required. A NiCad battery can be charged at a rather fast rate of 1.5 times it amp-hour rating. That is to say, if a NiCad battery has a rating of 500 mA-hours, then it can be brought up to full charge by a .75 amp charge rate for 1 hour. It is imperative that a NiCad battery is not over charged. Over charging will result in severe damage requiring replacement of the battery.

There are three methods commonly used for controlling the charge of a NiCad Battery. The first, and most common, is to monitor the voltage of the battery. When the battery achieves full voltage it is at maximum charge. The voltage of a NiCad battery will continuously rise until it reaches full charge and if the charging is continued the battery voltage will start to fall. Generally, a NiCad charger will incorporate an electronic monitoring circuit that watches the voltage rise of the battery. As long as the voltage continues to rise, the charge will continue. When the monitoring circuit detects a slight drop in voltage, the charging is stopped and the battery is at full charge.

The second, and less reliable method, is to monitor the temperature of the battery. NiCad batteries will remain cool until they reach their peak charge. At this point they will start to heat up. The temperature of the battery can be monitored and when there is a rise, the charging is stopped.

The third, and most straightforward method, is to charge the battery in reference to its amp-hour rating. In order to time charge a NiCad battery, it must be completely discharged first. Then 1.5 times its amp-hour rating can be applied for 1 hour and the battery will be up to full charge.

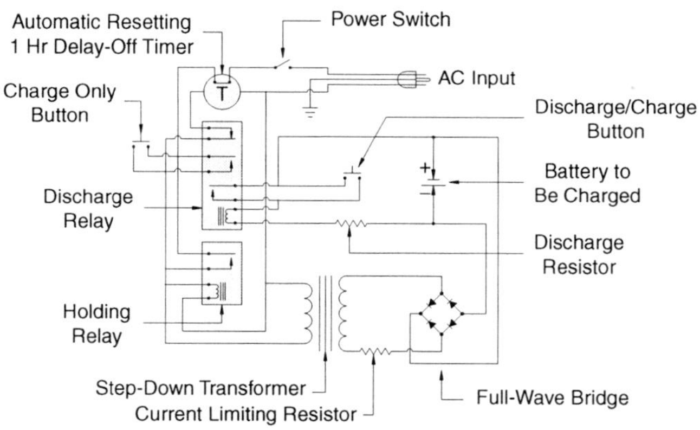

Figure 3-31 Discharge NiCad Battery Charger Schematic

Figure 3-31 shows a schematic representation of a timed NiCad battery charger. The system consists of a typical DC power supply that is controlled by two relays and a timer. The battery to be charged is placed into the socket and its remaining charge energizes the discharge relay. When the discharge relay energizes, it resets the timer, energizes the holding relay, and blocks power to the DC power supply. The coil of the discharge relay has a discharge resistor in series with the battery. This arrangement is intended to produce a controlled discharge rate on the battery. When the battery is fully discharged, it cannot provide enough power to the discharge relay. When the discharge relay resets, power is applied to the timer and DC power supply. The timer allows the power supply to operate for 1 hour, fully charging the battery. If a battery is fully discharged before it is placed into the charger, it will not energize the discharge relay. In these cases, the "charge only" button is pressed and the charge cycle initiates.

It should be noted that NiCad batteries will greatly benefit from a proper charging program. NiCad batteries should always be stored fully charged. They should be completely discharged and fully recharged on a periodic basis. They should never be subjected to over charging and under charging should be avoided. If a NiCad battery is properly handled, it will provide many years of excellent service. Inversely, if the batteries are mishandled, their useful service life will be dramatically shortened.

Nickel/metal hydride or NiHd batteries have become the rechargeable battery of choice for most rechargeable applications. This battery formulation provides some significant advantages over NiCad batteries, most noteworthy, an improved capacity. Charging these batteries is principally the same as for NiCad batteries, except complete discharge should be avoided with NiHd batteries. Charge rates vary with NiHd batteries but usually match mAh rating. The charge of battery is assessed by measuring voltage. If a NiHd battery is fully discharged, it should be placed on a trickle charger for a period of time specified by the manufacturer before receiving a higher charge rate. In any case, the batteries may be fully charged in 6 to 14 hours with a typical trickle charger.

Battery Holders

When using batteries it is necessary to mount them in a convenient and secure manner that is appropriate for the application. Commercial holders are available for most standard batteries. Figure 3-32 shows an inexpensive sheet metal holder for an ordinary "D" cell. Figure 3-33 shows a holder that arranges two batteries in series to produce a 3-volt output. Figure 3-34 shows a holder for four batteries to produce a 6-volt output.

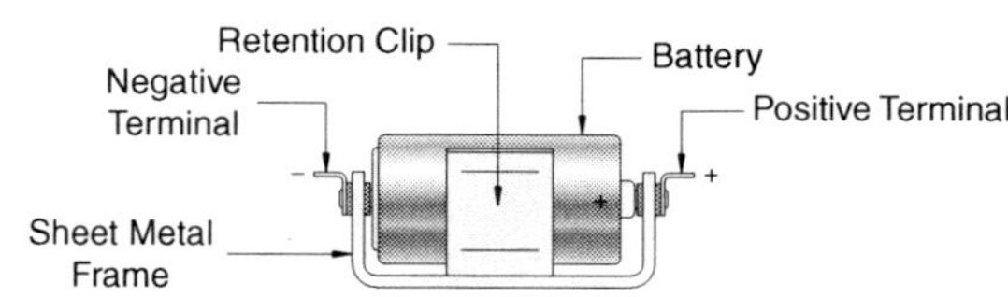

Figure 3-32 Battery Holder

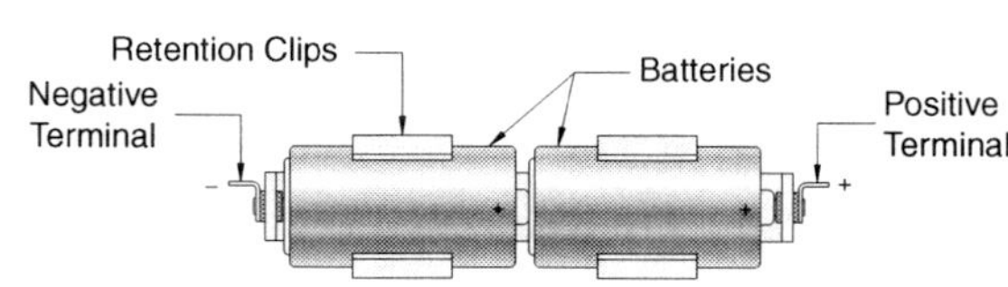

Figure 3-33 Series Battery Holder for 3-Volt Output

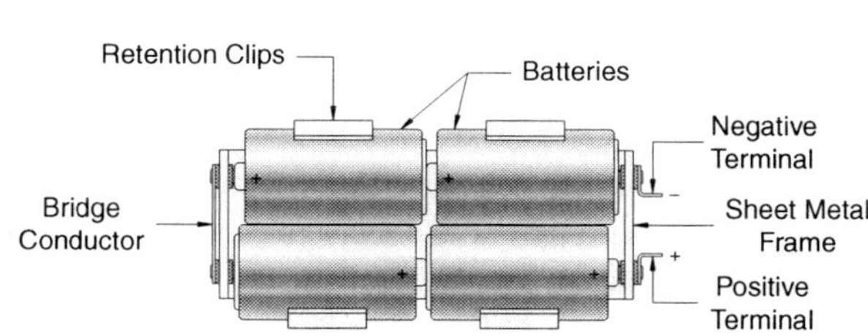

Figure 3-34 Series Battery Holder for 6-Volt Output

Automotive, marine, and deep-cycle batteries typically have a pair of slots or lips at the base of the battery for mounting purposes. Figure 3-35 shows a typical automotive style mounting tray for a lead/acid battery. These types of mounts are very secure and are appropriate for several applications. For lead/acid batteries that do not have clamping provisions a clamp frame as shown in Figure 3-36 is typically used. This mount generally consists of a battery tray and an upper frame. The frame is secured with two or four clamp bolts that run down to tabs on the tray. It should be noted that the clamp frame used in this type of mount is generally constructed of plated steel. Great care should be exercised to avoid shorting the battery terminals when installing the frame.

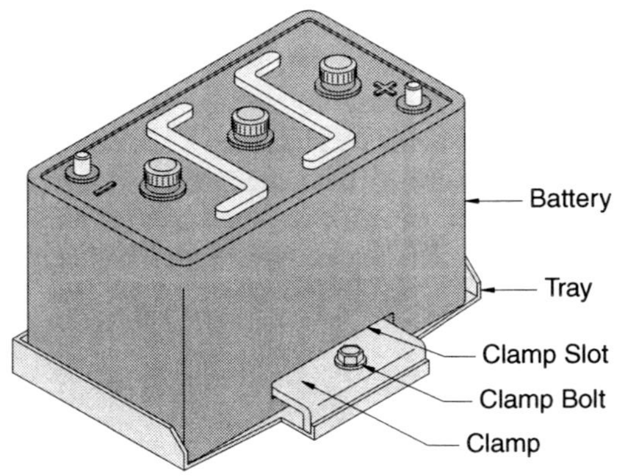

Figure 3-35 Lead/Acid Battery Mount

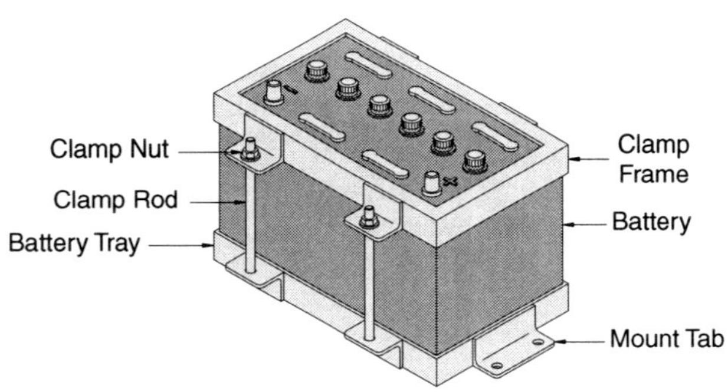

Figure 3-36 Clamp Frame Battery Mount

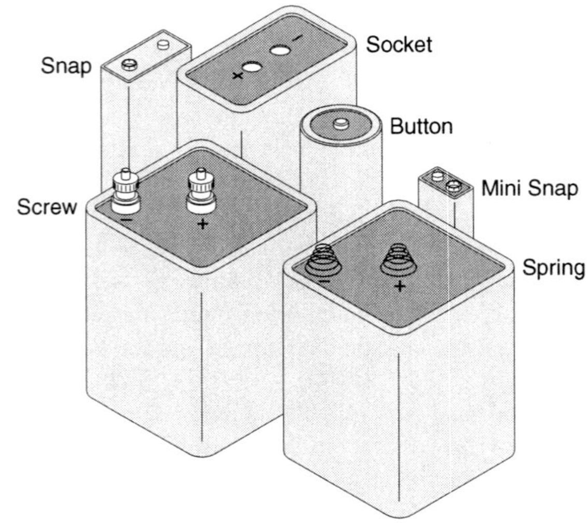

Figure 3-37 Common Dry Cell Battery Terminals

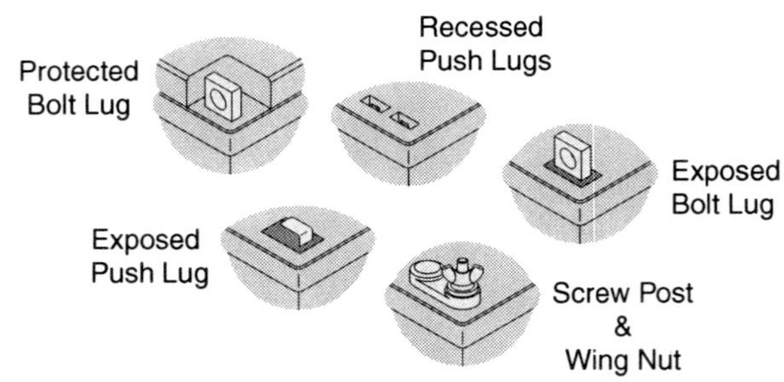

Figure 3-38 Common Lead/acid Battery Terminals

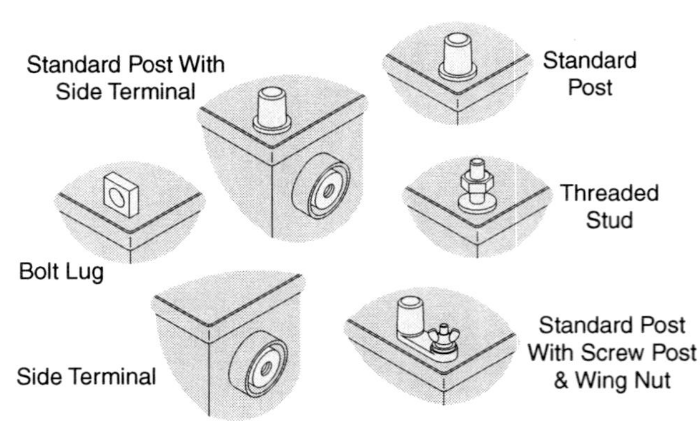

Figure 3-39 Automotive, Marine, and Deep-Cycle Terminals

Battery Terminals

There are nearly as many different types of battery terminals as there are battery types. Terminals can generally be grouped into three different categories, dry cells; sealed lead/acid batteries; and automotive, marine, and deep-cycle batteries. Figure 3-37 shows common terminals that are found on standard dry cells. These range from screw terminals to spring-loaded terminals. Figure 3-38 shows terminals that are generally found on sealed lead/acid batteries. These types of batteries are commonly found in computer and test equipment. Figure 3-39 shows terminals that are usually used on automotive, marine, and deep-cycle batteries. In this class of battery, the terminals are generally designed to carry high current discharge rates.

Battery Connectors

Most batteries will either accept standard electrical connectors or provide a screw terminal for connecting bare wire. Automotive, marine, and deep-cycle batteries are generally placed into high current service, making standard connectors inadequate. These batteries have either top posts or side terminals and require special connectors for interconnection. Figure 3-40 shows a selection of some of the more common

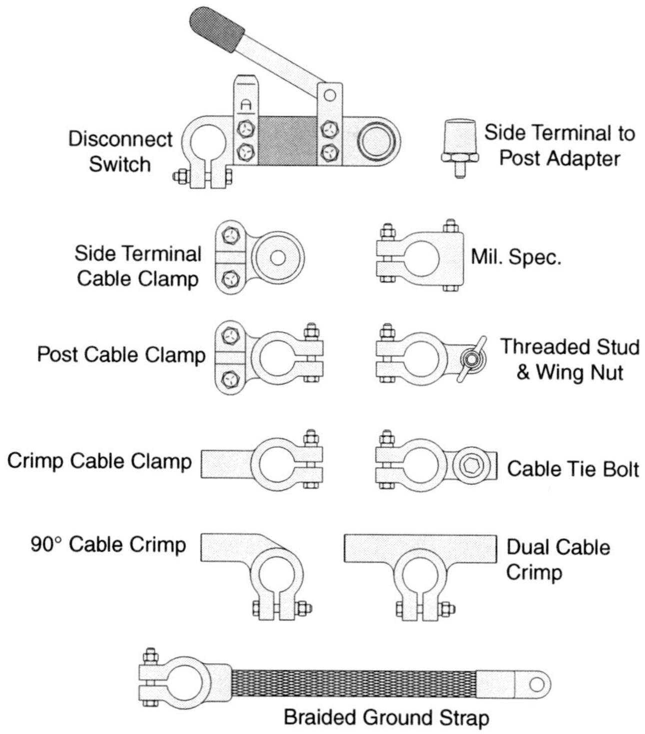

Figure 3-40 Automotive, Marine, and Deep-Cycle Battery Connectors

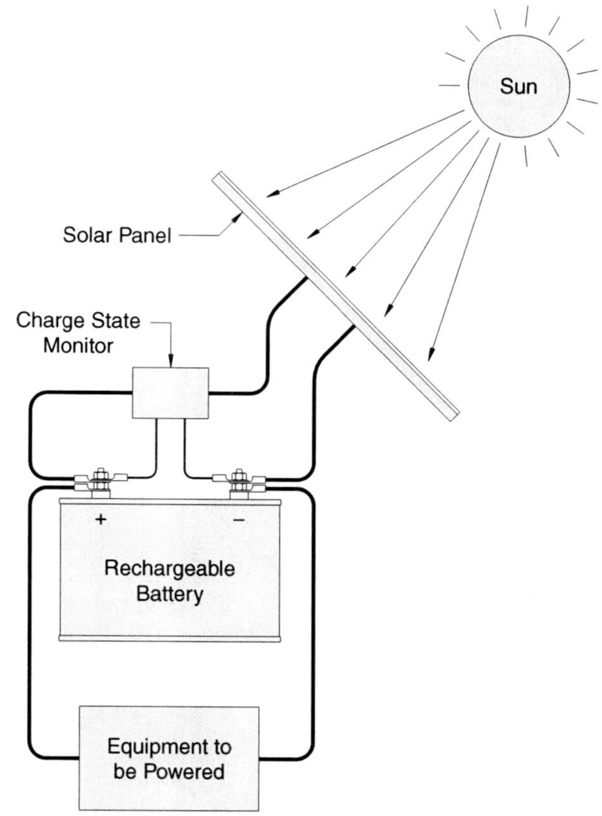

Figure 3-42 Solar Powered Trickle Charger

connectors available. The battery disconnect switch is particularly handy on equipment that requires regular maintenance. The braided ground strap is the classic method of grounding these types of batteries.

Solar Cells

Solar cells have gained a real niche within certain segments of industry. Although they have never been able to fulfill the promise of free electrical power for the home, they have provided us with an excellent charging system for remotely located equipments.

Figure 3-41 shows a typical solar panel. These panels are generally made up of many individual solar cells connected in series and/or parallel so that their voltage output is matched to the application. The undivided cells are normally mounted on some sort of backing plate. The plate is typically mounted into a frame which is appropriate for the environment where the panel will be installed.

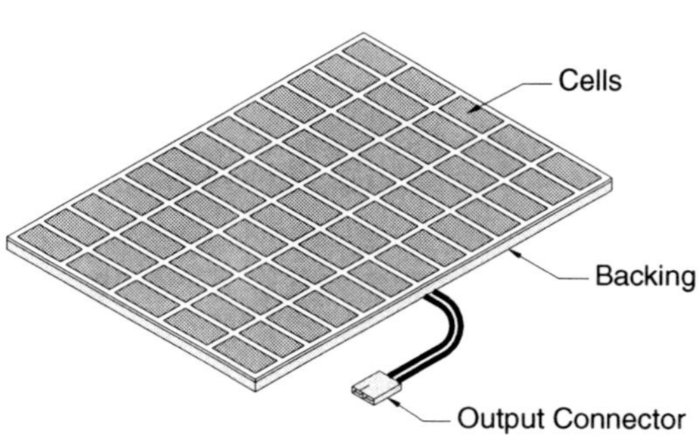

Figure 3-41 Solar Cell Panel

Figure 3-42 shows how a solar panel is used as a charger to replenish field batteries. The panel is configured to produce a voltage and charge rate that will bring the battery up to full charge with just a few hours of sunlight. The charge circuit includes a charge state monitor and electronic switch. When the battery is at full charge, the solar cell is disconnected. When the battery is at a reduced charge, the solar panel is connected. In this manner the battery is constantly maintained during the day, and even on overcast days. During the night the battery carries the entire load of the equipment.

An excellent example of this type of charger is a solar powered calculator, as shown in Figure 3-43. The solar cell maintains the battery almost indefinitely. If the calculator battery goes dead, just leave the unit out in the sun or under a lamp for a few minutes and you're back online.

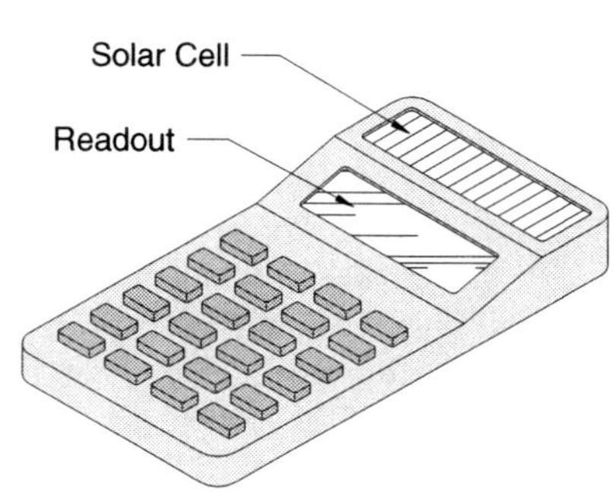

Figure 3-43 Solar Powered Pocket Calculator

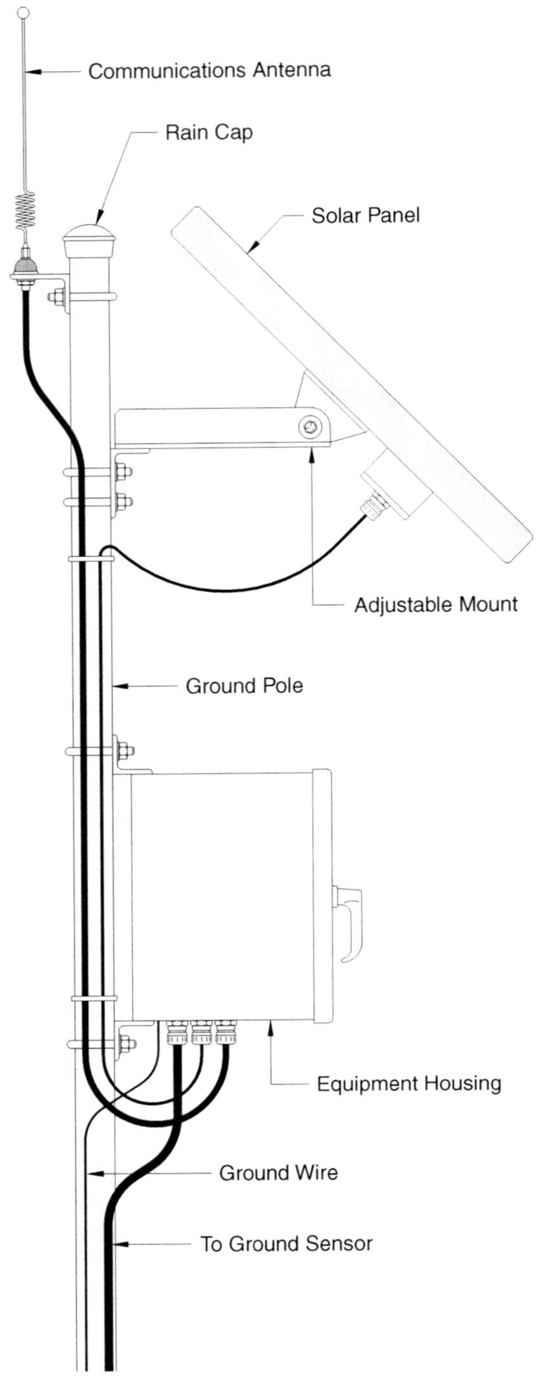

Figure 3-44 Solar Powered Monitoring Station

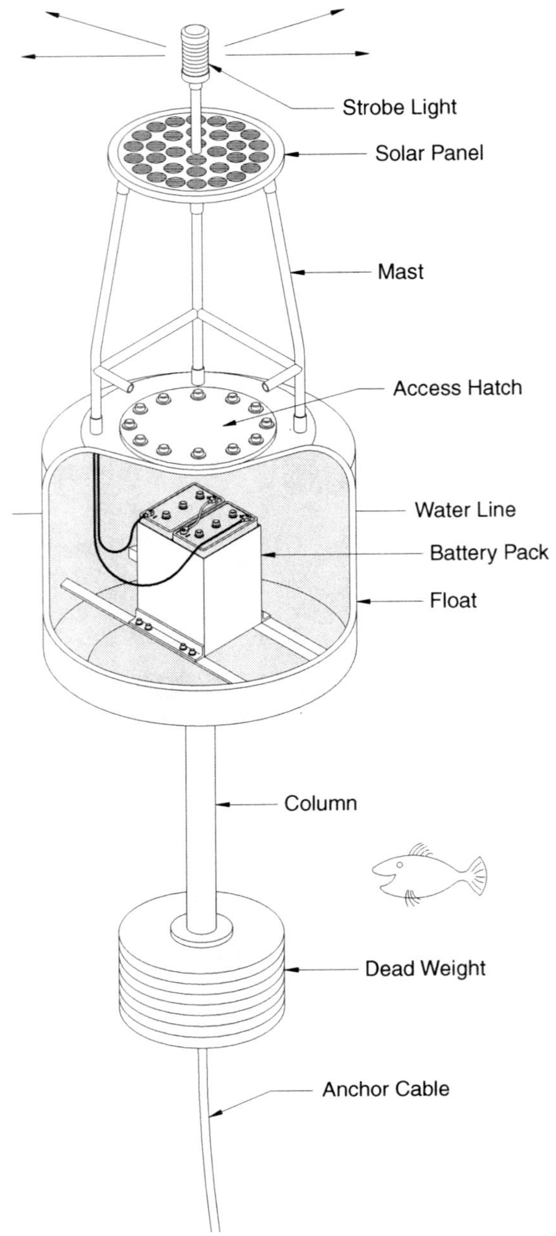

Figure 3-45 Flashing Marine Buoy with Solar Powered Battery Charger

Figure 3-44 shows a ground water monitoring station. These types of systems are set up to monitor water-and air-borne contaminates, flood waters, temperatures, weather data, seismic activity, and so forth. The system generally consists of a metal pole set in concrete on which all of the equipment is mounted. The equipment housing will typically contain the instruments, communication gear, charge monitor, and the battery used to power them. The solar panel and communication antenna are mounted at the top of the pole.

Solar charging systems have also found favor in marine buoy applications. Figure 3-45 shows a sectional view of a flashing marine buoy with a solar powered charger. The batteries are located in the buoy and the support equipment is located on the bottom of the solar panel.

Direct Current Power Supplies

For DC applications that have easy access to AC power, DC power supplies are the preferred power source. Direct Current power supplies can be as simple as a trickle charger or may be extremely complex, computer controlled units that are designed to supply specialized equipment.

Figure 3-46 shows a half-wave DC power supply. This is probably the simplest power supply design that is in common

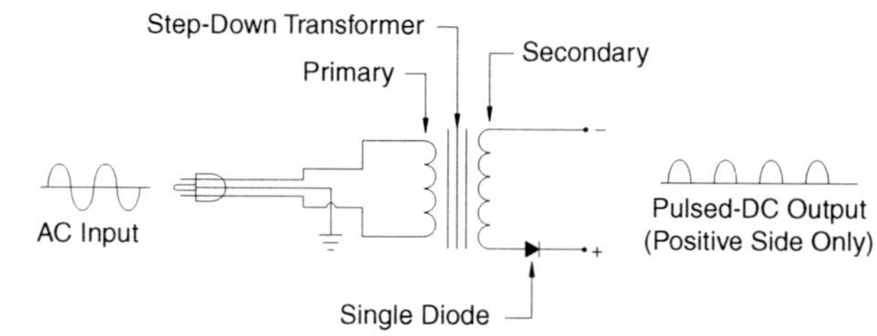

Figure 3-46 Half-Wave DC Power Supply Schematic

use. These supplies consist of a step down transformer and a single diode on one side of the secondary. The step-down transformer has a 120 volt primary and the secondary voltage is matched to the application. The diode acts as an electrical one-way valve and allows only the positive output through. Since the diode allows only the positive side of the wave to pass through, the output is referred to as pulsed DC. The illustration to the right shows a graphic representation of the output.

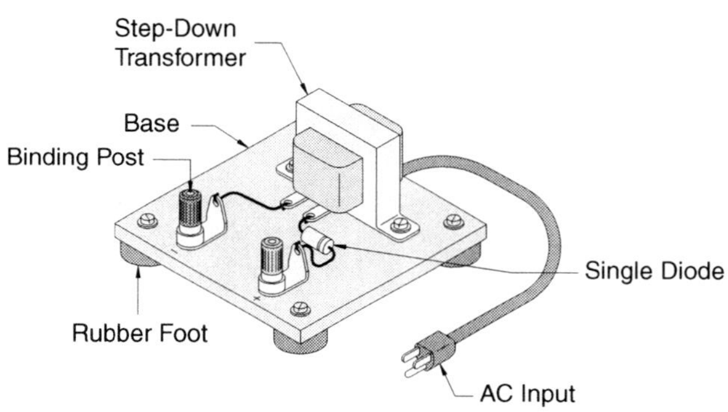

Figure 3-47 Half-Wave DC Power Supply Chassis

Figure 3-47 shows a view of a complete half-wave power supply chassis. Take note of just how simple this construction is.

The next progression in DC power supplies is the full-wave supply. A schematic representation of a full-wave DC power supply is shown in Figure 3-48. In this configuration four diodes are arranged so that they direct the positive and negative sides of the AC to either the negative or positive output terminals. In this manner the full output of the transformer is utilized. To improve the output, these power supplies usually have a filter capacitor that bridges the DC terminals. The

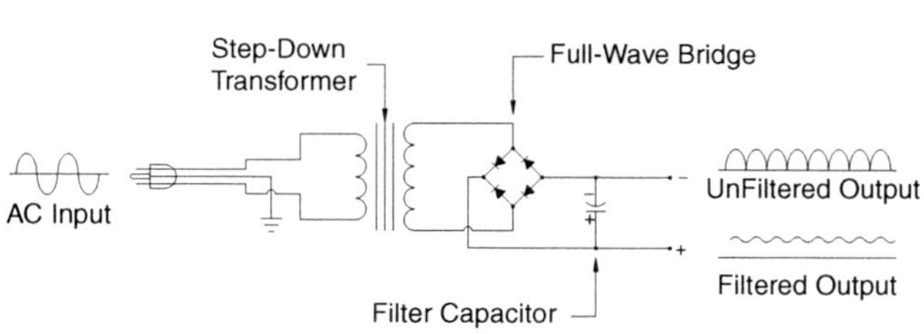

Figure 3-48 Full-Wave DC Power Supply Schematic

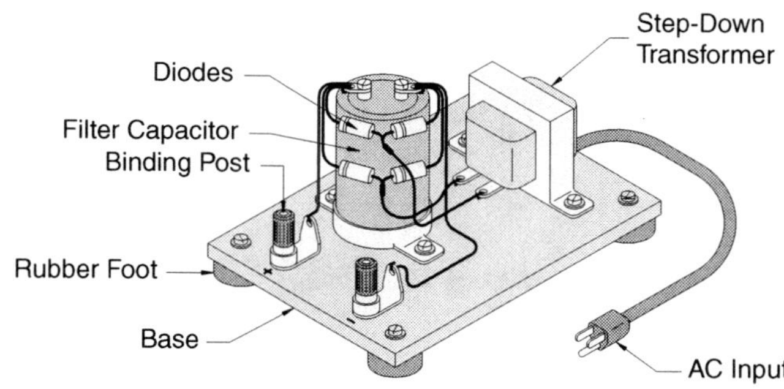

Figure 3-49 Full-Wave DC Power Supply Chassis

filter capacitor has the effect of smoothing the DC output. The illustration to the right shows a graphic representation of an unfiltered output and a filtered output.

Figure 3-49 shows a view of a complete full-wave power supply chassis. Notice the relative size of the filter capacitor. Generally speaking, the capacitor must have a substantial capacity to have the desired filtering effect.

For bench and test applications, a variable output DC power supply can be very useful. A variable output power supply is simply a full-wave unit with a variable autotransformer on the AC input. An autotransformer is a device that allows its output voltage to be adjusted by turning a knob. Figure 3-50 shows a schematic representation of a variable output DC power supply. The supply also has an output meter and is protected by fuses. The fuses are very important on a bench supply because the chance of inadvertently shorting the supply is greater in this environment. Without the fuses to protect the circuit, the supply could be severely damaged if shorted. The meter is a convenience to aid in adjusting the output voltage.

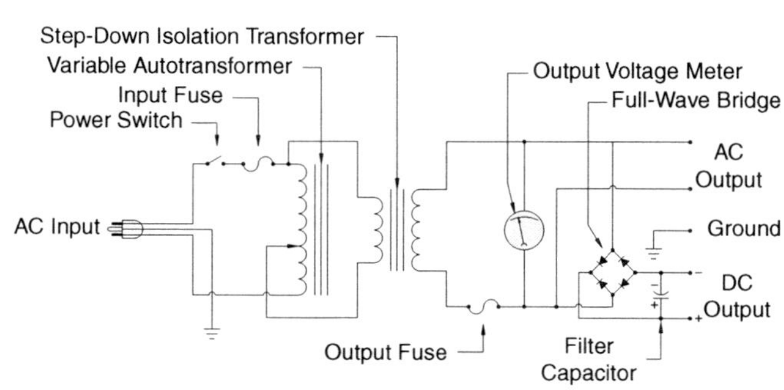

Figure 3-50 DC Power Supply with Variable Output

Figure 3-51 shows a view of a complete variable output DC power supply chassis. Take note that there are several common components with the half- and full-wave power supplies. It is clear, however, that the complexity of the supply is considerably higher than previous examples.

For clarity, Figure 3-52 shows an exploded view of the major components that make up the variable supply.

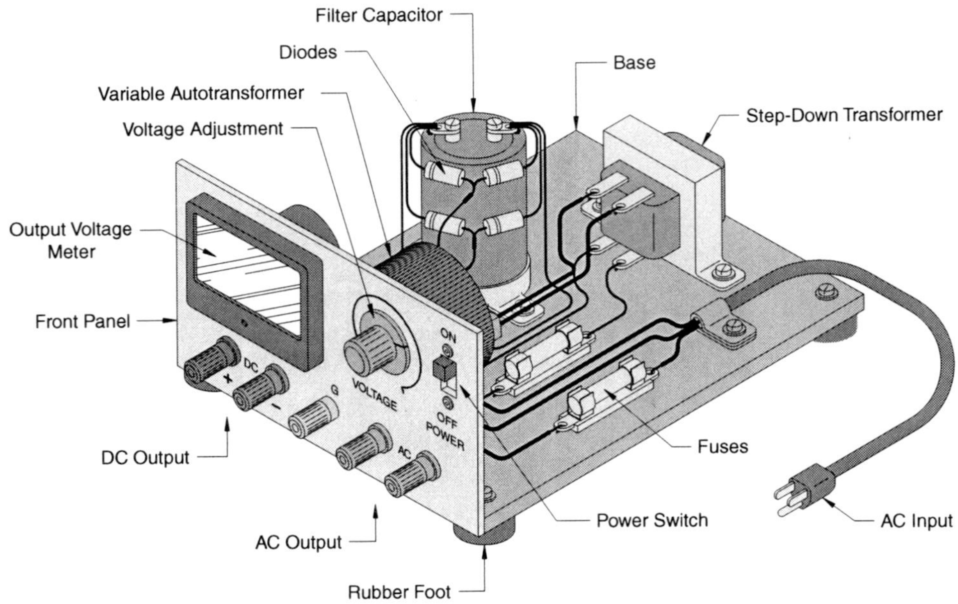

Figure 3-51 Variable Output Full-Wave AC/DC Power Supply Chassis

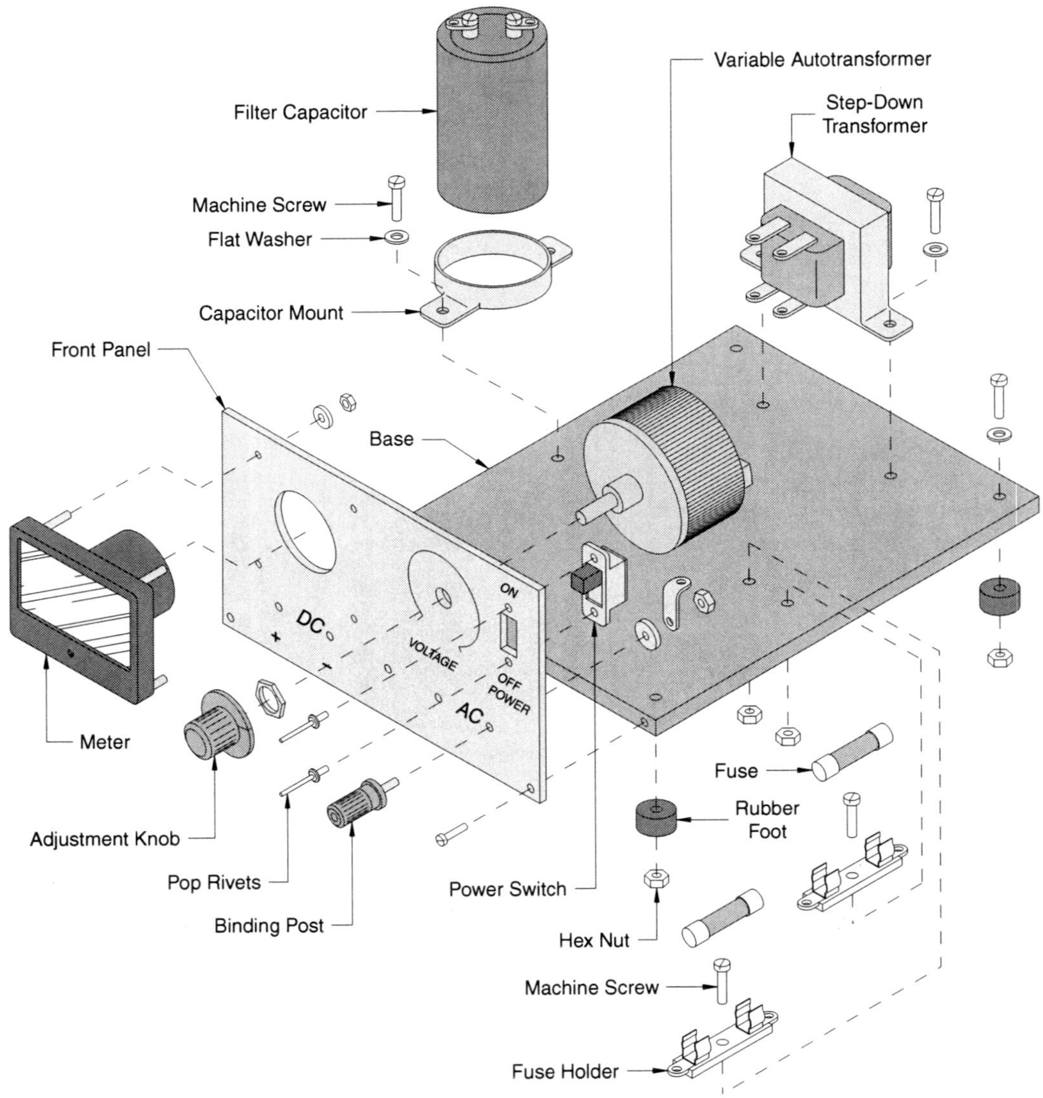

Figure 3-52 Power Supply Chassis Exploded View

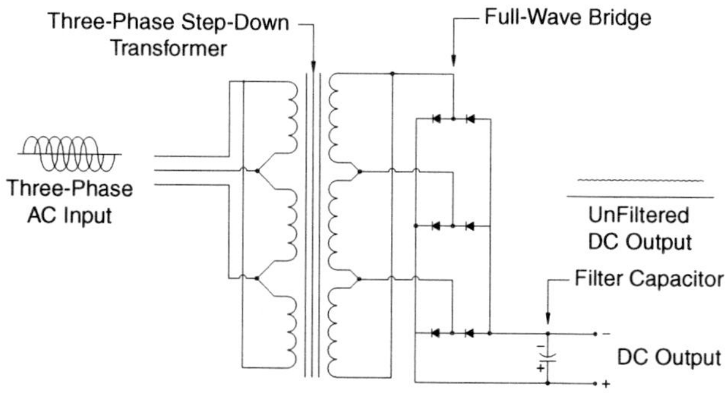

Figure 3-53 Three-Phase DC Power Supply Schematic

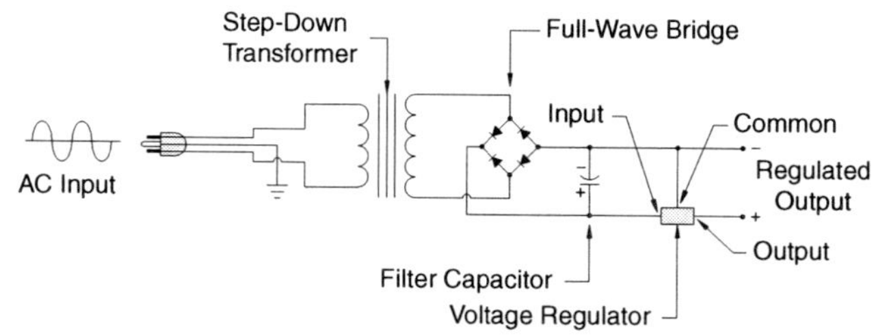

Figure 3-55 Regulated DC Power Supply Schematic

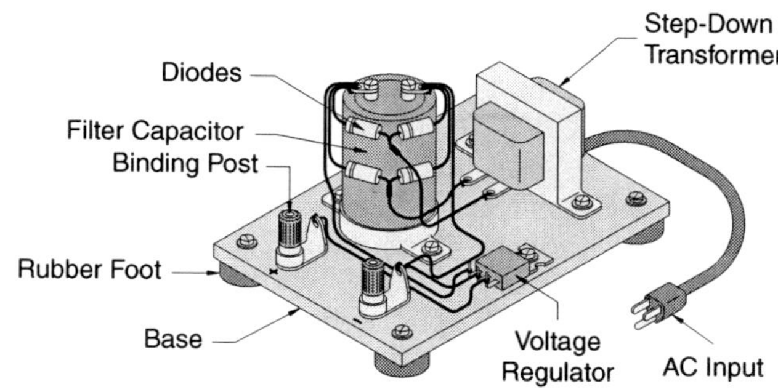

Figure 3-56 Regulated DC Power Supply Chassis

Figure 3-53 shows a schematic representation of a full-wave DC power supply that uses a three-phase input. Since a three-phase transformer has three output wires, six diodes are required, two per wire (one for positive and one for negative). The advantage of a three-phase power supply is that the unfiltered output is substantially smoother than a single-phase supply. This means that the filter capacitor can be much smaller for the same level of filtering, or that a larger capacitor will provide greatly improved filtering.

For many applications a simple capacitor cannot provide enough filtering. This is especially true for sensitive electronics, such as audio equipment and computers. This type of equipment usually requires a regulated power supply. Regulated power supplies provide power that is extremely precise. The heart of the regulated power supply is the voltage regulator. Figure 3-54 shows a typical voltage regulator. These devices generally have three terminals: input, output, and common. The regulator is selected based on the output voltage that is desired and is simply added to the output of a full-wave power supply.

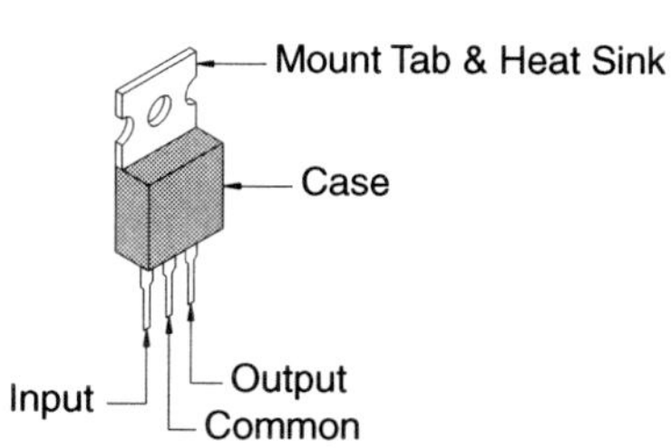

Figure 3-54 Voltage Regulator

Figure 3-55 shows a schematic representation of a regulated DC power supply. In this case, the regulator is placed between the filter capacitor and the positive output terminal. The common is connected to the negative terminal.

Figure 3-56 shows a view of a complete, regulated DC power supply chassis. Take note that the chassis is the same as the full-wave supply shown in Figure 3-49 except that a voltage regulator has been added.

Selecting Power Supply Components

When selecting different components for your DC power supply, it is first important to have a clear understanding of the application. What's the required voltage? How much current is necessary? What types of connections are needed? Where will it be mounted? Are there safety considerations? All of these questions and more must be answered before designing or building a power supply.

More often than not, available components dictate a great deal of a power supply's design. It does no good specifying a special transformer, if you can't get one. So let's take a look at what we really need to know.

The transformer is usually the most significant component of a power supply. Transformers usually have three different specifications that are of interest to the power supply designer, the primary voltage, secondary voltage, and the secondary current. Select the primary voltage to conform to the AC supply voltage, usually 120 volts and less frequently 240 volts. The output voltage should match the application. For a car battery charger, for instance, the output voltage should be 13 to 14 volts. The output current for an automotive trickle charger might be 2 amps.

Let's consider the bench supply shown in Figure 3-47. Since it's a bench supply, the transformer should have a 120-volt primary. We will want a 0- to 50-volt output, so the secondary should produce 50 volts. We also want a 5-amp output, so the secondary should be capable of producing about 10% higher current, or about 5.5 amps. The higher current is intended to provide a margin of safety to the finished supply.

Next we must consider the diodes to be used. If diodes are subjected to voltages higher than their maximum rating, they will fail. The voltage rating on a diode is given in peak inverted volts (PIV). Select diodes with a PIV rating of at least five times higher than the output voltage. This is intended to protect the diodes from high transient voltages that may be encountered while switching any equipment connected to the supply. The current rating should also be considered when selecting the diodes. The +10% rule is also applicable for this selection.

If we go back to the bench supply, we should select diodes with 400 PIV and 5.5 amp ratings. The higher PIV rating is because this is a bench supply and, at times, it may be subjected to very high transient voltages.

The filter capacitor size is based on the level of filtration desired. There are two parameters that are of interest to the power supply designer, the capacity, given in microfarads or μf, and the maximum voltage rating. As with diodes, overvolting a capacitor will cause severe damage. If enough current is available at the moment of over-voltage, some capacitors can actually burst because of excess internal temperatures. Direct current electrolytic capacitors are generally selected for the primary filter on any DC power supply. The filtration should be as high as possible. For most power supplies, the principal limit is the physical size of the capacitor. When selecting the filter capacitor, the highest μf rating that will fit into the physical limits should be selected.

In the case of the bench supply, a capacitor with a voltage rating between 150 and 250 volts at 2000 to 5000 μf should be selected. A capacitor of this size can be difficult to find, and when found can be rather cost prohibitive. It should be noted, however, that the surplus marketplace offers a wide selection of these types of capacitors at rock bottom prices.

If the filtration requirement is high enough to warrant a voltage regulator, then a device should be selected with the desired output voltage and current capabilities. It should also be noted that voltage regulators will have an input voltage range and a transformer should be selected that falls roughly in the middle of that range.

For a power supplies that are permanently installed into a piece of equipment, the output terminals will most likely be a soldered connection or some type of modular connector.

For our bench supply it is best to select a standard binding post for the terminals. The type of terminals that are most often used on bench supplies are combination banana plug and screw type binding post with an insulated thumb nut. These posts are generally supplied with insulating washers so that they may be installed into a metal panel.

Oftentimes power supplies do not even have their own chassis. They are commonly built into the chassis of the equipment that the supply serves. This technique provides a great deal of flexibility when selecting and arranging components.

Our bench supply uses a rather simple chassis. The base is a piece of $^1/_2$ inch thick, paper-based phenolic. The front panel is a piece of $^1/_8$ inch aluminum, which is attached to the base with two machine screws. The base has four ordinary rubber feet attached to the four corners. Before mounting, take a moment to lay out all of the components in a logical manner. Be sure to provide ample space for wiring and soldering.

For internal power supplies, the AC connection is generally a solder joint or modular connector. Desktop computers use a standard connector into which an inexpensive AC cord can be plugged. Be certain that the selected cord is large enough to carry the current that is required.

The bench supply has a 250-watt output and this translates to 2.1 amps at 120 volts. Almost any AC cord will carry this current. A more important consideration is the rough service that the cord will probably receive, so a heavy, durable AC cord with a molded plug is recommended.

Protecting a power supply from overloading is always a good idea. In the case of a nonregulated supply, a fuse is generally placed on the primary winding of the transformer. Transformers are generally hardy pieces of equipment, so a slow-blow fuse, with a current rating no higher than the maximum current draw of the transformer, is usually recommended. Voltage regulated supplies should also have a fast-blow fuse on the output of the regulator.

For our bench supply we use two fuses, one on the AC input and one on the AC output. The input fuse should be a 2 amp slow-blow fuse and the output should be a 5 amp fast-blow fuse.

Our bench supply also uses a variable autotransformer to adjust the output voltage. Most variable autotransformers have an output range from 0 volts to the line voltage, in this case 0 to 120 volts. Alternating current line voltage is connected to the input of the autotransformer and the output is connected to the input of the power supply's transformer. Using the autotransformer to vary the line voltage has a direct effect on the supply's output voltage.

The last significant item to consider is the panel meter. This item should be a 0- to 50-volt AC meter. The most significant item to consider when selecting a meter movement is how easy it is to read. It should have bold numbers and graduations and an easy-to-see pointer. It should also be large enough to comfortably read during operation.

Ulnterruptible Power Supplies (UPS)

One other type of power supply that warrants a brief discussion is the uninterruptible power supply (UPS). These supplies have exploded onto the market in the past decade and are primarily intended to provide emergency back-up power for computer systems. Figure 3-57 shows a block diagram of a typical UPS. The AC is passed through a transient suppressor to provide basic surge protection. The output of the suppressor is routed to a full-wave bridge, filter capacitor, and voltage regulator, which all act as a battery charger. The output of the suppressor is also routed to an electronic switch and utility outputs. During normal operation, the utility outlets receive power and the electronic switch directs AC power

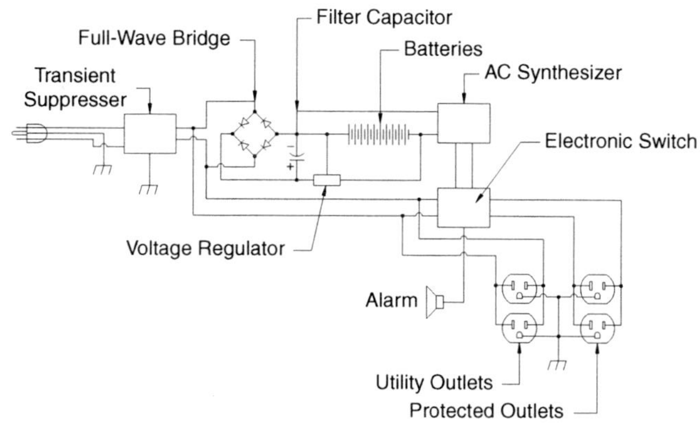

Figure 3-57 Uninterruptible Power Supply (UPS) Schematic

to the protected outputs. In the event of a power failure, the AC synthesizer automatically generates AC power from the batteries, which is directed to the electronic switch. The switch turns off the feed from the suppressor and turns on the feed from the synthesizer. The protected outlets are fed synthesized AC until the power is restored or the batteries are spent.

It should be noted that the utilities are intended to supply nonessential equipment and will lose power during a power failure.

Figure 3-58 shows a typical commercial UPS that is appropriate for desktop computers.

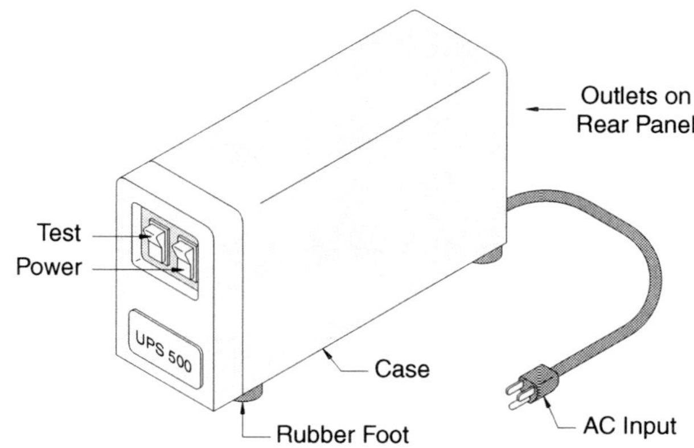

Figure 3-58 Commercial Uninterruptible Power Supply (UPS)

CHAPTER 4
ELECTRICAL CONTROLS

Electrical controls are the most common class of electromechanical devices. This arena, that comprises primarily switching devices, impacts virtually every aspect of our technical lives. We are all familiar with switching devices, every time we flip a light switch, we use one. When you turn the key to start your car, you use a switching device that, in turn, actuates a multitude of other switching devices. When you pick up the receiver on the phone, a switch activates the set. To better understand electromechanical devices, it is imperative that the reader understands control mechanisms. This chapter of the book provides a review of these mechanisms and their associated terminologies.

Manual Switches

Manually actuated switches are by far the most common electrical control devices. The simplest switch is the knife switch, as shown in Figure 4-1. The knife switch is simply a metal blade that can be rotated into a contact. The switch terminals are located at either end of the blade, on the pivot and on the contacts. To turn on the switch you simply push the blade into the contacts. To turn off the switch, lift the blade out of the contacts. In real life the basic knife switch is not very common. They are primarily used to switch high-power applications and for educational purposes.

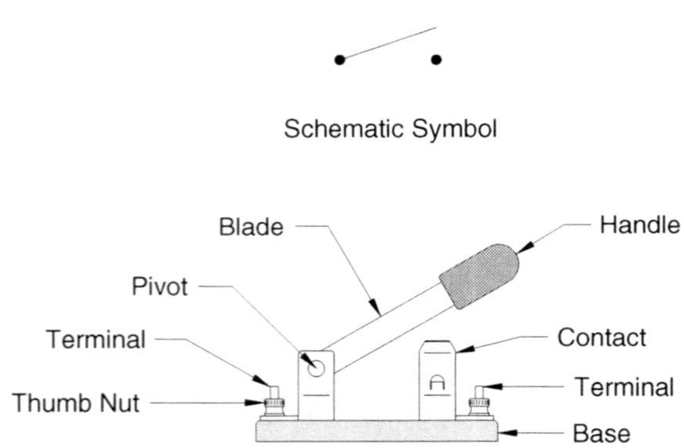

Figure 4-1 Single-Pole, Single Throw Knife Switch

A double throw switch is essentially a bidirectional valve for electricity. Power is connected to the common terminal and may be directed to one or another circuits. Figure 4-2 shows a single-pole, double throw knife switch. By throwing the blade to the right or left, the common terminal can be connected to either contact A or B.

Multipole switches are intended to switch two or more circuits simultaneously. Figure 4-3 shows a two-pole knife switch. This switch is simply two standard switches with a common handle and mounted on a common base.

Multipole switches are also used extensively in double throw applications. Figure 4-4 shows a double-pole, double throw knife switch. This arrangement is one of the most common switch configurations found.

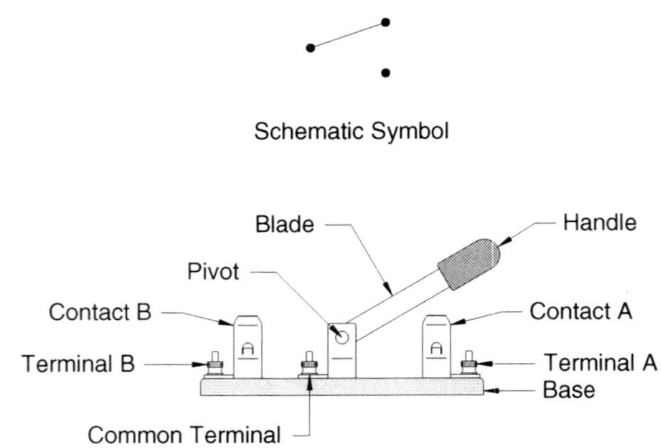

Figure 4-2 Single-Pole, Double Throw Knife Switch

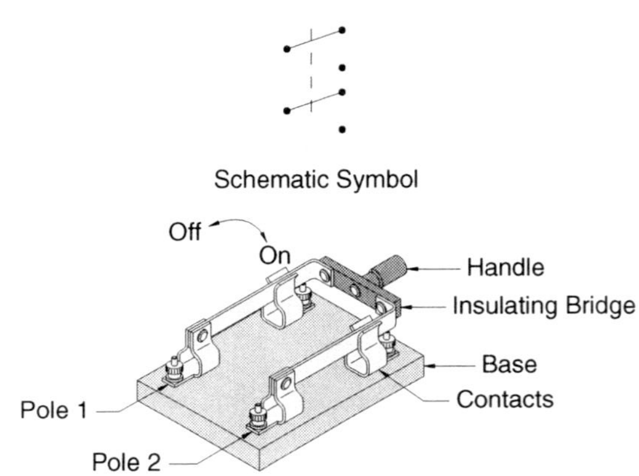

Figure 4-3 Double-Pole, Single Throw Knife Switch

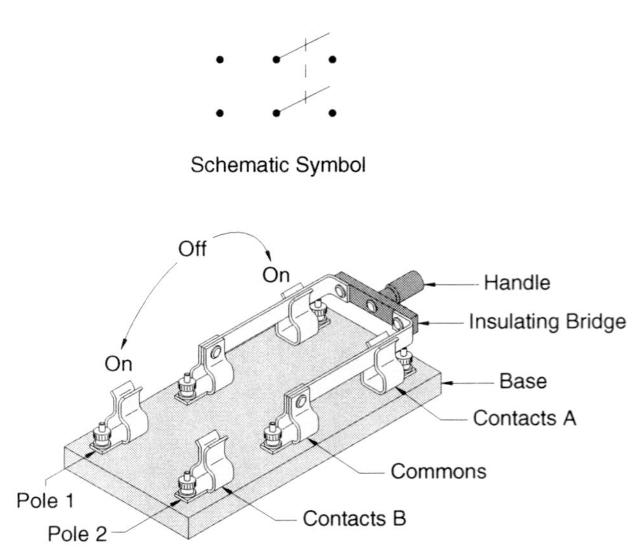

Figure 4-4 Double-Pole, Double Throw Knife Switch

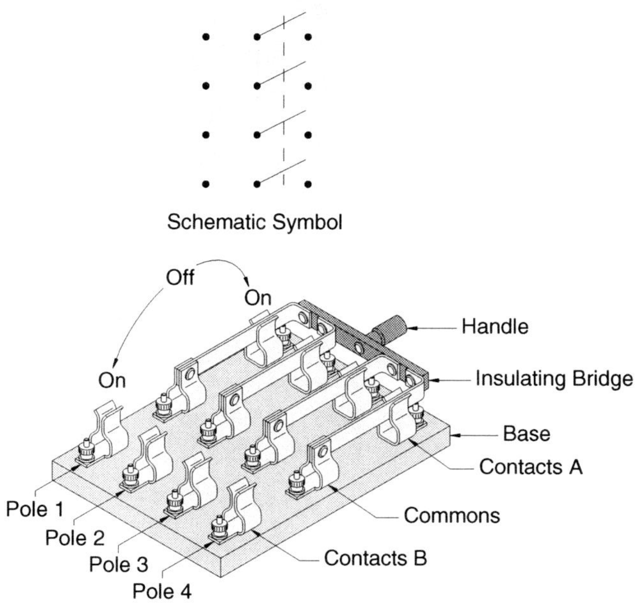

Figure 4-5 Four-Pole, Double Throw Knife Switch

Figure 4-5 shows a four-pole, double throw knife switch. The number of poles that a switch can have is only limited by practicality. Three-, four-, six-, and eight-pole switches are commonly available and can be utilized to solve a multitude of control applications.

One of the real world applications where knife switches excel, is high-power applications. Large, multi-blade knife switches as shown in Figure 4-6 can be found in power generation stations, as power disconnects for high-current applications and for switching auxiliary generation systems. Submarines of WWII used knife switch arrays to control their large DC electric propulsion motors. An order was relayed to the control station and the operator manually selected the voltage that corresponded with that speed.

Constructing a knife switch is a fairly simple process. Four copper angles are affixed to a nonconductive base, as shown in Figure 4-7. The blade is bolted or riveted into one set of angles while the opposite angles act as the contacts. An insulating handle can be made by wrapping the blade end with electrical tape or applying heat shrink. Figure 4-8 shows an exploded view of the bench built knife switch.

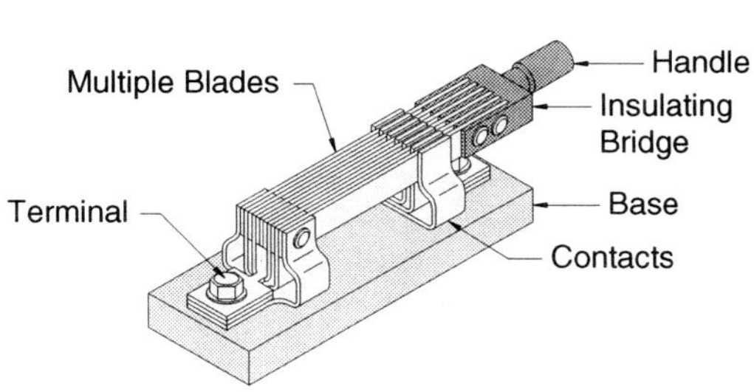

Figure 4-6 High-Current Knife Switch

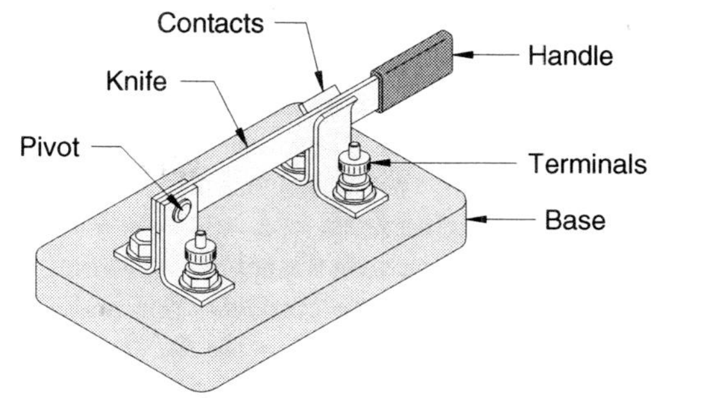

Figure 4-7 Bench Built Knife Switch

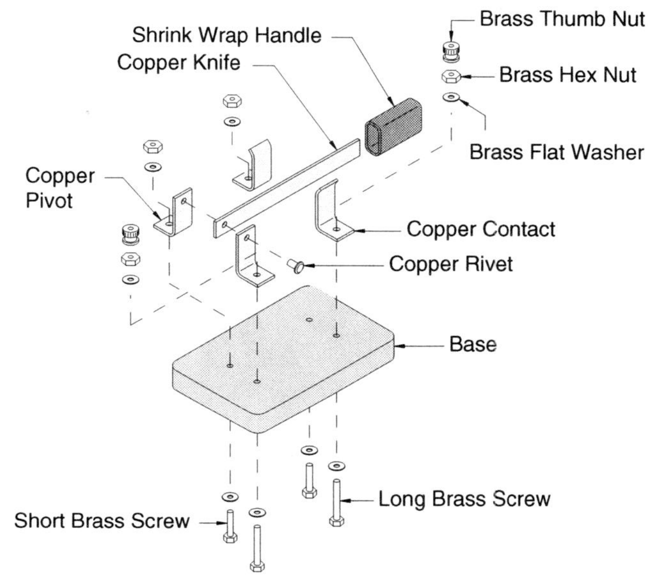

Figure 4-8 Knife Switch Exploded View

Another variation of the knife switch is the fused switch. In this arrangement the blade is interrupted with a section of insulating material. The blade has two snap sockets so that a fuse can bridge the insulator. The fuse assembly rides piggyback, as shown in Figure 4-9. These switches are available in multipole and double throw configurations.

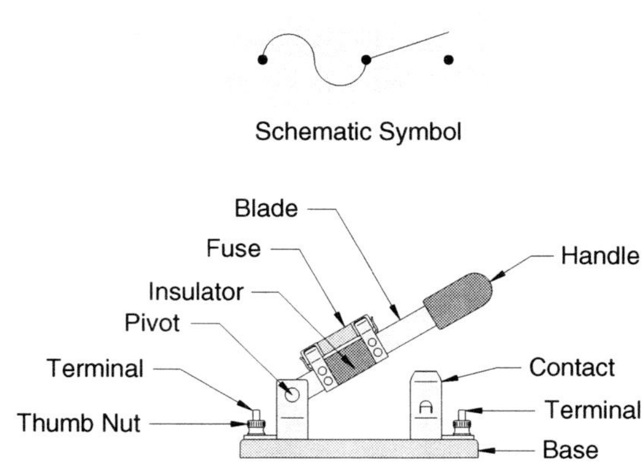

Figure 4-9 Fused Knife Switch

Switch Actions

The action of a switch is the mechanism by which the contacts are opened and closed. There are two basic actions that are commonly found in switches. The knife switch is considered a cam action. The contacts of a cam action switch open and close in direct relation with the actuator position. Because of the slow speed at which the contacts open and close, arcing can be problematic. To compensate for the damage that arcing may cause to the contacts, cam action switches generally use heavier duty construction.

Figure 4-10 shows a typical cam action switch mechanism. When the actuator is pushed to the left, the cam opens the contacts. When the actuator is pushed to the right, the cam allows the contacts to close. Many cam action switches have flats in the cam to provide some holding action for the position of the actuator.

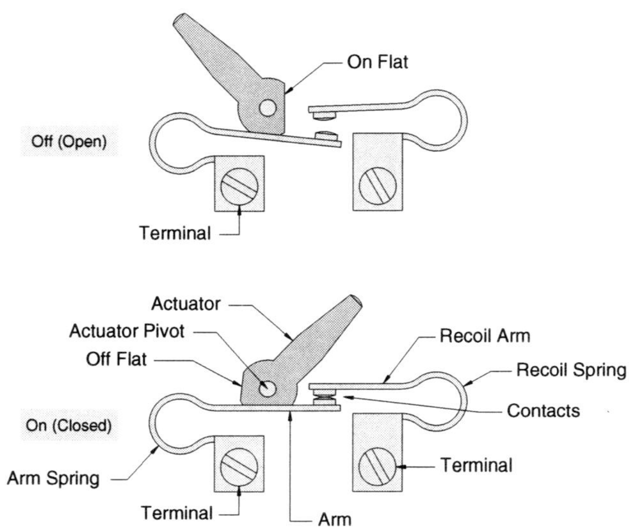

Figure 4-10 Cam Action Switch Mechanism

Snap action switches are designed to provide a very fast open/close cycle. Snap action switches incorporate a mechanism that stores the energy of the actuator and releases it to the contacts at a single moment. This system is primarily intended to minimize arcing between contacts. Snap action switches can be rather small in size in reference to their current carrying capabilities. In addition to their small size, the snap-action switch also provides excellent tactile feedback to the operator.

One drawback to snap action switches is contact bounce. When the switch is closed, the contacts are forced together at a very high rate of speed and the moving contact may recoil, or bounce, off the fixed contact. Generally, switch bounce is only a concern on circuits that have extremely sensitive switching requirements.

Figure 4-11 shows a snap action mechanism that might be found in a high-quality switch. The actuator energy is stored in the spring. As the spring crosses the pivot point, it pulls both the contact arm and the floating contact up into the fixed contact.

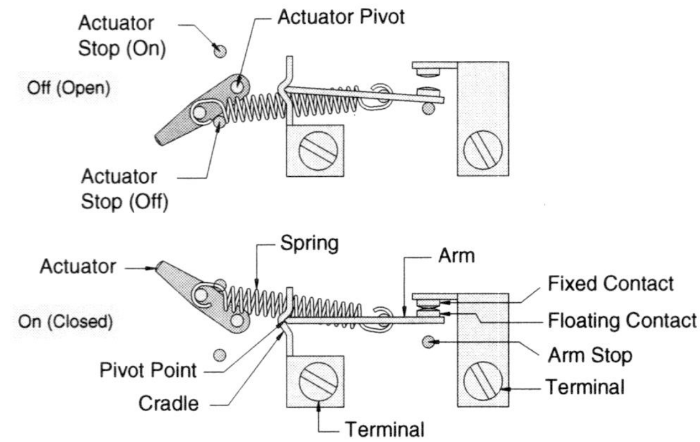

Figure 4-11 Snap-Action Switch Mechanism

The most common switch action is a hybrid of the cam and snap actions. The pseudo-snap-action switch is a cam action switch with a snap action actuator. This type of switch has a number of attributes that make it an attractive solution for many applications. The simplicity of the design provides relatively low manufacturing costs, minimized contact bounce, fast open/close cycle, and excellent tactile feedback.

Figure 4-12 shows a pseudo-snap action switch mechanism. Take notice that the contact and actuator design is very similar to the cam action shown in Figure 4-10. The principal difference is the actuator ball and spring, that are intended to provide the snap action.

A specialized type of pseudo-snap action switch is the drum switch. These switches are generally designed to provide a forward-off-reverse function for low horsepower, three phase motors. Figure 4-13 shows a typical commercial drum switch.

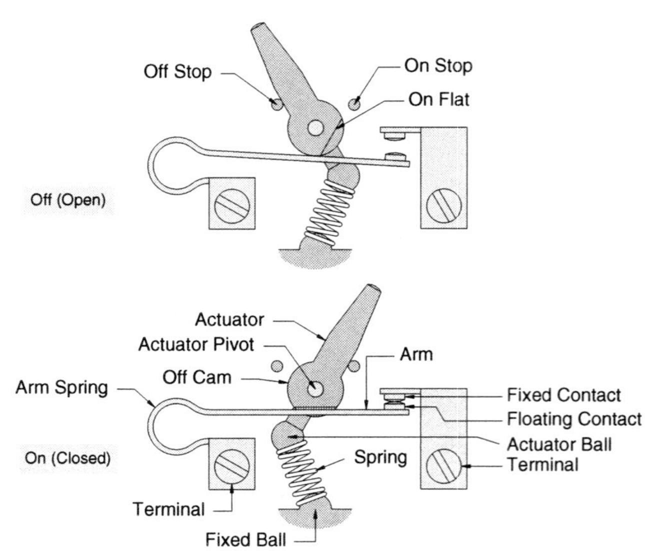

Figure 4-12 Pseudo-Snap Action Switch Mechanism

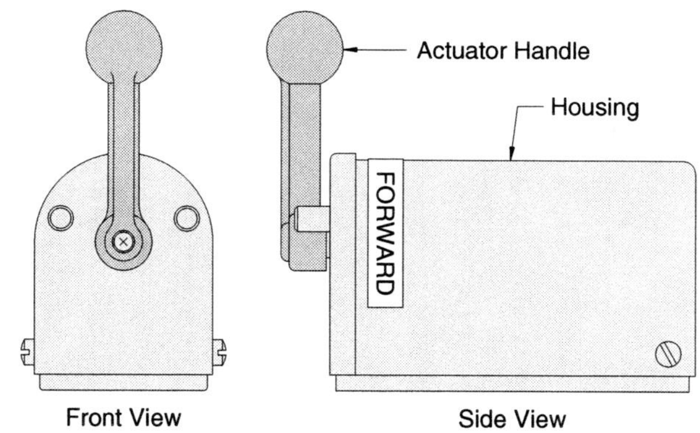

Figure 4-13 Drum Switch

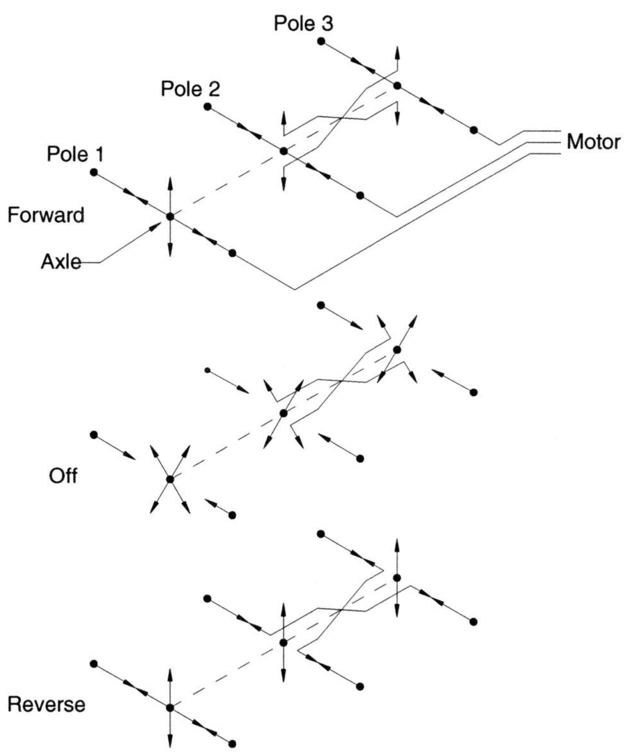

Figure 4-14 Drum Switch Schematic

Three-phase motors can be reversed by switching two of the three power feeds. The drum switch is a three position, three pole unit with the center position off. Pole one is on/off/on, poles two and three are configured for a reversing function. Figure 4-14 shows a schematic representation of a drum switch function. In the forward position the three power feeds are fed directly through to the motor. In the off position the three feeds are disconnected. In the reverse position, pole one is fed directly to the motor, while poles two and three are reversed.

Push Buttons

Push buttons, or momentary switches, are very common items. A perfect example is your door bell button. These switches should not be confused with on/off switches using push actuators. Momentary switches are available in multi-pole, normally open, normally closed, or a combination of both.

Figure 4-15 shows a simple leaf spring momentary switch. The leaf is simply a bent strip of copper with an insulating button affixed to one end and the opposite end screwed down to the base. The contact is a second strip, that is also screwed to the base. The terminals are at the ends of the copper strips.

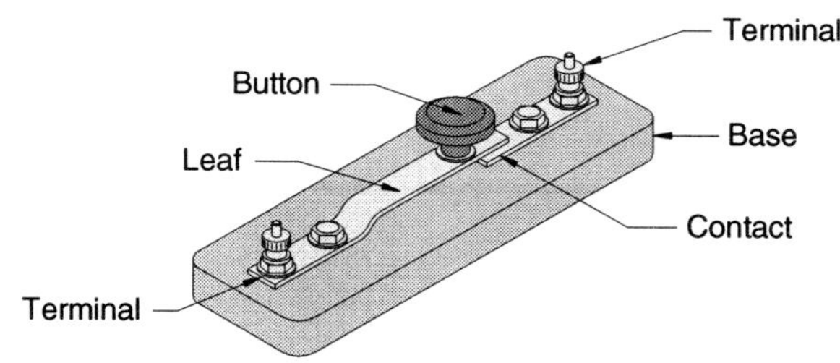

Figure 4-15 Leaf-Spring Push Button Switch

Figure 4-16 shows an exploded view of the leaf spring switch and just how simple the construction is.

Figure 4-17 shows a sectional view of a commercial, normally open, momentary switch. This particular unit incorporates two contact sets. The fixed contacts are set into the body of the switch and the floating contacts are affixed to the bridge. When the button is pressed, the bridge assembly is forced down to the fixed contacts and the circuit is closed.

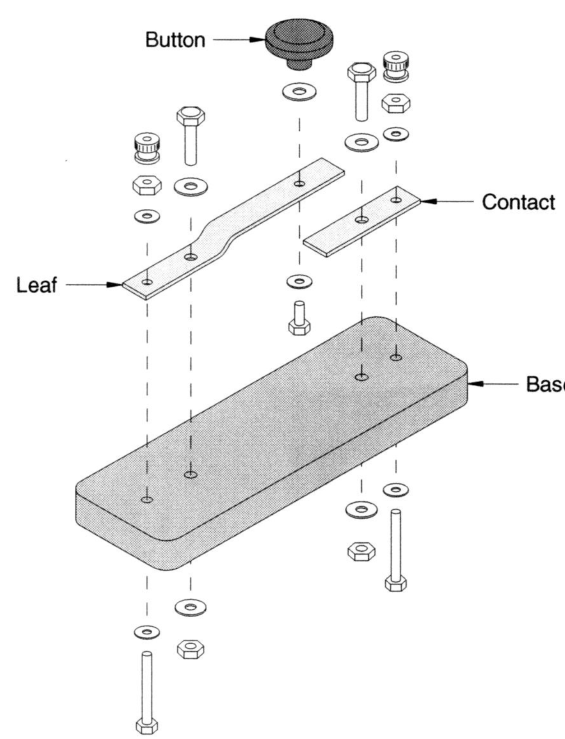

Figure 4-16 Leaf-Spring Switch Wxploded View

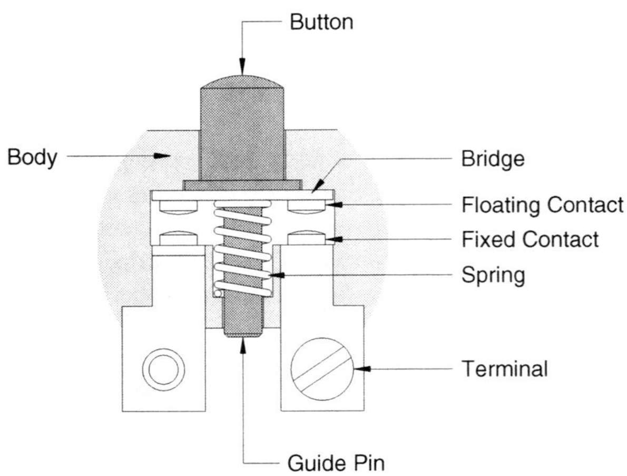

Figure 4-17 Dual-Contact Momentary Switch Mechanism

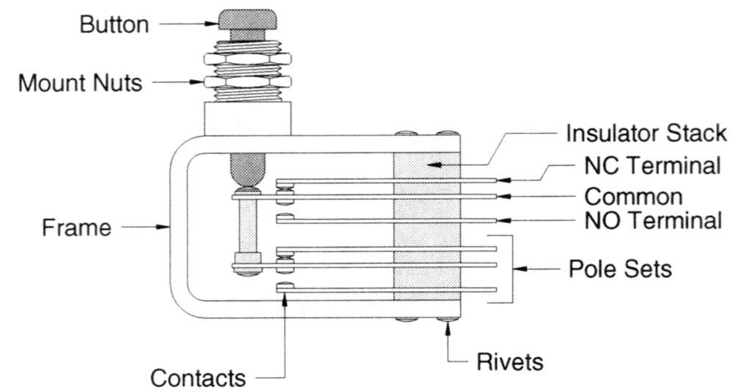

Figure 4-19 Two-Pole, Double Throw Reed Switch

For some applications push buttons must be snap action. To minimize size, a resilient dome is added to the configuration. As the button is pressed, the dome deflects until it snaps into a deflected position.

Figure 4-18 shows a sectional view of a typical dome-action momentary switch mechanism.

One common momentary switch configuration is the reed switch. These designs use a series of leaves stacked into an arrangement that suits the application. Because they have limited current carrying capabilities, the use of reed switches is generally limited to communication and test equipment.

Figure 4-19 shows a typical two-pole, double throw reed switch. Note how simple it would be to lengthen the frame and increase the elements in the stack. For this reason multipole reed switches are typically inexpensive.

The market offers virtually thousands of switches, available in every conceivable configuration. Figures 4-20 and 4-21 show just a small assortment of commercial switches. Most switch designs are available in cam, snap, or pseudo-snap action. They are commonly available in single-, double-, or multipole and one, two, or three position.

Power Disconnects

The next class of switches we will discuss is power disconnects. Power disconnects are large industrial switches specifically designed to switch high-current power feeds. Power disconnects fall into two principal categories. The first is a disconnect-only-role. These disconnects are not designed to switch the load, rather they are designed only to disconnect power for service, safety, and emergencies. Before disconnecting power, the machine being serviced should be completely shutdown. If a disconnect is used to switch power, extreme arcing can occur and damage to the contacts might result.

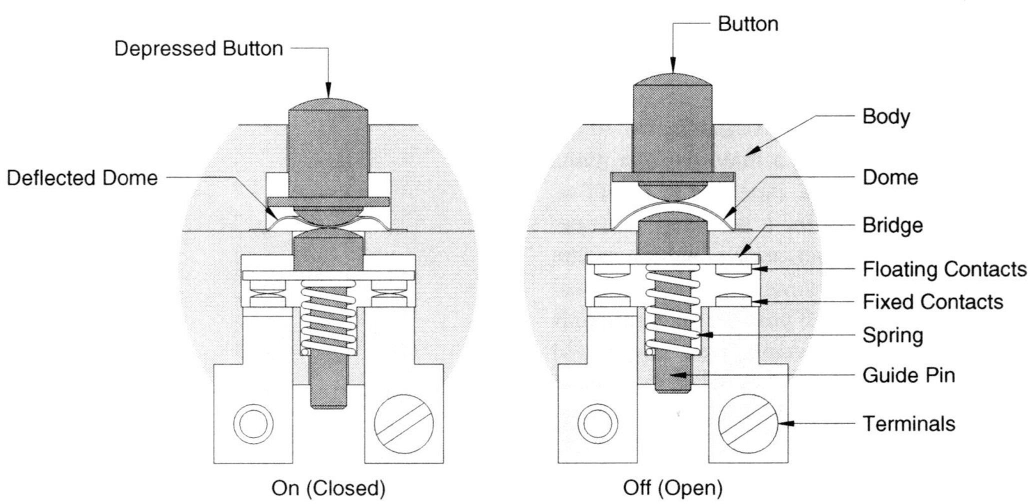

Figure 4-18 Dome-Action Momentary Switch Mechanism

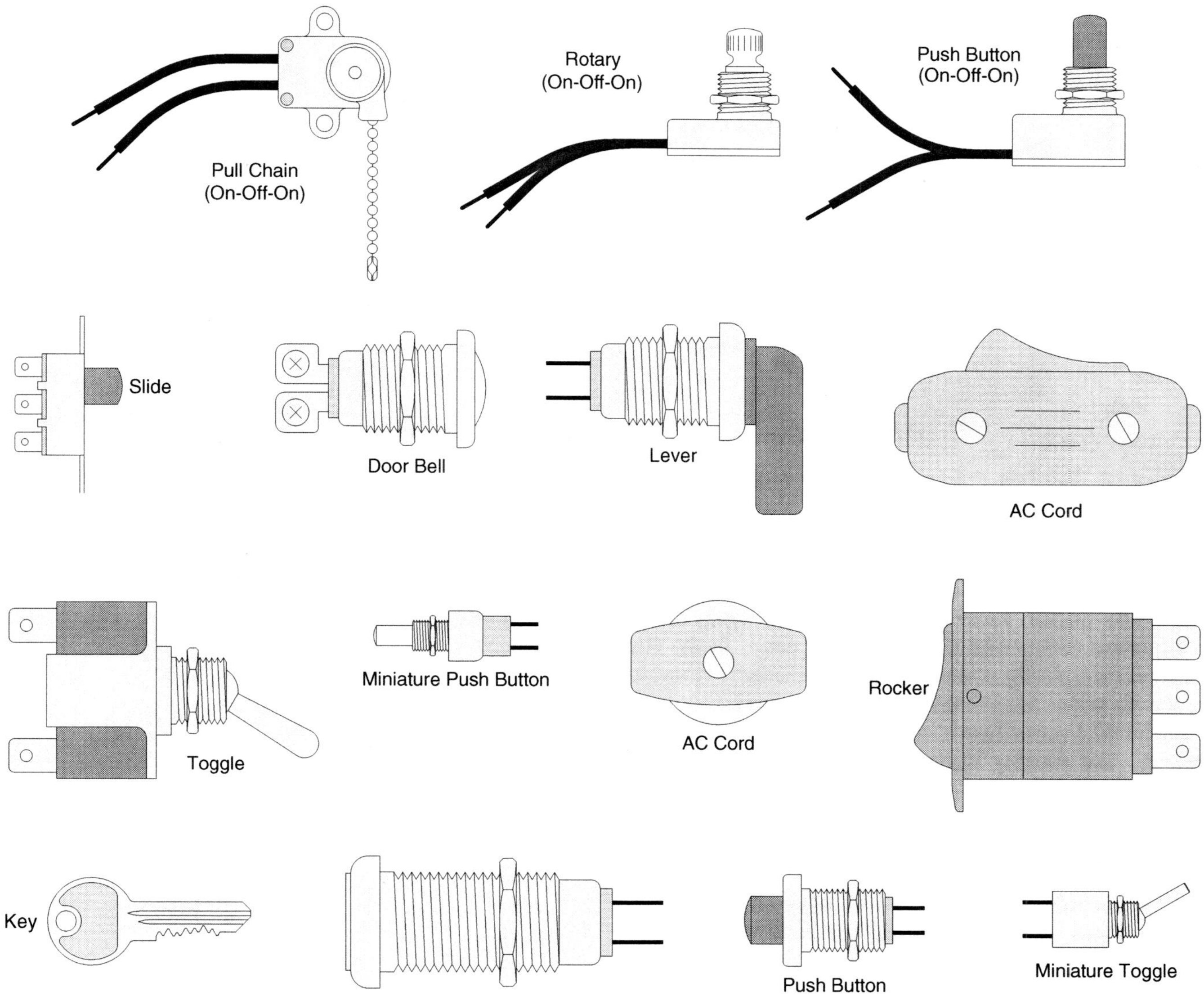

Figure 4-20 Commercial Switches

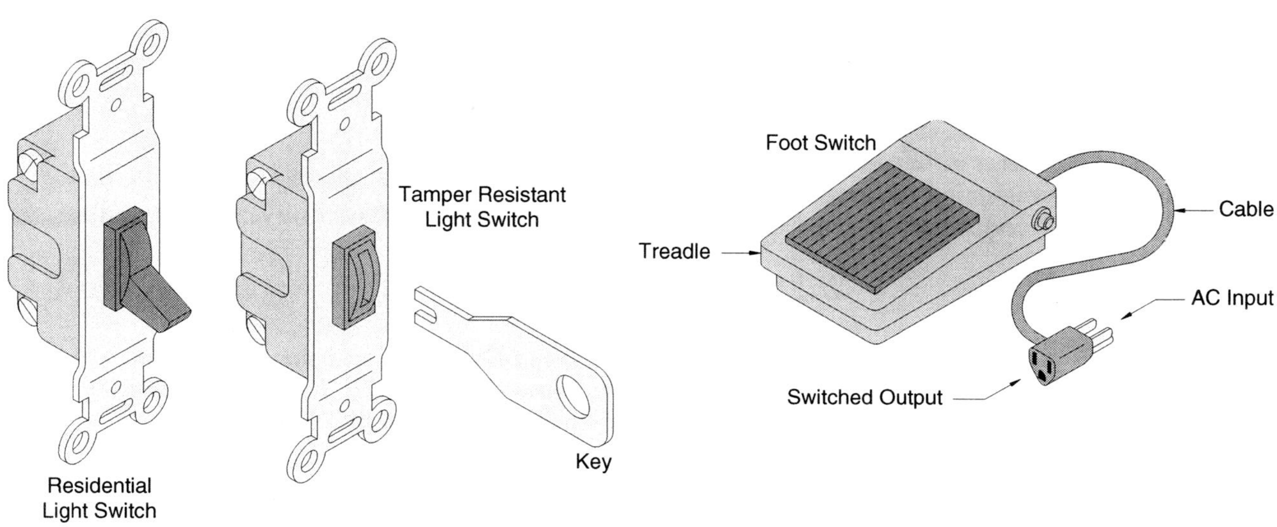

Figure 4-21 Commercial Switches

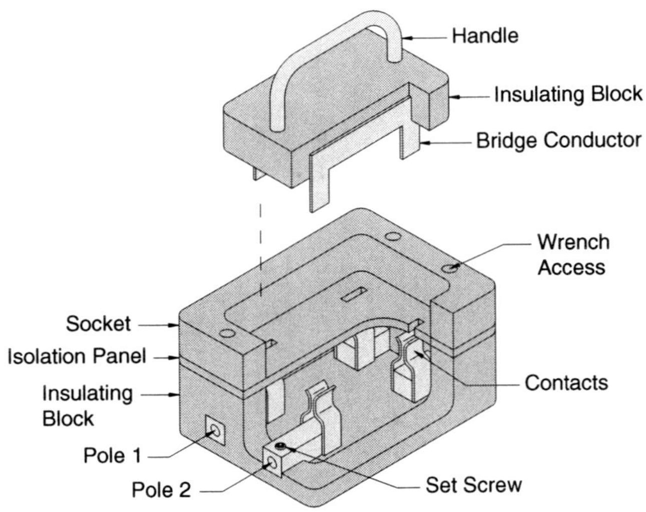

Figure 4-22 Two-Pole, Pull-Out, Power Disconnect

Figure 4-22 shows one of the most simple power disconnects on the market. These disconnects are commonly found on home and commercial air conditioning power feeds. The disconnect is typically mounted adjacent to the unit to provide the service technician ultimate control of the power.

Figure 4-23 shows how a rotating blade power disconnect operates. The rotating blade is a two-sided knife switch arrangement. A drive shaft connects an actuator handle to the pivot in the rotating blade. When the actuator is pulled through a 90° arc, the blades connect.

Figure 4-24 shows a three-pole, lever action power disconnect. The dashed lines show the blades in the off position.

The second category of power disconnects are those designed to switch live loads. These disconnects generally consist of a standard power disconnect with a snap action actuator added to the drive shaft.

Figure 4-25 shows a typical snap action actuator added to a standard power disconnect. This addition will minimize arcing and allow the disconnect to be used in switching live loads.

Figure 4-26 shows a typical snap action actuator mechanism. As the lever is pulled down, the actuator cam stores energy by compressing the follower spring. When the lever is pulled completely through its arc, the load assembly progresses past the pivot point and the energy of the spring forces the contact cam into position. Mechanisms of this nature can be used to safety and reliably switch several hundred amp loads.

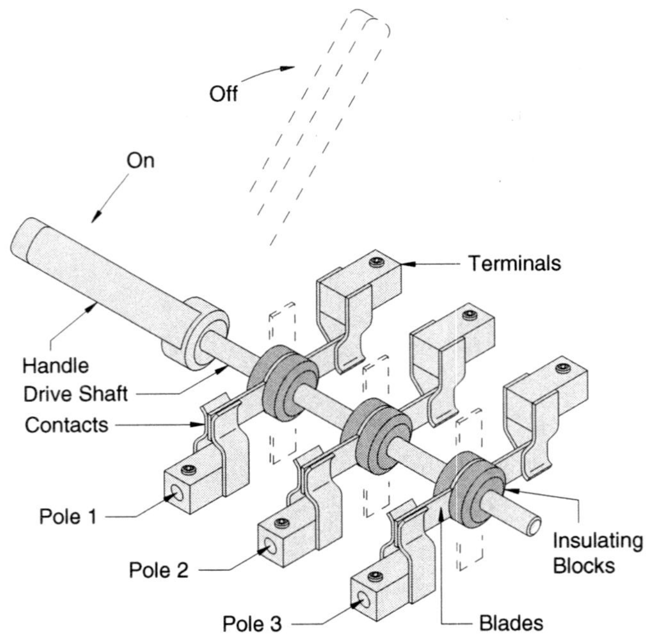

Figure 4-24 Three-Pole, Lever Action, Power Disconnect Mechanism

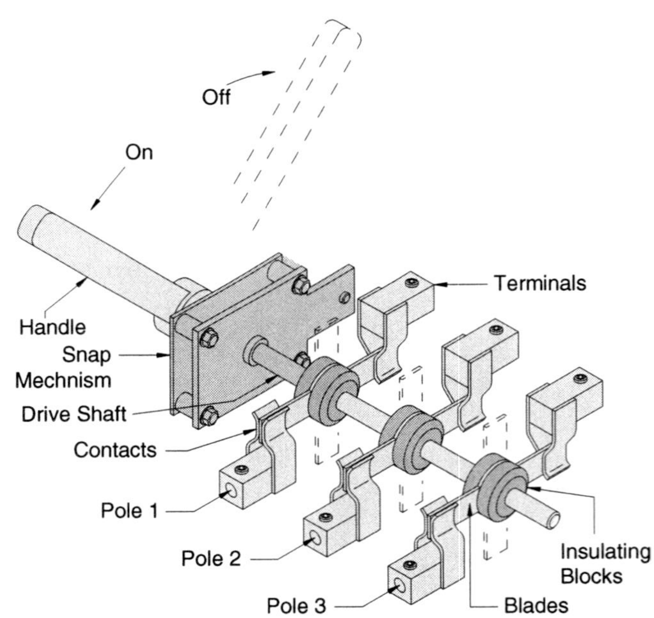

Figure 4-25 Power Disconnect with Snap-Action Actuator

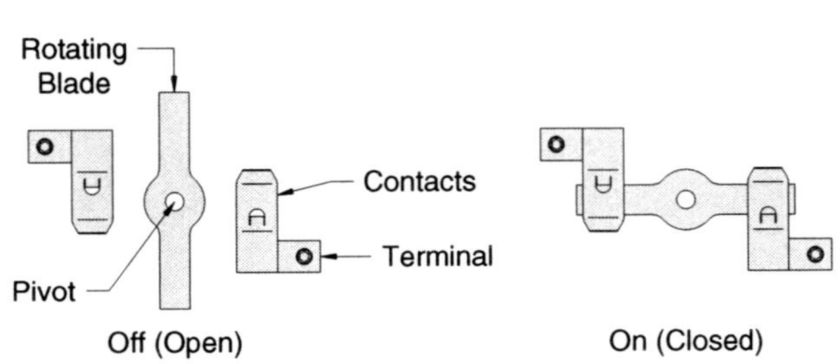

Figure 4-23 Rotating Blade Power Disconnect

Power disconnects are commonly supplied with fuse sets, as shown in Figure 4-27. The fuses are generally selected to protect the equipment the disconnect serves.

Selector Switches

For many applications there is a need to make selections between multiple circuits. This is the realm of the selector

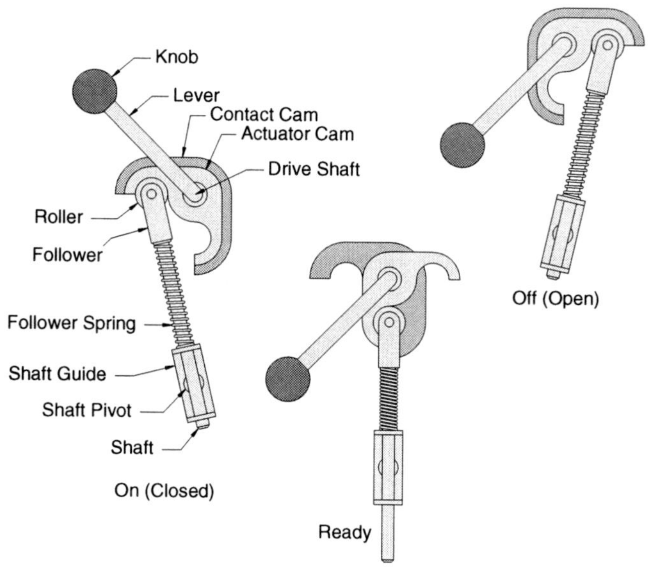

Figure 4-26 Snap-Action Mechanism for Power Disconnect

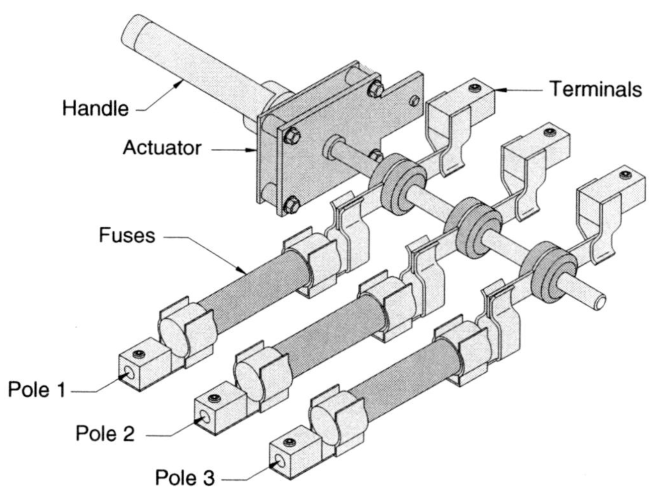

Figure 4-27 Power Disconnect with Fuse Set

switch. These switches typically have a common terminal that can be connected to several output terminals.

Figures 4-28 and 4-29 show a simple blade-type selector switch. The blade and contacts are simple strips of copper, that are screwed to an insulating base. Terminals are placed at the ends of the strips. An insulating handle is affixed to the blade. The blade may be adjusted to contact any one of the outputs.

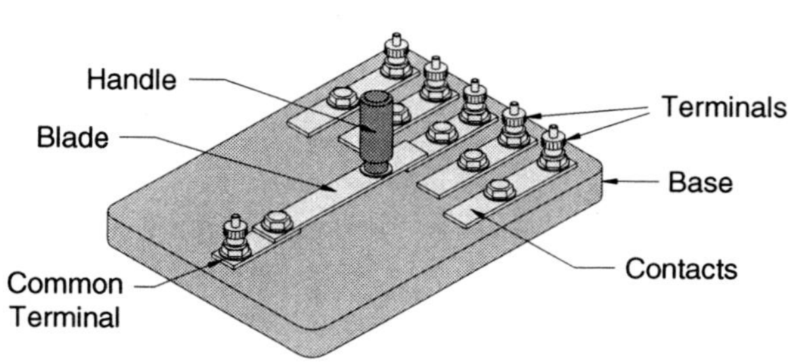

Figure 4-28 Blade-Type Selector Switch

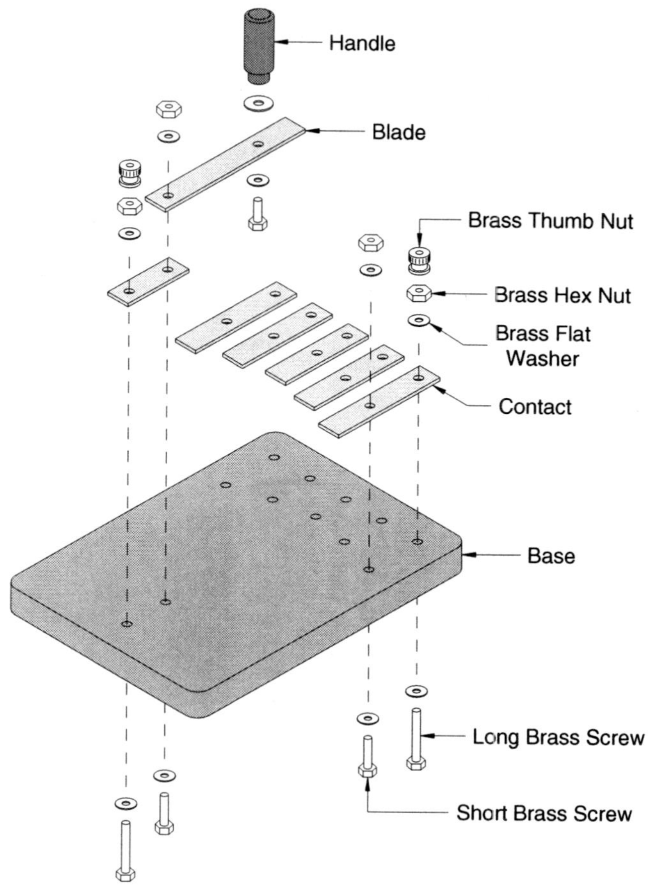

Figure 4-29 Selector Switch Exploded View

Another simple method of building a selector switch is to use a banana jumper configuration, as shown in Figures 4-30 and 4-31. The common jack is placed at the center of a circular array of jacks. The radius of the array is 0.75 inch, which is the center spacing of a standard dual banana plug. A shorting wire is added to the dual banana plug. The plug can be used to as the selector by simply pulling it out and reinserting into a different position. It should be noted that the center-to-center spacing of the jacks that make up the circular array should have a dimension other than 0.75 inch. This prevents the dual banana plug from being miss connected.

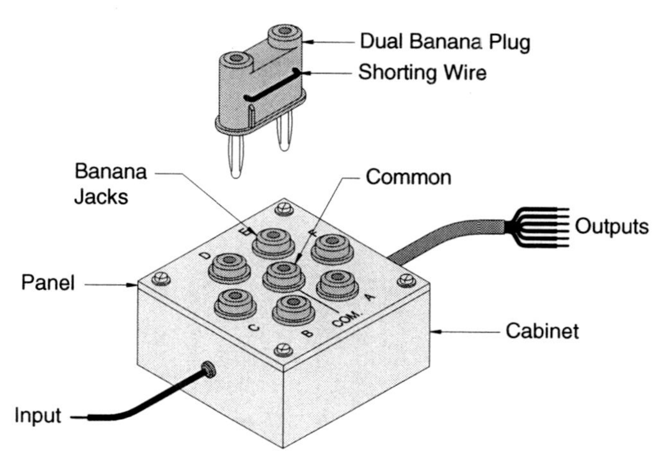

Figure 4-30 Banana Jumper Selector Switch

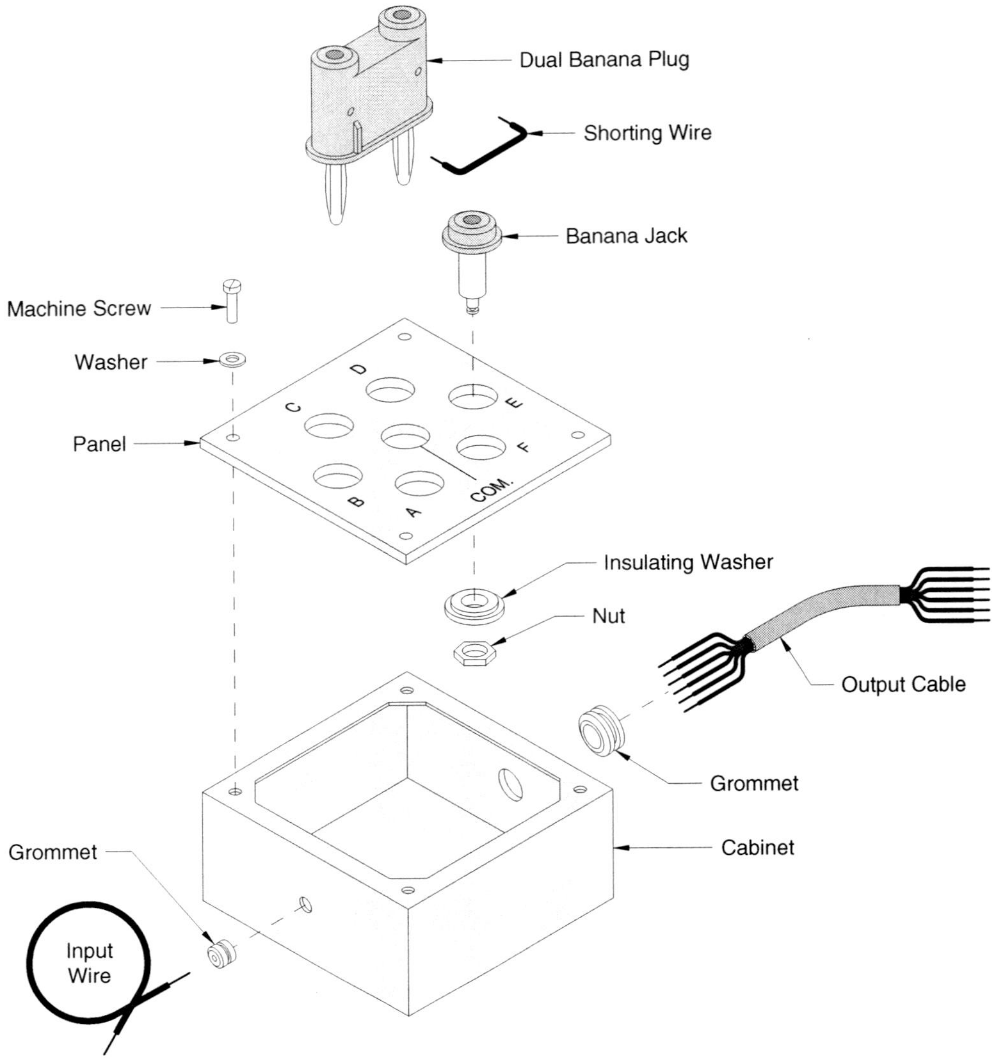

Figure 4-31 Jumper Selector Switch Exploded View

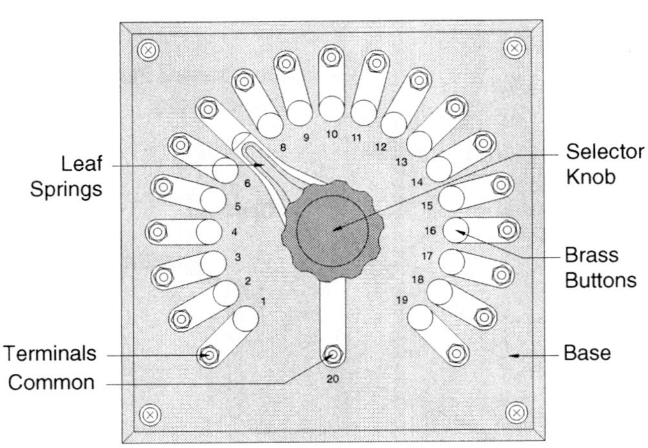

Figure 4-32 Button Selector Switch

Early test and radio equipment commonly used button selector switches, as shown in Figure 4-32. These switches offered excellent, low-resistance contacts at a time when switches were built entirely by the manufacturer of the equipment in which they were used. They are now found almost exclusively as educational aids.

Figure 4-33 shows a modern, open-frame, multipole, selector switch. These switches are available in a variety of configurations, positions, and pole counts. The decks are generally constructed from fiberglass insulating board with copper contact inserts. The switch decks are assembled onto the main deck with spacers and threaded rods. The main deck generally carries a detent mechanism that provides position accuracy and tactile feedback.

Figure 4-34 shows a typical enclosed selector switch design. These units are available with solder and screw terminals.

High-current selector switches, as shown in Figure 4-35, typically incorporate some type of snap action mechanism. These switches are not particularly comfortable to operate. The actuators are rather stiff and require rotation and loading until the mechanism snaps.

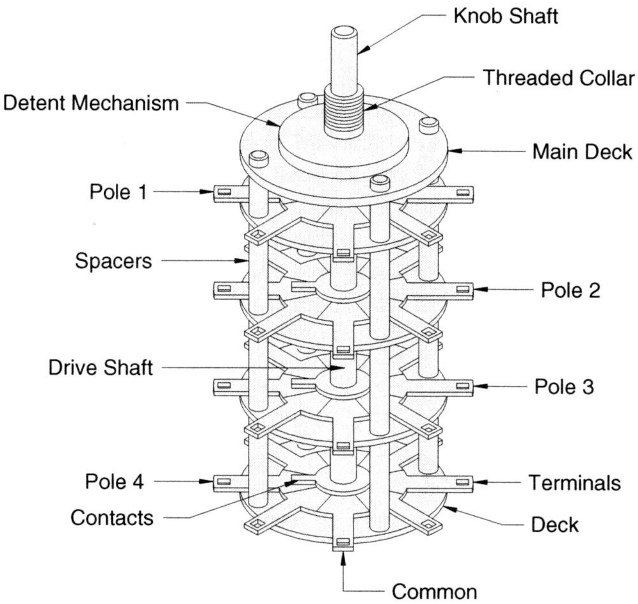

Figure 4-33 Multi-Deck, Open-Frame Selector Switch

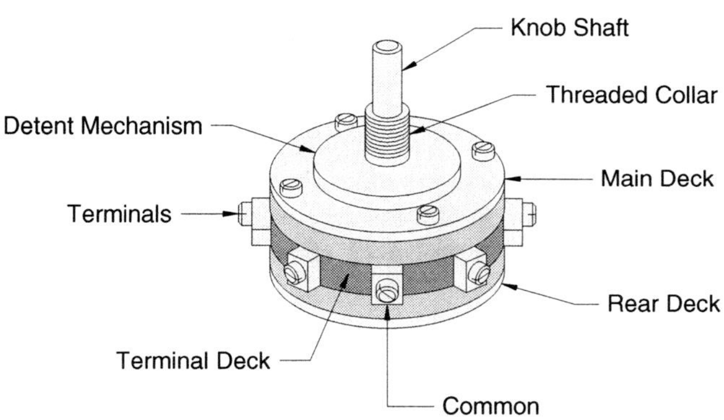

Figure 4-34 Single-Pole, Enclosed Frame Selector Switch

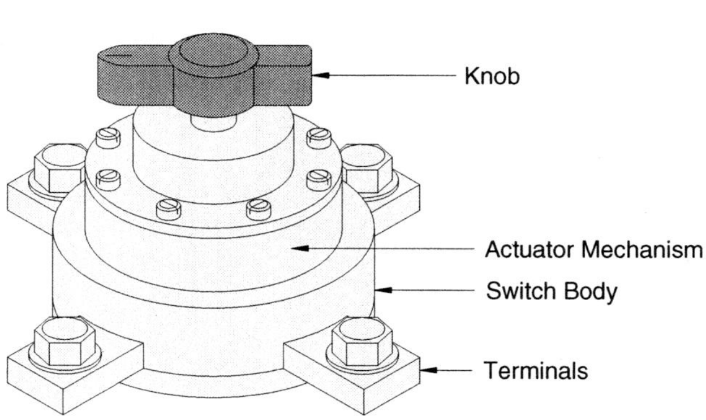

Figure 4-35 High-Current, Snap Action Selector Switch

Thumb wheel selector switches are normally configured to provide a base two output for use with microprocessor-based systems. A value is entered and a binary output that corresponds with that value is generated.

Figure 4-36 shows a typical thumb wheel selector switch array. The switches are single digit and several units can be arrayed together, as shown.

Figure 4-37 shows the outputs of octal (base 8), decimal (base 10), and hexadecimal (base 16) thumb wheel switches.

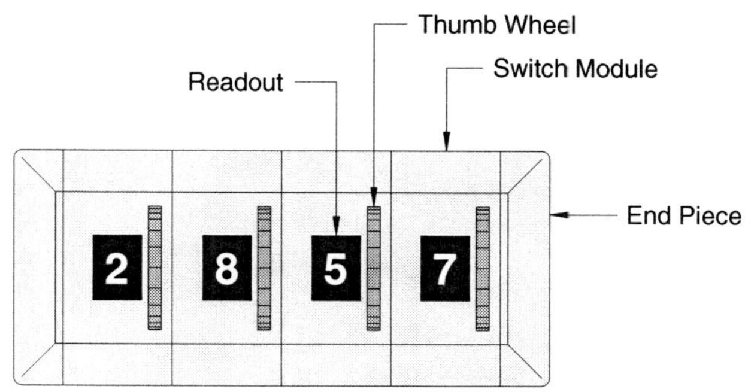

Figure 4-36 Thumb Wheel Selector Switch

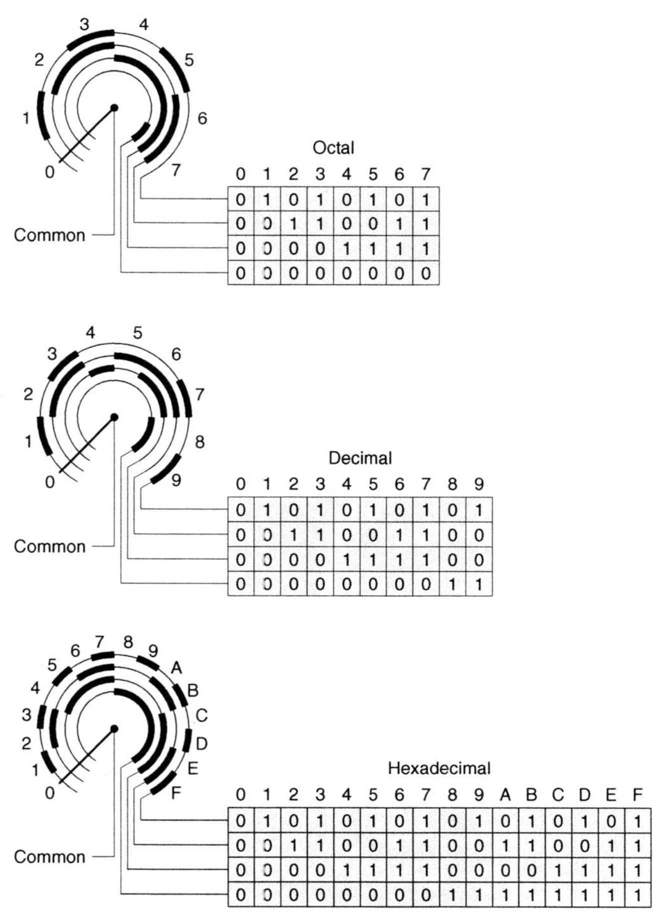

Figure 4-37 Thumb Wheel Encoding Patterns

A type of selector switch that most of us are familiar with is the automobile ignition distributor, as shown in Figure 4-38. The distributor is used to select the appropriate spark plug for ignition on internal combustion engines. It should be noted that a common distributor can be configured to serve as a high-voltage selector switch. Distributors are designed to switch tens-of-thousands of volts and have excellent, high-voltage terminals. Additionally, spark plug wires are a low cost source for high-voltage wire and connectors. When using a distributor for a selector switch, the housing should be grounded for safety. The contact in a standard distributor is a spark gap arrangement so a hard contact must be added to the rotor. A knob, detent, and indicator replace the timing gear.

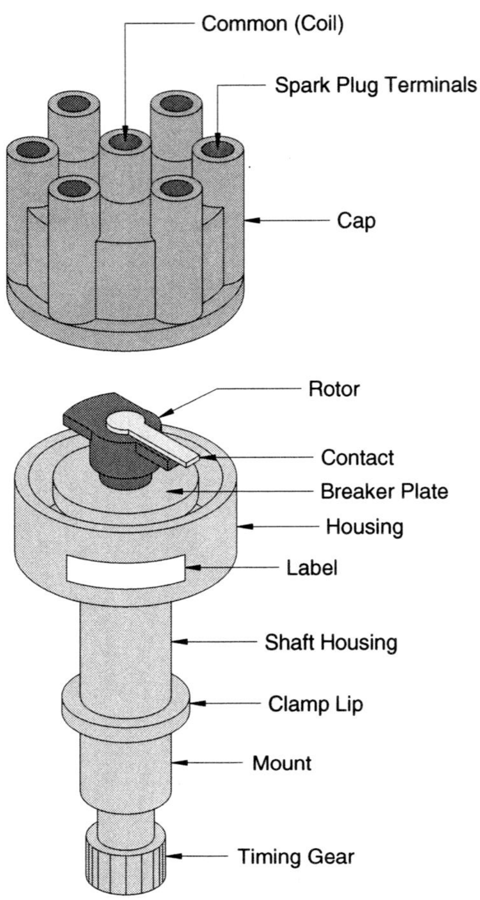

Figure 4-38 Automobile Ignition Distributor

Figure 4-39 shows an automobile distributor configured as a high-voltage selector switch. The timing gear is removed and replaced with a panel and selector knob with pointer. The housing is drilled and tapped for a ground wire, as shown.

Limit Switches

Limit switches are a type of switch that is specifically designed to detect machine motion. They are available in variety of sizes and configurations. They usually incorporate a single-pole, double-throw, snap action switch element. The principal difference from limit switches are their actuators. These range from micro buttons to rather sophisticated specialty actuators. A common example of a limit switch is the button that turns on the light in your refrigerator or car when you open the door.

Figure 4-40 shows an example of some of the common direct-acting actuators, the micro button and standard button being the most common. Figure 4-41 shows common lever arm actuators. Like the direct-acting actuators, lever arms are extremely common.

Different manufacturers offer a variety of specialty actuators and are configured for nearly every conceivable application. Figures 4-42 and 4-43 show just a few specialty actuators.

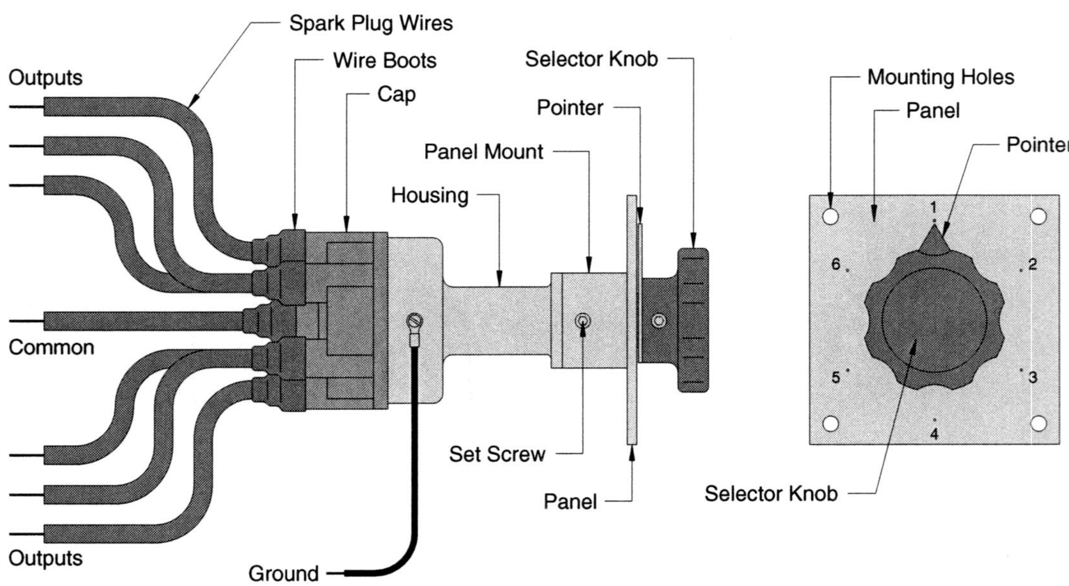

Figure 4-39 Distributor High-Voltage Selector Switch

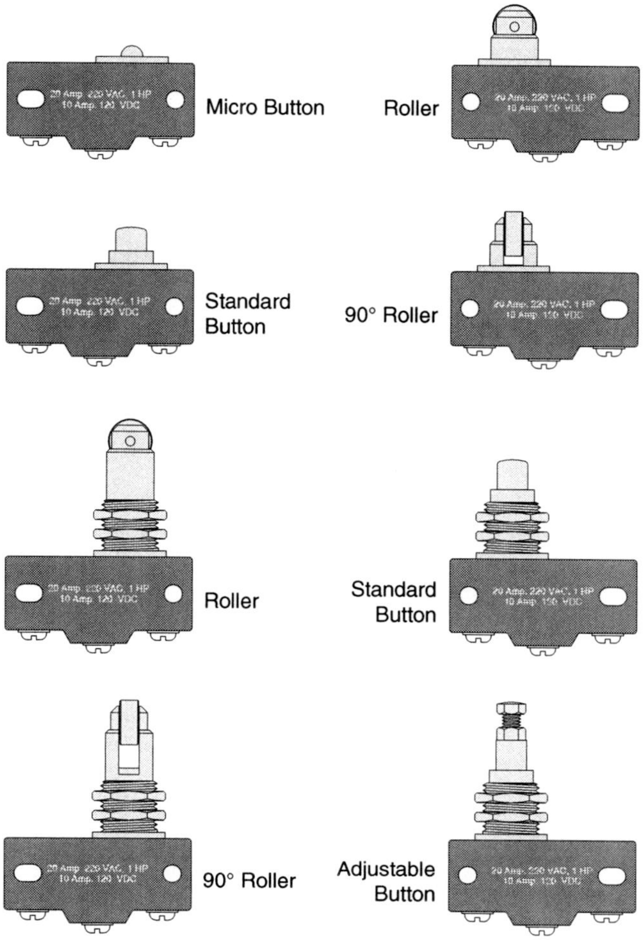

Figure 4-40 Direct-Acting Limit Switches

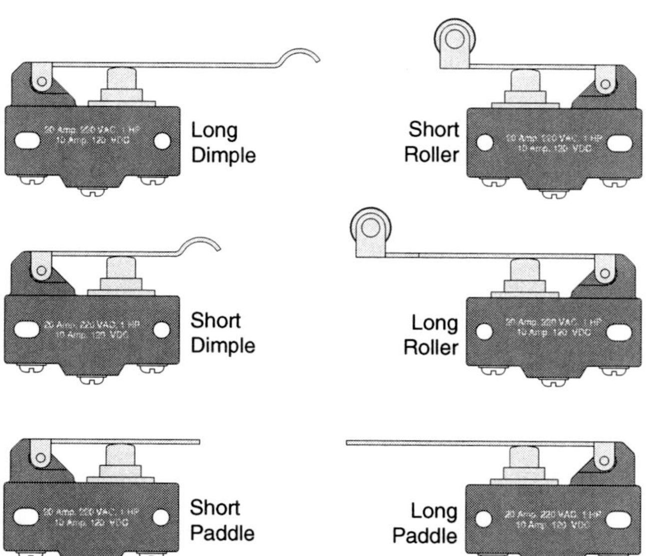

Figure 4-41 Basic Limit Switch Actuators

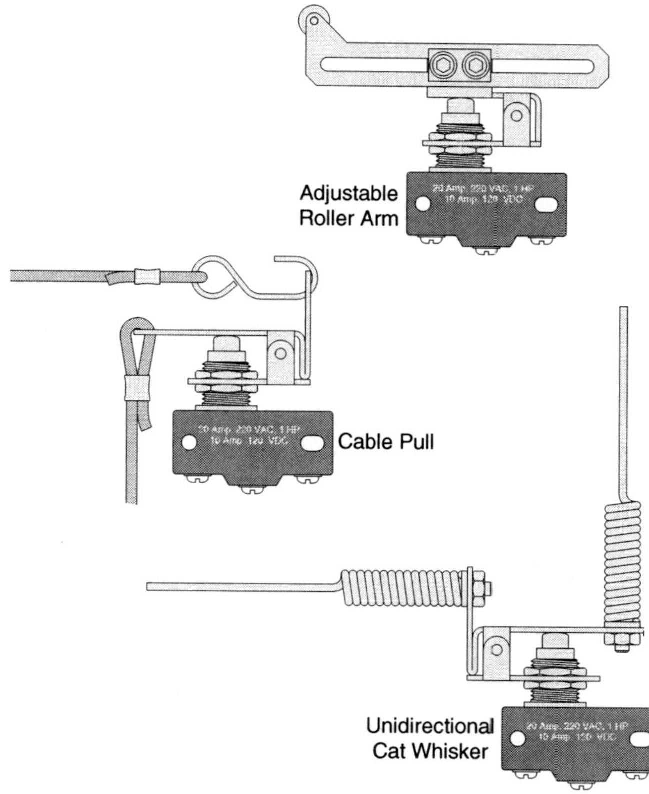

Figure 4-42 Specialty Actuators

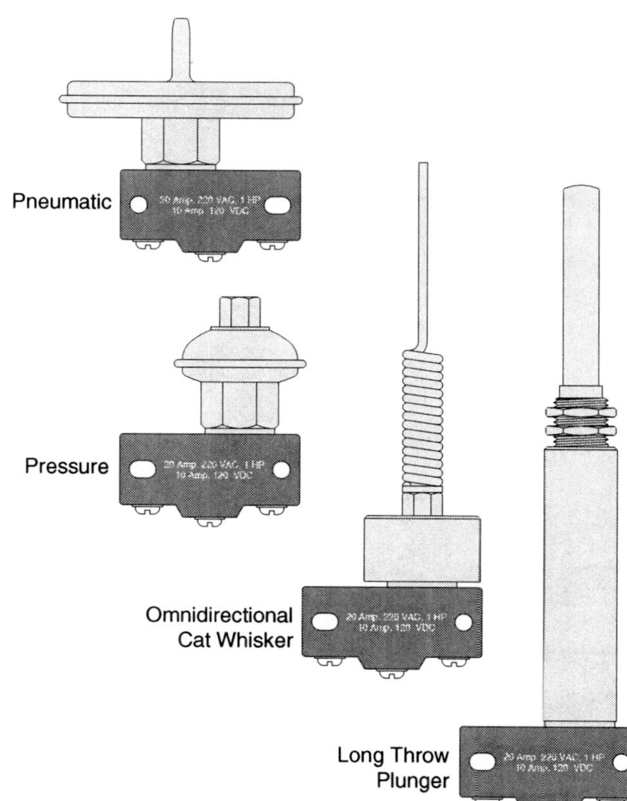

Figure 4-43 Specialty Actuators

Harsh environments, such as the plant floor, require especially rugged limit switches. These units are generally referred to as industrial limit switches. They generally have high-impact housings which are impervious to oil, water, and many chemicals. Figure 4-44 shows two industrial limit switches, with a center-loaded lever arm and one with latching lever arms.

For precise control of mechanical movement, a micrometer-adjustable limit switch may be configured, as shown in Figure 4-45. This arrangement provides control of the switch down to 0.001inch. Micrometer adjustable limit switches are commonly found in modern machine tools.

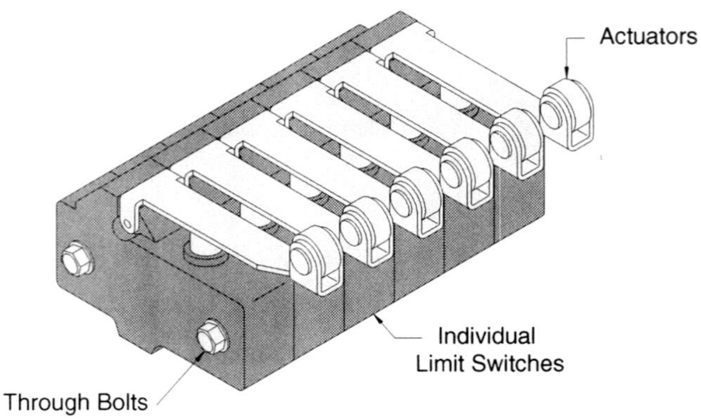

Figure 4-46 Ganged Limit Switches

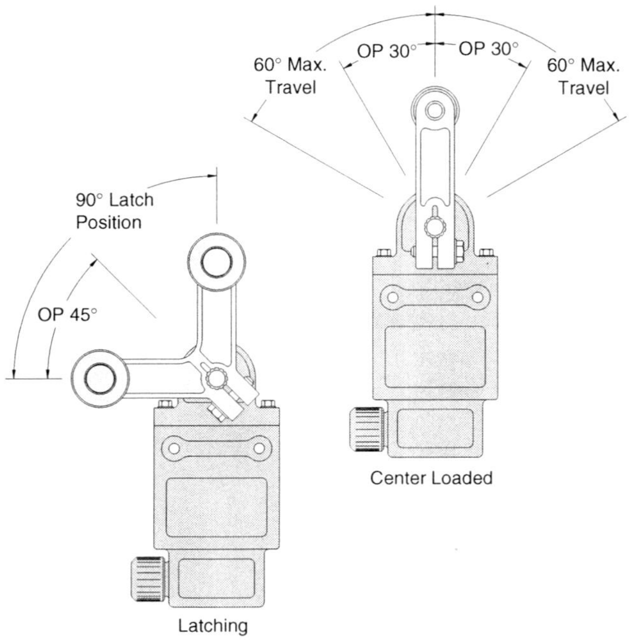

Figure 4-44 Industrial Limit Switches

Limit switches may be stacked, or ganged, together to detect a variety of motion. Figure 4-46 shows an array of limit switches with roller lever arm actuators.

Practically speaking, the application possibilities for limit switches are extremely broad. Limit switches are designed for, and have been placed into, virtually every conceivable situation. Figure 4-47 shows a few basic applications that limit switches can be applied to.

Slip clutches are the most common method to control torque. The drawback to a slip clutch is that it provides no visual indicator when the drive is applying too much torque. In some applications, the first indicator that there is too much torque being applied is that the slip clutch overheats. By using an expanding shoe and limit switch arrangement, as shown in Figure 4-48, the drive motor may be controlled or an alarm may be sounded. The drive bar acts against the shoes. The shoes are pulled together by a pair of extension springs. If the torque exceeds the rating, the drive bar forces the shoes out, tripping the limit switch. To adjust the torque rating, springs of different ratings may be selected.

To provide local control of a long conveyer belt, a limit switch travel control can be configured as shown in Figure 4-49. The lead screw rotates with the motor and drives the follower right and left. The follower trips the retract and extend limit switches and interrupts the motor power. The travel points can be adjusted by controlling the relative position of the switches. This is accomplished by rotating the adjustment screws, which carries the limit switch mounts.

Complex rotating machinery, such as printing presses, require a variety of control functions during their cycle. In many cases, the environment may be unsuitable for mounting a limit switch because of dirt, access, clearance, adjustment, and the like. In these cases a barrel limit switch array can be utilized, as shown in Figure 4-50. The limit switches can be located in a safe, clean, and easy to access location while the rotational information that they need can be provided via a timing belt. The belt drives a drum with preprogrammed cams which, in turn, trip the limit switches.

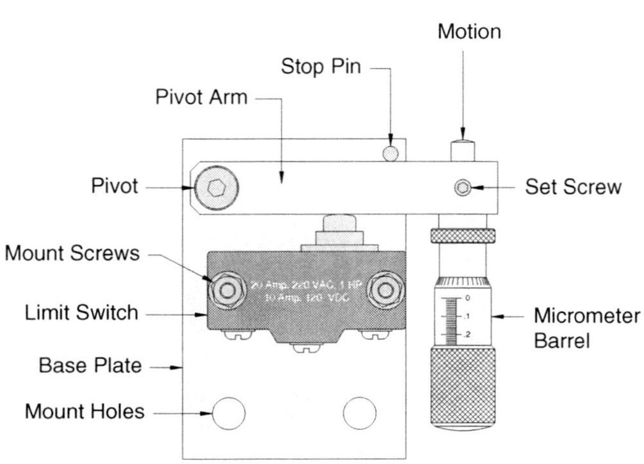

Figure 4-45 Micrometer Adjustable Limit Switch Assembly

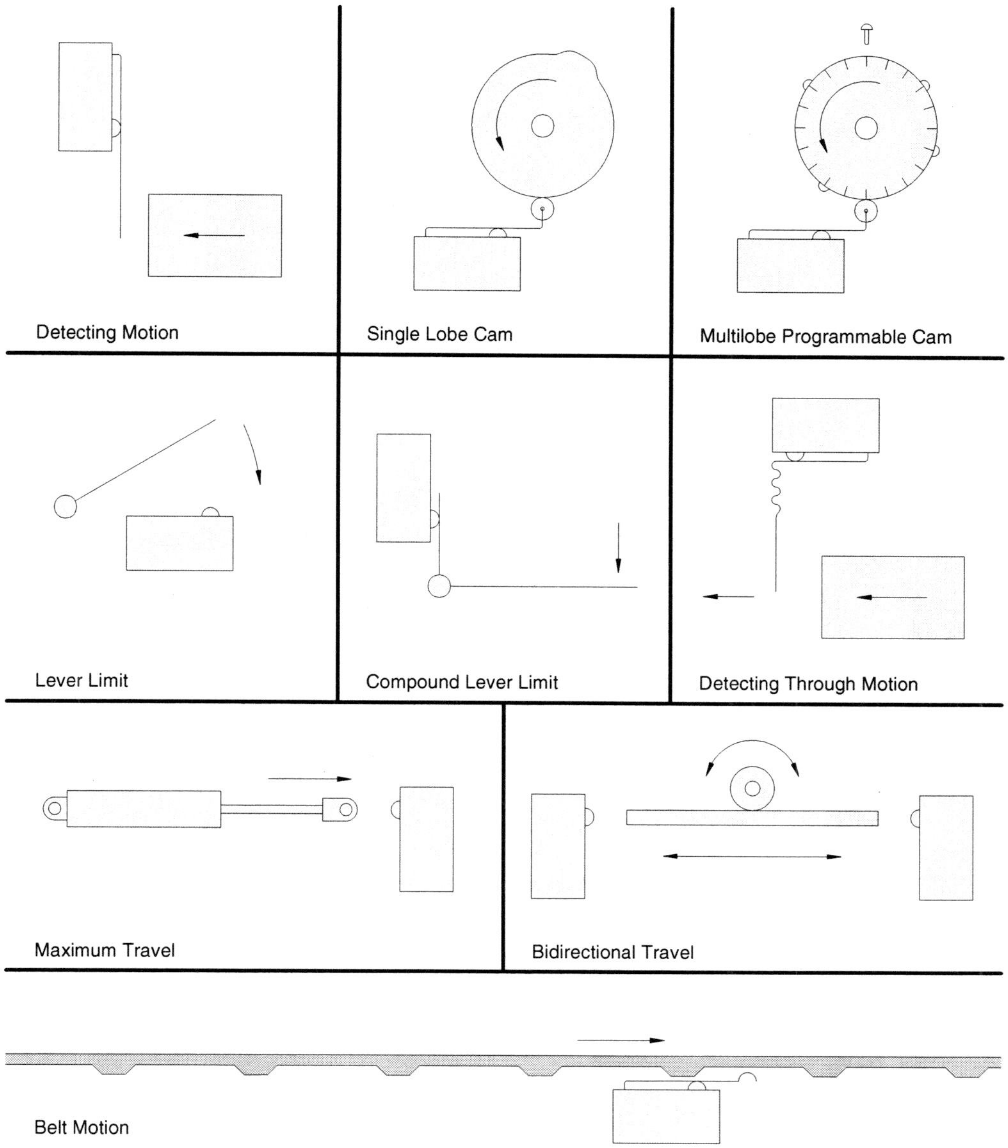

Figure 4-47 Limit Switch Applications

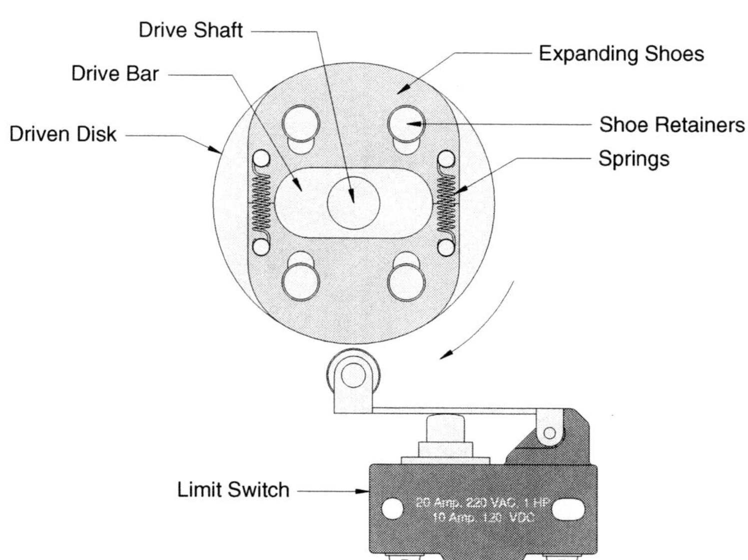

Figure 4-48 Limit Switch Over/Under Torque Detection

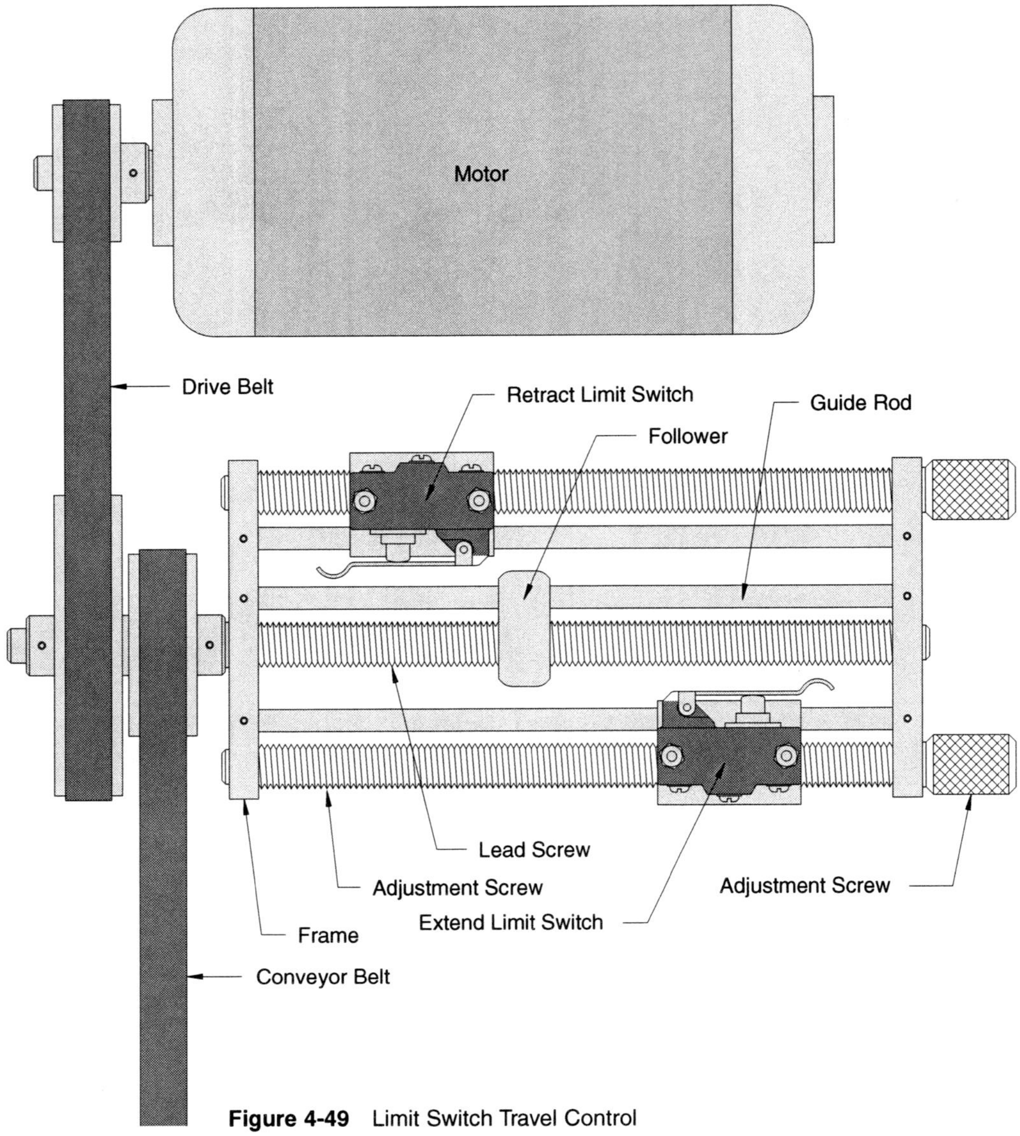

Figure 4-49 Limit Switch Travel Control

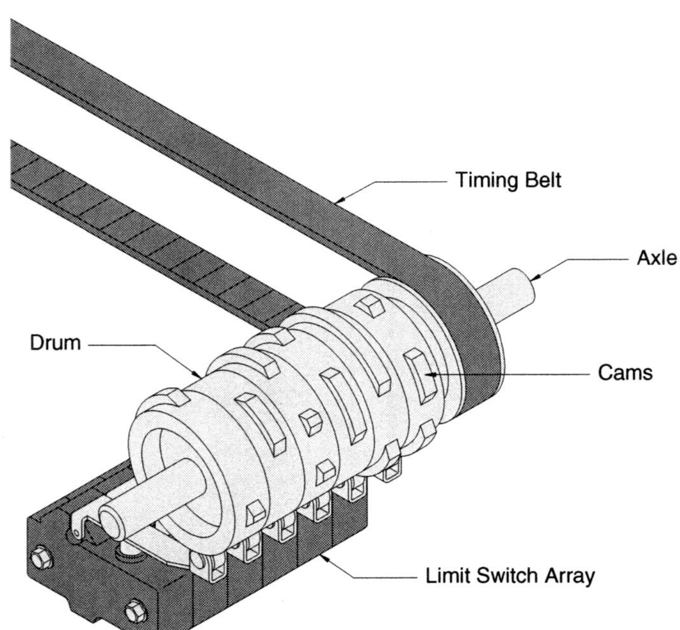

Figure 4-50 Barrel Limit Switch Array

Magnetic Switches

A variation of the limit switch is the magnetic switch. These switches are most commonly found as window sensors for alarm systems. The switch is quite simple in construction and therefore, inexpensive to purchase. The switch (Figure 4-51) is simply a reed with a contact and iron plug mounted into a plastic case. When a magnet is moved into position, the attraction to the iron deflects the reed and the switch closes. Because of the light nature of magnetic switches they are usually low-current devices and can only function as sensors.

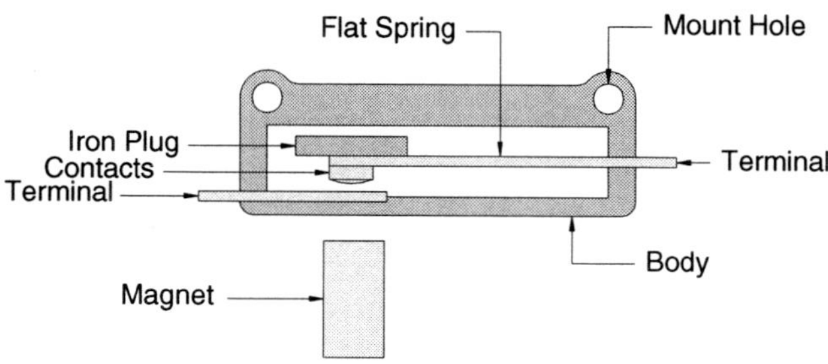

Figure 4-51 Magnetic Switch

Mercury Switches

A mercury switch is a type of limit switch that functions in reference to gravity. These switches have a cavity with a charge of mercury at the bottom and two terminals at the top. When the cavity is turned up-side-down, the mercury comes in contact with the terminals and the switch is closed. To open the switch simply turn the switch right-side-up. Figure 4-52 shows a simple mercury switch made from a test tube, rubber stopper and small brass rods.

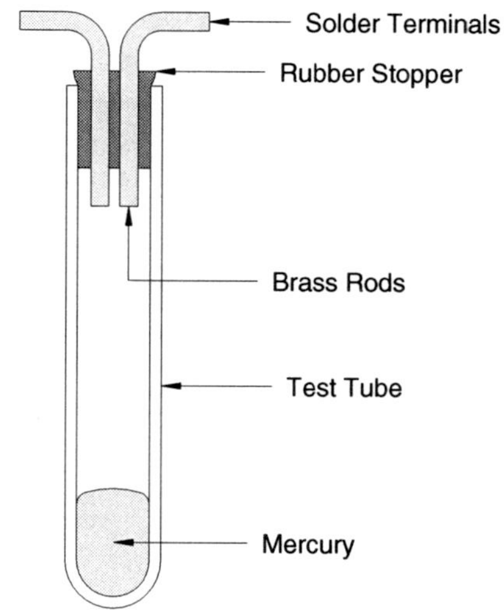

Figure 4-52 Simple Mercury Switch

Commercial mercury switches are generally double throw units. They are also rather compact and usually are supplied with wire leads preconnected. A common application for mercury switches is in mechanical home thermostats. The mercury switch is attached to a bimetal coil. When the temperature drops, the coil tilts the switch and contact is made. Figure 4-53 shows a typical commercial mercury switch.

Float Switches

Detecting fluid levels is another common requirement within industry. For these applications specialized float switches are manufactured in a variety of patterns.

Figure 4-54 shows how an ordinary limit switch may be configured to act as a float switch. The pivot arm carries a rod

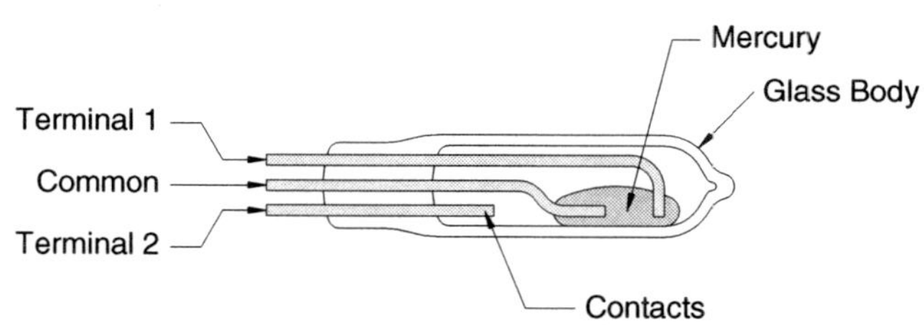

Figure 4-53 Commercial Double Throw Mercury Switch

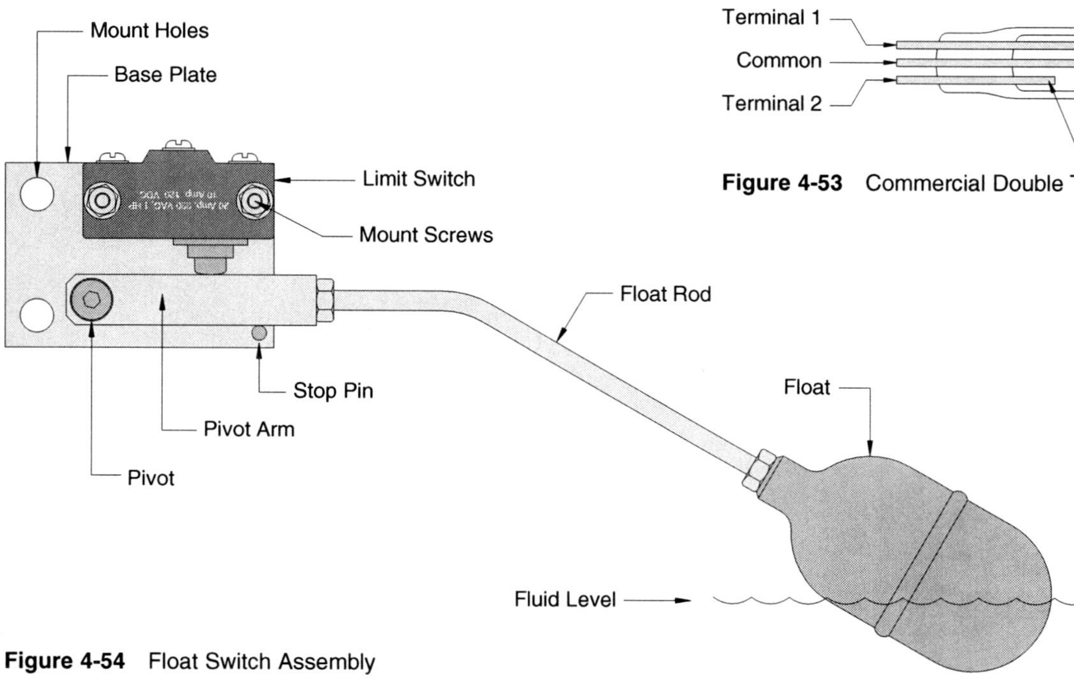

Figure 4-54 Float Switch Assembly

and float. The lower limit of the arm is controlled with a stop pin. As the liquid rises, the float lifts the arm which, in turn, trips the limit switch.

Figure 4-55 shows a simple free floating switch. A mercury switch is sealed into a rubber bladder and allowed to hang freely from its cable. When the fluid level rises, the bladder floats on its side and the mercury closes the switch.

Figure 4-56 shows a few common float switches. Through mount switches are used in tanks where the switch can only be installed from the outside. The float carries a magnet, which trips a magnetic switch in the body when it is aligned with the housing. By adjusting the float to the bottom or the top during installation, the switch can be set up as a normally open or a normally closed unit.

Top mount float switches are intended to be mounted in a vertical position, as shown. The float carries a magnet which trips a magnetic switch in the body when it is at the top of the housing. These switches can generally be converted from normally open to normally closed by flipping the float over.

Free floating switches are commonly used in sump applications. When the fluid level gets high enough, the float switch turns on a pump which discharges the contents of the sump. When the level is low enough, the pump shuts off. These switches are commonly available with a switched AC receptacle, which makes them very easy to install.

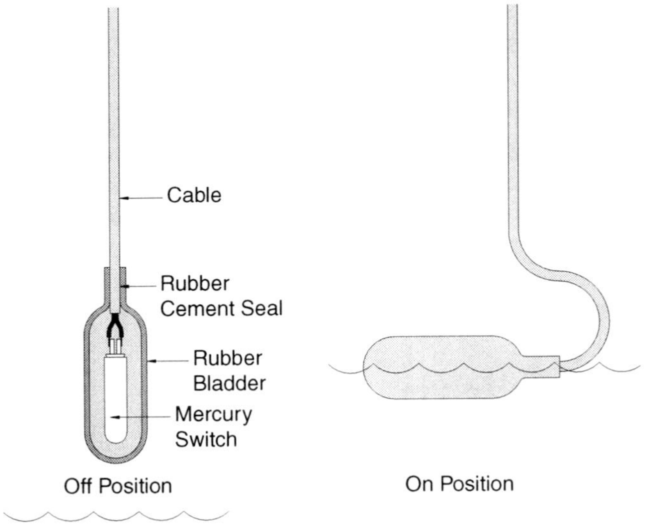

Figure 4-55 Free Floating Switch Assembly

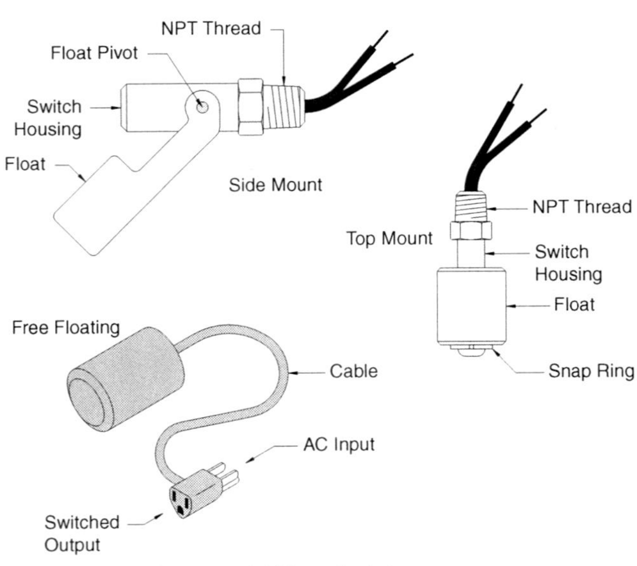

Figure 4-56 Commercial Float Switches

Contactors

For applications that require high-current switching, it is impractical to use manually-actuated switches. It is necessary to provide an interface between a small, operator-friendly switch and the high-current switching requirement. In addition to the current considerations, many loads must be switched from remote locations and it is impractical or too costly to install long runs of heavy gauge wire. Contactors are used for these applications.

A contactor is a set of high-current contacts that are actuated with a solenoid for the sole purpose of providing an on/off function. The solenoid typically requires a low voltage, low current signal and, therefore, can be actuated from remote locations with very light wire and a high degree of safety.

Figure 4-57 shows a knife switch contactor. When the solenoid is off, the return spring pulls the blade into an upright position and opens the contactor. When the solenoid is energized, the blade is pulled into the contacts and the contactor is closed.

Figure 4-58 shows a schematic representation of a basic contactor circuit. The control switch only controls the coil power. The main power is switched on by the heavy-duty contacts.

Commercial contactors, like the one shown in Figure 4-59, are readily available in many different configurations, voltages, and currents. Commercial contactors are commonly available with current ratings as high as 200 amps per pole. A four-pole contactor, with 125 amp contacts, can be wired in parallel and provide as much as 500 amps of switching capacity, all from a very compact and inexpensive package.

Figure 4-60 shows a sectional view of a typical commercial contactor. Take note of the visual indicator that can also double as a manual override. This is a particularly useful feature to service technicians.

Some contactors are supplied with auxiliary contacts to make setting up the controls a little easier. The schematic

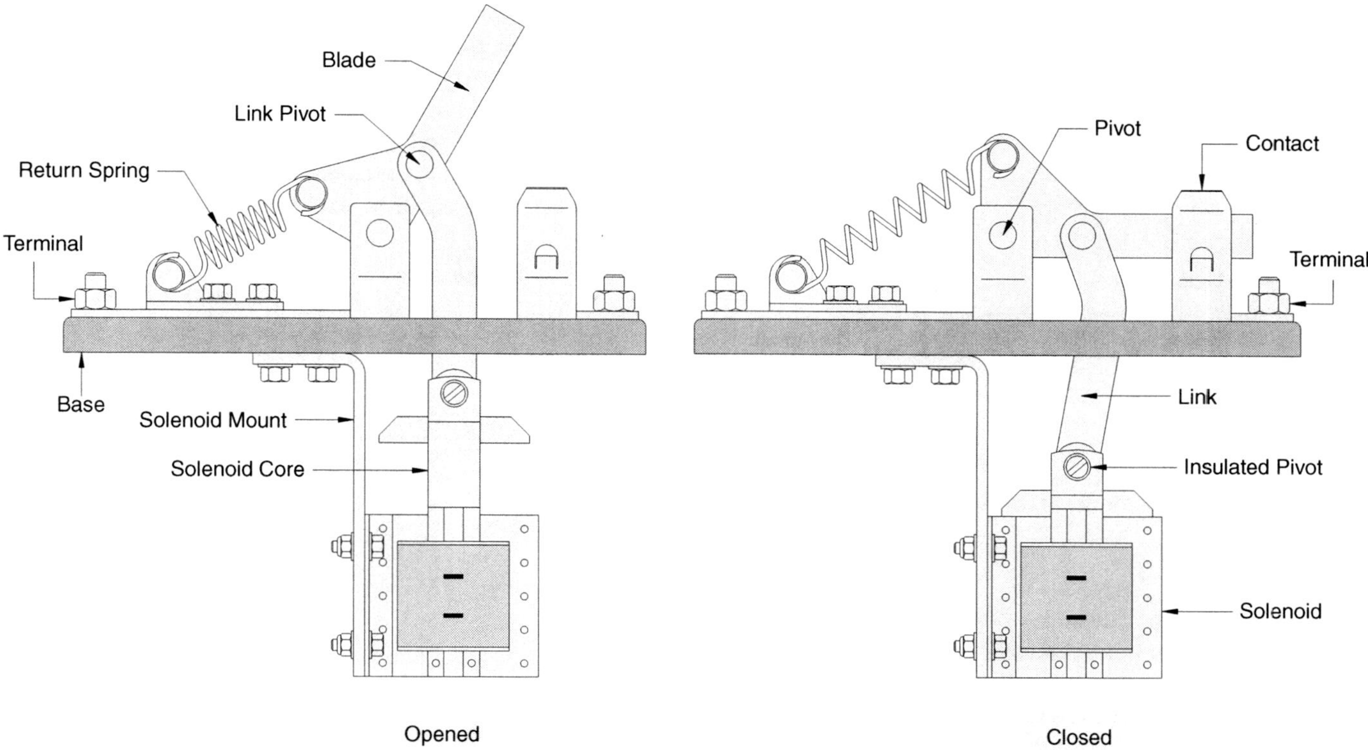

Figure 4-57 Knife Switch Contactor

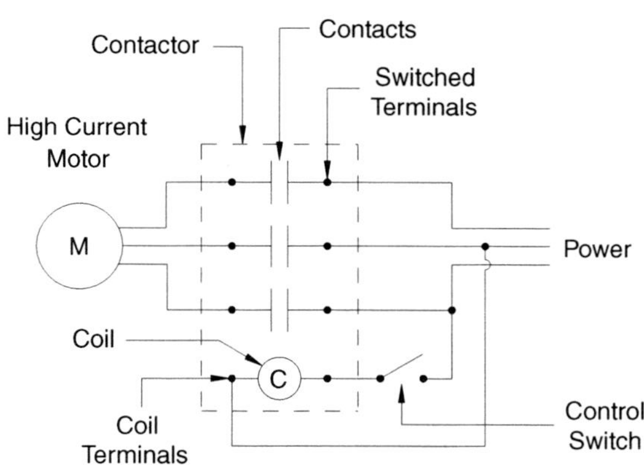

Figure 4-58 Contactor Schematic

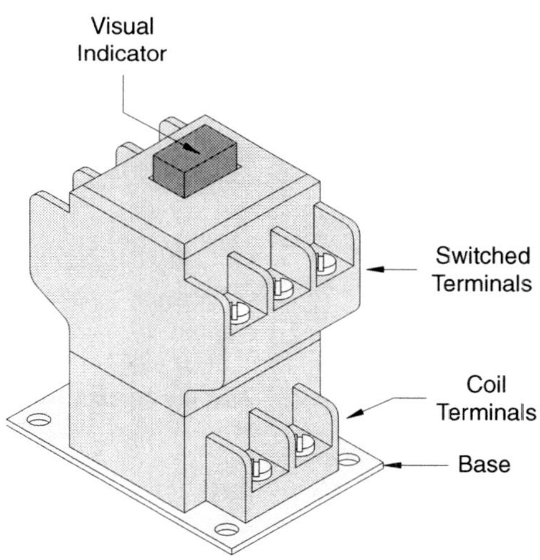

Figure 4-59 Commercial Three-Phase Contactor

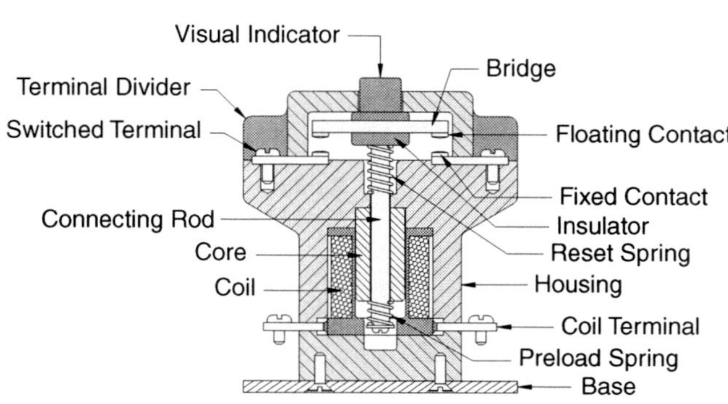

Figure 4-60 Contactor Sectional View

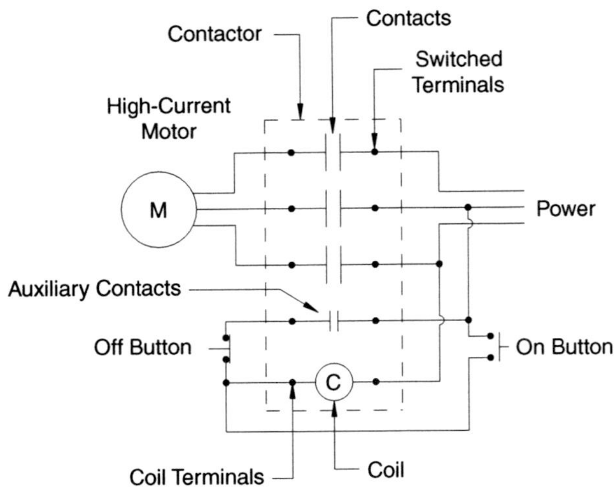

Figure 4-61 Contactor Schematic with Auxiliary Contacts

shown in Figure 4-61 illustrates how a set of push buttons could be wired to control a contactor with a single set of auxiliary contacts.

Figure 4-62 shows a commercial contactor with a two sets of auxiliary contacts. Generally, the auxiliary contacts are offered as an option for a standard contactor.

A basic motor controller, as shown in Figure 4-63, is an excellent example of how a contactor can be applied. In this case, the coil voltage is matched to the line voltage and a set of overloads are supplied on the outputs. The overloads will be discussed in greater detail in Chapter 8.

For greater safety, the circuit that controls the coil, or control circuit, uses a lower voltage than the line voltage. In these cases the contactor coil is a low-voltage unit and a step-down transformer is added to the controller. The low-control voltage is much safer and easier to work with. Figure 4-64 shows a

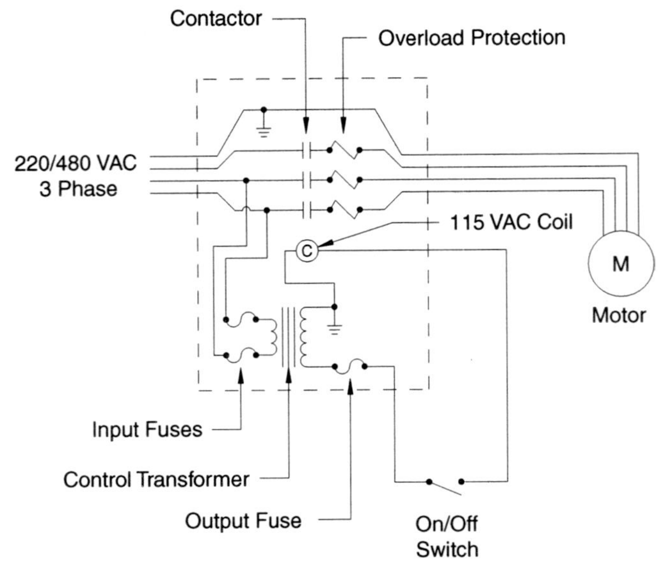

Figure 4-64 Three-Phase Motor Controller with 120 VAC Control Circuit

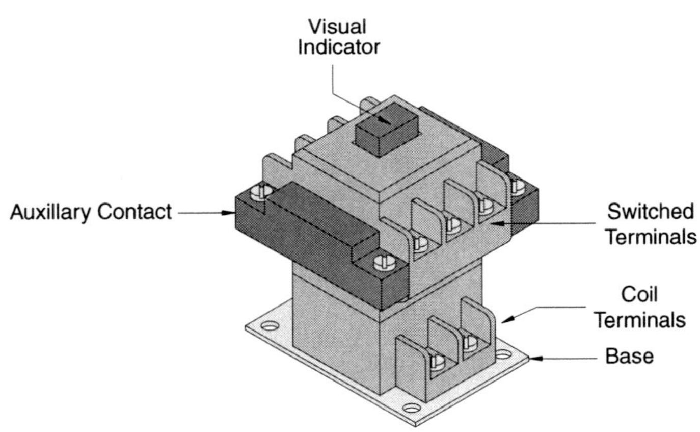

Figure 4-62 Commercial Three-Phase Contactor with Auxiliary Contacts

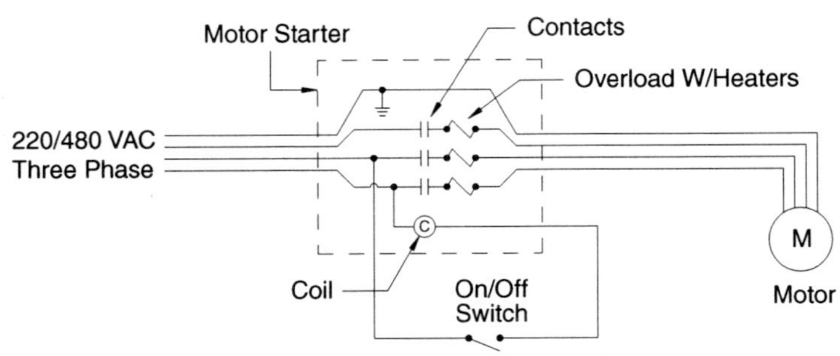

Figure 4-63 Three-Phase Motor Controller

schematic representation of a motor controller with a low voltage control circuit. Figure 4-65 shows how the finished assembly of this type of motor controller may appear. Note the fuses on the input and output of the control transformer. Because the control circuit is a stand-alone, system it must be protected separately from the rest of the system.

Figures 4-66 and 4-67 show how the control circuit can be interrupted to provide an interrupt in the event that unacceptable parameters are detected. The loop can be loaded with all manner of sensors. Notice that the low oil pressure sensor is a normally open switch. To start the machine this sensor must be overridden until the pressure builds to an acceptable level.

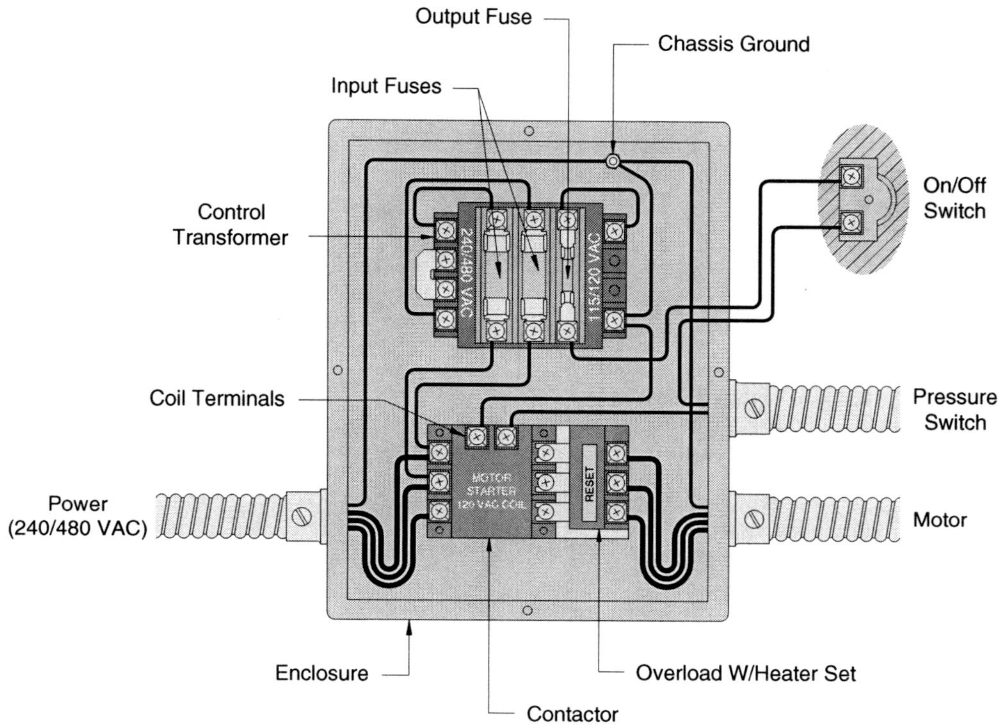

Figure 4-65 Three-Phase Motor Controller with 120 VAC Control Circuit

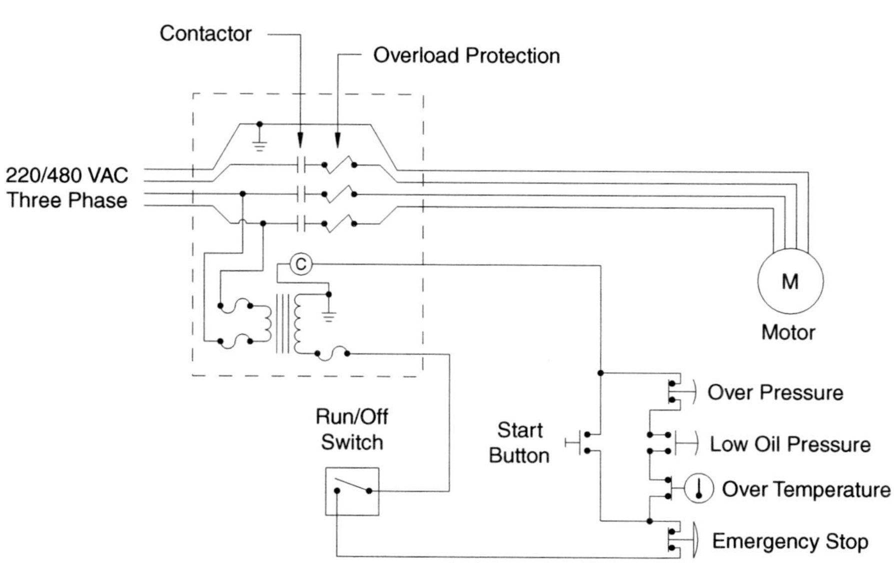

Figure 4-66 Three-Phase Motor Controller with Sensor Loop

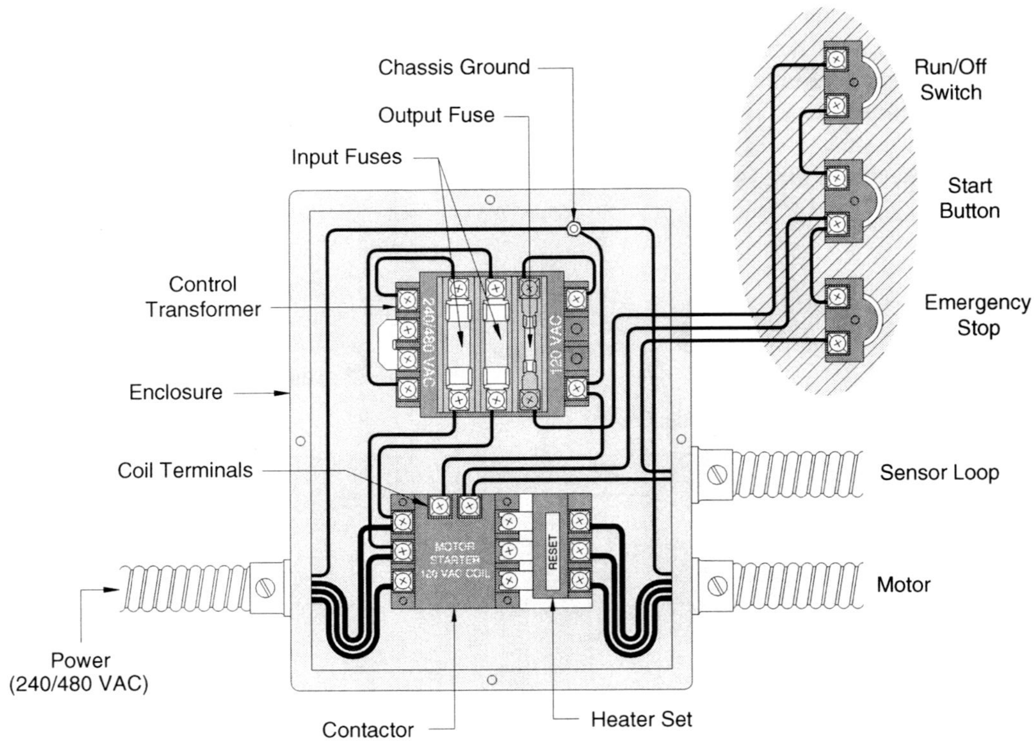

Figure 4-67 Commercial Motor Controller with Sensor Loop

This is the function of the start button, which bridges the sensor loop.

Figures 4-68 and 4-69 show a control circuit for a commercial screw compressor. This particular unit is designed to operate either continuously or in an automatic mode and provides a fault indicator in the event a sensor shuts down the system.

Another common use for contactors is in reversing circuits. A three-phase motor can be reversed by simply reversing two of the power wires. By using two contactors side-by-side, a

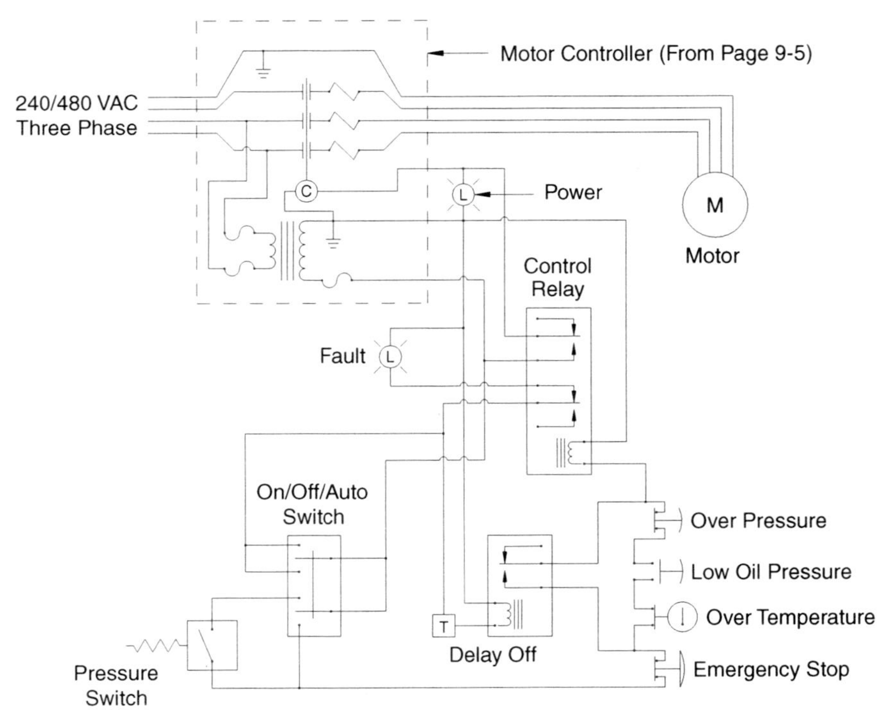

Figure 4-68 Three-Phase Motor Control Circuit Diagram with Run/Automatic Mode, Sensors, and Fault Indicator Lamp

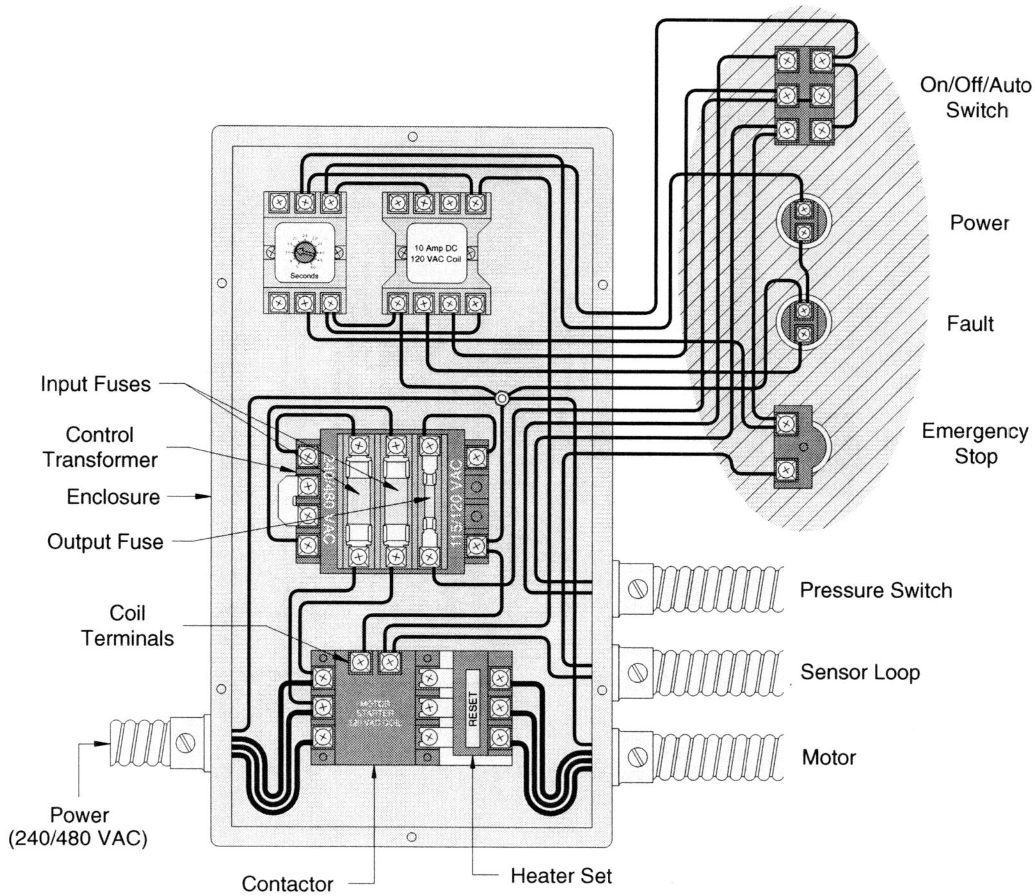

Figure 4-69 Motor Controller with Sensor Loop, Alarm, and Automatic Function

simple reversing circuit can be configured. The first, or forward, contactor is wired normally. The second, or reverse, contactor is wired with two conductors reversed. The contactors are controlled with a single-pole, double throw, center-off switch. When the forward contactor is energized by the control switch, the motor operates in forward. When the control is switched to reverse, the forward contactor is de-energized, the reverse contactor is energized, and the motor reverses. The center-offposition de-energizes both contactors. Figures 4-70 and 4-71 show how a simple, three-phase, reversing circuit can be constructed.

Delta/Wye Motor controllers typically use three contactors. The idea behind this configuration is to start the motor in a low-torque mode and then switch to a high-torque mode for run. This arrangement can be particularly useful for equipment with high inertial starting loads, such as punch presses.

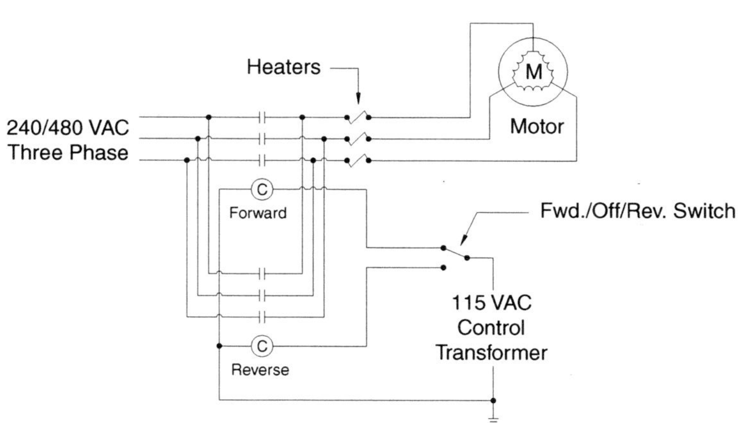

Figure 4-70 Three-Phase Motor Reversing Circuit

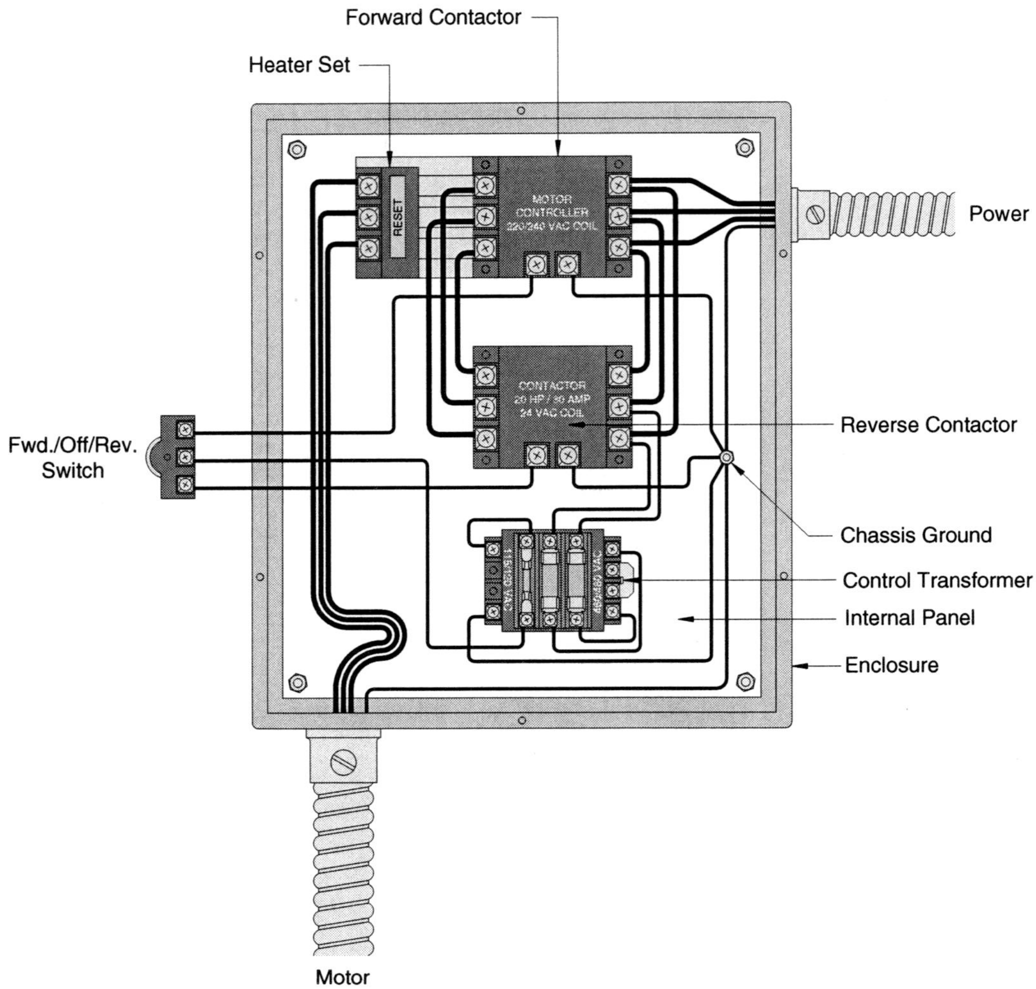

Figure 4-71 Three-Phase Reversible Motor Controller

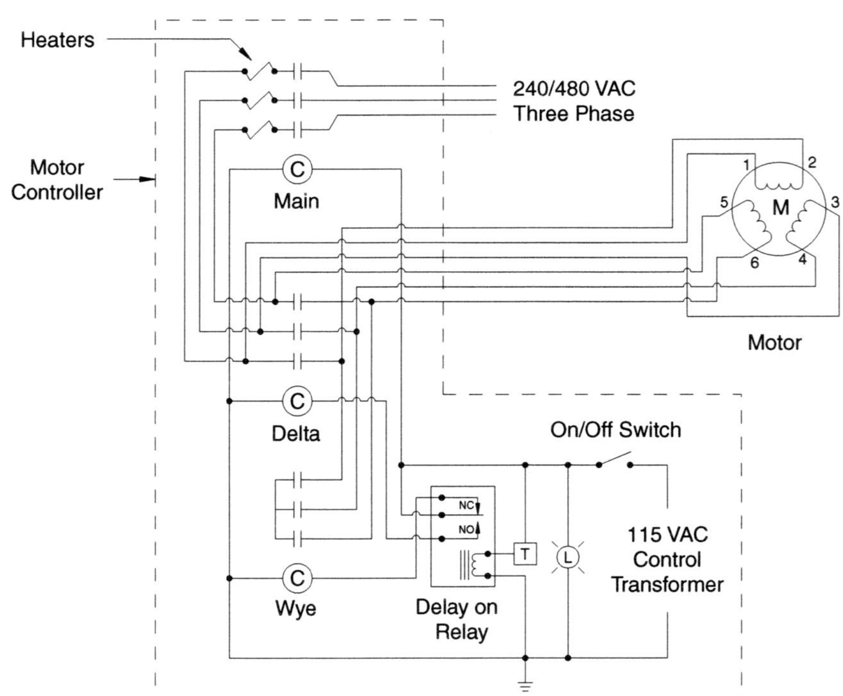

Figure 4-72 Delta/Wye Motor Controller Schematic

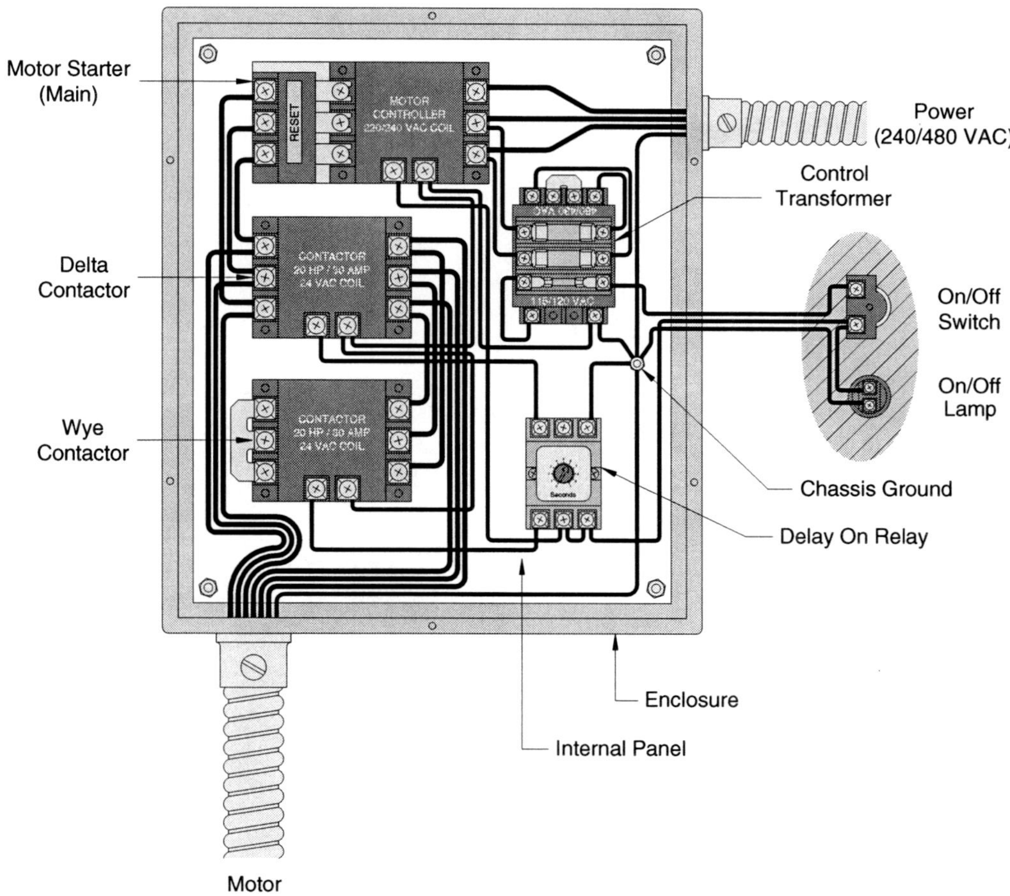

Figure 4-73 Delta/Wye Motor Controller

When the motor is turned on, the starter connects the motor in a Wye configuration which will produce lower torque characteristics. After a predetermined period of time, the Wye contactor opens and the Delta contactor closes. The controller remains in a delta configuration until the motor is restarted. Figures 4-72 and 4-73 show a basic Delta/Wye motor starter configuration.

Large three-phase resistive furnaces are typically controlled with contactors such as those shown in Figures 4-74 and 4-75. These furnaces use arrays of heating elements that are independently switched via contactors. The control circuit for the contactors is typically connected to a sector thermostat as shown in the schematic.

Relays

Relays are similar to contactors, except that they are generally designed to emulate higher-level switch functions. Relays are normally multipole, double throw devices and are usually designed for low-current switching. Relays are used extensively for control applications and are found in nearly every electromechanical appliance manufactured.

Figure 4-76 shows a single-pole, double throw knife switch relay. Take particular notice of its similarity to the

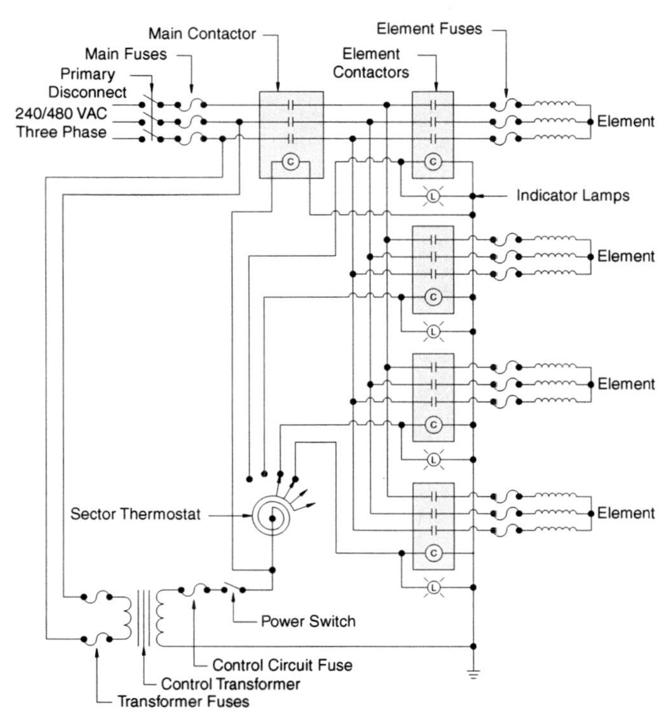

Figure 4-74 Electric Furnace Controller Schematic

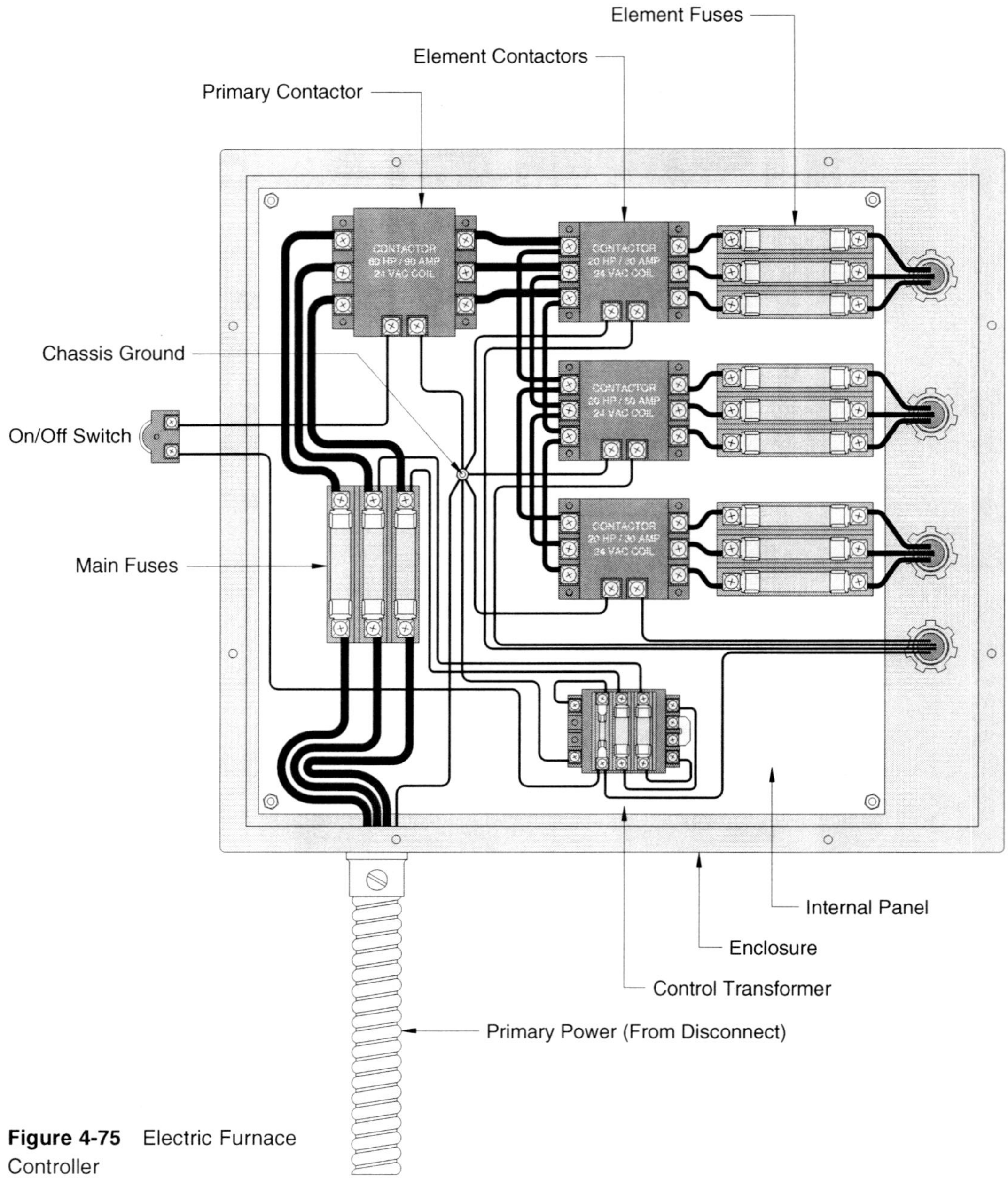

Figure 4-75 Electric Furnace Controller

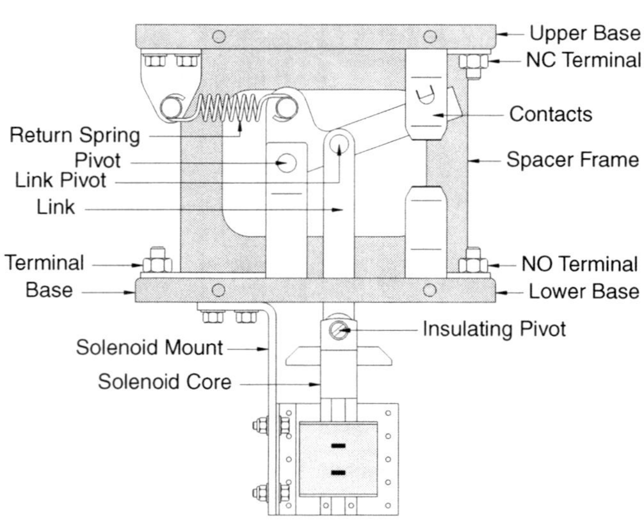

Figure 4-76 Double Throw Knife Switch Relay

knife switch relay shown in Figure 4-57. The principal difference is that the blade closes a contact when it is in the upright position. In this manner the relay has a normally open terminal and a normally closed terminal.

Figure 4-77 shows a typical double throw relay. A relay of this type may have as many as eight poles. These units are available in a wide range of configurations and capacities. They are often delivered with a protective plastic case which protects the mechanism from dust and dirt.

Reed relays are commonly found in communication and test equipment. These relays are essentially the same switch as shown in Figure 4-19, except a solenoid replaces the push button. They are often configured with a number of poles designed for specific applications. Figure 4-78 shows a typical commercial reed relay.

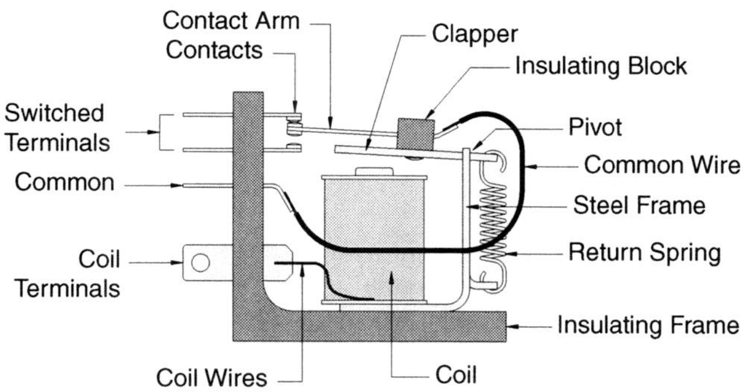

Figure 4-77 Double Throw Relay

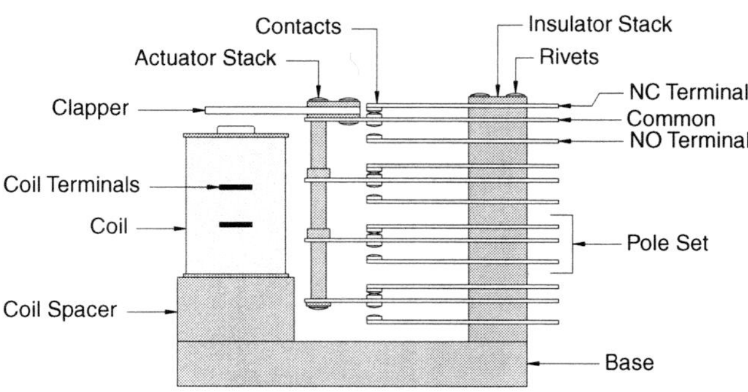

Figure 4-78 Four-Pole, Double Throw Reed Relay

Time delay relays offer an element of control that is critical for many applications. At one time pneumatic time delay systems dominated this arena. Figure 4-79 shows how the basic knife switch relay can be fitted with a pneumatic delay cylinder. The cylinder is equipped with needle valves that restrict how fast the piston can move. When the solenoid is energized the cylinder slows the switching action and provides a delay. This is also true when the solenoid is de-energized.

Figure 4-80 shows a commercial pneumatic time delay relay. Most of these units are assembled onto a common frame using standard limit switches and solenoids. The only specialty item is the delay diaphragm.

Sector Relays

Sector relays operate as a type of selector switch. They are typically single-pole, multi-position devices with a bidirectional control system.

Figure 4-81 shows a typical sector relay for general purpose applications. This particular unit has 10 switched contacts with a common wiper. The solenoid is a dual-coil unit that provides bidirectional control to the wiper. The dash pot provides a level of damping to control over-travel. The control terminals are intended as position sensors and are used within the control circuit.

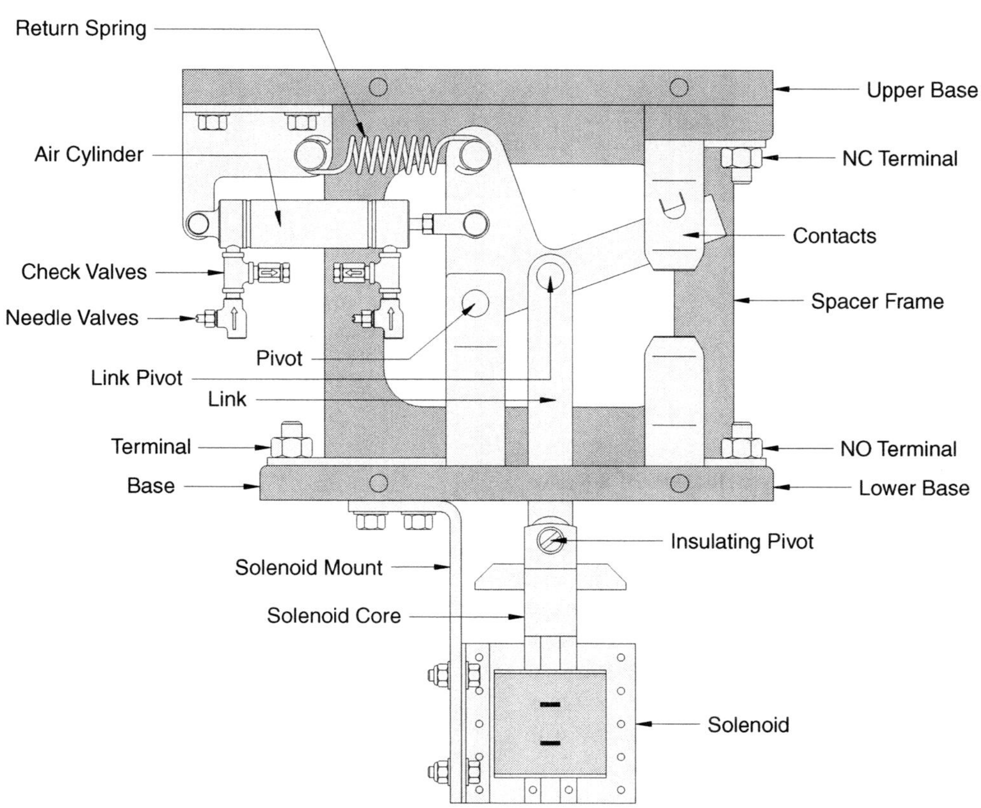

Figure 4-79 Double Throw Knife Switch Relay with Pneumatic Time Delay

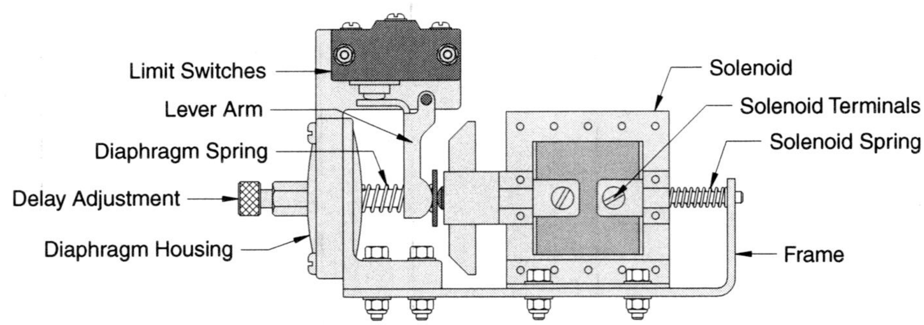

Figure 4-80 Commercial Relay with Pneumatic Time Delay

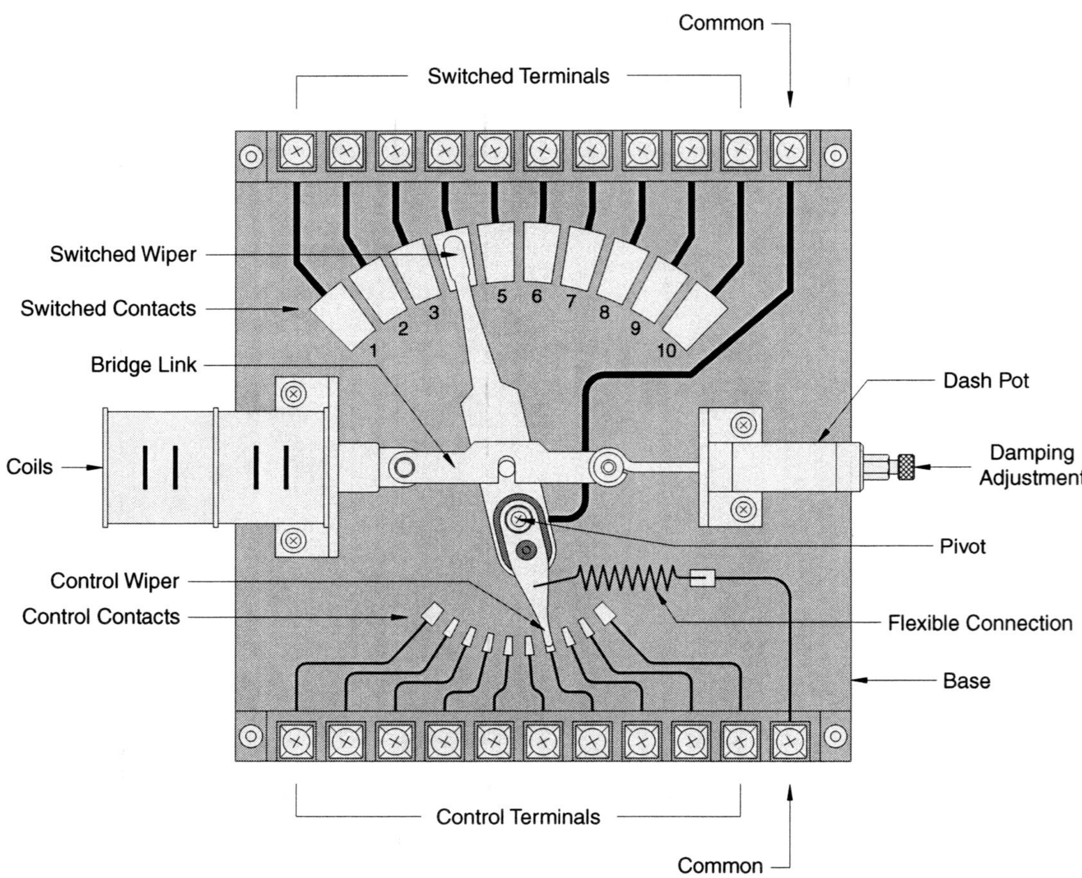

Figure 4-81 Sector Relay

Figure 4-82 shows a schematic representation of a sector relay control circuit. Each time the start button is pressed the relay advances one position either up or down, depending on the setting of the direction switch.

Figure 4-83 shows a sector relay configured as a voltage regulator for an engine-driven generator. As the output voltage of the generator climbs, it applies more power to the solenoid coil. The solenoid pulls the wiper to the right and switches more resistors into the field winding circuit. The rheostat is used to adjust the voltage regulation to the engine RPM. A schematic representation of this type of regulator is shown in Figure 4-84.

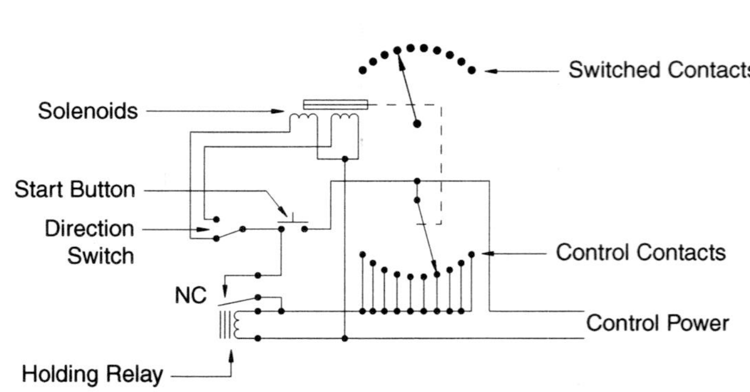

Figure 4-82 Sector Relay Control Schematic

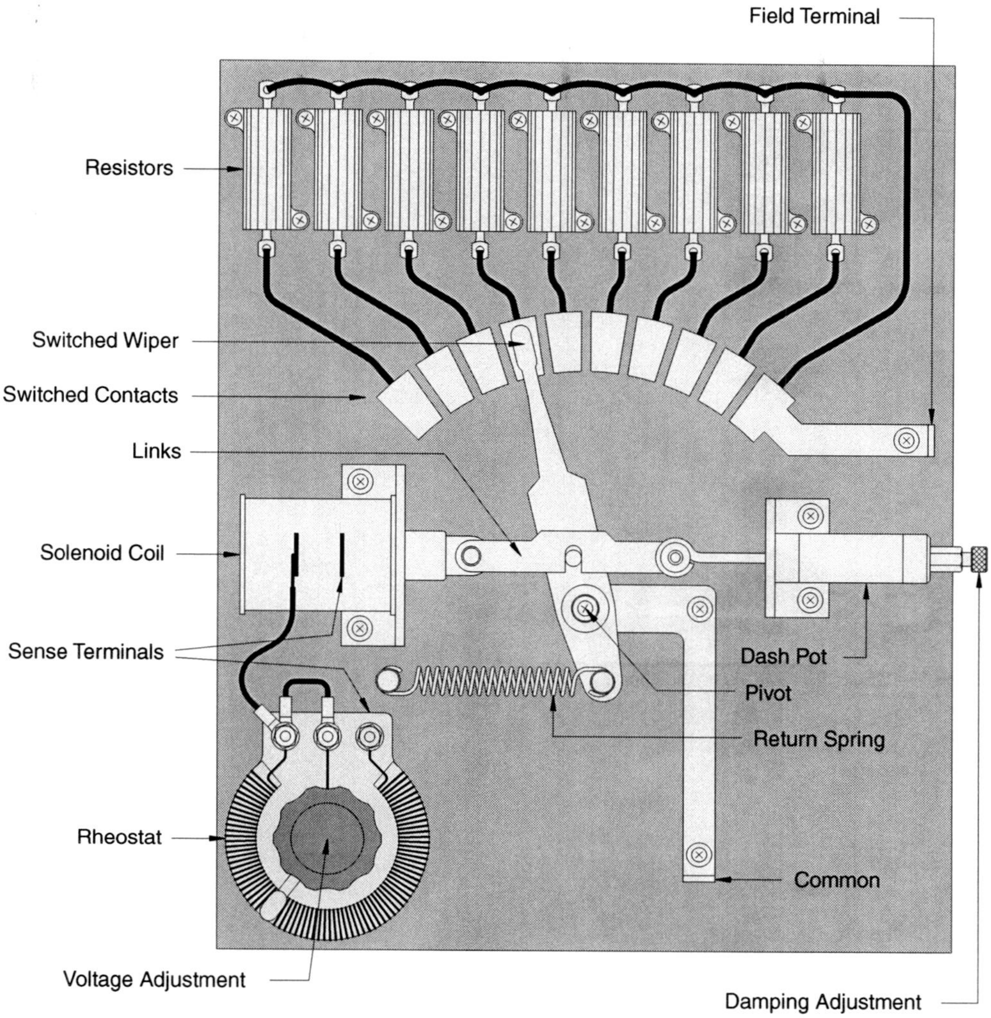

Figure 4-83 Sector Relay as a Voltage Regulator

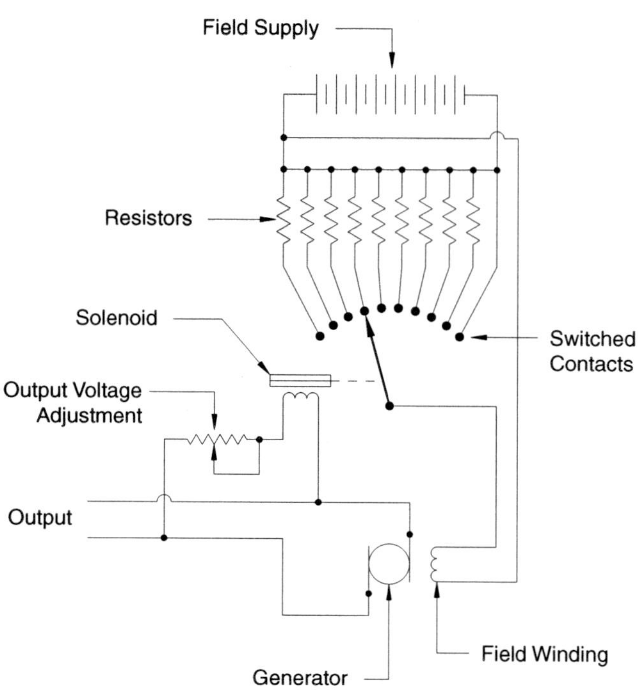

Figure 4-84 Voltage Regulator Schematic

Figure 4-85 shows a sector relay configured for communication applications. This type of relay was used on early dial telephone systems. Each time a pulse is sent to the advance solenoid, the wiper advances one position. If the number six was dialed, then six pulses would be sent to the relay and the wiper would advance six positions. These relays carry a reset solenoid that rotates the reset plate which, in turn, lifts the advance and hold ratchets. The return spring then resets the wiper back to zero.

Latching Relays

Latching relays are relays that can be switched and remain switched after power has been released. To switch the relay back, a second coil must be momentarily energized.

Figure 4-86 shows a latching knife switch relay. When the upper solenoid is energized it pulls the blade into the upper contacts. The contacts hold the blade in position even after the solenoid has been de-energized. To switch the relay the lower solenoid is energized and the blade is pulled into the lower contacts.

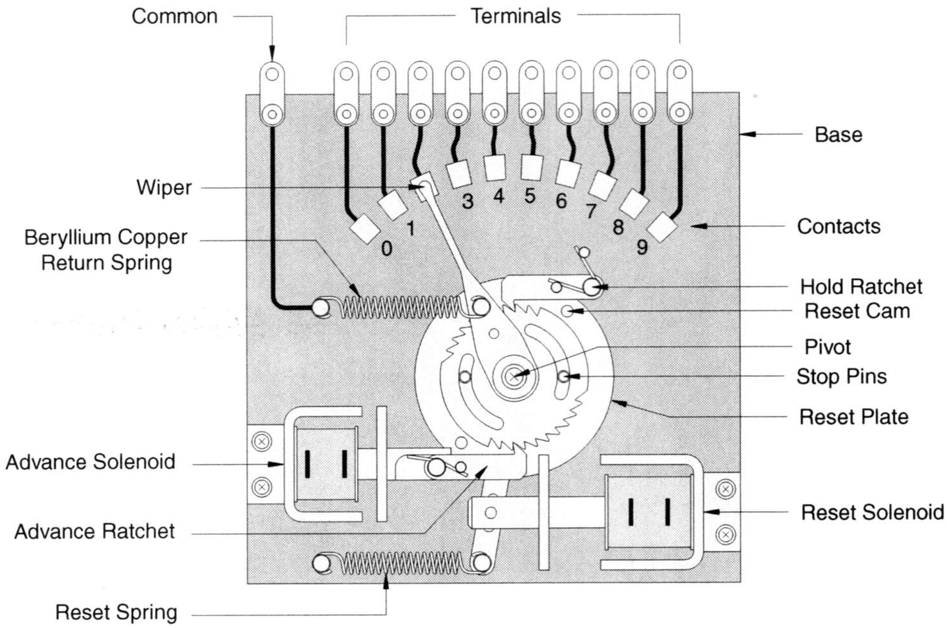

Figure 4-85 Communications Sector Relay

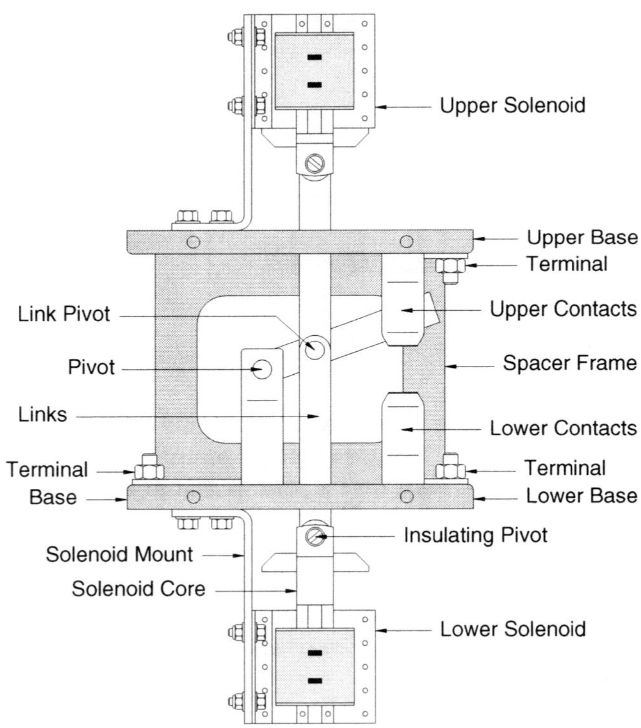

Figure 4-86 Double Throw, Latching Knife Switch Relay

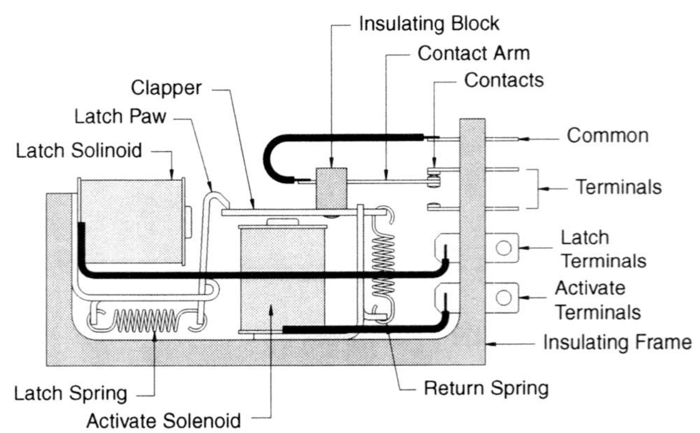

Figure 4-87 Commercial Latching Relay

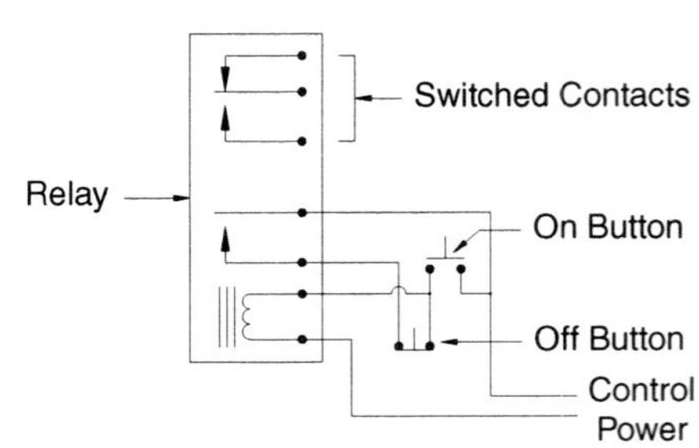

Figure 4-88 Holding Circuit

Commercial latching relays are similar in design to a standard relay, except that they incorporate a latching mechanism, as shown in Figure 4-87. When the relay coil is energized, the clapper is pulled down and captured by the latch paw. To reset the relay the latch solenoid is energized and the latch paw disengages.

Another method to latch a relay is to use one of the normally open set of contacts in a standard relay. Figure 4-88 shows a schematic representation of a typical holding circuit. When the on button is depressed, the coil is energized and the relay closes. The closed contacts feed power to the coil and the relay remains closed after the on button has been released. When the off button is pressed the power to the coil is broken and the relay resets.

Mercury pool relays are switching devices that are intended for very high currents. These types of relays are more like contactors in their function. The schematic representation shown in Figure 4-89 shows two pools of mercury that are in contact with the terminals. A solenoid pulls a conductor bridge down into the pools and the switch is closed. Figure 4-90 shows a typical commercial mercury pool relay.

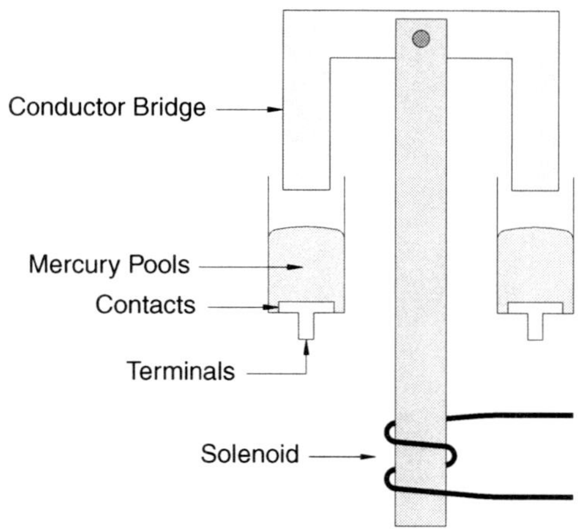

Figure 4-89 Mercury Pool Relay

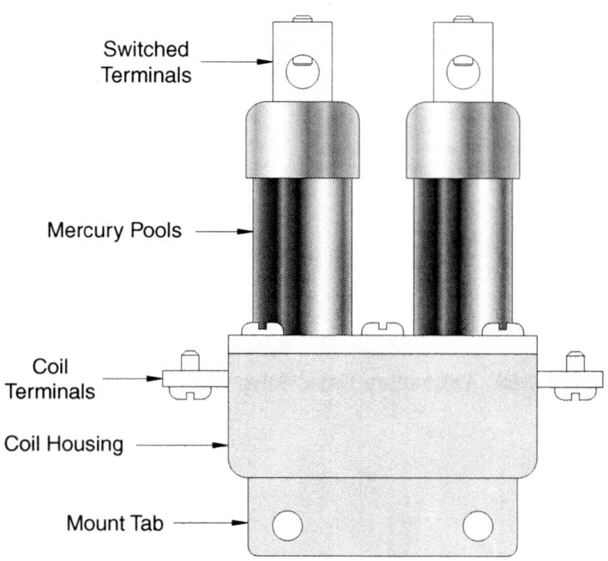

Figure 4-90 Commercial Mercury Pool Relay

Relay Sockets

Most small relays are designed to be used with some type of standard socket. Sockets are available in standard octal patterns or in square patterns. Octal pattern relays must be used with a socket; however, the square pattern relays are often dual purpose. These units can be placed into a socket or the terminals may serve as solder connections. Figure 4-91 shows a few of the more common commercial relay sockets.

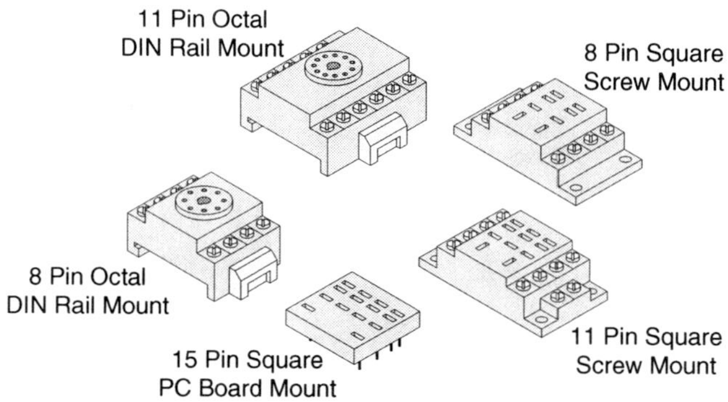

Figure 4-91 Commercial Relay Sockets

Motor Relays

Motor relays are similar to sector relays, except they are typically high-current devices and are driven by a gear motor. The control circuits that are typically used are similar to the sector relay controls. Figure 4-92 shows a 10 position, two-pole motor relay. The switch elements are standard limit switches actuated with a rotating cam.

Timers

Timers are devices that reference either a preset time interval or the 24-hour time cycle. In either case the timer typically trips a limit switch at the end of a time interval or at different times during the day.

The most common electromechanical timer is the ordinary clock. We all have experience with these devices. The typical wall clock uses a synchronous motor which operates in reference to the utility company's 60 Hz AC power. The motor usually drives a gear box with a 1 RPM output. The second hand is driven at a 1:1 ratio, the minute hand at a 60:1 ratio, and the hour hand at a 720:1 ratio. The hour hand on a 24-hour clock is driven at a 1440:1 ratio. Figure 4-93 shows a phantom view of a 12-hour synchronous motor wall clock.

Much like wall clocks, lab timers generally utilize synchronous gear head motors. The most common time interval for lab timers is 60 minutes (1 hour); however, these timers are available in a variety of other intervals ranging from 60 seconds to 48 hours. The mechanism shown in Figure 4-94 uses a 1 RPM synchronous gear head motor, which drives the pointer and trip cam at a 60:1 ratio. The pointer and trip cam are connected to the driven gear through a slip clutch. The operator sets the pointer to the desired interval by turning the knob, while at the same time the trip cam rotates along with the pointer. When the timer motor is energized, it runs until the cam trips the limit switch. Figure 4-95 shows a typical schematic for a synchronous motor lab timer. These units normally have one switched AC outlet and an audible alarm that can be turned on or off.

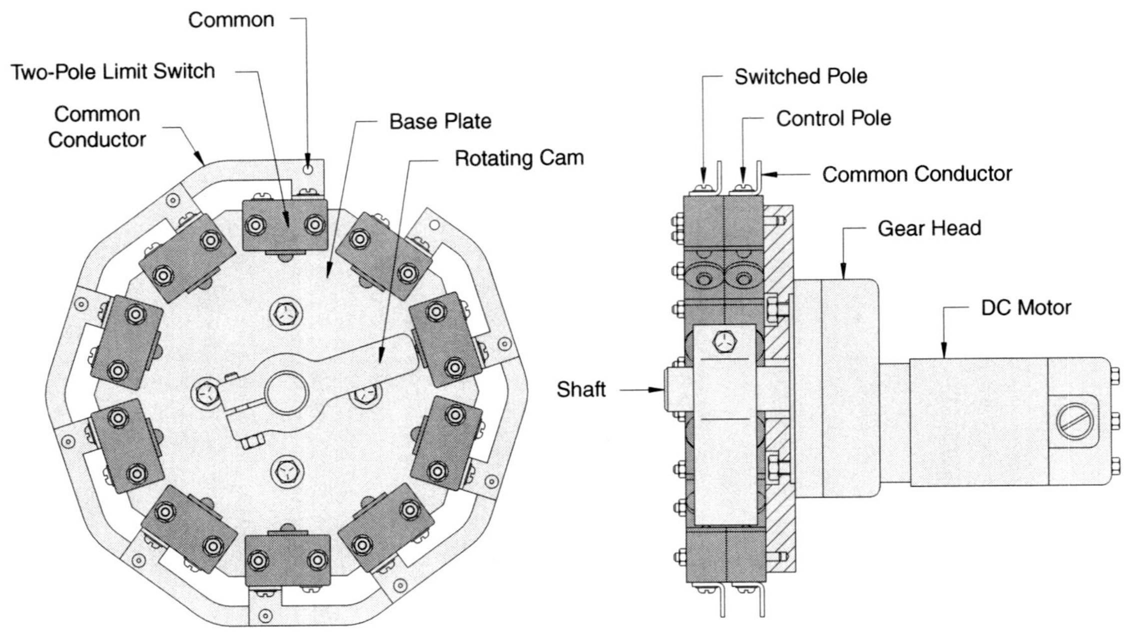

Figure 4-92 Motor Powered Relay

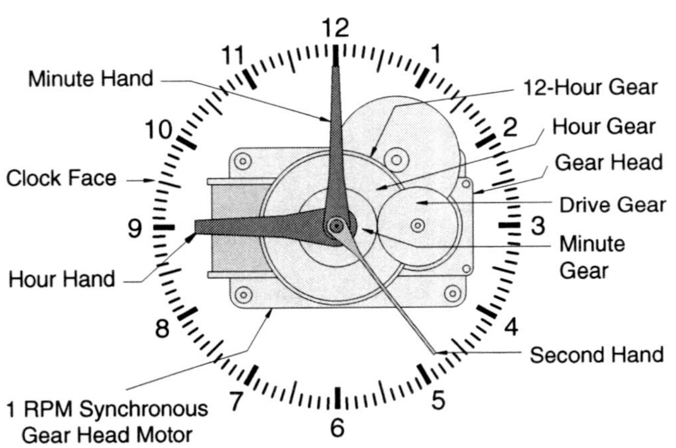

Figure 4-93 12-Hour Synchronous Motor Clock Drive

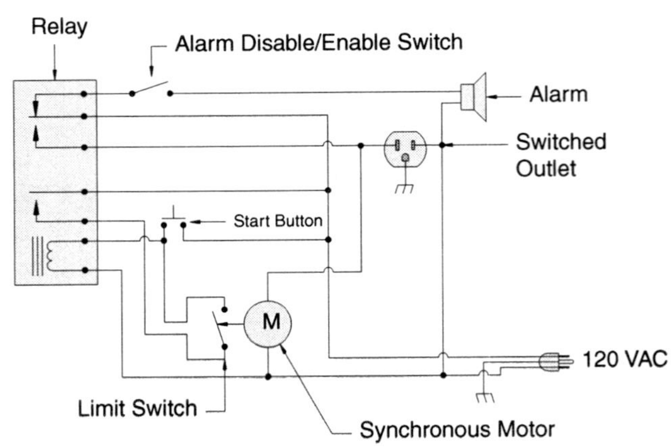

Figure 4-95 Laboratory Timer Schematic

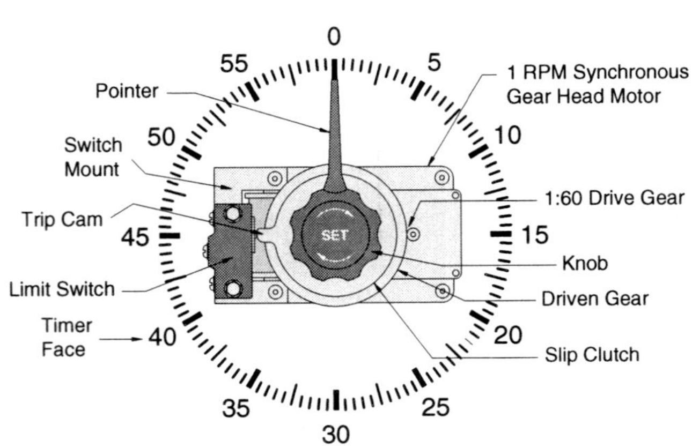

Figure 4-94 60-Minute Laboratory Timer Drive

For panel applications, spring return timers are particularly useful. These timers consist of a spring motor which can be set by rotating a knob. The operator rotates the knob to the desired interval. During rotation a spring is compressed and released to drive the timer. At the same time the time interval is set, the trip cam rotates along with the knob. Figure 4-96 shows a typical spring return timer.

For all practical purposes, digital time delay relays have become the preferred choice for most timer applications. These relays are available is a wide variety of configurations and, in most instances, are less expensive than their mechanical counterparts. Figure 4-97 shows just a few of the time delay relays that are commonly available in the market. Take note that these units are available in standard packages as well

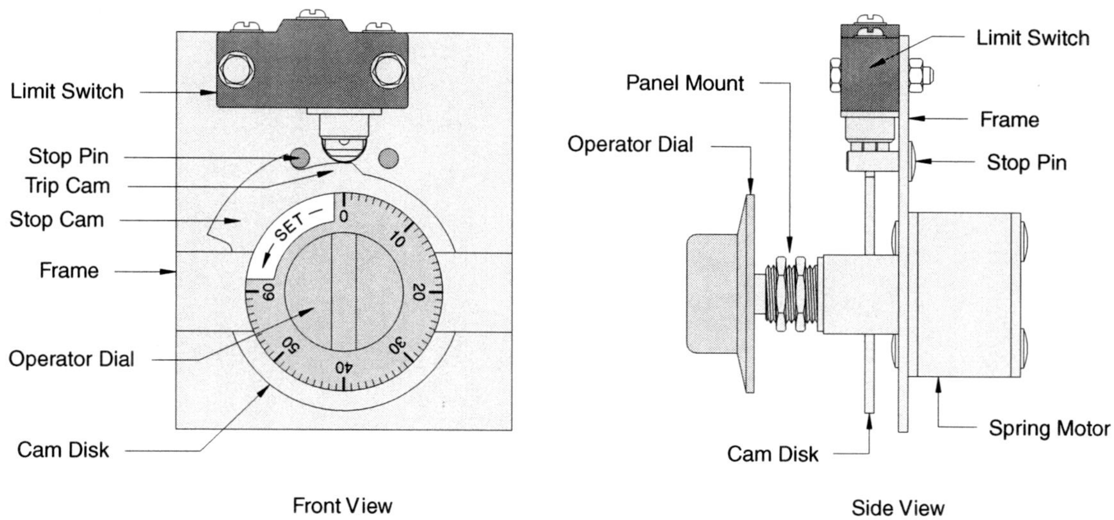

Figure 4-96 0- to 60-Second Spring Return Timer

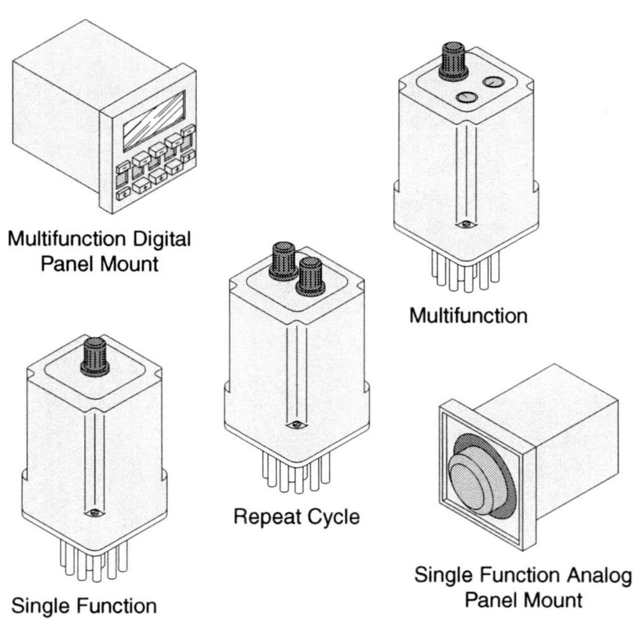

Figure 4-97 Commercial Time Delay Relays

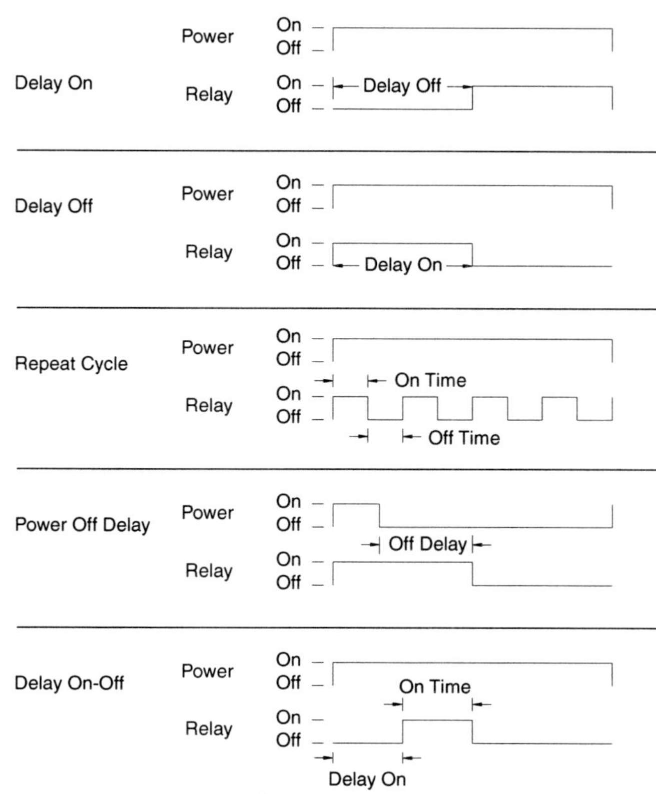

Figure 4-98 Timing Relay Functions

as panel mount configurations. Figure 4-98 shows graphical representation of the most common timing functions that these relays can provide.

Some lab timers are designed to use a ratchet drive with a repeat cycle relay for the clock. Figure 4-99 shows a typical ratchet drive lab timer. The ratchet has 60 teeth and is connected to the solenoid via the ratchet paw. The solenoid is momentarily energized once a second until the limit switch is tripped. Each time the solenoid is energized, the pointer and trip cam advances 1 second. Figure 4-100 shows a schematic representation of the control circuit that might be used on a ratchet drive timer. Like its synchronous counterpart, these timers generally have a switched AC outlet and an audible alarm.

Figure 4-101 shows a high-accuracy digital lab timer that uses a commercial multifunction timing relay. The relay provides excellent timing functions while the balance of the

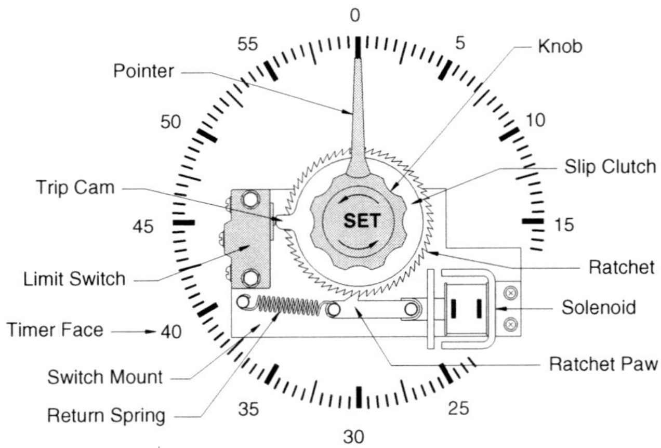

Figure 4-99 60-Second Ratchet Drive Timer

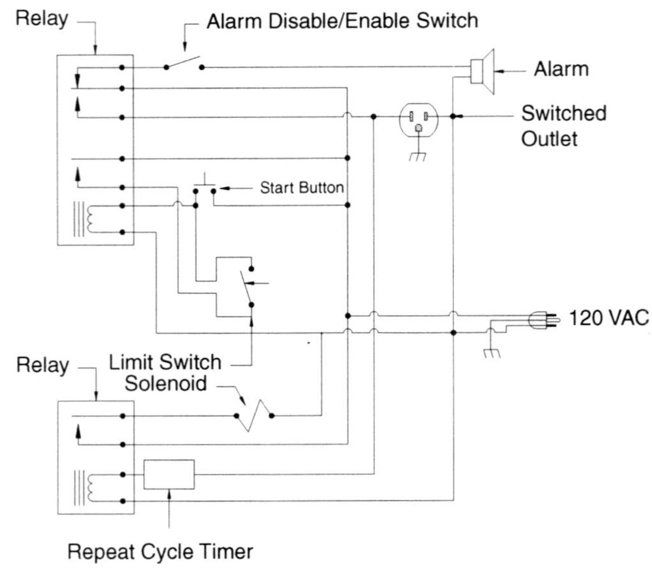

Figure 4-100 Ratchet Timer Schematic

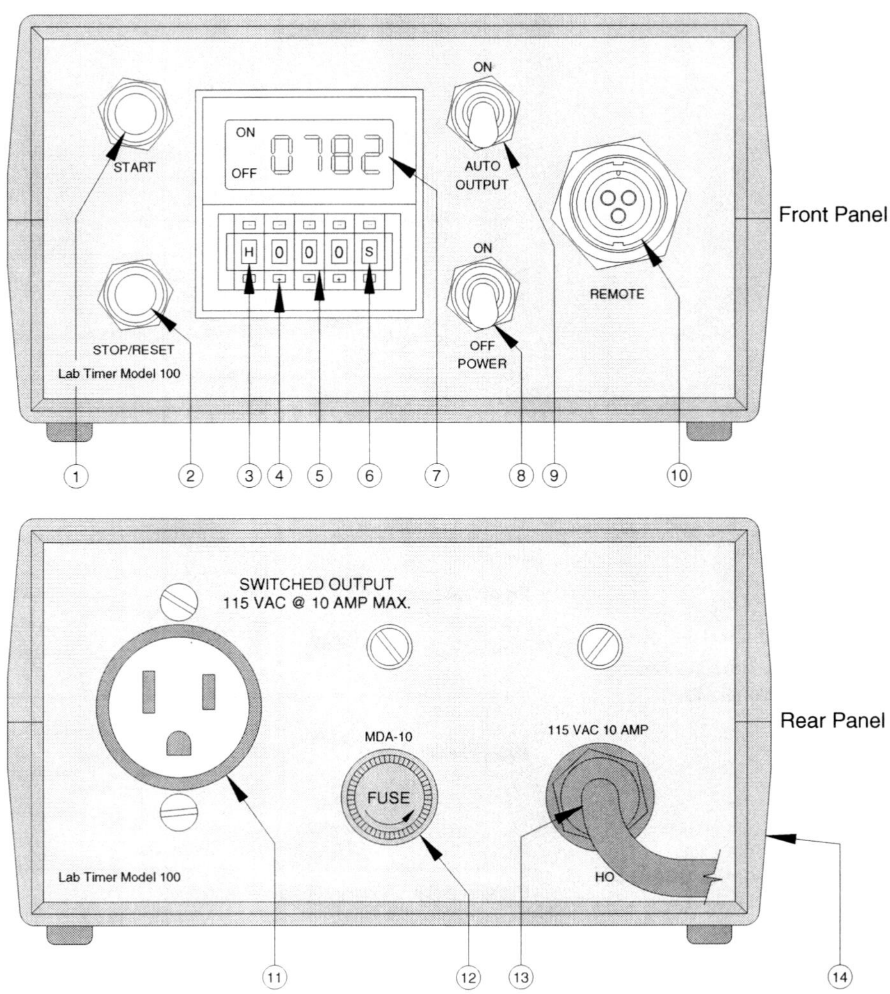

1. Start Button
2. Stop/Reset Button
3. Function
4. Selector Buttons
5. Time
6. Multiplier
7. Read Out
8. Power Switch
9. Automatic/Manual Switch
10. Remote Trigger Input
11. Switched Output
12. 10 amp Slow Blow Fuse
13. AC Cord
14. Cabinet

Figure 4-101 Bench Timer Using a Commercial Multifunction Relay

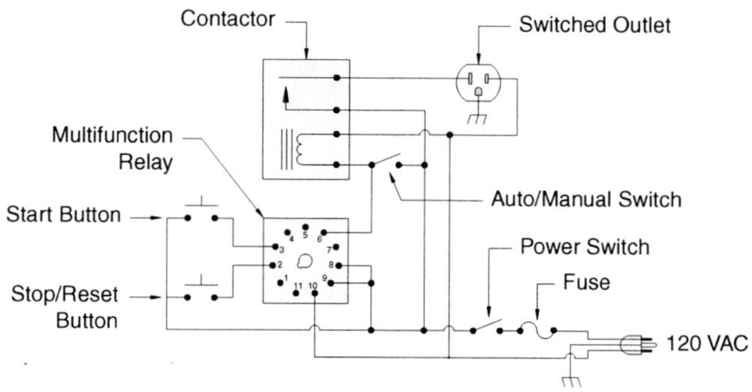

Figure 4-102 Bench Timer Schematic

timer components are simply support electrics. Figure 4-102 shows a schematic for the digital bench timer. The pin-out of the relay is dependent on the particular unit that is selected for the project. Figure 4-103 shows an exploded view of the chassis assembly.

One of the most common control timers in the market is the lamp timer. These units are available in most hardware stores and many drug stores and markets. The timer is a synchronous motor unit with a timer dial carrying multiple actuators. Generally these timers have 15 minute resolution. To trip the switch on the actuators which correspond to the desired "on" time are pulled out, as shown in Figure 4-104. These timers usually have a single switched AC outlet and a manual on-off switch.

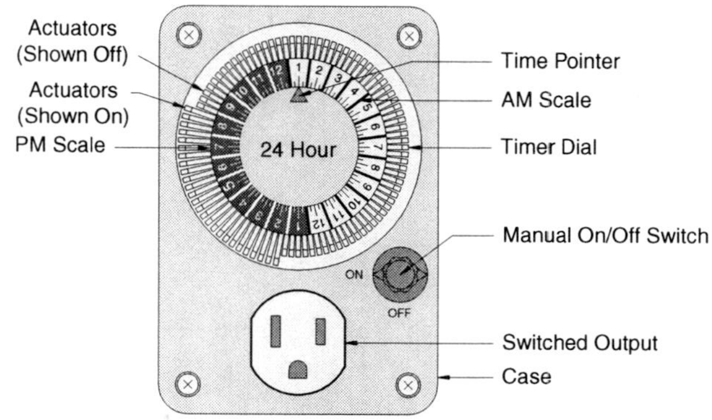

Figure 4-104 24-Hour AC Receptacle Timer

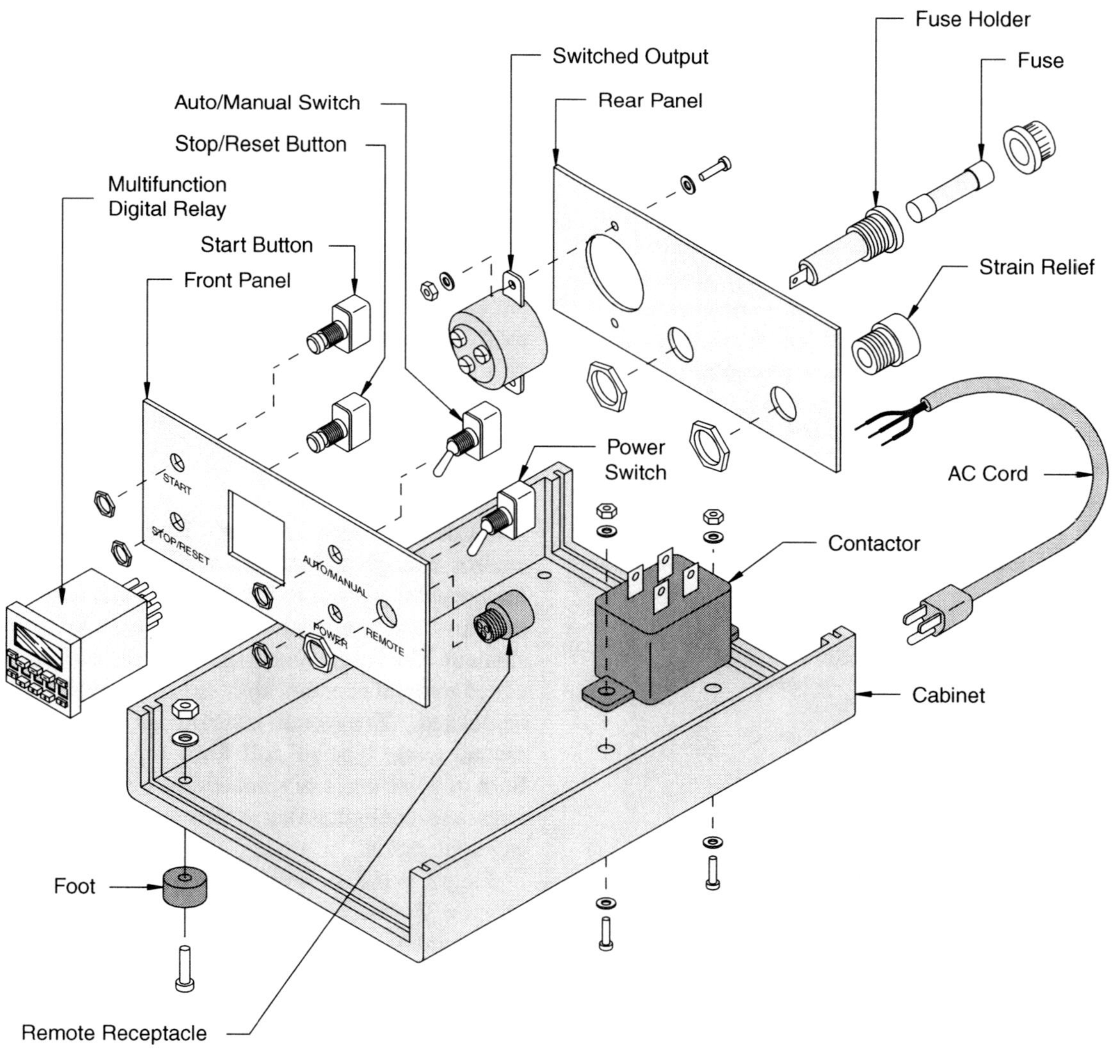

Figure 4-103 Bench Timer Exploded View

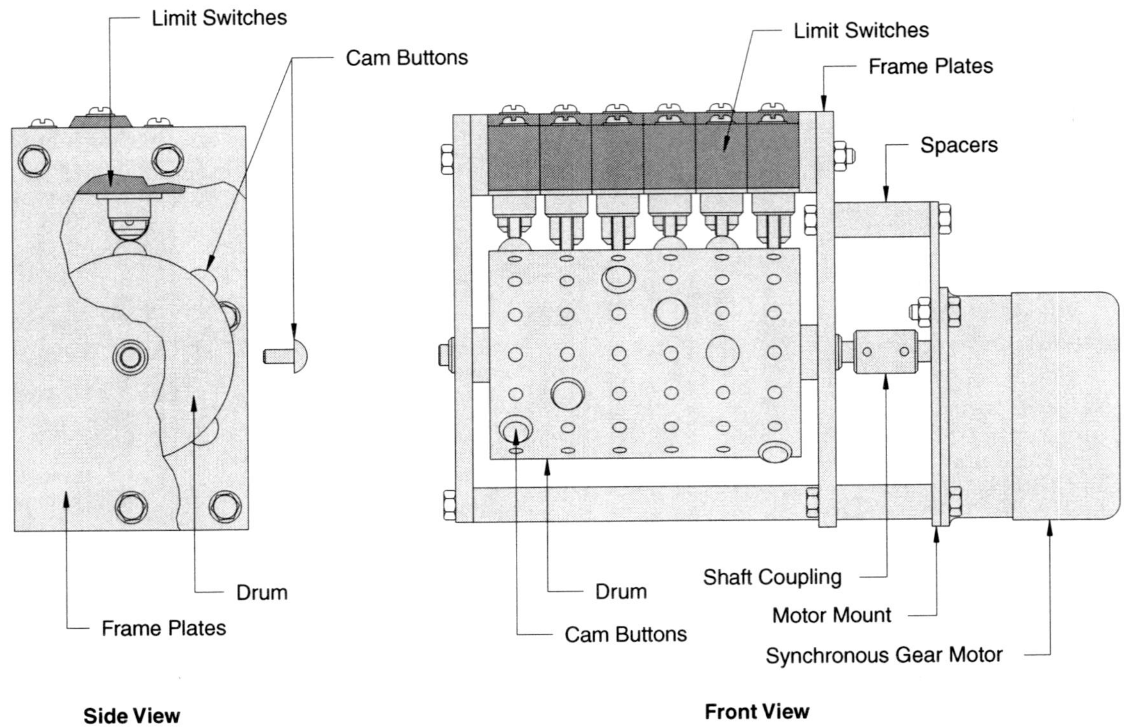

Figure 4-105 Cam Programmable Synchronous Motor Barrel Timer

Taking the ordinary wall timer to the other end of the spectrum is the cam programmable barrel timer. These timers can provide very sophisticated, complex, and precise control for all types of industrial equipment. A typical barrel timer has a drum with a series of buttons that trip limit switches as the drum rotates. The drum may have fixed buttons or may carry an array of holes so that the buttons can be placed in any location that is appropriate. The drum is generally driven by a synchronous gear head motor. Figure 4-105 shows a cam programmable barrel timer.

An inexpensive barrel timer can be constructed as shown in Figure 4-106. The drum is mounted on the shaft of the gear motor. A fine mesh brass screen is wrapped around the drum and insulating tape, plastic sheeting, or heat shrink covers the screen. A series of brushes are constructed from brass strips and mounted so that they come in contact with the outside diameter of the drum. To program the "on" cycles, holes are cut through the insulating cover. A timer of this nature can be easily constructed to have as many on/off cycles and poles as necessary for the application.

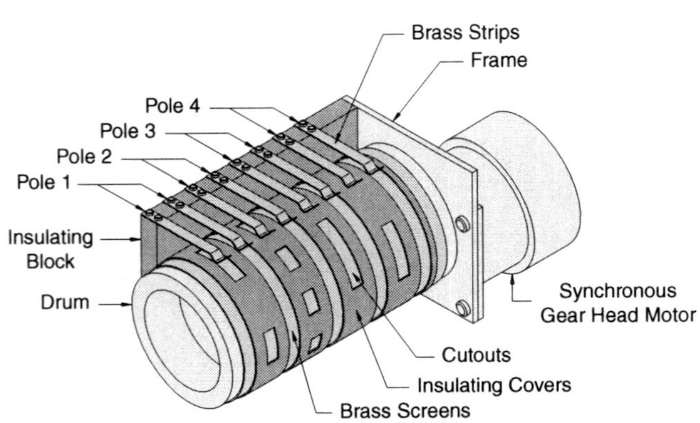

Figure 4-106 Four-Pole, Brush Contact Programmable Barrel Timer

Resistors

As is demonstrated in Ohm's law, resistors can be used to control voltage and current within a circuit. There are two fundamental types of resistors, carbon and wire wound. Carbon resistors use a measured plug of carbon for the resistive element. The length and diameter of the carbon can be manipulated to build resistors with virtually any resistance that may be desired. Wire wound resistors use a length of wire coiled around some type of coil form for their resistive element. Both of these types of resistors are available in a wide variety sizes and configurations which are appropriate for virtually any application.

Figure 4-107 shows a sectional view of a typical carbon resistor. The electrodes are impeded into the carbon plug and the assembly is encased into a plastic housing. These are very inexpensive items and are readily available.

The values of smaller-size resistors are generally identified with a four- or five- band color coding system. The color code system will clearly indicate the ohm rating of the resistor and

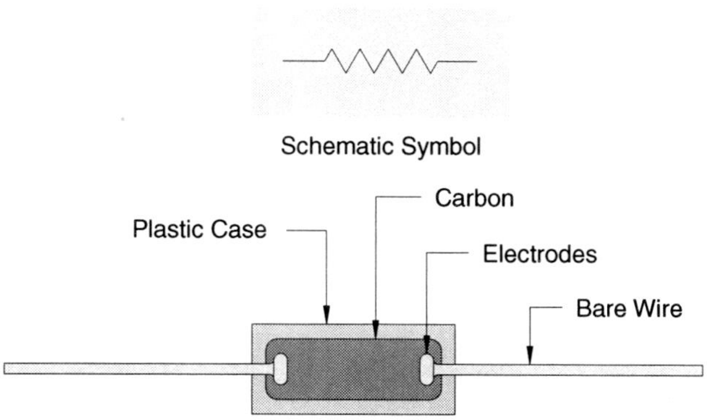

Figure 4-107 Carbon Resistor

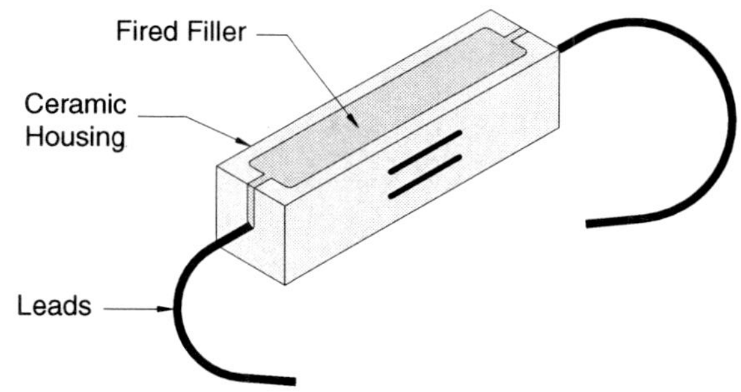

Figure 4-109 High-Wattage Ceramic Resistor

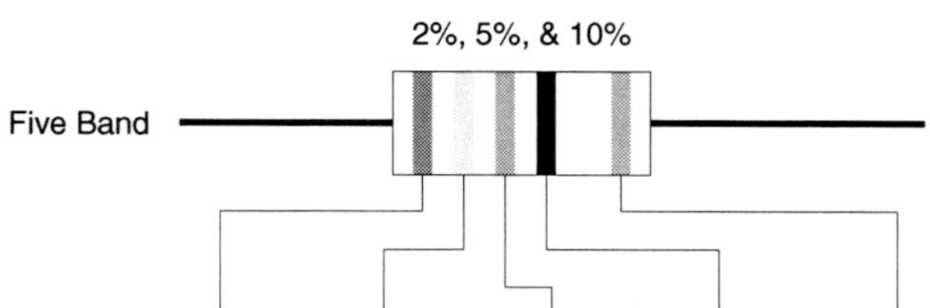

Color	1st Band	2nd Band	3rd Band	Multiplier	Tolerance
Black	1	1	1	1	1%
Brown	2	2	2	10	2%
Red	3	3	3	100	-
Orange	4	4	4	1K	-
Yellow	5	5	5	10K	-
Green	5	5	5	100K	0.5%
Blue	6	6	6	1M	0.25%
Violet	7	7	7	10M	0.10%
Gray	8	8	8	-	0.05%
White	9	9	9	-	-
Gold	-	-	-	0.1	5%
Silver	-	-	-	0.01	10%

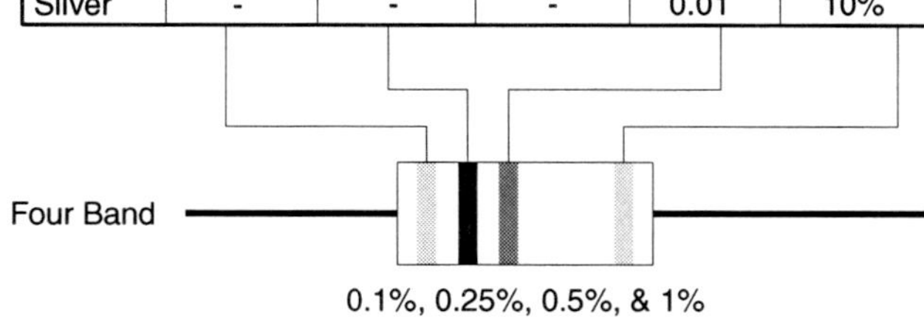

Figure 4-108 Four-and-Five Band Resistor Color Codes

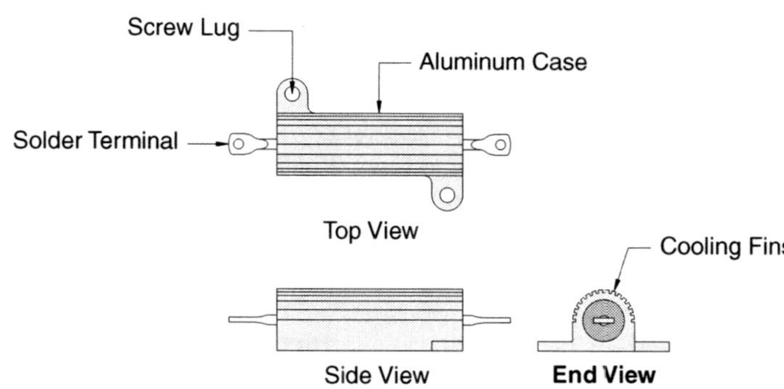

Figure 4-110 Screw Mount High Wattage Resistor

its precision. Figure 4-108 shows the standard color code chart for four- and five- band coding systems. Notice that the tolerance band is spaced further then the value bands. This provides a clear indicator of how to orient the resistor so that it may be read left-to-right.

Resistors that are required to carry a higher-current load are generally embedded into a ceramic housing, as shown in Figure 4-109. These types of resistors can become very hot during operation, so care should be exercised when working with live circuits using ceramic resistors.

For very high wattages, resistors are mounted into extruded aluminum housings that can be mounted to a heat sink. Figure 4-110 shows a high-wattage resistor which is appropriate for heat sink mounting. When mounting these devices it is a good idea to use a thermally conductive paste between the mounting surface of the resistor and the heat sink.

Figure 4-111 shows a wire wound resistor. The terminal wires are molded into the coil form. The resistive element is wound around the coil form and the ends are soldered to the terminal wires. Wire wound resistors are more expensive and less available than carbon resistors, however, they are commonly available through normal supply channels.

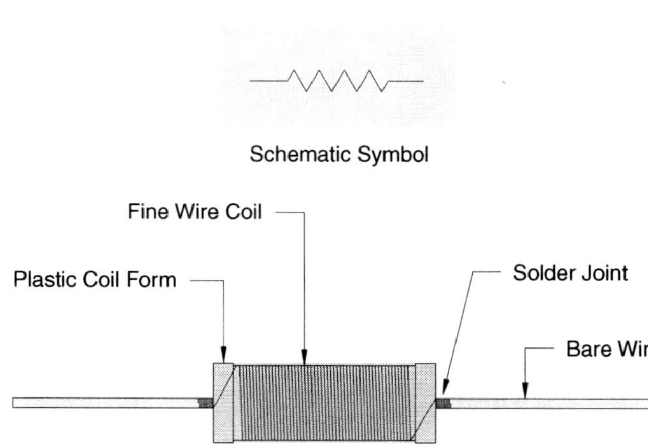

Figure 4-111 Wire Wound Resister

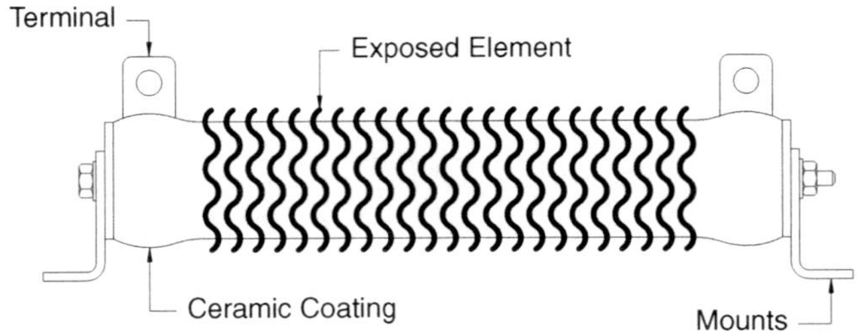

Figure 4-112 High-Wattage Industrial Resistor with Exposed Element

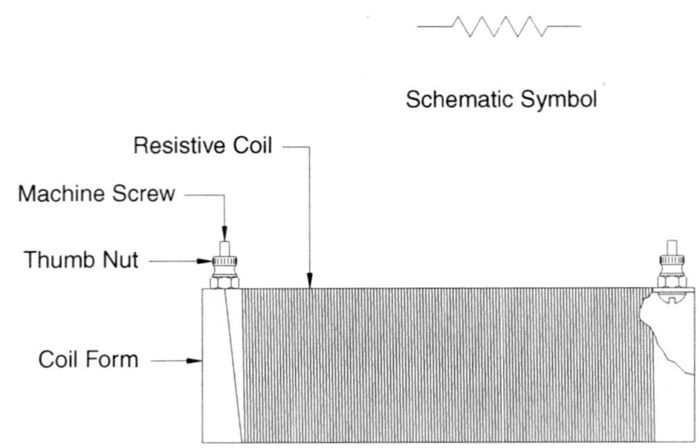

Figure 4-113 Bench Built, Wire Wound Resister

Wire wound resistors are the preferred choice for extremely high-current applications. Figure 4-112 shows a high-wattage resistor which uses an exposed nichrome element for improved heat dissipation. Resistors like this can become so hot during operation that the solder joints can melt. Therefore, these resistors are often installed with the lead wires bolted on.

Figure 4-113 shows a bench built, wire wound resistor. A coil form is wound with a length of wire sufficient to provide the appropriate resistance. A terminal is added to either end of the form where the element is connected.

To determine the resistance of a wire wound resistor, the resistance per length of selected wire must be known. The resistance is directly related to the cross sectional area and the overall length of the wire. Figure 4-114 provides resistance, in ohms per 1000 feet, for common copper wires. If a 25-ohm resistor is to be wound using 32 gauge wire then 152 feet of wire must be used.

The formula below is used to calculate the resistance of a given wire:

$$R = (\rho \times L) \div A$$

Where ρ = Resistivity
L = Length of the conductor
A = Cross sectional area of the conductor.

AWG	Dia."	Cm**	Resistance*
0000	.4600	211600	0.049
000	.4096	167810	0.062
00	.3648	133080	0.078
0	.3248	105530	0.098
1	.2893	83694	0.124
2	.2576	66373	0.156
3	.2294	52634	0.197
4	.2043	41742	0.249
5	.1819	33102	0.313
6	.1620	26250	0.395
7	.1443	20816	0.498
8	.1289	16509	0.628
9	.1144	13094	0.792
10	.1019	10381	0.999
11	.0907	8234	1.260
12	.0808	6529	1.588
13	.0720	5178	2.003
14	.0641	4107	2.525
15	.0571	3257	3.184
16	.0508	2583	4.016
17	.0453	2048	5.064
18	.0403	1624	6.385

AWG	Dia."	Cm**	Resistance*
19	.0359	1288	8.051
20	.0320	1022	10.150
21	.0285	810	12.800
22	.0254	642	16.140
23	.0226	509	20.360
24	.0201	404	25.670
25	.0179	320	32.370
26	.0159	254	40.810
27	.0142	201	51.470
28	.0126	160	64.900
29	.0113	127	81.830
30	.0100	101	103.200
31	.0089	80	130.100
32	.0080	63	164.100
33	.0071	50	206.900
34	.0063	40	260.900
35	.0056	32	329.000
36	.0050	25	414.800
37	.0045	20	523.100
38	.0040	16	659.600
39	.0035	12	831.800
40	.0031	10	1049.000

* Ohms per 1000 Feet at 68°F

** Circular mills. 1 Cm = .001" Dia.

Figure 4-114 Resistance of Common Copper Wires (AWG)

Material	Resistivity (p) (ohm/inch)
Aluminum	.0673
Carbon	.0762 - 1.5240
Copper (Pure)	.04267
Copper (Common Wire)	.0437
Gold	.0607
Iron	.2466
Lead	.5588
Mercury	2.4890
Nichrome	2.5400
Nickel	.1895
Platinum	.2692
Silver	.0404
Tungsten	.1422

Figure 4-115 Resistively of Common Materials

Figure 4-115 provides a list of resistivities that correspond to different metals commonly used as conductors and/or to manufacture wire.

Variable Resistors

For many applications, there is a need to adjust the parameters of the circuit during operation. The easiest parameter to adjust is resistance. This is accomplished using variable resistors. One of the most common uses for variable resistors is a volume control. When you turn the knob to raise or lower the volume on your stereo, you are actually adjusting the resistance of the internal circuit.

Figure 4-116 shows a classic, laboratory type, wire wound, variable resistor. The primary element is a coil of wire with a terminal on both ends. A third wiper can slide up and down the length of the coil. At any given position, the resistance is equal to the length of the wire between one of the terminals and the wiper.

To construct a variable resistor, a coil is wound onto a coil form and connected to a terminal at both ends. A wiper is made from a strip of brass and an insulating handle. The insulation must be sanded off the wire at the apex of the coil where the wiper contacts the wire. Figure 4-117 shows a bench built, variable, wire wound resistor.

For applications that require setup or periodic adjustments, there are a wide variety of center-tap resistors available. These resistors act as a fixed assembly, but allow a degree of adjustment during setup or de-energized times. Figure 4-118 shows a wire wound ceramic resistor with a clamp-on center tap.

For low-power applications, PC board mount 10-turn precision wire wound resistors are used. These devices are typically built into a compact plastic housing and have an adjustment screw on one end, as shown in Figure 4-119.

Rheostats are a type of resistor that is designed to be adjusted while the circuit is energized. They are typically found in high-current applications such as motor field, reactor, and heater control. Typically, rheostats are wire wound

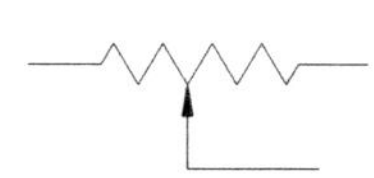

Schematic Symbol

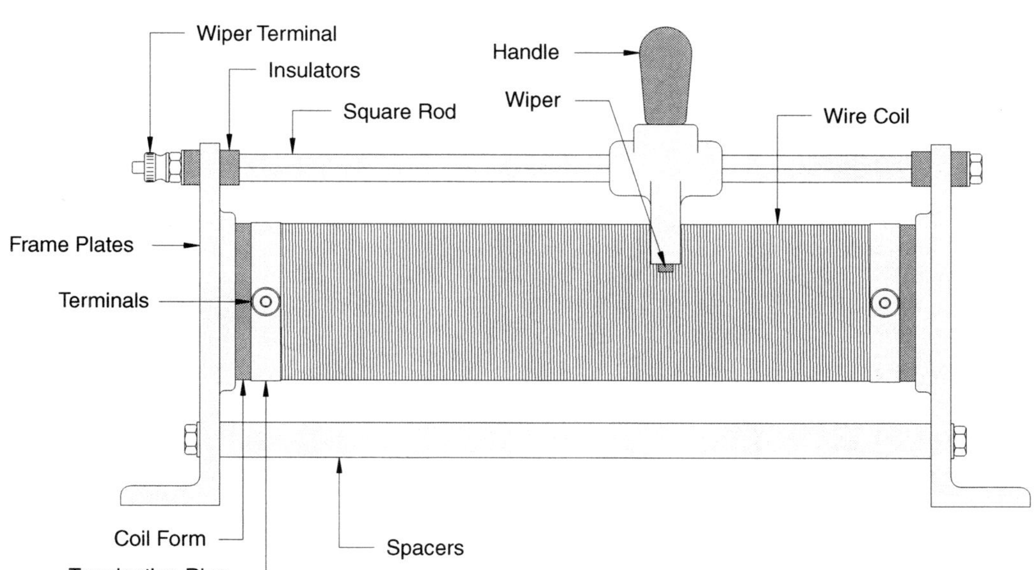

Figure 4-116 Lab Type, Wire Wound, Variable Resistor

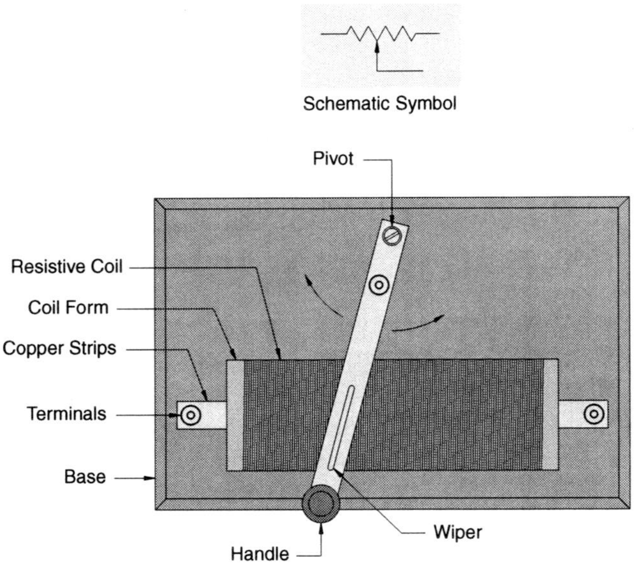

Figure 4-117 Bench Built, Variable, Wire Wound Resister

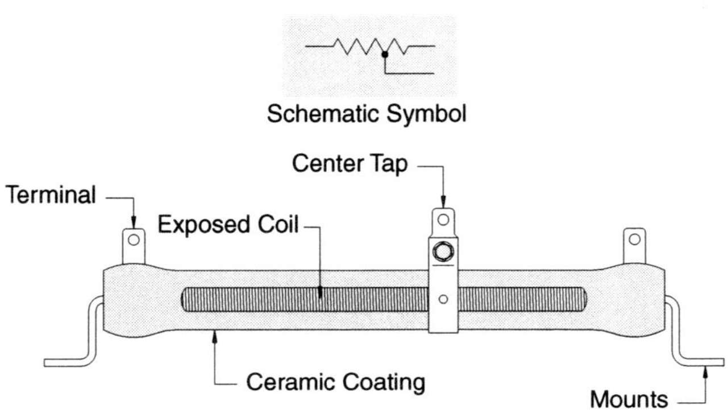

Figure 4-118 Center-Tap Resistor

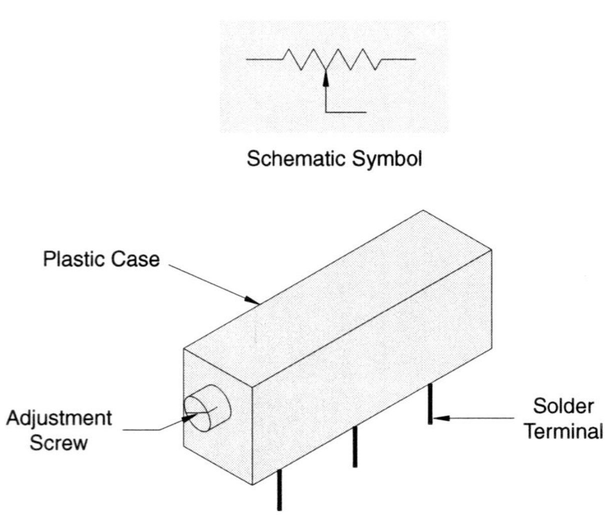

Figure 4-119 Ten Turn PC Board Mount Wire Wound Potentiometer

resistors. Figure 4-120 shows a high-current rheostat. The coil form is in the shape of a horseshoe with a wiper mounted on an axle that allows it to be adjusted to any position on the coil. Normal rheostat configuration calls for the wiper to be electrically connected to one of the coil terminals. In this manner the unit acts as a simple variable resistor.

Potentiometers are variable resistors that are generally used for more sensitive applications. Just about any electronic control knob that you might encounter is connected to a potentiometer. Much like standard resistors, potentiometers are generally available with carbon or wire wound elements. Figure 4-121 shows a typical carbon film potentiometer and Figure 4-122 shows a wire wound unit.

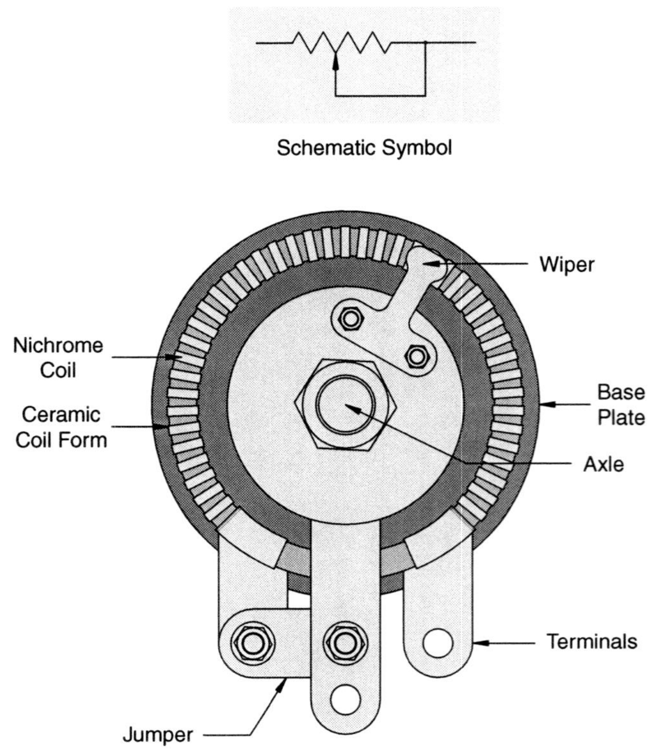

Figure 4-120 Rheostat

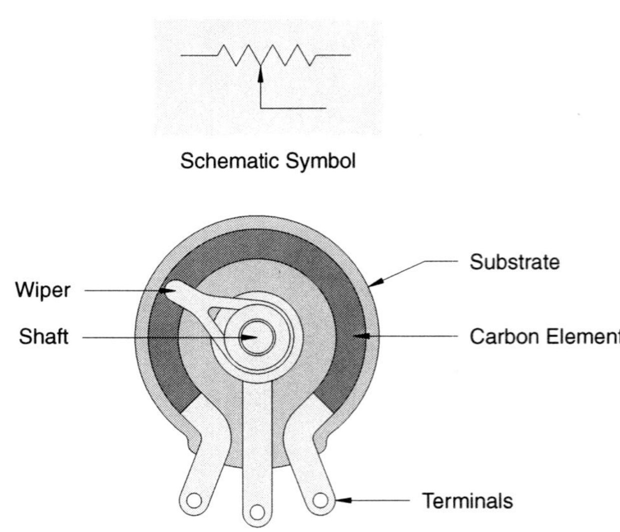

Figure 4-121 Carbon Film Potentiometer

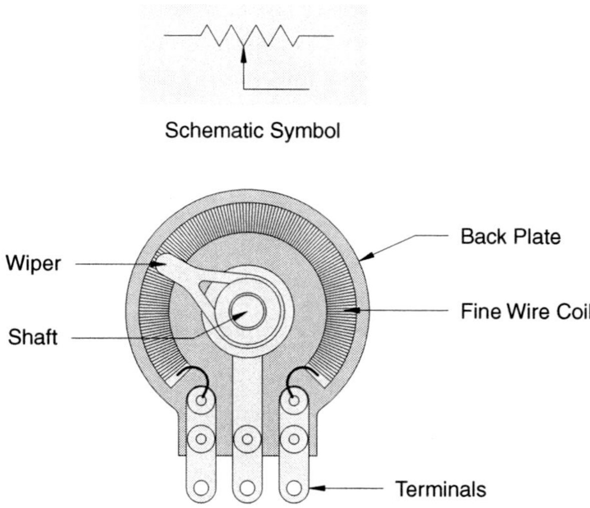

Figure 4-122 Wire Wound Potentiometer

Some select applications require potentiometer with center taps. Although these units are available for general purpose applications, most center-tap potentiometers are custom built for the specific circuit on which they are installed. Figure 4-123 shows a carbon film potentiometer with a center tap.

Because potentiometers are generally used as the human interface, there are times that the operator's perception of the change is important. For this reason potentiometers are normally supplied with one of three different tapers. The term taper refers to the change in resistance as the knob is rotated. Figure 4-124 shows a graph that illustrates the different tapers. A standard potentiometer uses a linear taper and this is the most common version. For audio applications, audio tapers are used. This taper is intended to match the volume perception of the average human. Log tapers are used when a logarithmic progression is necessary.

Carbon pile resistors are generally used for extremely high-current applications. These units are variable resistors that clamp a stack of carbon plates together to form their element. As the clamping force of the stack is increased, the overall resistance lowers. When the clamping force is decreased, the resistance increases. These resistors are often-

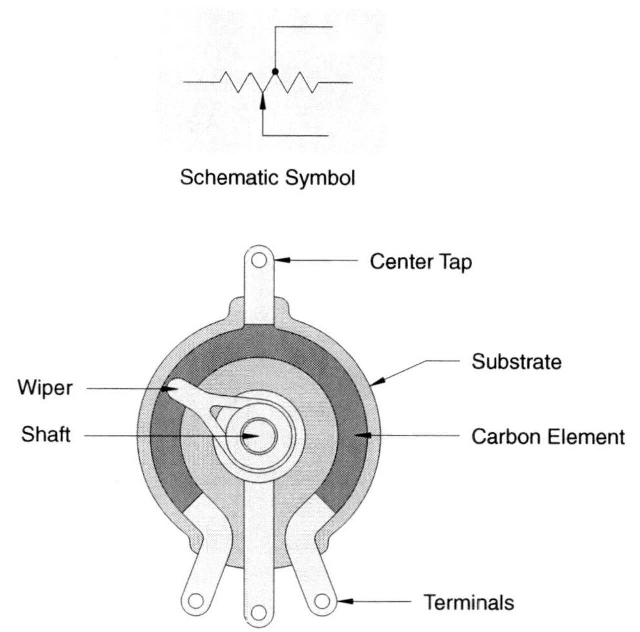

Figure 4-123 Carbon Film Potentiometer with Center Tap

times found as load dumps for testing generators and large power supplies. It is not uncommon for one of these units to glow red hot during operation. Figure 4-125 shows a small benchtop carbon pile resistor.

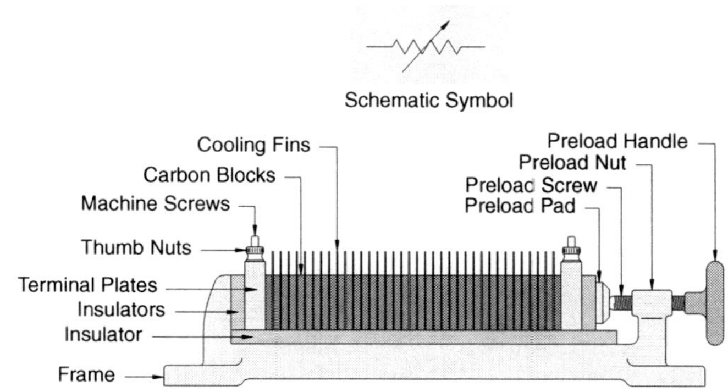

Figure 4-125 Carbon Pile Resister

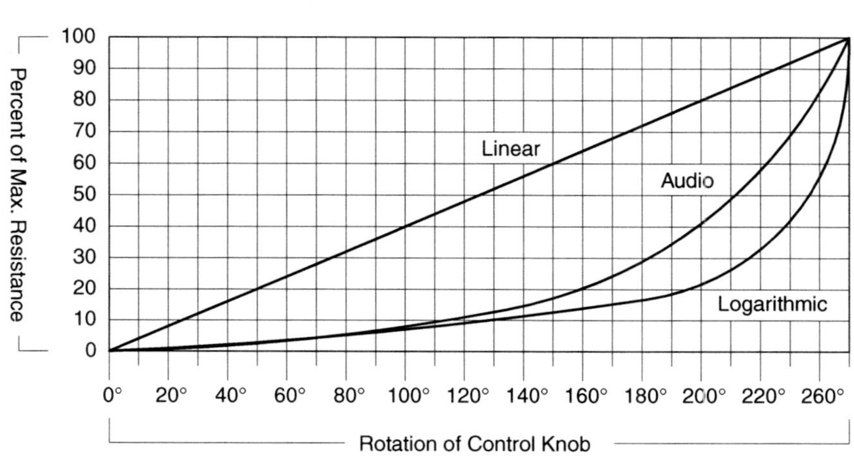

Figure 4-124 Liner, Audio, and Log Potentiometer Tapers

Decade Resistance Boxes

For laboratory applications, decade resistance boxes are very useful pieces of equipment. They are generally supplied with a 1 megs-ohm maximum resistance which is switchable in 1-ohm increments. Figure 4-126 shows a typical commercial decade resistance box. Figure 4-127 shows a schematic representation of the unit.

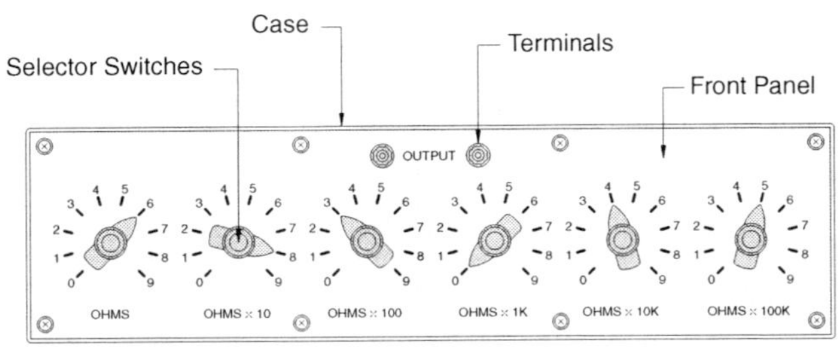

Figure 4-126 1-Mega-Ohm Decade Resistance Box

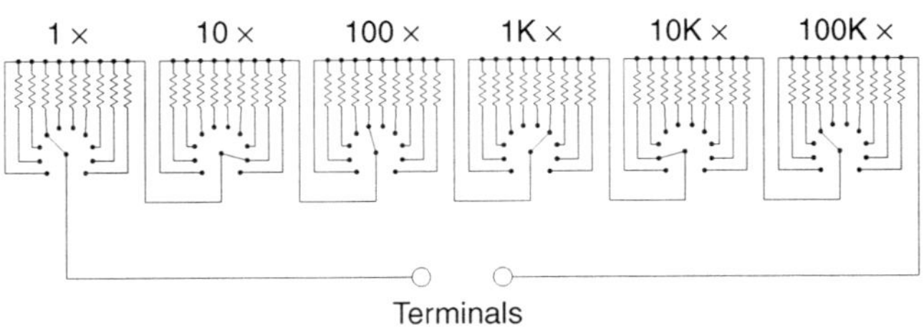

Figure 4-127 Decade Resistance Box Schematic

A low cost decade resistance box may be constructed using a banana jacks and jumpers, as shown in Figures 4-128 and 4-129. This unit has the same functionality as the commercial unit but can be built to carry much higher currents.

Voltage Dividers

A very common use for resistors is as voltage dividers. By selecting two different value resistors and arranging them as shown in Figure 4-130, a lower output voltage can be generated. The output voltage is based on the ratio of R_1 and R_2. To calculate the output voltage use the following formula:

$$V_2 = V_1 \times [R_1 \div (R_1 + R_2)]$$

$$\text{Example: } V_2\ (3 \text{ volts}) = V_1\ (12 \text{ volts}) \times [R_1\ (75\ \text{k}\Omega) \div (R_1\ (75\ \text{k}\Omega) + R_2\ (25\ \text{k}\Omega)$$

Figure 4-131 shows a typical voltage divider arrangement. Low wattage versions of this configuration can be constructed and heat shrunk directly into a feed cable.

Figure 4-132 shows a potentiometer set up as an adjustable voltage divider.

Another common use for resistors is as current limiting devices. A resistor is added to a circuit to prevent the possibility of a short-circuit situation. Figure 4-133 shows how a

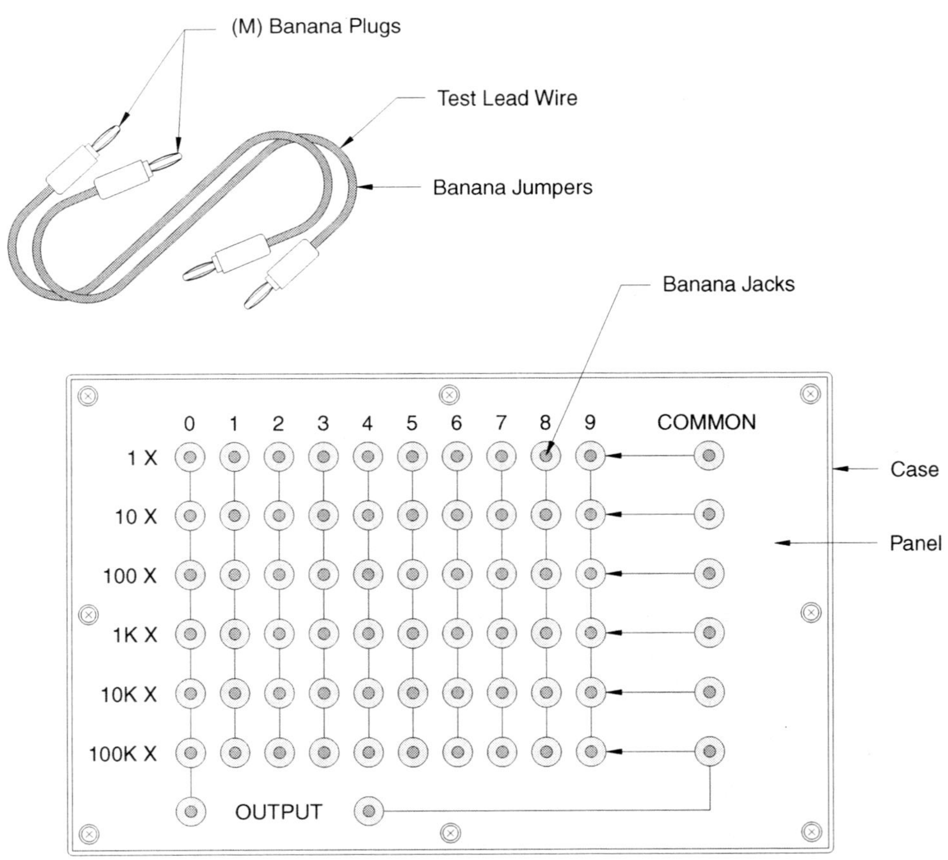

Figure 4-128 Decade Resistance Box Using Banana Jumper Plugs

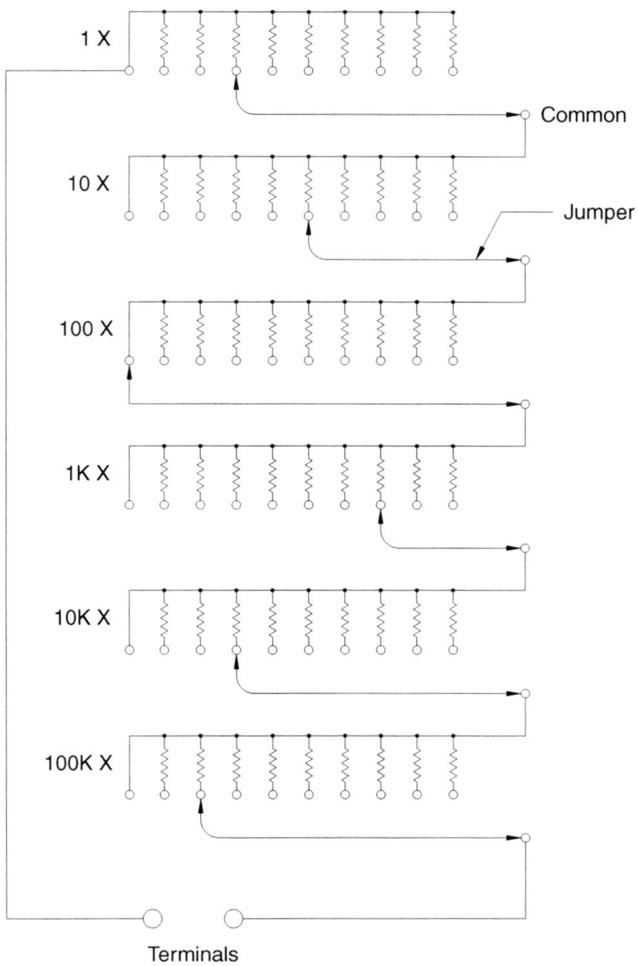

Figure 4-129 Jumper Decade Resistance Box Schematic

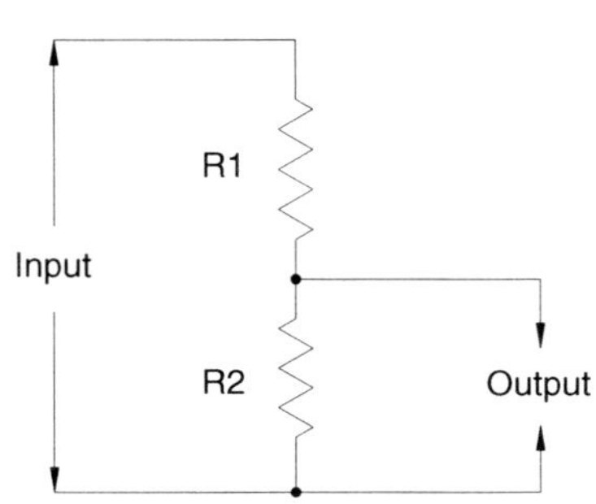

Figure 4-130 Voltage Divider Schematic

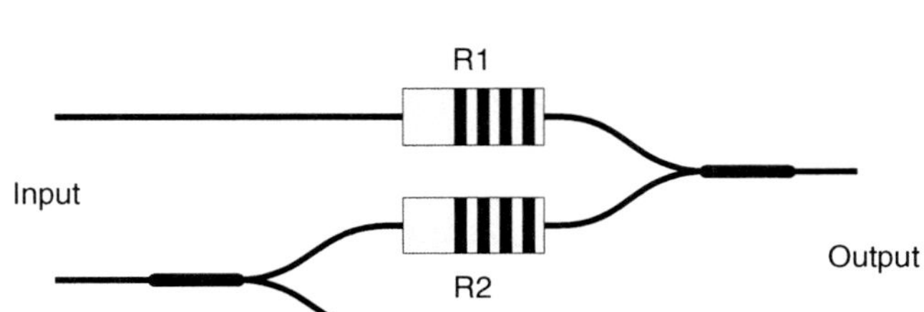

Figure 4-131 Voltage Divider

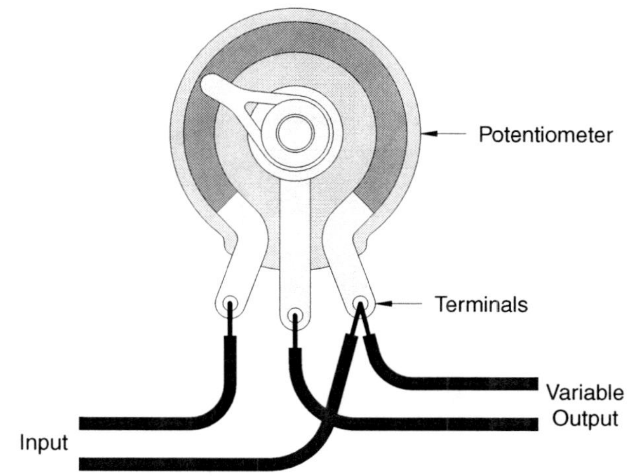

Figure 4-132 Potentiometer Voltage Divider

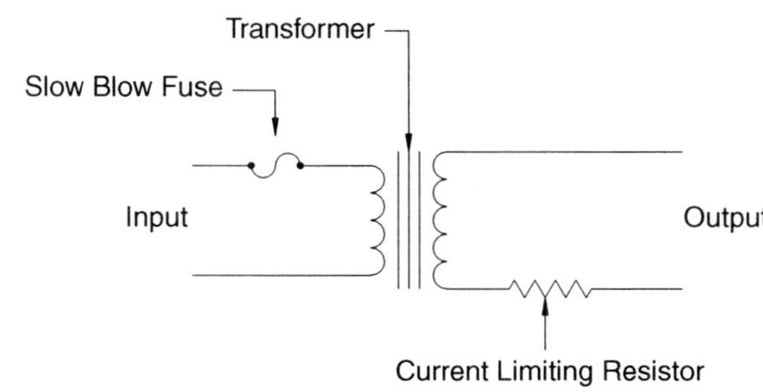

Figure 4-133 Current Limiting Resistor Schematic

current limiting resistor would be applied to protect a transformer. The resistor provides enough resistance to the output of the transformer to slow the overload process during a complete short. This arrangement is very common on battery chargers where a fully discharged battery may present a "dead" short for the first couple of seconds after the battery charger is turned on.

Capacitors

Capacitors function very much like a type of electrical shock absorber. They can receive and dump their charge in reference to their capacity and the specific application. A very common use for capacitors is as a filter, as shown in the power supplies of Chapter 3.

The first capacitor was invented by Professor Musschenbrock of the University of Leyden, Holland, in 1746. The Leyden jar is simply a glass jar that is lined on the inside and outside with foil, as shown in Figure 4-134. The terminal, which is mounted in a rubber stopper, is connected to the inner liner with a hanging chain. A charge is applied to the terminal and the inner liner builds up a surplus of electrons. If the terminal is connected to the outer liner, the charge flows between the liners and is neutralized.

The Leyden jar was used as the standard capacitor well after Benjamin Franklin invented the improved glass plate capacitor. The glass plate capacitor uses alternating glass

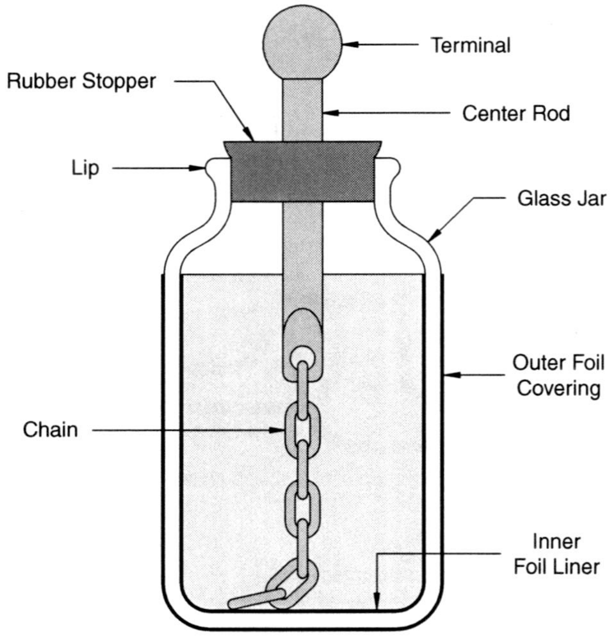

Figure 4-134 Leyden Jar

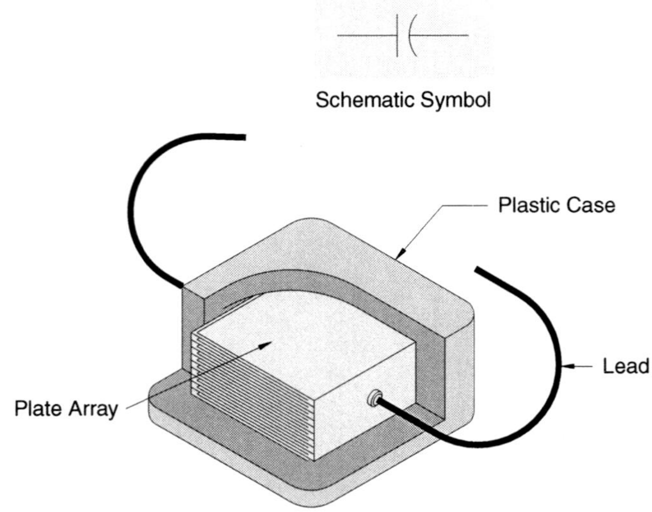

Figure 4-136 Plate Capacitor

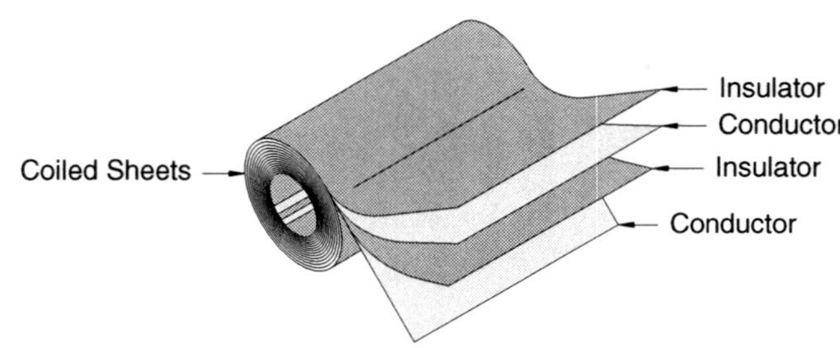

Figure 4-137 Coil-Wound Capacitor

plates and conductor sheets arranged in a stack. This design has a profound effect on the size and capacity of these devices. Figure 4-135 shows a bench built glass plate capacitor. Notice that there are separate positive and negative stacks of conducting sheets. These stacks form the storage elements of the assembly.

Many commercial capacitors are miniaturized versions of the glass plate design. Plate arrays are molded into a plastic case, as shown in Figure 4-136. Other designs use coils of conductive sheet sandwiched with insulating sheet, as shown in Figure 4-137.

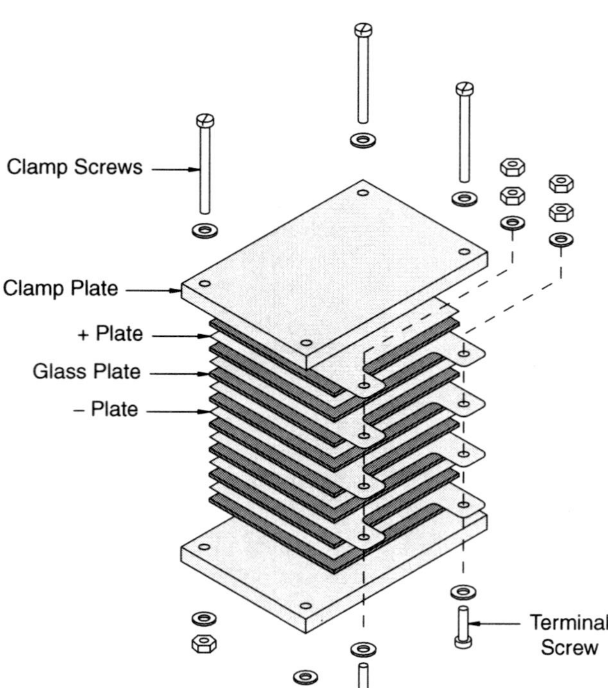

Figure 4-135 Bench Built Glass Plate Capacitor

Commercial capacitors are available in nearly any size, voltage, and capacity imaginable. Figure 4-138 shows just a small sampling of commercial capacitors, that are designed for a variety of applications.

Variable capacitors are not particularly effective because the insulation between plates is generally air. Using air as the insulator requires that the spacing between the plates must be sufficient to prevent arcing and to provide required production tolerances for manufacturing. To improve the performance of variable capacitors, some of these units are immersed into a high dielectric oil bath.

Figure 4-139 shows a typical variable capacitor that may be found in radio equipment to tune the resonant frequency of the circuit. These assemblies are typically a set of fixed plates and a set of moving plates that are set up to mesh together in reference to the operator input.

Diodes

A diode, as shown in Figure 4-140 is essentially a one-way valve for electricity. The device is commonly used for rectifying AC current into DC current. An example of this application is shown in Chapter 3.

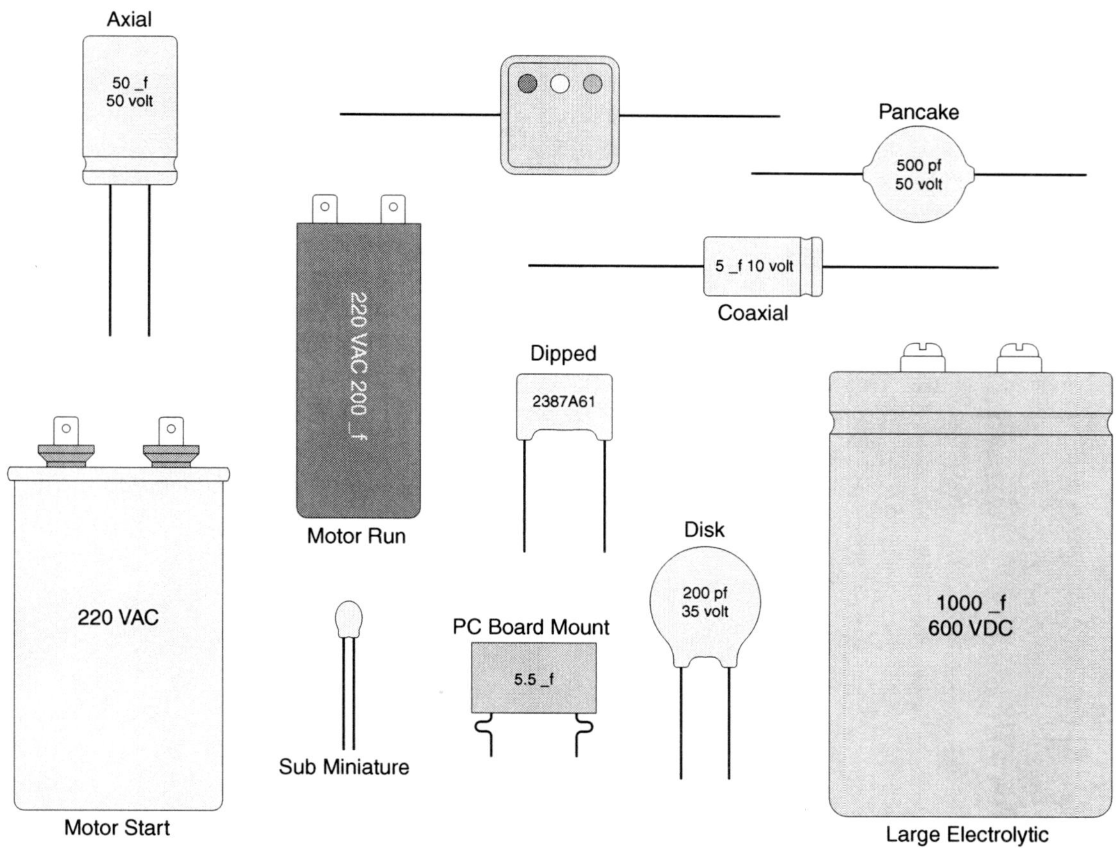

Figure 4-138 Commercial Capacitors

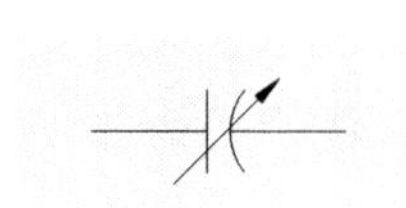

Schematic Symbol

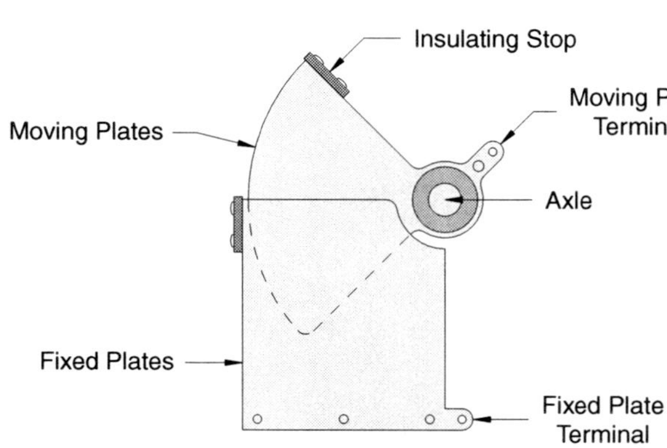

Figure 4-139 Variable Capacitor

Silicone Controlled Rectifiers (SCR)

A silicone controlled rectifier or SCR, as shown in Figure 4-141, is essentially a diode that provides a trigger function. The SCR can be turned on by applying a voltage to the gate and therefore makes these devices particularly popular among

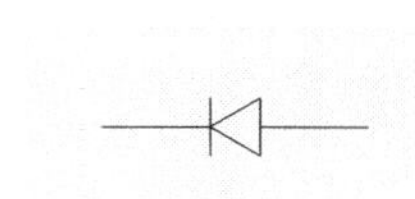

Schematic Symbol

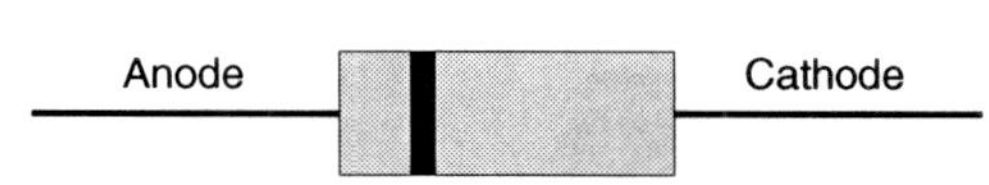

Figure 4-140 Diode

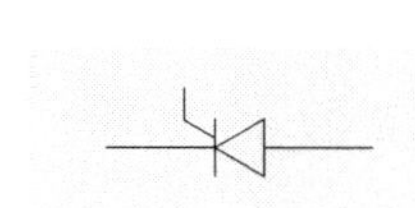

Schematic Symbol

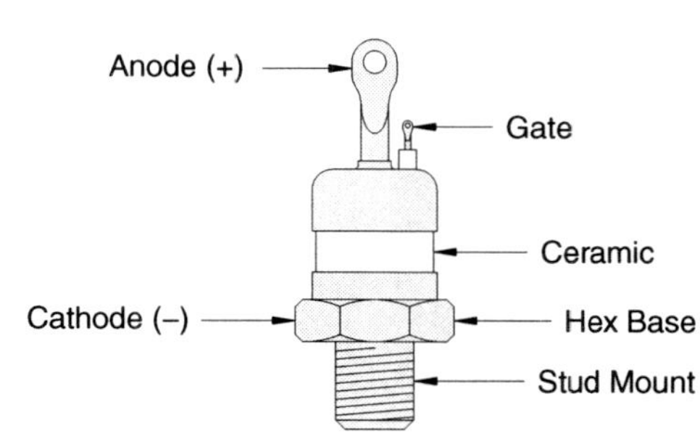

Figure 4-141 Silicone Control Rectifier

power supply and motor controller designers. Once the SCR is turned on it will remain conductive until the voltage across the anode and cathode reaches zero, even after the gate voltage has been removed. This attribute makes these devices ideal for switching AC because the voltage in an AC signal drops to zero 120 times per second.

Triacs

A triac is simply a pair of SCRs that are placed back-to-back to form a solid state AC switch. Triacs are commonly found in AC motor controllers, such as ceiling fan speed controls. Figure 4-142 shows a typical triac package.

Transistors

A transistor is essentially a semiconducting diode that has the ability to control the current flow throughput. These devices have a base, emitter, and collector. The emitter and collector make up the primary terminals and the base acts as a variable trigger. The amount of current that flows through the device is proportionally dependent on a lesser current applied to the base. Figure 4-143 shows a typical commercial transistor that may be found in ordinary electronic assemblies.

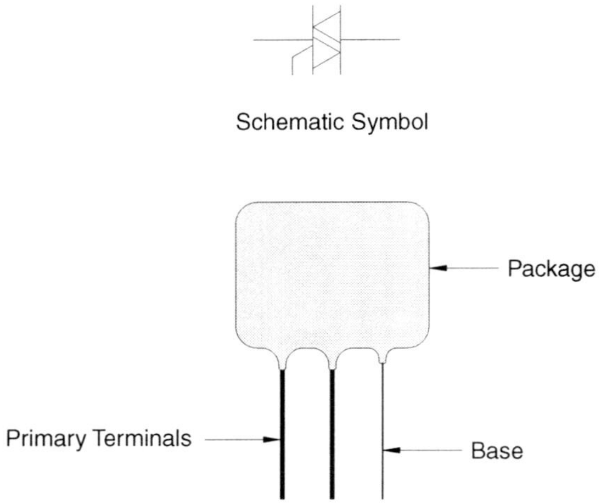

Figure 4-142 Triac

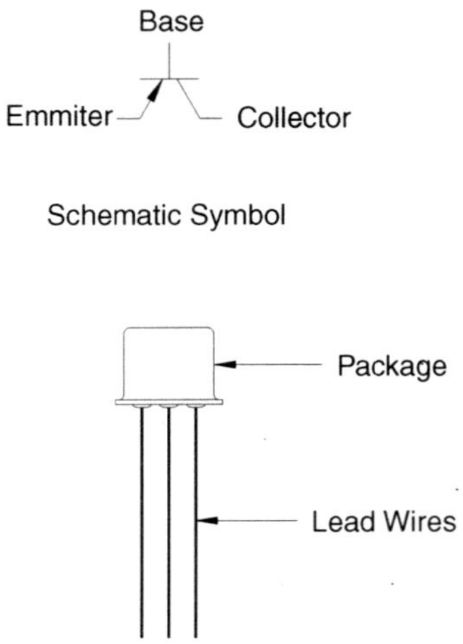

Figure 4-143 Transistor

CHAPTER 5
MAGNETIC COMPONENTS

Magnetics play an important role in the field of electromechanical devices. The manipulation of magnetic fields provides an interface between electricity and mechanical components. Most transformers, for instance, don't have an actual electrical connection between their inputs and outputs. They are instead connected with a magnetic field. Solenoids rely on a magnetic field, generated by an electrical signal, to produce motion; while motors derive their rotation through the variation of magnetic fields.

Electromotive Force

The key to understanding the relationship between electricity, magnetics, and mechanics is gained through the understanding of electromotive force. Simply stated, electromotive force is the electrical energy generated when a conductor is passed through a magnetic field.

Electrical energy can be generated in two principal ways, electrochemical and induction. Batteries, as discussed in Chapter 3, generate electrical energy through the electrochemical process. Generators and transformers utilize induction to generate electrical energy. Figure 5-1 shows how electrical energy may be generated with a simple horseshoe magnet and a wire. A wire loop is connected to a +/− voltmeter, as shown. The wire is then moved up and down through the magnetic field. As the wire is moved up, positive polarity is produced and as the wire is moved down, a negative polarity is produced. When the wire is repeatedly moved up and down, the needle on the voltmeter will swing to the positive and negative in direct reference to the direction of the wire movement.

To produce higher voltages, multiple loops of wire are moved within the magnetic field. If a coil is connected to the +/− voltmeter, as shown in Figure 5-2, and moved in the magnetic field, the needle will deflect much further than with the single-wire setup. The increase in voltage is directly proportional to the number of turns on the coil. If the coil has 10 turns, it will produce 10 times the voltage of a single wire.

In the same way that a voltage can be produced from a magnetic field, a magnetic field can be produced with electrical power. If a coil is connected to an electrical power source, a magnetic field will be generated, as shown in Figure 5-3. The flux lines show how the field balances in reference to the coil. When power is connected to the coil, a field is generated. When the power is disconnected, the field collapses.

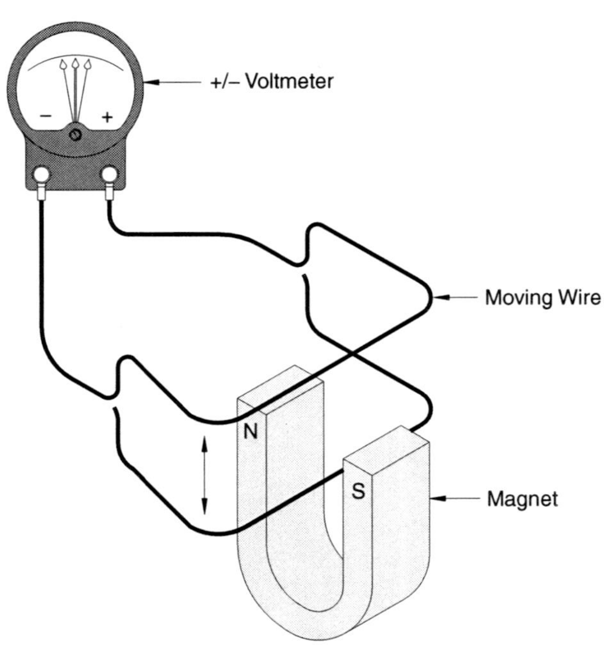

Figure 5-1 Inducing Electrical Current

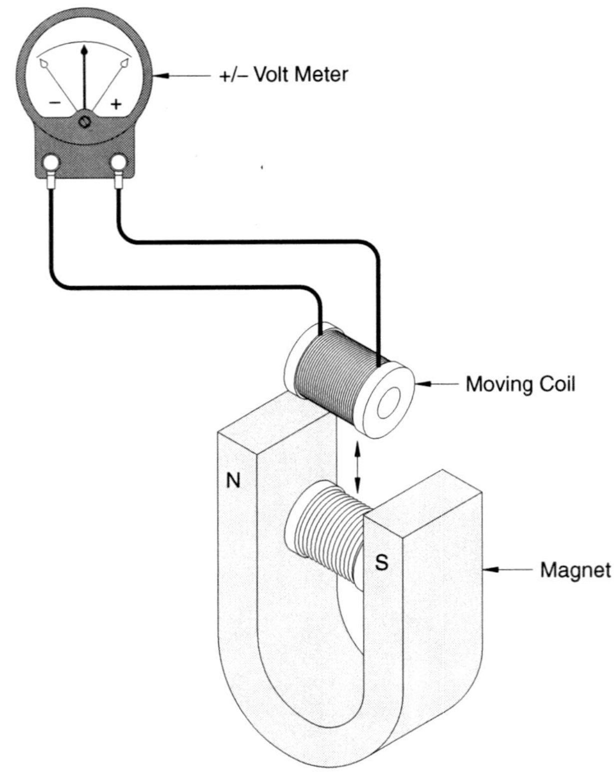

Figure 5-2 Increasing Voltage

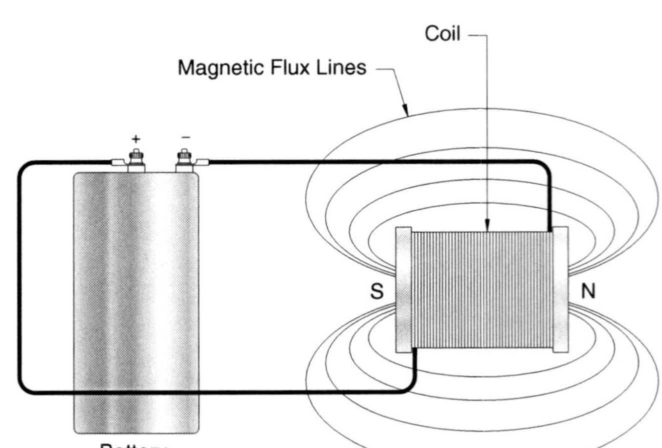

Figure 5-3 Flux Lines Surrounding a Solenoid Coil

Transformers

By placing two coils in close proximity to one other it is possible to achieve magnetic coupling. The first coil (primary) is connected to a power source and the second coil (secondary) is connected to a +/−voltmeter. Each time the power to the primary is turned on and off, the magnetic field is raised or collapsed. As the magnetic field is raised and collapsed, a voltage is induced in the secondary coil and the meter will deflect. Figure 5-4 shows two coils that are inductively coupled. Coils set up in this manner constitute a transformer.

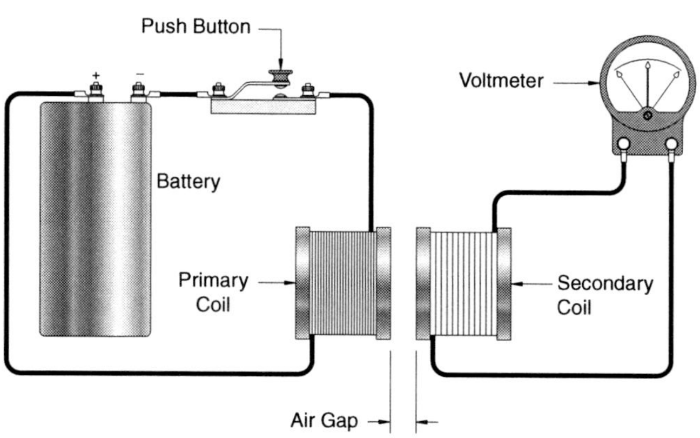

Figure 5-4 Magnetic or Inductive Coupling

To better control the magnetic field, the coils can be assembled onto an iron core. The iron can sustain a significantly higher flux density than air, so the coupling of the two coils is considerably more efficient. Figure 5-5 shows an iron core transformer.

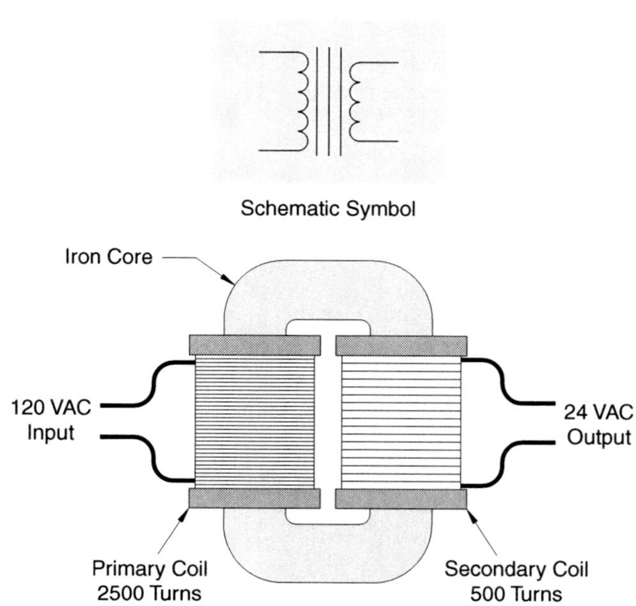

Figure 5-5 Step-Down, Iron Core Transformer

The output voltage of a transformer is a function of the ratio of the number of windings of the primary to the secondary. By adjusting the number of turns, a transformer can produce either lower or higher output voltages. The transformer in Figure 5-5 has a 2500-turn primary and a 500-turn secondary. This arrangement will produce a 24-volt output with a 120-volt input. Inversely, if roles of the coils are reversed the output would produce 600 volts for a 120-volt input. To calculate the output voltage of a simple transformer divide the input voltage by the ratio of the coils.

$$\text{output volts} = \text{input volts} \div (\text{number of primary turns} \div \text{number of secondary turns})$$

As an example, if a transformer has a 10,000-turn primary, a 500-turn secondary, and is receiving 480-volt input the output will be 24 volts.

$$24 \text{ volts} = 480 \text{ volts} \div (10{,}000 \text{ turns} \div 500 \text{ turns})$$

Most commercial transformers are built around a laminated "E" frame core. The laminations are stamped pieces of sheet metal in the shape of the letter E. The laminations are stacked to produce the necessary mass of iron for any given design. The two "E" sections are assembled with the coil around the middle leg, as shown in Figure 5-6. To improve magnetic coupling between the two "E" sections, some designs overlap the laminations on the outside legs.

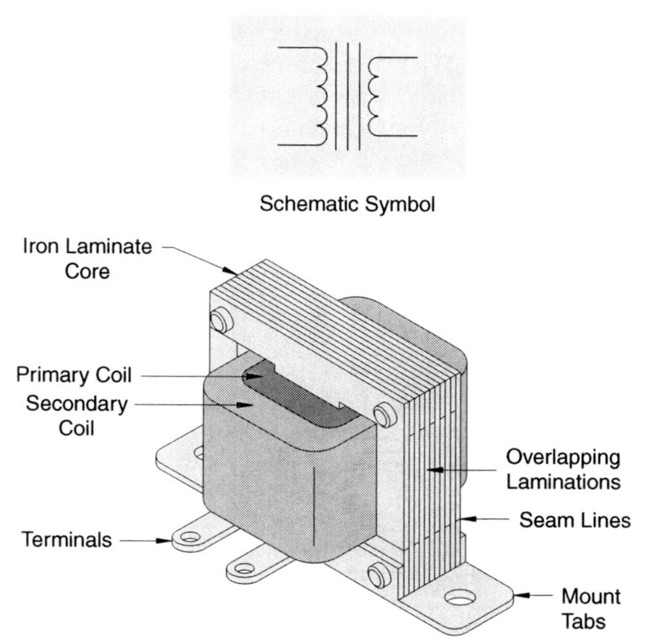

Figure 5-6 Commercial E-Frame Transformer

Center Taps

Many transformers provide facilities for several output voltages, the most common being transformers with a center tap on their secondary. The center tap is connected to the middle point of the secondary coil and, therefore, produces half of the

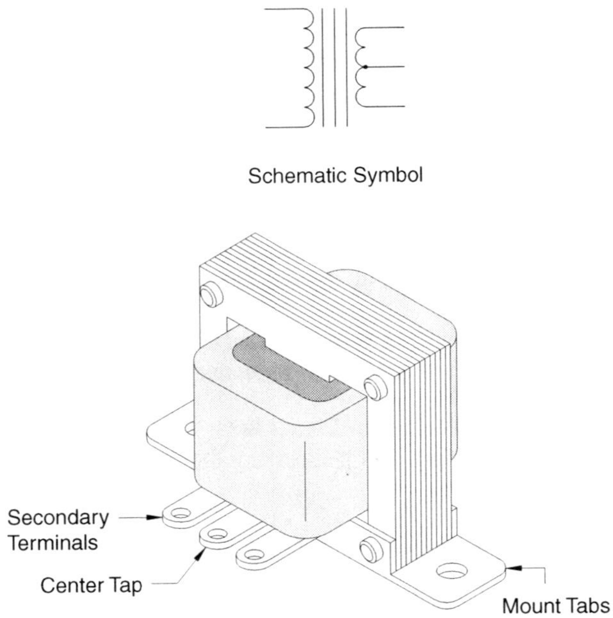

Figure 5-7 Commercial Transformer with Center Tap

output voltage. Specialized transformers are commonly manufactured that have a number of voltage taps, which can provide all necessary voltages for a given design. Figure 5-7 shows an "E" core transformer with a center tap on its secondary.

Control transformers typically have a dual-voltage primary. A dual-voltage primary is usually two independent coils that can be wired in series or parallel. If the transformer is set up for a low-voltage input, the primary coils are wired in parallel and the number of turns is half of the total turns. If the transformer is set up for high-voltage input, the primary coils are wired in series and the full number of turns is used. In either case the output voltage is the same. Figure 5-8 shows

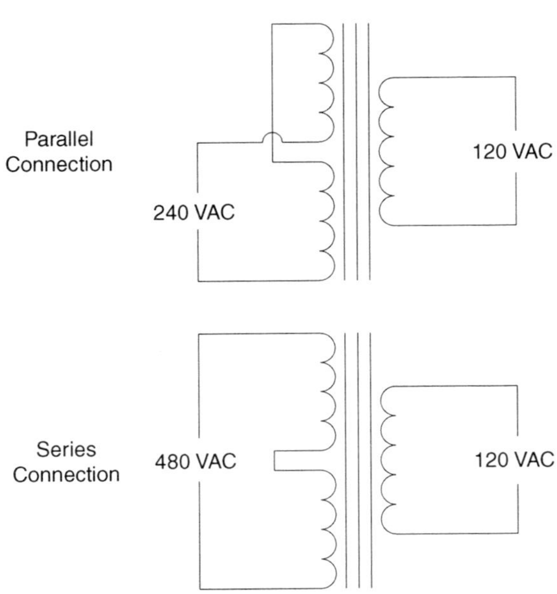

Figure 5-8 Transformer Schematic for Selectable Input Voltages

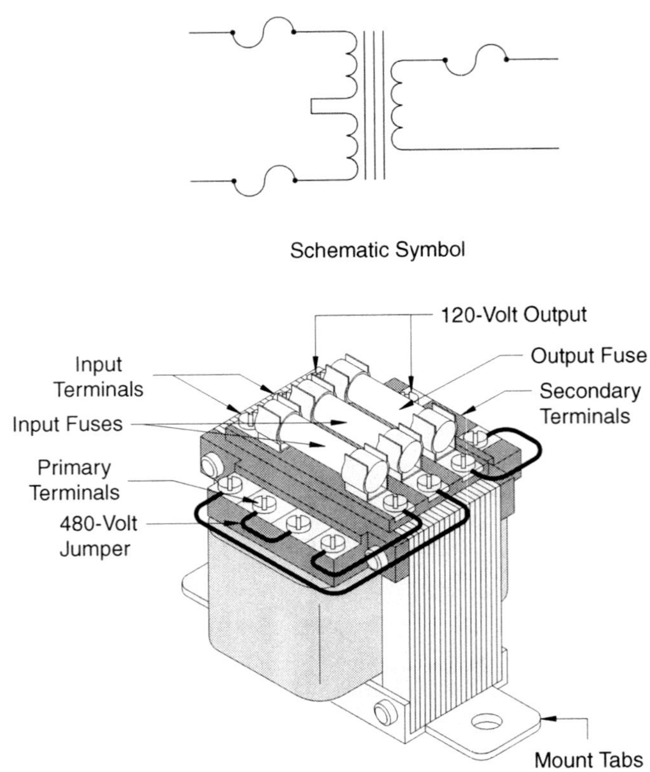

Figure 5-9 High-Inrush-Control Transformer with Fuse Set

control transformer schematics for low-and high-voltage inputs. Figure 5-9 shows a commercial high-inrush-control transformer complete with integral fuse set. Normally, these transformers carry two input fuses and one output fuse.

To produce a more compact package many transformers are constructed using a toroidal configuration. These types of transformers are generally designed for original equipment manufacturers (OEM) applications and are commonly used in switching power supplies. Figure 5-10 shows a typical toroidal core transformer.

In addition to changing voltages, transformers may also be configured for impedance matching. For circuits whose inputs and outputs are resistively mismatched, an impedance matching transformer can be configured. These transformers typically have the same number of turns in the primary and the secondary. The wire in the primary is selected to have a resistance that matches the output of the source circuit. The wire size of the secondary is selected to have a resistance that matches the input of the receiving circuit. In this manner the input and output voltages are the same, while the input and output resistances, or impedances, are different. Figure 5-11 shows a stylized schematic of an impendence-matching transformer.

Power transformers are generally constructed around two "C" cores, as shown in Figure 5-12. The primary and secondary coils are independent from and adjacent to one another. The two coils are set side-by-side and the "C" cores are assembled from the top and bottom. To minimize vibration, the coils are wedged with shims and the entire assembly is banded together.

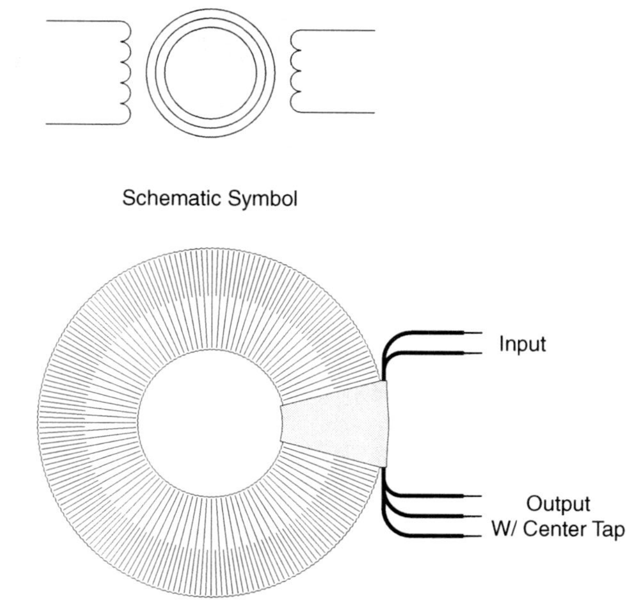

Figure 5-10 Toroidal Core Transformer

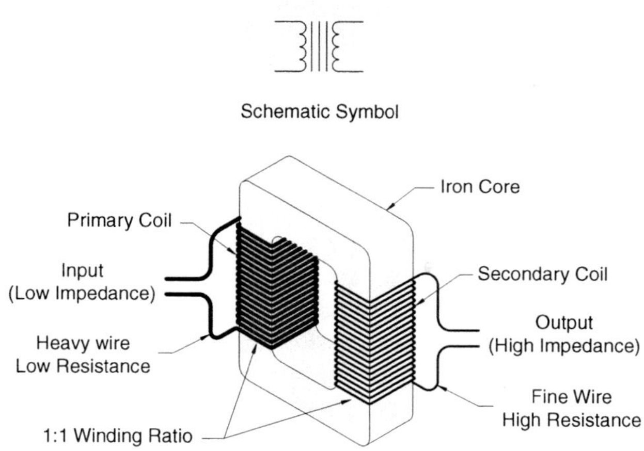

Figure 5-11 Impedance-Matching Transformer

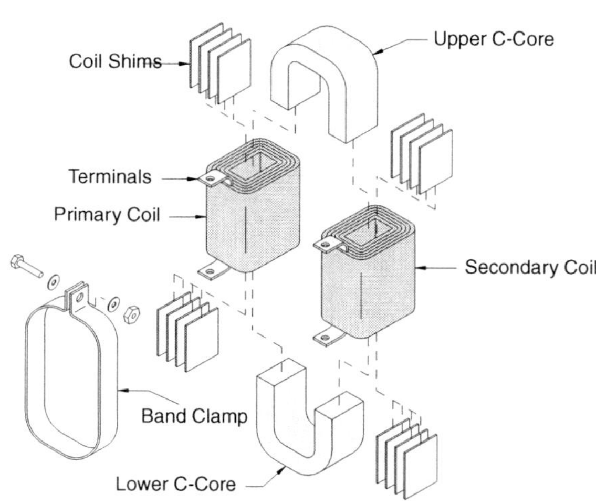

Figure 5-12 C-Frame Power Transformer Assembly

Isolation Transformers

The transformers discussed thus far are constructed with two separate coils, therefore the inputs and the outputs are electrically isolated. This type of design is generally referred to as an isolation transformer and the level of isolation is a function of the insulation between the coils and/or the core. Usually, the level of isolation is given in volts. Isolation transformers are particularly useful for electrically detaching a sensitive application from a noisy power source. The most notable use of isolation transformers is for home and business power distribution. Power transformers mounted on the poles electrically isolate power drops from the distribution grid and therefore protect homes and businesses from the high-voltage transients that routinely occur.

Autotransformers

Autotransformers are a type of transformer that uses only one winding. These units do not have an isolation function and should only be used when isolation is not warranted. These transformers are generally used for voltage-matching applications. If a 208 VAC machine must be placed in a location that only provides access to 240 VAC, an autotransformer can be placed to step-down the voltage. In most instances the installation of an autotransformer is considerably less expensive than installing a special power drop specifically for the machine. Figure 5-13 shows schematics of step-up and step-down autotransformers. Figure 5-14 shows a commercial voltage-matching autotransformer.

Another common use for autotransformers is as a variable voltage source. By replacing the fixed center tap with a sliding tap, any output voltage within the range of the transformer can be adjusted. Figure 5-15 shows a schematic representation of a variable autotransformer.

Commercial variable autotransformers are generally manufactured in a stand-alone cabinet, as shown in Figure 5-16.

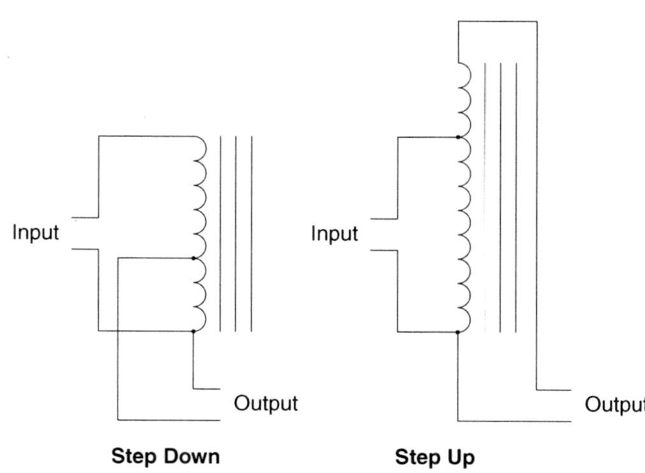

Figure 5-13 Autotransformer Schematic

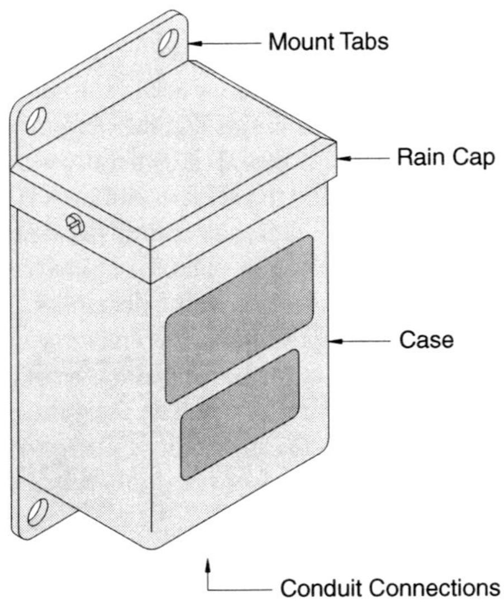

Figure 5-14 Commercial Voltage-Matching Autotransformer

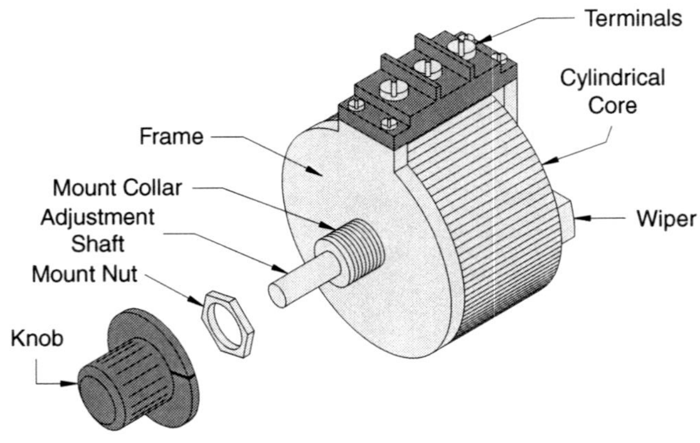

Figure 5-17 Commercial Variable Autotransformer

These units are excellent pieces of equipment for a test bench or for temporary installations. They generally have a standard AC cord for the input and a fused AC receptacle for the output. Some models are even supplied with output voltmeters.

Variable autotransformers are also supplied in panel mount versions, as shown in Figure 5-17. This arrangement makes them particularly suitable for custom or OEM installations.

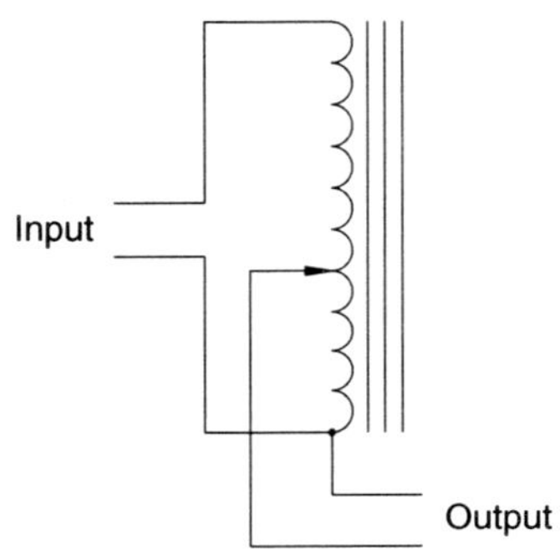

Figure 5-15 Variable Autotransformer Schematic

Three-Phase Transformers

Transformers are also used with three-phase power. A three-phase transformer is three, single-phase transformers that share a common core. They can be wired to accept and to output either Delta or Wye configurations (see Chapter 2). Figures 5-18, 5-19, 5-20, and 5-21 show the four basic configurations for three-phase transformers.

Commercial three-phase transformers have the same general appearance as a single-phase unit, except that there are

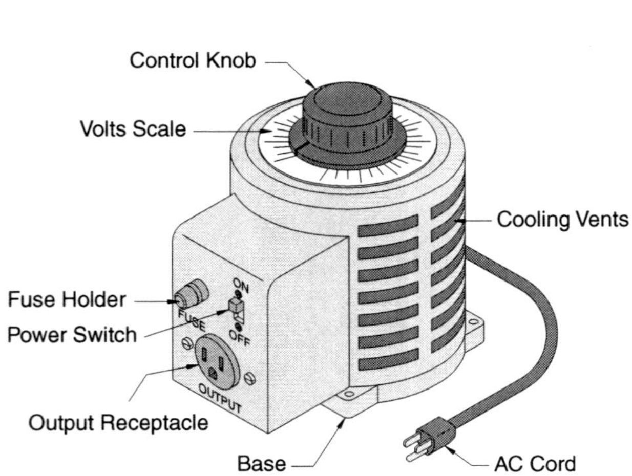

Figure 5-16 Packaged Commercial Variable Autotransformer

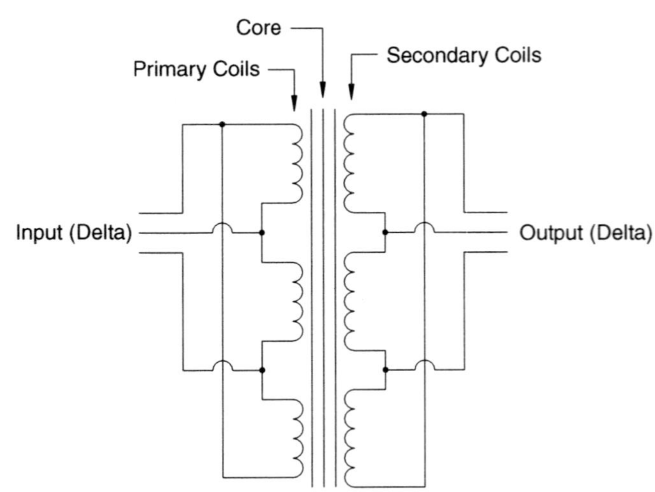

Figure 5-18 Delta-Delta Configured Three-Phase Transformer Schematic

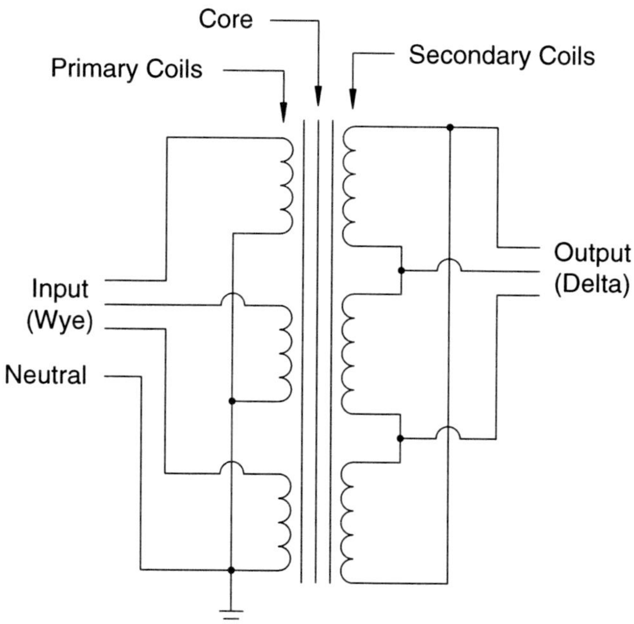

Figure 5-19 Wye-Delta Configured Three-Phase Transformer Schematic

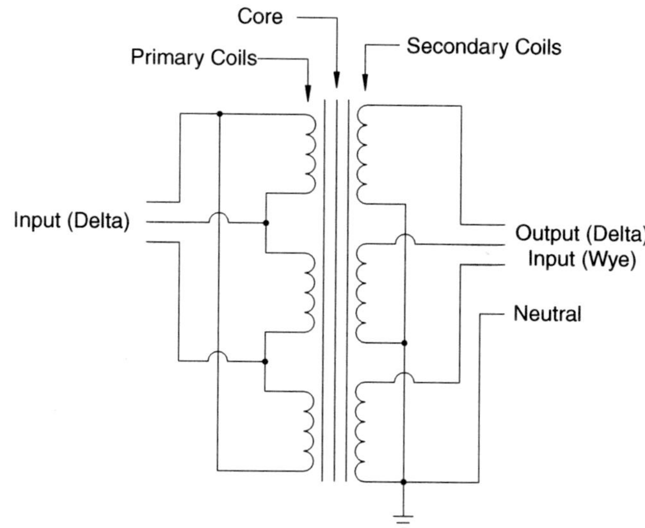

Figure 5-21 Delta-Wye Configured Three-Phase Transformer Schematic

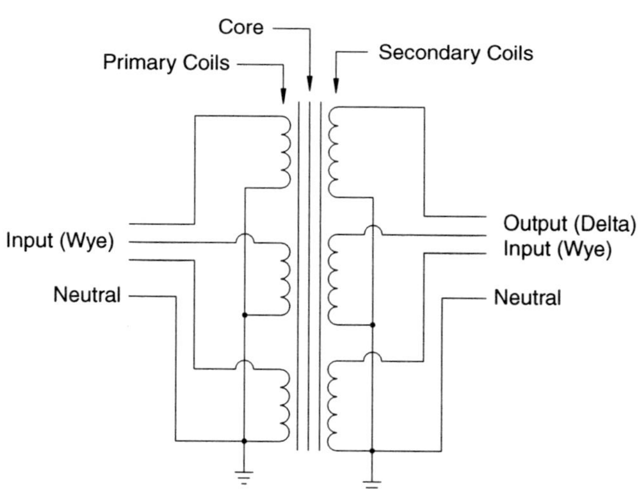

Figure 5-20 Wye-Wye Configured Three-Phase Transformer Schematic

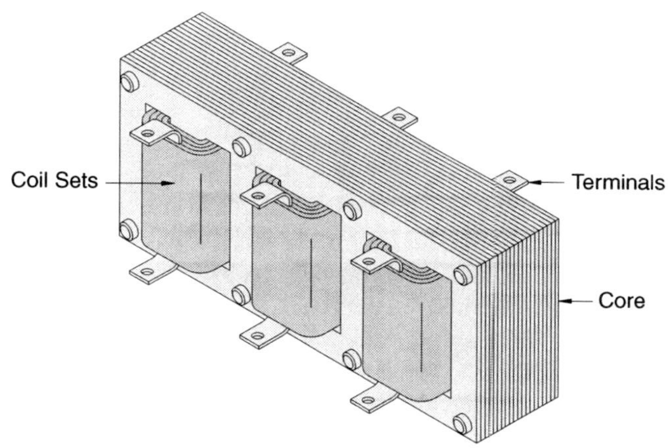

Figure 5-22 Commercial Three-Phase Transformer

three coil sets instead of one. Figure 5-22 shows a typical commercial three-phase transformer. Note that the terminals for each coil set are independent from the other coils. This allows the input and output to be configured to either Delta or Wye.

Three single-phase transformers may be configured to operate as a three-phase unit if they are wired correctly. Figure 5-23 shows three single-phase transformers configured for Delta input and output.

Large power distribution stations use large three-phase transformers. These units may be isolation or autotransformers,

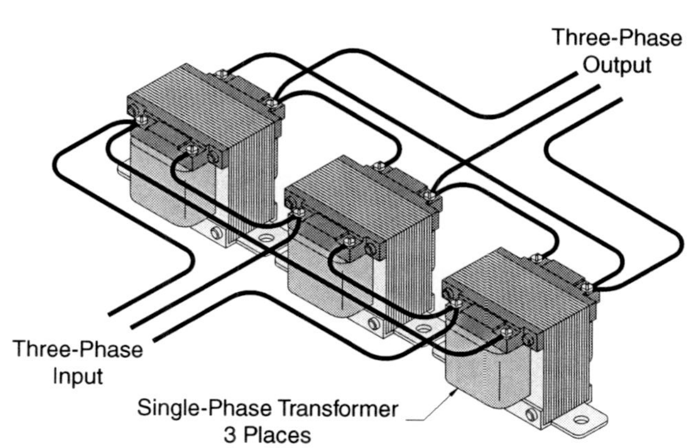

Figure 5-23 Three-Phase Transformer Configured Using Three Single-Phase Units

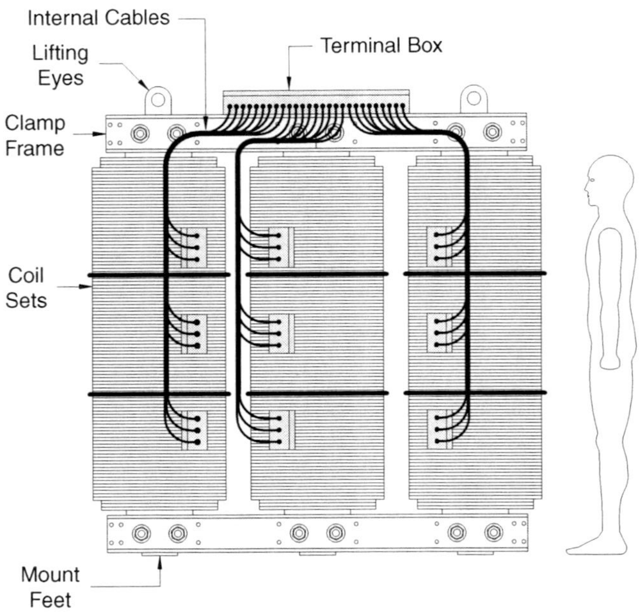

Figure 5-24 Large Three-Phase Power Distribution Transformer

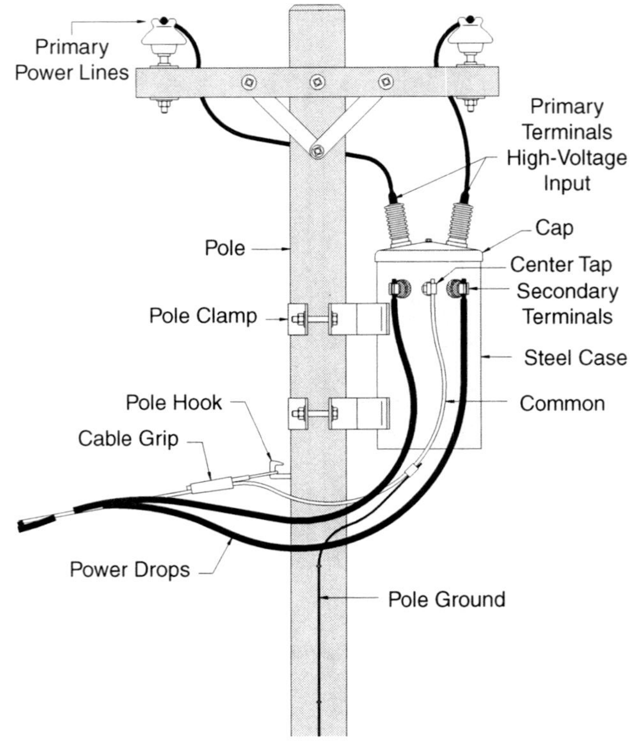

Figure 5-25 Pole Transformer

depending on the requirements of the specific application. Figure 5-24 shows a large three-phase power distribution transformer. Transformers like this are generally immersed in a high dielectric oil to improve insulation and cooling. Larger units may have forced air heat exchangers as an integral part of the overall assembly.

Most of us have noticed the pole transformers that dot our community. These are typically power transformers immersed in high dielectric oil. The steel case is designed to provide a high level of protection against almost any weather condition. It is common to find these transformers providing excellent service even 50 years after they had been installed. The large terminals on the top are the primary and the side terminals are the secondary. The secondary typically has a center tap, which is connected to ground. Figure 5-25 shows a typical single-phase pole transformer and associated wiring. Note that the center tap of the secondary is connected to the common leg, which is grounded at the pole as well as the building it is serving.

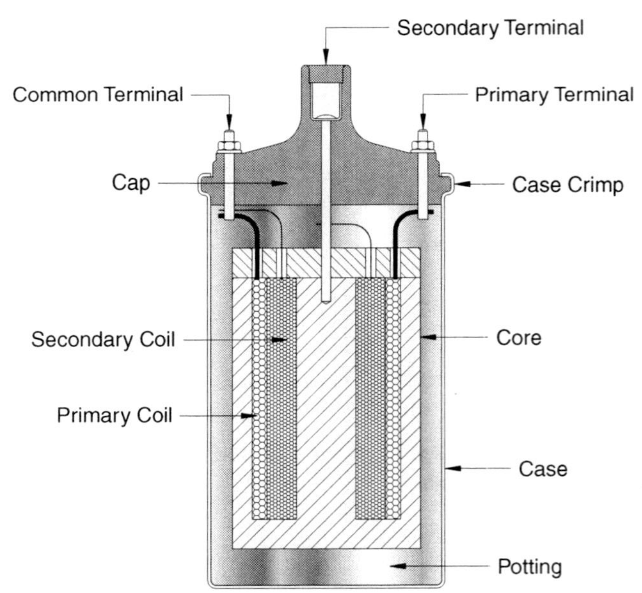

Figure 5-26 Automobile Ignition Coil

Ignition Coils

Another common transformer that most of us are aware of is the automobile ignition coil, as shown in Figure 5-26. These transformers are designed to provide a high-voltage pulse to generate a spark. The input is usually 12 volts while the output is between 30,000 and 70,000 volts! They typically have two solenoid coils placed into a cup-style core. The core/coil assembly is potted into a steel case and a plastic cap with the high-voltage terminal is crimped onto the top.

Figure 5-27 shows a schematic representation of an automobile ignition coil. The primary is the smaller coil shown with the heavy line, and the secondary is the larger coil shown with the fine line. One side of both the primary and the secondary is connected to the common terminal and the core is typically connected to the high-voltage terminal.

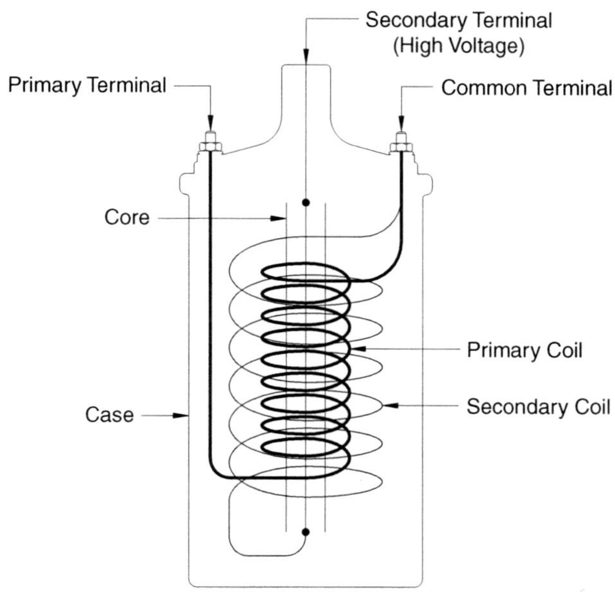

Figure 5-27 Ignition Coil Schematic

Because automobiles use DC power, some type of interruption is required to activate the ignition coil. The common terminal is usually connected to ground through a set of contact points, as shown in Figure 5-28. Each time the points open, the magnetic field in the coil collapses and a high voltage pulse is generated at the secondary terminal. The points are synchronized with the rotor, which directs the high-voltage pulse to the cylinder requiring ignition. The primary terminal of the ignition coil is connected to the positive terminal of the battery through an ignition switch. The capacitor bridging the contacts points is intended to minimize arcing and extend the life of the points.

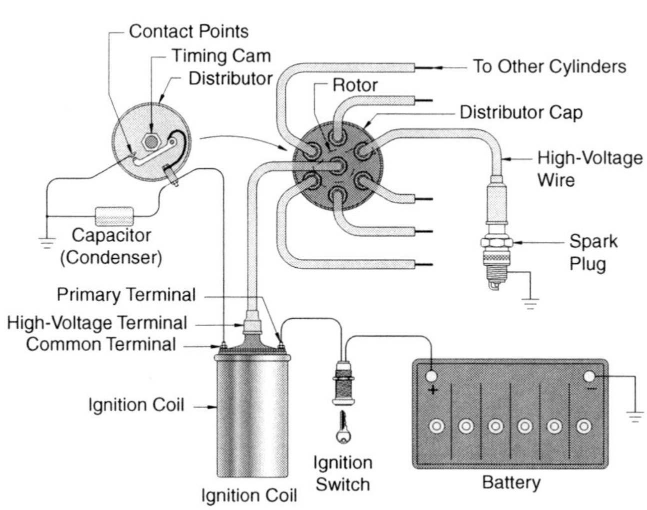

Figure 5-28 Automobile Ignition System

Saturatable Core Transformers

Limiting the output current of a transformer has many applications. Among the most noteworthy are battery chargers and arc welders. In these situations the load is essentially 0 ohms. If connected to a standard transformer, either the circuit protection will trip or the coils will be irreparably damaged. For these applications a saturatable core transformer is generally specified.

The output current of any transformer is dependent on the magnetic capacity of the core. Once the core reaches its full magnetic capacity, or saturation, the output current is maintained at a level that reflects the magnetic condition of the core. Therefore, by manipulating the core's magnetic capacity the output current can be controlled or limited.

There are two approaches to controlling the saturation of a transformer. The first is by changing the amount and location of the iron. Figure 5-29 shows a moving-core saturateable core transformer such as might be found in a small AC arc welder. The core is a typical "E" core design, except that the center leg can be retracted. As the center leg is retracted, the magnetic capacity of the core is reduced and reaches saturation at a lower current level. To increase the current, the leg is inserted into the core. To decrease current, the leg is retracted.

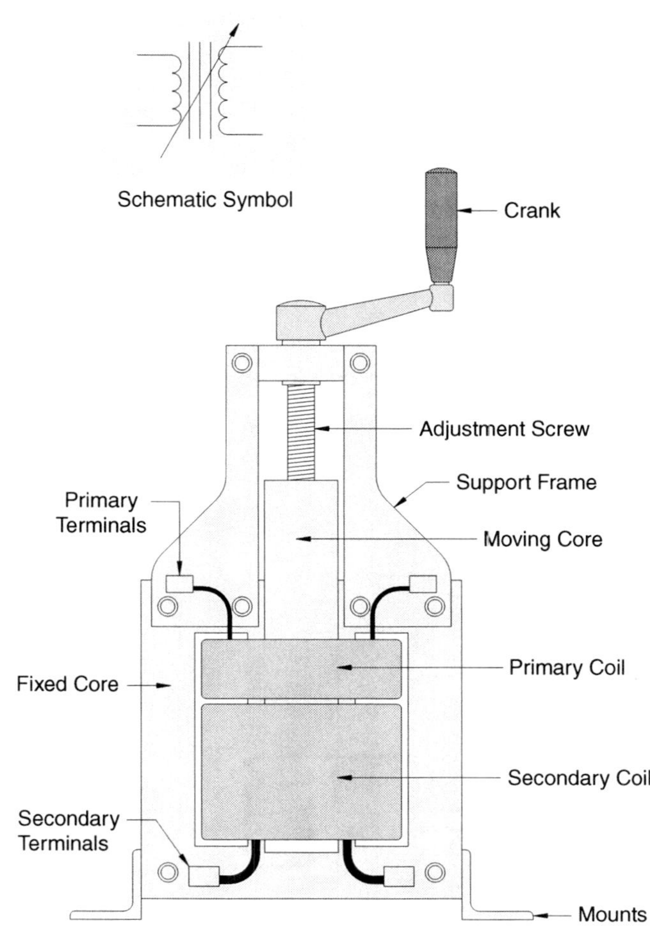

Figure 5-29 Moving-Core Saturatable Transformer

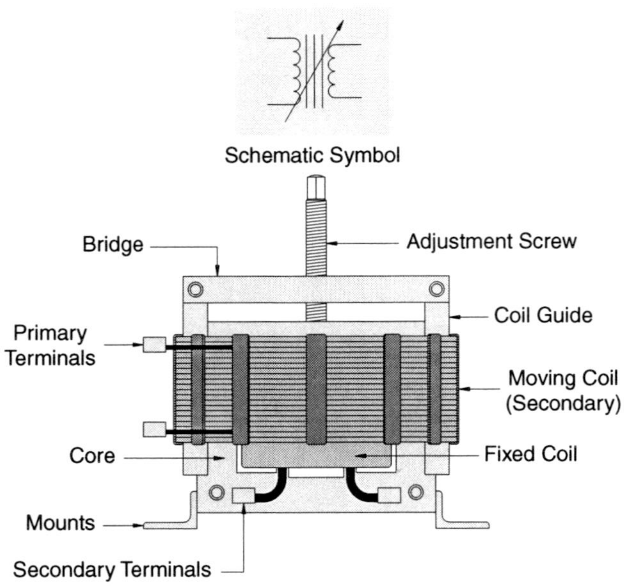

Figure 5-30 Moving-Coil Transformer

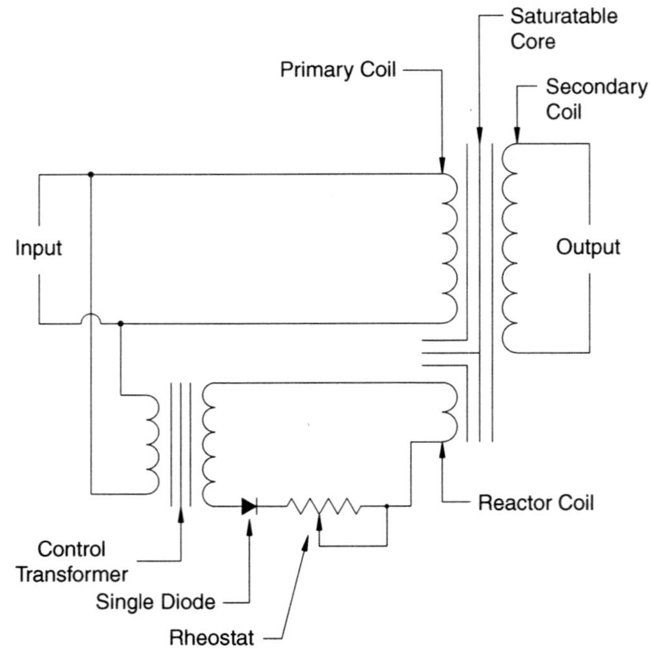

Figure 5-32 Saturatable Transformer Control Schematic

The same current limiting effect can be achieved by moving one of the coils into a position that will reduce the magnetic coupling. Figure 5-30 shows a moving-coil transformer. In this case, the secondary is raised or lowered to adjust the coupling efficiency and therefore limiting the output current.

The second current-limiting method is to electrically induce additional magnetic flux into the core. This is done by adding a third coil to the core, as shown in Figure 5-31. The additional coil is generally referred to as a reactor. When power is applied to the reactor it magnetizes the core and, in effect, uses up some of its magnetic capacity. This leaves less capacity for the function of the transformer, limiting the output current.

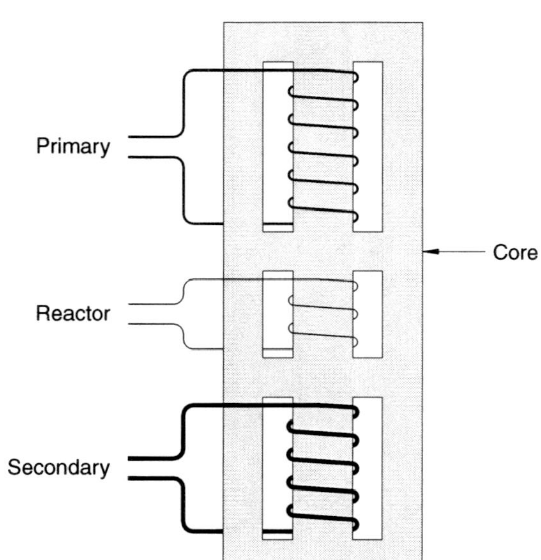

Figure 5-31 Saturatable Core Transformer with Reactor

Figure 5-32 shows a schematic representation of a control circuit that might be used with a saturatable core transformer. Take note of the simplicity of the circuit. This allows for a low-cost and robust design which finds favor in the demanding role of electric arc welding supplies.

Figure 5-33 shows a small, commercial saturatable core transformer with integral reactor. These units are not commonly available and are normally custom-made for specific applications.

A common use for saturatable transformers is as power supplies for neon lights. A neon light requires very high voltage to start (8000 to 15,000 volts) and a considerably lower voltage to operate (400 volts). Neon sign transformers are designed to have a high open-circuit voltage and a low current capacity. When the neon lamp is off, its internal resistance is very high and a very high voltage is required to ionize the gas particles in the tube. However, once the lamp turns on, the internal resistance drops to a low level and effectively shorts the transformer. At this point, the output voltage of the

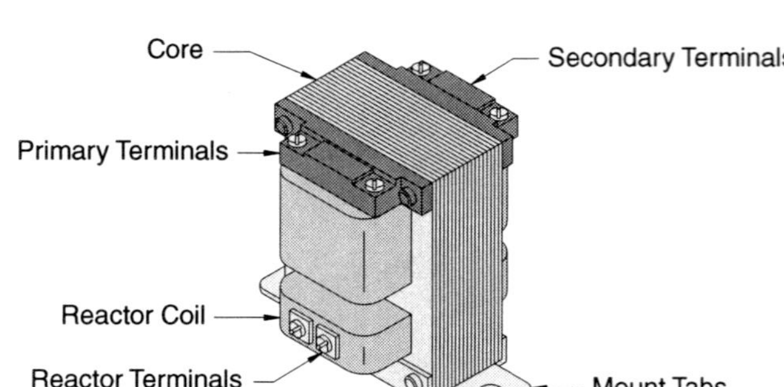

Figure 5-33 Saturatable Core Transformer

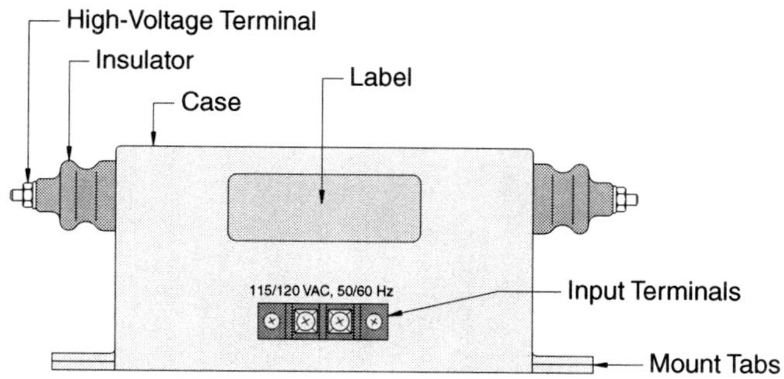

Figure 5-34 Neon Light Transformer

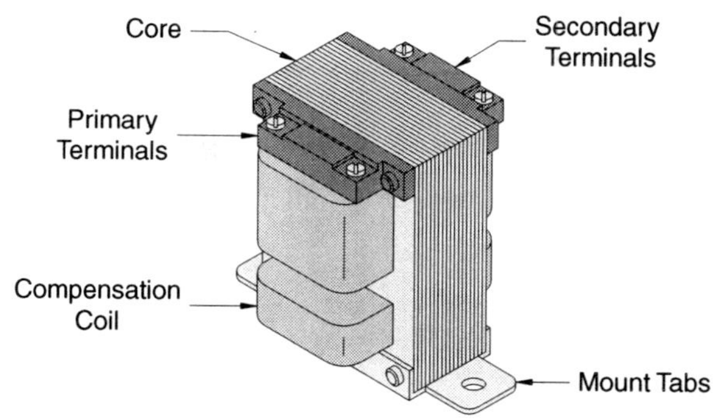

Figure 5-36 Constant Voltage Transformer

transformer drops to a level that matches the current and resistance operating the tube on this lower voltage. Figure 5-34 shows a typical neon light transformer.

Constant Voltage Transformers

Constant voltage transformers are generally used in applications that have precise power requirements, yet only have access to a poor quality power distribution system. These units are particularly popular in third world nations where the uniformity of the power distribution system is, at best, variable. They also find favor at remote installations that generate on-site power.

Constant voltage transformers produce a regulated output by taking advantage of ferro-resonance. A compensation coil is added to the core and connected to the output of the secondary in series with a capacitor. The capacitor is selected to match the magnetic resonance of the core. If the input voltage varies, then the capacitor/compensation coil set adjusts the saturation level of the core to produce a constant voltage output. Figure 5-35 shows a schematic representation of a constant voltage transformer.

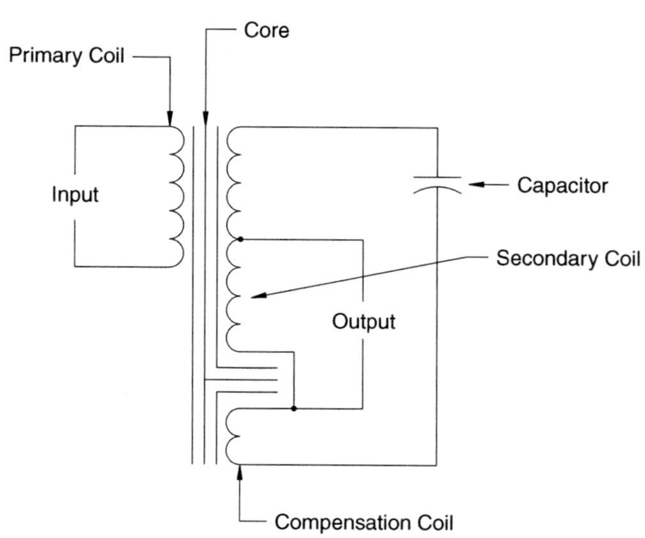

Figure 5-35 Constant Voltage Transformer Schematic

Figure 5-36 shows a commercial constant voltage transformer. Notice that the unit has a similar appearance to the saturatable core transformer shown in Figure 5-33.

For small point-of-use applications, constant voltage transformers are available in a stand-alone package, as shown in Figure 5-37.

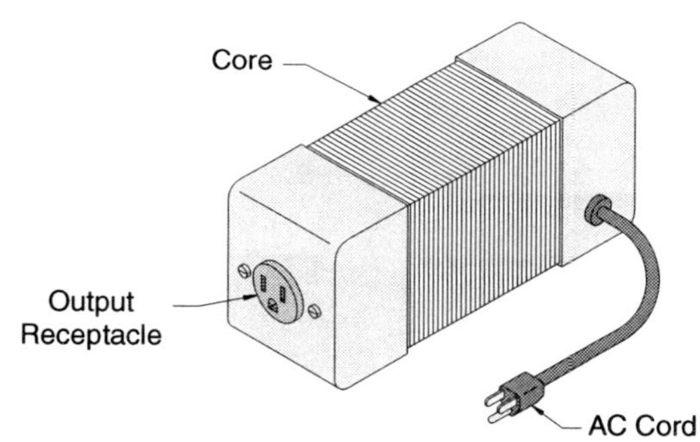

Figure 5-37 Stand-Alone Constant Voltage Transformer

Effects of Frequency on Transformer Design

The frequency of the AC power must be taken into consideration when designing transformers. In effect, the core volume must be large enough to store the magnetic flux generated by half of the AC cycle. Therefore, transformers that operate at higher frequencies will require less iron than their lower frequency counterparts. Figure 5-38 shows a comparison of the storage requirements between a 60 Hz wave and a 400 Hz wave. The 400 Hz wave has 0.15 times less area and therefore the iron required would be approximately seven times smaller than its 60 Hz counterpart. For this reason 400 Hz AC power is typically used on aircraft. The total weight of 400 Hz equipment is about $^1/_7$ that of 60 Hz equipment.

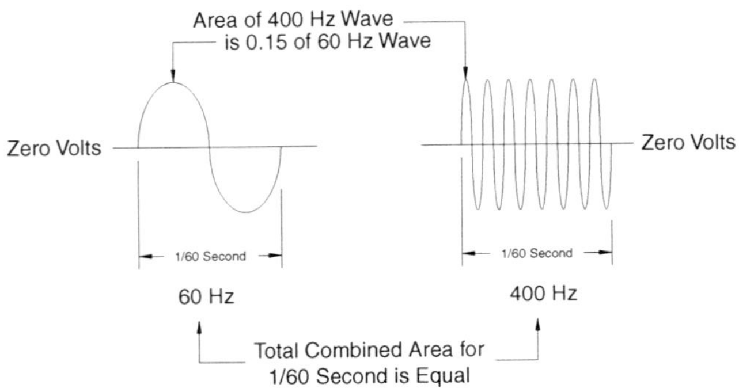

Figure 5-38 Effects of Frequency on Core Mass

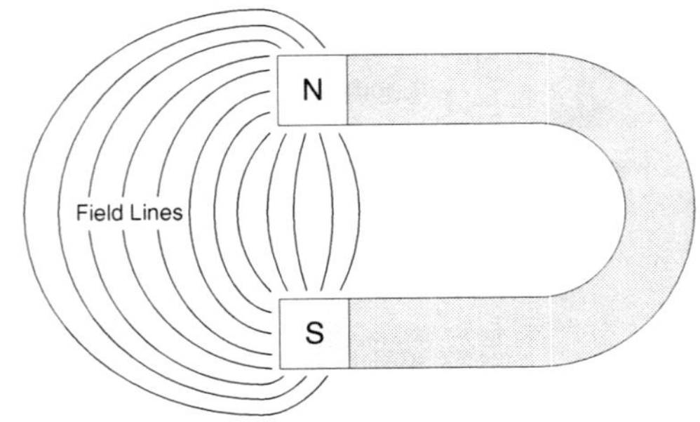

Figure 5-41 Horseshoe Magnet and Field Lines

Figure 5-39 shows a size comparison between a 60 Hz and a 400 Hz transformer with the same voltage and current carrying capabilities. More often then not, the size of a high frequency transformer is dictated by the physical size of the coil and terminals instead of the core size.

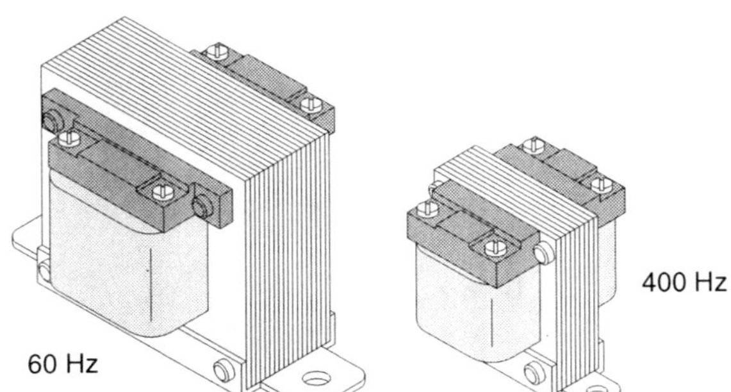

Figure 5-39 60 Hz and 400 Hz Transformer Size Comparison

Permanent Magnets

When considering electromagnetic components, it's a good idea to have a basic understanding of permanent magnets and their field characteristics. Figure 5-40 shows a common bar magnet and its associated field lines. Notice that in the absence of any outside influences the field lines are balanced. This is the natural state of the magnet.

Figure 5-41 shows a horseshoe magnet and associated field lines. Like the bar magnet the field lines are balanced between the poles, except the flux density is higher because the pole spacing is less.

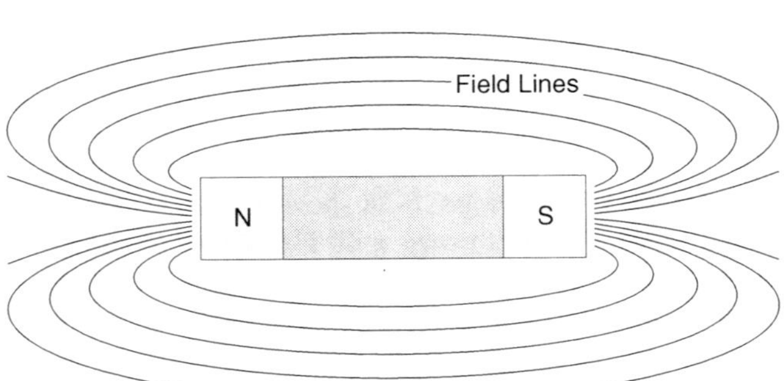

Figure 5-40 Flat Bar Magnet and Field Lines

Electromagnets

An electromagnet can be produced by simply feeding a current through a coil of wire that is wrapped around an iron bar. When the current is turned on, the iron bar becomes magnetized, and when the current is turned off, the magnetism is lost. In this manner a magnet can be constructed that may be turned on and off at will. This phenomenon has profound implications for electromechanical devices. Figure 5-42 shows a simple electromagnet, power source, and field lines. Notice the field lines are similar to the lines of a common bar magnet.

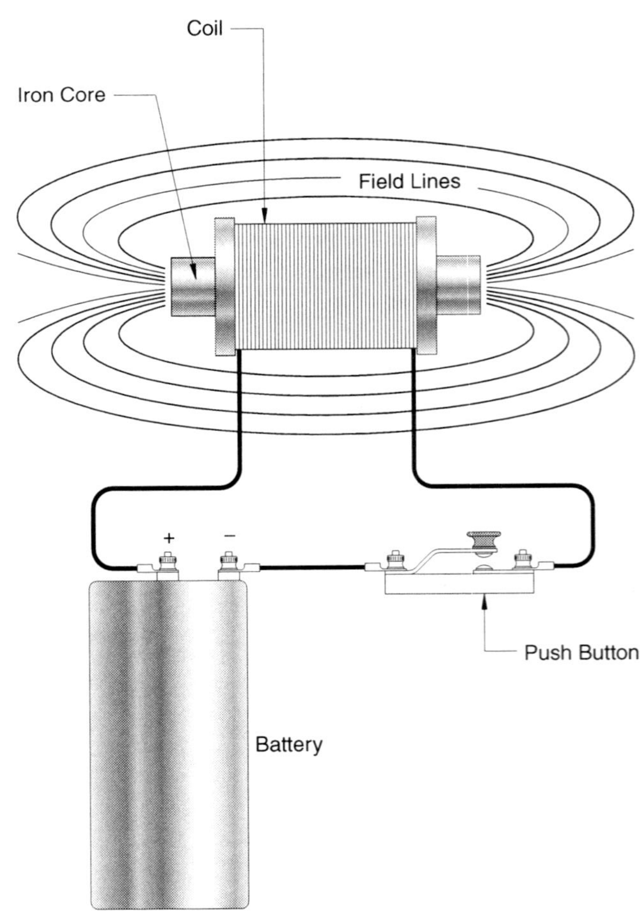

Figure 5-42 Electromagnet

My father used to tell a story of how he and his friend built an electromagnet in their high school industrial arts class. The idea was that they would place the magnet on top of the glass of a pinball machine and move the ball in order to rack up points. The sandwich shop down the road from the school would give a free sandwich to anyone who got a score above a certain level. My father and his friend thought that this would be a good way to get free sandwiches. They hauled the 19 pound electromagnet down to the sandwich shop and, somehow, slipped in unnoticed. His friend carried a book bag with six dry cell batteries connected in series. They placed the magnet on top of the pinball machine, slid it over the ball and connected it to the batteries. The magnet was so strong that it sucked the ball up and shattered the glass. It took them six weekends to work off the repair costs and their shop-teacher never gave the magnet back.

Cup-style electromagnets can produce very high flux densities at their poles. For this reason they are commonly used in mechanisms that require mechanical high loads. A common use for these magnets is as door locks. The locking force is so strong that the door cannot be opened, even with considerable force, until the power is turned off. Figure 5-43 shows a cup-style electromagnet.

Figure 5-44 shows a "C" style electromagnet. By manipulating the gap spacing extremely high fields can be produced with relatively low power and coil sizes.

Figure 5-45 shows a typical commercial application of a cup-style electromagnet for scrap recycling. These magnets can be seen in action at nearly every scrap yard in the nation.

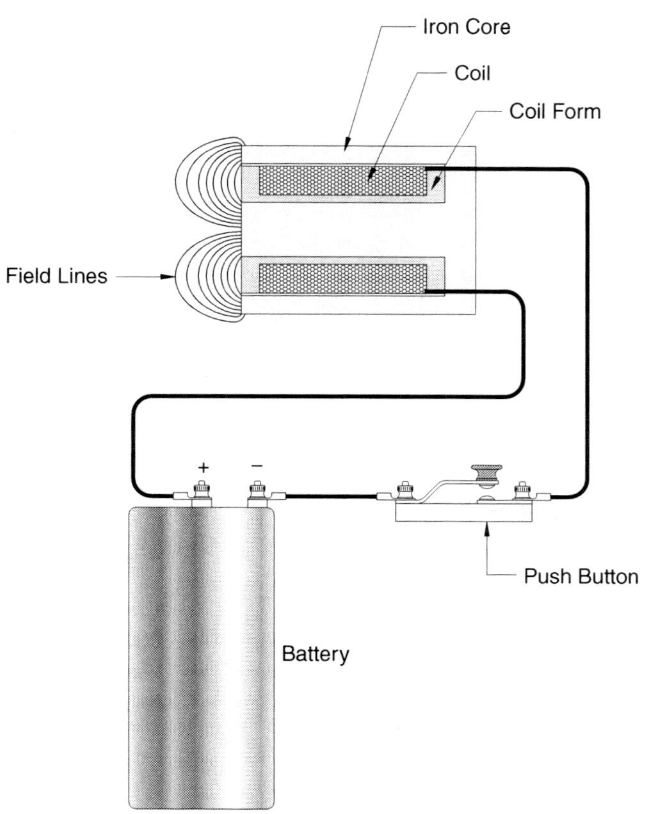

Figure 5-43 Cup-Style Electromagnet

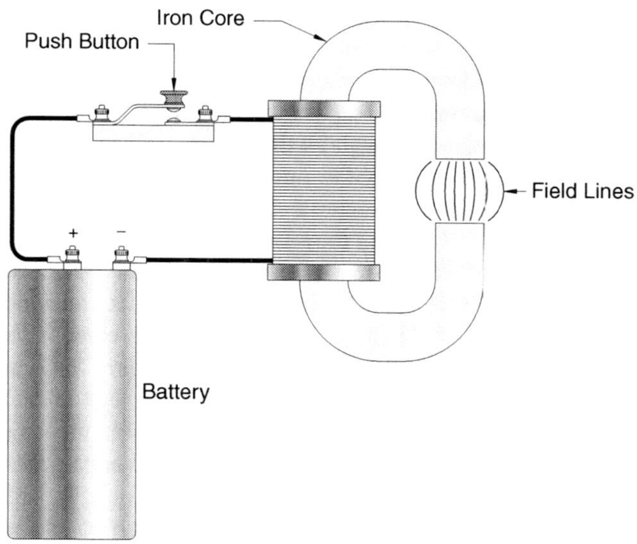

Figure 5-44 "C" Style Electromagnet

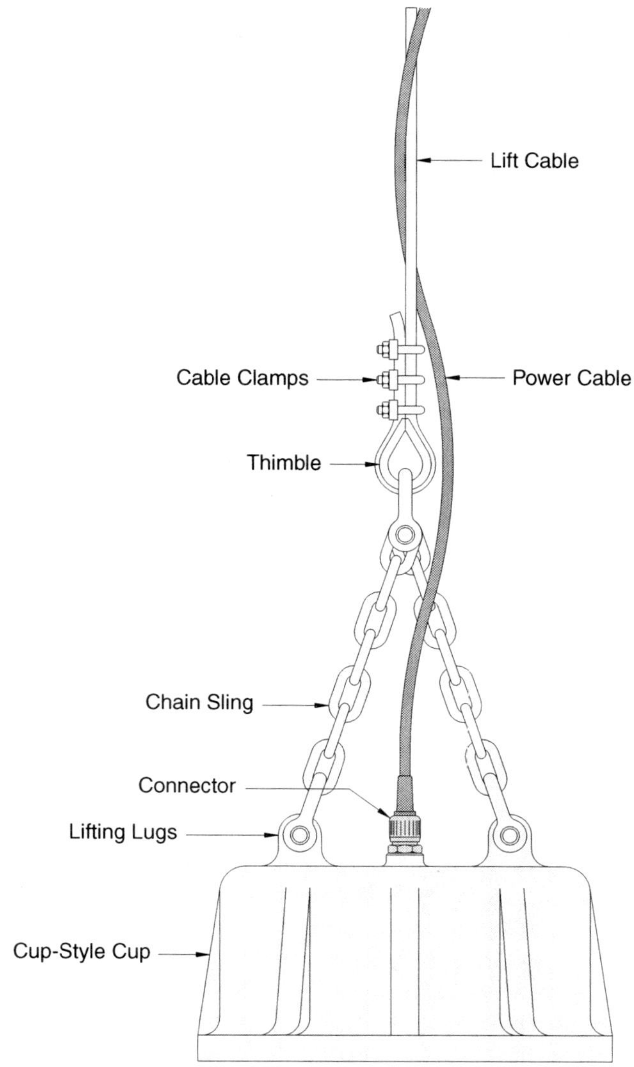

Figure 5-45 Electromagnet for Scrap Recycling

Solenoids

Solenoids represent an extremely important use of magnetics in the mechanical world. They are used in literally millions of different applications throughout the world.

When a current is applied to a coil, a magnetic field is produced in air, as shown in Figure 5-46. The field lines are similar to a bar magnet, except they are in the shape of a toroid. In the absence of any outside influences the field is balanced.

When an iron core is placed into the coil, an electromagnet is formed and the field lines are coupled to the core. If the core is not centered in the coil, as shown in Figure 5-47, an asymmetric load is generated. The asymmetric load tries to force the core into the center of the coil and balance the magnetic field. In this way usable mechanical force can be generated at the push of a button.

Figure 5-48 shows a paddle-style solenoid. The "L"-frame core and paddle represent the magnetic circuit. When the coil is energized, the paddle is pulled onto the core with considerable force. "C"-frame solenoids provide more travel than a paddle type. The magnetic circuit is in the form of a C-shaped frame with the coil mounted in the center. A guide rod and keeper made from iron complete the circuit. Figure 5-49 shows a "C"-frame style solenoid. These types of solenoids are generally inexpensive and can be found in all manner of equipment.

Laminated-core AC solenoids are generally used for high-load applications. They are similar in geometry to "C"-frame units, except that they use a riveted laminated core and the

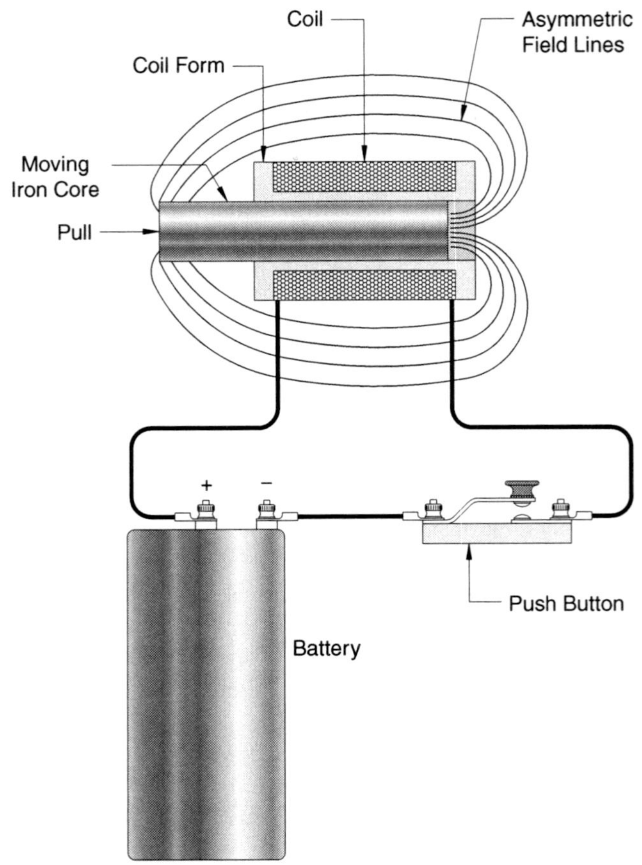

Figure 5-47 Solenoid

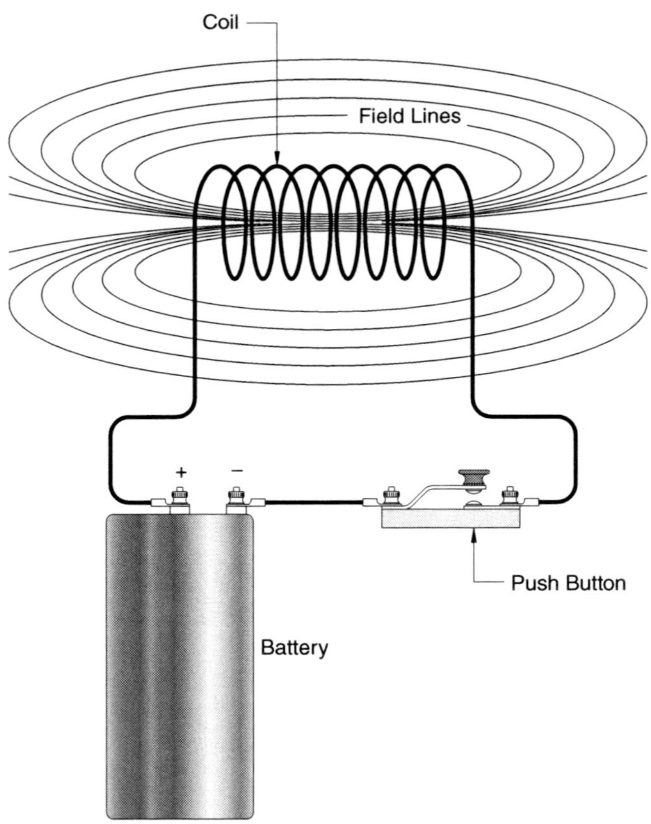

Figure 5-46 Solenoid Coil

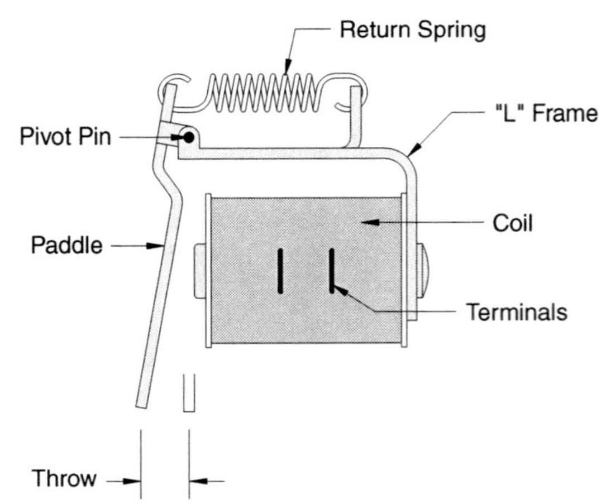

Figure 5-48 Paddle-Style Solenoid

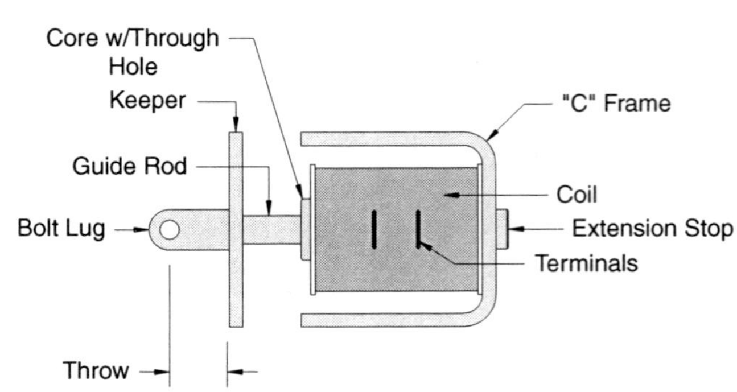

Figure 5-49 C-frame Style Solenoid

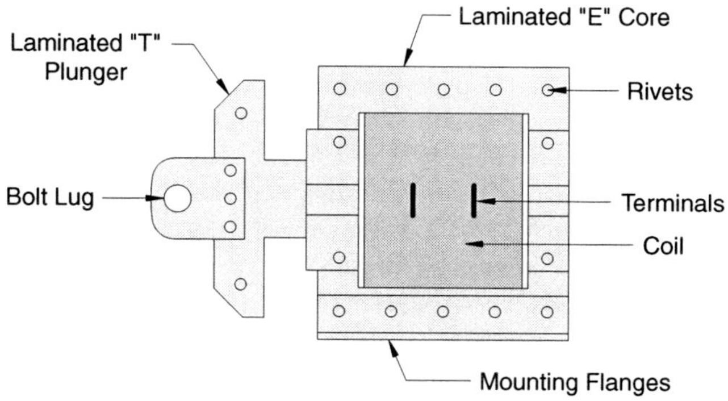

Figure 5-50 Laminated Core AC Solenoid

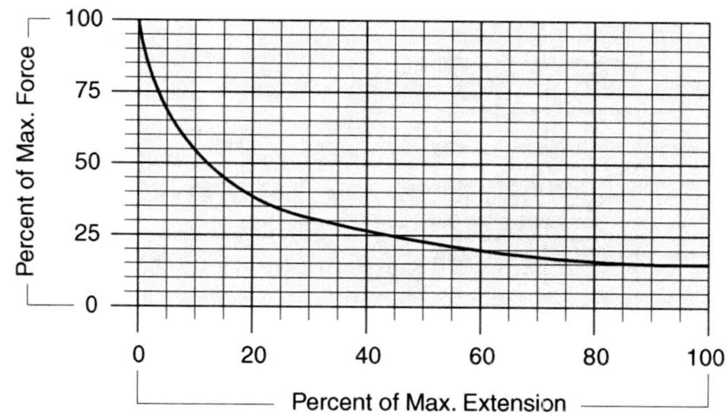

Figure 5-52 Solenoid Force Profile

coil is designed to operate on AC power. These units are manufactured in particularly large sizes that can produce hundreds of pounds of force. Figure 5-50 shows a commercial laminated core AC solenoid.

Cylindrical solenoids are used extensively for applications that do not require particularly high force. They are typically small units that have an integral nose mount. They will usually have two wires as their terminals and are available in AC or DC versions. Figure 5-51 shows a commercial cylindrical solenoid.

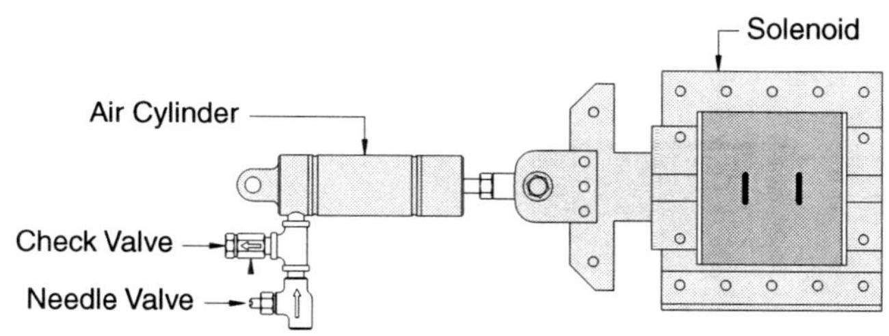

Figure 5-53 Damping Solenoid Motion

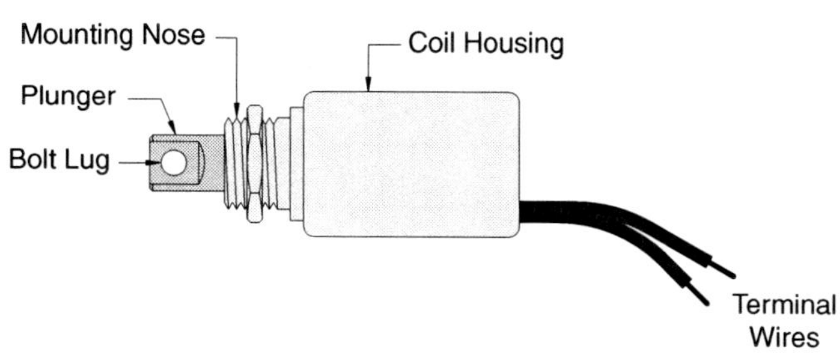

Figure 5-51 Cylindrical Solenoid

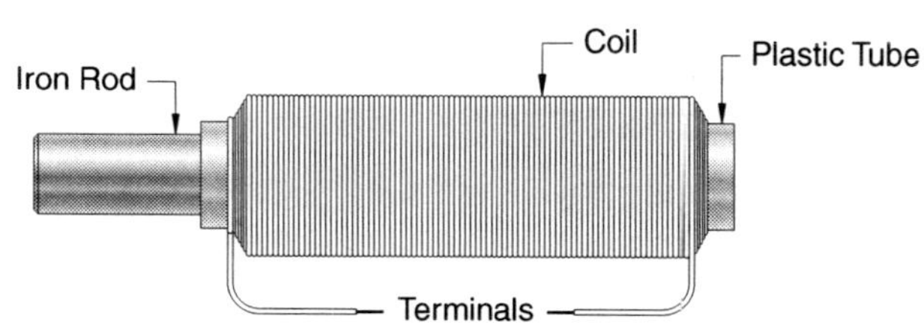

Figure 5-54 Bench Built Solenoid

The pulling force that any solenoid generates is dependent on the position of the core in relation to the coil. Solenoids will always produce their maximum force when the core is completely drawn in. Inversely, they will produce their weakest force when the coil is at full extension. Figure 5-52 shows a force profile of a typical solenoid. Notice that the highest forces are generated in the first 20% of extension.

Typically, solenoids operate at very high speeds. In many applications the speed of actuation must be dampened. For smaller solenoids, a simple dash pot can be applied to the plunger. For larger applications, a pneumatic damper can be constructed using an air cylinder equipped with check and needle valves, as shown in Figure 5-53. The needle valve can be used to control the rate at which the solenoid retracts. The check valve allows the core to extend with no restriction.

A simple solenoid can be constructed as shown in Figure 5-54. A plastic tube is wrapped with bell wire and an iron rod is inserted in the center. When the coil is connected to a battery, the core will be pulled into the center of the core.

One very common use of solenoids is as valve actuators. Solenoid valves are available in a wide variety of designs. Figure 5-55 shows a typical commercial solenoid valve.

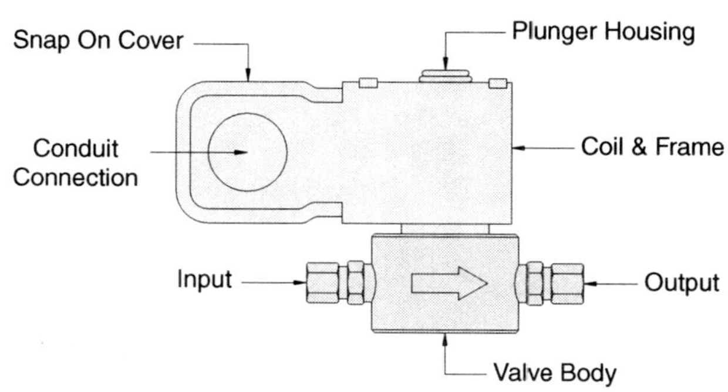

Figure 5-55 Solenoid Valve

Eddy Currents

If a conductor is moved through a magnetic field, an electrical charge is induced. Any conductor will harbor eddy currents. Eddy currents are complete electric circuits that reside entirely in the conductor, as shown in Figure 5-56. Because the currents are contained within the conductor, they represent a short circuit. Two things result from eddy currents, first the conductor will heat up in direct proportion to the amount of power being dissipated. Second, the energy that goes into the eddy currents represents a loss. In wires, eddy currents do not represent a significant problem. However, in the cores of magnetic components such as transformers, solenoids, and motors, this loss can be problematic. To minimize the path that eddy currents can be generated in, most AC magnetic components use laminated cores. The laminated plates present a single magnetic mass and, at the same time, are electrically isolated. The thin plates reduce the effective path and eddy currents are minimized.

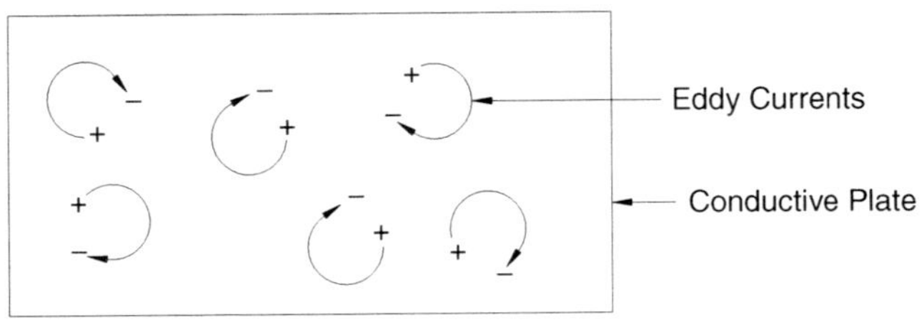

Figure 5-56 Eddy Currents

An interesting demonstration of the effects of eddy currents is shown in Figure 5-57. Place a nonmagnetic, conductive plate (aluminum or brass) on a table. Set a horseshoe magnet on the center of the plate. Slide the magnet back-and-forth very slowly. You will notice that there is no restriction to the movement of the magnet. Now move the magnet very fast and you will notice that there is a great deal of restriction to the movement. In fact you will most likely have to clamp the plate down to keep it from moving with the magnet.

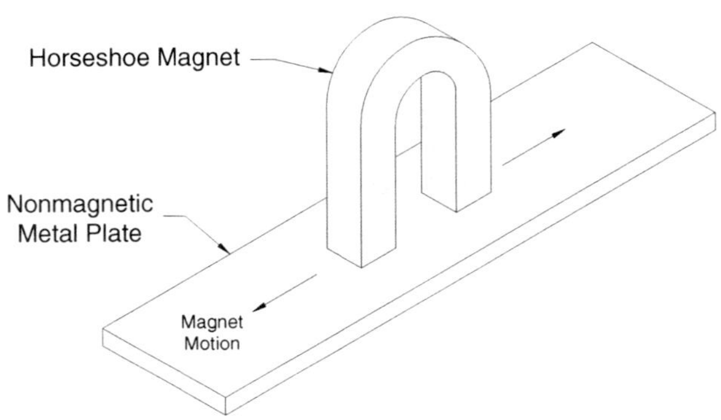

Figure 5-57 Eddy Current Demonstration

As the magnet moves, it induces eddy currents into the plate. When the currents are present, they can work against the magnet's field and the plate becomes momentarily magnetic. As soon as the magnet motion stops, the eddy currents diminish and the plate resumes its nonmagnetic characteristics.

One common use of eddy currents is as shock absorbers or dampers. Figure 5-58 shows a permanent magnet eddy current damper. The horseshoe magnet may be hung from a pendulum and the plate is fixed. If the pendulum motion is slow, it can move freely. As its speed increases, progressively stronger eddy currents are induced into the plate and the motion is damped.

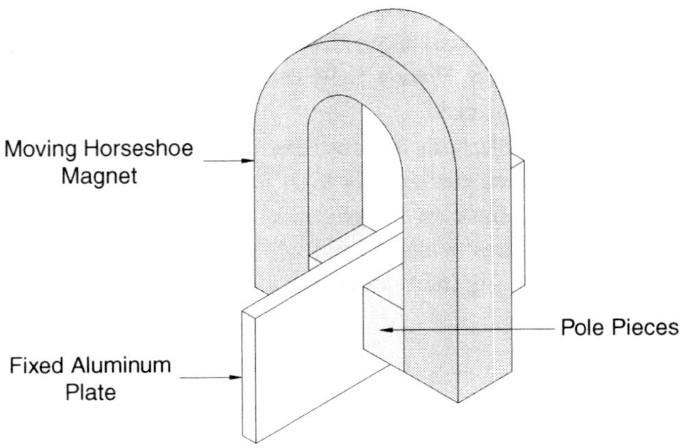

Figure 5-58 Permanent Magnet Eddy Current Damper

Figure 5-59 shows an electromagnet eddy current damper. In this manner the effects of the damper can be switched on and off. Additionally, by varying the magnetic field, the strength of the eddy currents can be controlled and variable damping can be utilized.

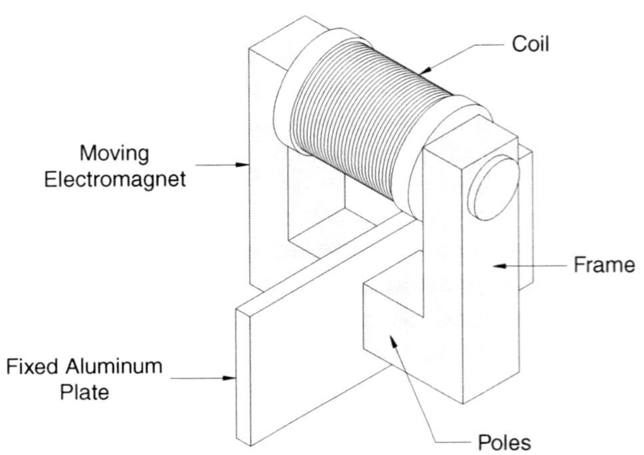

Figure 5-59 Electromagnet Eddy Current Damper

Inductors

An inductor is a device that is intended to limit current based on the rise and collapse of a magnetic field. Most inductors are coils of wire and any coil is an inductor. Transformers, solenoids, motors, and the coiled filament of a light bulb are all inductors. Inductors are rather simple in operation. An AC signal is sent through the coil and an oscillating magnetic field is set up. As the field rises and collapses, it induces currents back into the coil that are out of phase with the line signal. These out-of-phase currents cancel out a part of the line signal and, in effect, change the resistance of the coil. As the resistance of the coil rises, less current can pass. The unit of measure for an inductor is the *hennery* and inductors are generally found rated in milli-henneries.

Figure 5-60 shows a common air core inductor. These devices are simply a coil of wire wrapped onto a coil form. Much like transformers, inductors can benefit from magnetic cores. Figure 5-61 shows a common iron core inductor design with a solid "C" frame. Figure 5-62 shows an iron core inductor with a laminated "E" core. The laminations are used to limit eddy current losses within the core. Notice that this unit is very similar in appearance to a typical transformer.

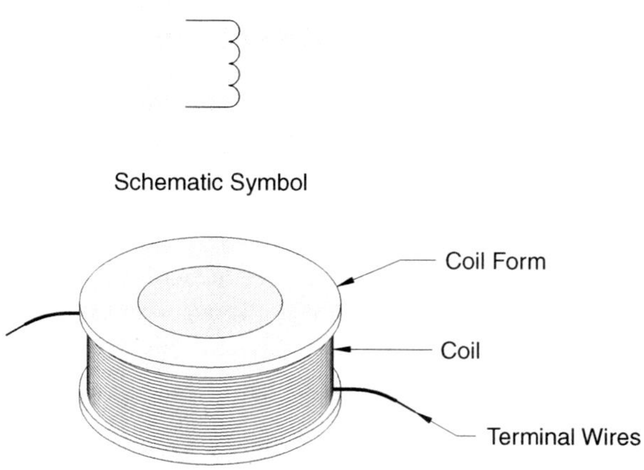

Figure 5-60 Air Core Inductor

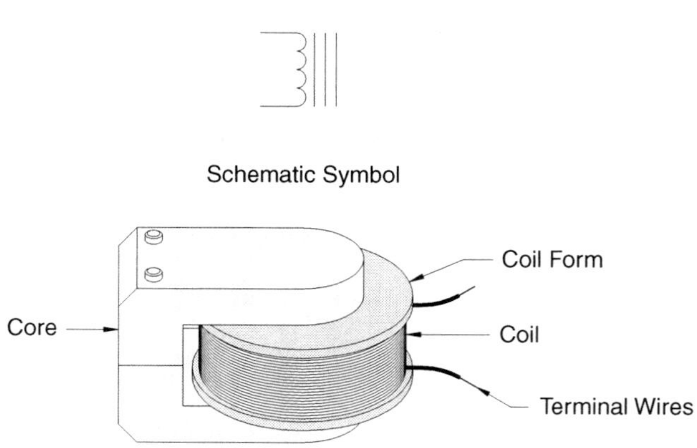

Figure 5-61 Iron Core Inductor with "C" Frame

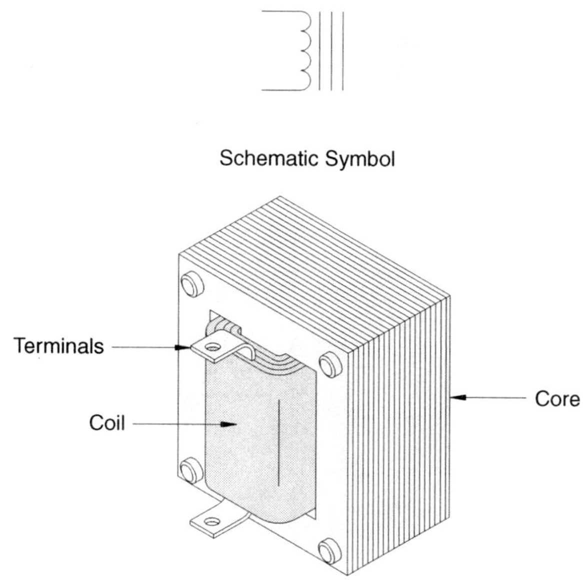

Figure 5-62 Iron Core Inductor with "E" Frame

To further limit eddycurrent losses, sintered metal cores are manufactured from ferrite. In this process powdered metal is cold welded into shapes through a pressing process. Each particle of metal has an extremely short electrical path and, therefore, eddy current losses are extremely low. Ferrite-core inductors are commonly found in high-frequency circuits. Figure 5-63 shows a typical ferrite core inductor, while Figure 5-64 shows an exploded view of the unit.

Changing the magnetic characteristics of an inductor will affect its performance. Tunable inductors typically have an adjustable core, as shown in Figure 5-65. As the screw is turned, the core moves in and out of the coil and the inductance is affected.

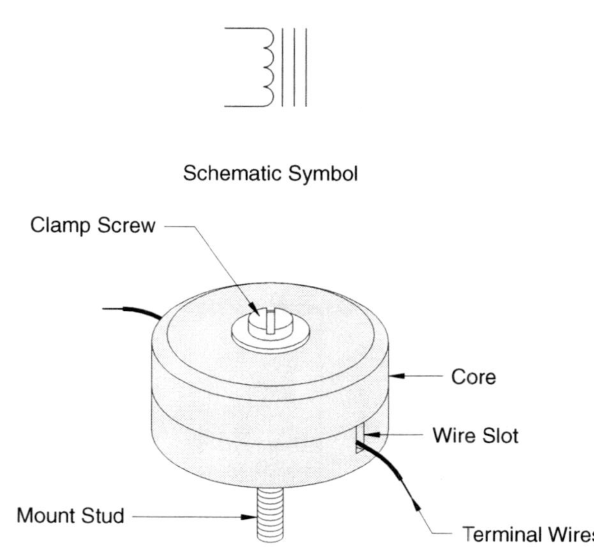

Figure 5-63 Ferrite Core Inductor with Toroidal Frame

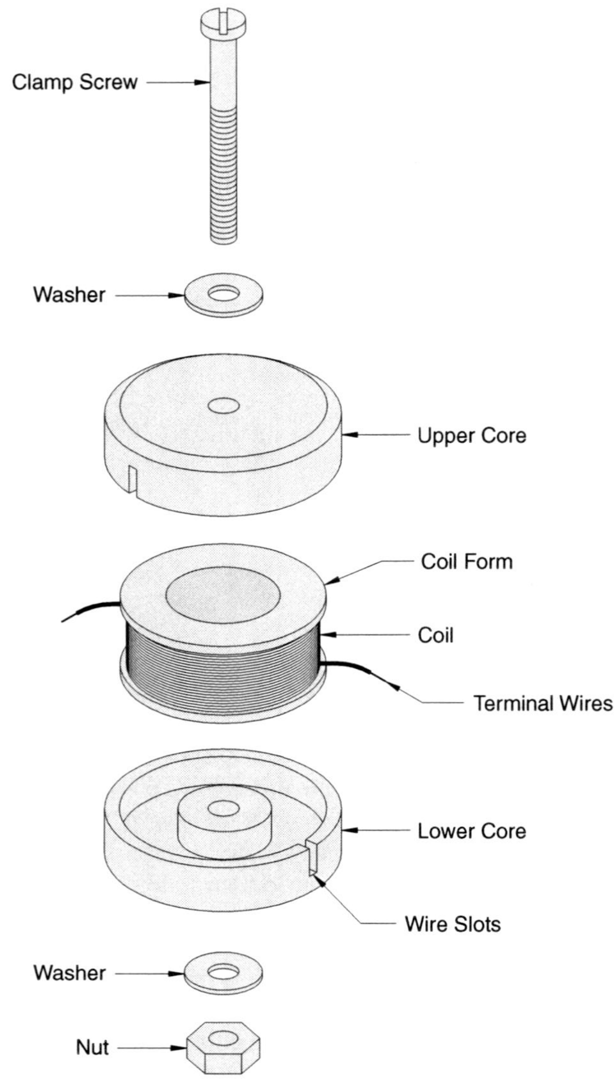

Figure 5-64 Exploded View of a Ferrite Core Inductor with Toroidal Frame

Magnetic Amplifiers

Magnetic amplifiers, or saturatable core reactors as they are sometimes referred to, were first developed in the early 1900s as an alternative to vacuum tube technology. At this time vacuum tubes could not handle particularly high power and were quite fragile. The magnetic amplifier was extremely tough and could be built to handle virtually any power level. For this reason these units found favor in industrial and marine applications and were quite common well into the 1990s. High-power transistor technologies have replaced most magnetic amplifier applications, however, in extreme conditions these units are still used. One area where magnet amplifiers are still used with great effect, is as extremely high-current regulated power supplies.

Figure 5-66 shows a basic magnetic amplifier schematic. The unit is an iron-core inductor with an additional reactor coil. The input signal is used to manipulate the level of core saturation and, in turn, the resistance of the inductor. Figure 5-67 shows a typical commercial magnetic amplifier. Notice the similarity in appearance to a saturateable core transformer.

Magnetic Recording Devices

Magnetic recording technologies play an important role in our day-to-day lives. Most notably as the little magnetic strips on the back of our credit cards and the hard drives in our computers. Although digital technologies have displaced most magnetic recording systems, most of us have had dealings with either video or audio cassette decks in our lives.

Magnetic recording was invented by Valdemar Poulsen, in 1898. Considering the impact that magnetic recording has had on society, it seems a shame that this man doesn't occupy a noteworthy position in the history of technology. Rather, he has slipped into obscurity and, with the exception of his occasional mention in texts such as this one, his name is completely unknown.

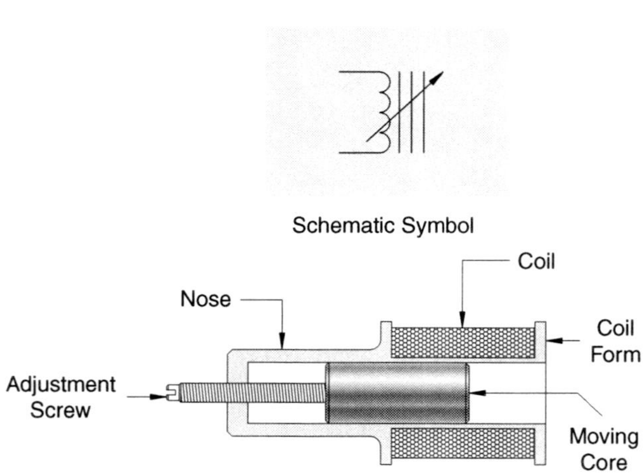

Figure 5-65 Moving Core Inductor

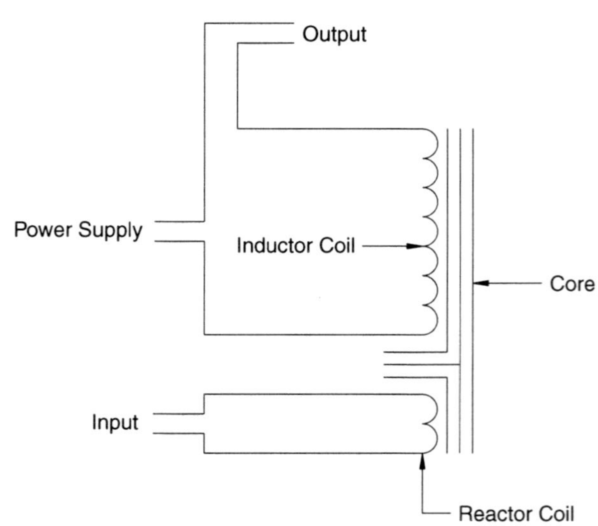

Figure 5-66 Magnetic Amplifier Schematic

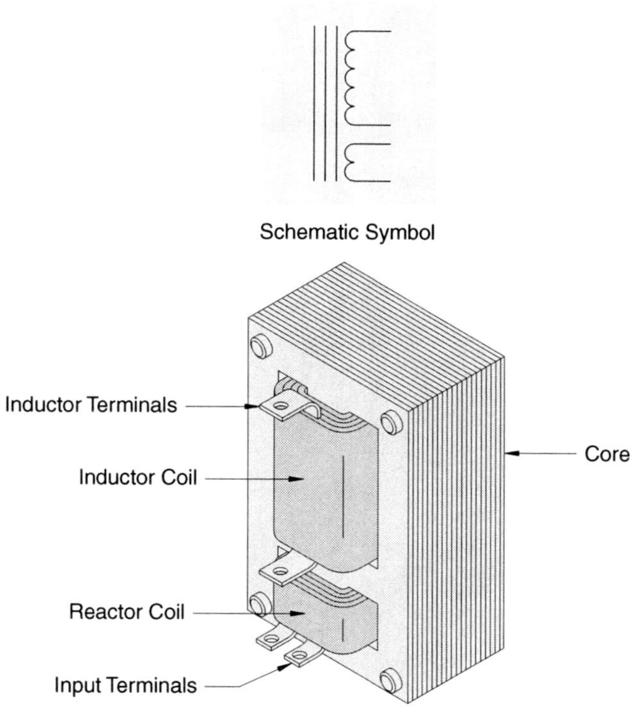

Figure 5-67 Commercial Magnetic Amplifier

The first magnetic recording machine was introduced to the market around 1920 by the American Telegraphone Company. The machine used a wire traveling at 7 feet per second for its recording media. It was poorly marketed and difficult to use, and the company soon went out of business.

It wasn't until about 1935 that a German company introduced a paper-backed magnetic recording tape which fostered significant advancements in the technology. From the paper tape came plastic tapes, cassettes, floppy disks, hard drives, and the like.

Figure 5-68 shows a typical magnetic recording head. The unit is simply an electromagnet with small gap (0.001 to 0.0005 inch). The gap is placed onto a moving tape and a signal is applied to the coil. As the tape passes, varying magnetic signals are imprinted on the tape. When the tape is played back, the magnetic signals induce a voltage in the coil, which is fed to an amplifier.

Figure 5-69 shows stylized schematic of a three-head tape recording system. The system has an erase head, record head, and a play head. Three-heads are used because the gap and magnetic characteristics are optimized for their specific duty. The system also has a tape reel, cap stand, and take-up reel. The cap stand is intended to provide very precise speed control for the tape. This assures accurate recording and playback signals. Figure 5-70 shows a block diagram of a typical magnetic recording system.

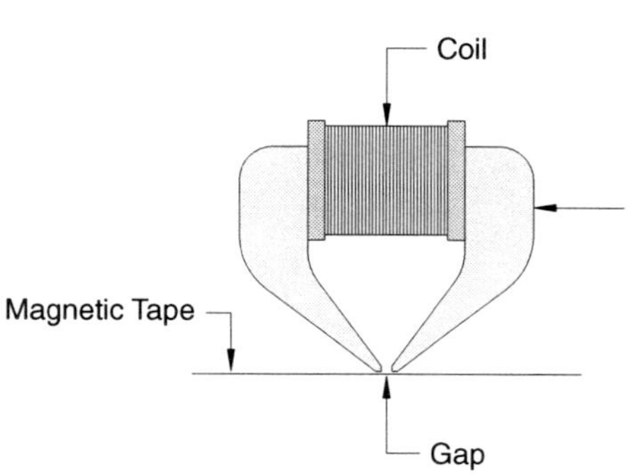

Figure 5-68 Magnetic Recording Head

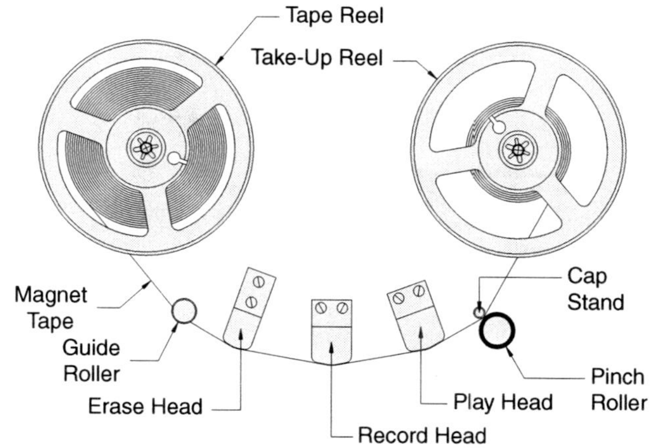

Figure 5-69 Three-Head Magnetic Recording System

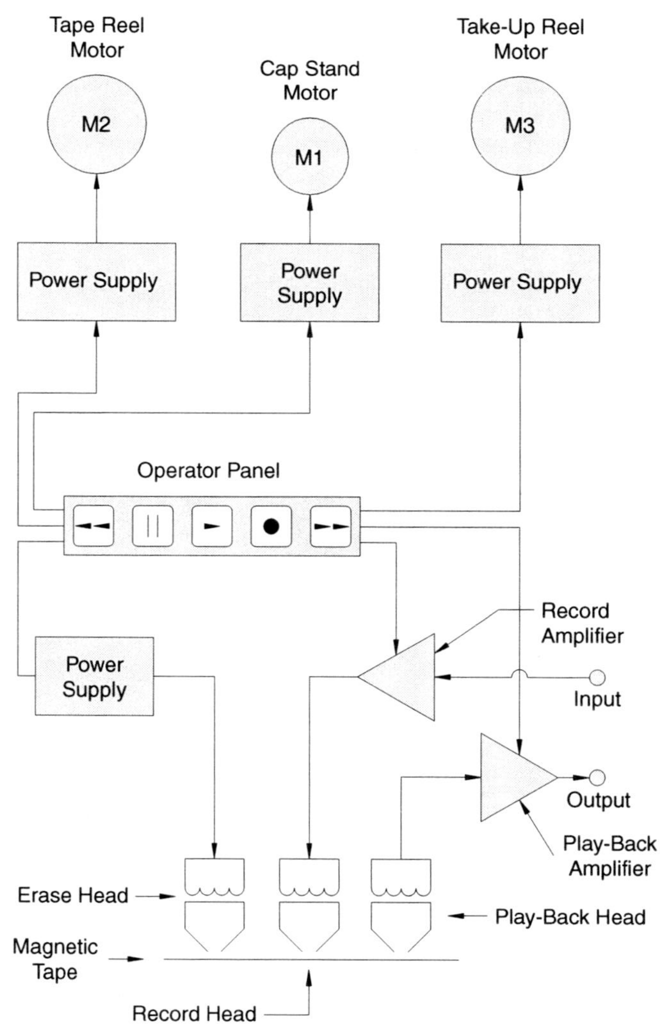

Figure 5-70 Magnetic Recording System Block Diagram

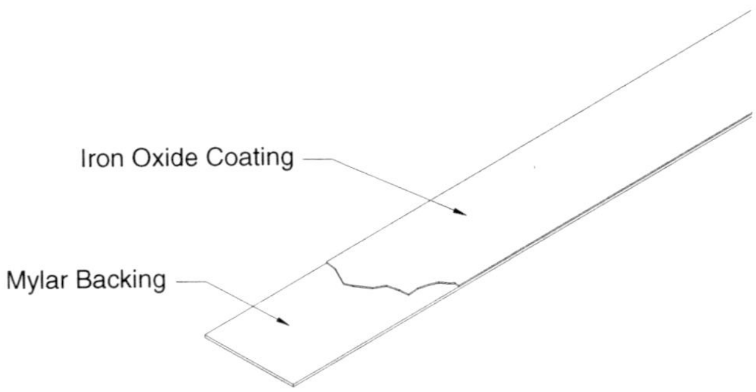

Figure 5-71 Magnetic Recording Media

The magnetic recording media is typically a plastic backing with magnetically permeable coating. Figure 5-71 shows a typical recording tape with a Mylar backing and iron oxide coating. Figure 5-72 shows various types of magnetic recording media.

Disk drives are probably the most common use of magnetic recording media today. Both floppy drives and hard drives use advanced forms of this technology.

Figure 5-73 shows the internals of a typical floppy drive. The floppy disk is inserted into the disk slot and the locking lever is rotated down. As the lever is rotated down, the clamp hub locks in the disk and activates the drive. The head rides on a transport mechanism which is driven by a stepper motor, rack, and pinion. All of the control functions and the digital interface are carried on the controller board.

Figure 5-74 shows the internals of a typical hard drive. The disk spins at very high speed in an effort to minimize data

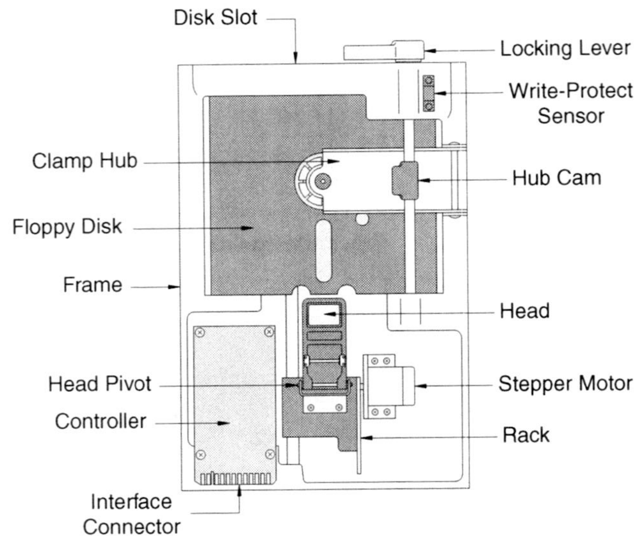

Figure 5-73 Floppy Disk Drive

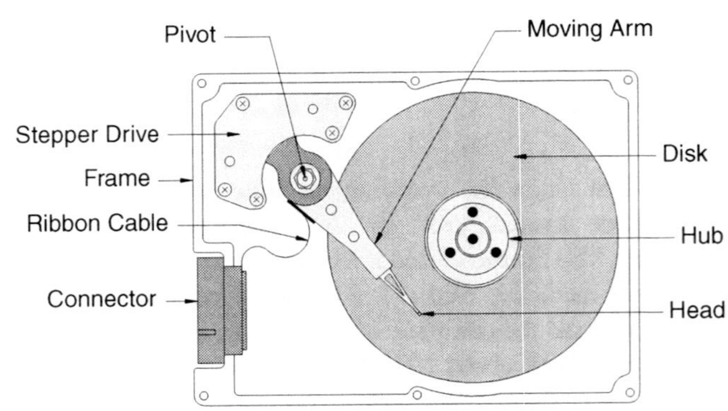

Figure 5-74 Personal Computer Hard Drive

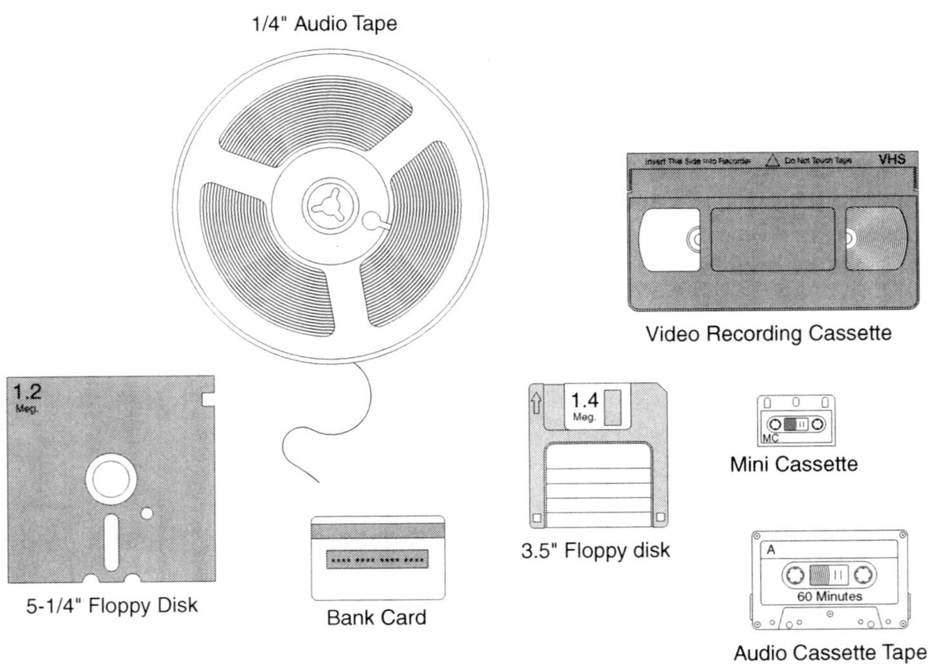

Figure 5-72 Various Types of Magnetic Recording Media

access times. The head is mounted to the end of a moving arm and its position is controlled by a special stepper drive. The controller electronics and drive motor are located on the backside of the frame.

Figure 5-75 shows a magnetic core memory. These were used in early computers instead of the solid-state memory that is used today. The core is made from a magnetic permeable material. Depending on the orientation of the magnetic poles, the computer will interpret either a 1 or a 0. The pole state is controlled by wires Z and X. Wires S and Y are used to sense the magnetic polarity.

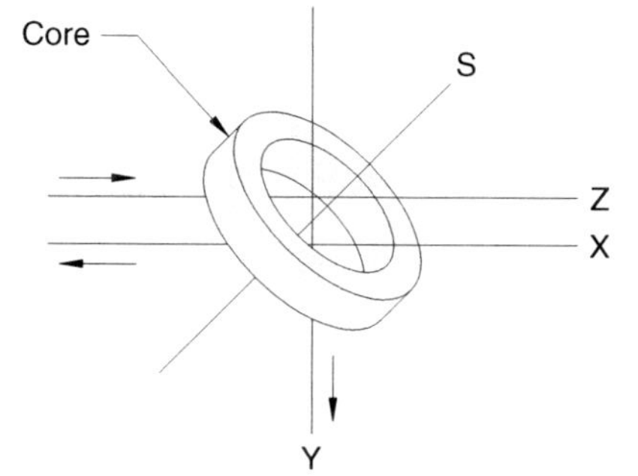

Figure 5-75 Magnetic Core Memory

CHAPTER 6
ROTATING COMPONENTS

One of the most important divisions in electromechanics is rotating equipment. The two most noteworthy pieces of rotating equipment are the motor and the generator. We encountered electric motors virtually everywhere we go. Our kitchens have a variety of motors inside the appliances that dot our countertops and cabinets. Our automobiles use electric motors to drive the windshield wipers, heater fan, power windows, automatic doors, and fuel pumps. Every time we ride an elevator we experience the work generated by an electric motor. The air-conditioning and heat we enjoy would not be possible without the motors that drive the air handlers and compressors within the systems.

The electric motors that make our lives so comfortable must have a ready source of power; this is where generators come into play. Although most of us do not encounter generators in our day-to-day lives, they are closer than you might think. All of the electric power that you rely on at home and work is produced by generators. Most people take their access to reliable electric power for granted and never really consider where it comes from or how it's produced. Even the common terms that are associated with power generation like coal, hydroelectric, and nuclear have no real meaning to the average person.

Probably the closest generator to most people is the alternator in their automobile. Without this critical piece of equipment, the electrical system of your car would quickly fail. Many of us have experienced a failed alternator while driving. When the alternator stops producing power, the car's electrical system starts to operate on the battery alone. It doesn't take long before the load depletes the battery and the car shuts down.

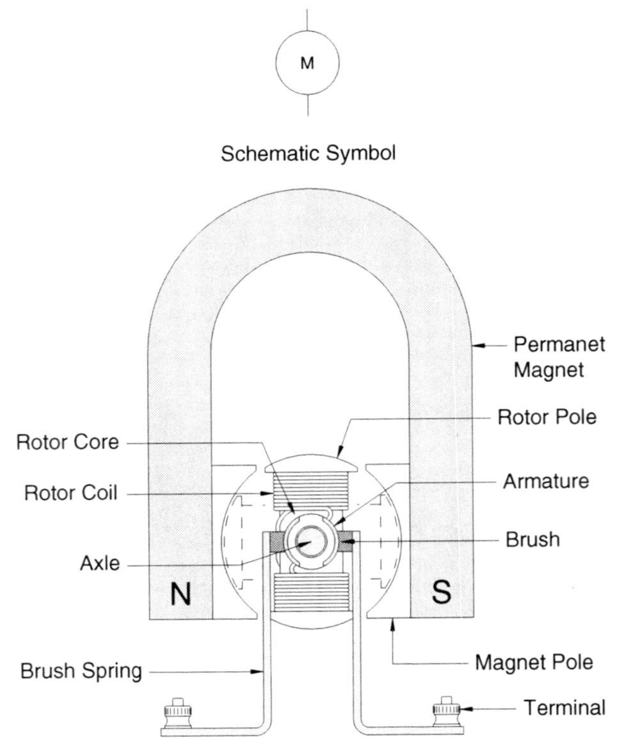

Figure 6-1 Permanent Magnet DC Motor

Permanent Magnet DC Motors

The basic permanent magnet DC motor, as shown in Figure 6-1, is fairly easy to understand. It generates rotation by manipulating the interaction of the fields between a permanent magnet and an electromagnet. When the rotor poles are vertical, power is connected to the rotor coil, which, in turn, generates a magnetic field in the core of the rotor. The two magnetic fields attract one another and the rotor aligns to the permanent magnet (dotted lines). When the rotor spins into the horizontal position, the rotating armature disconnects the power to the coils and the rotor freewheels toward the vertical. As the rotor approaches vertical, the armature reconnects the coil and a reverse field is generated within the core. In this manner the magnetic field in the rotor is reversed every half revolution and a spinning motion is created. Figure 6-2 shows a schematic representation of a two-pole permanent magnet electric motor. Figure 6-3 shows a typical permanent magnet DC motor. Note that most of these motors have facilities to easily replace the brush set, the part of a DC motor that most often requires maintenance.

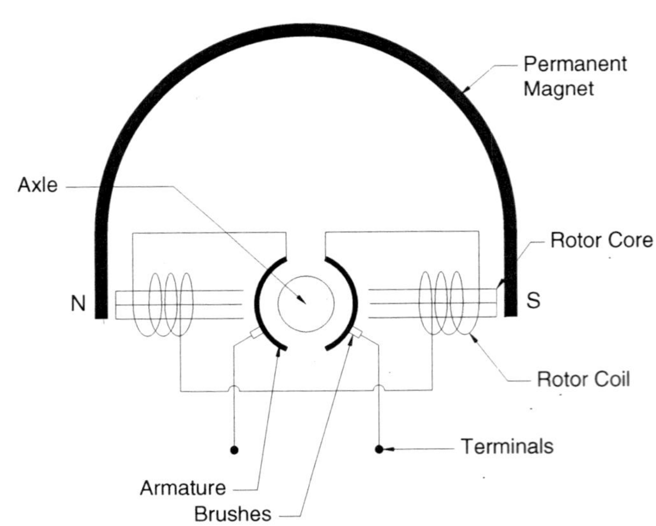

Figure 6-2 Permanent Magnet DC Motor Schematic

Shunt Wound DC Motors

A common variation of the permanent magnet DC motor is the shunt wound motor. A shunt wound motor is the same as a permanent magnet motor, except that an electromagnet replaces the permanent magnet. Shunt wound motors are generally used for higher horsepower motors because the electromagnet can supply a much higher field strength than a permanent magnet can. Figure 6-4 shows a two-pole shunt wound motor. Notice that the only difference between this unit and a permanent magnet unit is that this design uses an electromagnet. Otherwise the two motors operate in the same fashion. Figure 6-5 shows a shunt wound motor schematic.

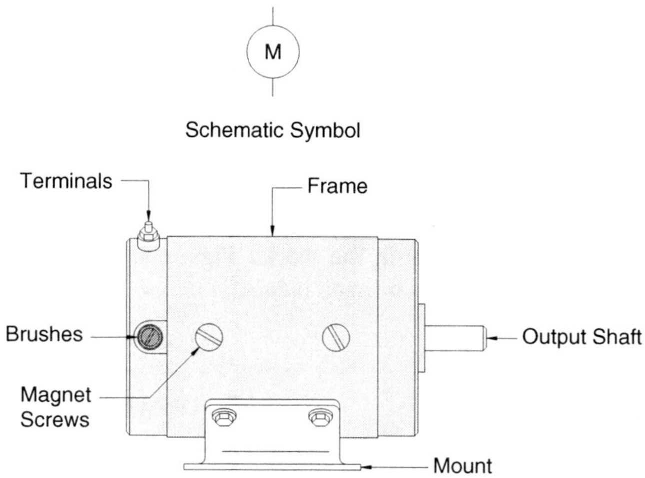

Figure 6-3 Permanent Magnet DC Motor

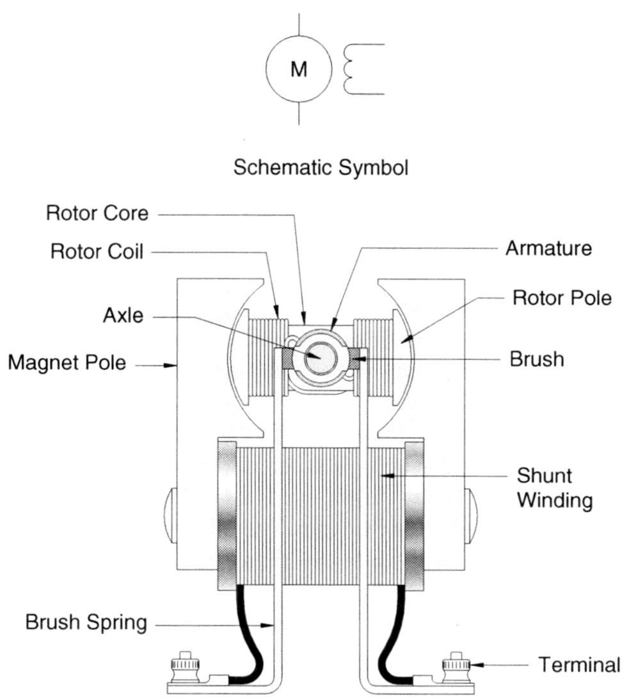

Figure 6-4 Shunt Wound DC Motor

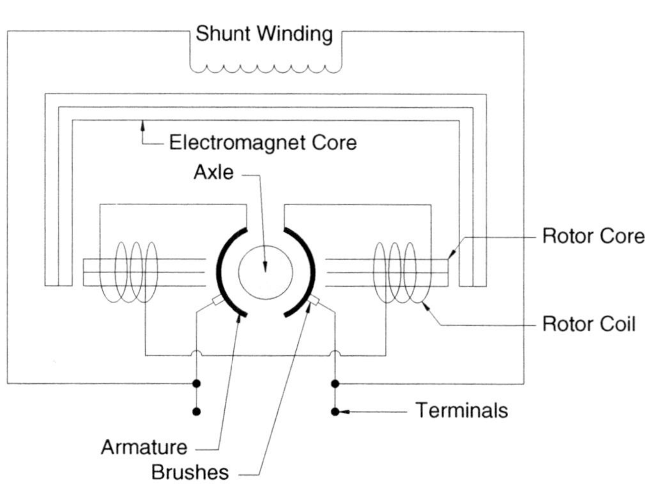

Figure 6-5 Shunt Wound DC Motor Schematic

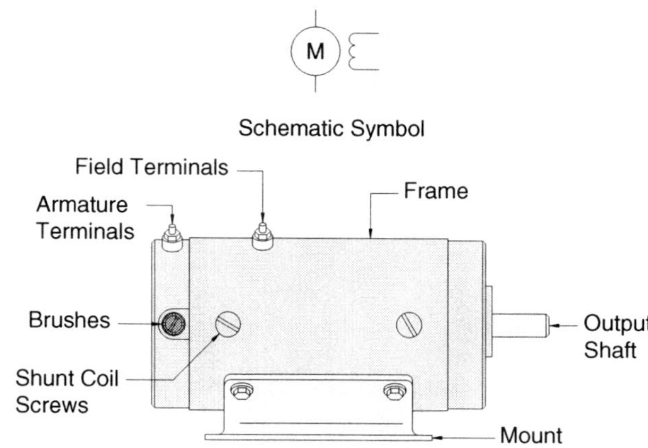

Figure 6-6 Shunt Wound DC Motor

Figure 6-6 shows a typical commercial shunt wound motor. Note that most of these motors also have facilities to easily replace the brush set. Like the permanent magnet motor, this is the part of a shunt wound motor that most often requires maintenance.

The speed of a shunt wound motor can be controlled by limiting the current that the field receives. By lowering the current, field strength lowers and the motor produces a lower speed. Figure 6-7 shows a schematic representation of how a shunt wound motor can be configured for speed control.

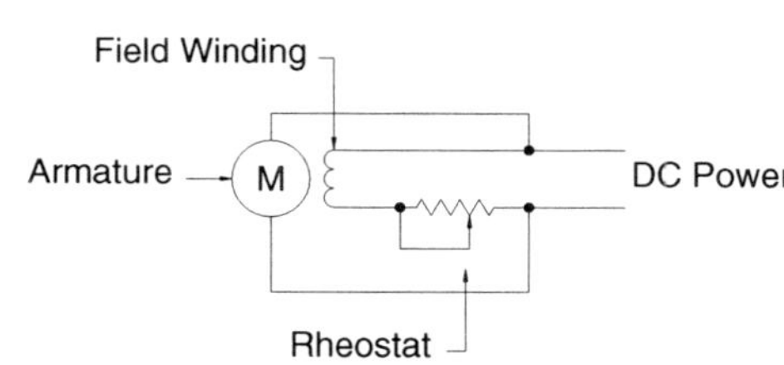

Figure 6-7 Shunt Wound Motor Speed Control Schematic

Universal Motors

A universal motor is a shunt wound motor that is designed to operate on either AC or DC power. These motors are commonly found on sewing machines as they provide excellent speed control at a low cost per unit. Figure 6-8 shows a typical universal motor. The appearance of a universal motor differs very little from most DC motors and usually the difference can't be determined without reading the nameplate.

Controlling the speed of a universal motor is accomplished by varying the input voltage. Variable autotransformers are ideal for this service. Figure 6-9 shows a schematic of a universal motor speed control utilizing a variable autotransformer. Figure 6-10 shows a high/medium/low speed control using a multi-tap transformer. Figure 6-11 shows speed control accomplished using an electronic SCR (silicone-controlled rectifier) speed controller.

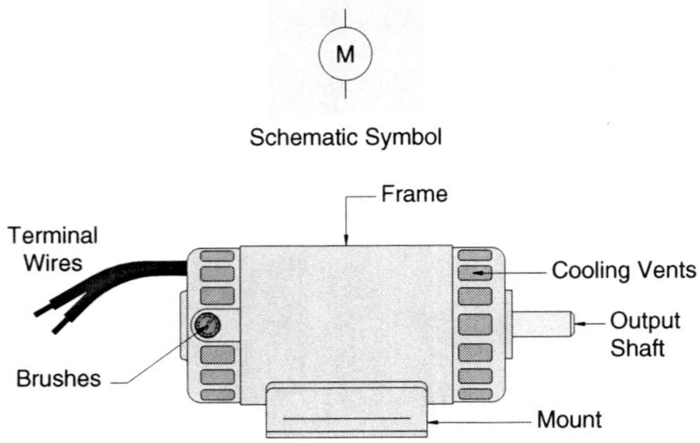

Figure 6-8 Universal Motor

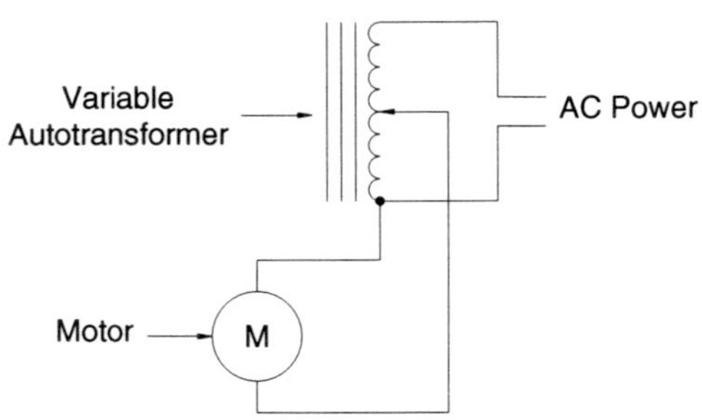

Figure 6-9 Universal Motor Speed Control Schematic

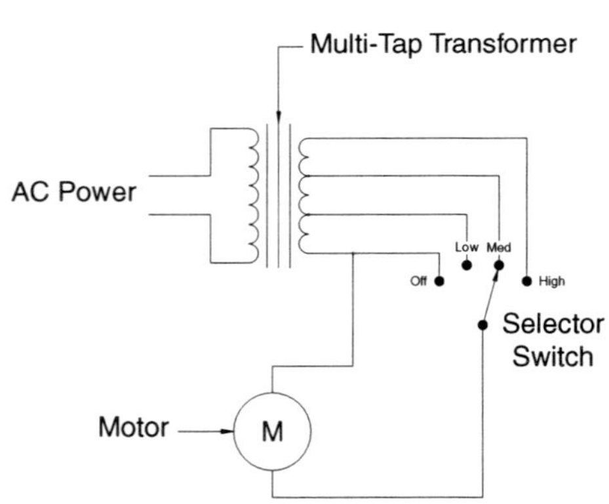

Figure 6-10 Three-Speed Universal Motor Control Schematic

Figure 6-11 Universal Motor Speed Control with SCR Controller

Induction Motors

By far, the largest class of electric motors are induction motors. These motors represent the most efficient use AC power and are the least expensive class of motor to manufacture. They can be designed to produce outputs from low fractional horsepower to tens of thousands of horsepower. Induction motors are found in virtually every home, office, and industrial facility in the world. Figure 6-12 shows a typical commercial open-frame induction motor.

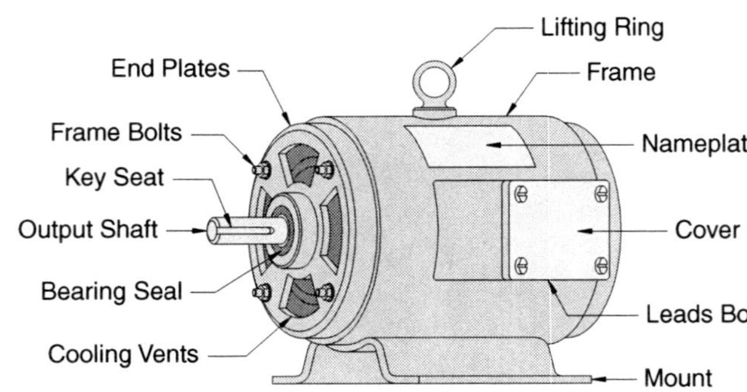

Figure 6-12 Commercial Open-Frame Induction Motor

Induction motors operate by inducing a current into their rotors. The induced rotor current then produces a magnetic field, which is attracted by the field generated in the stator. Because of the continuously reversing poles of AC power, the stator field rotates and drags or pulls the rotor into a spinning motion. Figure 6-13 shows a schematic representation of an induction motor. When the voltage rises and falls in the stator, a current is induced into the rotor. The induced rotor field acts against the field in the stator and rotary motion is produced.

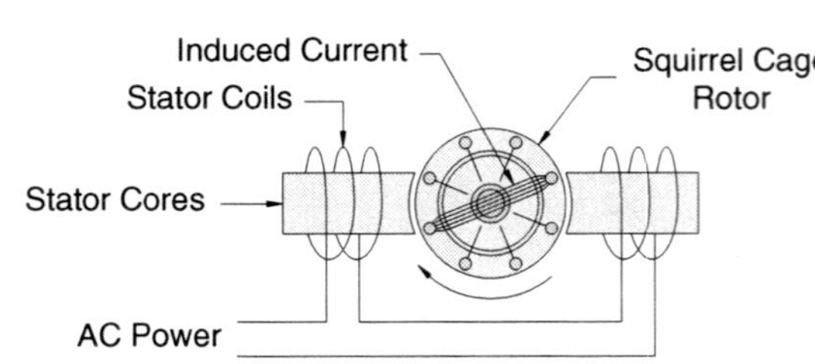

Figure 6-13 Stylized Induction Motor Schematic

Most induction motors use what is termed a "squirrel cage rotor." A squirrel cage rotor is a stack of circular iron laminations clamped between two end plates, which are connected with a series of nonmagnetic conductors. The end plates and the conductors form closed electrical circuits that a current may be induced into. The iron lamination creates a magnetic core that is designed to act against the stator field. Figure 6-14 shows a typical squirrel cage rotor found in many induction motors.

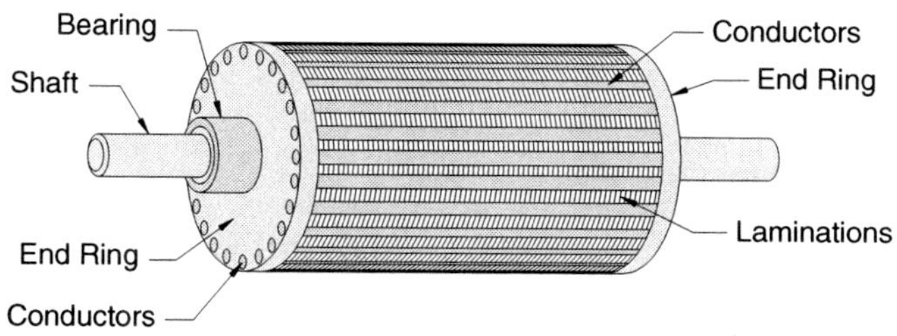

Figure 6-14 Squirrel Cage Rotor

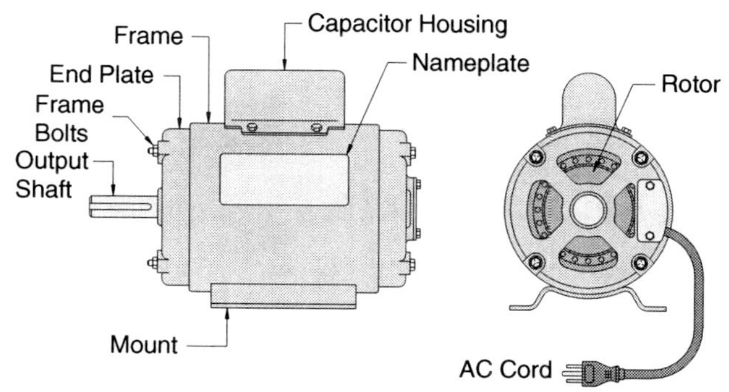

Figure 6-16 Commercial Capacitor Start Motor

A basic induction motor will not start on its own. The magnetic circuit will lock and the rotor will not rotate. Therefore some type of start mechanism must be introduced into an induction motor. Simply spinning the motor by hand and connecting the power is generally enough to start a typical induction motor. However, for obvious reasons, hand starting an induction motor is impractical. Because of this, induction motors are equipped with some type of starting circuit.

Capacitor Start Motors

Capacitor start motors are commonly found in small equipment. These motors produce good starting torque and excellent efficiency. They are the motors of choice for most small equipment that requires $^1/_2$ through $1^1/_2$ horsepower. Figure 6-15 shows a schematic representation of a capacitor start motor. In addition to the run winding, these motors have a start winding. The start winding is connected to the power source through a capacitor and centrifugal switch. When the rotor is at rest and power is supplied, the capacitor generates a phase shift and the start winding creates asymmetry in the field. This, in turn, starts the rotation of the rotor. As the speed of the rotor increases, the centrifugal switch opens and disconnects the start winding. At this point, the motor operates on the run winding only. Figure 6-16 shows a typical capacitor start induction motor.

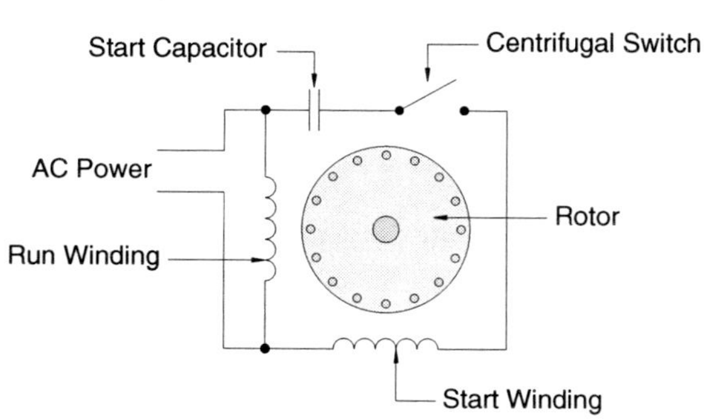

Figure 6-15 Capacitor Start Motor Schematic

Split Phase Motors

A split phase motor is similar to a capacitor start motor, except that there is no capacitor in the circuit. The asymmetry in the field is accomplished by adjusting the position of the start winding in reference to the run winding. Figure 6-17 shows a schematic representation of a split phase motor. Split phase motors are generally supplied in the $^1/_4$ through $^3/_4$ horsepower range. They do not produce as high of a starting torque as a capacitor start motor and are generally used for applications that have minimal starting requirements. These motors are commonly found in the air handling equipment of homes and small businesses. Figure 6-18 shows a commercial split phase motor. It should be noted that most of these

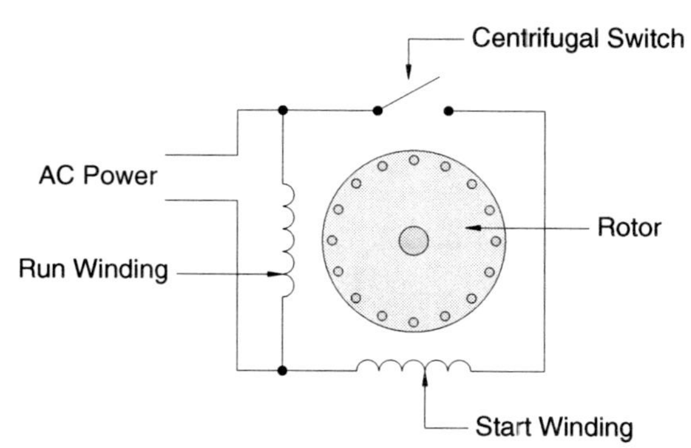

Figure 6-17 Split Phase Motor Schematic

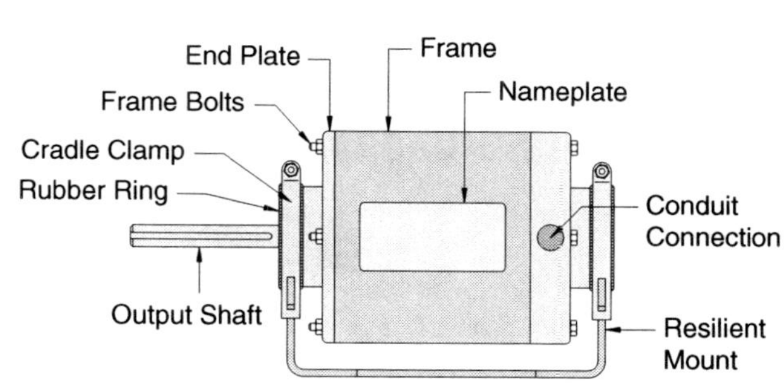

Figure 6-18 Commercial Split Phase Motor

motors are supplied on what is termed a resilient mount. These mounts are specifically designed to minimize noise and vibration.

Split Capacitor Motors

Split capacitor motors are similar in design to a capacitor start motor, except that the centrifugal switch is eliminated from the assembly. The start winding is connected to the power source through a capacitor and is designed to be continuously energized. This sets up a continuous asymmetry in the field and, therefore, the power of the start winding must be minimal. Figure 6-19 shows a schematic of a split phase motor. These motors have poor starting torque and generally provide poor efficiency. They are usually used in low fractional horsepower applications. One real advantage of these motors is that they have very few moving parts and are extremely reliable. They are excellent choices for small equipment for which regular maintenance is difficult or impossible. Figure 6-20 shows a typical split capacitor motor.

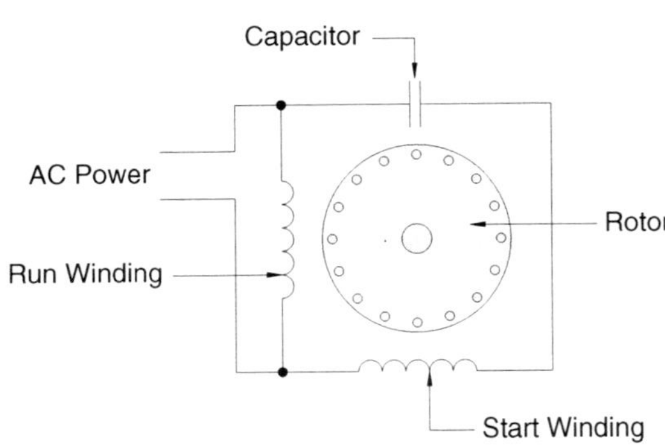

Figure 6-19 Split Capacitor Motor Schematic

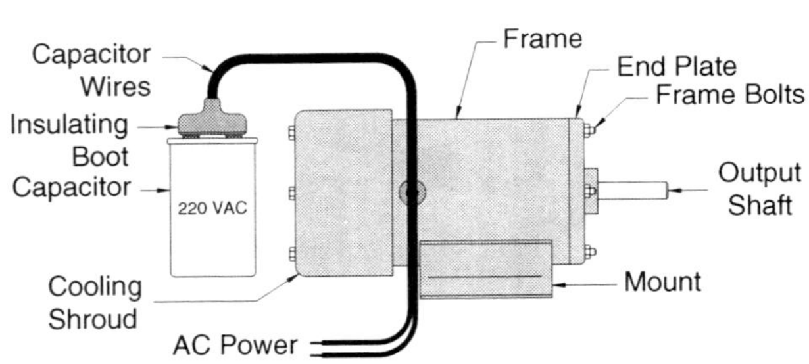

Figure 6-20 Split Capacitor Motor

Capacitor Start/Capacitor Run Motors

For applications in the $^3/_4$ through 10 horsepower range, the capacitor start/capacitor run motor is the preferred choice. In this design the start winding is connected to the power source through two capacitors. The start capacitor is connected

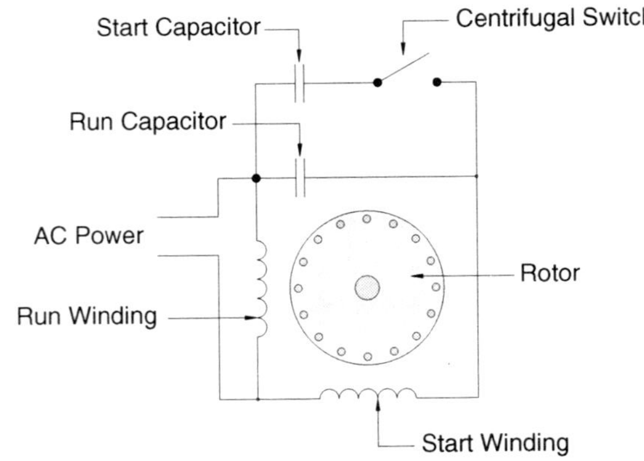

Figure 6-21 Capacitor Start/Capacitor Run Motor Schematic

through a centrifugal switch. During start-up, the start capacitor creates a gross asymmetry in the fields and very high start torques are generated. As the motor speed increases, the centrifugal switch opens and the run capacitor creates a moderate asymmetry in the field. Figure 6-21 shows a capacitor start/capacitor run motor schematic. This arrangement is typically configured to produce high horsepower from relatively small packages. These motors also exhibit excellent efficiencies. They represent a very good choice for small equipment that requires high starting torques, such as compressors, machine tools, and conveyor systems. Figure 6-22 shows a commercial capacitor start/capacitor run motor. The telltale of these motors are the two cylindrical capacitor housing that are normally mounted on the top of the frame.

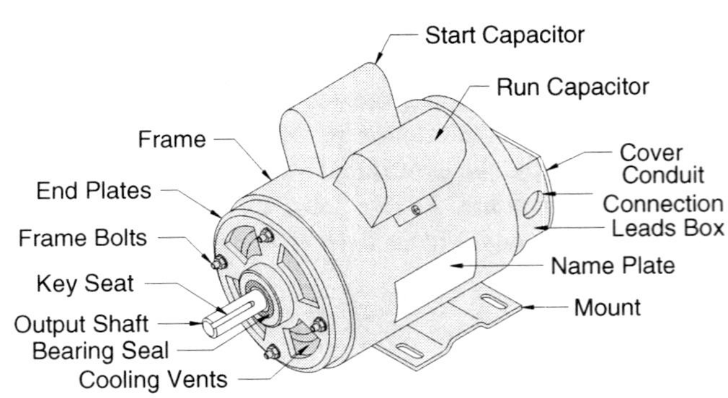

Figure 6-22 Capacitor Start/Capacitor Run Motor

Shaded Pole Motors

Shaded pole motors are very common in small appliances and the like. These motors are generally supplied in very low fractional horsepower, typically in the $^1/_{16}$ through $^1/_{120}$ horsepower range. The start mechanism consists of a copper ring added to a limited section of the pole face. This conductor loop sets up an asymmetry in the field and produces the necessary starting

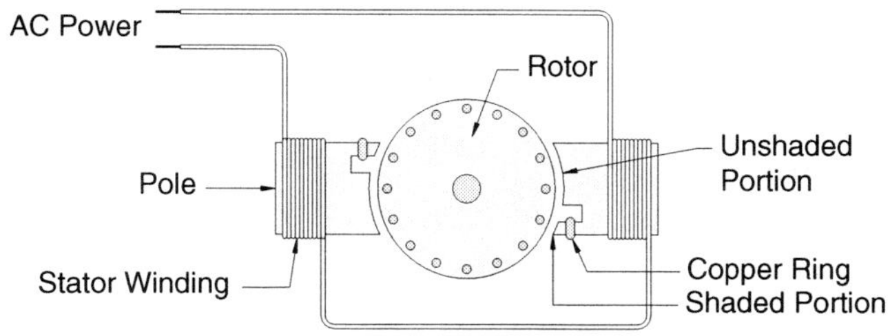

Figure 6-23 Shaded-Pole Induction Motor Schematic

force. Figure 6-23 shows a stylized schematic representation of a shaded pole motor. These motors are typically very inexpensive and, when they fail, are usually treated as a disposable item. Their efficiency is very poor, usually less than 20%, and provide extremely low starting torques. Figure 6-24 shows a typical shaded pole motor. Note the copper rings on the upper left corner of the core.

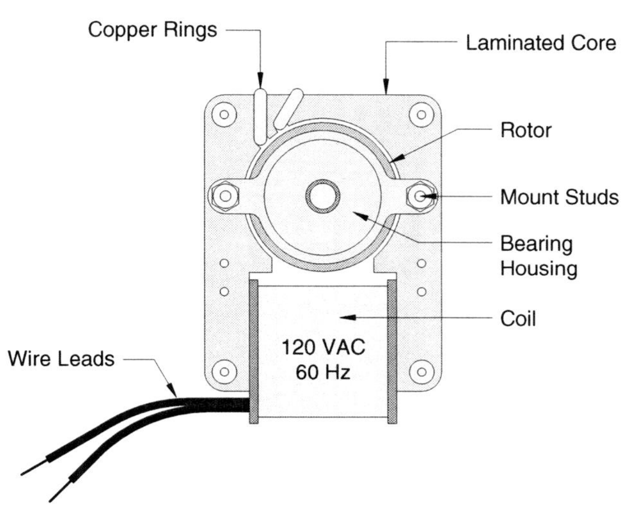

Figure 6-24 Shaded-Pole Induction Motor

Induction Starting Torques

Figure 6-25 shows starting torque curves for common induction motor types. All induction motors will produce starting torques above their full load torque. Capacitor start/run and capacitor start motors should be selected for applications that have high start loads. Split phase motors provide good general purpose starting torque. Split capacitor and shaded pole motors should only be used in applications with low starting loads.

Three-Phase Induction Motors

Three-phase motors are used almost exclusively in industrial and commercial applications. These motors provide excellent starting torque, high efficiency, compact packaging, and low maintenance. In addition, the motor rotation can be reversed by simply switching two of the power legs and the only maintenance item is the shaft bearings.

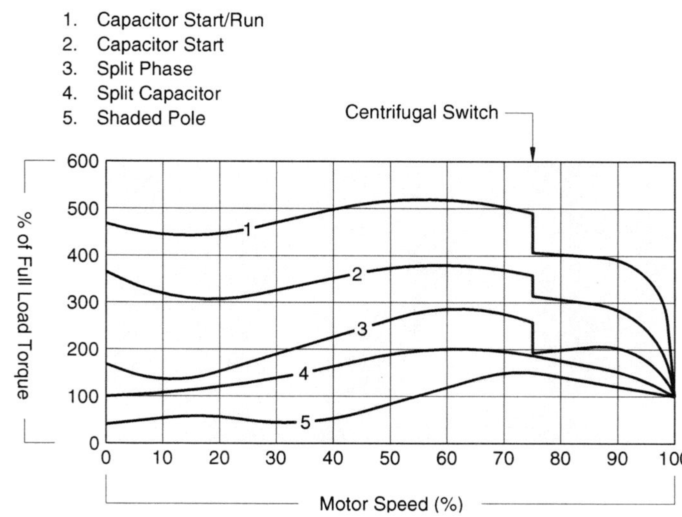

Figure 6-25 Induction Motor Torque-Speed Curves

Because of the winding geometry, three-phase motors are self starting and do not require any special starting facilities. Figure 6-26 shows a schematic of a three-phase induction motor. Figure 6-27 shows a typical totally enclosed, fan cooled (TEFC) three-phase induction motor. TEFC motors generally have cooling fins protruding from the frame to provide higher cooling efficiency. The back of the motor has a cooling shroud that encloses a fan which forces air across the cooling fins.

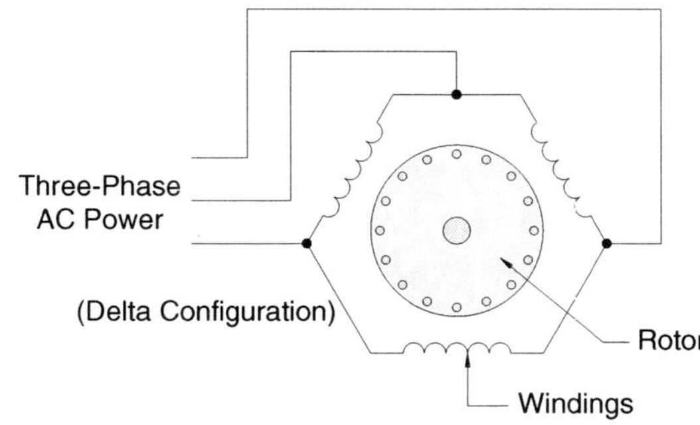

Figure 6-26 Three-Phase Induction Motor Schematic

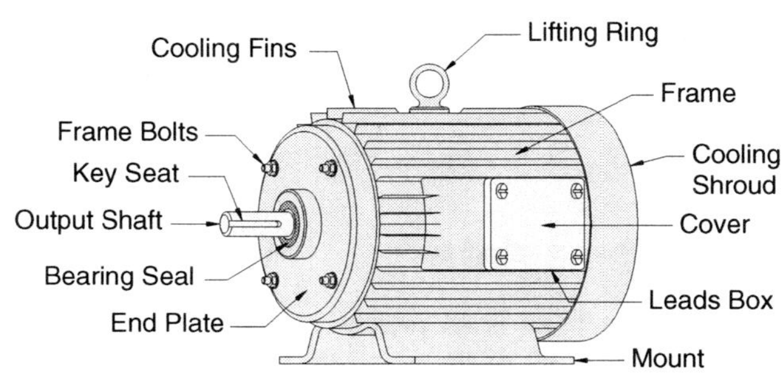

Figure 6-27 Totally Enclosed, Fan Cooled, Three-Phase Induction Motor

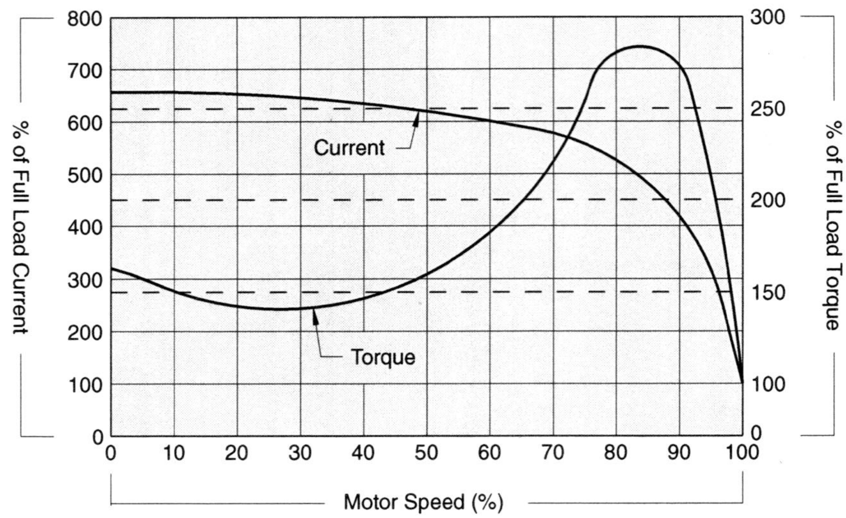

Figure 6-28 Three-Phase Induction Motor Torque/Current Curves

The starting torque of a typical three-phase motor is generally between 250% and 750% of the full load torque. The initial torque profile of these motors makes them ideal for applications that have high starting loads. Figure 6-28 shows the starting torque and current curve for a typical three-phase induction motor. The starting current may be as high as 700% of the full load current. This high starting current must be considered when installing these motors. Generally, commercial motor starters that are rated at the full load current are designed to deal with the high start current profile of these motors.

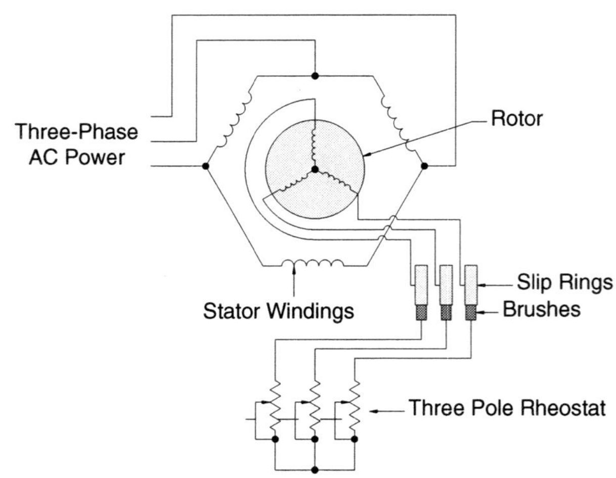

Figure 6-29 Three-Phase Wound Rotor Schematic

Wound Rotor Three-Phase Induction Motors

A wound rotor motor is a more traditional method to provide speed control. Instead of a squirrel cage rotor, these motors have a wound rotor and brush set more like a DC motor. The rotor coils are connected to a three-pole rheostat, which is used to adjust the resistance. As the resistance of the rotor is increased, the speed of the motor decreases, as the resistance is lowered, the speed increases. Figure 6-29 shows a schematic of a wound rotor three-phase motor. These types of motors are particularly expensive and their slip rings require considerable maintenance. Coupled with the introduction of low cost, electronic speed controls, the wound rotor motors have all but vanished from the industrial landscape.

Motor Nameplate

The motor nameplate is an extremely important part of the motor. It provides all of the pertinent information about the motor and without it the motor is practically useless. Nameplates used on commercial induction motors will provide standard information about the motor including: voltage(s), full load current, RPM, horsepower, frequency, power type (single- or three-phase), service factor, frame type, duty cycle, insulation class, and efficiency rating. The plates will also carry the manufacturers name, contact information, and manufacturing information (ORD. No.). Figure 6-30 shows a typical induction motor nameplate.

Name of Manufacturer

ORD. No.	216B347-H					
TYPE	High Efficiency			Frame	286T	
H.P.	5			Sev. Fac.	1.10	3 PH
AMPS	13/6.5			Volts	240/480	Y
R.P.M.	1430/1724			Hertz	50/60	4 Pole
DUTY	Cont.			Date	03-11-57	
Class Insul	F	NEMA Design	B	NEMA Eff.	95	

Manufacturers Contact Information

Figure 6-30 Typical Induction Motor Nameplate

- **Type-**Motor type, that is capacitor start, reversing, high temperature, TEFC, compressor rated, wound rotor, and so on.
- **H.P.-**Full load horsepower.
- **Amps**-Full load current (not starting current). If the motor is a dual-voltage unit the first number is for low voltage and the second is for high voltage.
- **RPM-**Full load revolutions per minute. If the motor is a dual-frequency unit the first number is for the low frequency and the second is for high frequency.
- **DUTY-**Duty cycle (Cont. means continuous or 100%)
- **Class Insul-**Insulation classification.
- **Frame-**NEMA Frame type.
- **Service factor-**The overload capacity. As an example, a motor that has a service factor of 1.25 is able to produce 25% more continuous horsepower then the nameplate states without damage.
- **Volts-**Line voltage.
- **Hertz-**Line frequency to produce rated RPM. If the motor is a dual-voltage unit the frequency is displayed with the lower frequency first (50/60)
- **Phase-**Labeled single or three phase. With three phase there is generally a D or Y to designate Delta or Wye configuration.
- **NEMA Eff.-**Efficiency rating.

Synchronous Motors

A synchronous motor is a unit whose output RPM is dependent on line frequency. Although any AC induction motor may be considered a synchronous motor, the accuracy of the output RPM varies because of a certain amount of "slip." A typical synchronous motor is designed to have an extremely accurate RPM output. These motors are typically used for applications where timing is critical, such as a wall clock or strip chart recorder. Figure 6-31 shows a typical synchronous motor.

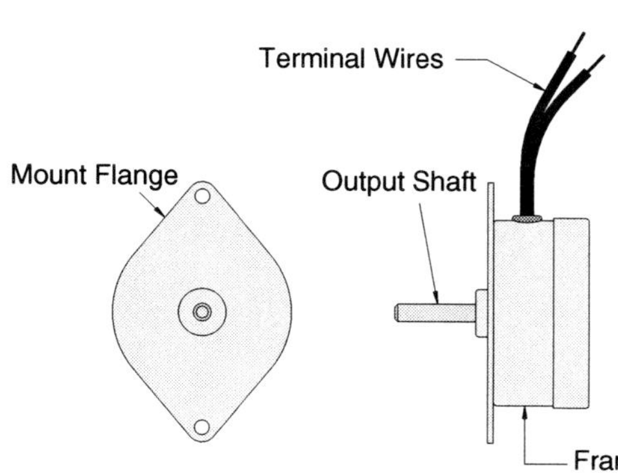

Figure 6-31 Synchronous Motor

Stepper Motors

Stepper motors are generally used for motion control applications. These motors are excellent for equipment that has low or fairly consistent loads. The stepper motor is a multipole design that allows extremely precise positioning of the rotor, even at 0 RPM. This attribute make them very friendly to the motion control engineer and these motors are found in most computer equipment such as disk drives and printers.

Figure 6-32 shows a schematic representation of a stepper motor. The motor has six sets of opposing stator windings that can be controlled independently from one another. When two opposing poles are energized (poles five) a magnetic field is generated and the rotor is forced into a position that corresponds with the field. If poles five are turned off and poles four are turned on then the rotor will jump into the new position. By carefully manipulating these pole sets, the position and RPM of the rotor can be closely controlled.

To provide further resolution, stepper motors can be operated in a half-step mode. In this operation four poles are energized and the rotor takes up a position between the two pole sets. This effectively doubles the rotational resolution of the motor. Figure 6-33 shows the motor in a half-step position.

Carrying this concept further is microstep control. By controlling the field strength of the two pole sets the position of the rotor between the poles can be placed anywhere within the arc. Figure 6-34 shows the motor in microstep mode.

Figure 6-35 shows a typical commercial stepper motor. These motors are usually supplied with a mounting flange which includes an alignment boss. The position of the output shaft in reference to the alignment boss is very precise and is appropriate for mounting the motor directly into a gear box.

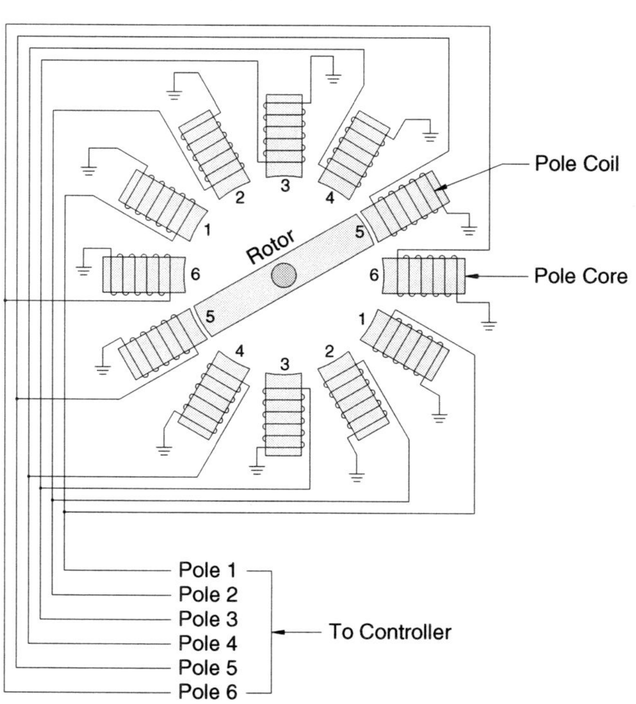

Figure 6-32 Six-Pole Stepper Motor Schematic

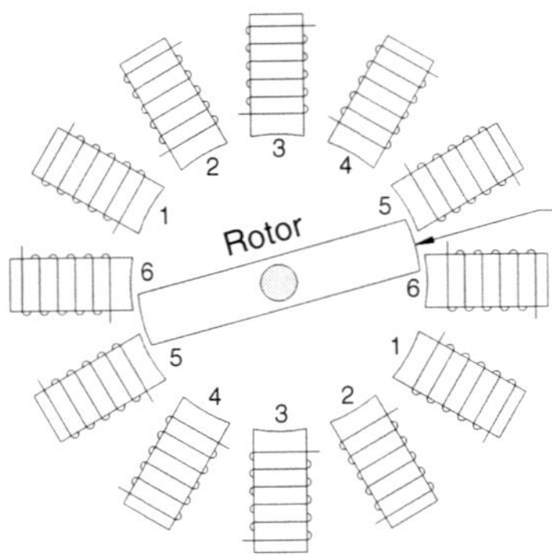

Figure 6-33 Half-Step Position

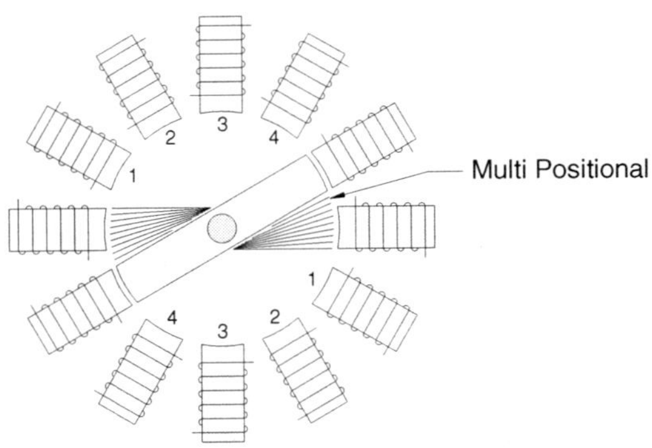

Figure 6-34 Microstep Positioning

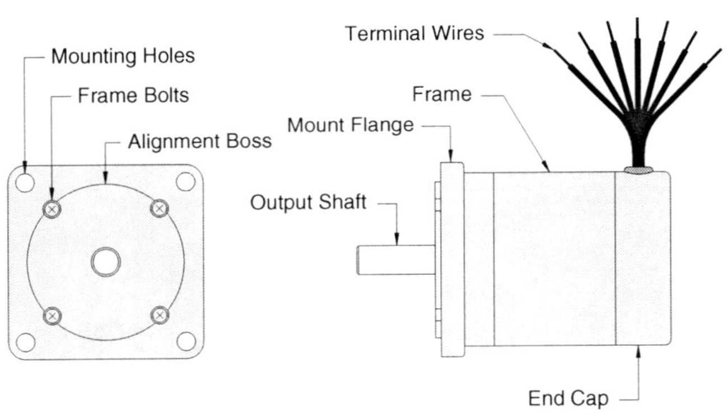

Figure 6-35 Commercial Stepper Motor

Stepper motors require a special controller to operate. The controller generally consists of a series of switches that turn the poles on and off. These switches are usually in the form of power transistors, which direct the output of a DC power supply to the appropriate pole set. The switches are connected the stepper controller, which provides speed, step and rotation control. Figure 6-36 shows a block diagram of a typical stepper motor control system. It should also be noted that this arrangement is referred to as an open-loop system. In an open loop system the motor does not provide any feedback to the controller.

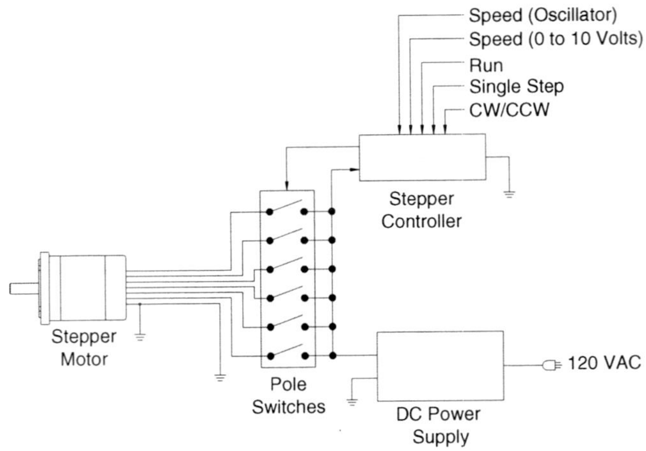

Figure 6-36 Stepper Motor Control

Servo Motors

A servo motor is simply a DC motor with a positional feedback resolver attached to its shaft. These motors are the preferred choice for motion control applications that have high or varying loads, such as machine tools and material-handling systems. Figure 6-37 shows a commercial DC servo motor. Note the resolver housing on the back side of the motor. The housing contains a sensor that can feed speed and position information back to the controller.

Figure 6-38 shows a block diagram of a typical servo motor control system. The motor resolver feeds position and

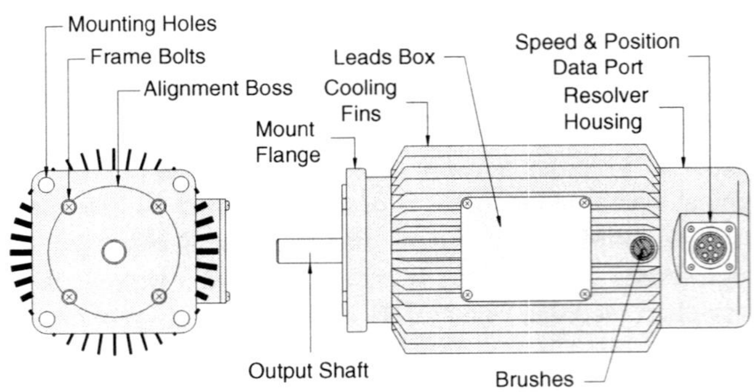

Figure 6-37 Commercial DC Servo Motor

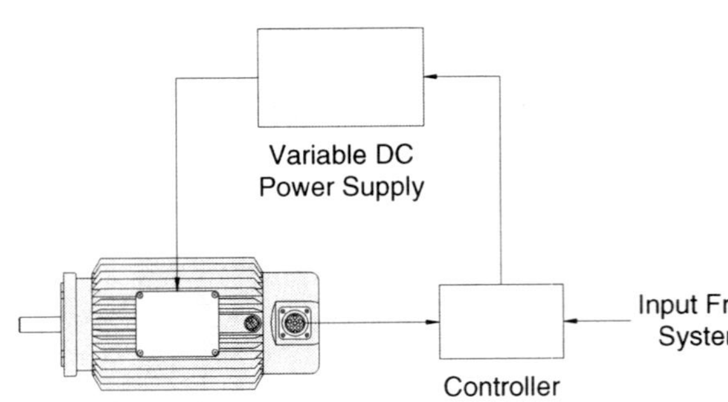

Figure 6-38 Closed-Loop Servo Motor Control System

speed information back to the controller. The controller uses this information to adjust the output of the power supply, assuring the motor operates within the parameters that the system requires. This type of system is referred to as a closed-loop system.

Solenoid/Piston Motors

A type of motor that is generally found in demonstration roles is the solenoid/piston unit. These motors operate in roughly the same manner as a piston engine, except that instead of combustion, they utilize a magnetic field. Figure 6-39 shows a schematic representation of a solenoid/piston motor. The crank shaft carries a cam that operates a double throw, double-pole reversing switch. The piston is a magnet and when the polarity attracts the magnet, it is pulled into the cylinder. When the polarity is reversed, the magnet is pushed out of the cylinder. The motion of the magnet is coupled to a crank shaft and rotation is generated. Figure 6-40 shows a single-cylinder solenoid/piston motor.

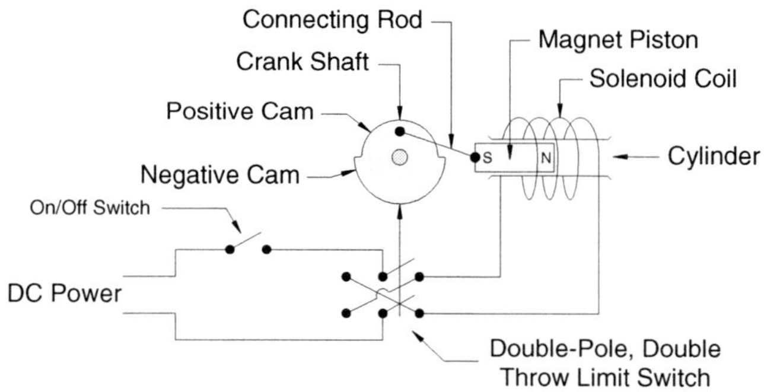

Figure 6-39 Solenoid/Piston Motor Schematic

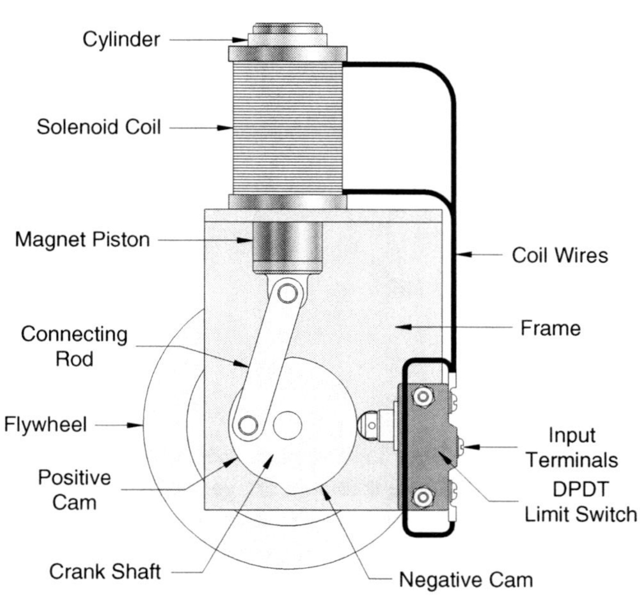

Figure 6-40 Solenoid/Piston Motor

Speed Reduction

Motor speed reduction is necessary for a variety of applications. There are three basic power transmission systems in common use today—gear, belt, and chain. However, there are a myriad of variations within each of these categories, which makes power transmission a broad topic that is too extensive for the scope of this book.

Gear drives can be acquired as standalone units or as an integral part of a motor. The latter are referred to as gear head motors. Gear head motors offer a compact package that can simplify a design at a very attractive price. Figure 6-41 shows a few typical gear head motors. Gear heads are available in single- or multistage designs that can provide virtually any output RPM desired.

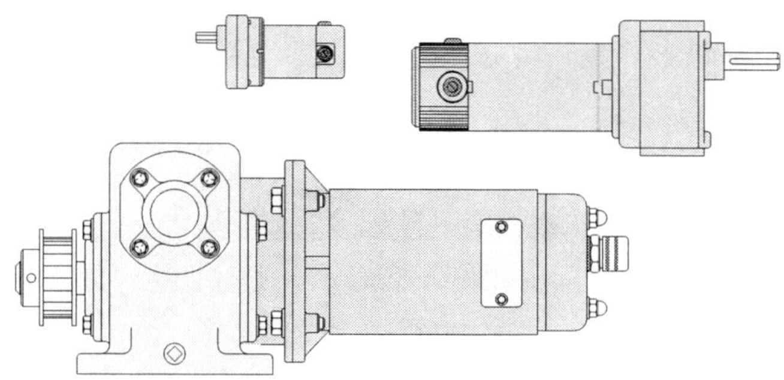

Figure 6-41 Various Gear Head Motors

Belt drives are very common as the primary drive element for electric motors. They can be found in all manner of equipment from home washing machines to massive industrial equipment. They can be configured as single- or multistage systems; however, they are most commonly found in single-stage applications. Figure 6-42 illustrates the elements of a typical two-stage V-belt drive system.

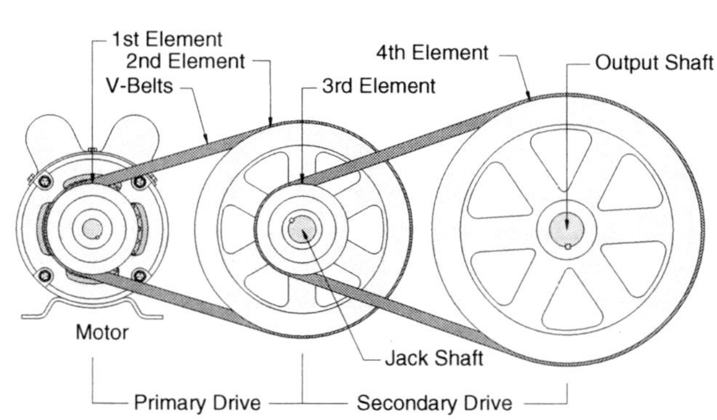

Figure 6-42 Two-Stage V-Belt Reduction Drive

Translating Rotary Motion to Linear Motion

Motion control most often means controlling linear motion. Since motors produce rotary motion, some sort of linear conversion must take place. The most common method is to couple the output of the motor to a threaded shaft, or lead screw.

For high load applications, recirculating ball screws are usually used. The nuts used in these types of screws use ball bearings to engage the threads. The balls are free to rotate and are fed back into the front of the nut via an external conduit as they roll out of the back. In this way a threaded shaft and nut can be produced with minimal friction and, therefore, minimal wear. These types of screws are commonly found in computer controlled machine tools. Figure 6-43 shows a powered recirculating ball screw with a fixed nut.

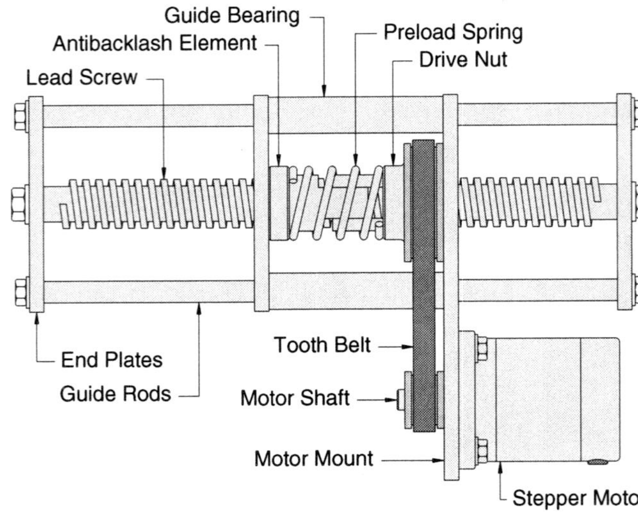

Figure 6-45 Rotating Antibacklash Nut

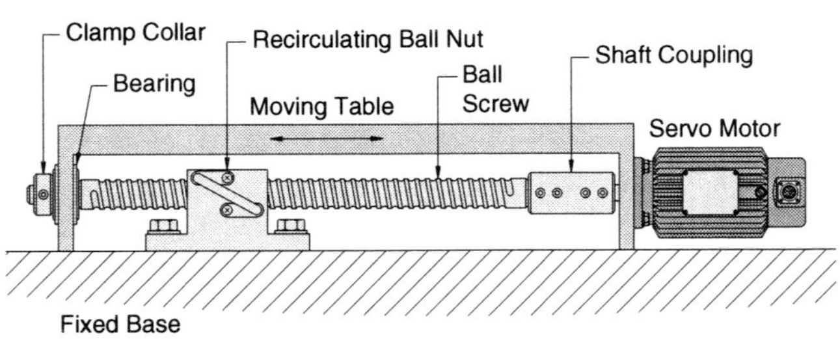

Figure 6-43 Powered Lead Screw

Motorized linear actuators, as shown in Figure 6-44, are common devices for adding motion control to all sorts of equipment. These units can be found on retractable awnings, convertible automobile tops, satellite antennas, and hospital beds. They are typically a cylinder fitted with a hollow piston. The bottom of the piston has a bronze nut with a threaded steel shaft. The threaded shaft is powered by a DC motor. Switching the motor to forward or reverse either extends or retracts the actuator.

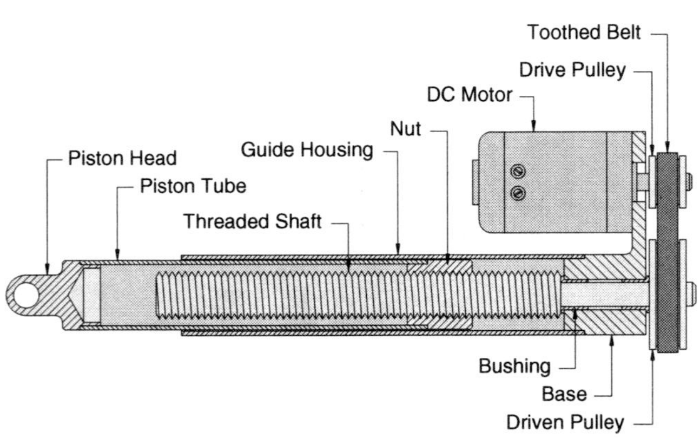

Figure 6-44 Motorized Screw Thread Actuator

For applications that have low loads, an acme lead screw with an antibacklash nut can be specified. Figure 6-45 shows a linear actuator that uses an acme lead screw and antibacklash nut. In this configuration the lead screw is fixed and the nut is rotated. This arrangement is very popular for low load applications that incorporate stepper motors.

Another method to achieve liner motion is to use a toothed belt, as shown in Figure 6-46. This is a very common way to control computer equipment such as disk drives and printer heads. In a belt arrangement the positional resolution of the belt is dependent on the circumference of the drive pulley and the accuracy of the drive motor.

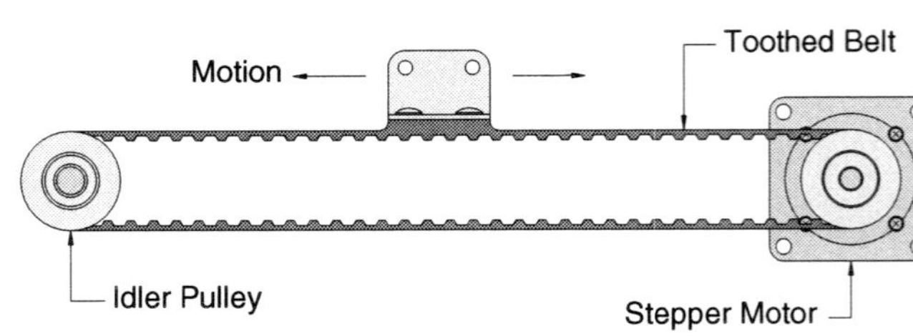

Figure 6-46 Linear Belt Drive

For extremely high load applications a rotary actuator may be specified. These actuators are generally built using a worm gear drive. Worm drives can produce very high mechanical advantages and, as such, are used in the most rigorous applications. Figure 6-47 shows a worm drive rotary actuator.

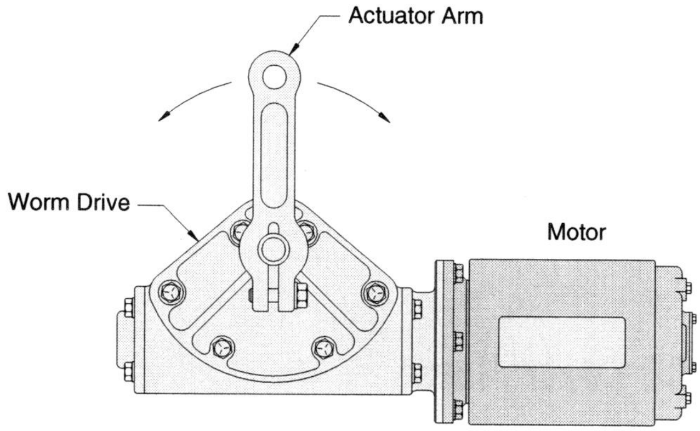

Figure 6-47 Rotary Actuator

Variable Frequency Drives

Variable frequency motor speed controllers, or variable frequency drives as they are commonly referred to, are an excellent method to control the speed of induction motors. Figure 6-48 shows a typical variable frequency drive. These units will generally allow down to 70% reduction and up to 100% increase of the output of a standard induction motor. If the selected motor is rated at 1725 RPM, then its slowest speed would be 518 RPM and its highest speed would be 3450 RPM. The reason that most variable frequency drives are limited in the low setting is that the torque of a typical induction motor will start to fall to unacceptable levels below 18 to 20 Hz. Figure 6-49 shows how a variable frequency drive is set up with a standard induction motor. Note that there is no motor contactor in the circuit. The function of the motor starter is incorporated into the variable frequency drive.

Motor Speed Control

Controlling the speed of DC motors is generally accomplished by simply adjusting the input voltage. Controlling the speed of an induction motor is not so simple. Induction motors operate from an AC input and their speed is generally dependent on the supply frequency. If the voltage is lowered, the coils pull progressively higher current and the motor will eventually overheat. All the while the motor operates at the same speed.

To change the speed of most induction motors the input frequency must be adjusted. Changing the frequency does not affect the voltage and current, and the motor will continue to operate well within its temperature limits. If a motor is rated at 1725 RPM at 60 Hz, then supplying power at 30 Hz will halve the RPM. Inversely, supplying the motor with 120 Hz power will double the RPM.

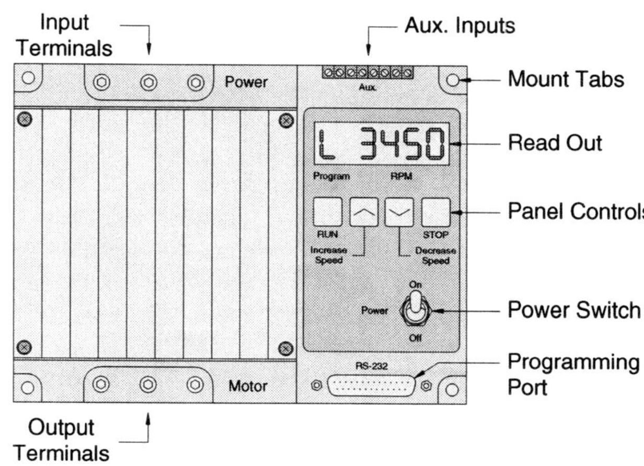

Figure 6-48 Variable Frequency Motor Speed Controller

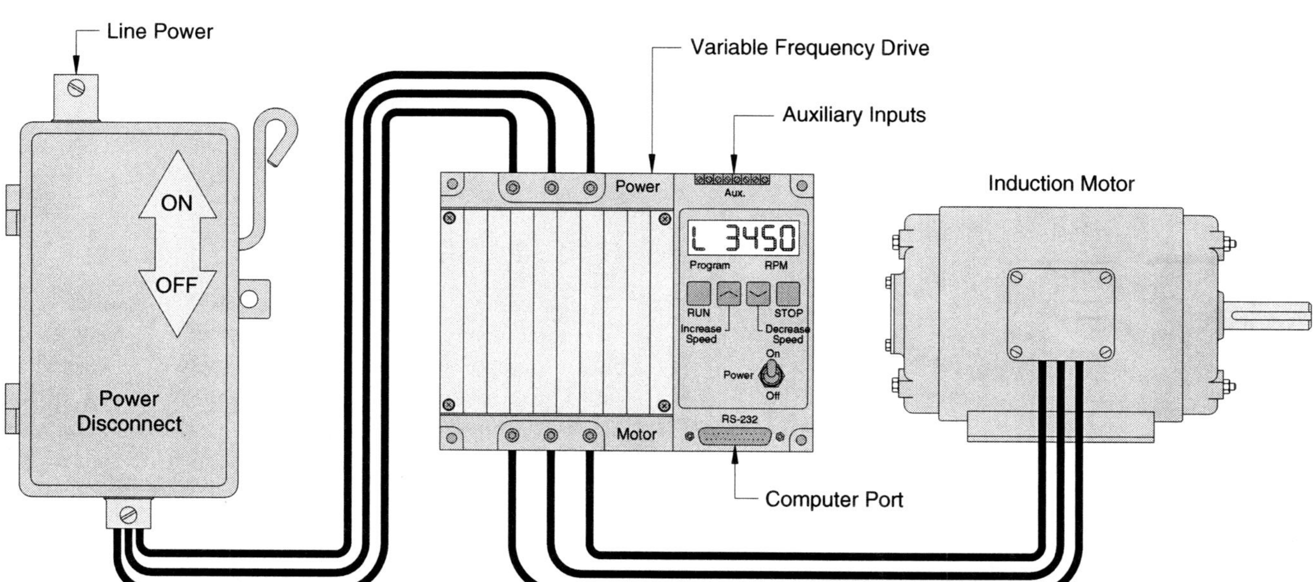

Figure 6-49 Variable Frequency Drive Circuit

Another function that variable frequency drives can provide is soft starting. Soft starting is useful for applications that have high inertial loads, such as heavy flywheels and cable spools. If a motor is exposed to an excessively long start cycle it may overheat and can be severely damaged. By starting a motor at a lower RPM, the starting load may be mitigated. Figure 6-50 shows the progression of a variable frequency drive during a soft start cycle.

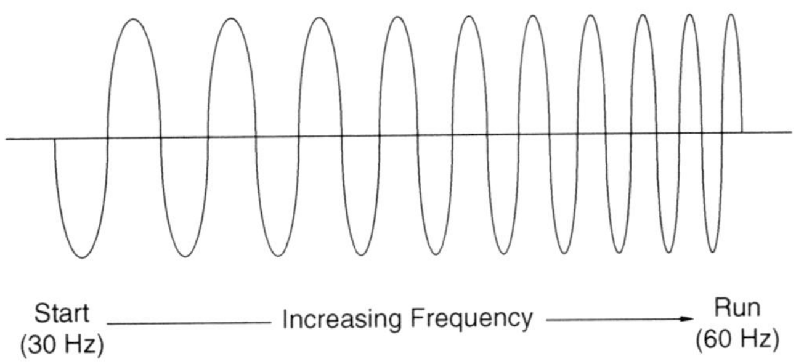

Figure 6-50 Variable Frequency Soft Start Cycle

Soft Starters

For applications that do not require speed control, but will benefit from soft starting, a dedicated soft starter may be used, as shown in Figure 6-51. These units are considerably less expensive than a variable frequency drive and provide better soft start control. Figure 6-52 shows a soft starter switching cycle. These units do not rely on frequency control, rather they switch the AC cycle on at a controlled delay time. Each cycle the delay time is shortened until the full cycle is supplied to the motor.

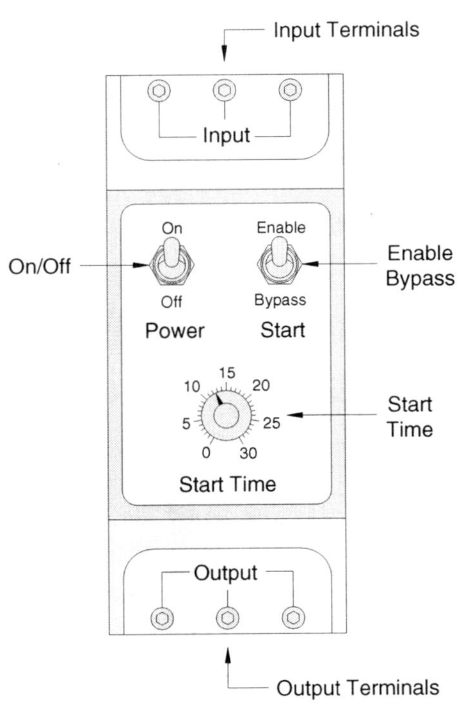

Figure 6-51 Commercial Soft Starter

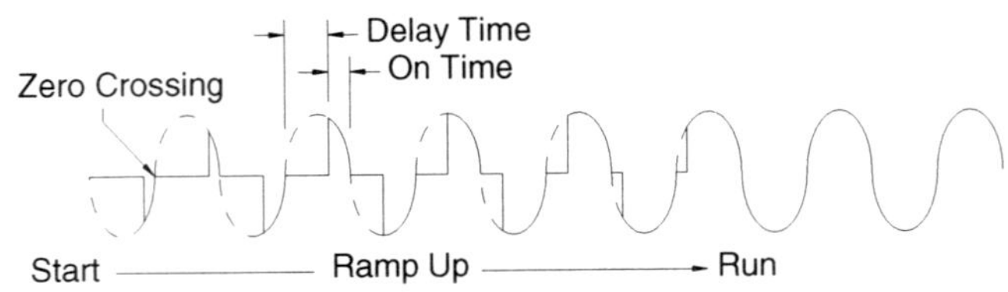

Figure 6-52 Soft Starter Switching Cycle

SCR Controllers

Universal motors can be controlled with an SCR controller. These controllers switch the AC cycle to vary the total power that the motor receives. Figure 6-53 shows an SCR controller. These units are often supplied with an AC cord and receptacle so that they may be simply plugged in.

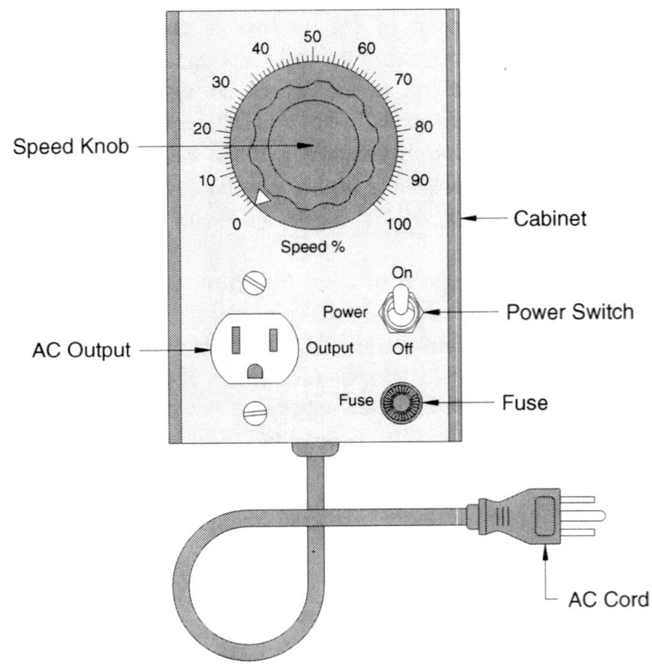

Figure 6-53 Commercial SCR Motor Speed Controller

Torque Converters

Variable pulleys, or torque converters as they are often referred to, are the most popular mechanical method of controlling the output speed of an induction motor. Figure 6-54 shows a typical torque converter. The spacing of the sheaves on the motor pulley is controlled with a screw and hand crank. The spacing of the sheaves of the output pulley is spring loaded to the center. As the input sheaves are forced together, the effective diameter of the pulley increases. This, in turn, pulls the sheaves apart on the output pulley, effectively decreasing the diameter. In this manner the ratio of the two pulleys can be varied to produce any output speed within the range of the drive.

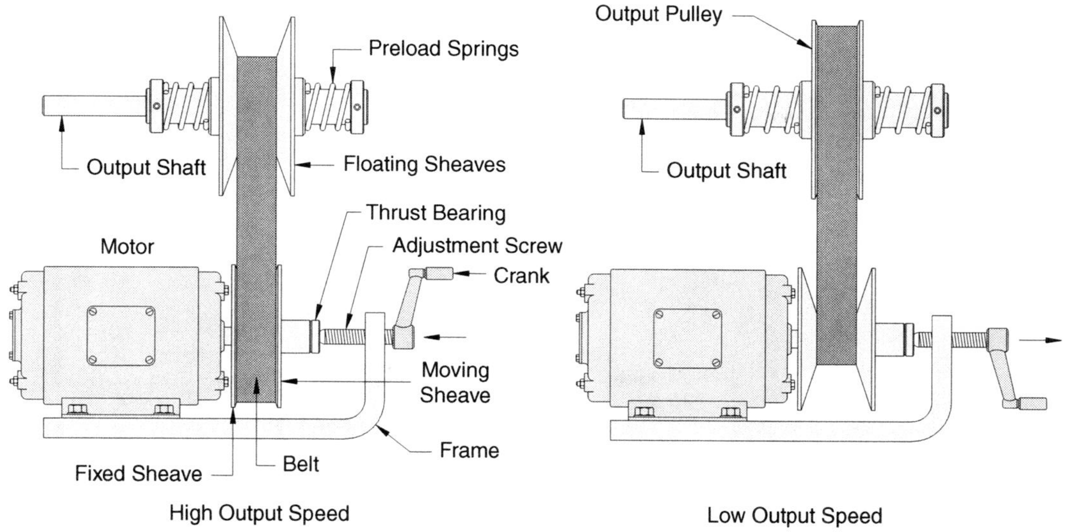

Figure 6-54 Torque Converter

Generators

As we reviewed in Chapter 1, moving a coil of wire through a magnetic field will generate a voltage, as shown in Figure 6-55. Generators use this principal to create a usable voltage. Figure 6-56 shows a simple AC generator. A magnet is rotated between two coils. Each time a pole of the magnet passes the coils, a voltage in generated. Every half-revolution the polarity of the output is reversed because the poles of the magnet are reversed by the spinning action. In this manner AC voltage is generated on a continuous basis.

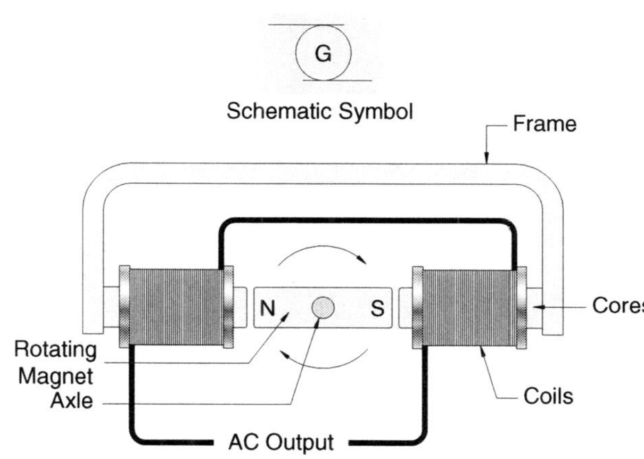

Figure 6-56 Rotating Magnet AC Generator

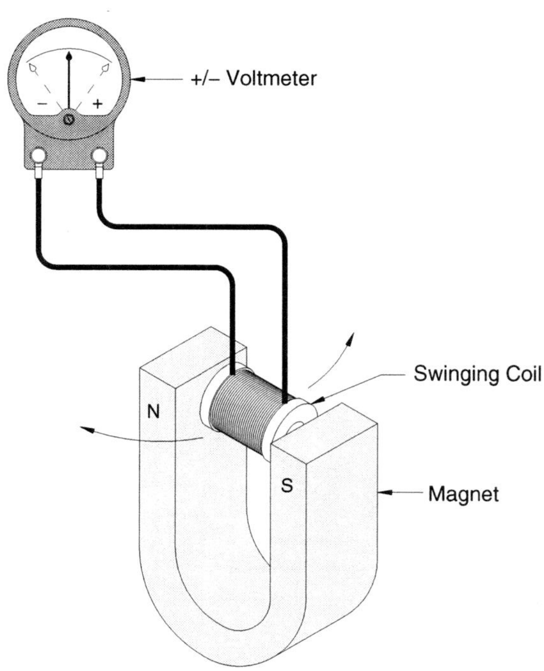

Figure 6-55 Generating Voltage Through Induction

Figure 6-57 shows a fixed magnet AC generator. In this arrangement the coils are carried on the rotor and the magnet is fixed. This allows the magnet to be considerably larger and, therefore, have a considerably higher field than its spinning magnet counterpart. The AC output is taken off from two slip rings on either end of the rotor.

Because generators are inherently AC in nature, it is necessary to use reversing brushes to produce a DC output. Figure 6-58 shows a permanent magnet DC generator. Notice that the only real difference in the design is that the rotor carries an armature instead of a pair of slip rings. Each half-revolution the coils generate a reverse polarity, which is countered by the reversing armature.

Alternators

Alternators are principally the same as a moving coil generator, except that the rotor has an electromagnet instead of a fixed magnet. This allows a greater field strength than could

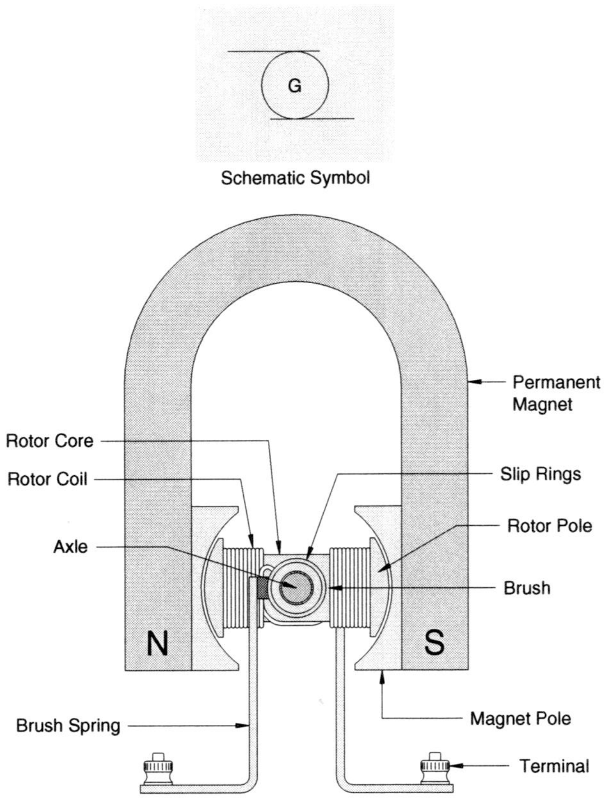

Figure 6-57 Fixed Magnet AC Generator

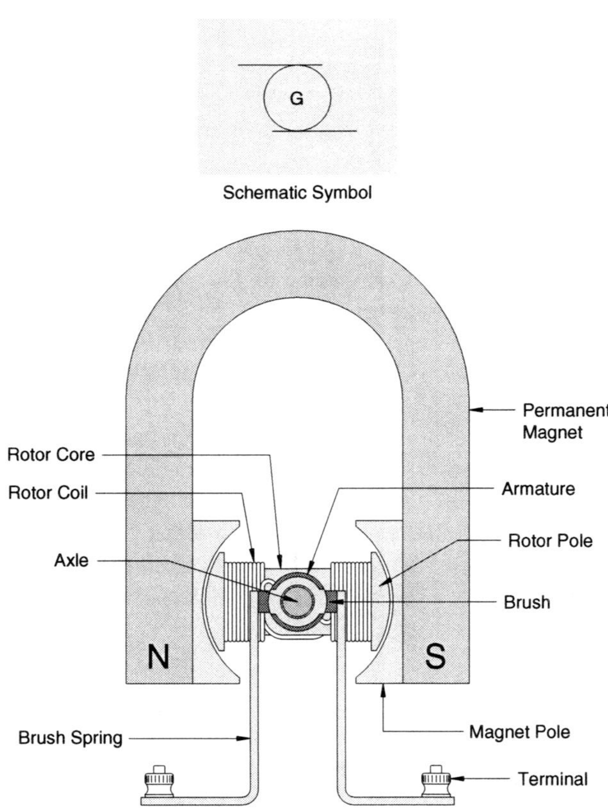

Figure 6-58 Fixed Magnet DC Generator

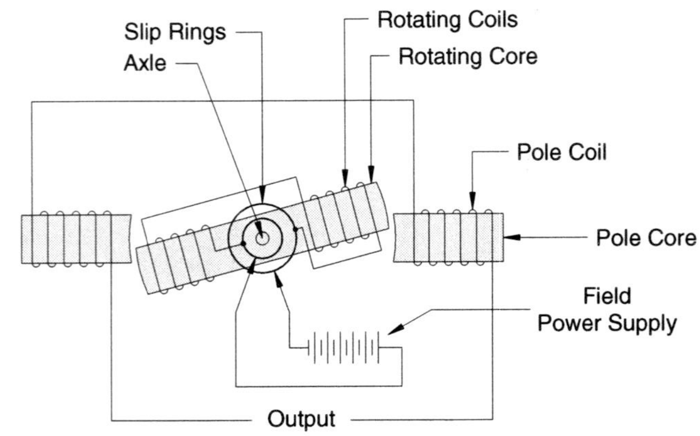

Figure 6-59 Alternator

be gained with a permanent element. Alternators are the most common type of generator in existence. Figure 6-59 shows a schematic representation of an alternator. Figure 6-60 shows a typical commercial power generator. These units are manufactured in sizes that range from small, portable generators to huge units that are installed in hydroelectric and coal-fired power plants.

Figure 6-61 shows a schematic of a three-phase alternator with a full wave DC bridge output and voltage regulation circuit. This arrangement can be found in nearly every automobile manufactured today. Figure 6-62 shows a typical automobile alternator.

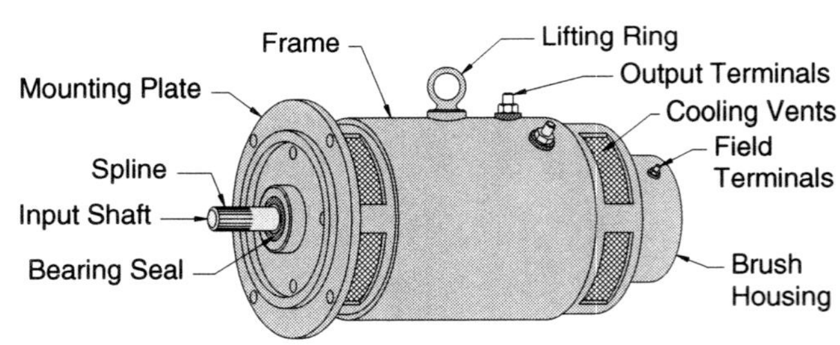

Figure 6-60 Commercial 60 Hz Three-Phase Power Generation Alternator

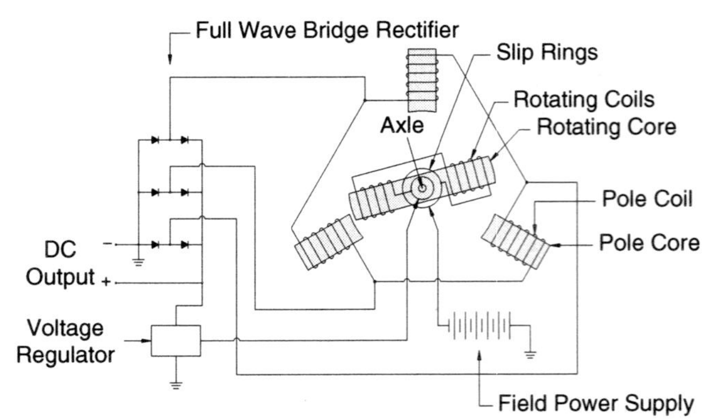

Figure 6-61 Three-Phase Alternator with Regulated DC Output

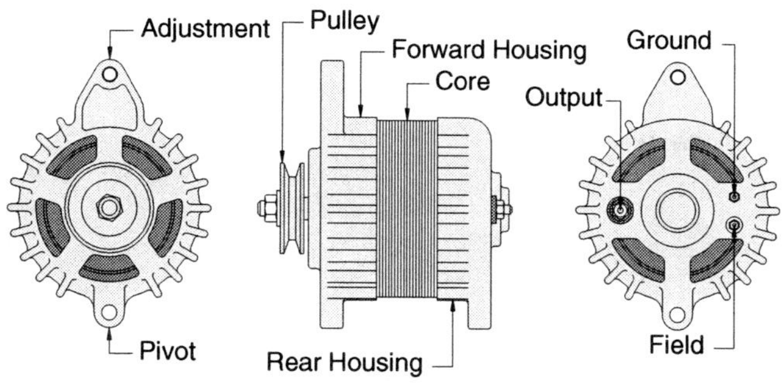

Figure 6-62 Typical Automobile Alternator

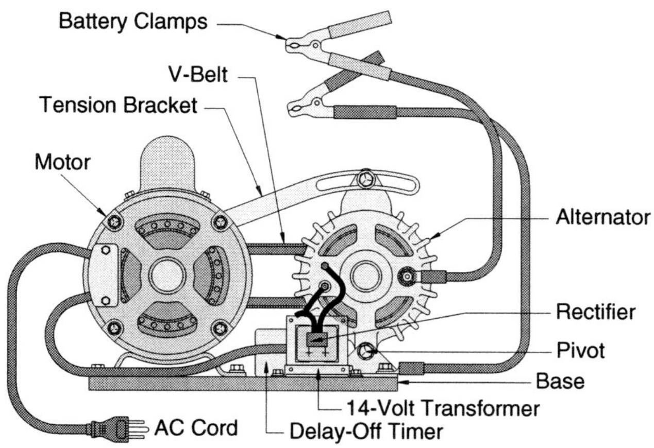

Figure 6-63 Motor/Generator Automobile Battery Charger

Figure 6-63 shows a unique use for an automobile alternator as a standalone battery charger. This arrangement makes an excellent, high power battery charger and can be constructed for little or no money. Figure 6-64 shows a schematic of the battery charger.

Engine-driven generators make excellent emergency or remote location power sources. These units may use a single cylinder engine, as shown in Figure 6-65, or have multi cylinder diesel engines that can provide backup power for large industrial facilities.

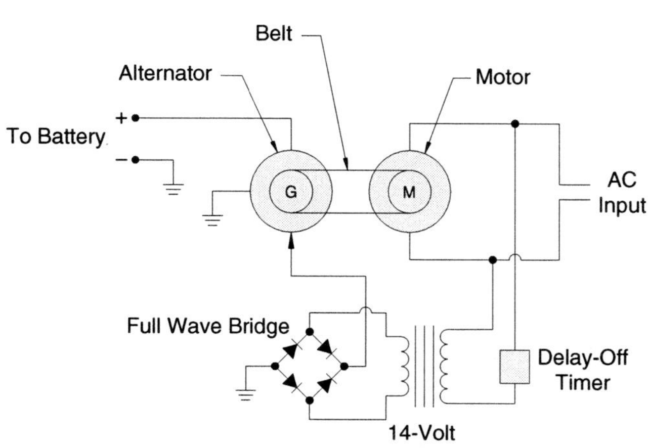

Figure 6-64 Motor/Generator Automobile Battery Charger Schematic

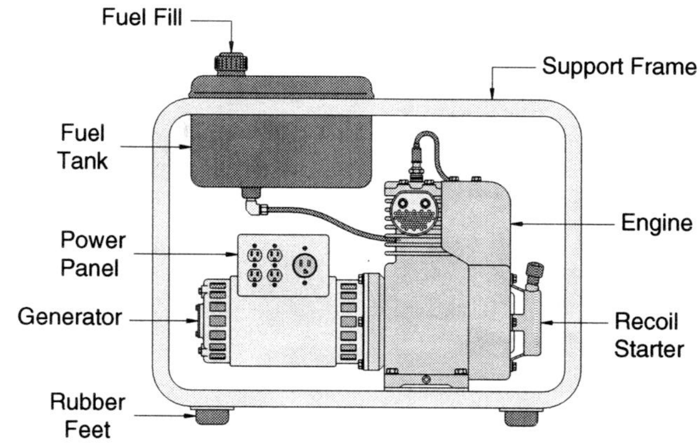

Figure 6-65 Single Cylinder Engine Driven Portable Generator Set

Magnetos

Magnetos generally have a spinning magnet and are set up to generate pulse outputs. They are most commonly found in small, single cylinder engines. Figure 6-66 shows a typical magneto for a commercial single cylinder engine. The magnet is embedded into the flywheel and the point set is used to time and initiate the high-voltage pulse for the spark plug.

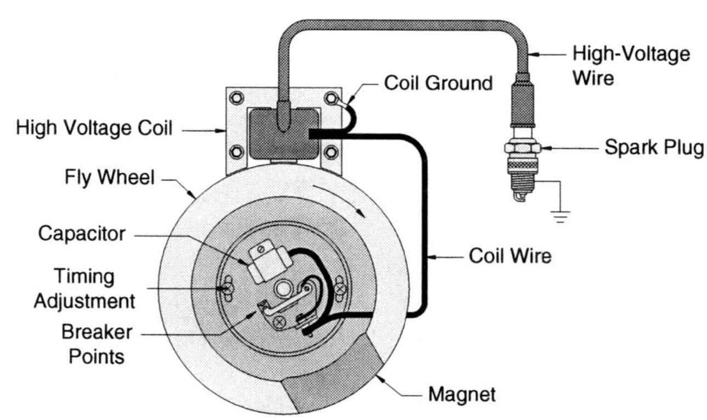

Figure 6-66 Single Cylinder Engine Magneto

Dynamometers

Dynamometers are systems that are used to measure the output of rotating equipment like motors and engines. These systems typically consist of a generator, load resistor, and ammeter, as shown in Figure 6-67. The engine being tested is coupled to the generator. After the engine is started, the field is slowly increased until the engine can no longer provide enough power to sustain a higher load. At this point the voltage and current being generated is read and the horsepower can be calculated. As an example, 1 horsepower equals 746 watts. Suppose that we had an engine that produces 15.9 amps at 237 volts.

$$15.9 \text{ amps} \times 237 \text{ volts} = 3768.3 \text{ watts} \div 746$$
$$= 5.05 \text{ horsepower}$$

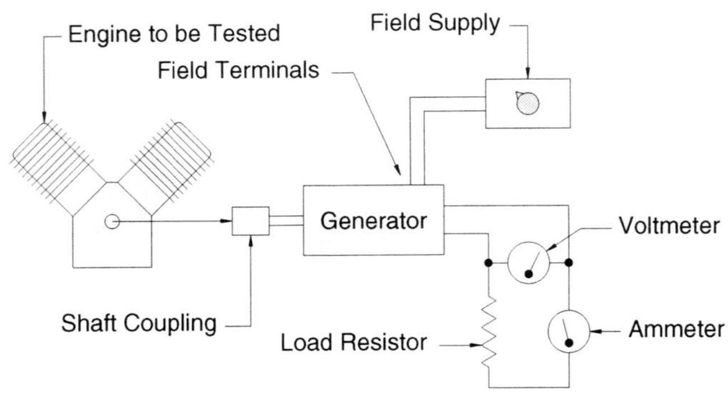

Figure 6-67 Electric Dynamometer Schematic

High-Voltage Generators

Some special applications require extremely high voltages, at times in the millions of volts range. This is particularly true in the field of high-energy physics research. Generating these types of voltages is impractical with conventional generator technologies. Over the years many high-voltage generator designs have been produced; however, one in particular has emerged as the most effective. This is the Van de Graff generator, named after its inventor. These are rather simple devices, which, when carefully designed and constructed, can produce extremely high voltages. There are Van de Graff generators in existence that will easily produce 25 million volt outputs!

Figure 6-68 shows a simple Van de Graff generator. A unit like this will only produce 50,000 to 100,000 volts. Even so, these machines can be quite dangerous to be around while operating. In suitable climate conditions, a 50,000 volt charge can easily jump as much as 5 inches! The unit has a nonconductive flat belt connecting a drive pulley and the upper pulley. The two pulleys are mounted within an insulated tube or column. The bottom of the belt is in constant contact with a grounded brush and the top is in close proximity to a set of needles. The needles are mounted to a high-voltage pick up assembly, which is connected to a large globe, or high-voltage terminal. The belt is driven by an ordinary electric motor. As the belt rotates it picks up free electrons from the grounded brush. The free electrons are carried up the belt and jump off onto the high-voltage pickup. As the unit runs, a significant accumulation of electrons build up on the terminal and a high-voltage potential is generated.

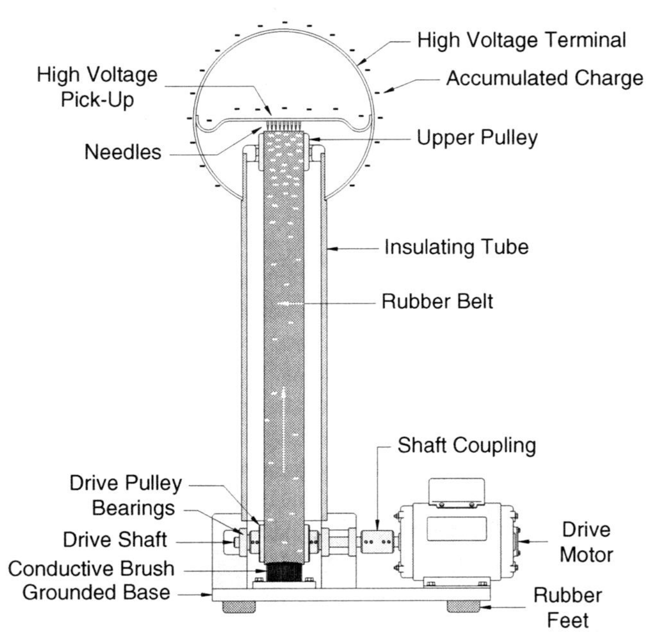

Figure 6-68 Van de Graff Generator

Rotary Converters

Rotary conversion is an established method to convert an electrical signal to a grossly different signal. This is principally accomplished by driving a generator with a motor, as shown in Figure 6-69. The motor uses the available power, such as 12 DC from an automobile battery, and drives a generator that will produce 240 VAC, three phase. Figure 6-70 shows a motor generator set that is intended to accept 48-VDC power from a marine engine and supply 120/240 VAC to operate appliances in the crew's compartment. Figure 6-71 shows a motor generator set designed to produce a 50 Hz output from a 60 Hz input. These sets are commonly used by

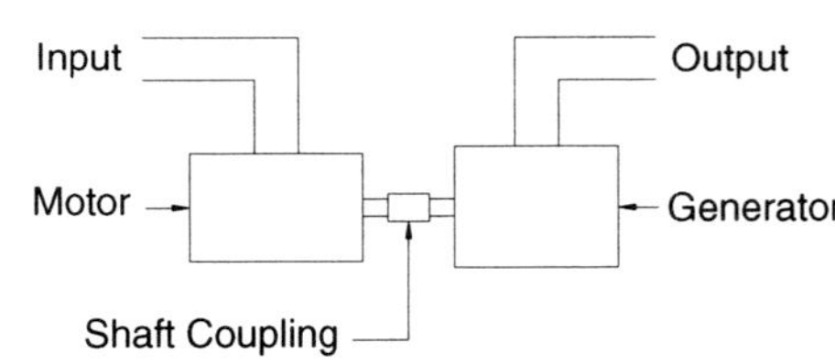

Figure 6-69 Basic Rotary Converter

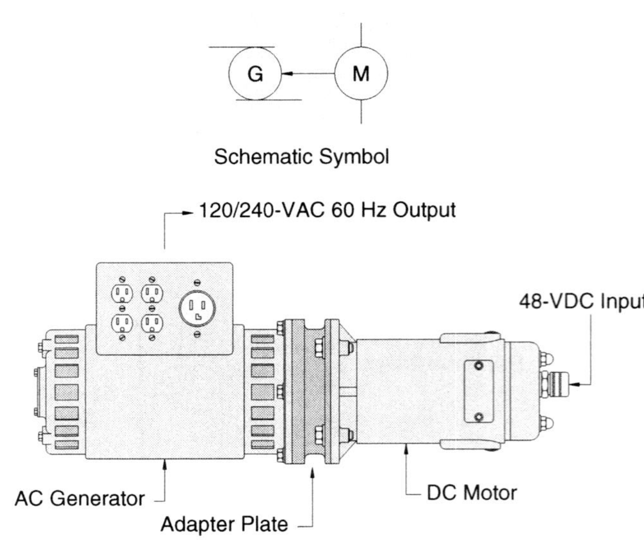

Figure 6-70 Motor/Generator Set

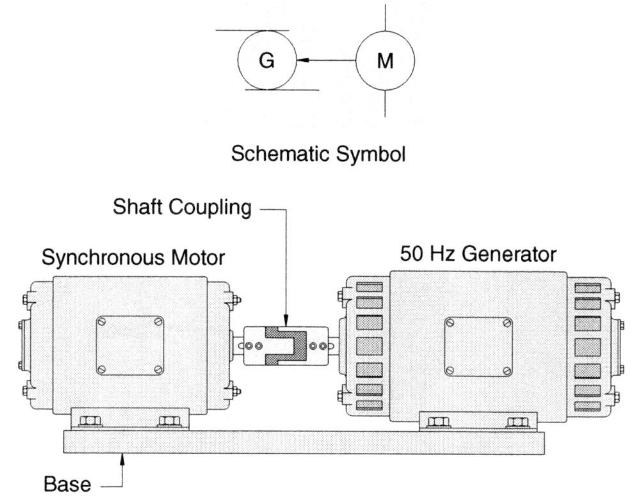

Figure 6-71 Motor/Generator Set to Produce 50 Hz Power

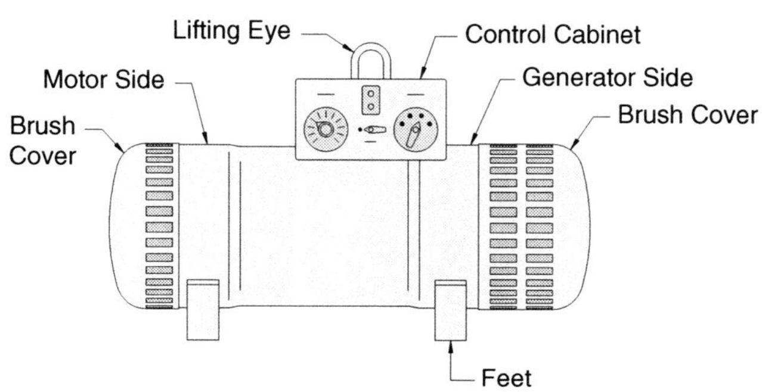

Figure 6-72 Motor/Generator DC Arc Welding Machine

American companies producing electrical equipment for markets that use 50 Hz power. Figure 6-72 shows a motor/generator arc welder. These units were very common before the availability of high-power rectifiers. The motor operates from standard AC power and the generator produces low-voltage, high-current DC power.

Dynamotors

Small, packaged Motor/generator sets, as shown in Figure 6-73, are commonly referred to as dynamotors. These units were very common in aircraft built from the 1930s through the 1970s.

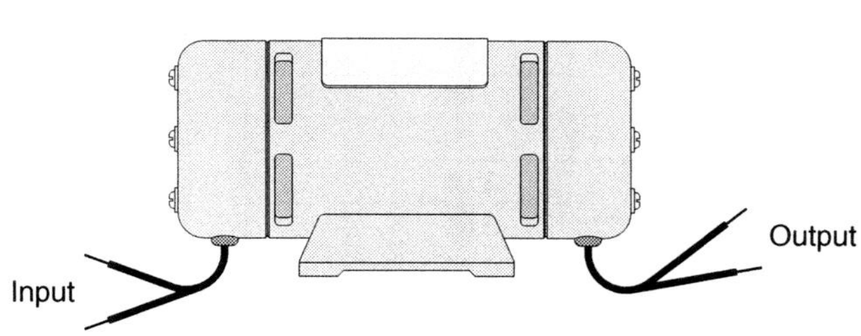

Figure 6-73 Dynamotor

The unit may use a 24-VDC input and produce a 600 VAC, 400 Hz output that is specified for a special piece of equipment. Dynamotors have been almost completely replaced by solid state power supplies.

Single- to Three-Phase Converters

In some applications it is necessary to operate a three-phase piece of equipment in a building or locations that are only served with single-phase power. In these cases a single- to three-phase converter can be used to generate the appropriate power. Figure 6-74 shows a typical single- to three-phase rotary converter. The unit is constructed from a three-phase motor that is set up to operate on single phase. The single-phase power is wired to one of the three coils. As the rotor turns, it induces power into the two unused coils. Three-phase power can be taken off of the three coils.

Figure 6-75 shows a schematic representation of a three-phase rotary converter. The unit must be set up with a starting circuit. In this case a start capacitor is connected to the second coil through a delay off relay. When the converter is energized, the start capacitor is connected for a predetermined time and then automatically disconnected. The run capacitors are primarily used to tune the phase angle of the two generated phases.

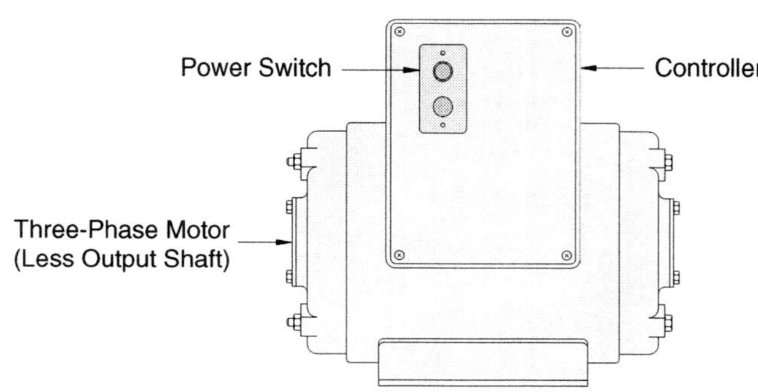

Figure 6-74 Standalone Single- to Three-Phase Converter

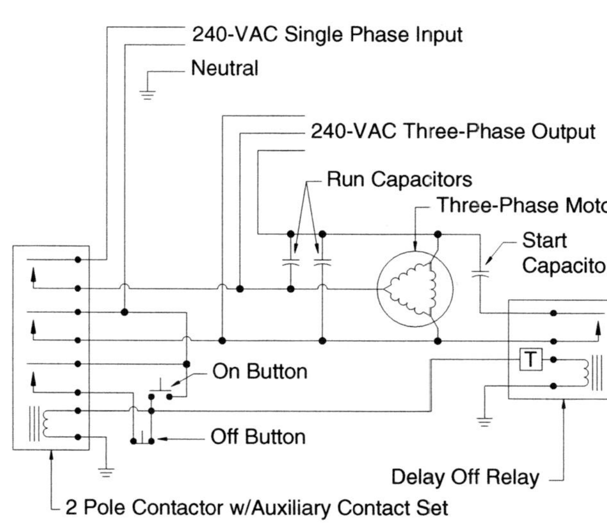

Figure 6-75 Single- to Three-Phase Converter Schematic

A three-phase motor may be converted to a single-phase motor with a loss of approximately one-third of its power. Figure 6-76 shows a commercial single- to three-phase adaptor used to convert motors. These units operate in the same manner as a rotary converter, except that the power is taken from the motor shaft.

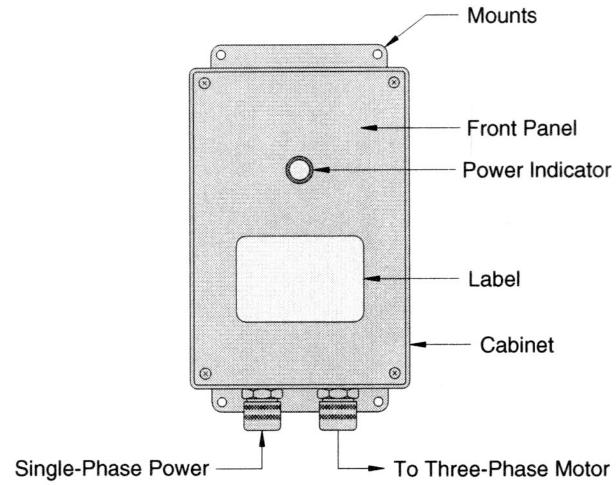

Figure 6-76 Commercial Single- to Three-Phase Converter

NEMA Motor Frame Dimensions

The National Electric Manufactures Association (NEMA) publishes standard dimensions that are used for electric motor manufacturing. Most motor manufactures adhere to these standards. If a motor is encountered that does not conform to these standards, then it was, most likely manufactured for a special application. Figure 6-77 is a table that provides standard NEMA motor frame dimensions.

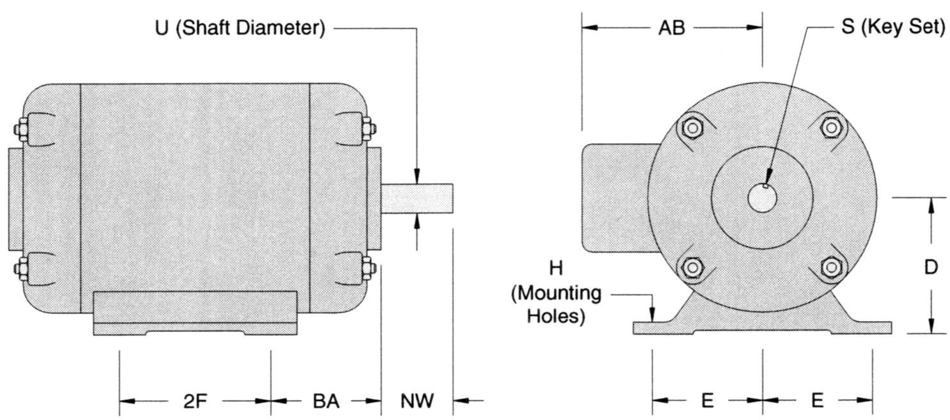

FRAME	D	E	2F	H	U	BA	NW	S
48	3.00	2.12	2.75	0.30	0.50	2.50	1.50	Flat
56	3.50	2.44	3.00	0.30	0.63	2.80	1.88	0.19
143	3.50	2.75	4.00	0.30	0.75	2.30	2.00	0.19
143T	3.50	2.75	4.00	0.30	0.88	2.30	2.25	0.19
145	3.50	2.75	5.00	0.30	0.75	2.30	2.00	0.19
145T	3.50	2.75	5.00	0.30	0.88	2.30	2.25	0.19
182	4.50	3.75	4.50	0.40	0.88	2.80	2.25	0.19
182T	4.50	3.75	4.50	0.40	1.13	2.80	2.75	0.25
184	4.50	3.75	5.50	0.40	0.88	2.80	2.25	0.19
184T	4.50	3.75	5.50	0.40	1.13	2.80	2.75	0.25
203	5.00	4.00	5.50	0.40	0.75	3.10	2.25	0.19
204	5.00	4.00	6.50	0.40	0.75	3.10	2.25	0.19
213	5.30	4.25	5.50	0.40	1.13	3.50	3.00	0.25
213T	5.30	4.25	5.50	0.40	1.38	3.50	3.38	0.31
215	5.30	4.25	7.00	0.40	1.13	3.50	3.00	0.25
215T	5.30	4.25	7.00	0.40	1.38	3.50	3.38	0.31
224	5.50	4.50	6.75	0.40	1.00	3.50	3.00	0.25
225	5.50	4.50	7.50	0.40	1.00	3.50	3.00	0.25
254	6.30	5.00	8.25	0.50	1.13	4.30	3.37	0.25

Figure 6-77 NEMA Motor Frame Dimensions

FRAME	D	E	2F	H	U	BA	NW	S
254U	6.30	5.00	8.25	0.50	1.38	4.30	3.75	0.31
254T	6.30	5.00	8.25	0.50	1.63	4.30	4.00	0.38
256U	6.30	5.00	10.00	0.50	1.38	4.30	3.75	0.31
256T	6.30	5.00	10.00	0.50	1.63	4.30	4.00	0.38
284	7.00	5.50	9.50	0.50	1.25	4.80	3.75	0.25
284U	7.00	5.50	9.50	0.50	1.63	4.80	4.88	0.38
284T	7.00	5.50	9.50	0.50	1.88	4.80	4.62	0.50
284TS	7.00	5.50	9.50	0.50	1.63	4.80	3.25	0.38
286U	7.00	5.50	11.00	0.50	1.63	4.80	4.88	0.38
286T	7.00	5.50	11.00	0.50	1.88	4.80	4.62	0.50
286TS	7.00	5.50	11.00	0.50	1.63	4.80	3.25	3.75
324	8.00	6.25	10.50	0.70	1.63	5.30	4.87	0.38
324U	8.00	6.25	10.50	0.70	1.88	5.30	5.62	0.50
324S	8.00	6.25	10.50	0.70	1.63	5.30	3.25	0.38
324T	8.00	6.25	10.50	0.70	2.13	5.30	5.25	0.50
324TS	8.00	6.25	10.50	0.70	1.88	5.30	3.75	0.50
326	8.00	6.25	12.00	0.70	1.63	5.30	4.87	0.38
326U	8.00	6.25	12.00	0.70	1.88	5.30	5.62	0.50
326S	8.00	6.25	12.00	0.70	1.63	5.30	3.25	0.38
326T	8.00	6.25	12.00	0.70	2.13	5.30	5.25	0.50
326TS	8.00	6.25	12.00	0.70	1.88	5.30	3.75	0.50
364	9.00	7.00	11.30	0.70	1.88	5.90	5.62	0.50
364S	9.00	7.00	11.30	0.70	1.63	5.90	3.25	0.38
364U	9.00	7.00	11.30	0.70	2.13	5.90	6.37	0.50
364US	9.00	7.00	11.30	0.70	1.88	5.90	3.75	0.50
364T	9.00	7.00	11.30	0.70	2.38	5.90	5.88	0.63
364TS	9.00	7.00	11.30	0.70	1.88	5.90	3.75	0.50
365	9.00	7.00	12.30	0.70	1.88	5.90	5.62	0.50
365S	9.00	7.00	12.30	0.70	1.63	5.90	3.25	0.38
365U	9.00	7.00	12.30	0.70	2.13	5.90	6.37	0.50
365US	9.00	7.00	12.30	0.70	1.88	5.90	3.75	0.50
365T	9.00	7.00	12.30	0.70	2.38	5.90	5.88	0.63
365TS	9.00	7.00	12.30	0.70	1.88	5.90	3.75	0.50
404	10.00	8.00	12.30	0.80	2.13	6.60	6.37	0.50
404S	10.00	8.00	12.30	0.80	1.88	6.60	3.75	0.50
404U	10.00	8.00	12.30	0.80	2.38	6.60	7.12	0.63
404US	10.00	8.00	12.30	0.80	2.13	6.60	4.25	0.50
404T	10.00	8.00	12.30	0.80	2.88	6.60	7.25	0.75
404TS	10.00	8.00	12.30	0.80	2.13	6.60	4.25	0.50
405	10.00	8.00	13.80	0.80	2.13	6.60	6.37	0.50
405S	10.00	8.00	13.80	0.80	1.88	6.60	3.75	0.50
405U	10.00	8.00	13.80	0.80	2.38	6.60	7.12	0.63
405US	10.00	8.00	13.80	0.80	2.13	6.60	4.25	0.50
405T	10.00	8.00	13.80	0.80	2.88	6.60	7.25	0.75
405TS	10.00	8.00	13.80	0.80	2.13	6.60	4.25	0.50
444	11.00	9.00	14.50	0.80	2.38	7.50	7.12	0.63
444S	11.00	9.00	14.50	0.80	2.13	7.50	4.25	0.50
444U	11.00	9.00	14.50	0.80	2.88	7.50	8.62	0.75

Figure 6-77 (*Continued*)

FRAME	D	E	2F	H	U	BA	NW	S
444US	11.00	9.00	14.50	0.80	2.13	7.50	4.25	0.50
444T	11.00	9.00	14.50	0.80	3.38	7.50	8.50	0.88
444TS	11.00	9.00	14.50	0.80	2.38	7.50	4.75	0.63
445	11.00	9.00	16.50	0.80	2.38	7.50	7.12	0.63
445S	11.00	9.00	16.50	0.80	2.13	7.50	4.25	0.50
445U	11.00	9.00	16.50	0.80	2.88	7.50	8.62	0.75
445US	11.00	9.00	16.50	0.80	2.13	7.50	4.25	0.50
445T	11.00	9.00	16.50	0.80	3.38	7.50	8.50	0.88
445TS	11.00	9.00	16.50	0.80	2.38	7.50	4.75	0.63
447TS	11.00	9.00	20.00	*	*	*	*	*
449TS	11.00	9.00	25.00	*	*	*	*	*
504U	13.00	10.00	16.00	0.90	2.88	8.50	8.62	0.75
505	13.00	10.00	18.00	0.90	2.88	8.50	8.62	0.75
505S	13.00	10.00	18.00	0.90	2.13	8.50	4.25	0.50

* Per manufacturer's specification

Figure 6-77 (*Continued*)

CHAPTER 7
HEATING

Uses for electrical heating are nearly as numerous as motor applications. We find electrical heating elements wherever we go. Our kitchens have toasters, stove tops, crock pots, coffee makers, and various other appliances, all with electrical heating elements. Our hair dryers and curlers have electrical heating elements. Many of our hot water heaters and home heating systems rely on electrical heating. Just by looking around as you go through your day-to-day life, you'll see hundreds of electrical heating applications.

Most heating elements are resistive in nature. That is to say that the heating element represents a high-power resistor. When a current is passed through the element, it glows red hot and emits heat. Figure 7-1 shows a typical ribbon-type heating element. The unit is constructed with two threaded rods that pinch a series of ceramic insulators into a column. The two columns are separated by a ceramic frame. The element is a nickle-chromium (Ni-Chrome) ribbon that is wound around the insulators. The ends of the ribbon are terminated at the clamp rods, which also serve as the terminals.

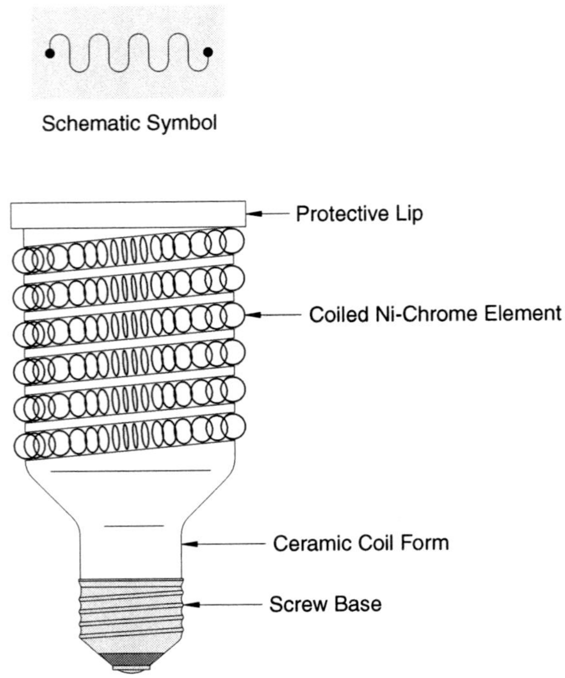

Figure 7-2 Screw-in Heater Element

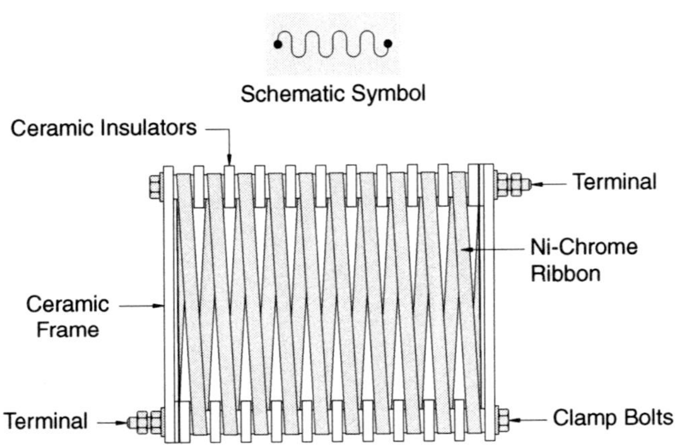

Figure 7-1 Ribbon Heater Element

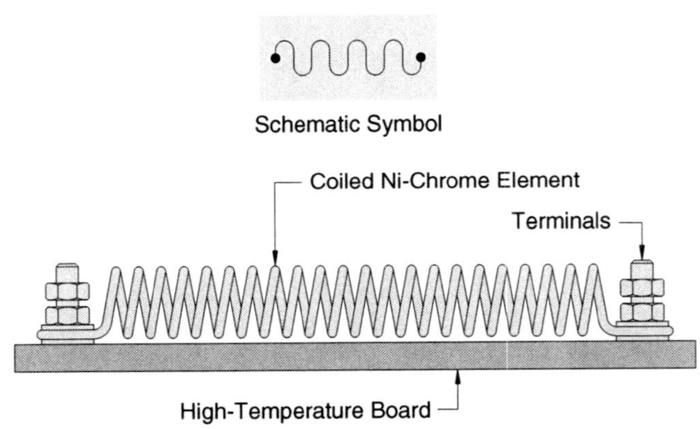

Figure 7-3 Coiled Heater Element

Figure 7-2 shows a heating element that is constructed onto an ordinary light bulb base. These elements can be screwed directly into a lamp base to produce spot heating. Care should be taken to assure that the lamp base has a sufficient current rating for the selected heating element. This particular element uses a coiled Ni-Chrome wire.

One of the most common geometries for resistive heating elements is the coil. These elements have the same general appearance as a spring. Figure 7-3 shows a typical coiled heater element. The element is generally bolted to some sort of insulating, high-temperature board.

Small, two-stroke engines and diesel engines use a heating element to initiate combustion during start-up. These elements are referred to as glow plugs. There are two basic varieties of glow plugs, open element and shrouded element. Figure 7-4 shows a typical example of both types. Open elements are generally used for small applications such as model airplane engines. Shrouded glow plugs are used for automobile and industrial applications.

Nickel-chromium wire is used almost exclusively for open air heating elements because it is particularly resistant to oxidation at elevated temperatures. This provides an exceptional life expectancy. Most of us probably know someone who uses a toaster that was manufactured in the 50s or 60s. Now that's a good service life.

The other factor that makes Ni-Chrome an excellent element is its resistive characteristics. Ni-Chrome has a high resistance when compared to other common conductors. This allows the conductor to dissipate a great deal of energy during start-up. As the temperature of the conductor increases, its resistance increases until the temperature/resistance reaches equilibrium with the power source. Figure 7-5 shows the temperature/resistance profile of three common gauges of Ni-Chrome wire.

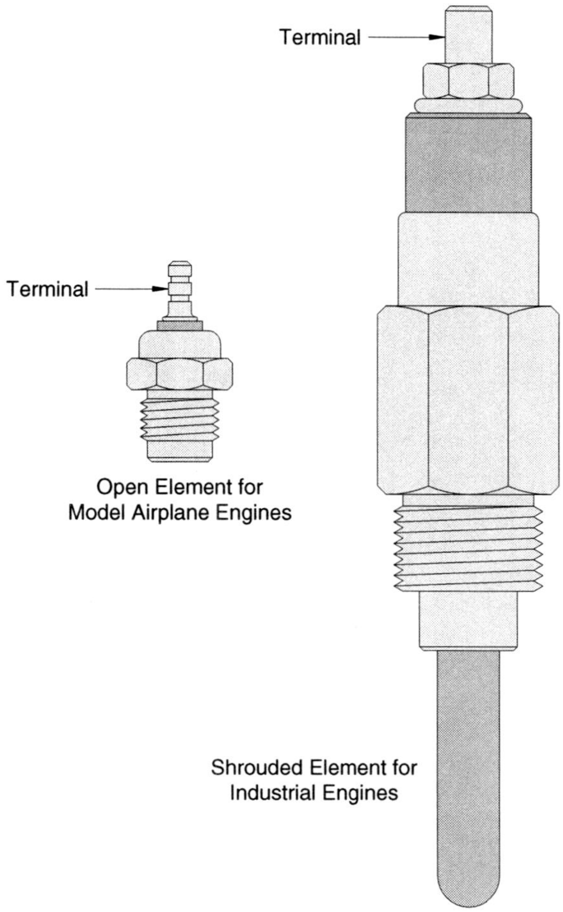

Figure 7-4 Commercial Glow Plugs

Inductive Heating

Metal components can be readily heated by manipulating the eddy currents that naturally form when the piece is exposed to an AC signal. This is a common practice in industry and is frequently used in forging and heat-treating operations. The basic concept is rather simple. The part that needs to be heated is placed into a coil and a high frequency signal is fed to the coil. Eddy currents are set up in the piece and resistive

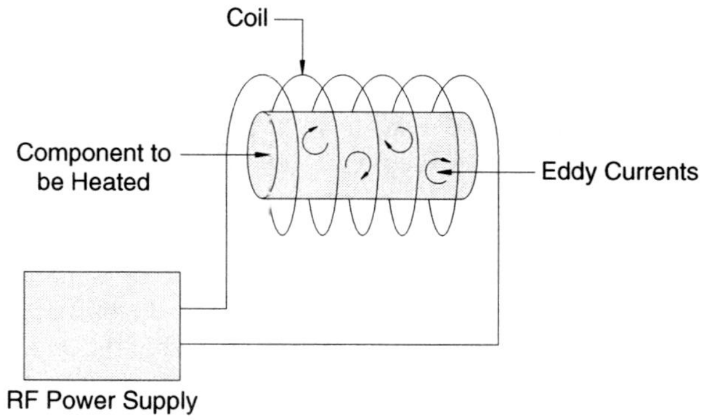

Figure 7-6 Induction Heating Schematic

heating occurs. Figure 7-6 shows schematic representation of an induction heating system. The frequency of the AC signal is generally between 22,000 and 100,000 Hz. The frequency is dependent on the geometry of the part and the type of thermal patterns that are required.

Figure 7-7 shows a typical induction coil for heating a section of tubing. The coils are normally constructed from copper tube so that continuous cooling can be applied during the operation. Electrical terminals are simply tabs of copper silver soldered to the coil ends.

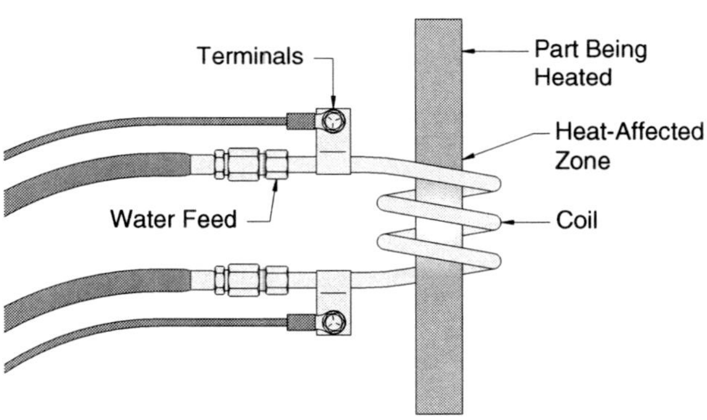

Figure 7-7 Induction Heating Element

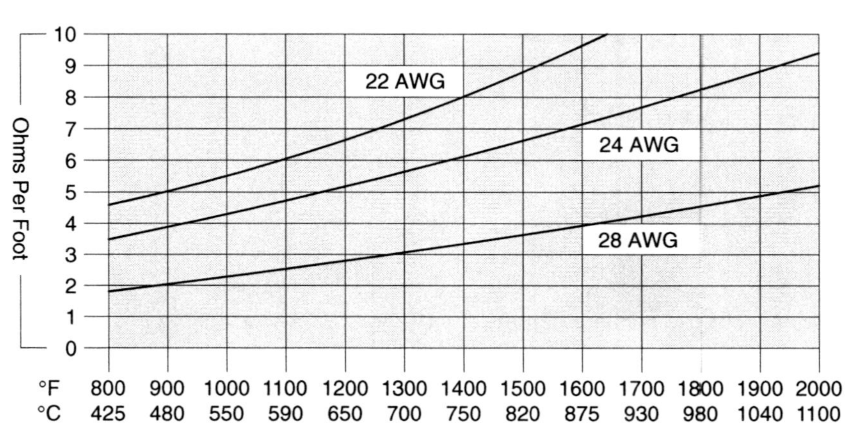

Figure 7-5 Resistance Change Due to Heating of Ni-Chrome Wire

Resistive Heating

Resistively heating a part is using the part itself as the heating element. Generally, the component to be heated is connected to a high-current power supply, as shown in Figure 7-8. When the power supply is turned on, the component will be heated between the two terminals.

The most common use for resistive heating is to thaw frozen water pipes, as shown in Figure 7-9. In this case, a small AC arc welder is connected to either end of the exposed pipe and then switched on. The pipe slowly warms, and the internal ice melts.

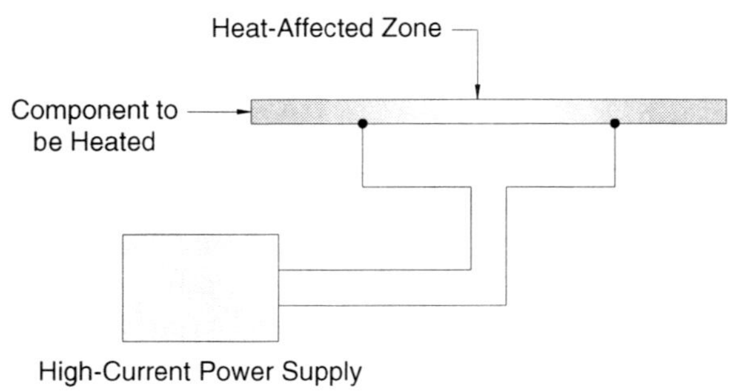

Figure 7-8 Resistive Heating Schematic

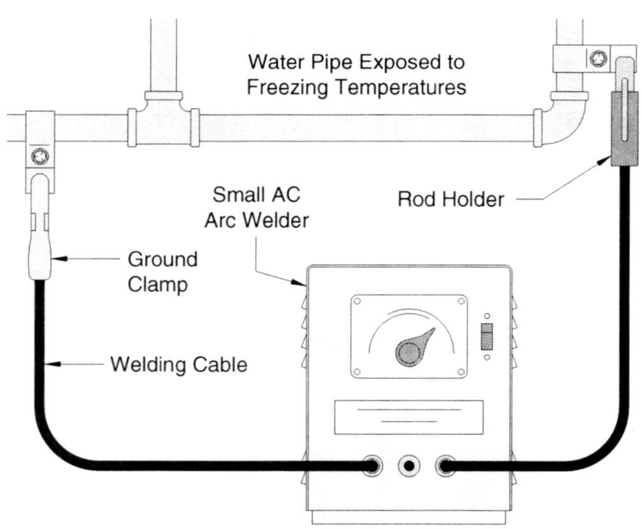

Figure 7-9 Thawing Pipes with Resistive Heating

Arc Heating

For extreme heating applications, electric arcs are deployed. An electric arc will produce very high temperatures and are commonly used for melting and alloying metals. Most of the recycled steel made in the world is melted with electric arc furnaces.

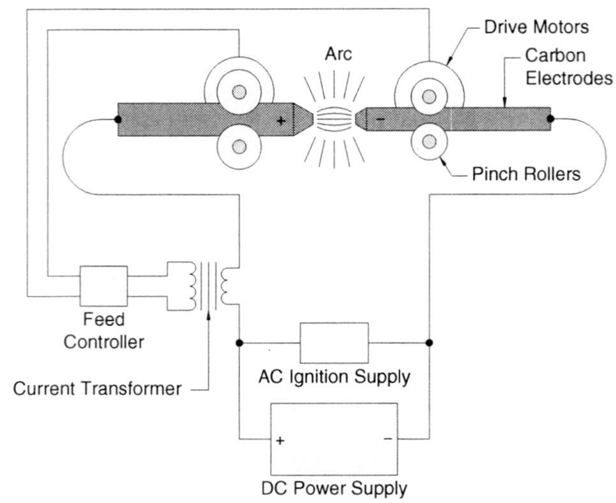

Figure 7-10 Carbon Arc Schematic

Figure 7-10 shows a schematic representation of a carbon arc heater. The arc is fed with a high-current DC power supply. The AC ignition supply is used to generate a spark to start the arc. The electrodes are mounted into roll feeders that automatically compensate for their loss due to erosion. The feeders are controlled by monitoring the current that the arc is drawing.

Arc Gouging

Arc gouging is a process that is used extensively in the steel fabrication industries. In this process a carbon electrode is used to produce an arc to the base metal. The arc generates an intensely high temperature zone that locally melts the steel. The electrode holder carries two air jets that are connected to a compressed air source. As the metal melts, the air jets blow the molten material clear of the workpiece. Figure 7-11 shows a typical arc gouging operation.

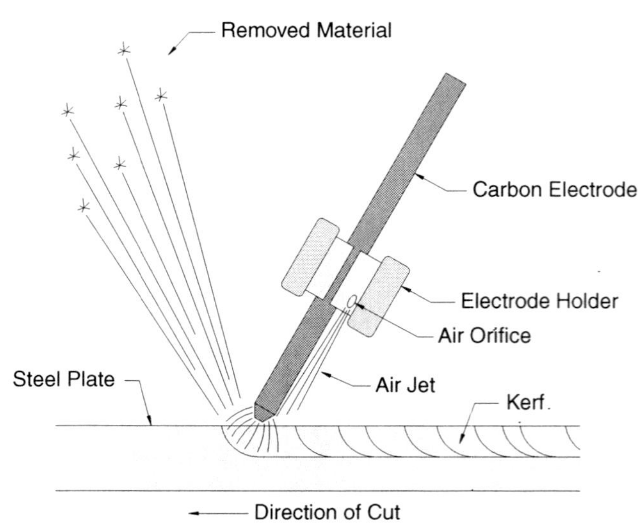

Figure 7-11 Carbon Arc Gouging

Arc Welding

Probably the most common use for arc heating is electric welding. This is a very common process that is used in all types of manufactured products. There are many different types of electric arc welding processes; however, there are three that are the most common.

Stick Rod

Figure 7-12 shows a stick rod welding operation. The electrode is connected to a power supply, either AC or DC, depending on the type of electrode used. The electrode is a steel wire with a hard flux cover. The arc produces extremely high localized heating and melts both the electrode wire and the base metal. The molten wire precipitates to the base metal and solidifies. The hard flux cover is vaporized and forms a cloud that protects the arc from inclusion of the atmosphere.

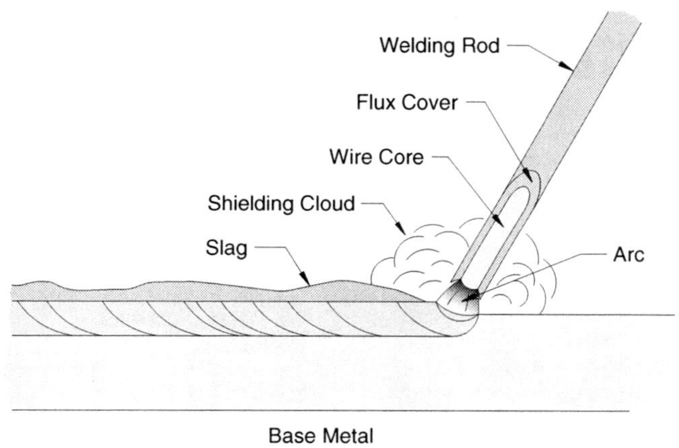

Figure 7-12 Stick Rod Welding

Tungsten Inert Gas (TIG)

Tungsten inert gas welding uses a tungsten electrode to precipitate the arc to the base metal, as shown in Figure 7-13. A flow of inert gas, usually argon, is flowed around the arc to prevent inclusion of the atmosphere. The filler is added from a separate rod. This process is very precise and is applicable to almost any type of metal.

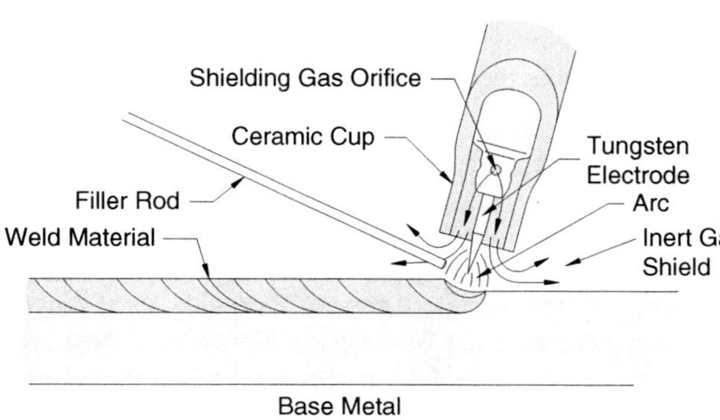

Figure 7-13 Tungsten Inert Gas Arc Welding (TIG)

Metal Inert Gas (MIG)

Metal inert gas welding is similar to the TIG process, except the tungsten electrode is replaced with a continuous feed wire, as shown in Figure 7-14. The wire acts as both the electrode and the filler material. Because the wire cannot carry particularly high currents, MIG welding is generally relegated to general purpose applications that do not require critical weld strength.

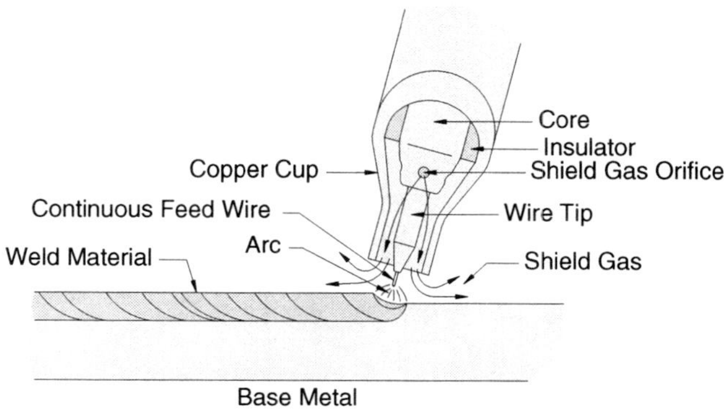

Figure 7-14 Metal Inert Gas Arc Welding (MIG)

Atomic-Hydrogen

One of the highest temperature arc heating processes is also a welding application. The atomic-hydrogen cycle produces base metal temperatures as high as 6000°F. An arc is formed between two tungsten electrodes. Hydrogen is flowed through the arc. As the hydrogen passes through the heat of the arc, it disassociates itself into singular atoms and absorbs a great deal of energy. When the hydrogen leaves the arc, it recombines and releases the stored energy in the form of heat. Figure 7-15 shows an atomic-hydrogen arc welding process.

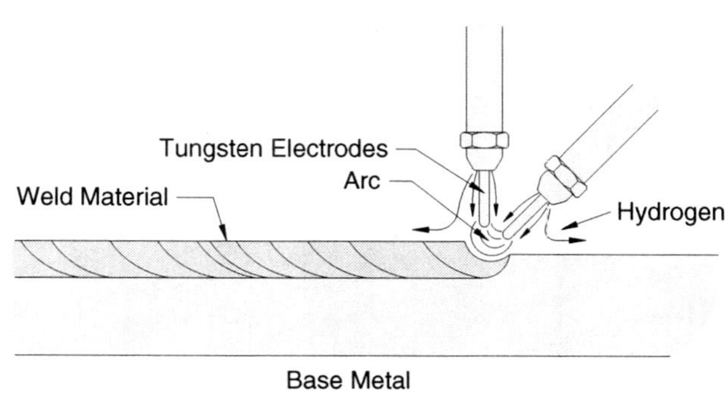

Figure 7-15 Atomic-Hydrogen Arc Welding

Arc Furnaces

Electric arcs provide an excellent heat source for melting metals. Figure 7-16 shows a small laboratory vacuum arc furnace. These units are used to alloy small amounts of metals for testing and prototyping. The system is constructed inside of a bell jar. The base is the negative terminal and the electrode holder is the positive terminal. The bases are generally water-cooled and carry a port to bleed in reaction gases. The vacuum port is connected to an appropriate vacuum system.

Large scrap steel processing is done almost exclusively with electric arc furnaces. Figure 7-17 shows a schematic representation of a commercial arc furnace. These units use a three-phase arc set and the electrodes may be as large as 36 inches in diameter.

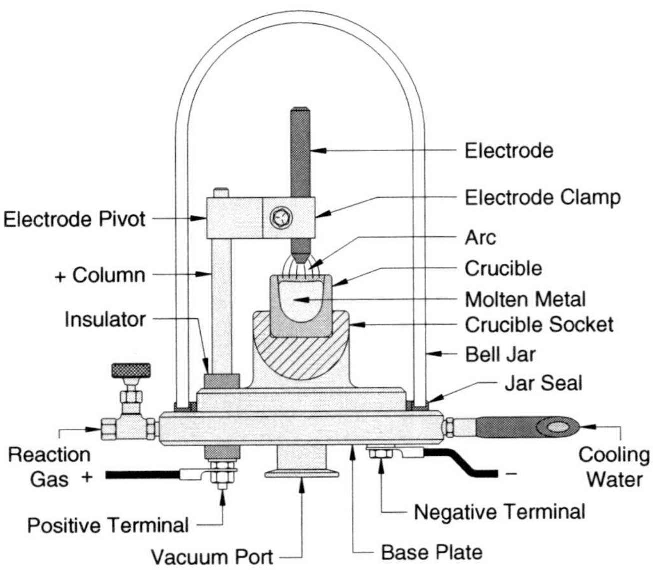

Figure 7-16 Vacuum Arc Furnace

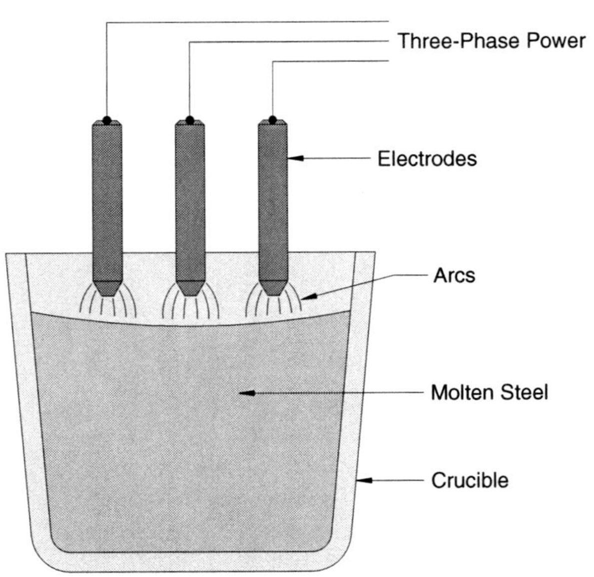

Figure 7-17 Three-Phase Arc Furnace Schematic

Thermostats

Thermostats are the most common device used to control a heater. These units are extremely common and can be found in nearly every home and office in the world. Figure 7-18 shows a typical coiled bimetal strip thermostat. A mercury switch is mounted to the outside end of a coiled bimetal strip. The inside end is attached to a fixed mount. As the temperature changes, the bimetal strip deflects and the mercury switch rotates into an actuation angle (dotted lines).

Figure 7-19 shows a typical control circuit for a high-power heater controlled with a thermostat. The thermostat operates on a 24-VAC signal and controls the contactor that switches the heater element.

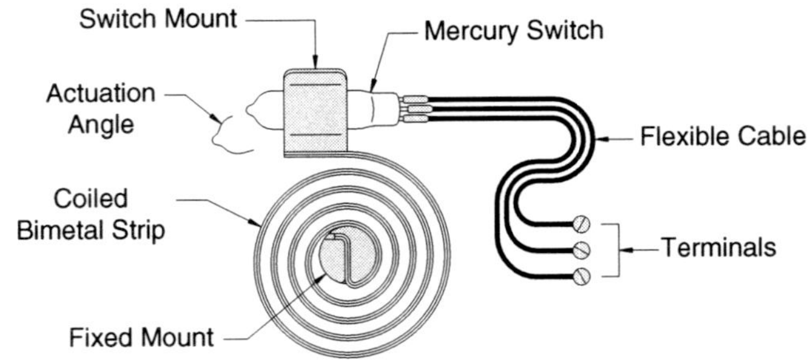

Figure 7-18 Coiled Bimetal Strip Thermostat

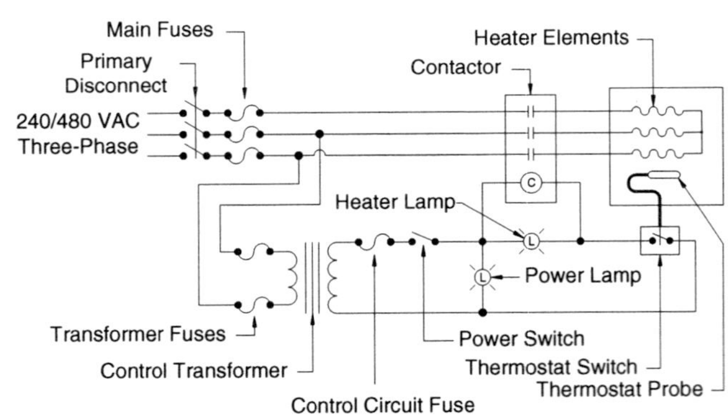

Figure 7-19 Heater Control Schematic

Figure 7-20 shows a few commercial thermostats. Thermostats are available for nearly any control environment and installation imaginable. They are typically inexpensive and highly reliable pieces of equipment.

Temperature Controllers

Temperature controllers perform the same basic function as a thermostat, except that they provide a higher degree of control. Temperature controllers generally provide some control over the temperature the heater generates. Mechanical controllers provide fairly rudimentary control while digital controllers provide a high degree of control. Figure 7-21 shows two common commercial temperature controllers.

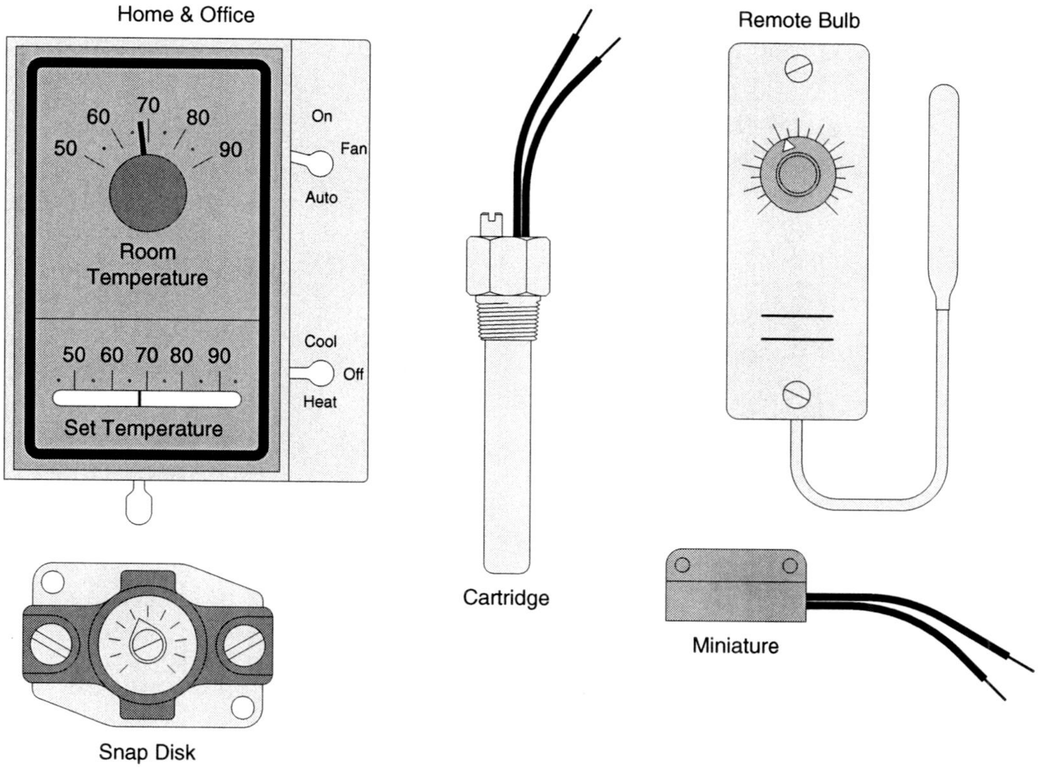

Figure 7-20 Various Commercial Thermostats

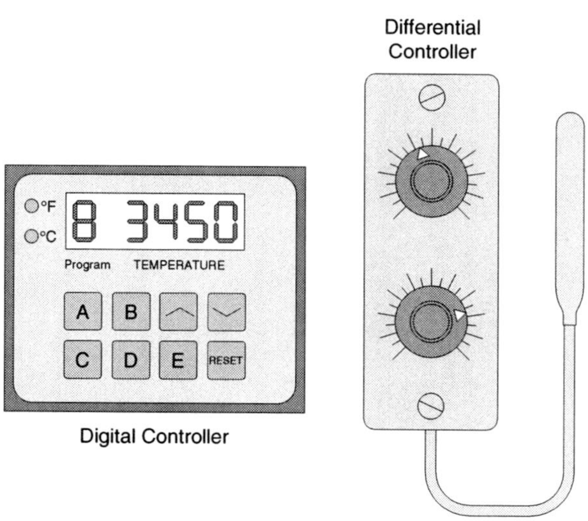

Figure 7-21 Temperature Controllers

Microwave Heating

Microwave heating was discovered in 1946 by Percy Spencer, an engineer who was engaged in the development of the magnetron tube. When working with a magnetron tube he noticed that the candy bar in his top pocket had melted. Out of curiosity he placed some popping corn in front of the tube and was delighted to see that it would pop almost immediately when the tube was powered up. The next day he placed an egg in front of the tube and in very short order it was heated to a point where it exploded. At that point in time, the microwave oven was born.

The heart of the microwave oven is the magnetron tube. This is a vacuum tube that emits microwave radiation in the 2.5 GHz range when it is excited with a high voltage. Figure 7-22 shows a sectional view of a magnetron tube. Water, fat, and sugar molecules absorb microwave radiation and convert it to atomic motion, or heat. When food is placed into a cavity that is fed by the output of the tube, it is effectively heated. Magnetron tubes will be discussed in greater detail in Chapter 14.

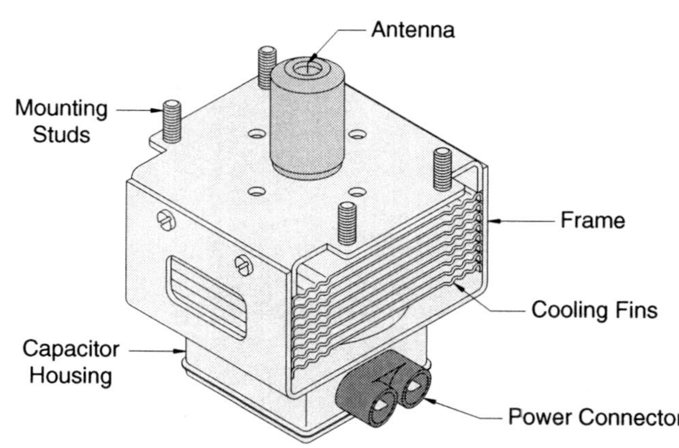

Figure 7-22 Commercial Magnetron Tube

CHAPTER 8
CIRCUIT PROTECTION

Most electrical circuits require some type of protective device that will limit the current and/or voltage. In the home, circuit breakers protect from plugging too many appliances into a single circuit. Without the circuit breaker, the distribution wires would carry too much current, overheat and eventually melt. The combination of arcing and high-temperature metal creates a severe fire hazard. Most of us have heard of a home fire that was caused by an electrical malfunction.

In smaller circuits, fuses protect the device from power line transients, miss-connections or the malfunction of a component in another section of the system. Most multimeters have fuses to protect against incorrect connection of the test leads. Testing a 240-volt circuit while the meter is set to ohms (a mistake that is easily made) would instantly destroy the unit. Instead of replacing the entire meter, the fuse fails and the problem can be corrected for a few cents.

Fuses

The basic fuse is a rather simple device. It is a small metal link with a central element that is designed to fail if the current rises above a certain level. Figure 8-1 shows a typical fuse element. The center portion of the link has a reduced width element. If the current rises above the limit of the element, it melts and the circuit is broken.

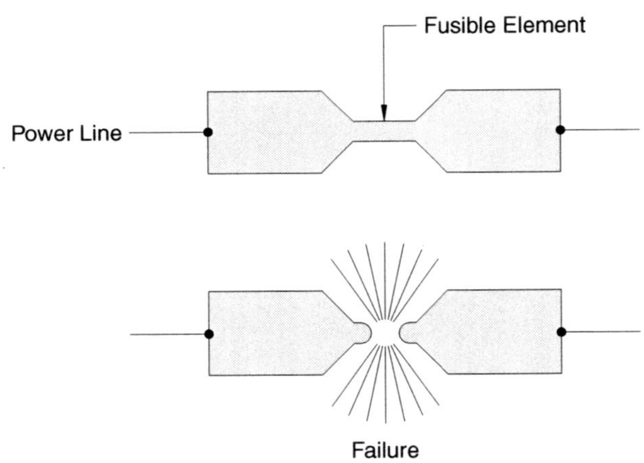

Figure 8-1 Fuse Element

For applications that have a high in-rush current, a delayed action or slow-blow fuse must be used. These links have a reduced width element that is connected in-line with a coil or delay element. The delay element allows the fuse to carry a higher current load for a short period of time. If the current surges for a longer time, the element will fail. Figure 8-2 shows a slow-blow fuse element.

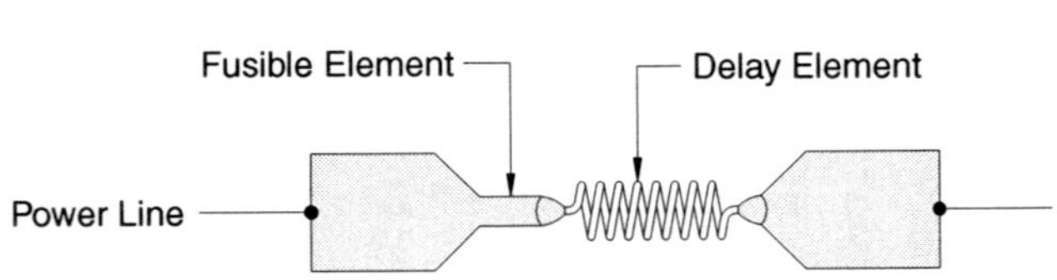

Figure 8-2 Slow-Blow Fuse Element

Fuses are available in every size, voltage and current rating imaginable. Fuse ratings generally include a maximum voltage, current, and in-rush current. In-rush current is the amount of current the fuse can carry during starting operations. Figure 8-3 shows just a few types of standard fuses that are commonly used in different applications. Figure 8-4 shows the different types of fuse holders that are readily available in the market.

For removing large cartridge fuses, always use a fuse puller, as shown in Figure 8-5. These tools will make pulling a fuse very easy, protect the fuse block from damage, and allow the operator to avoid inadvertent electrocution.

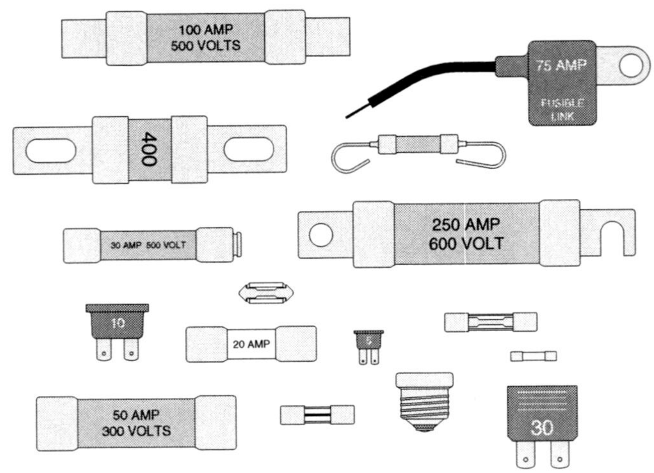

Figure 8-3 Various Commercial Fuse Types

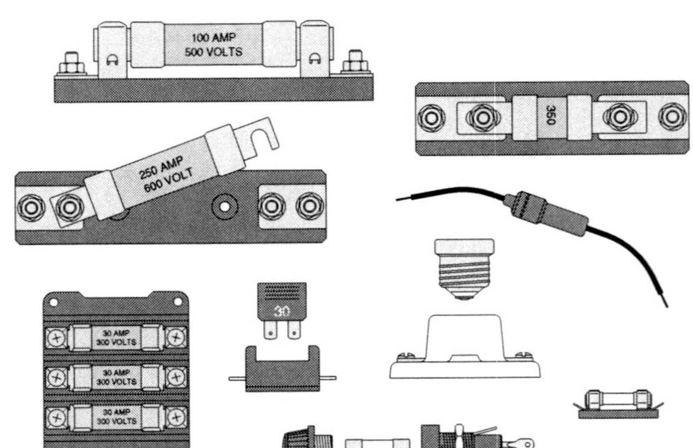

Figure 8-4 Various Commercial Fuse Holders

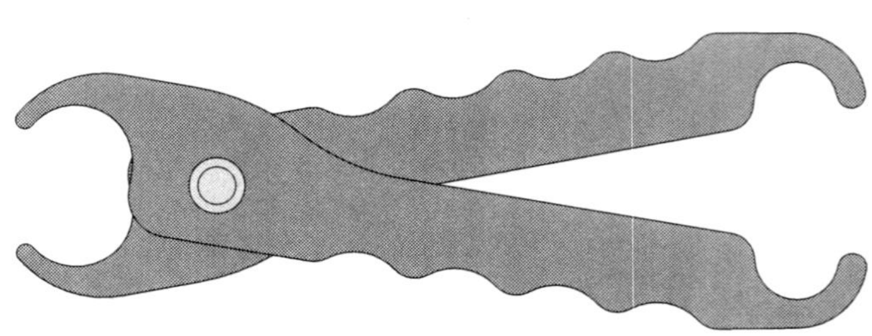

Figure 8-5 Commercial Fuse Puller

Circuit Breakers

Circuit breakers are functionally resettable fuses. Most of us are familiar with the circuit breakers in our homes and have reset a few over the years. Circuit breakers are also found in virtually every type of electrical circuit made. Circuit breakers can be found in the smallest appliances, with ratings as low as a few milliamps, all the way through to huge power distribution systems with ratings in the million amp range. Circuit breakers provide a level of convenience that a fuse cannot match; however, this convenience comes at a cost. Circuit breakers are typically much more expensive than the equivalent fuse.

Figure 8-6 shows a stylized schematic of a thermal circuit breaker. Power is fed through a bimetal strip, flexible cable, flat spring, and a set of contacts. If the current exceeds the rating of the breaker, the bimetal strip heats and curves up, releasing the latch and allowing the contacts to open. After the bimetal strip cools the contacts can be reset.

For panel mount applications, small circuit breakers are available that have the same general appearance as a push button. Figure 8-7 shows a panel mount unit. When the breaker trips, the button extends out and provides a visual indicator. A panel mount breaker should always have the current rating clearly printed on the face of the button and the side of the body.

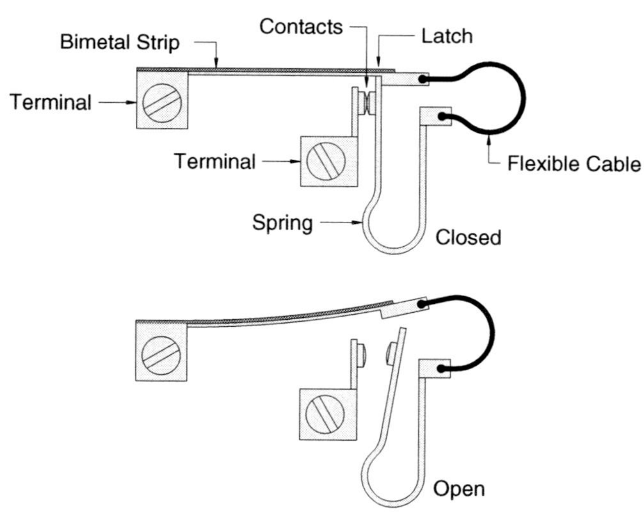

Figure 8-6 Thermal Circuit Breaker Schematic

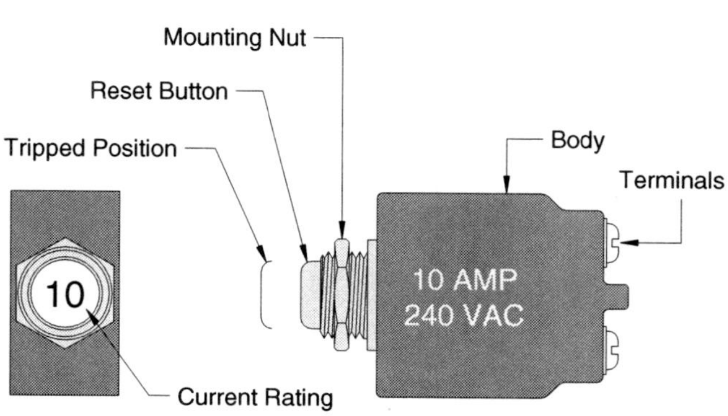

Figure 8-7 Push Button, Panel Mount Circuit Breaker

Multipole, panel mount circuit breakers are typically ganged, single-pole units with flip-type actuators. The actuators are pinned together so that if one breaker trips it will trip the other two at the same time, thus protecting the entire circuit. Figure 8-8 shows a multipole, panel mount circuit breaker assembly.

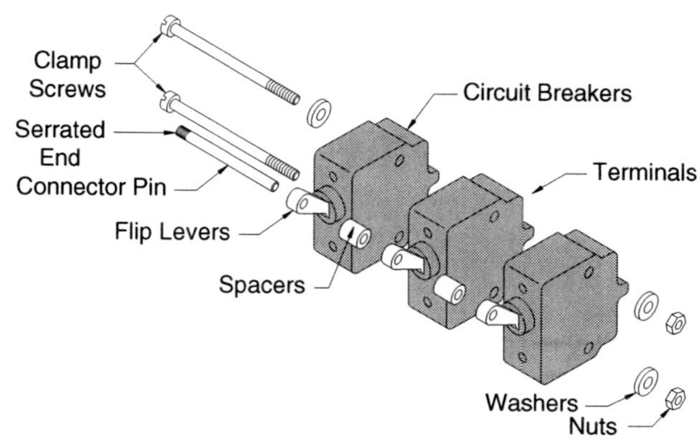

Figure 8-8 Multipole, Panel Mount Circuit Breaker Assembly

Older light bulb base fuses are available in circuit breaker versions. Although these units are not very common anymore, they are convenient units for bench work. A panel of light bulb sockets can be set up and an assortment of breakers can be kept on hand. Simply selecting the appropriate rating and screwing it into a socket provides ready protection for any given project. Figure 8-9 shows a light bulb base circuit breaker.

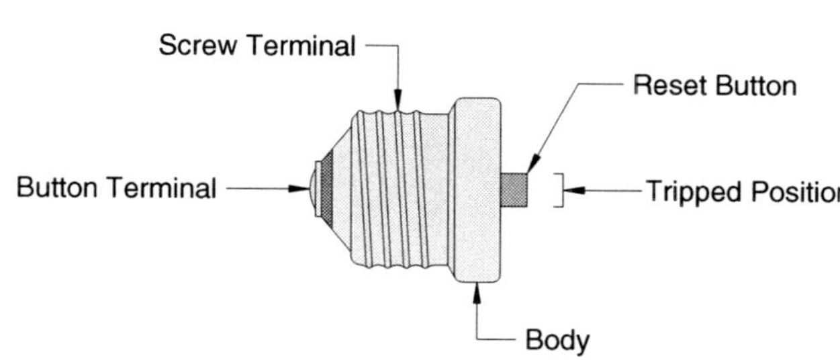

Figure 8-9 Light Bulb Base Circuit Breaker

The power centers or breaker boxes in homes and offices use specially designed circuit breakers that are easily inserted and removed. Figure 8-10 shows a typical power center circuit breaker. The mounting socket is clipped onto a rail and then the top of the breaker is rotated in until the input terminal engages the power buss. The output is generally a screw

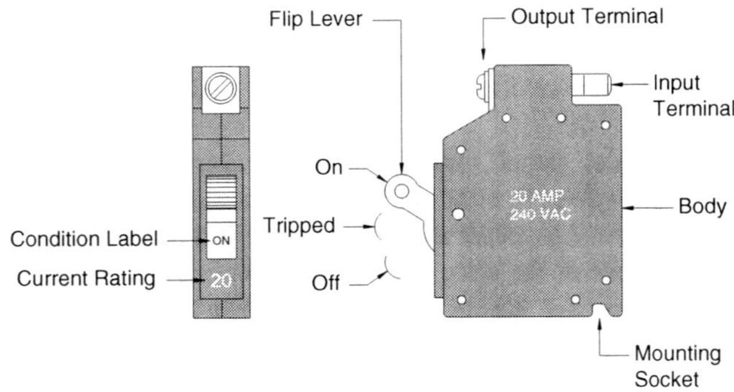

Figure 8-10 Power Center Circuit Breaker

terminal that allows a wire to be easily connected. These breakers have a flip-type actuator that is up when on, down when off, and in the center when tripped. Figure 8-11 shows a typical home power center. Always select a power center with a main breaker as shown. It is also a good idea to label the breakers. Labeling can make de-energizing a circuit a simple matter instead of an arduous chore.

Circuit breakers are manufactured in extremely large sizes to provide circuit protection for large power distribution systems. Figure 8-12 shows a 240,000 volt, 40,000 amp circuit breaker which is used to protect cross-country transmission lines.

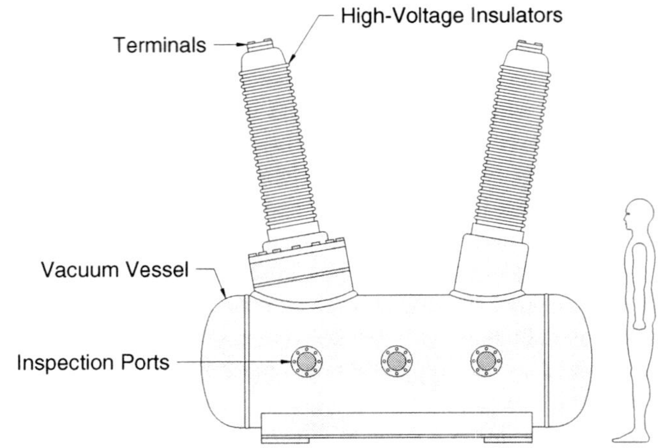

Figure 8-12 240,000-Volt, 40,000-Amp Power Transmission Circuit Breaker

Arc Suppression

The formation of arcs during the opening phase of contacts can be extremely damaging. This is the principal reason for snap and pseudo-snap action switches. The contacts are opened very fast, mitigating the damaging effects of the arc. As voltages and currents get higher, special provisions must be made to limit arc damage.

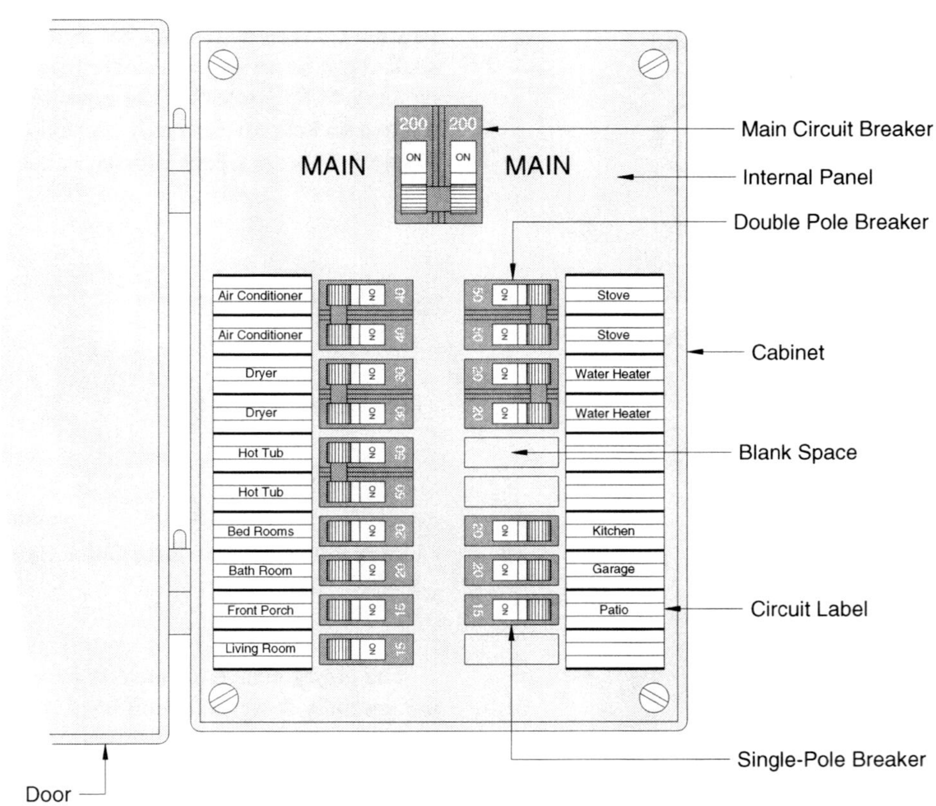

Figure 8-11 Home Power Center

Long Throw Interrupters

Figure 8-13 shows a fast acting, long throw interrupter. The unit has a moving contact that is spring loaded for fast opening. When the trigger is pressed, the moving contact is pulled into the open position. The distance between the contacts is too wide to allow the arc to sustain. The speed at which the contact retracts limits the duration of the arc and mitigates the damage that it causes. Long throw interrupters are only practical to a certain upper limit. As the voltages become higher, the size and speed of these units becomes a factor.

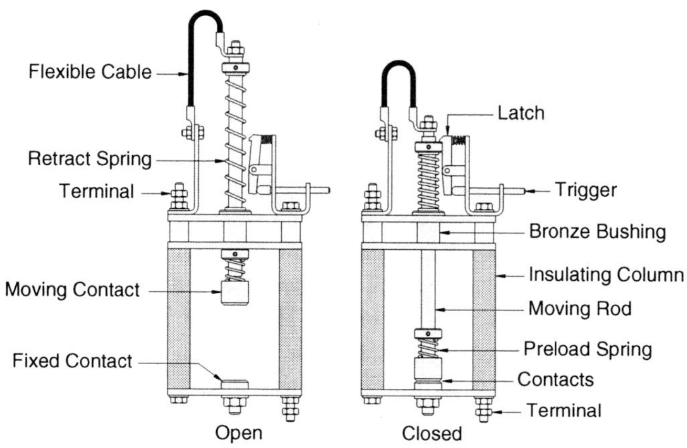

Figure 8-13 Fast Acting, Long Throw Interrupter

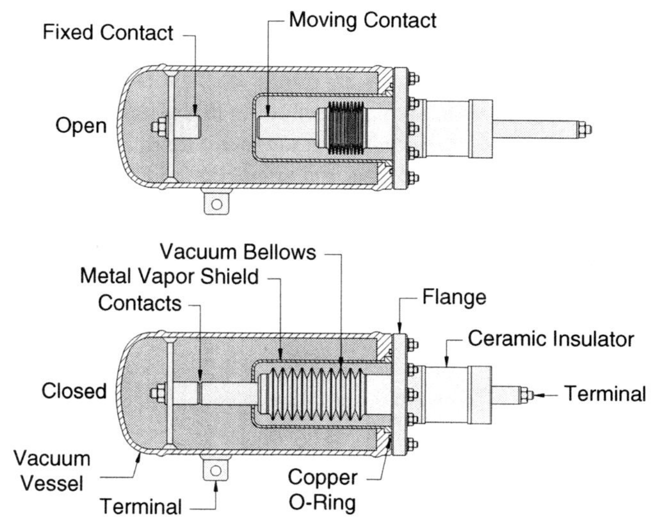

Figure 8-14 Vacuum Interrupter

Vacuum Interrupters

To reduce size, interrupters are often constructed so that the contacts can operate in a vacuum. The lack of ionizing gas surrounding the contacts provides considerably improved arc control. Vacuum interrupters are not typically preferred because maintaining a suitable vacuum over extended periods of time can be problematic. If the vacuum leaks down, the interrupter could open and maintain a continuous arc, severely damaging the interrupter and providing almost no protection. A variation of the vacuum interrupter uses a high dielectric oil in place of the vacuum. The oil provides better insulating qualities and draws the heat from the arc with great efficiency. These interrupters have better long-term reliability and do not require the rigorous monitoring of a vacuum unit. Figure 8-14 shows a sectional view of a typical vacuum interrupter.

Pneumatic Interrupters

Pneumatic, or air blast, interrupters are the most common high-power units used. These units can be seen at many of the high-power switching stations that dot the country. These units use a blast or jet of compressed gas to blow the arc away from the contacts. Figure 8-15 shows a typical pneumatic suppression interrupter. The contacts are built into a pair of bolted clam shells. A diaphragm is sandwiched between the flanges. The center of the diaphragm is fixed to the moving contact. When the contacts are opened, the cavity behind the diaphragm is pressurized and a jet of gas is forced through the center orifice. The gas jet blows the arc away from the contacts and it is extinguished. To improve the performance of these interrupters, the air can be replaced with a gas having a higher dielectric, such as sulfurhexifloride.

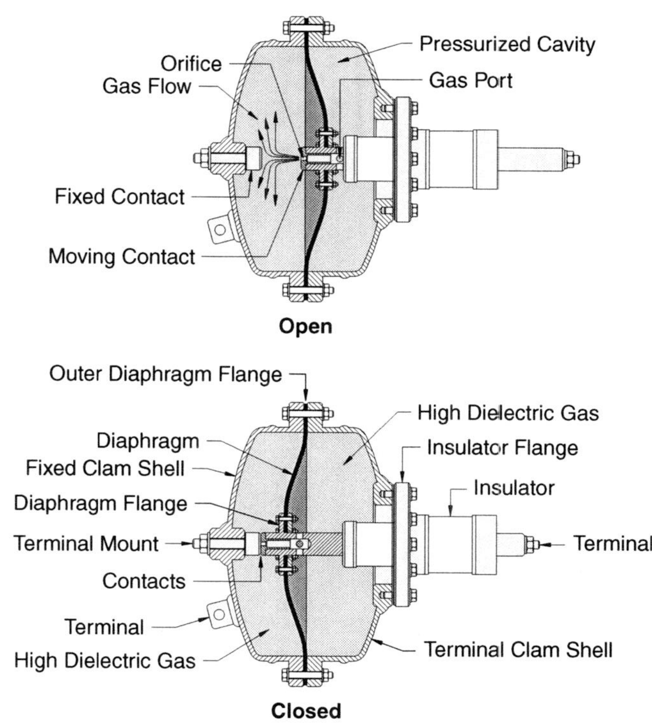

Figure 8-15 Pneumatic Suppression Interrupter

Magnetic Arc Suppression

Smaller applications, such as motor contactors, will often incorporate magnetic arc suppression. In this arrangement the arc is exposed to a perpendicular magnetic field, as shown in Figure 8-16. As the arc forms, the magnetic field forces it into a long, curved path and the arc extinguishes.

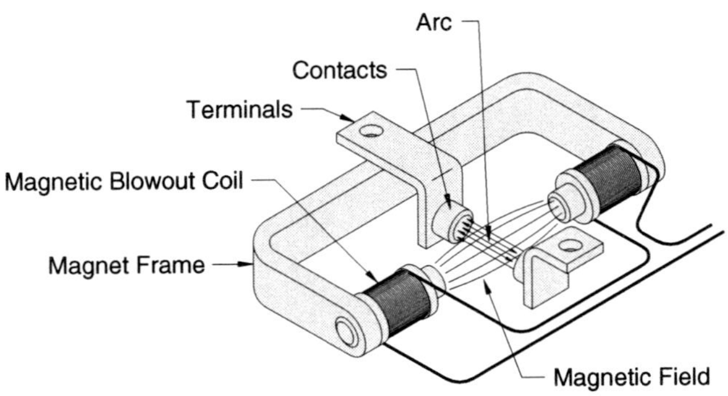

Figure 8-16 Magnetic Arc Suppression

Arc Dividers

An arc divider is a series of nonconductive plates that are arranged in close proximity to the formation area of the arc, as shown in Figure 8-17. As the arc forms, the heat forces it to rise and it enters the plate array. Once the arc has entered the array, the path is broken.

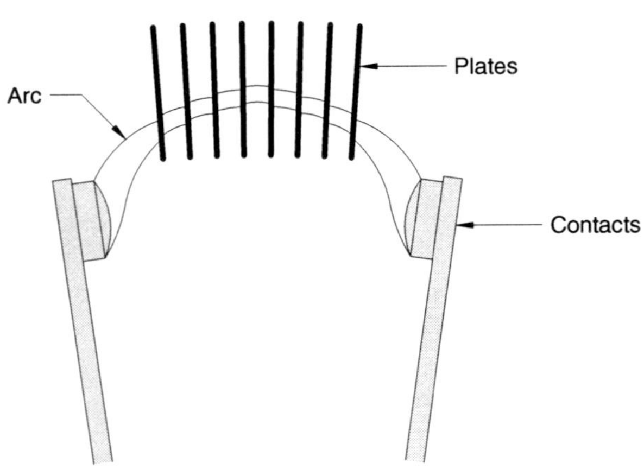

Figure 8-17 Arc Divider

Motor Heaters

Protecting induction motors generally requires a special type of system based on a heating profile as well as current. The motor should be connected through a set of standard fuses and a motor starter. There are two basic types of heaters used on motor starters, inductive and thermal. The first type that we will discuss is the inductive heater. Each pole of a commercial motor starter is equipped with a thermal kick-out mechanism, as shown in Figure 8-18. The heart of the system is the inductive heater, which is matched to the operating voltage and horsepower of the motor. The heater is connected in series with the motor and in conjunction with the core, mimicking the basic heat profile of the motor. The heater is installed in close proximity to an axle mounted within a housing (the core). The head of the axle and housing are joined with a low-temperature solder joint. The opposite end of the axle has a serrated wheel. A ratcheted rack engages the wheel and is designed to close and hold the contact set. The opposite end of the ratchet rack has a reset button. Under normal conditions the inductive heater produces very little heat in the core and the motor circuit operates without incident. Under overload conditions the heater induces enough heat to melt the solder joint, allowing the axle and serrated wheel to freely rotate. At this point, the trip spring forces the ratchet rack back and the contacts open. The contacts are wired in series to the contactor coil and when open, the contactor opens. Figure 8-19 shows a schematic representation of a motor starter and Figure 8-20 shows a typical commercial motor starter.

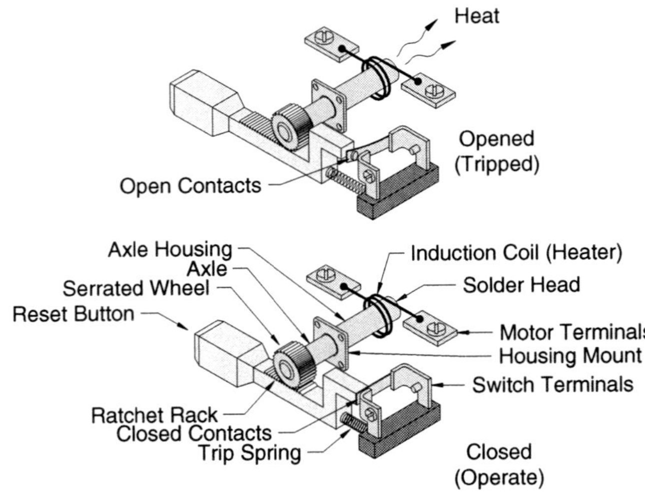

Figure 8-18 Inductive Heating Motor Protection Assembly

The second type of motor protection is the thermal heater. Like the inductive unit, each pole of the motor starter is equipped with a thermal kick-out mechanism, as shown in

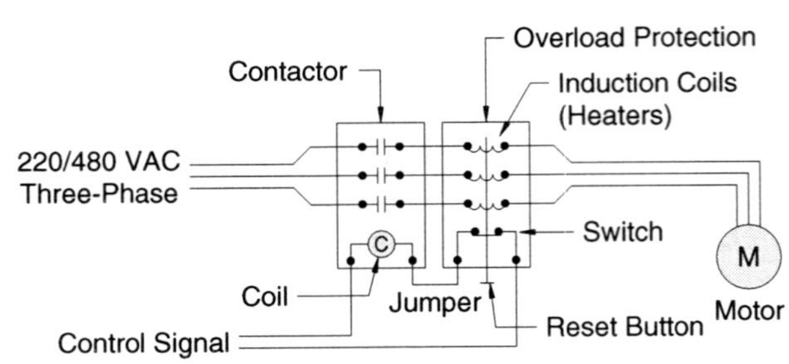

Figure 8-19 Schematic for a Motor Starter with Inductive Heaters

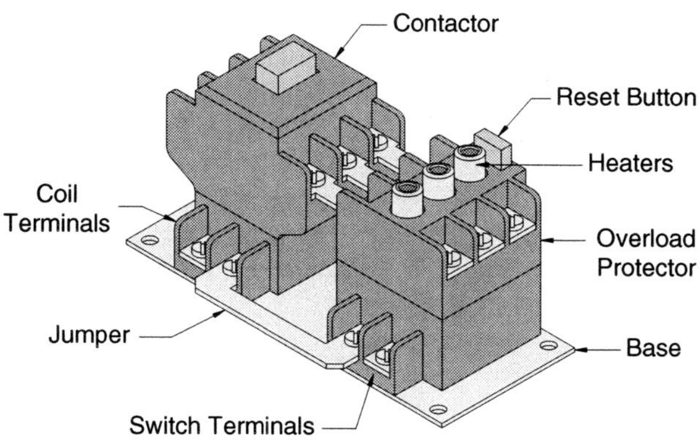

Figure 8-20 Commercial Motor Starter

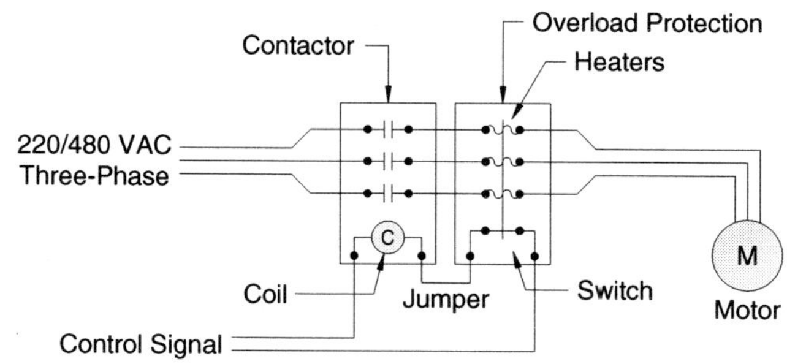

Figure 8-22 Schematic for a Motor Starter with Thermal Heaters

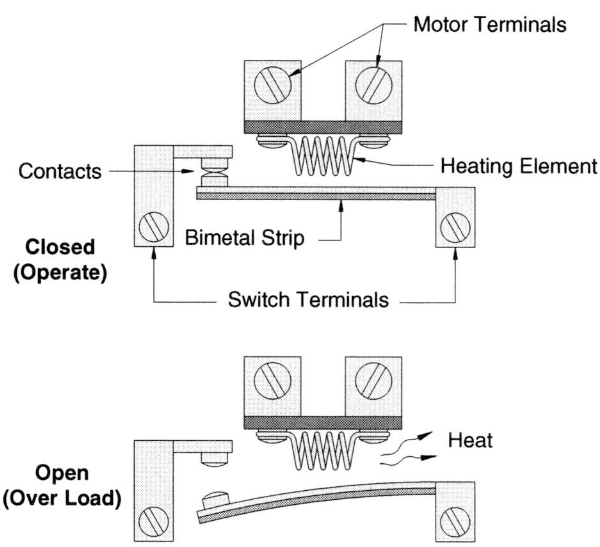

Figure 8-21 Thermal Heating Motor Protection Assembly

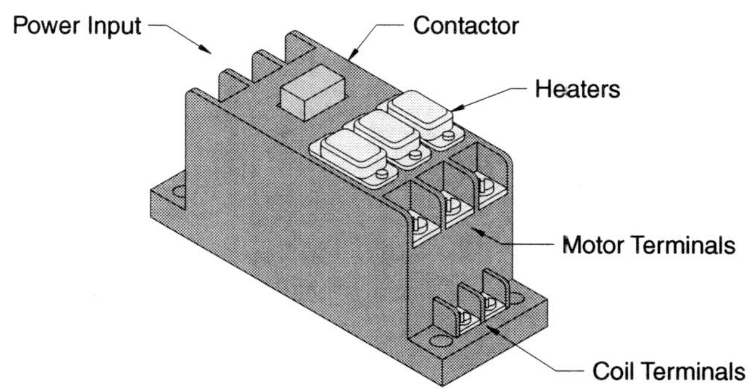

Figure 8-23 Commercial Motor Starter with Integral Thermal Heater Protection

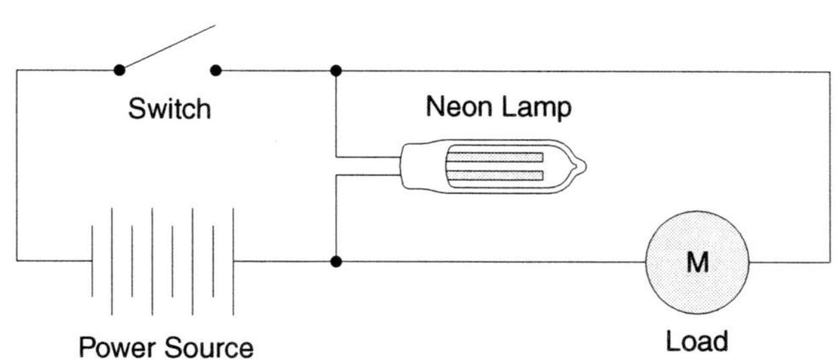

Figure 8-24 Glow Discharge Protection

Figure 8-21. A thermal heater is matched to the operating voltage and horsepower of the motor. This heater is also connected in series with the motor and mimics the basic heat profile of the motor. The heaters are installed in close proximity to a bimetal strip which operates a contact set. Under normal conditions the heaters produce very little heat and the contacts remain closed. Under overload conditions the heater generates enough heat to cause the bimetal strip to deflect and open the contacts. The contacts are wired in series to the contactor coil and when open, the contactor opens. Thermal starters will automatically reset when they have sufficiently cooled. Figure 8-22 shows a schematic representation of a motor starter and Figure 8-23 shows a typical commercial motor starter with thermal protection.

Glow Discharge Protection

For applications that need a modest amount of transient protection, a neon lamp may be used, as shown in Figure 8-24.

The neon lamp will turn on at a specific voltage and while operating has a very low resistance. If a transient is experienced, the lamp will flash on and suppress the excess voltage.

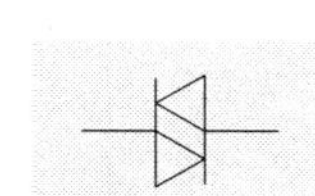

Schematic Symbol

Metal Oxide Varistor (MOV)

To provide more precise transient control a metal oxide varistor (MOV) as shown in Figure 8-25, is generally used. These devices short at a specific voltage and are quite effective in suppressing the inductive

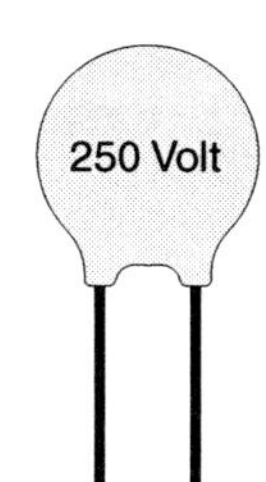

Figure 8-25 Metal Oxide Varistor (MOV)

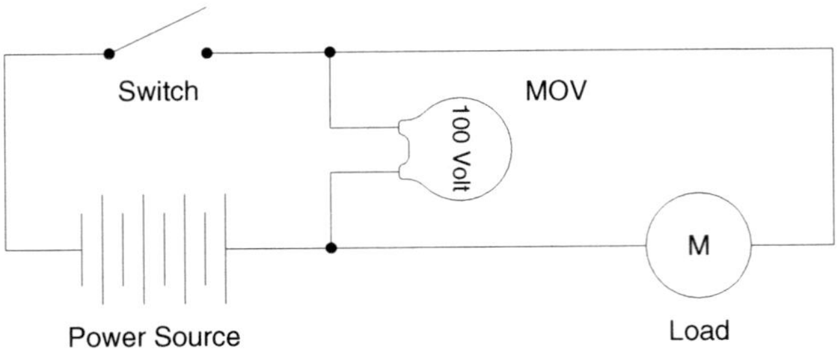

Figure 8-26 MOV Protection

kickback of coils and motors. The voltage is printed on the side of the device as shown. Figure 8-26 shows how a MOV is applied in a switching circuit. Note the similarity of this circuit to the neon lamp circuit shown in Figure 8-24.

Spark Gaps

Spark gaps can be effectively used to limit high-voltage transients. The gap is set to a spacing that will arc when a certain voltage is reached. Figure 8-27 shows three different spark gaps. Different electrode shapes play a large role in determining the performance of the spark gap. The precision of a spark gap is generally not very good because they typically operate in air. The atmospheric conditions affect the standoff voltage to a significant degree.

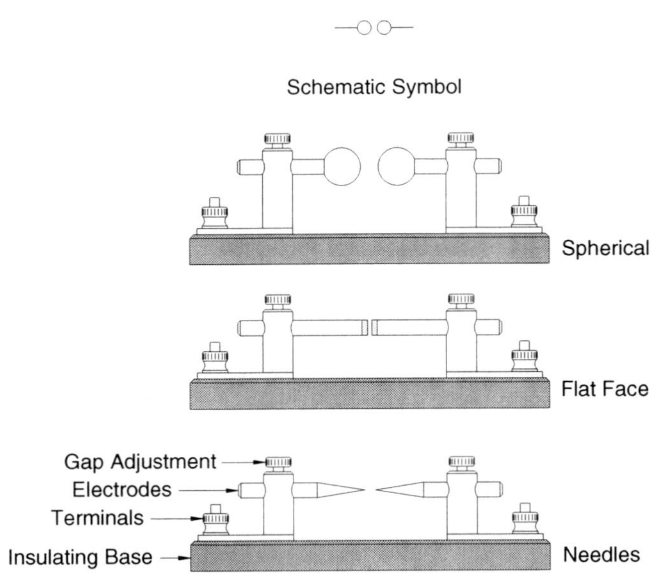

Figure 8-27 Spark Gaps

Grounds

Grounding is the single most effective method to protect an electrical circuit. In addition, grounding will provide a significant margin of safety to personnel working with or around a circuit. Grounds and grounding methodologies are the base circuit for virtually all electrical controls, electromechanical devices, and equipment. When working with electrical equipment a good rule of thumb is "when in doubt, ground." A good example of this is the use of grounding or shorting shunts on power lines during service activities. Figure 8-28 shows the application of shorting shunts to protect workers from accidental electrocution. The power is turned off and a bolt-on wire clamp is attached to each of the primary power lines. The clamps are connected to a third clamp that is connected to the ground wire. If the power is inadvertently turned on, the line breakers will immediately trip and the workers will be protected.

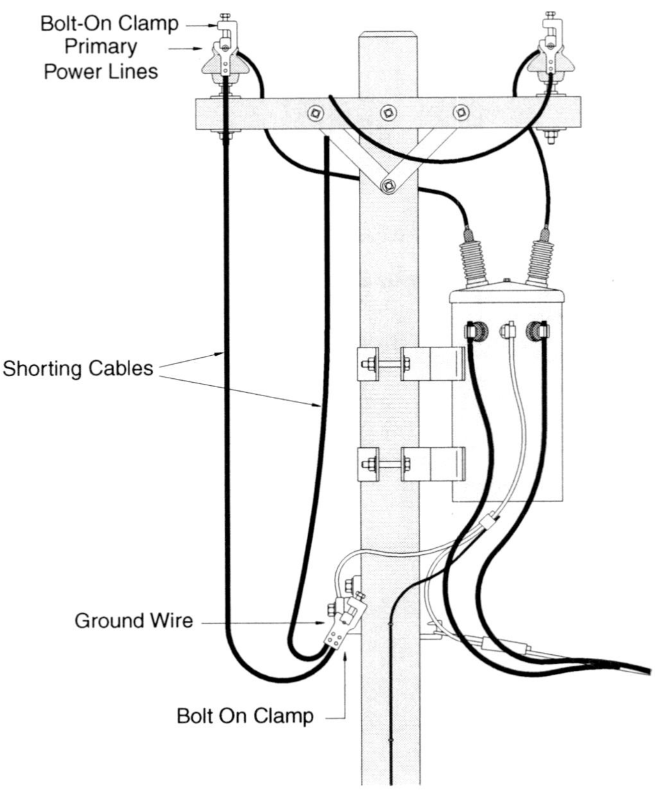

Figure 8-28 Shorting Shunts

Ground Connections

The term ground literally means "the ground." Any ground loop must be ultimately connected to an earth ground. An earth ground is typically a bronze rod that is embedded into the ground at least 6 feet, as shown in Figure 8-29. The top of the rod has a bolt-on clamp with a ground terminal. This arrangement will provide a suitable ground for most instruments and equipments. Another suitable ground is a metal cold water pipe. A cold water system that is constructed with metal pipe is electrically connected to an earth ground. Figure 8-30 shows a typical cold water pipe ground connection. It should be noted that polyvinyl chloride (PVC) pipe is nonconductive and, therefore, cannot be used for grounding purposes.

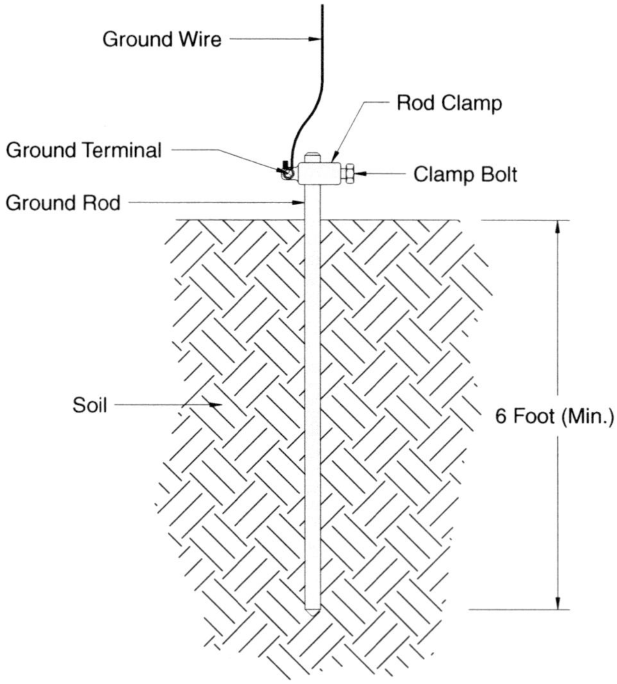

Figure 8-29 Typical Ground Rod Installation

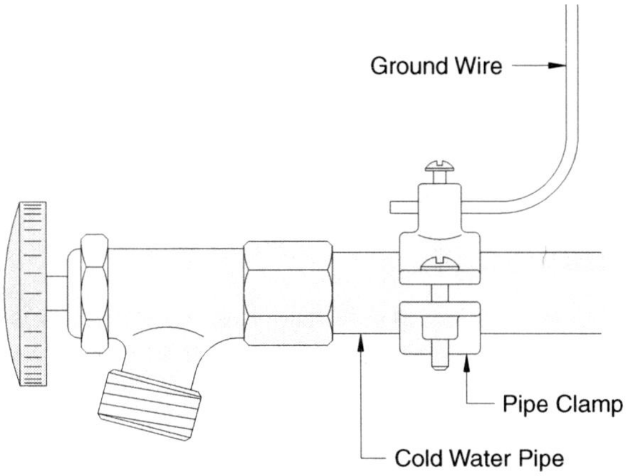

Figure 8-30 Cold Water Pipe Ground Connection

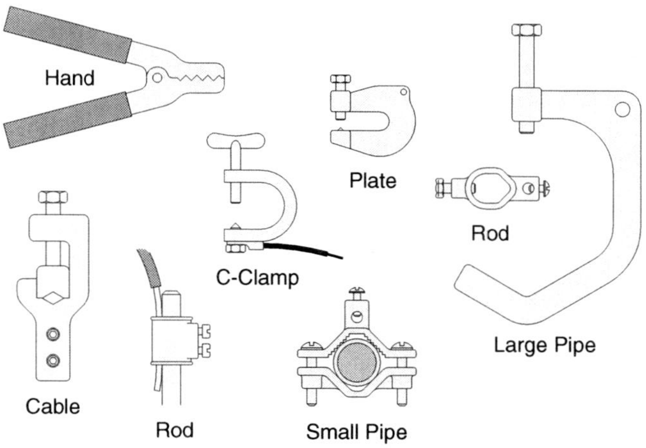

Figure 8-31 Various Commercial Ground Clamps

Control Cabinet Grounds

Most control boxes provide grounding facilities, as shown in Figure 8-32. The door and internal panel(s) should have a hardwired ground loop that is connected to the main cabinet ground. The cabinet should be connected to a suitable earth ground.

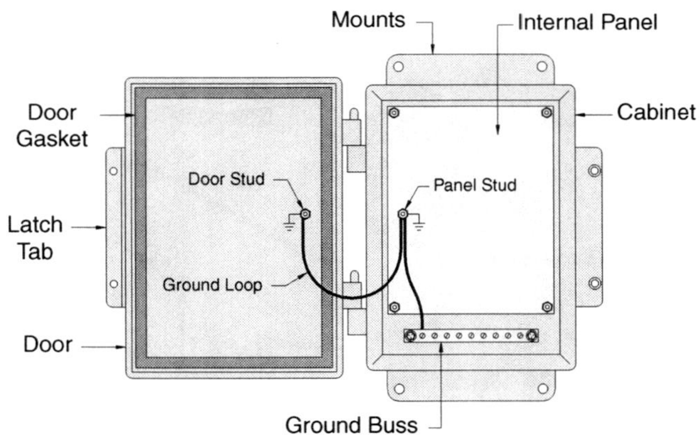

Figure 8-32 Control Cabinet Grounding Facilities

Ground Clamps

Ground clamps can take the form of almost any connector or clamp that will provide an electrically conductive junction. However, there are a number of clamps that are specifically designed for grounding applications. Figure 8-31 shows just a few common ground clamps that are available in the market today. Some of these clamps are designed for temporary applications, such as the hand and C-clamp configurations. Some are designed for semipermanent applications, such as the cable clamp. Some are designed for permanent applications, such as the plate, rod, and pipe clamps.

Grounding Conduit and Junction Boxes

Electrical conduit and associated junction boxes are specifically designed to facilitate grounding. All screws and fittings are designed to produce a high-quality ground connection. In addition, a ground wire is routed along with all conductors and connected to the conduit system whenever possible. Figure 8-33 shows typical grounding procedures in reference to conduit and junction boxes.

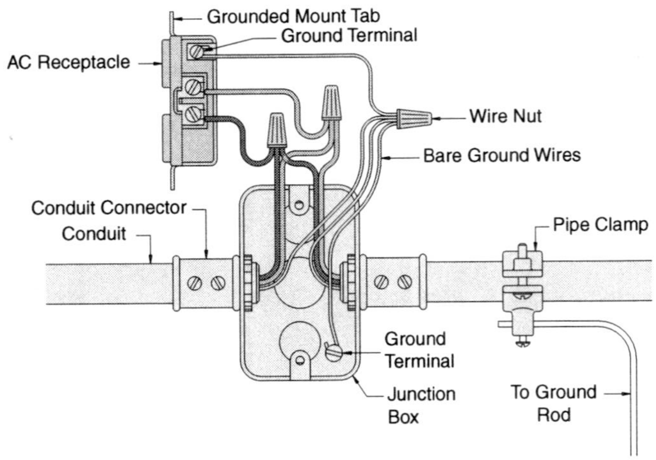

Figure 8-33 Grounding Conduit and Junction Boxes

Static Protection

When working with or servicing sensitive electronics, the slightest discharge may severely damage the sensitive circuitry. Most service technicians wear a wrist ground during these periods. The wrist ground, as shown in Figure 8-34, is simple a brass rivet held against the wrist by a hook and loop strap. The brass rivet is connected to an earth ground. Any charge that may build up on the service technician is harmlessly shorted to ground. In this manner sensitive electronics are protected from stray transients.

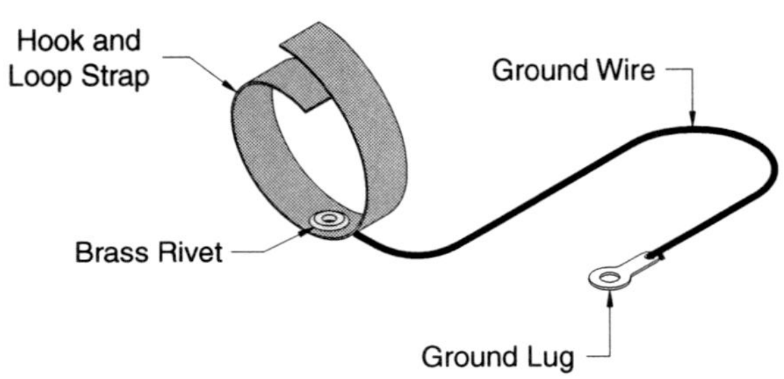

Figure 8-34 Wrist Ground

Grounding Hooks

Grounding hooks, or Jesus sticks as they are sometimes called, are important safety devices when dealing with high-voltage equipment. When you ask a technician why it's called a Jesus stick, he'll respond, "When working with this equipment, it's the only thing that keeps you from meeting Jesus." Figure 8-35 shows a typical grounding hook. The ground clamp should be a semipermanent or permanent connection so that it cannot be inadvertently pulled off and disconnected. To assure that a conductor is at zero potential, the metal hook is looped over the wire and any existing charge will be shorted to ground. These devices should be inspected regularly for damage and dirt. If found to be in a poor state of repair, they should be discarded.

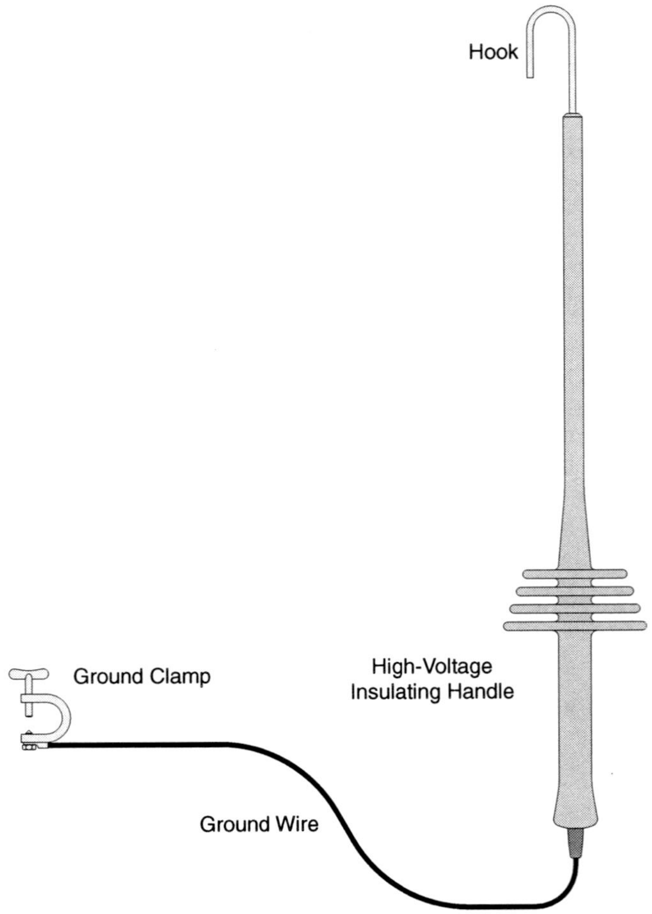

Figure 8-35 Grounding Hook or "Jesus Stick"

Faraday Cage

Protecting delicate equipment from external transients is the task of the faraday cage. A faraday cage is simply a conductive enclosure that is grounded. Any transient that reaches the cage is immediately conducted to ground. The equipment in the cage is completely protected by the cage. Figure 8-36

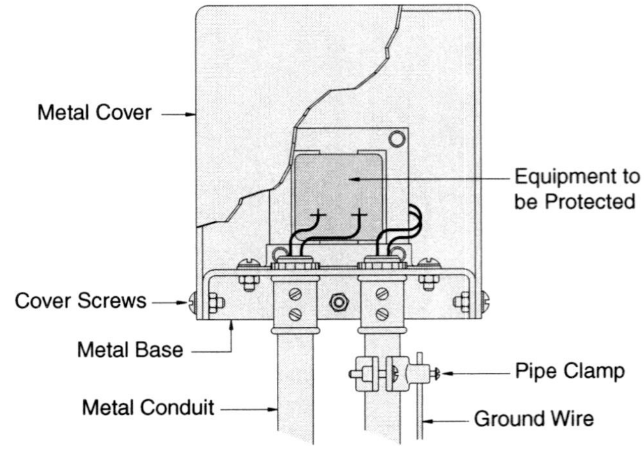

Figure 8-36 Faraday Cage

shows a typical faraday cage used to protect equipment exposed to high transient signals. Notice that the equipment is 100% enclosed within the housing. The pipe clamp and ground wire are connected to an earth ground.

When testing highly sensitive electronics, any external signals may produce erroneous results. Therefore, lager faraday cages can be built that shield the equipment from these stray signals. The cage, as shown in Figure 8-37 is constructed with a brass frame mounted to a grounded steel deck plate. The panels are brass screens that are soldered to the brass frame work. The door carries a special conductive seal. The inside of a cage like this is an extremely clean electronic environment.

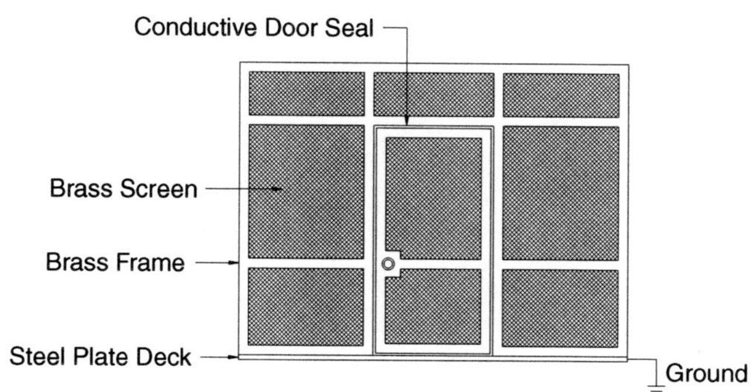

Figure 8-37 Faraday Cage for Servicing High Frequency Equipment

Lightning Protection

Lightning carries exceptionally high energies and the resulting strike damage can be severe. Buildings can burst into flames, electrical and electronic devices can be completely destroyed, metal fixtures can be melted, and a person can be instantly killed. When trees are struck by lightning, so much energy is absorbed that the water inside the trunks will flash vaporize and the trees will explode.

Lightning Rods

Most of us are familiar with lightning rods. We have seen them on top of buildings, barns, and radio towers. The idea behind the lightning rod is that the lightning strike is conducted to ground over a predetermined and, therefore, harmless path. Figure 8-38 shows a typical lightning rod that may be found on top of any modern building. The top of tall parking garages is an excellent place to inspect a lightning rod installation. Needless to say, inspections should only be carried out on a clear day with no atmospheric electrical activity. The rods are usually mounted along the perimeter of the highest part of the building. The rods are connected to a heavy ground wire which is connected to an earth ground. Do not connect a lightning rod to the internals of the building. The ground wire should be routed down the outside of the building.

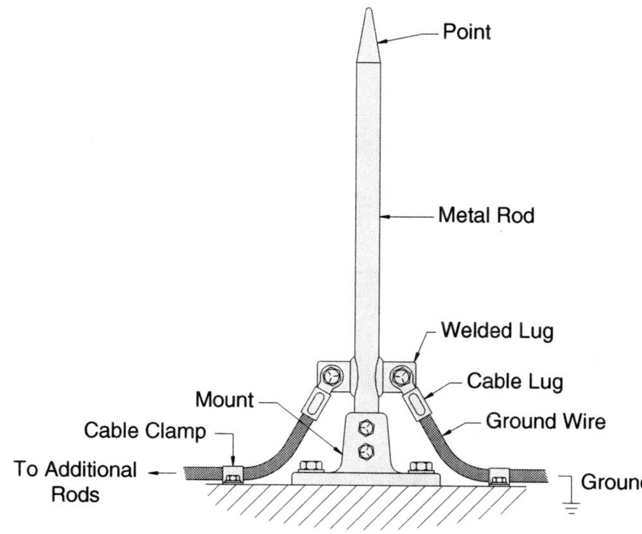

Figure 8-38 Lightning Rod

A lightning rod approximates a faraday cage. The volume that is protected is in the form of a 90° cone, referred to as the cone of protection. Figure 8-39 shows a lightning rod cone of protection. For some applications the pole height may be restricted by the area that requires protection. In these cases, multiple lightning rods are incorporated. Figure 8-40 shows a house that is protected with three rods.

When protecting timber structures great care should be taken in placing the ground wire and rod. Because of the extreme energies involved, lightning will oftentimes travel

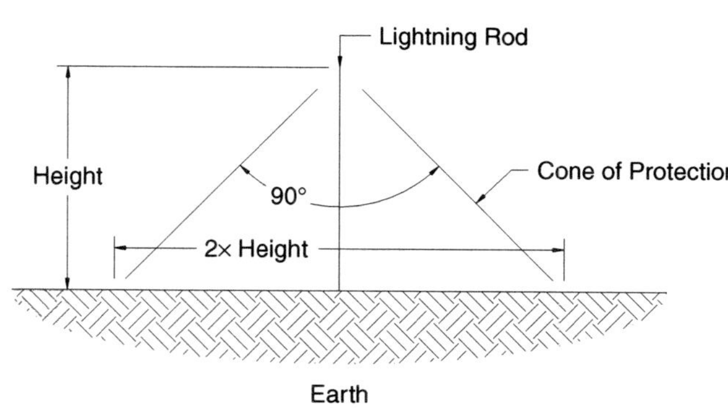

Figure 8-39 Lightning Rod Cone of Protection

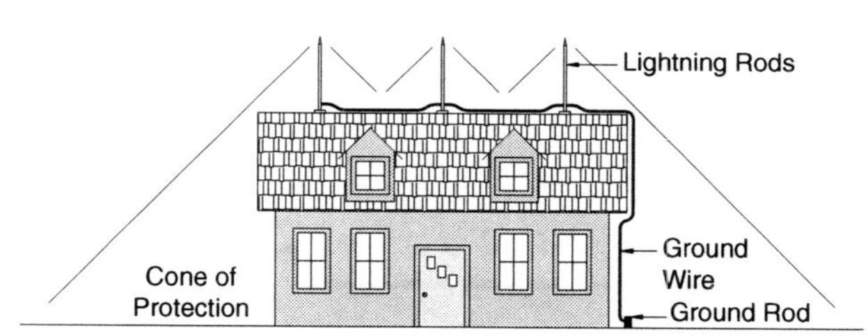

Figure 8-40 Multiple Lightning Rod Protection

outside and adjacent to the ground wire. In these cases the ground wire operates as an arc path and not necessarily as a conductor. If the wire is laid directly on a wooden surface, as shown in Figure 8-41, the arc can create a flash fire and the protection that the rod provides is principally negated. Notice that the arc path in the illustration is partly within the wooden structure. Figure 8-42 shows the proper placement of the ground wire. Notice that the arc path is isolated from the structure.

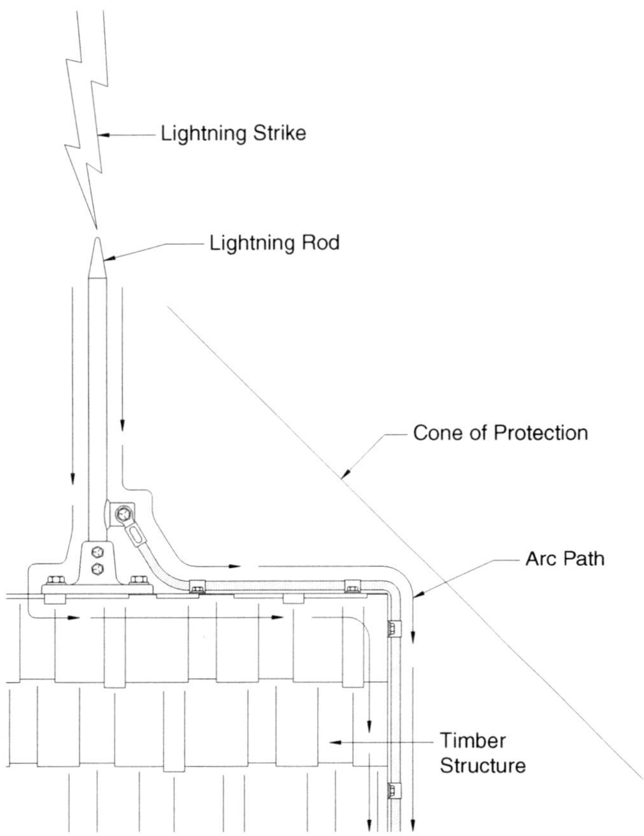

Figure 8-41 Incorrect Placement of Ground Wire

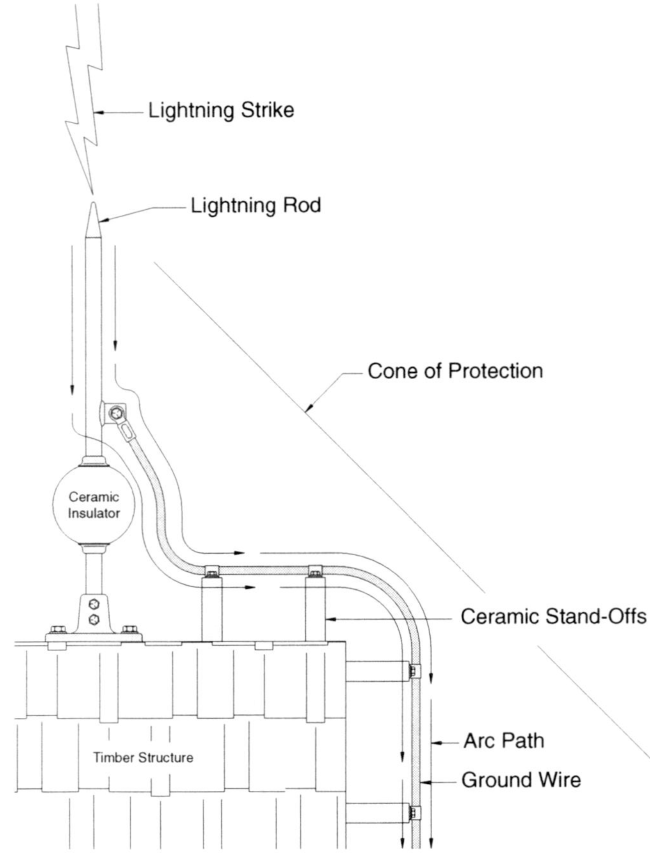

Figure 8-42 Correct Placement of Ground Wire and Rod

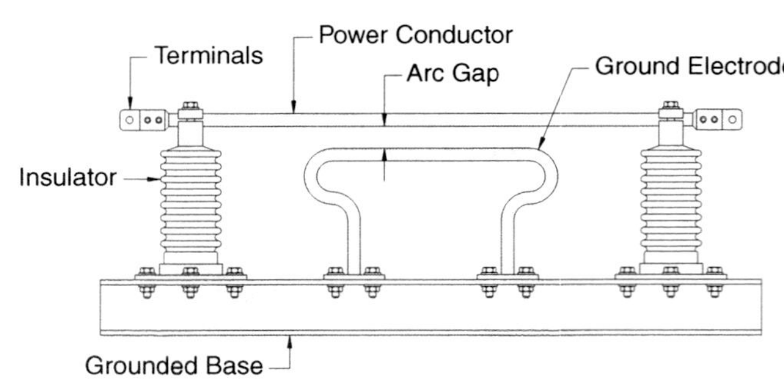

Figure 8-43 Lightning Arrestor

Lightning Arrestor

For large power distribution systems lightning strikes are a constant problem. On the occasion that lightning strikes a conductor, a larger portion of the energy can be controlled by using a lightning arrestor. Figure 8-43 shows a typical power transmission lightning arrestor. The power line is placed in close proximity to a parallel ground electrode. If lightning strikes the power line, it will arc to the ground electrode and the damage can be greatly diminished.

Protecting transmission lines is accomplished by running a top ground wire, as shown in Figure 8-44. The top wire is placed high enough to provide a cone of protection that encompasses the transmission lines. In some cases two top wires may be used. The top wire is connected to an earth ground at every pole.

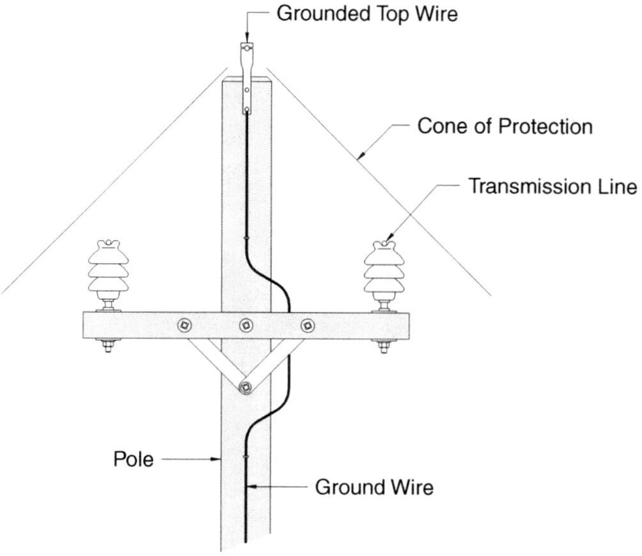

Figure 8-44 Lightning Protection on Power Transmission Pole

Ground Fault Interrupter

Ground fault interrupters are solid-state devices that monitor the current flow between the power line and the ground. If the current flow is above a preset limit, the interrupter automatically switches off the power. The acceptable current bleed rate to ground is extremely low and in this way the interrupter can monitor the condition of isolation in the circuit. Figure 8-45 shows a typical 120-VAC receptacle with an integral ground fault interrupter.

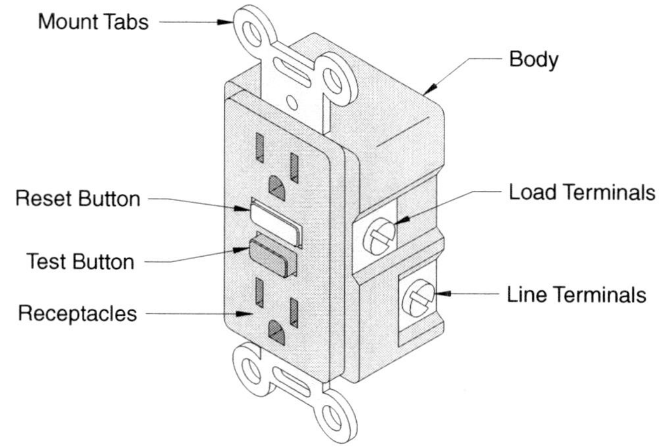

Figure 8-45 Ground Fault Circuit Interrupter Receptacle

CHAPTER 9
CONNECTORS

When considering the type of electrical connectors to use in any system, one is faced with a dizzying array of different designs to select from. In the absence of any real knowledge about what applications different connectors are designed for, it can become a daunting task to sift through the literally thousands of different configurations currently on the market. This chapter of the book reviews the different methods and the most common connectors used for connecting electrical circuits. It will provide base level knowledge and make the task of selecting connectors and connection types a little less daunting.

Twisted Connections

Figure 9-1 shows the most common way to join two conductors, the twist splice. There are two basic methods to prepare a twisted splice, the basic twist and the Western Union splice. Both of these splices should be soldered after they are complete. The Western Union splice was originally designed for joining telegraph lines and was usually made up without solder. After either splice is soldered, a protective insulating coating should be applied, generally electrical tape or heat shrink.

Wire Nuts

Wire nuts are used to make the basic twist splice quicker, more reliable, and solder free. Wire nuts, as shown in Figure 9-2, are extremely common in home, commercial, and industrial power distribution systems. A wire nut will typically have three elements for joining the wires, a thread insert, a transition, and insulation threads. The thread insert is normally a diamond-shaped piece of spring steel coiled into a taper and molded into the body. The sharp edges of the shape cut into the wire and the taper forces the wires together. This assures a high quality, low resistance connection. Larger wire nuts have grip wings molded into the body to provide the necessary torque to join heavier wires. The transition is a cone that is intended to guide the wires into the wire threads. The lower threads are intended to engage the wire insulation and prevent it from sliding back and exposing the conductor. The body of the wire nut is made from high-impact plastic.

Although electricians will typically allow the wire nut to twist the conductors together, this is not recommended. To properly make up a wire nut connection the conductors should be pretwisted as shown in Figure 9-3 and the wire tightened on.

Wire nuts are also available with an additional pigtail, as shown in Figure 9-4. These units can be very handy for

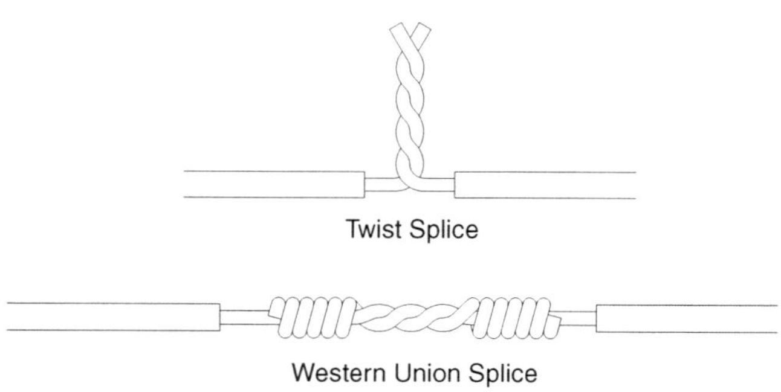

Figure 9-1 Wire Splices

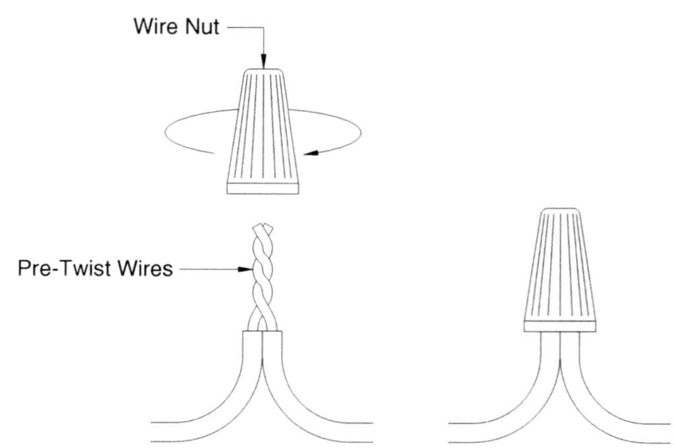

Figure 9-3 Wire Nut Splice

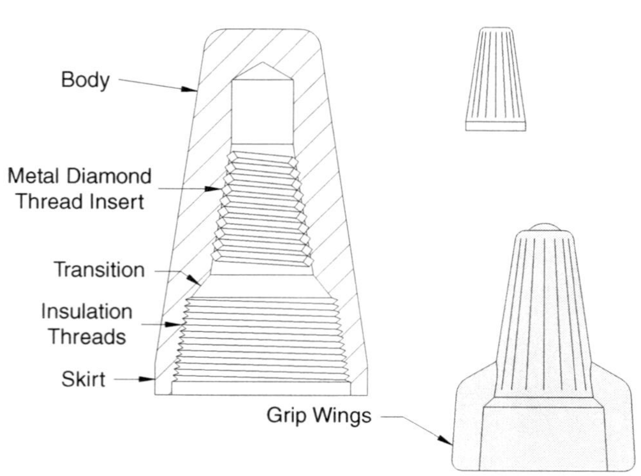

Figure 9-2 Wire Nuts

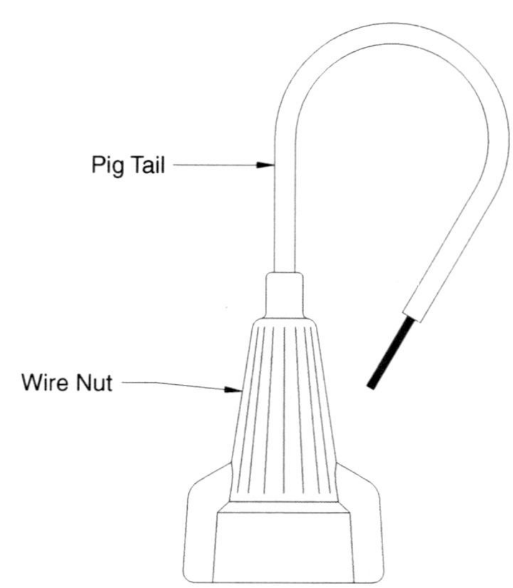

Figure 9-4 Wire Nut with Pig Tail

adding a circuit to a connection that is already at its maximum wire count. These types of wire nuts are also useful for connecting ground loops to receptacles, switches, and conduit.

Wire nuts are usually color-coded so that their capacity is easily gauged. The chart in Figure 9-5 shows standard wire nut colors and their wire capacities. It is generally recommended that a selection of wire nuts be kept on hand at all times. Many home improvement stores sell wire nut kits that have a supply of the various sizes arranged in individual compartments in a handy plastic box.

Color	Range (AWG)	Wire (Min.)	Wire (Max.)
Gray	22-16	(2) 22	(2) 16
Blue	22-14	(3) 22	(3) 16
Orange	22-14	(3) 22	(2) 14 with (1) 18
Yellow	18-10	(1) 14 with (1) 18	(1) 10 with (1) 14
Red	18-10	(2) 14	(4) 12 or (2) 10

Figure 9-5 Wire Nut Color Chart

For joining larger conductors or large bundles of smaller wires, bolted wire nuts are used. Figure 9-6 shows a typical bolted wire nut. The unit is made from a U-shaped saddle that is threaded on both sides. A clamp block and nut are used to force the conductors together. These wire nuts provide an excellent, high-current connection and are commonly found in industrial applications.

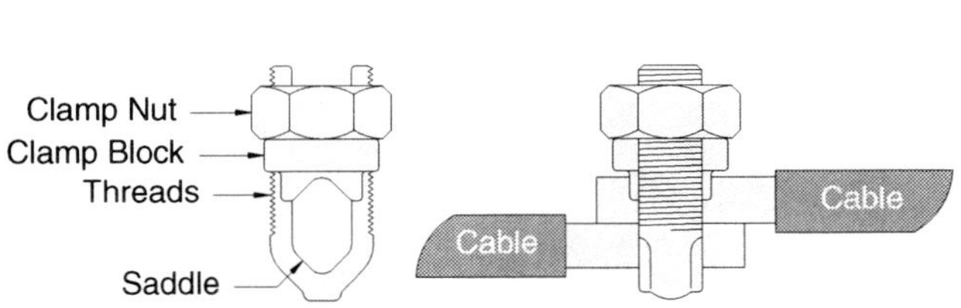

Figure 9-6 Bolted Wire Nut

Because the bolted wire nut is not insulated, it must be coated after make up. The standard method is to thoroughly wrap the connection with friction or "tar" tape. After the connection is fully insulated, the friction tape should be wrapped with electrical tape. Figure 9-7 shows a properly wrapped connection.

For low-voltage, high-current applications, such as arc welding, set screw connectors are a convenient way to join two conductors. Figure 9-8 shows a set screw wire connector. The cables are stripped and inserted into the brass body. The set screw is tightened down and an excellent high-current connection is made. After the cables are in place, a cover is slid over the body and a lock screw is inserted.

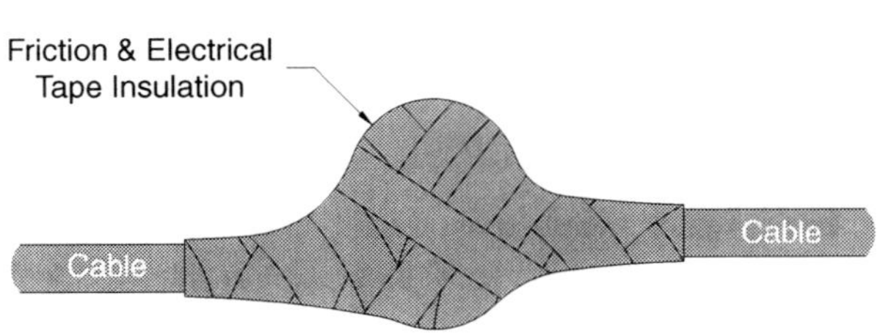

Figure 9-7 Taped Wire Nut

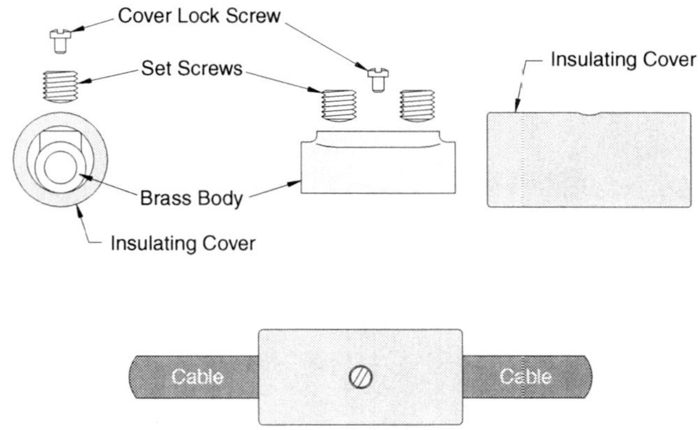

Figure 9-8 Set Screw Wire Connector

Crimp Connections

Crimping is the defamation of a conductor in order to force a connection with a wire. A crimp connection is typically a cylinder of metal that a wire is inserted into. The cylinder is crushed and permanently captures the wire. Figure 9-9 shows a typical crimp lug. Crimping is preferred because of the speed at which the connection can be made. A typical crimp can be made in just a few seconds and most crimp lugs also have an integral insulator. A solder joint, on the other hand, requires considerably more effort and the insulator can only be applied after the joint has cooled down.

Figure 9-10 shows a variety of crimping tools. Most of us have seen electrical utility pliers at the local hardware store. These are very handy tools and are recommended for any tool box. My personal favorite is the aviation crimper. These pliers only crimp bare lugs, but provide a superior small wire crimp. Ratchet crimpers are typically found in production applications. These crimpers are designed for small wire crimping and usually carry an adjustable crimp force.

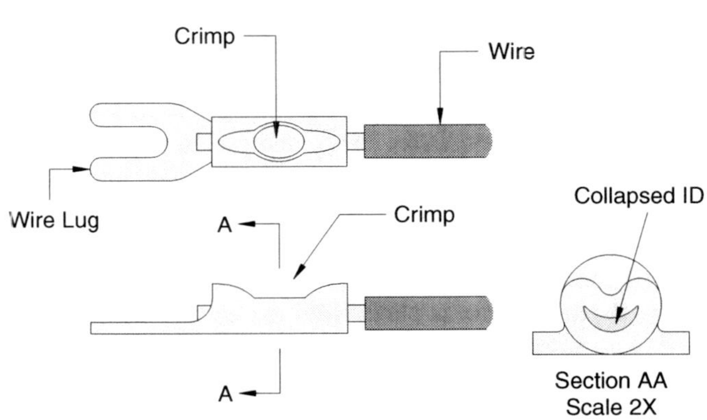

Figure 9-9 Crimp Lug

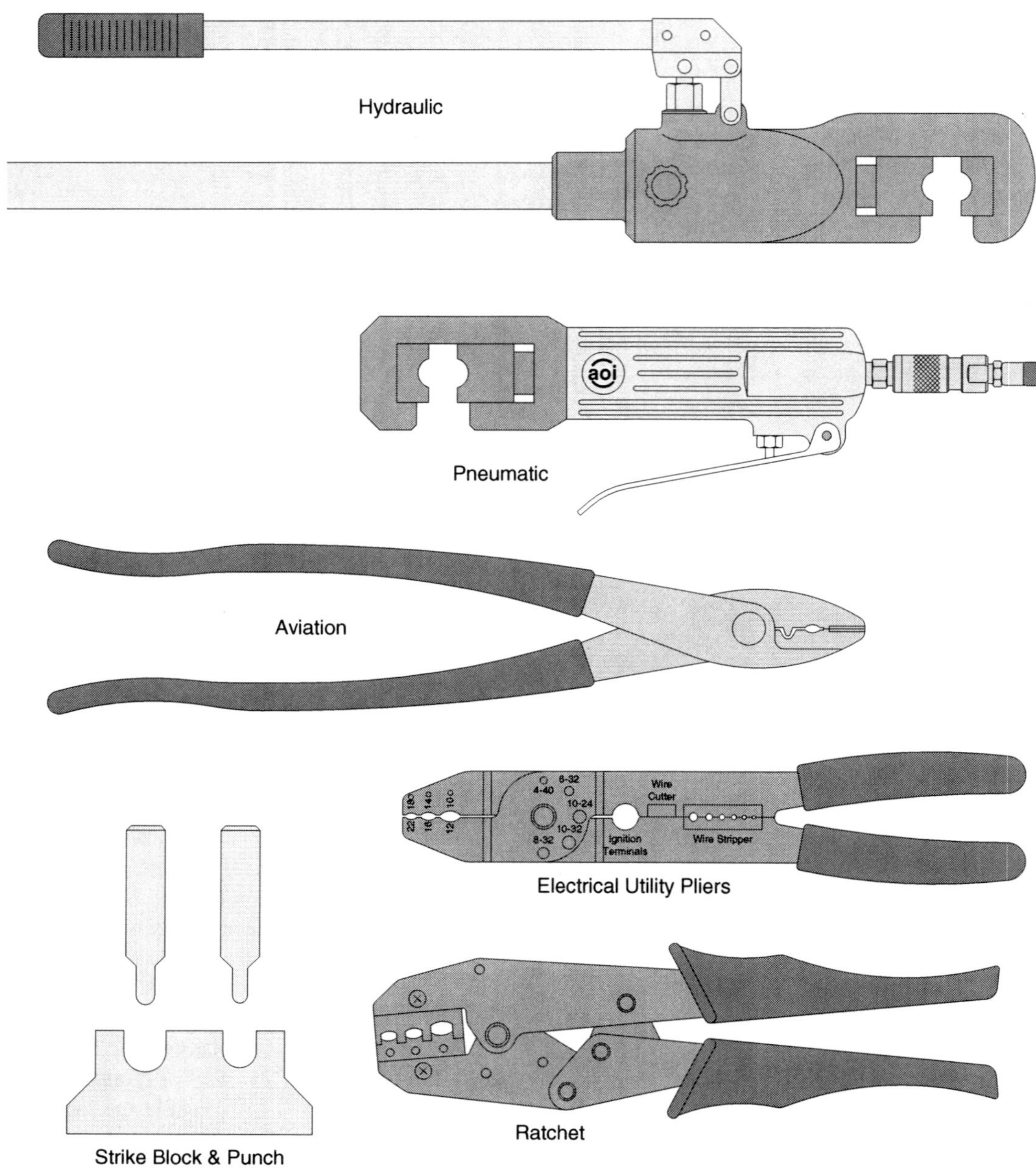

Figure 9-10 Crimping Tools

Pneumatic crimpers are generally found in production facilities that manufacture high-current equipment, where they are usually set up for crimping large cables. Hydraulic crimpers are used for crimping the largest conductors. These units are used in field service and installation applications. They are not particularly quick; however, they can produce a suitable large wire crimp from a small, light weight tool. For large, fine wire conductors, strike block and punch sets are used. These consist of a block that the lug is placed into and a punch, which is hammered into the lug to form the crimp.

For large, fine wire conductors, primarily welding cables, lugs are manufactured with integral crimp blocks as shown in Figure 9-11. The wire is placed into the lug and a hammer is used to complete the crimp. These types of lugs are very popular among industrial maintenance personnel.

Crimp lugs are available in virtually any style. Figure 9-12 shows a few of the more common lug types that are manufactured.

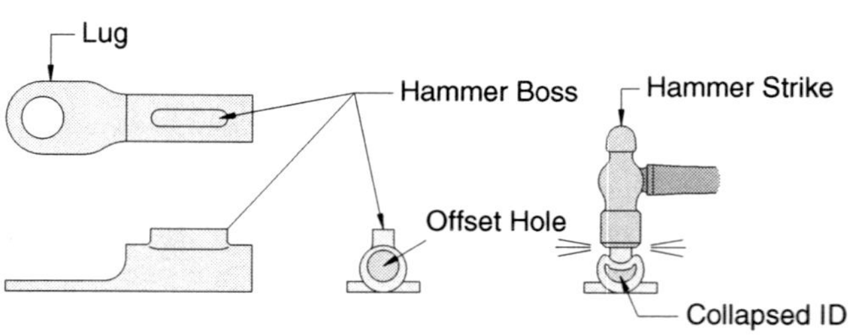

Figure 9-11 Cable Lug with Integral Crimp Block

Solder Connections

The basic solder connection should have two elements, a mechanical joint and a soldered joint. This is intended to provide redundancy in the event the solder bond fails. Figure 9-13 shows the two steps for making a solder joint. A good rule of

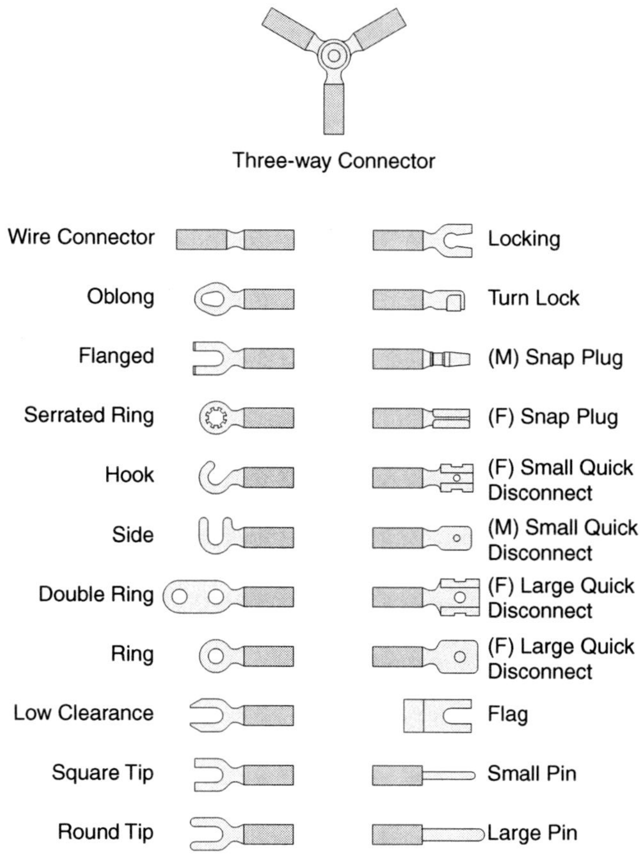

Figure 9-12 Various Crimp Lugs and Connectors

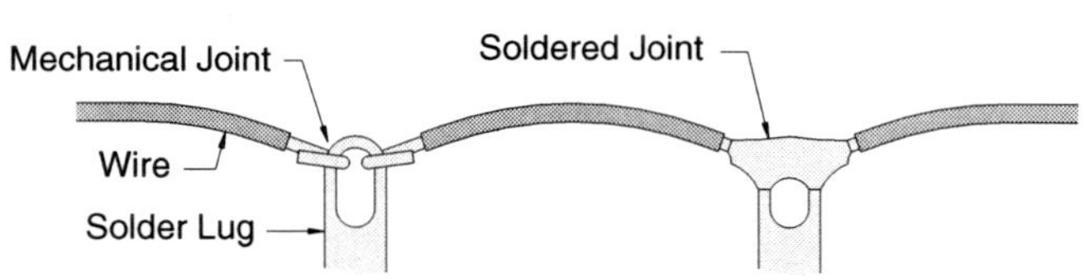

Figure 9-13 Solder Connection

thumb is if the wire will fall off the joint before solder is applied, it's not a suitable mechanical connection.

Figure 9-14 shows a typical solder terminal strip. These units were the mainstay in electronic manufacturing during the vacuum tube era. These units are still available today; however, printed circuit boards have all but eliminated them in modern electronics.

Solder lugs differ very little from crimp lugs. If the insulator is removed from a crimp lug, you have a solder lug. Figure 9-15 shows a typical solder lug. The wire is inserted

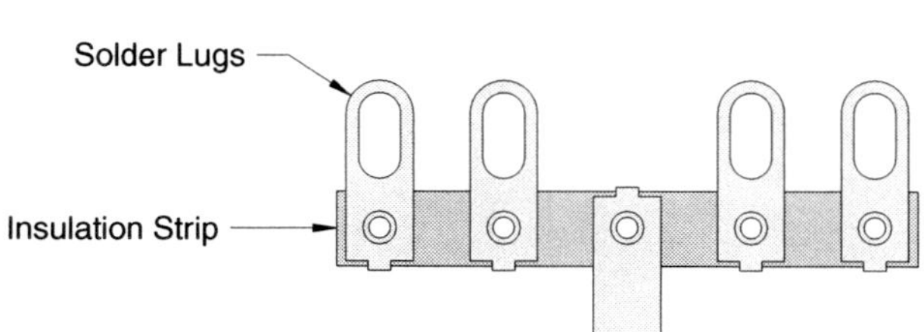

Figure 9-14 Solder Terminal Strip

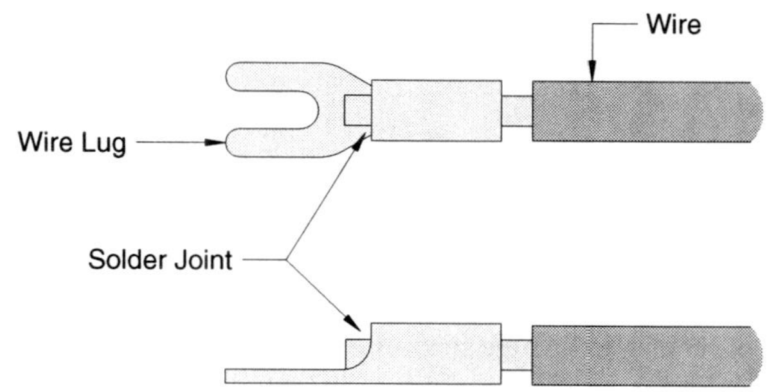

Figure 9-15 Solder Lug

into the barrel and the joint is flooded with solder. After soldering it is important to remove all of the excess flux before insulating the joint.

Printed circuit (PC) board connections are usually a copper pad that is bonded to an insulating board. A wire is simply placed into the hole and the solder is flooded into the joint. Edge pads are also provided for connecting wires, as shown in Figure 9-16. Edge pads do not provide a mechanical element to the connection and should only be used where redundancy is not critical.

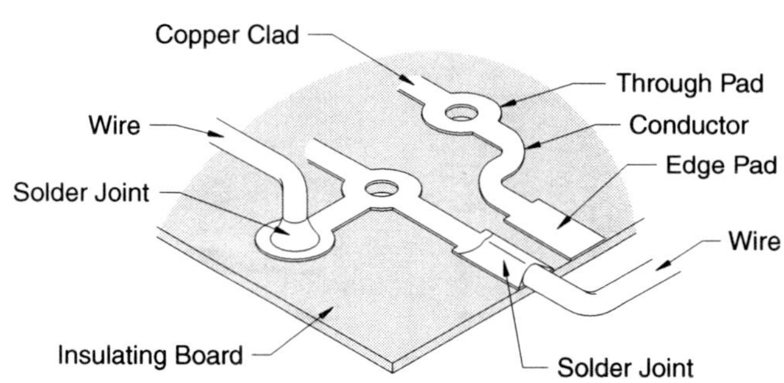

Figure 9-16 PC Board Solder Connection

Socket solder joints are typically found in pin connections. They consist of a hollow cylinder that the wire is inserted into and then flooded with solder. Because of the depth of the joint some type of air relief is required. Some pins have a hole in the tip that the wire protrudes from and some have a small vent hole on the side, as shown in Figure 9-17.

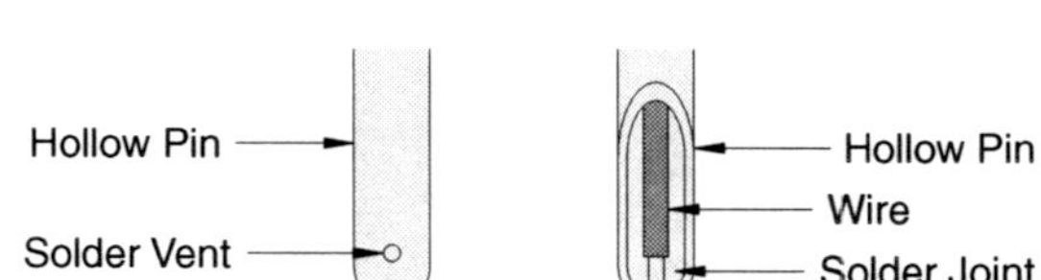

Figure 9-17 Socket Solder Joints

Ground lugs should always use solder joints. A good mechanical/solder joint should be applied to a loop-type lug. This provides the necessary redundancy that any ground connection should have. Figure 9-18 shows a typical lug made specifically for grounding applications. Note the serrated screw hole is to provide an improved connection to the chassis. It is best to rivet these lugs to the chassis.

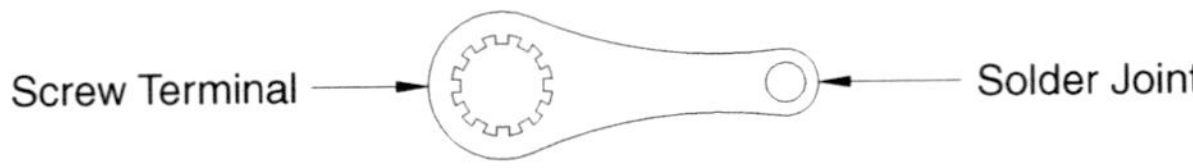

Figure 9-18 Ground Lug

Soldering irons are generally rated in watts. Lower wattage irons are used for delicate electronics that may be damaged by applying too much heat. These units also have small tips that allow the iron to be used in cramped applications. Figure 9-19 shows two typical low-wattage, electronic soldering irons. The top illustration shows a basic, general purpose unit. The bottom illustration shows a precision unit with temperature control.

Medium sized units in the 60 to 150 watt range are used for general applications, such as automotive, machine, and appliance wiring. Figure 9-20 shows two typical solder irons in the medium size range. The top illustration shows a dual-

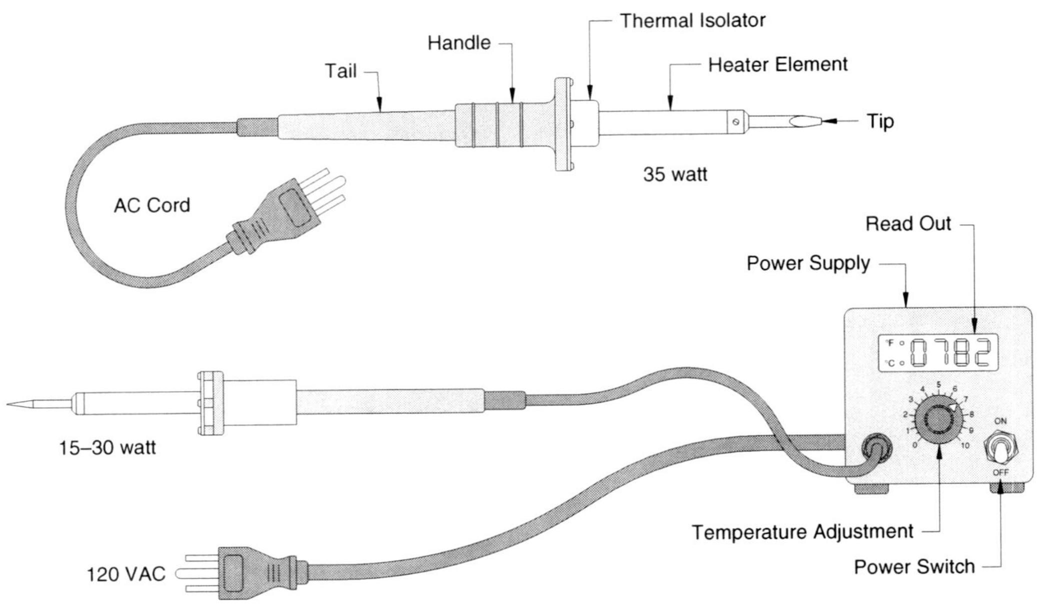

Figure 9-19 Electronic Soldering Irons

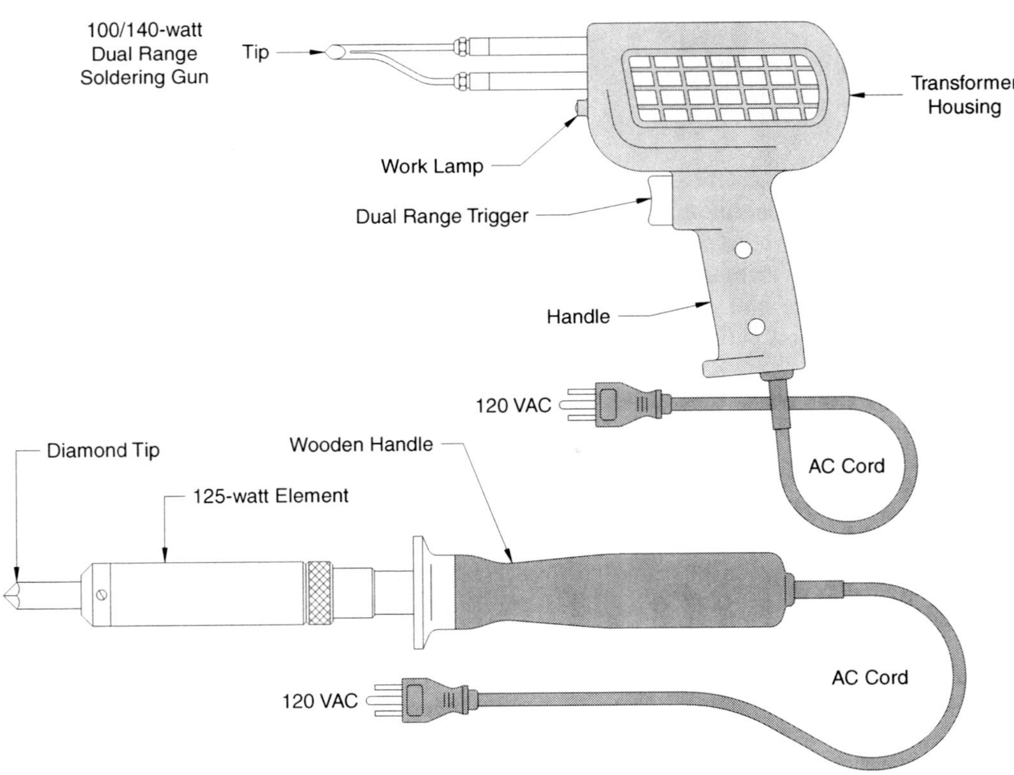

Figure 9-20 General Purpose Soldering Irons

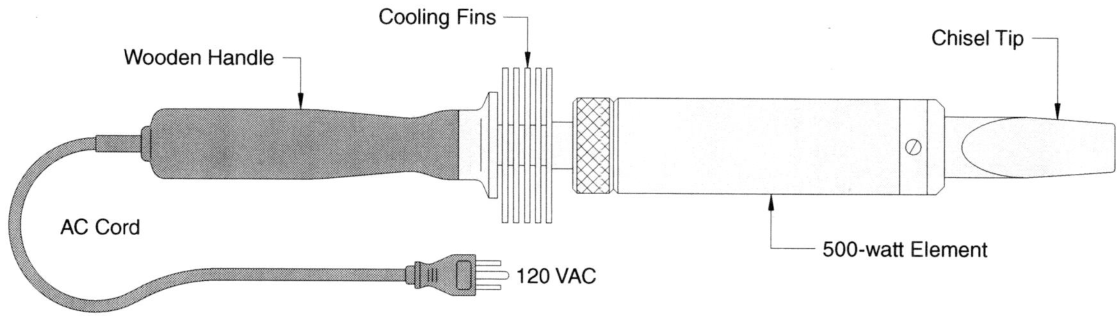

Figure 9-21 Heavy Duty Soldering Iron

range soldering gun. These units are very handy tools. They have the attribute of heating up only when the trigger is pushed, which makes them very friendly. They usually have an integral light so the work area is properly illuminated whenever the iron is in use. The bottom illustration shows a more traditional 125-watt soldering iron. A unit like this can provide a powerful heat source and produce very rapid solder joints.

Soldering irons are also available with very large heating elements. Figure 9-21 shows a 500-watt soldering iron with a broad chisel point. These irons can produce solder joints on very large cable. They can also be used to solder buss bars and brass blocks for high-current applications.

For the largest applications, resistive soldering machines are specified. These machines are available in sizes as high as 2500 watts. The work piece is clamped between the electrodes of the hand piece and low voltage, high current is flowed through the item to be soldered. As the work piece heats up, solder is applied directly to the joint. Resistive soldering machines are generally used for the largest cables and buss bars. Figure 9-22 shows a commercial resistive soldering machine.

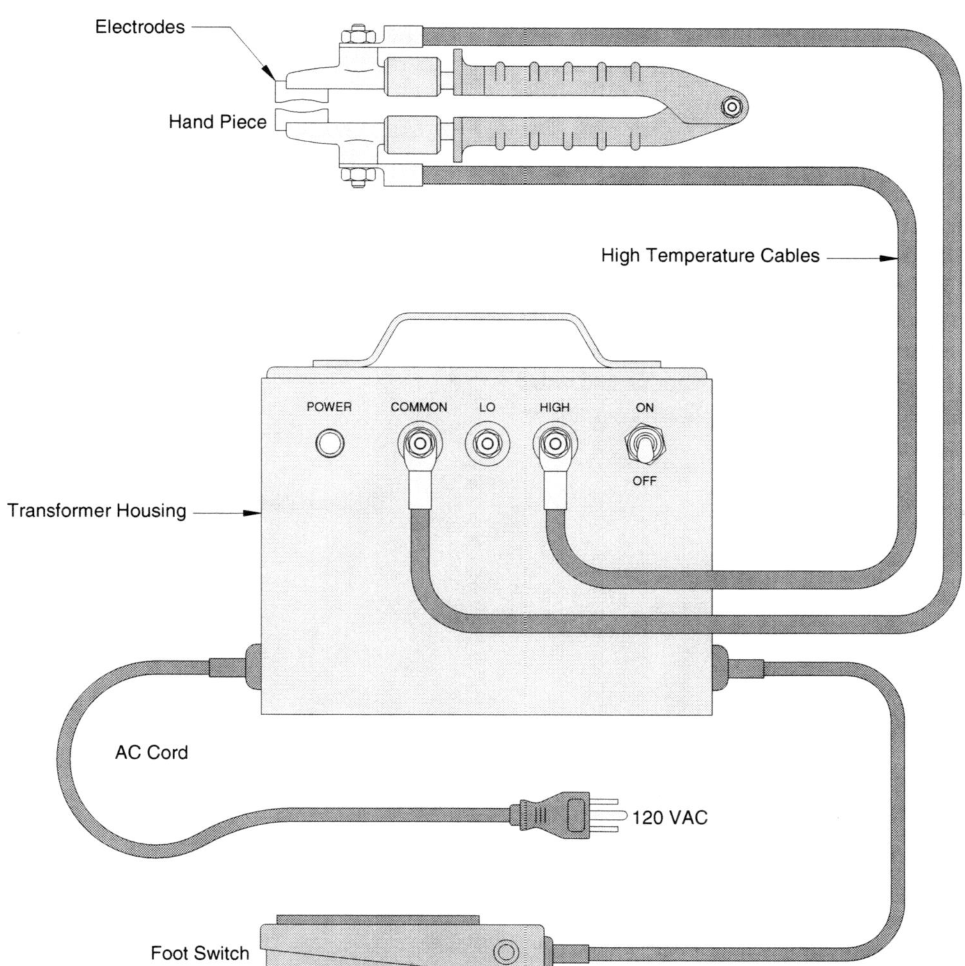

Figure 9-22 Resistive Soldering Machine

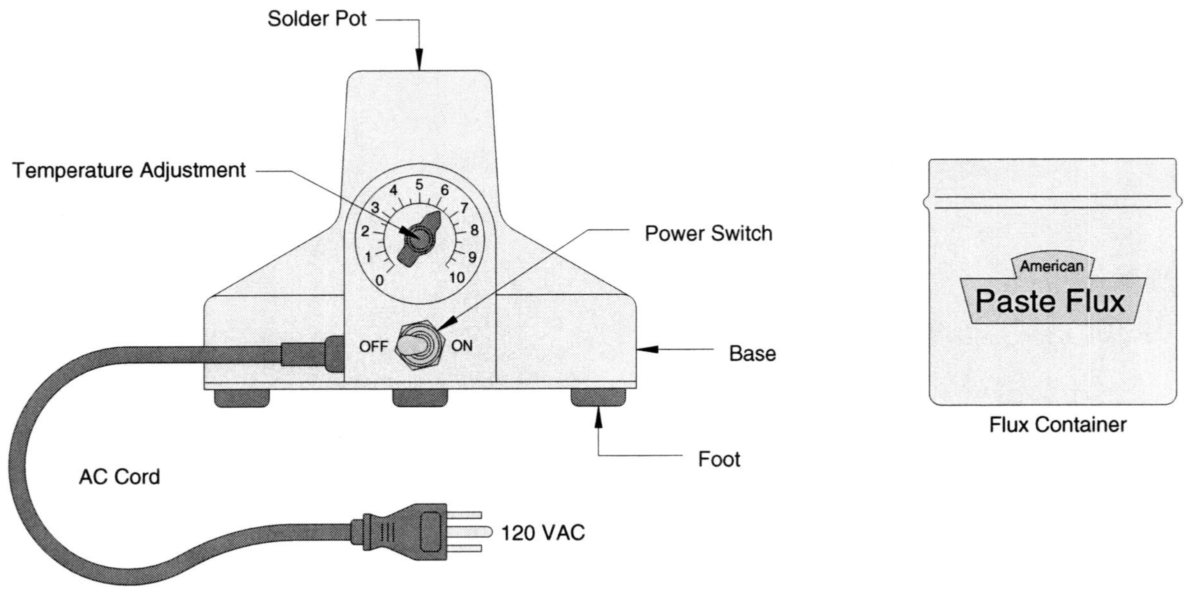

Figure 9-23 Solder Pot

Tinning is the process of coating a component with solder. This is a common step during production soldering operations. Solder pots, as shown in Figure 9-23, are used to heat a pool of solder. A flux container is placed adjacent to the solder pot. A stripped wire is dipped into the flux and then into the solder and is instantly tinned. The tinned wire is now prepared for a production solder joint.

Most solder intended for electrical use is $^{60}/_{40}$ flux core. The $^{60}/_{40}$ refers the ratio of tin to lead that makes up the alloy. $^{60}/_{40}$ solder is 60% tin and 40% lead. The flux, which is sometimes referred to as rosin, is carried in a circular core at the center of the solder. Solder that does not have a flux core requires that the flux be applied before soldering the joint. Figure 9-24 shows three different solder cores. Multicore is generally found in very fine solder intended for electronic work. Flux core is usually used for general wiring applications and solid core is principally specified for heavy applications such as large cables and buss bars.

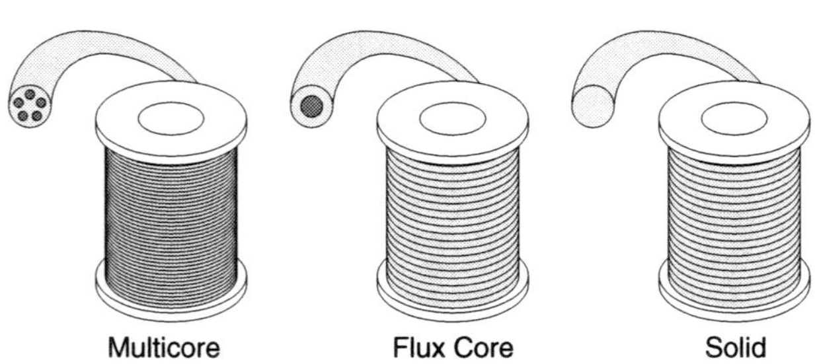

Figure 9-24 Solder Cores

Disassembling a solder joint can be rather difficult, especially if the joint was done correctly. Some technicians heat the joint and forcibly blow the molten solder off. This works if a source of compressed air is available and you don't mind spraying down adjacent components with molten solder. Sometimes, this is the only effective way to remove solder from large joints, but is not recommended for smaller, high-density assemblies. Figure 9-25 shows three common solder removal tools. A solder wick is used to draw the solder from the joint. As the joint and wick are heated, the solder flows into the tinned wire braid and is drawn away from the joint. When the wick cools the solder-filled end is cut off and discarded. A solder sucker is a heavy rubber ball with a high temperature tip. The bulb is squeezed and the tip is placed at the

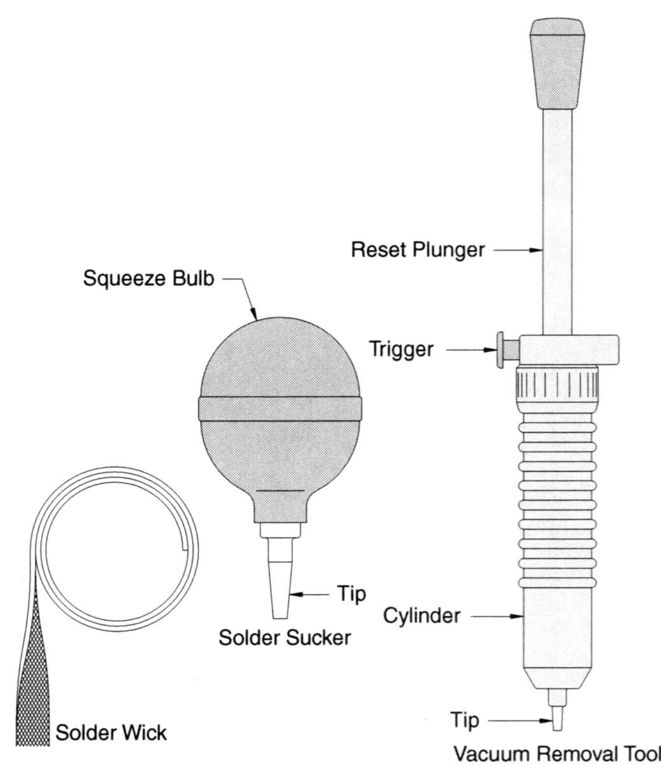

Figure 9-25 Solder Removal Tools

joint to be desoldered. After the joint is heated, the bulb is released and the solder is sucked off the joint. Solder suckers can be difficult to use and they don't provide enough power for some joints. The preferred solder removal tool is the vacuum removal tool. These units operate in the same fashion as the solder sucker, except that they have a spring loaded cylinder. The piston is cocked into the down position. When the trigger is pressed the piston is forced up and a high vacuum flow is generated at the tip. These tools are very inexpensive and do an excellent job of removing solder from electronic and electrical solder joints.

Delicate components can be severely damaged during soldering operations due to overheating. To protect these components, heat sinks are attached to the wire between the component and the solder joint. As the joint is soldered, the heat travels up the wire until it encounters the heat sink. It then flows into the heat sink instead of the delicate component. Figure 9-26 shows two examples of commercial heat sinks designed for soldering operations.

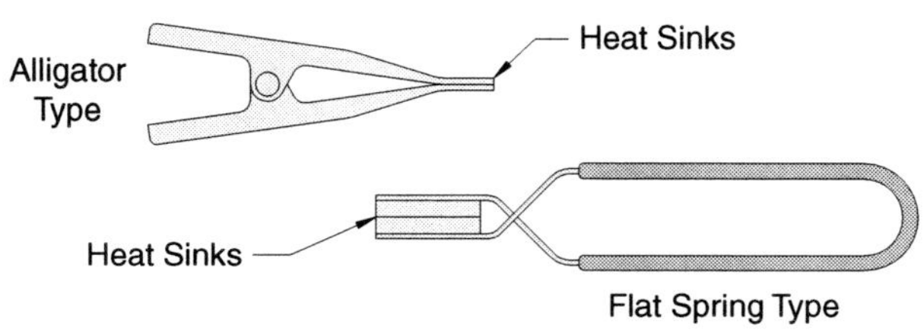

Figure 9-26 Soldering Heat Sinks

Binding Posts

The most fundamental connector is the binding post. Binding posts have been around for as long as electricity has been utilized by man. Binding posts may not represent the most elegant method of connecting conductors; however, they do represent the most reliable. Nothing produces a detachable electrical connection like an old fashioned nut and bolt.

Figure 9-27 shows a basic binding post arrangement. A brass screw is attached to an insulating board and wire lug with two flat washers and a nut. A thumb nut is installed onto the top of the screw. To attach a wire, simply wrap it around the screw and tighten the thumb nut. For a more permanent connection, the thumb nut can be replaced with a hex nut and tightened with a nut driver.

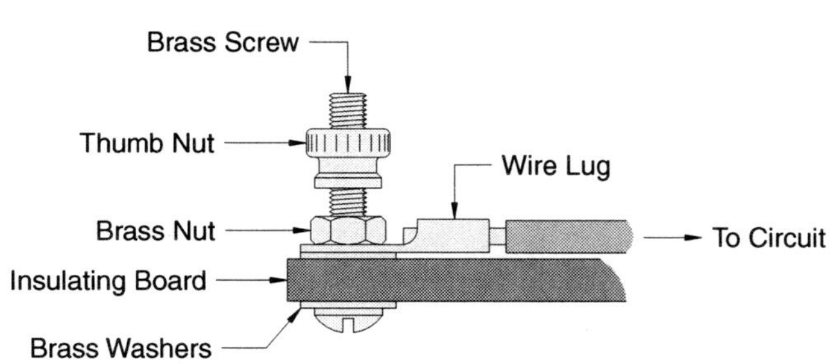

Figure 9-27 Brass Screw Binding Post

The flat spring binding post, as shown in Figure 9-28, is a very old standard. The idea is to press down on the tab until the wire hook protrudes through the element. A wire is placed under the hook and the tab is released. The spring forces the wire against the hook and a solid electrical connection is made.

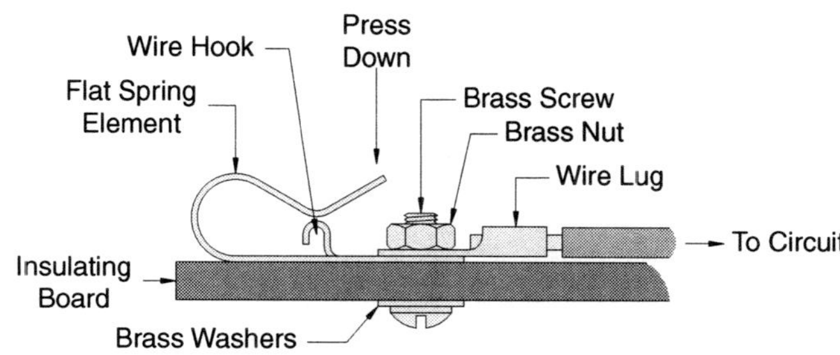

Figure 9-28 Flat Spring Binding Post

The coil spring binding post takes the flat spring unit a step further. A plastic button is pressed down, which, inturn, compresses the coil spring below the through hole. A wire is inserted through a slot in the button collar and through the wire connection. When the button is released, the wire is forced into contact with the inside of the hole and a quality electrical connection is made. Figure 9-29 shows a coil spring binding post.

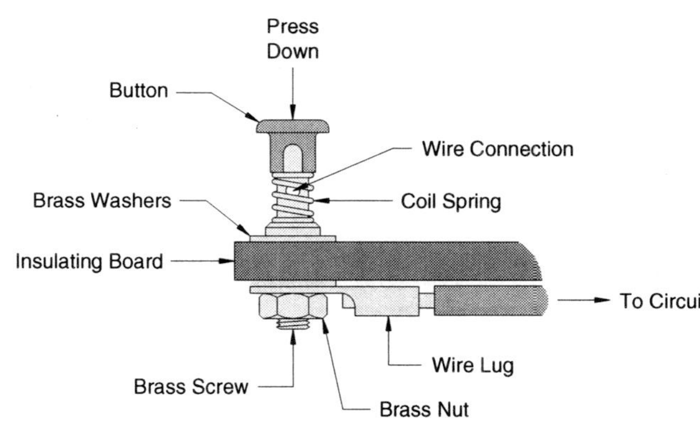

Figure 9-29 Coil Spring Binding Post

Screw tight binding posts are intended to provide the convenience of a spring post and the connection quality of a nut and bolt post. Figure 9-30 shows a typical screw tight binding post. The thumb nut is screwed open, which lifts the clamping block and opens the wire connection. A wire is placed in and the nut is tightened.

The combination binding post, shown in Figure 9-31, carries the best features of all posts. It has a screw tight thumb nut with a wrap-around and through wire connection. The nut is insulated and the top of the post carries a standard banana jack. These units are generally supplied with a set of isolation washers so that they can be mounted in a metal panel. Permanent connection can be made with a wire lug or to the

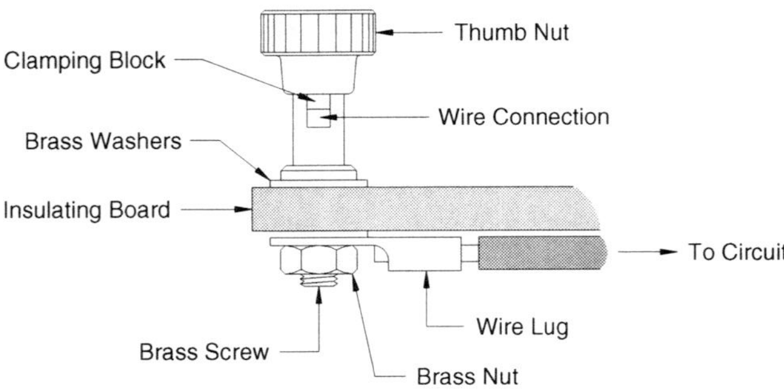

Figure 9-30 Screw Tight Binding Post

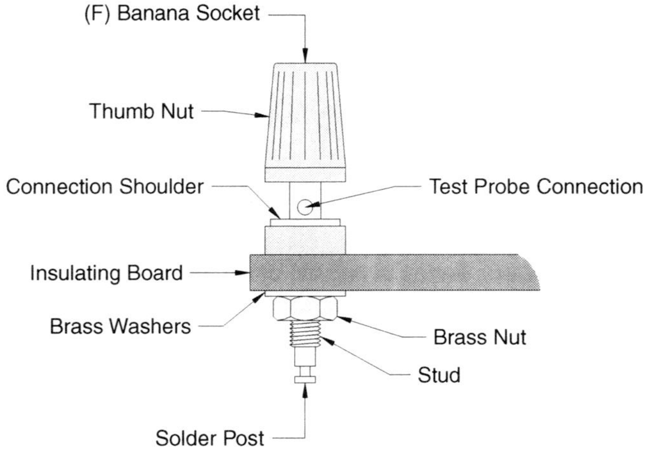

Figure 9-31 Combination Binding Post with Banana Socket

solder post. These units are very inexpensive and are, definitely, the preferred choice for binding posts.

Instrument and Test Connectors

The two most common connectors in the test and instrumentation world are the banana plug and the BNC connector. These two connectors are found on nearly every piece of test equipment and on most instrumentation. They both provide a broad operational range, coupled with extremely durable construction, at a very low cost per unit.

Banana Plugs

Banana plugs are primarily a single conductor plug and jack arrangement that provide a high-current, low-resistance connection in an easy-to-connect package. Figure 9-32 shows a few banana plug and jack arrangements. Most banana plugs have a no-solder wire connection that is optimized for standard test lead wire. Most jacks are supplied with an integral solder post. Grounding jacks are all metal construction. Dual banana plugs are set up on 0.75 inch centers and the plug will carry a polarity tab, as shown.

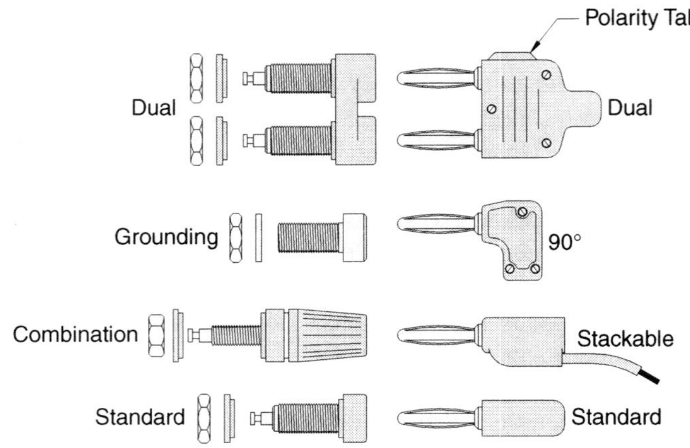

Figure 9-32 Banana Plugs

BNC Connectors

BNC stands for *Bayonet Neill Concelman* after the man who designed the connector in the 1940s. It was originally developed as a miniature version of the type CRF (type "C" radio frequency) connector. Over the years it has become the fundamental connector for test equipment and instrumentation. It provides excellent RF characteristics, a 500-VDC rating and particularly good shielding from stray electrical signals. It does not, however, have a very good current-carrying capability. The connector set has male and female sides and is mated by pushing the male connector onto the female and rotating the collar one quarter turn. The collar provides a tactile feedback when the locking point is achieved. Figure 9-33 shows a few BNC configurations and adaptors. Because the BNC connector is so common, adaptors are readily available for almost every standard connector made.

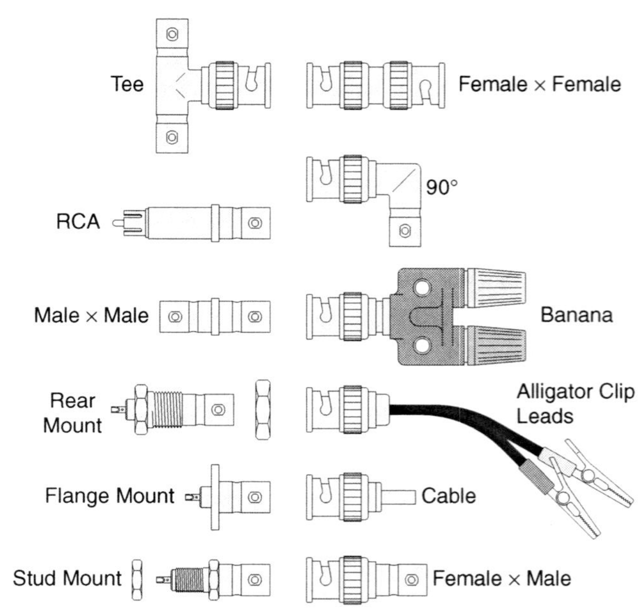

Figure 9-33 BNC Connectors and Adapters

Variations on the BNC are the MHV and the SHV connectors. These are high voltage versions of the BNC and stand for *miniature high voltage* and *safety high voltage* respectively.

The MHV has two significant shortcomings. First, if enough force is applied it can be made to mate with a standard BNC connector. Unfortunately, forcing these two connectors to mate will severely damage both units. The only recourse is to replace the damaged connectors. The second drawback in a safety issue is when using these connectors with a live circuit, high voltage is exposed to the operator and electrocution is a very real hazard. Although there many MHV connectors on all types of test equipment and instrumentation, they should not be used unless absolutely necessary.

To solve the shortcomings of the MHV, the SHV connector was developed. These connectors will not mate with either BNC or MHV units and provide voltage protection when working with live circuits. Safety high voltage connectors are easily identified by the circular spring set that protrudes from the center of the male connector. The female SHV is considerably longer than a BNC or MHV connector. For any high-voltage applications, the SHV should be exclusively selected. Figure 9-34 shows a comparison between MHV and SHV connectors.

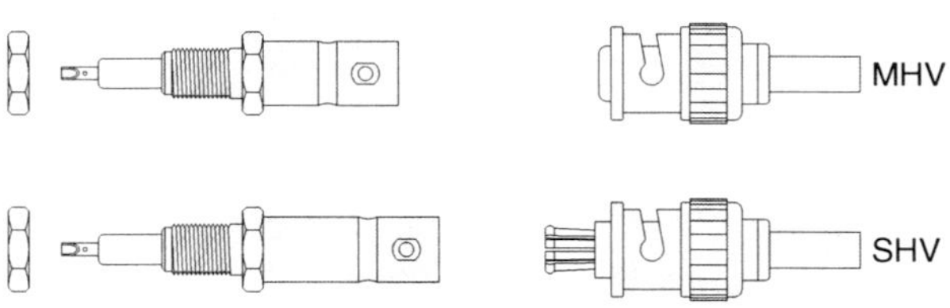

Figure 9-34 MHV and SHV Connectors

Radio Frequency (RF) Connectors

When dealing with RF power in such applications as radio and television, special connectors must be used. These connectors are specifically designed to deal with the unique problems associated with RF energies, such as leakage and stray signals.

The most common RF connector is the type F. These connectors are used extensively on cable TV connections. They are a small threaded connector that is specifically designed to mate with RG-59-U cables, discussed in Chapter 10. Figure 9-35 shows a few common configurations for type F connectors. A push-on version is available for applications that require frequent connect/disconnect operations.

Figure 9-36 shows a few *subminiature* size RF connectors. This class of connector is generally utilized on the internals of RF equipment.

Figure 9-37 shows an assortment of *medium* size RF connectors. This size range of connectors is commonly found on amateur, commercial, and marine radio communications equipment. The ultra high frequency (UHF) design is the connector of choice for citizens band (CB) radios.

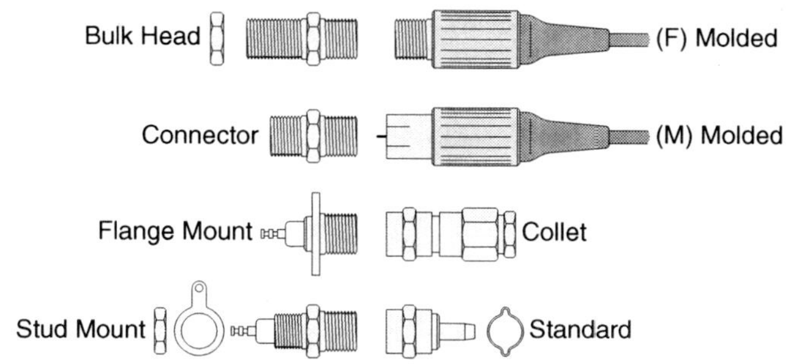

Figure 9-35 Type F Connectors

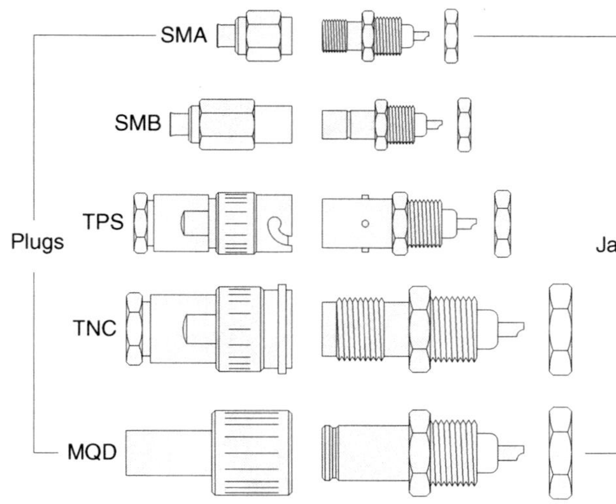

Figure 9-36 Miniature and Subminiature RF Connectors

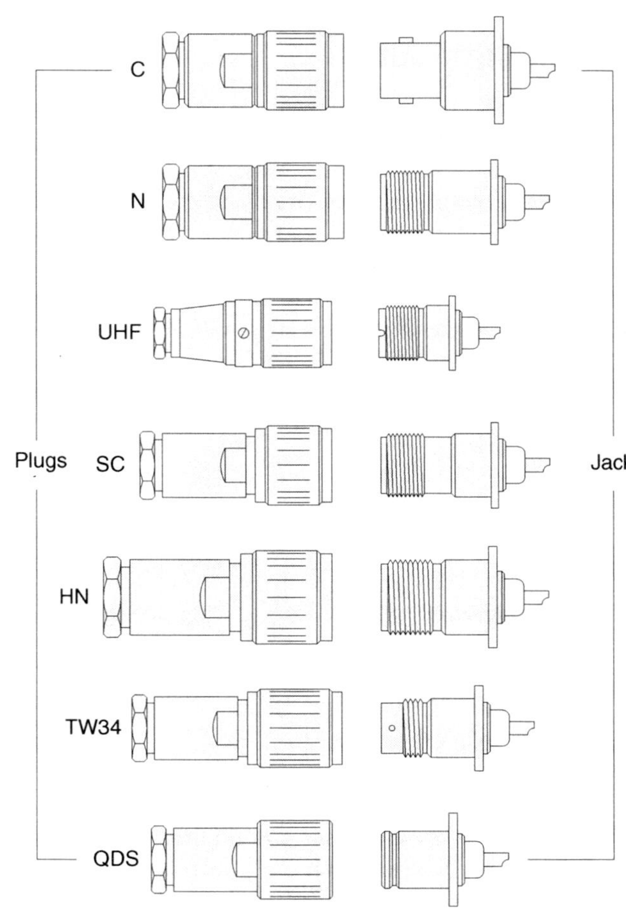

Figure 9-37 Medium Size RF Connectors

Radio frequency connectors in the *large* size range are designed for higher frequencies and power levels. These connectors are found on high-power radio transmitters and military equipment. The G874 connector is unique because it is the only RF connector that is a unisex design. Figure 9-38 shows various large RF connectors.

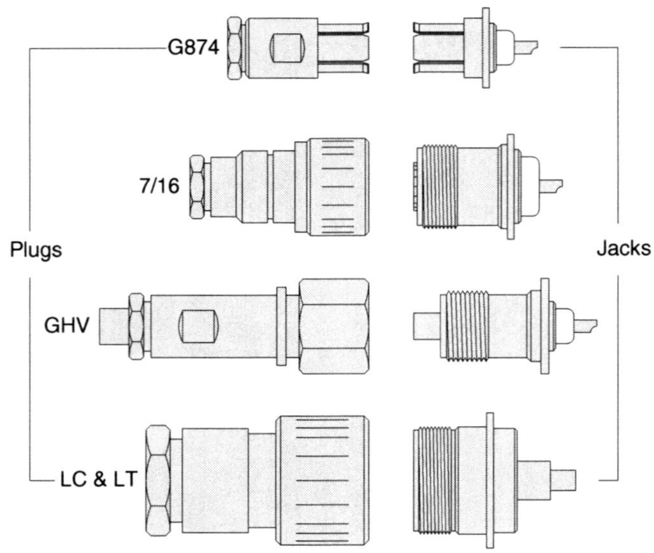

Figure 9-38 Large Size RF Connectors

Audio Connectors

Within the audio community there are only four connectors commonly used. These are the RCA, $^1/_4$-inch phone, $^1/_8$-inch phone, and the XLR connectors. Most of us have experience with the RCA and phone connectors in reference to our home stereos. The XLR connectors are primarily found on professional recording and public address systems. XLR stands for Type **X** connector, with **L**atch and **R**ubber surrounded terminals. Figure 9-39 shows a few common RCA connector configurations. This style of connector is very inexpensive and provides excellent performance with the sensitive signals found in typical audio equipment.

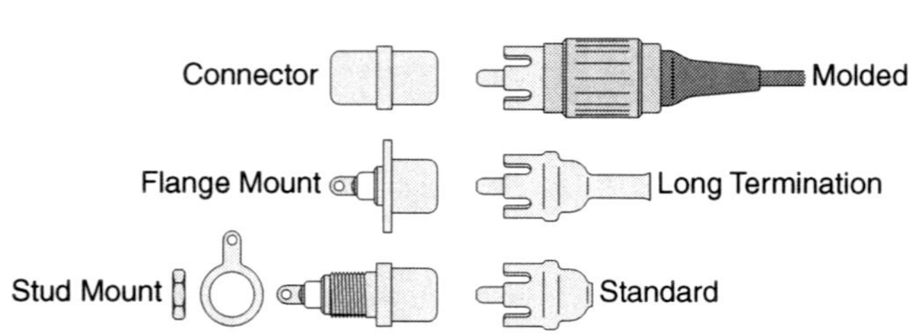

Figure 9-39 RCA Connectors

The $^1/_4$-inch phone connector was originally designed for switch board applications with early telephone systems. It has, however, proven to be a very adaptable design and finds favor in a broad range of audio applications. The original

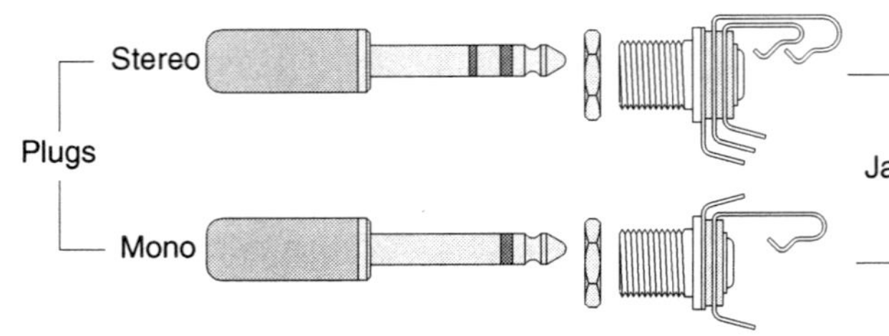

Figure 9-40 $^1/_4$-Inch Phone Connectors

$^1/_4$-inch phone jack was a two-pole unit. With the advent of stereo equipment a third pole was added and the design lives on. Figure 9-40 shows both mono and stereo $^1/_4$-inch phone plugs and jacks.

As audio equipment was miniaturized, the $^1/_4$-inch phone jack became a little unweilding in size. To accommodate the smaller equipment format, the basic $^1/_4$-inch phone jack was shrunk to half of its original size and designated the $^1/_8$-inch phone jack. This is the connector that is used on most portable cassette and CD players. Figure 9-41 shows both mono and stereo $^1/_8$-inch phone plugs and jacks.

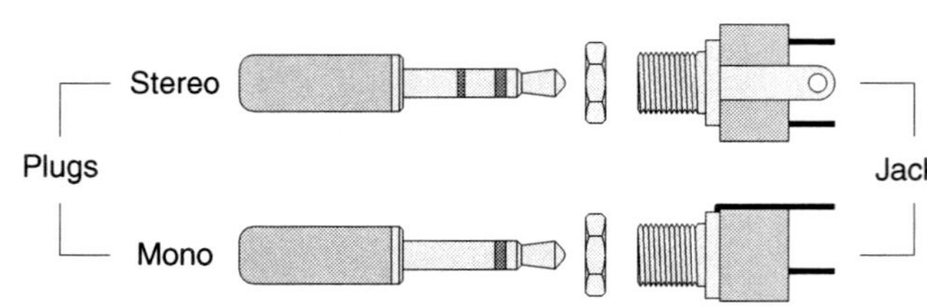

Figure 9-41 $^1/_8$-Inch Phone Connectors

The XLR connector, as shown in Figure 9-42, is a particularly versatile audio connector. It is a three-pin design with a fully shielded housing. The plug incorporates an automatic locking mechanism which must be manually released to disconnect the connector. Male and female versions are available for panel mount or cable assemblies. These connectors are an excellent choice for public address (PA) equipment as they are very durable and stand up well to years of service. They are appropriate for low-level signals (microphone), intermediate signals (preamp outputs, tone controls, and so forth), and low-power amplifier outputs.

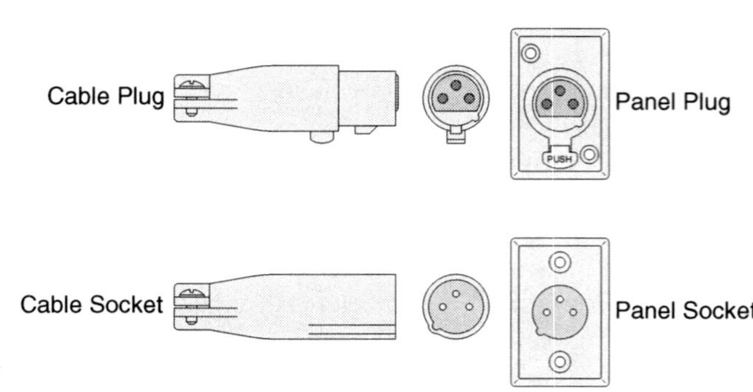

Figure 9-42 XLR Phone Connectors

Data Connectors

In our digital lives, data connectors have become omnipresent. Most notably they are used throughout our personal computers and the telephone system. They are also found on all manner of equipment that relies on digital controllers.

The DB (type **D**-su**B**mintire) series connectors are one of the most common connectors within the digital world. The DB designation is followed by a number that is representative of the number of pins in the connector. That is a DB 9 has 9 pins, a DB 25 has 25 pins, and so on. The HD 15 is a special version that is normally used to connect VGA computer monitors. The DB series connectors also have a set of locking screws on either end of the connector. The plug will have a set of screws and the jack will have a set of matching nuts. The plugs and jacks are available in either male or female versions. These connectors also have applications for low-level signal processing, test equipment, and instrumentation. Figure 9-43 shows a side view and pin arrangements for the most common DB connectors.

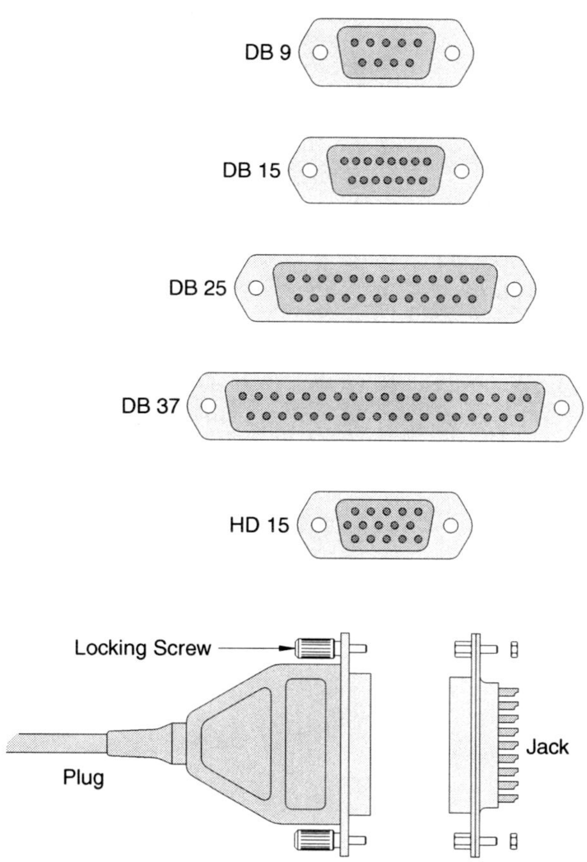

Figure 9-43 DB Connectors

The Centronics 36 connector is most commonly found as a parallel connector on printers. The male connector carries two cutouts on either end which correspond with a pair of locking clips on the female connector. The plugs are mated and the clips are snapped in place. These connectors also have applications for low-level signal processing, test equipment and instrumentation. Figure 9-44 shows a Centronics 36 connector set.

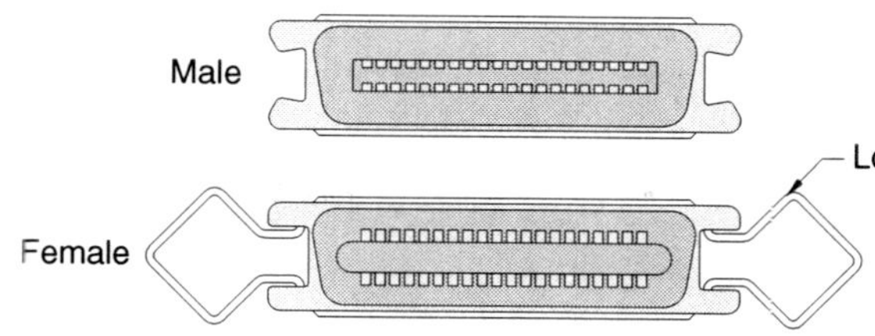

Figure 9-44 Centronics 36 Connector

Universal serial bus (USB) connectors have become very popular with personal computers. This port has two different connectors associated with it. Figure 9-45 shows the type A and B USB connectors and a pin out chart.

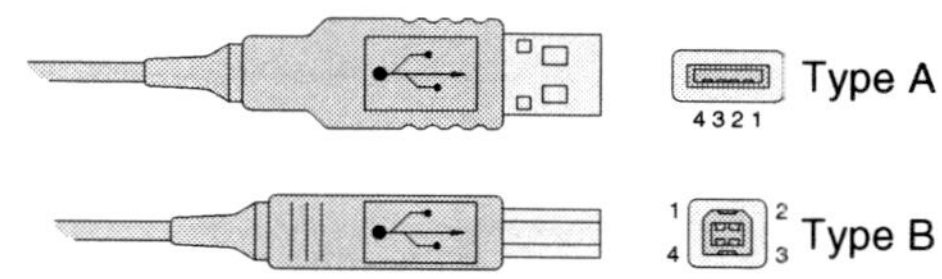

Pin	Name	Description
1	VBUS	+5 VDC
2	D–	Data–
3	D+	Data+
4	GND	Ground

Figure 9-45 USB Connectors

DIN connectors (**D**eusches **I**nstiut fur **N**ormung) are most commonly found as the plug on your mouse and keyboard. They are, however, used in all types of control environments including audio, test equipment, and instrumentation. Oftentimes a manufacturer will replace a standard plug with a DIN connector just to maintain a proprietary design. Figure 9-46 shows examples of both standard and mini DIN connectors.

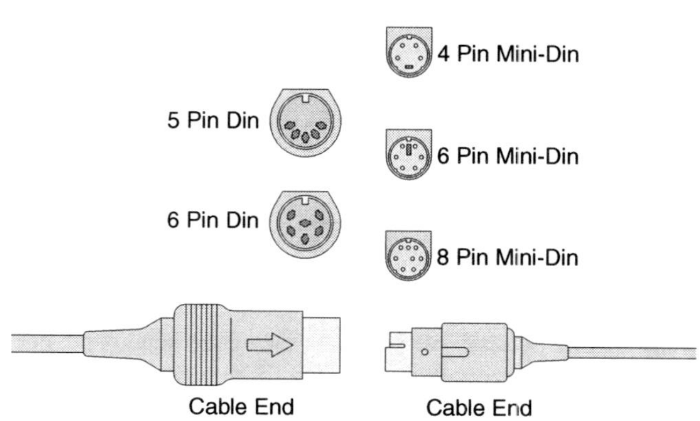

Figure 9-46 DIN Connectors

Registered jacks (RJ) connectors, are commonly found connecting our telephones. The RJ-10-2 is used to connect the headset to the receiver and the RJ-11/14 and RJ-12 are used to connect the receiver to the wall panel. RJ-48 is commonly used for Ethernet connections. These connectors have poor current-carrying capabilities and are only useful for low-level signals. Figure 9-47 shows standard RJ connectors and pin configurations.

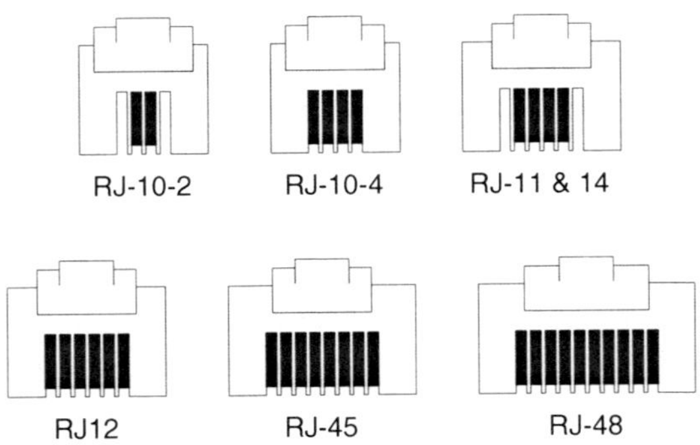

Jack	Pins	Pins Installed
RJ-10	4	All or 2,3
RJ-11 & 14	6	2,3,4,5
RJ-12	6	All
RJ-45	8	All
RJ-48	10	All

Figure 9-47 RJ Series Connectors

PC Board Connectors

Edge connectors are an excellent method of interfacing to digital and control electronics. The PC board is designed with a series of pads along one edge, as shown in Figure 9-48. A key slot is cut into the board to assure proper alignment of the connector.

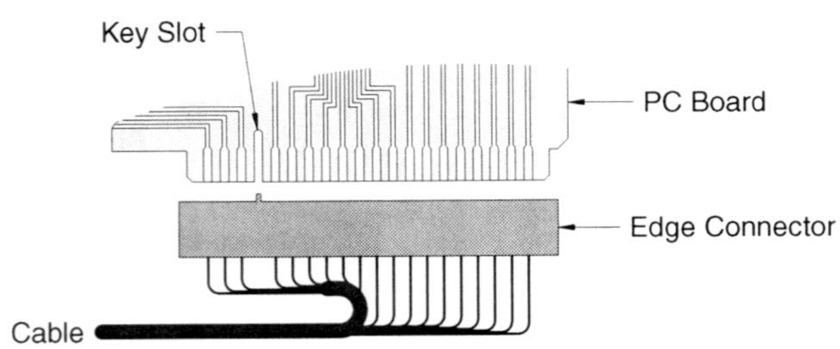

Figure 9-48 PC Board Edge Connector

Many edge connectors are designed with ribbon cable terminators. The ribbon cable is inserted into the connector and the clamp head is pressed into place. As the clamp head is pressed, it forces the ribbon cable into the pin edges, which in turn cuts through the insulation and forms a connection with the conductors. Figure 9-49 shows a typical ribbon cable snap connector.

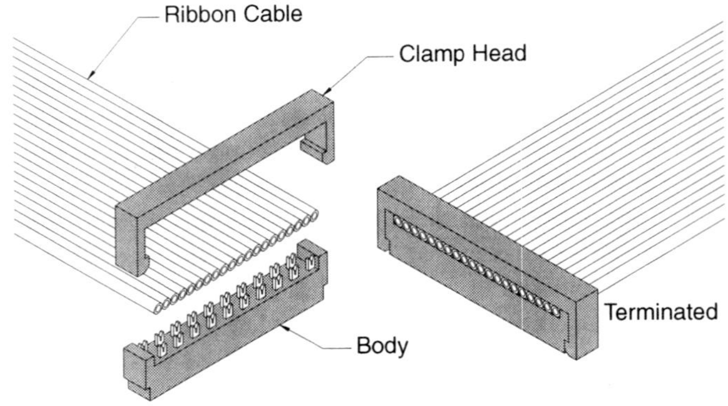

Figure 9-49 Ribbon Cable Snap Connector

General Purpose Connectors

Multipin connectors can be found in nearly every piece of electromechanical equipment ever manufactured. The judicious application of connectors can make final assembly and service a very simple task. Connectors also provide facilities to easily test, tune, and troubleshoot sub-assemblies. An excellent example of connectors in this role is the electrical systems in modern automobiles. Virtually every component in these systems is connected via a multipin connector. This "black box" approach makes manufacturing and service very friendly.

Collar lock connectors generally represent the highest quality connectors in this category. These connectors are available in every conceivable pin configuration and with screw-on collars or bayonet collars. They are available with plastic or metal housings and in waterproof versions. Figure 9-50 shows a few multipin collar lock connectors. It should be noted that nearly every collar lock connector on the market is also available in a nonlocking version as shown.

Probably the most common multipin connectors are the modular series. These are the white plastic connectors that are commonly found in computers and home appliances. They are available in a number of pin configurations and current

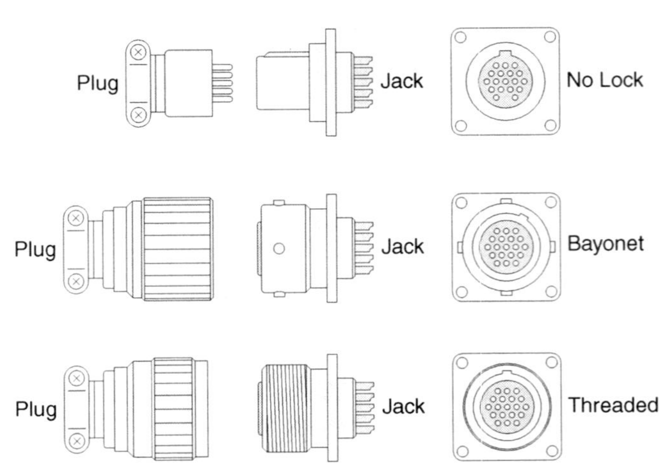

Figure 9-50 Multipin, Collar Lock Connectors

ratings. These connectors are designed for use on the internals of machines or appliances that are not subjected to harsh environmental conditions. The connector is supplied with the pins separate. The wires are connected either with a crimp or solder joint and snap inserted into the body. A special tool is required to remove the pins. Figure 9-51 shows an example of an eight-pin modular connector.

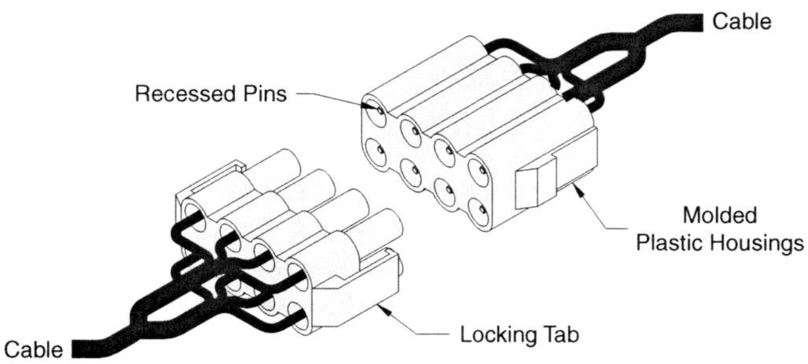

Figure 9-51 Modular Connector

Jones connectors are one of the age old standards in multipin connectors. They have been used in all manner of equipment and are still commonly found, however, modern designs have principally replaced these connectors. The design is based around a bake-a-lite base with flat "spade" style pins. A metal housing is affixed to the base via two small screws or rivets. The housing also carries a cable clamp and strain relief. Figure 9-52 shows a typical example of a six-pin Jones connector.

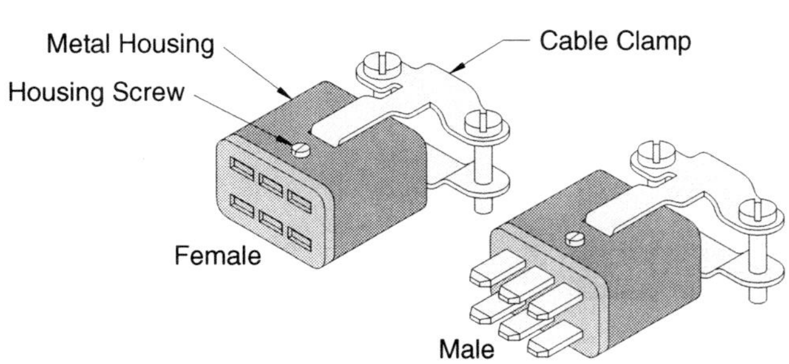

Figure 9-52 Jones Connector

Another vintage standard is octal connectors. These connectors were primarily used for tube sockets in the 40s, 50s, and 60s. Unlike the Jones plug, the octal plug has become one of the standard sockets for modern control relays. These connectors are available in 8- or 11-pin versions. They are particularly durable connectors and are easy to plug and unplug. Another useful item that the market offers are small cabinets with integral octal connectors. These allow small subassemblies to be constructed and simply plugged into a standard relay socket. Figure 9-53 shows both an 8-pin and 11-pin octal connector set. Also shown is an octal cabinet assembly.

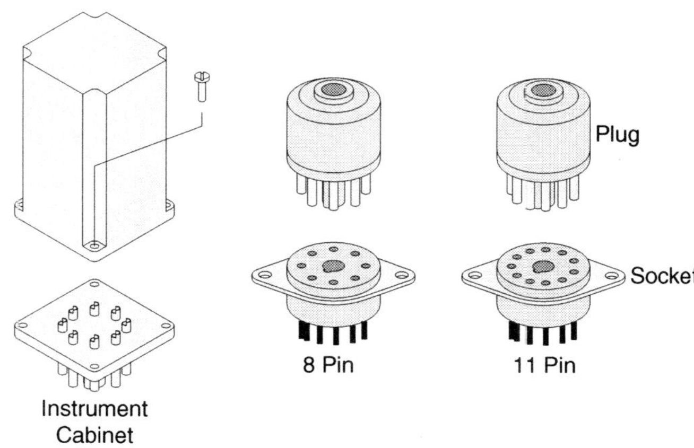

Figure 9-53 8- and 11-Pin Octal Connectors

AC Connectors

Most of us are familiar with the standard 120-VAC connector. We are aware that there are two-prong versions and three-prong versions, with the third prong providing a ground. Most modern 120-VAC equipment is supplied with a three-prong plug unless the appliance is *double insulated.* Figure 9-54 shows a few standard 120-VAC connectors. The wall receptacle is the same unit that you would find in your bedroom. The panel unit is intended for equipment applications.

Figure 9-54 Standard 120-VAC Connectors

Figure 9-55 shows the standard 240-VAC connector. These plugs are less known because 240 volts is not common for small appliances. The most common use for these receptacles is to provide power for window-mounted air-conditioning units.

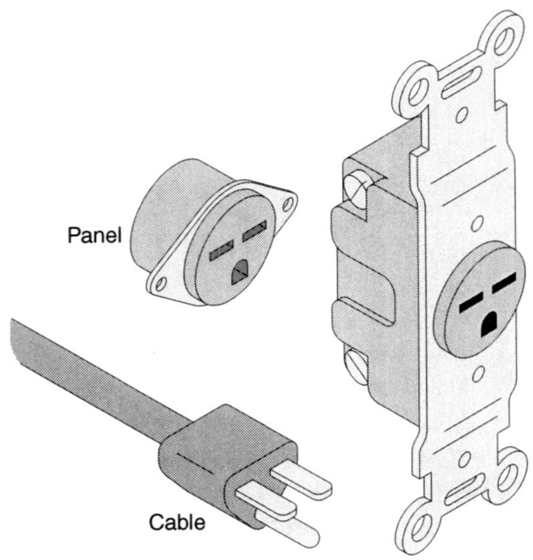

Figure 9-55 Standard 220-VAC Connectors

Most 240 volt power is used for major appliances such as stoves, dryers, hot water heaters, home welding machines, and the like. These units use higher current receptacles, as shown in Figure 9-56. Receptacles in this size range can have current ratings anywhere between 25 to 100 amps.

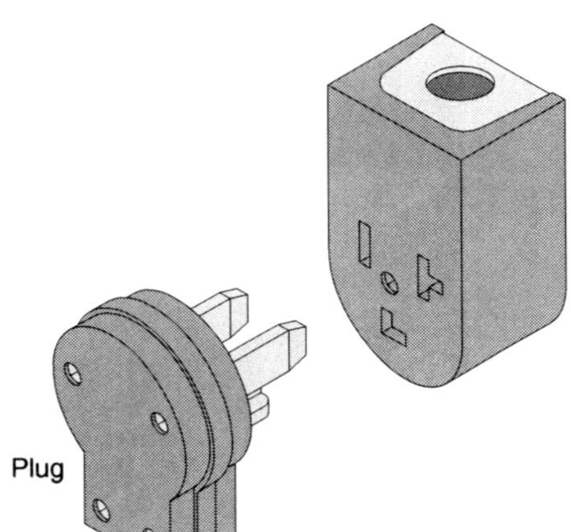

Figure 9-56 High-Current 220-VAC Connector

Turn lock AC connectors, as shown in Figure 9-57, are commonly used in environments where accidental disconnection is a possibility. To mate, the two connectors are pushed together and twisted in their locked position. These connectors are often used in manufacturing facilities where electrical power tools are used at the end of a long extension cord. The locking action prevents the connectors from being pulled apart when the cord is pulled by a worker. Another attribute is that the connector is nonstandard, which means that a power tool with a twist lock connector can only be used at a facility that has matching receptacles. This feature greatly reduces equipment theft because the tool can't be used anywhere else and a pawn shop won't pay any thing for a tool with an odd connector.

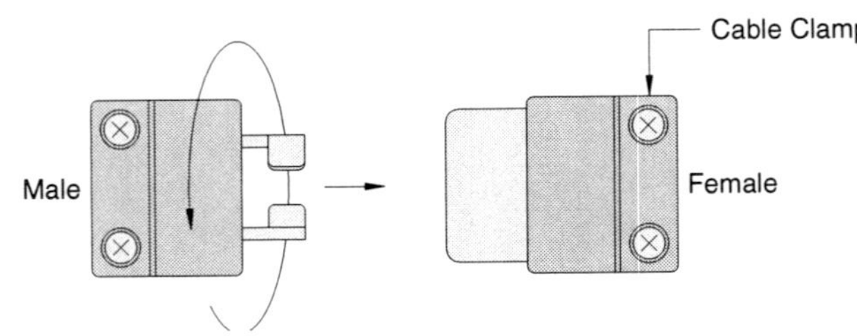

Figure 9-57 Turn-Lock AC Connector

Automotive Connectors

Within the automotive community there are three common connectors used. These are the barrel, flat or "spade," and hook or "twist lock" connectors. All three of these connectors seem to perform well in the harsh automotive environment.

The barrel connector is simply a cylinder-shaped plug and matching barrel. The barrel is split and spring-loaded closed. The plug has a tapered nose and a locking groove. When the plug is pushed into the barrel it springs open and a detent snaps into the locking groove. Figure 9-58 shows a crimp-on barrel connector.

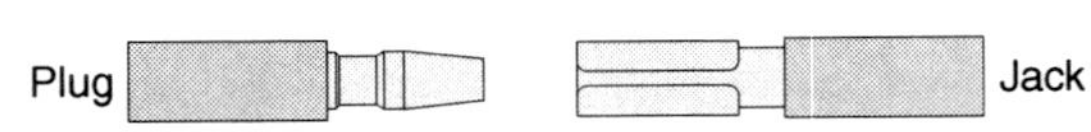

Figure 9-58 Barrel Connector

Flat, or "spade," connectors consist of a flat male plug and a formed female receptacle. The receptacle has the edges rolled in a fashion that pinches the outer edges of the male plug when it is inserted. These connectors are available in uninsulated versions, as shown in Figure 9-59, as well as fully insulated configurations.

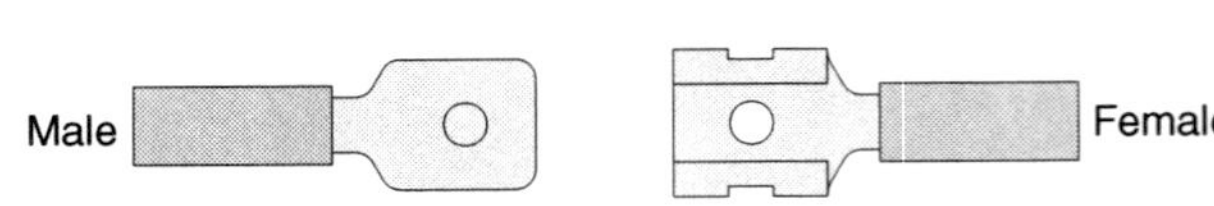

Figure 9-59 Flat or Spade Connectors

Turn-lock, or hook, connectors form an exceptionally solid connection. They are ideal for permanent and semipermanent applications. They are not insulated and require wrapping with electrical tape or heat shrink after connection. Figure 9-60 shows a typical turn-lock connector.

Terminal Strips

Terminal strips are the preferred connection for permanently wired subassemblies and controls within an electromechanical system. Terminal strips are offered in a wide variety of designs, configurations, and terminal counts.

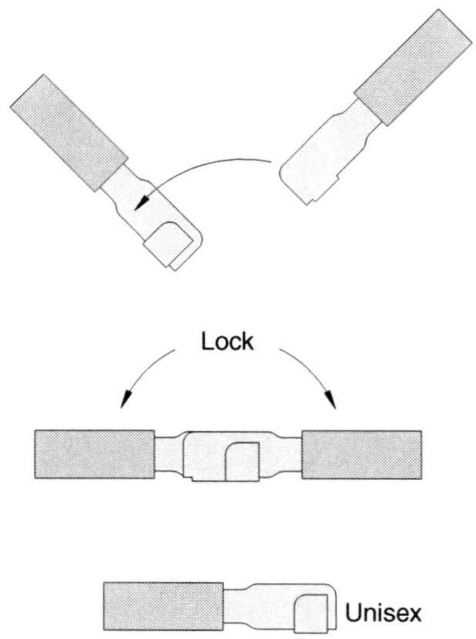

Figure 9-60 Turn-Lock or Hook Connectors

Figure 9-61 shows a typical terminal strip. The base is black bake-a-lite and the terminals are number eight plated brass screws. Subassembly wires are attached to one side and interface wires are attached to the opposite side. Terminal strips like this are a convenient way to terminate all sorts of electrical control and interface requirements.

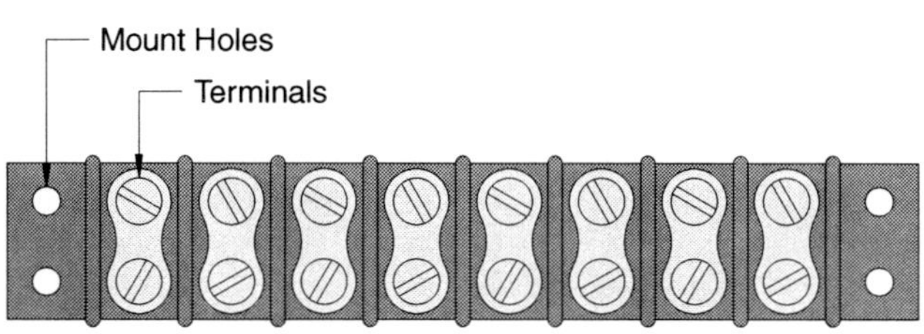

Figure 9-61 Terminal Strip

A standard terminal strip has exposed conductors, which may represent a shock hazard in some installations. To protect personnel from this hazard, a plastic plate is mounted above the strip, as shown in Figure 9-62. Two extra-long studs are

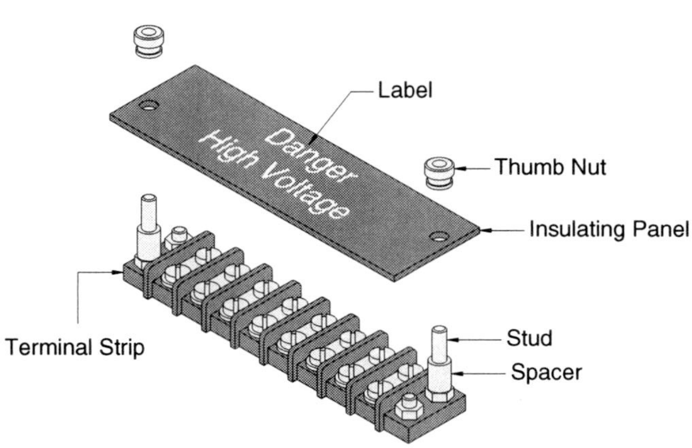

Figure 9-62 Terminal Strip Insulating Panel

used to mount the strip. Spacers are fitted to the studs and thumb nuts are used to secure the protective plate. The plate can also serve to identify the function of the device by printing a label on the top as shown.

To use a terminal strip as a multipin connector, a series of screw lugs are mounted to an insulating board, as shown in Figure 9-63. The screws in the terminal strip are loosened and the plug assembly is inserted. The terminal strip screws are tightened and a high-quality connection is made.

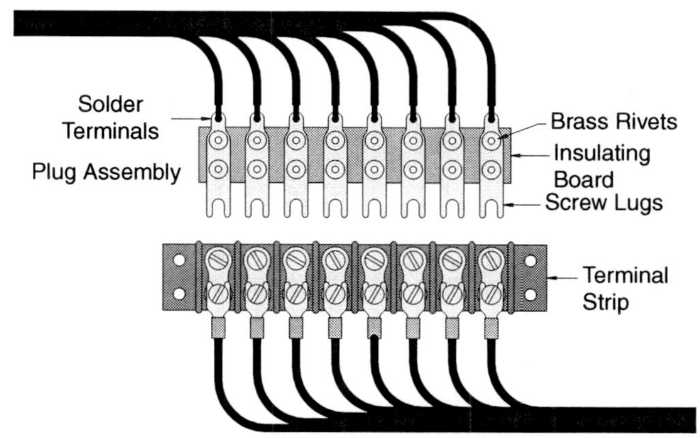

Figure 9-63 Terminal Strip as a Connector

Fully insulated terminal strips are available from a number of commercial sources. These strips are generally a molded insulating block with wire sockets and clamp screws. The wire is stripped and inserted into the socket. When the screw is tightened a high-quality connection is made. Figure 9-64 shows a typical insulated terminal block.

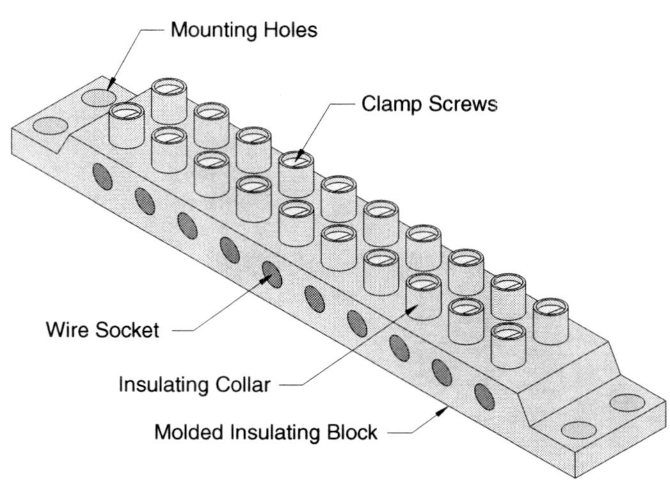

Figure 9-64 Insulated Terminal Strip

For quicker assemblies, push-in terminal blocks are available. These blocks simply require that the wire is stripped and pushed into the socket. To release the wire, a small screw driver is pushed into the release hole and the wire is removed. Figure 9-65 shows a typical push-in terminal block.

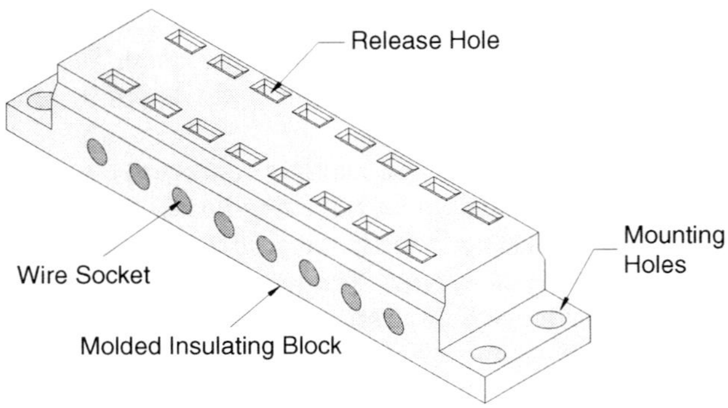

Figure 9-65 Push-in Terminal Strip

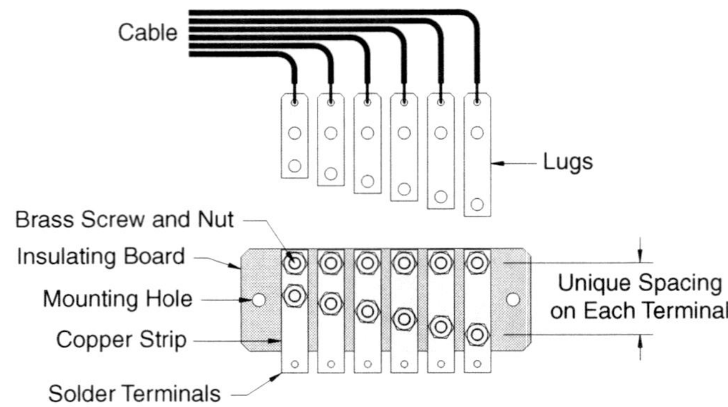

Figure 9-67 Bench Built Keyed Terminal Strip

Terminal strips can be easily constructed using a variety of different components. Figure 9-66 shows just a few bench built terminal strips. The value in constructing your own terminal strips is that they can be designed specifically for the application at hand.

Building a keyed terminal strip, as shown in Figure 9-67, is an excellent method to assure that wires are not miss-connected. Each pole of the strip has two binding posts with a spacing that is unique to that pole. Lugs are built that have a corresponding hole spacing. In this manner the lugs cannot be connected to the incorrect terminal.

Power Distribution Busses

In some assemblies it may be convenient to provide a central power source. A power distribution buss, as shown in Figure 9-68, is used for this purpose. This particular unit is simply a copper bar with a large stud acting as an input terminal and 12 smaller studs as outputs.

For obvious reasons it may be beneficial to provide full insulation to the power buss. Figure 9-69 shows a commercial power buss with a molded insulating jacket and insulating

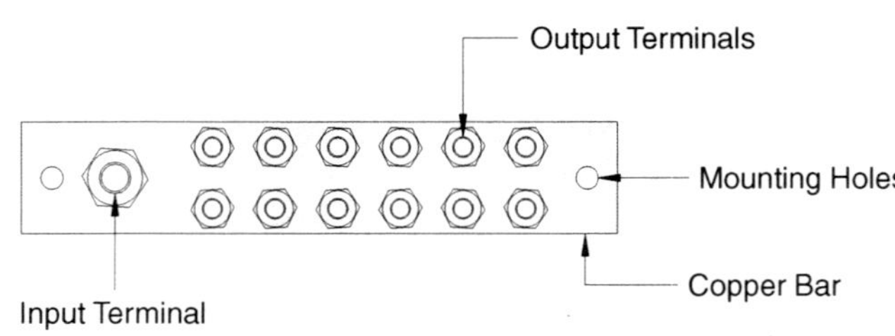

Figure 9-68 Stud-Type Power Distribution Buss

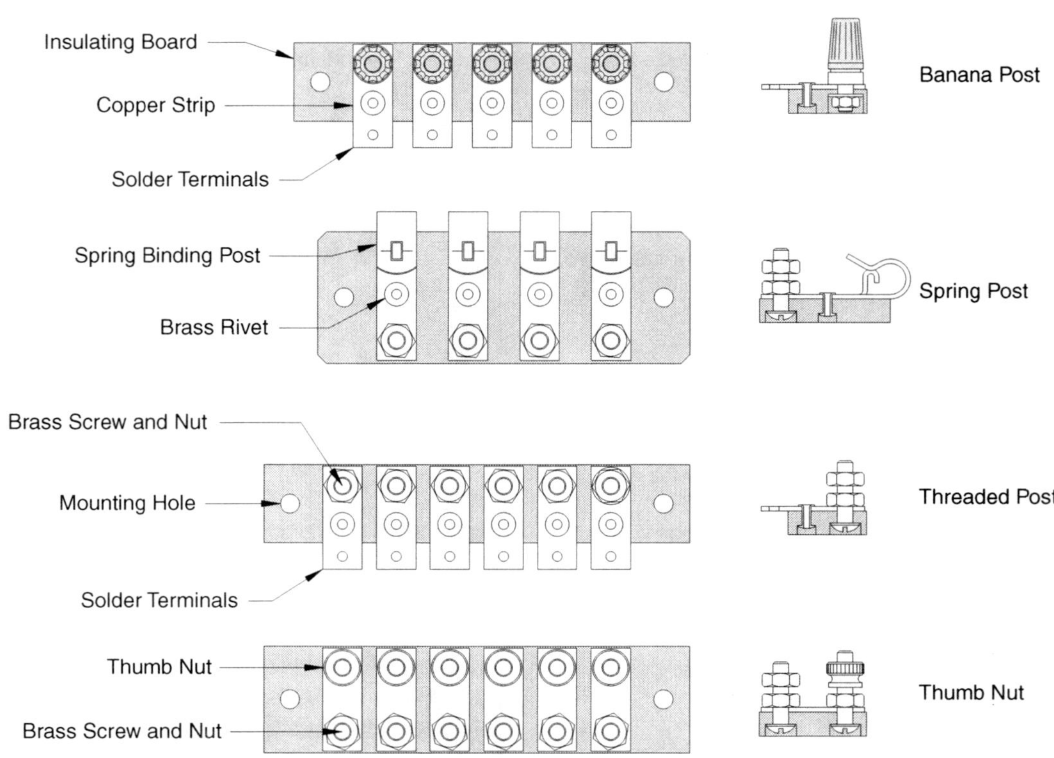

Figure 9-66 Bench Built Terminal Strips

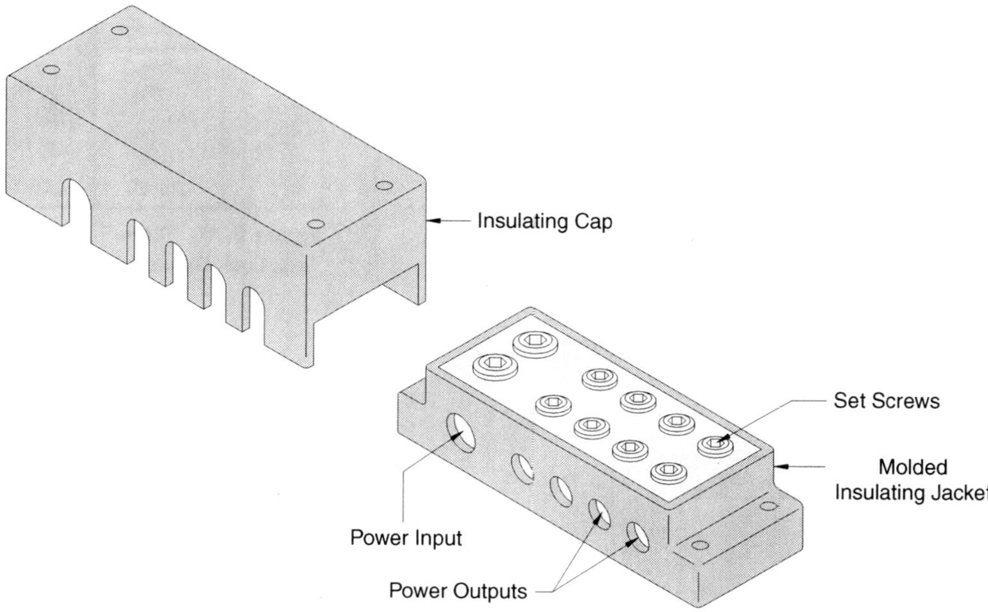

Figure 9-69 Socket-Type Power Distribution Buss

cap. This type of buss is popular in industrial distribution systems and equipment.

NEMA Connectors

The National Electric Manufacturers Association (NEMA) publishes standards for the construction of a wide range of AC power connectors. These connectors are universally accepted by the electrical community and are available in virtually every hardware and home improvement store in the United States. The more obscure patterns are readily available from any electrical or industrial supply house. NEMA connectors fall into two basic categories, straight blade and twist lock. Figures 9-70 through 9-73 show the straight blade patterns and their associated voltage and current ratings. Figures 9-74 through 9-78 show the twist lock patterns and their associated voltage and current ratings.

	15 Amp	20 Amp	30 Amp
125 V	1-15		
250 V	2-15	2-20	2-30

Figure 9-70 Two-Pole NEMA Straight Blade Connectors

	15 Amp	20 Amp	30 Amp	50 Amp
125 V	5-15	5-20	5-30	5-50
250 V	6-15	6-20	6-30	6-50
277 V	7-15	7-20	7-30	7-50
347 V	24-15	24-20	24-30	24-50

Figure 9-71 Two-Pole, Three Wire Grounding NEMA Straight Blade Connectors

	15 Amp	20 Amp	30 Amp	50 Amp
125 V 250 V		10-20	10-30	10-50
250 V 3 Phase	11-15	11-20	11-30	11-50

Figure 9-72 Three-Pole, Three Wire NEMA Straight Blade Connectors

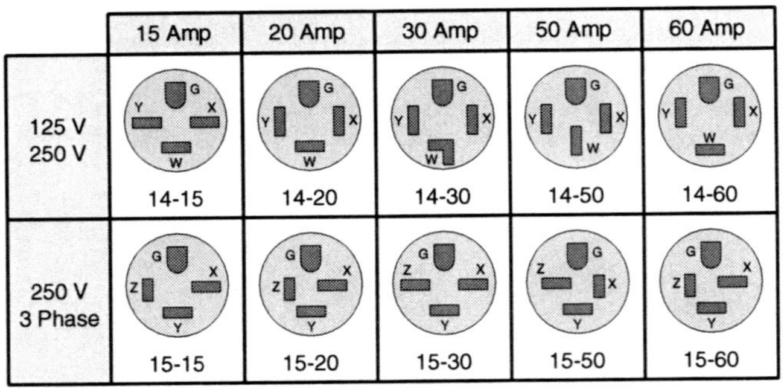

Figure 9-73 Three-Pole, Four Wire Grounding NEMA Straight Blade Connectors

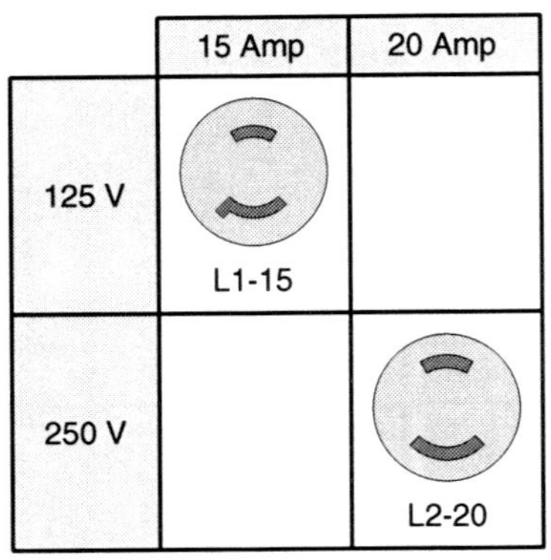

Figure 9-74 Two-Pole NEMA Twist Lock Connectors

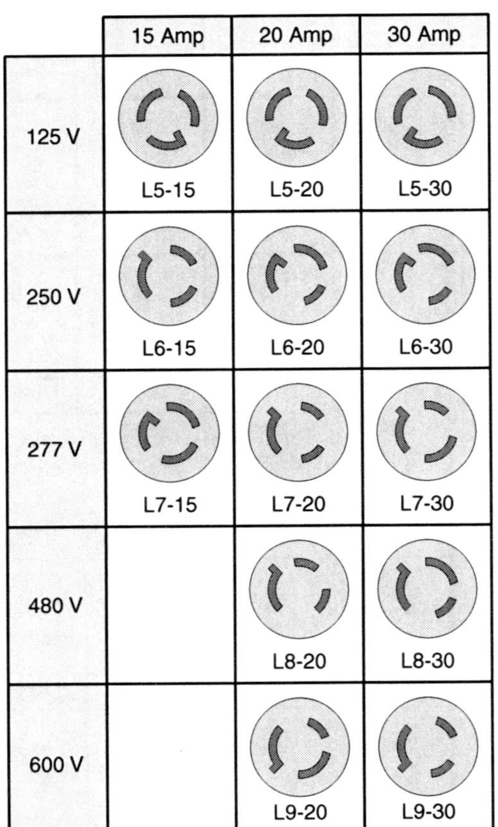

Figure 9-75 Two-Pole, Three Wire Grounding NEMA Twist Lock Connectors

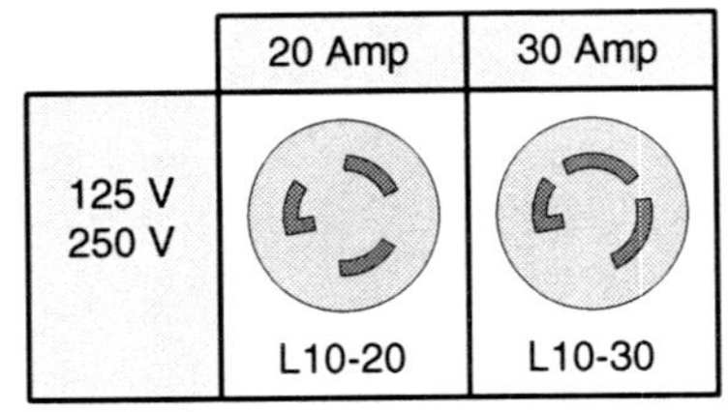

Figure 9-76 Three-Pole NEMA Twist Lock Connectors

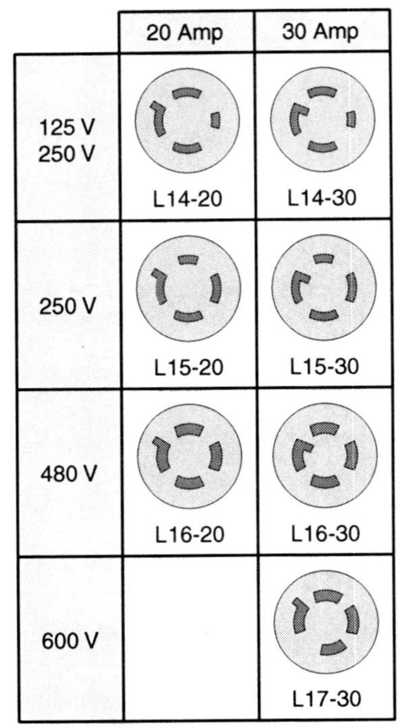

Figure 9-77 Three-Pole, Four Wire Grounding NEMA Twist Lock Connectors

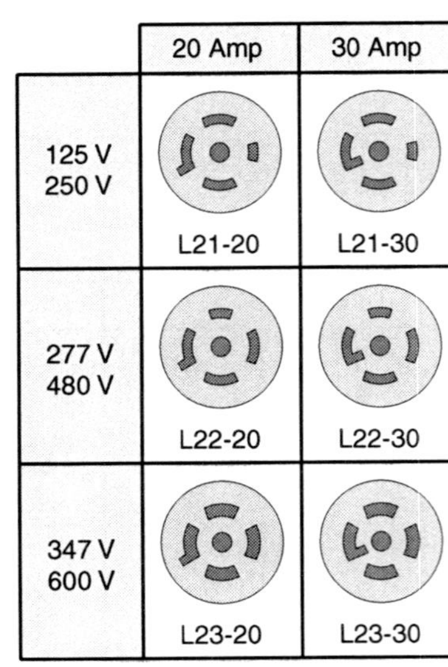

Figure 9-78 Four-Pole, Five Wire Grounding NEMA Twist Lock Connectors

CHAPTER 10
WIRE AND CONDUCTORS

Without a doubt, wire is the most important electromechanical device ever conceived. Without it, electrical and electromechanical devices could not exist.

Most of us pay very little attention to wire until we need some for a project. When we go to the hardware store, we are presented with a dizzying array of different types of wire and conductors. The store attendant is typically of little value and we are left to try to decipher the wire types and their intended purpose by reading the little tags on the spools. This is a very poor method to educate yourself about something as important as wire. To make matters worse, most hardware stores carry at least two specialty wire types for a special customer and you have no way of knowing if the particular conductor you are looking at is common to all stores.

Although there are only about 20 different types of wires commonly used, the different gauges and conductor configurations provide us with hundreds of different wires to choose from.

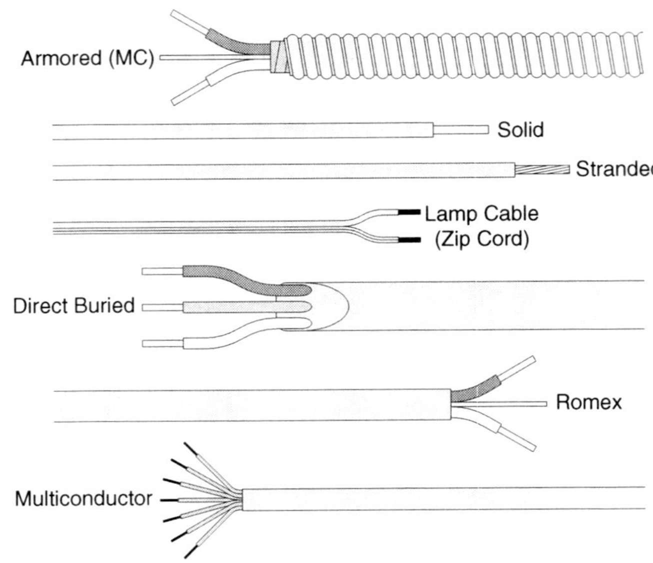

Figure 10-1 Various Commercial Wire Types

Common Wire Types

Figure 10-1 shows a few of the most common general purpose wire types. Armored and Romex are commonly used for home and office wiring. Lamp cord or, as it is commonly referred to, zip cord, is used for household lamps and light appliances. It also makes an excellent speaker wire. Single-strand wire, both solid and stranded, is generally used for commercial and industrial wiring. Single-strand wire is also commonly used in wiring the internals of home, commercial, and industrial equipment. Direct buried cable is used when it is necessary to route a conductor underground in the absence of buried conduit. Multiconductor cable is used in a wide range of control applications.

Figure 10-2 shows a few common specialty wires. Self-retracting wires are commonly found in telephone receiver sets

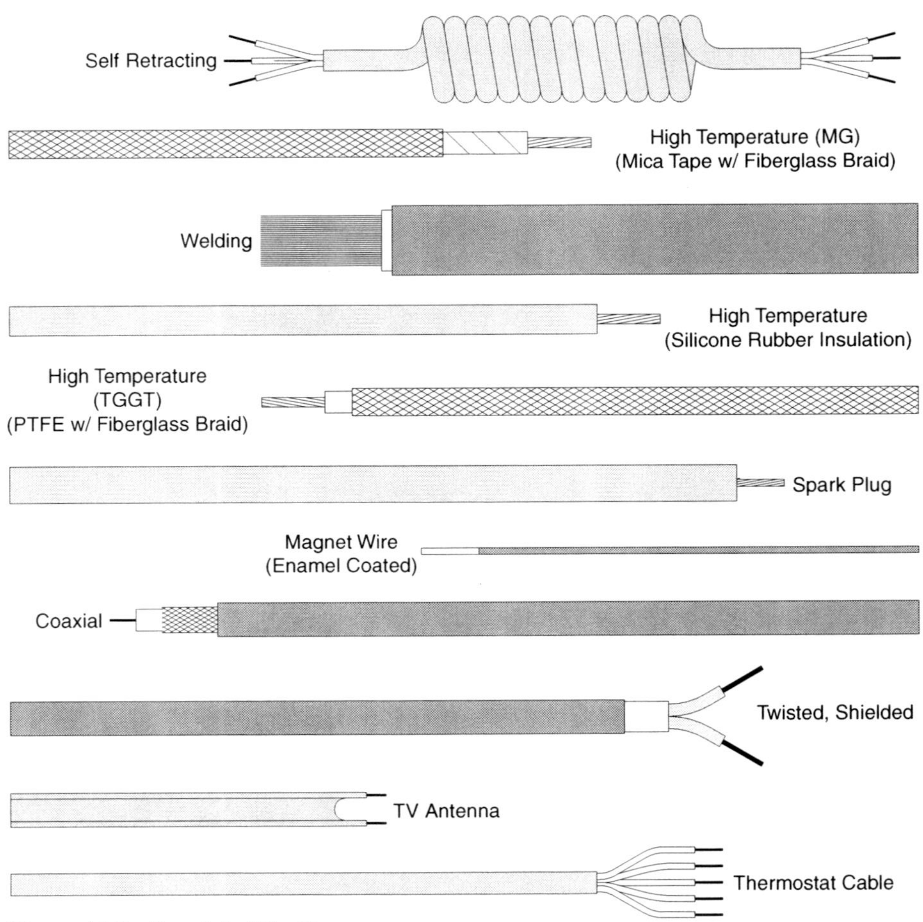

Figure 10-2 Specialty Wire Types

Wire Type	Common Application
Romex Coaxial Lamp cord Speaker cable Outdoor lighting cable	Home wiring Cable TV, Radio, High frequency, RF, Low signal Lamps, Home appliances Home and business speakers Low voltage outdoor lighting.
Bell wire Fixture wire (TFFN) Machine tool wire (MTW) Building Wire (THHN) Silicone rubber	Low voltage, Telephone, Hobby Commercial light fixtures Internal wiring for commercial and industrial equipment Commercial industrial power distribution, Conduit High Temperature (300°F Max.)
Hook-up (PTFE) High temperature Spark plug Ultra high temperature (MG) Continuous flex	Broad temperature range (–67° to +390°F) High temperature (480°F Max.) Automotive ignition High temperature (840°F Max.) Motion control
Audio cable Shielded Thermostat cable Fire alarm cable Thermocouple wire	Low-voltage, shielded, Audio, Security, Control Application that are sensitive to stray signals HVAC, Low-voltage control High reliability Connecting thermocouples
Multi conductor Aluminum armored Plastic armored Direct buried Marine	High and Low-voltage control Home and commercial power distribution Machine hook-up, Splash protection, Vibration isolation Burying with out conduit Resistant to fresh and salt water
Reinforced Magnet wire Ribbon cable Irrigation Chemical resistant	Applications with high pulling loads Motor, Coils, Solenoids, Electromagnets Low Voltage, Digital Direct bury, Low voltage Chemically active environments
Service cable Battery cable Severe environment Self retracting	Portable equipment, Vibration isolation, Temporary Low voltage, High current Mining, Construction, Tunneling, Field operations Manufacturing, Motion control, Telephone, Radio, PA

Figure 10-3 Common Wire Types and Applications

and are also available for higher-power applications. High-temperature conductor is used inside ovens and heaters or for environments where cooling is limited. Welding cable is designed to provide an extremely flexible cable for high-current, low-voltage applications. Spark plug wire is a low-cost conductor that is appropriate for voltages as high as 50 kilo volt. Magnet wire is used to wind the coils in most electric motors and solenoid coils. Coaxial cable is most commonly found for cable TV hook-up and computer networking. Twisted, shielded pairs are intended for applications that are sensitive to stray signals. We've all seen the flat antenna wire used on TVs of the 50s, 60s, and 70s. For the most part, this type of wire has been replaced with coaxial cable. Thermostat cable is found in nearly every home and office in North America.

Figure 10-3 shows common wire types and their common applications.

Another common conductor is service cable. Service cable has a rubber jacket and is designed for portable commercial and industrial applications. Figure 10-4 provides a guide to common service cables. These cables are available from two conductors through multiconductor. They are an excellent choice for machinery and equipment in harsh environments.

The single most common term applied to wire is gauge. The gauge of a wire is an indication of the cross section of the conductor. In North America, the American wire gauge (AWG) is the standard that is commonly used. Figure 10-5 shows a size reference comparison of wire gauges between 4/0 through 32 AWG.

Any wire used should be clearly marked with its various ratings and specifications, as shown in Figure 10-6. Generally a wire is marked with its insulation type, maximum voltage, size, and number of conductors and its maximum operating temperature.

Conductors are commonly supplied in either solid or stranded configurations. Solid wire is generally used in homes, offices, solenoids, motors, inductors, and resistors. Stranded wire provides greater flexibility and is generally used in equipment, industrial, and control cables. Figure 10-7 shows a comparison of solid and stranded wire cross sections.

Standard Service Cables

Type	Description	volts	Duty
SO	Thermoset Rubber	600	Heavy
SOW	Thermoset Rubber	600	Heavy
SOOW	Thermoset Rubber	600	Heavy
SEOW	Elastomer	600	Heavy
SEOOW	Elastomer	600	Heavy
ST	Thermoplastic	600	Heavy
STO	Thermoplastic	600	Heavy
STOOW	Thermoplastic	600	Heavy
STW	Thermoplastic	600	Heavy
SJEW	Elastomer	300	Medium
SJO	Thermoset Rubber	300	Medium
SJOW	Thermoset Rubber	300	Medium
SJOOW	Thermoset Rubber	300	Medium
SJEO	Elastomer	300	Medium
SJEOW	Elastomer	300	Medium
SJEOOW	Elastomer	300	Medium
SJT	Thermoplastic	300	Medium
SJTO	Thermoplastic	300	Medium
SJTW	Thermoplastic	300	Medium
SVO	Thermoset Rubber	300	Light
SVT	Thermoplastic	300	Light
SPT	Thermoplastic	300	Light

S: Standard (600 volts)	OO: Oil Resistant Inner & Outer Jacket
SJ: Junior (300 volts)	P: Parallel Construction
E: Elastomer	W Outdoor Use
T: Thermoplastic	V: Lightweight
O: Oil Resistant Outer Jacket	

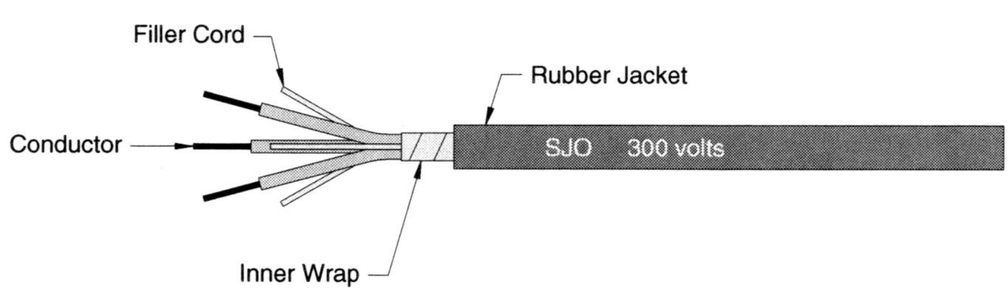

Figure 10-4 Service Cable

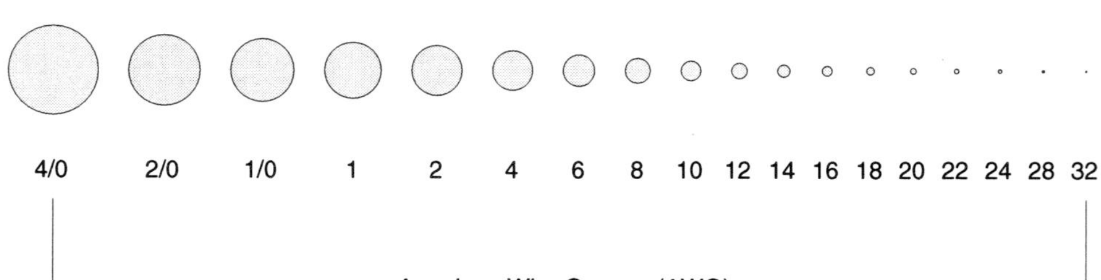

Figure 10-5 American Wire Gauges (AWG)

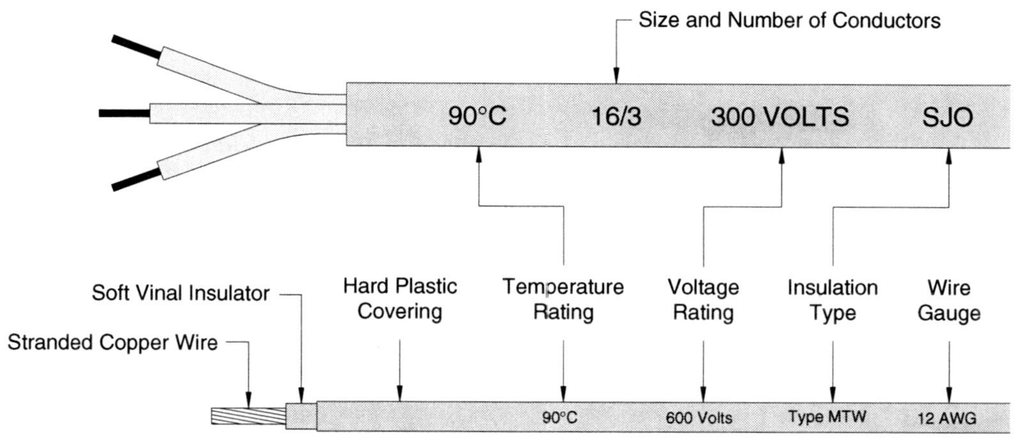

Figure 10-6 Wire Markings

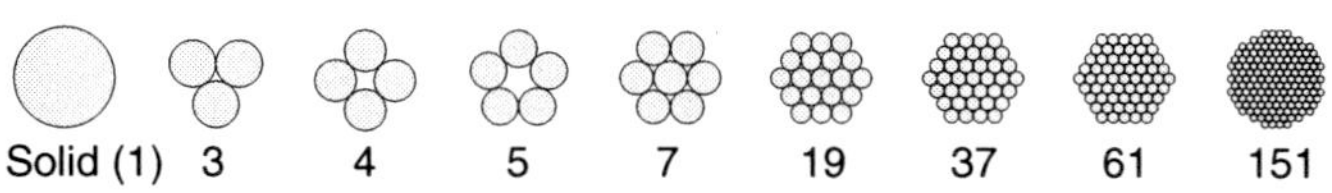

Figure 10-7 Stranded Wire

Figure 10-8 shows the type of cable that is typically used for home and office power drops. The bare center wire provides support as well as a ground reference. The two or three power lines (single or three phase) are wrapped around the center wire.

Armored cable, as shown in Figure 10-9, is used in applications where conduit is required but can't be easily installed.

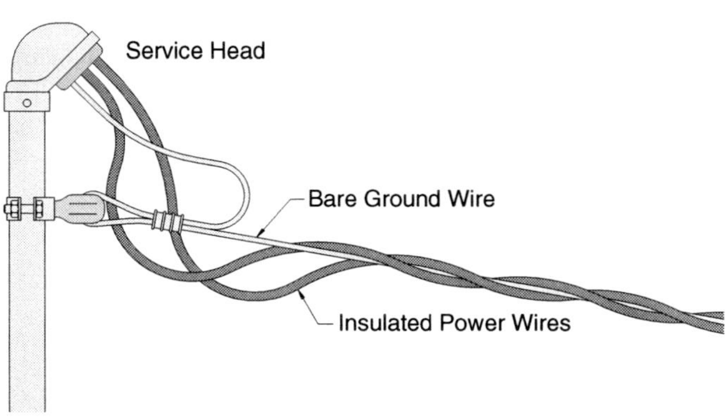

Figure 10-8 Power Connection Cable

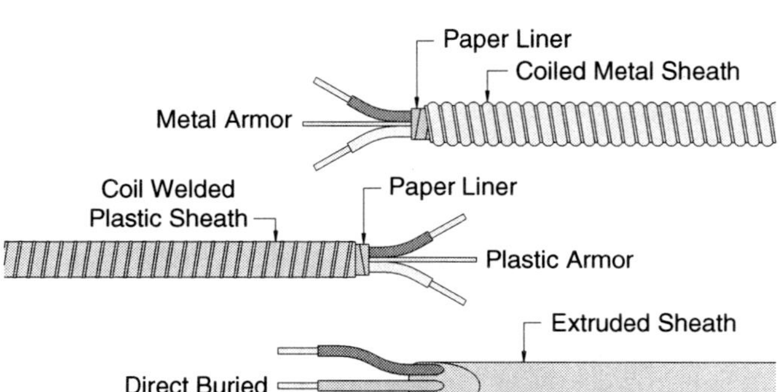

Figure 10-9 Armored Cable

Coiled, metal sheath is common in home, office, and commercial wiring. Coiled welded plastic cable is used in outdoor applications and in applications where minimal exposure to harsh environmental conditions may be encountered. Direct buried cable is used in applications that require underground installations.

Shielded Cable

Shielding a wire or cable is particularly useful when the signal being carried is sensitive to stray magnetic fields, or noise as it is referred to. Any wire with a current passing through it will generate a magnetic field around it. In the case of AC or switching DC, these fields can couple with other conductors and produce noise. This is particularly problematic in circuits that rely on very low signals, such as test, audio, radio frequency (RF), and digital equipment. To combat this tendency, cables use a variety of methods to provide some type of shielding stray signals.

The twisted pair cable, as shown in Figure 10-10, is two wires twisted together. The twist places both conductors in continuously reversing polarity in reference to one another. In doing so, any stray signal that the cable may pick up is canceled by opposing twist. Twisted pair cable is also commonly supplied with a metal shield which is connected to ground. This grounded shield further reduces the effects of stray electromagnetic fields.

Figure 10-11 shows the three basic shielding methods that is commonly available in the market today. The conductors may be jacketed in a wire braid, which is covered with a smooth plastic cover. Lower-cost cables use a metal foil that is either wrapped around the wires like a tape or rolled like a blanket. In either case a bare wire is embedded into the cable to provide a ground connection.

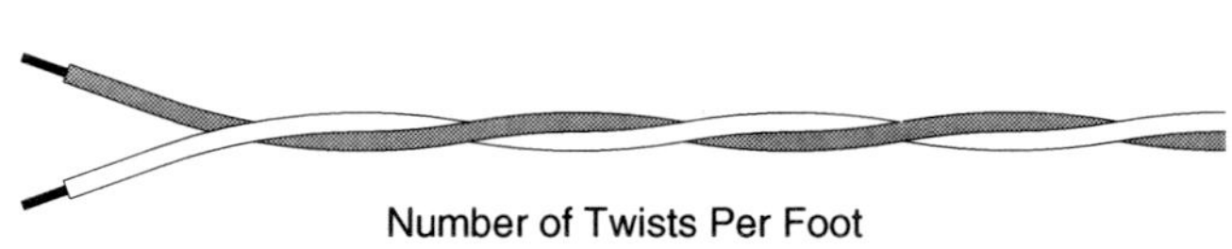

Figure 10-10 Twisted Pair

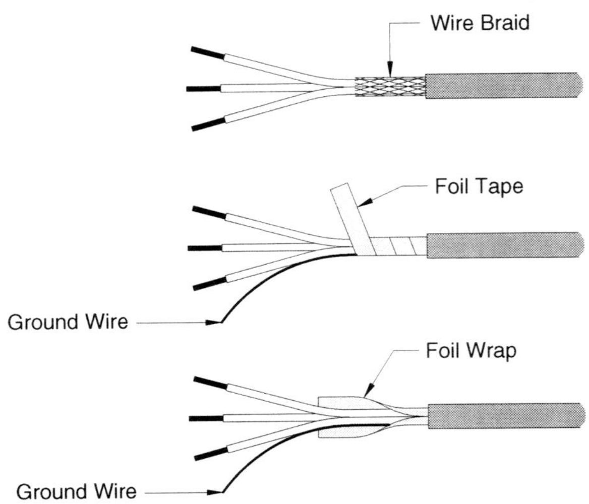

Figure 10-11 Shield Types

Coaxial cable is a type of conductor that is specifically designed for RF signals. These signals are particularly susceptible to noise and must be shielded at all times. A coaxial cable has a single center conductor surrounded by a thick insulator. The insulator is jacketed with a braided wire shield. Figure 10-12 shows a typical coaxial cable.

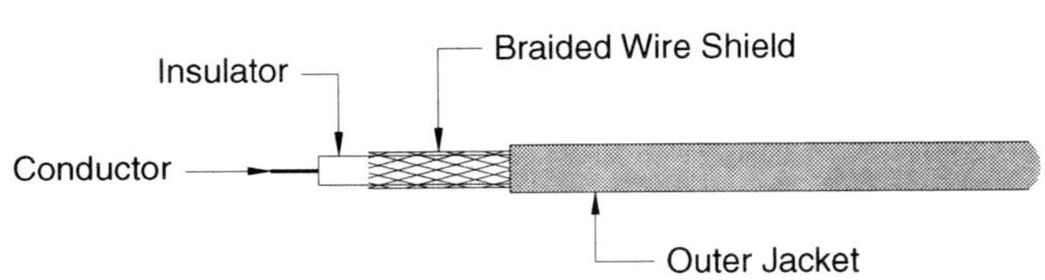

Figure 10-12 Coaxial Cable

High-voltage wire is generally constructed in a coaxial configuration, as shown in Figure 10-13. A piece of test lead is embedded into a thick silicone insulating core. The silicone insulator is jacketed with a fine braid shield and the braid is covered with a soft rubber jacket. The shield is grounded to provide protection in the event that the silicone insulation fails. The silicon insulation, fine shield braid, and soft rubber jacket are selected to maximize flexibility of the finished cable.

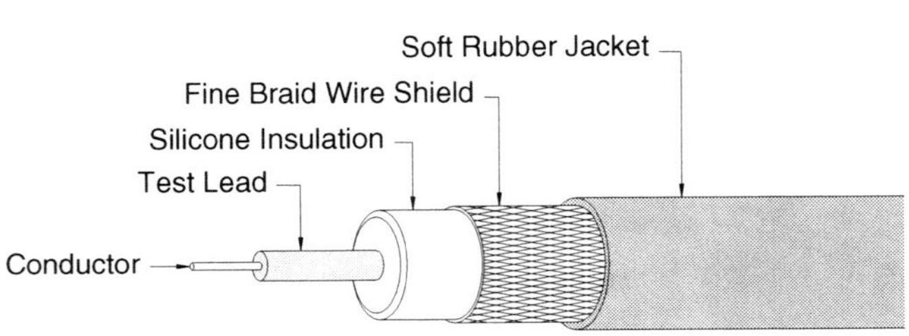

Figure 10-13 High Flexibility High Voltage Wire

Welding cable requires high flexibility and high current-carrying capacity. These cables are generally made with hundreds of fine wires in a central bundle. The bundle is jacketed with a silicone liner, which is covered with an oil resistant rubber jacket. This method of construction produces a cable that is quite flexible while providing a rugged assembly that will stand up well to tough environments. Figure 10-14 shows a typical welding cable.

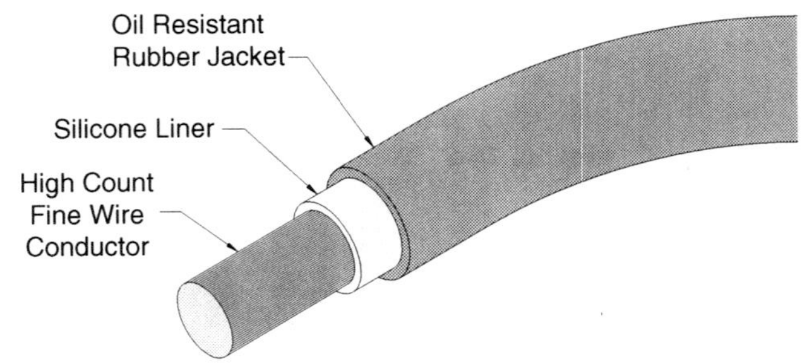

Figure 10-14 Welding Cable

Extension Cords

Most of us have had experiences with extension cords. We use them throughout our houses, offices, and shops. Life with our various appliances would be very difficult without extension cords.

Figure 10-15 shows a few of the more common extension cords for AC service. These types of cords are readily available in a variety of lengths, gauges, and voltages. Two conductor cords used for 120-VAC service are normally used for lamps, small appliances, and double insulated equipment. The grounded 120-VAC is the most common and is universally

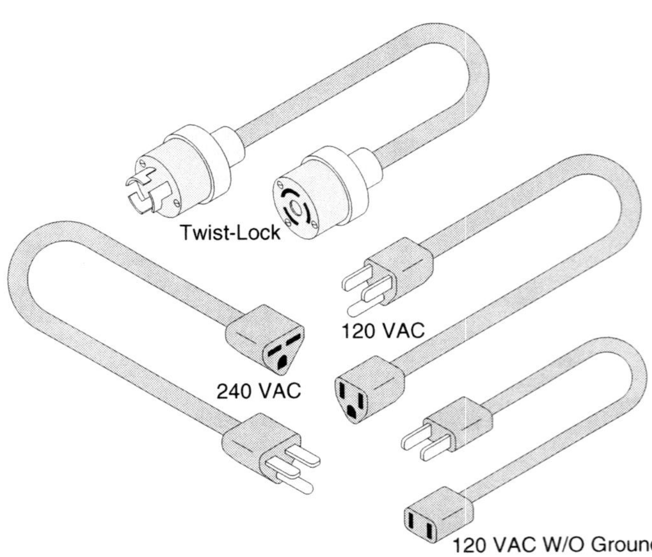

Figure 10-15 AC Extension Cords

adaptable to all appliances and equipment that use that voltage. Cords used for 240-VAC service are generally used to provide a little more reach for major appliances, such as window air conditioners. Twist-lock cords are common in the shop environment to reduce accidental disconnections and reduce theft.

Plug strips, as shown in Figure 10-16, are tucked in behind nearly every computer and stereo system in the country. These handy assemblies provide a convenient method to connect several low-current appliances to a central location. They are typically supplied with a main power switch and line fuse.

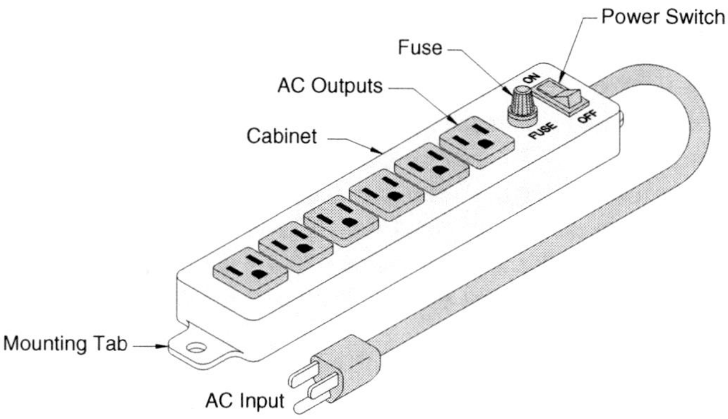

Figure 10-16 Plug Strip

In the shop environment overhead extension cord spools, as shown in Figure 10-17, are extremely handy devices. They provide convenient access to electrical power, keep the shop floor free from tripping hazards, and are easily retracted when not in use.

Extension cords can be very difficult to keep rolled up. To this end, an extension cord reel is a very useful tool. Figure 10-18 shows a simple extension cord reel intended to provide a power

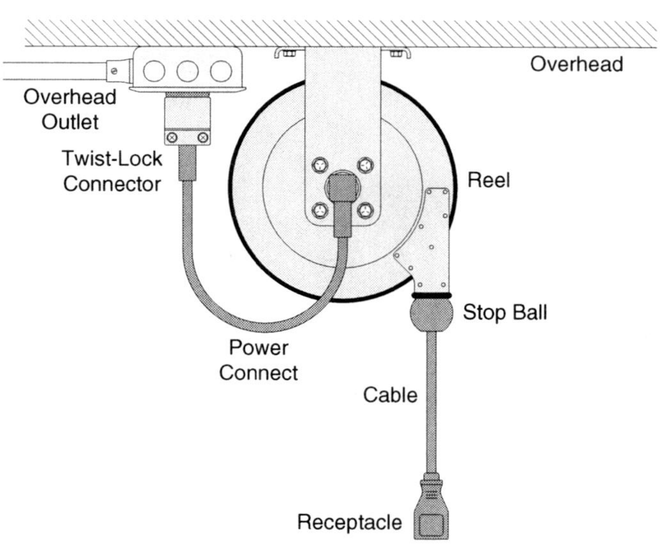

Figure 10-17 Overhead, Self Retracting Extension Cord Reel

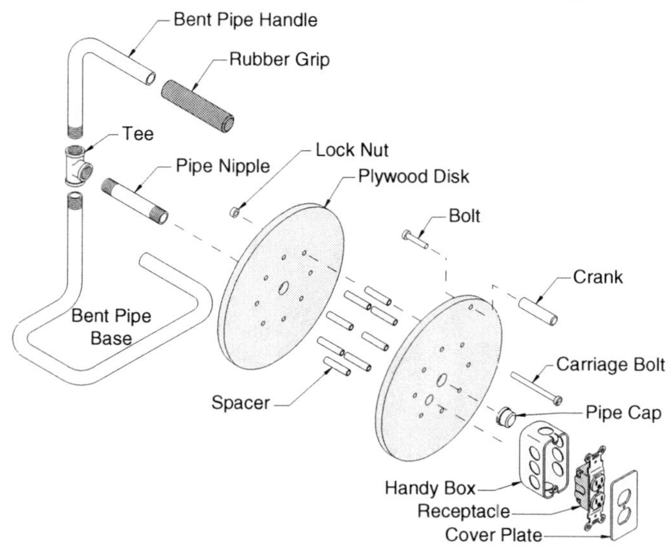

Figure 10-18 Shop Built Extension Cord Reel

drop to a remote work site. The disks are made from ordinary plywood and the frame is constructed from water pipe, a tee, and cap. The reel spacers are cut from small water or gas pipe while the fasteners are hardware store carriage bolts. The output is a handy box that is screwed to the outside disk.

Welding cables are also difficult to keep coiled and stored. A simple four post cable reel can be constructed as shown in Figure 10-19. The mount is an ordinary pipe flange and the axle is a nipple with screw on cap.

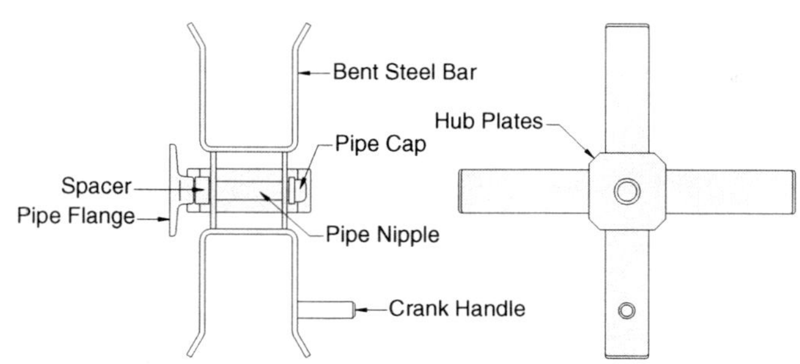

Figure 10-19 Four Post Cord Reel for Heavy Duty Cable

Wiring Harness

Inside of almost all electromechanical equipment, there exists a need for custom cable or wiring harnesses. Wiring harnesses are simply cables that are made up with wires, gauges, shielding, and terminations specific to the equipment that they serve. A perfect example of wiring harness is found under the hood of any modern automobile. Wiring harnesses are made up in advance and even a complex wiring system can be easily installed in just few minutes.

Joining wires is a necessity in almost any wiring harness. Standard crimp connectors can be used for this application, but the preferred method is to splice and solder the wires and protect them as shown in Figure 10-20. After the wires are soldered and the flux is thoroughly cleaned off, a piece of heat shrink is slid over the splice. The heat shrink is then heated causing it to shrink tightly around the joint.

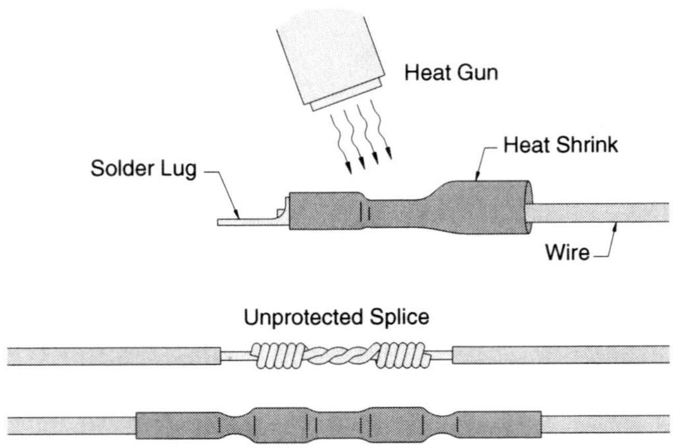

Figure 10-20 Heat Shrink

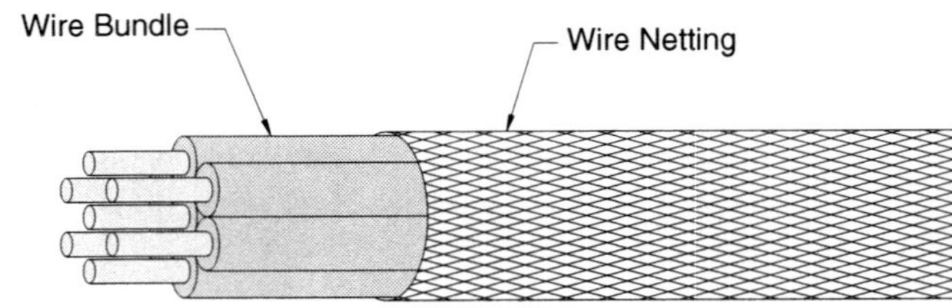

Figure 10-21 Wire Netting

There are a variety of methods used to jacket wiring harness. One of the most attractive jackets is wire netting, as shown in Figure 10-21. The netting is easily slid over the wire bundle and constricts tightly to form a neat cable. Wire netting is available in a variety of colors and color codes, which makes identifying different harnesses very friendly.

Wire netting is rather slippery and thus has a tendency to slip back up the bundle if pulled. This creates an unsightly assembly and can cause the integrity of the harness to deteriorate over time. To combat this tendency, the ends of the netting must be properly terminated, as shown in Figure 10-22. The wire bundle is wrapped with a layer of friction tap. The wire netting is placed over the tape and a second layer of tape is wrapped around the netting. The friction tape is sticky and

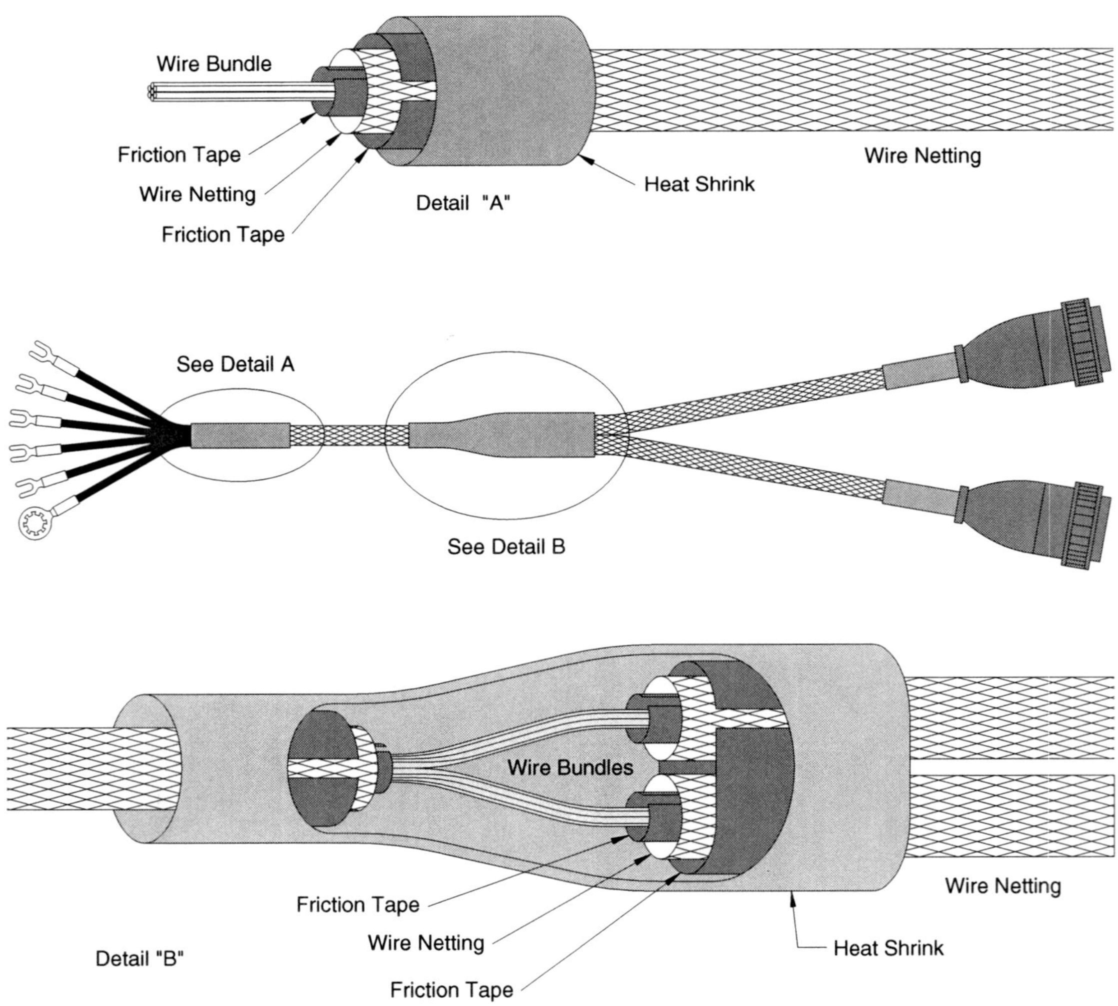

Figure 10-22 Wire Netting Construction

adheres to the wire and netting. The taped joint is then covered with heat shrink. This method of harness construction produces a particularly clean assembly and greatly enhances the internal appearance of any finished equipment. This method of wiring harness construction was originally designed by the author.

Wire lacing is the age-old method of building wiring harness. The wire is bundled and is tied with a bee's wax impregnated cord, called lacing cord. This type of construction is not in common practice anymore. Other, more efficient methods have all but eliminated wire lacing. Any military or aviation electrical equipment of the 30s, 40s, and 50s will use wire lacing in their construction. Figure 10-23 shows a wire bundle using lacing.

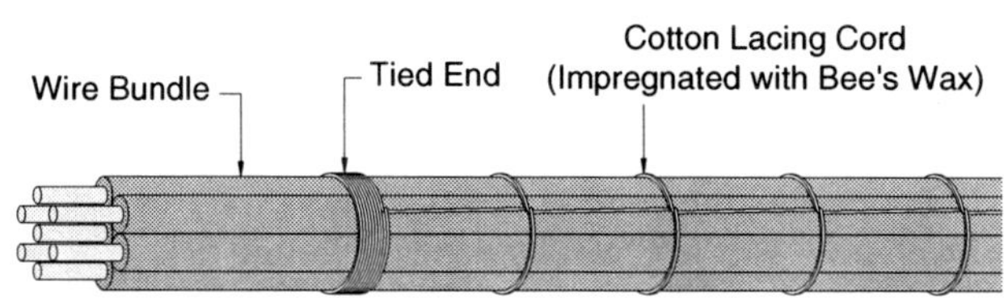

Figure 10-23 Wire Lacing

Loose-fitting plastic sleeving is often used for low-performance wiring harness construction. The wire bundles are pushed through the sleeving and the ends and joints are generally sealed off with electrical tape. This produces a functional, although unsightly, assembly. Figure 10-24 shows an example of sleeve construction.

Coil sleeve is a continuous spring of flat plastic material, as shown in Figure 10-25. After the harness is complete, the coil sleeve is wrapped around the bundles to form the finished assembly.

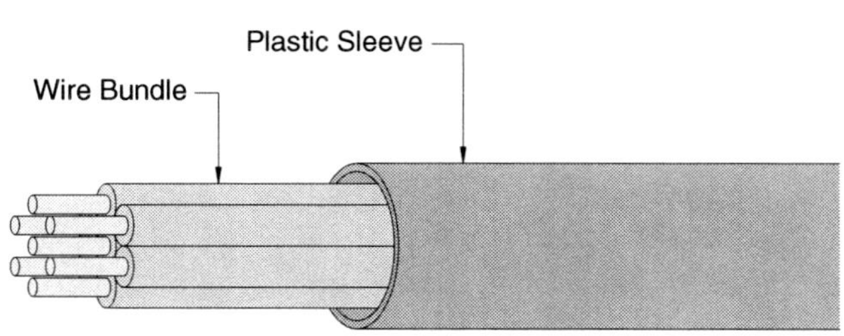

Figure 10-24 Loose Fit Plastic Sleeving

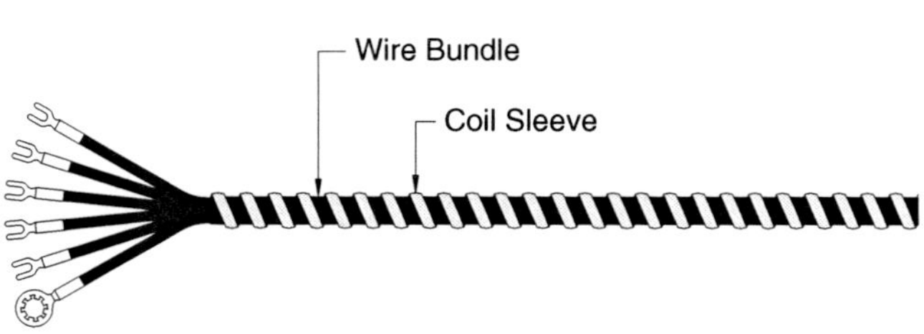

Figure 10-25 Coil Sleeve

Split sleeve is commonly found in automobile wiring harnesses. The finished bundle is simply pushed into the sleeve through a continuous split that runs the entire length of the sleeve. Wires can also be placed individually, which makes it very easy to add conductors after the assembly is complete. To aid in flexibility, split sleeve is generally made with alternating ribs and valleys, as shown in Figure 10-26.

Ordinary tie wraps are a great way to construct one-off or prototype wiring harnesses. They may also be deployed to clean up an unsightly wiring job. Figure 10-27 shows tie wrap construction.

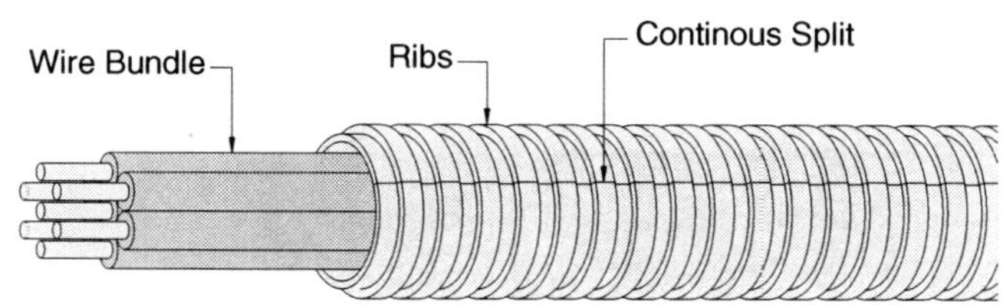

Figure 10-26 Split Sleeve

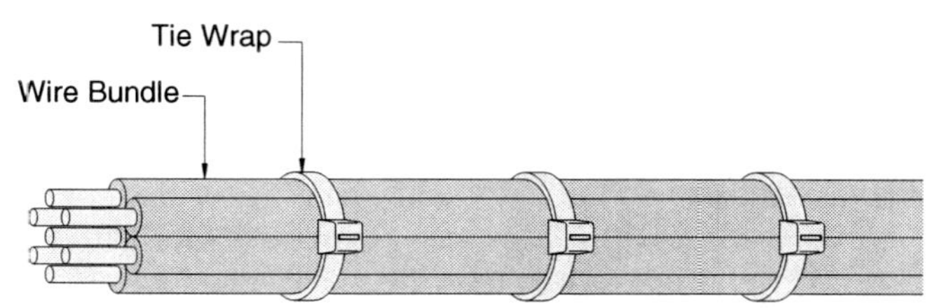

Figure 10-27 Plastic Tie Wraps

Extreme environments can represent significant problems when designing a wiring harness. One method is to build the harness inside standard plumbing fittings and hose, as shown in Figure 10-28. The hose can be selected to withstand abrasion and chemical environments, giving the wires superior protection.

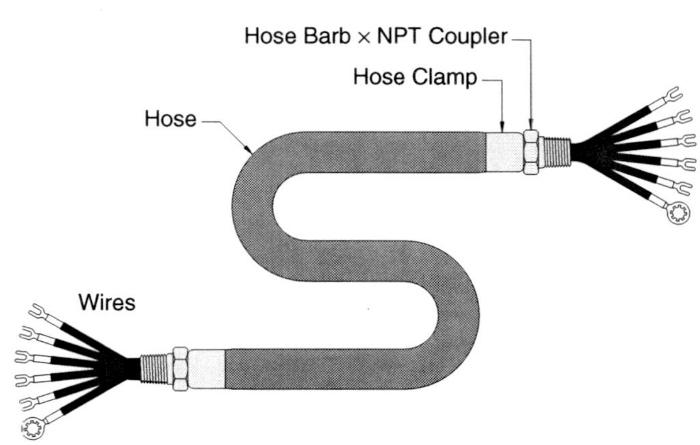

Figure 10-28 Hose and Hose Barb Cable

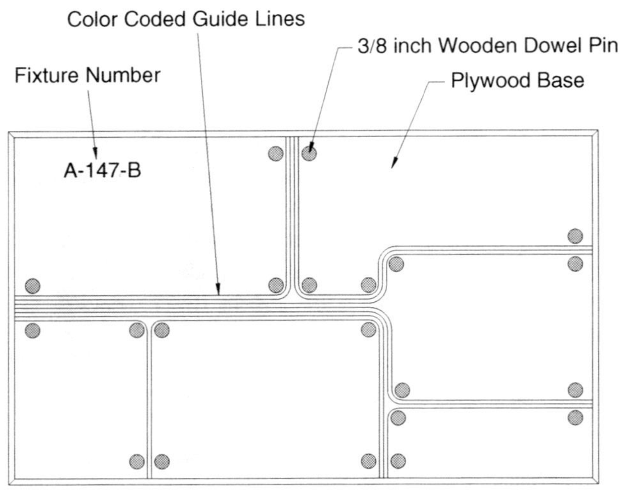

Figure 10-29 Wiring Harness Fixture

For building wire harnesses on a production basis, a simple fixture can be constructed, as shown in Figure 10-29. A board is laid out with a series of wooden dowel pine and colored guidelines are drawn on the board to guide the construction process. Each board is given a fixture number and can be easily stored when not in use.

For wiring harness labs, a configurable table can be set up as shown in Figure 10-30. The table is a 4 inch × 8 inch sheet of plywood with a Formica top. Guide holes are drilled at 1 or 2 inch intervals and grid lines are engraved onto the surface. Each horizontal line is labeled with a letter and the vertical lines are numbered. Written instructions can be provided to the technician for setting any given wiring harness design.

Cutting and Stripping Wire

Cutting wire is a simple process aided by a variety of cutters that are commonly available. Figure 10-31 shows a few wire cutters that may be found in any tool box or electrical shop. The side cutter is the anchor of the wire cutting world. These tools are available in very small units for delicate work through large, heavy duty units for cutting thick conductors. End cutters are similar to side cutters, except that the cutting edges are oriented on the end. These provide easy access for locations that have limited surrounding clearance. A wire cutting element is routinely added to all sorts of pliers. The most common are the lineman and needle nose pliers. For cutting heavy wires and cables, special hook nose cutters are used. These prevent the cable from slipping out of the jaws during cutting operations.

To connect a conductor it is usually necessary to strip off the insulation. This is generally done by cutting or scoring the insulation and then pulling off the end piece. Larger conductors can be scored with an ordinary razor blade knife, then pulled off by hand. Great care should be taken to avoid damaging the conductor when stripping wire with this method. For smaller wire and production stripping, plier type strippers are preferred, as shown in Figure 10-32. The V-notch strippers are clamped around the insulation and then the end is pulled off. The diameter is adjusted with a sliding screw stop that limits how far the jaws can close. Stripping pliers work with

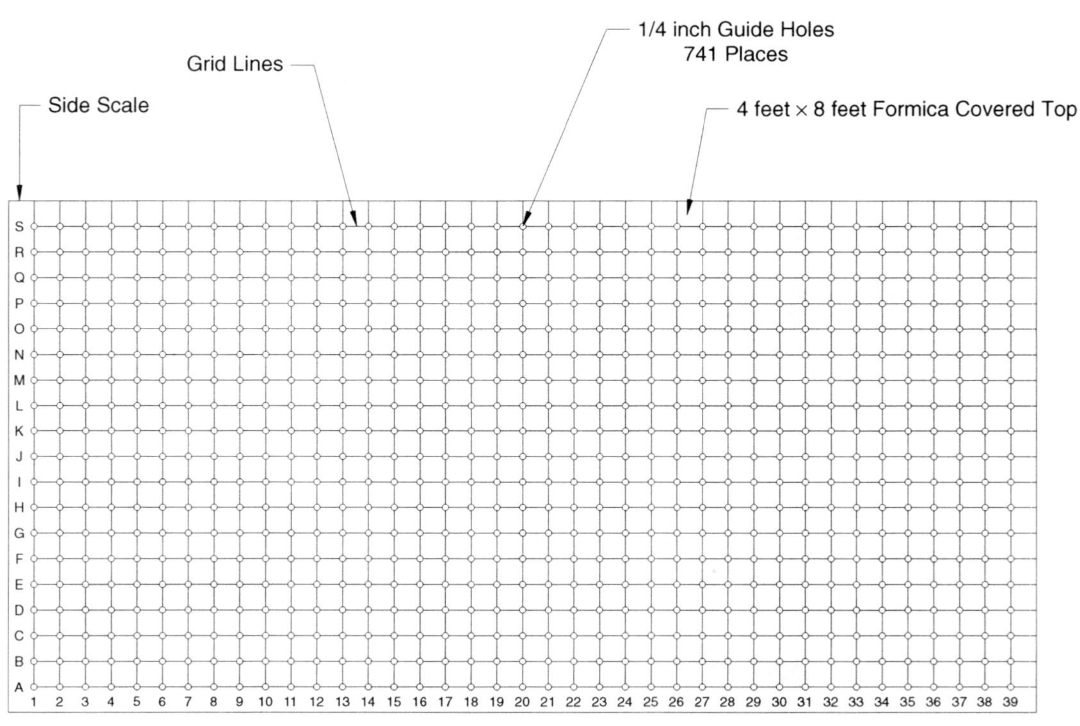

Figure 10-30 Wiring Harness Table

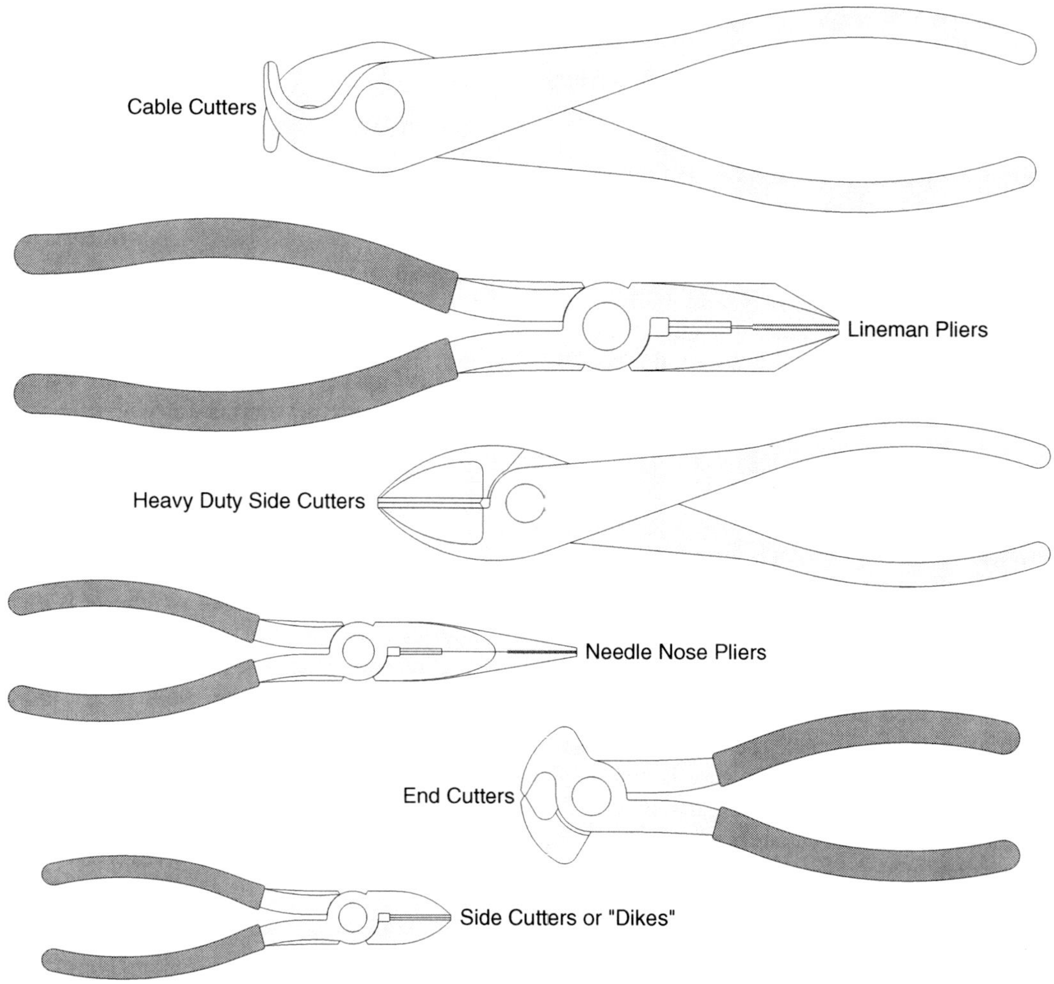

Figure 10-31 Various Wire Cutters

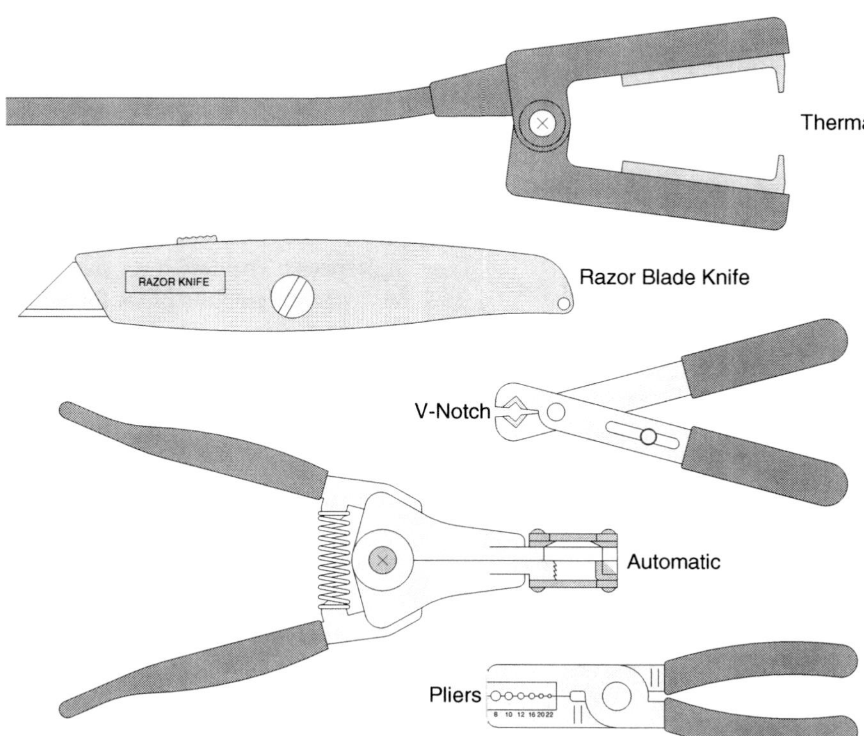

Figure 10-32 Wire Strippers

the same method as the V-notch unit, except that they have several specific wire size locations in the jaws. These are particularly useful in situations where a technician is working with several different sizes of wires. Automatic wire strippers are generally used in production stripping. The wire is placed into the jaws, a quick squeeze of the handle, and the wire is stripped and ejected in one quick operation. Thermal wire strippers are generally used on outer jackets of multiconductor cables or in applications where it is critical that the conductor is not nicked. The jaws are closed onto the insulator, the plastic is weakened, and the insulation is pulled off. The temperature of the stripper is adjusted to match the material from which the insulation is made.

Rotary Conductors

Making an electrical connection through a rotating axle is a difficult proposition. However, there are many applications for such connections, the most noteworthy being the armature on a DC motor.

Figure 10-33 shows a high-current, low-speed, slip ring rotary conductor. These units consist of a rotating brass hub with a preloaded brass block in contact. The interface is lightly oiled to minimize wear. This type of slip ring is commonly used to connect welding currents to a rotating assembly.

Multipole plate rotary conductors as shown in Figure 10-34, can be used for moderately high currents and medium speeds. The preload spring is adjusted to assure contact. The disk plates are lightly oiled for low speeds and use conductive grease for higher speeds.

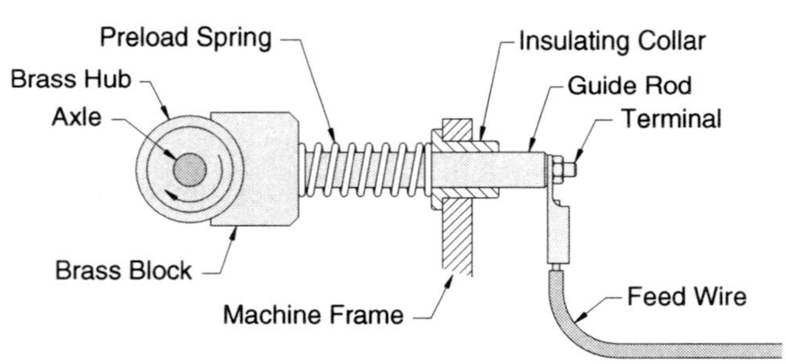

Figure 10-33 Slip Ring Rotary Conductor

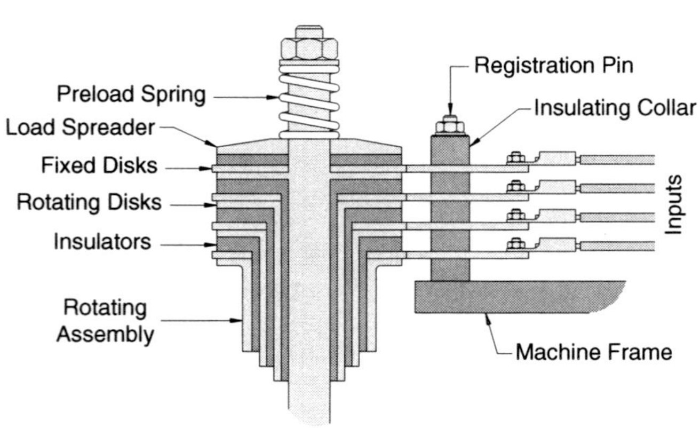

Figure 10-34 Multipole Plate Rotary Conductor

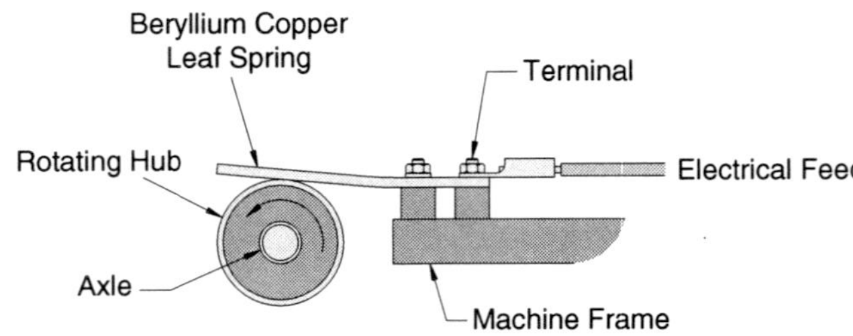

Figure 10-35 Leaf Spring Rotary Conductor

Leaf spring rotary conductors are the classic design. They provide moderate current-carrying capacity at low to medium speeds and they are simple to build and service. Multipole arrangements can be set up by simply adding additional hub/spring sets. Figure 10-35 shows a typical leaf spring rotary conductor.

Wire brush rotary conductors are similar to their leaf spring counterpart. The principal difference is that the leaf spring is replaced with a brass wire brush. These conductors provide low current capacity at medium to high speeds. Because of the multiple contacts that are made by the individual wires, these conductors have very consistent conductivity. Figure 10-36 shows a view of a wire brush rotary conductor.

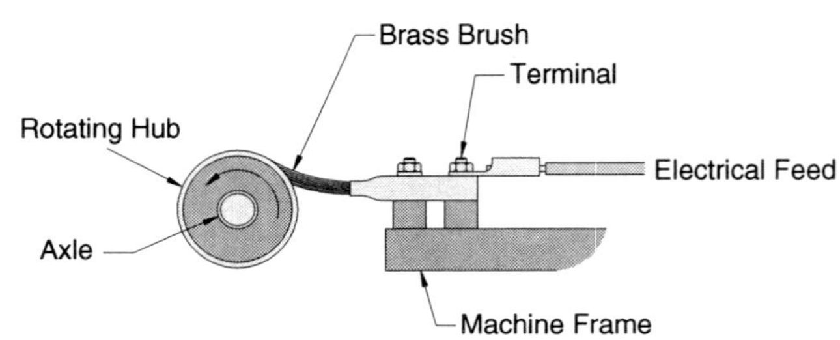

Figure 10-36 Wire Brush Rotary Conductor

One of the most common rotary conductors is the carbon brush set. This type of conductor is found in nearly every universal and DC motor manufactured today. These conductors provide moderate current-carrying capabilities at medium to high speeds. The design consists of a rotating brass hub which has a carbon/graphite block forced into it by a preload spring. The block rides in a brass housing that doubles as the electrical terminal. Figure 10-37 shows a sectional view of a typical carbon brush rotary conductor.

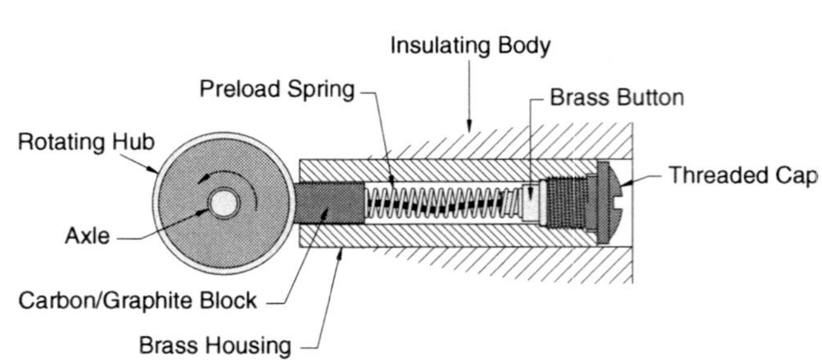

Figure 10-37 Carbon Brush Rotary Conductor

Buss Bars

Buss bars are typically used to provide a central distribution point for power or a common ground point. Figure 10-38 shows two of the most common buss bars. The flat bar is simply a piece of flat copper with a number of studs and mounting holes. The block and set screw type are commonly found as neutral/ground strips in power distribution boxes.

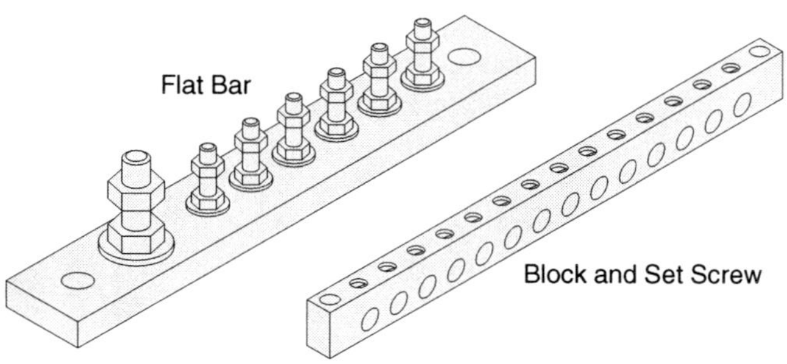

Figure 10-38 Typical Buss Bars

Electrical Construction Methods

There are a number of methods for assembling electrical circuits. Among these are printed circuit boards, wire and post, terminal strip, buss wire, and point-to-point. The following provides a brief review of these methods.

Printed Circuit Boards

The most common electronic construction method is the use of printed circuit boards. This is an excellent method for low current electronics and control circuits. Figure 10-39 shows a typical circuit board. The insulating board is usually a fiberglass panel with the conductors and solder pads laminated to its surface.

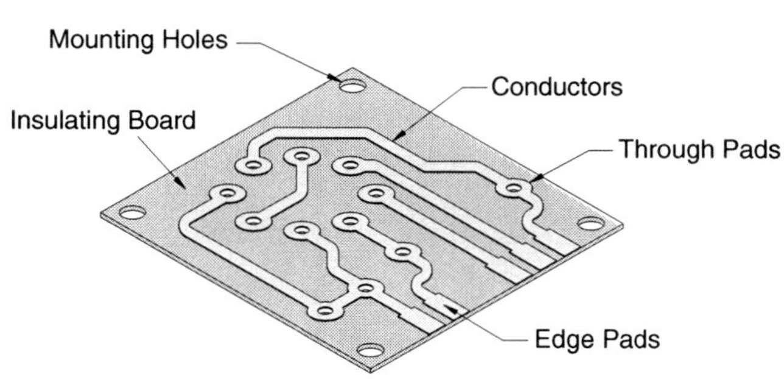

Figure 10-39 Printed Circuit Board

The Wire and Post Construction

Early radio, test, and electrical equipment were principally constructed using the wire and post method. This practice uses an insulating board with a number of brass screws which protrude up from the surface. Stiff, solid copper wire is soldered to the top of the screws and a circuit can be laid out. Components are then soldered to the wires as necessary. This type of construction has all but been forgotten, but is still an excellent method to construct prototype and one-off circuits. Figure 10-40 shows a typical wire and post assembly.

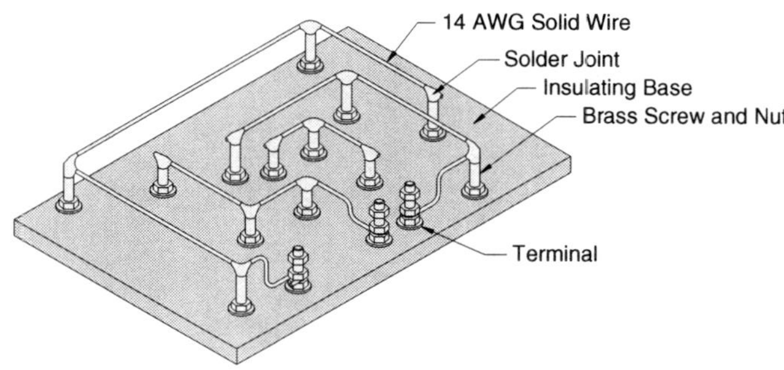

Figure 10-40 Wire and Post Construction

Solder Strip Construction

Electronics of the 40s, 50s, and 60s commonly used solder strip construction, as illustrated in Figure 10-41. Like wire and post, this type of construction is principally unused today. However, it still remains a good method for prototyping and one-off circuits.

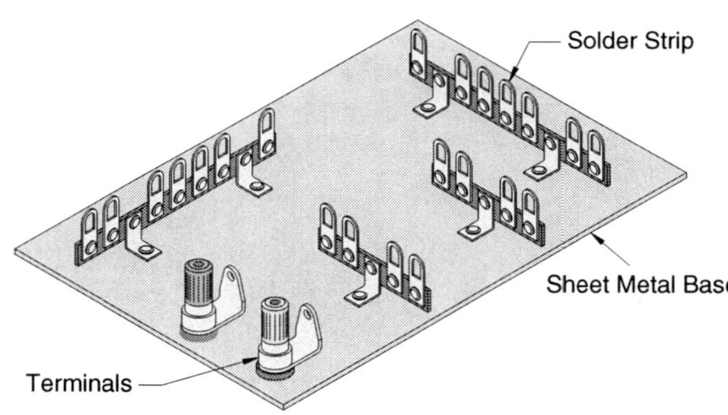

Figure 10-41 Solder Strip Construction

Lead Wire Construction

Lead or buss wire construction is a type of free form assembly. Components are soldered together using only the lead wires. For obvious reasons the method is also referred to as ball construction. After the circuit is complete and tested, it is generally potted into some sort of case. This produces a special circuit with the appearance of a production unit, but without the associated tooling costs. Figure 10-42 shows an example of lead wire or *ball* construction.

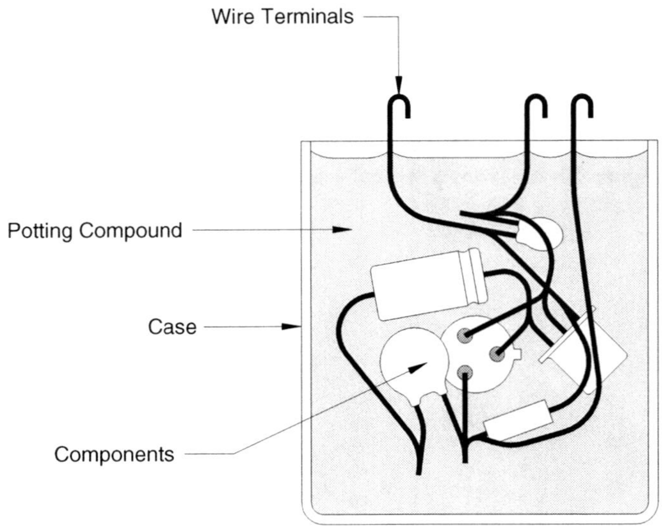

Figure 10-42 Lead Wire Construction

Point-to-Point Construction

Industrial controls, in particular, are generally constructed using the point-to-point method. In this case major components are mounted to a base or chassis and interconnected with wire to secondary components. Almost any industrial control cabinet is assembled using this method and can easily be inspected by simply opening the front panel. Figure 10-43 shows a small control panel using point-to-point wiring.

Cable Clamps and Strain Reliefs

Fixing a cable to a cabinet is critical to the long-term operation of any circuit. The movement of a loose cable can fray the conductors or damage the components to which it is connected.

Figure 10-44 shows typical collet style cable clamp. A cable is placed through the center of the clamp and a threaded cap is tightened down. As the cap is tightened a collet is compressed around the outside diameter (OD) of the cable, clamping it firmly. These units are available in a wide variety of styles and materials that are suitable for almost any situation.

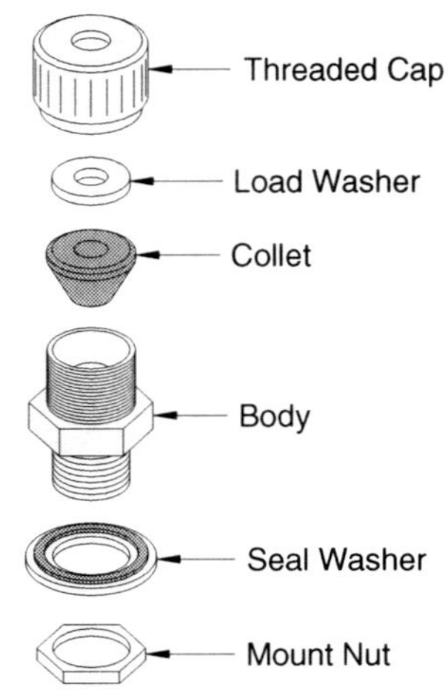

Figure 10-44 Typical Cable Clamp

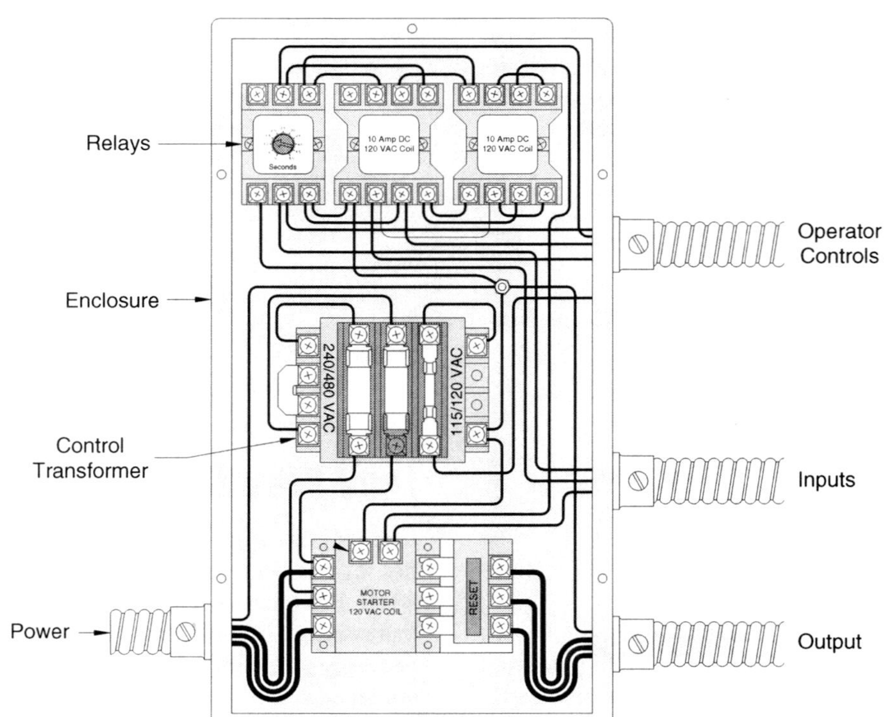

Figure 10-43 Industrial Control Construction

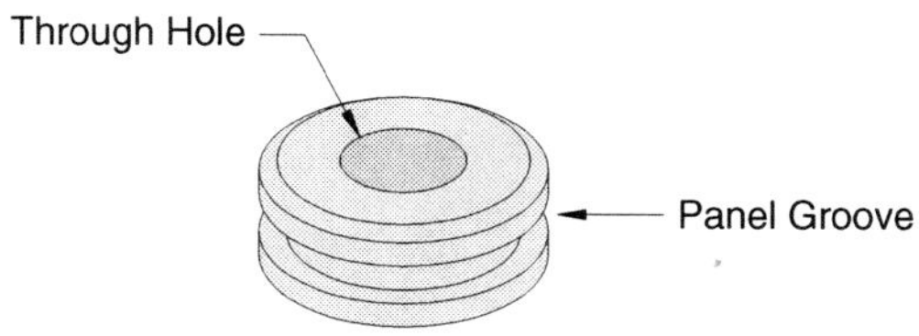

Figure 10-45 Rubber Grommet

Rubber grommets, as shown in Figure 10-45, are the classic wire strain relief. The chassis is punched with a hole, which matches the ID of the panel groove, and the grommet is inserted. Although it is not a generally recommended practice, many companies simply tie a knot in the cord to prevent it from being pulled through.

Tab style cable clamps can be used as strain reliefs or as internal cable supports. They are an inexpensive and effective method to provide moderate service in this regard. Figure 10-46 shows an example of a table style cable clamp.

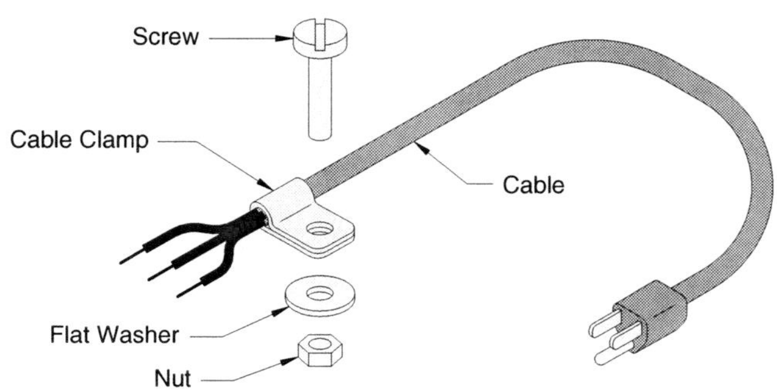

Figure 10-46 Tab Type Cable Clamp

Romex cable clamps are very versatile devices. They are designed to clamp Romex cable in home and office wiring systems; however, they are useful in a variety of other applications. Figure 10-47 shows a typical Romex cable clamp.

To extend the service life of a cable in heavy use, a bend relief is applied to the cable at the clamp. Figure 10-48 shows a few common bend reliefs. Simply adding a length of rubber hose to a cable and Romex clamp provides an excellent, low cost solution for the problem. Many manufactured cables come with a molded bend relief as an integral part of the assembly. For industrial applications a braided cable bend relief is generally preferred.

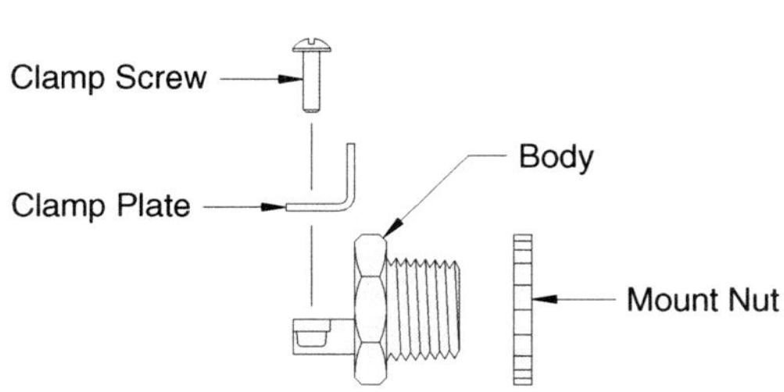

Figure 10-47 Romex Cable Clamp

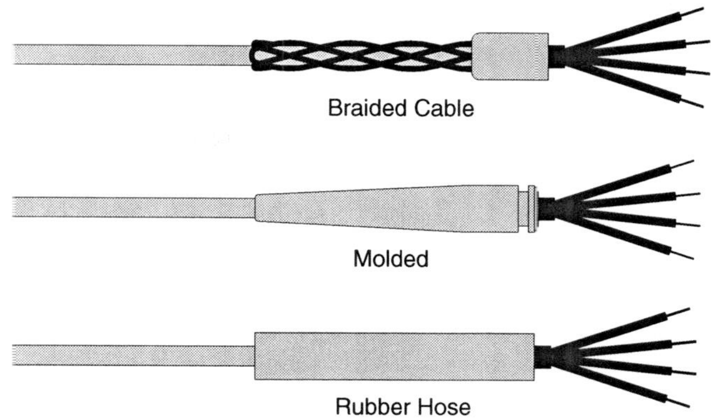

Figure 10-48 Common Bend Reliefs

Insulators

Electrical isolation is imperative to minimize losses and maintain a high degree of safety. There are a variety of insulators on the market that are designed to accomplish just that. An insulator is a piece of nonconductive material shaped in a fashion that makes it convenient for electrical isolation applications.

Figure 10-49 shows a basic wire and post ceramic insulator. These insulators were very common in the early days of electrification. They are still available; however, they become less and less common as time goes on.

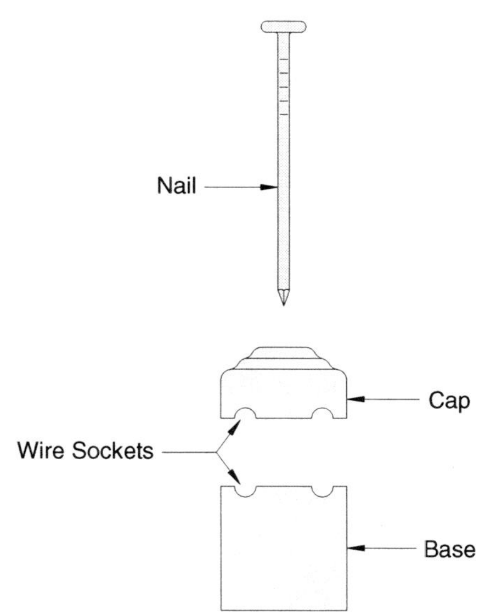

Figure 10-49 Wire and Post Insulator

Also used in conjunction with the wire and post insulators are ceramic feedthrough insulators, as shown in Figure 10-50. A hole is drilled through a joist and the insulator is pushed in. The wire is then fed through the insulator.

Insulators used on early telegraph and power poles were made of soda lime glass. These insulators were clear green

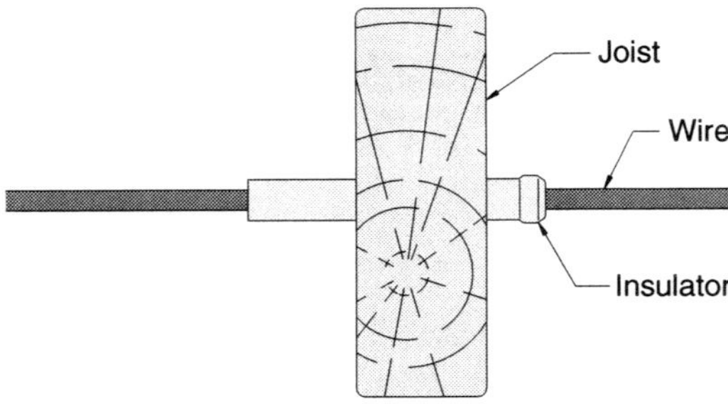

Figure 10-50 Ceramic Feedthrough Insulator

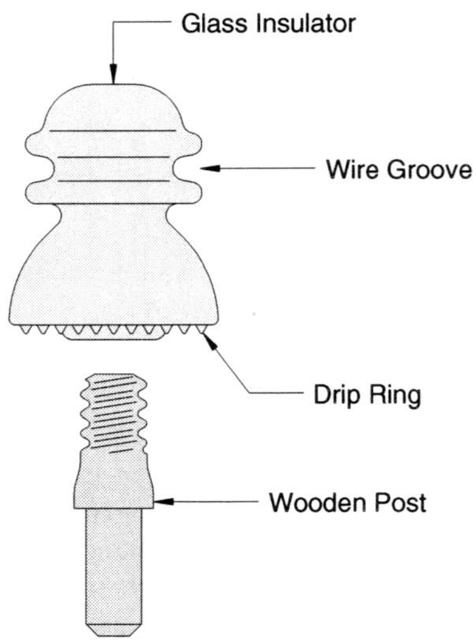

Figure 10-51 Glass Insulator

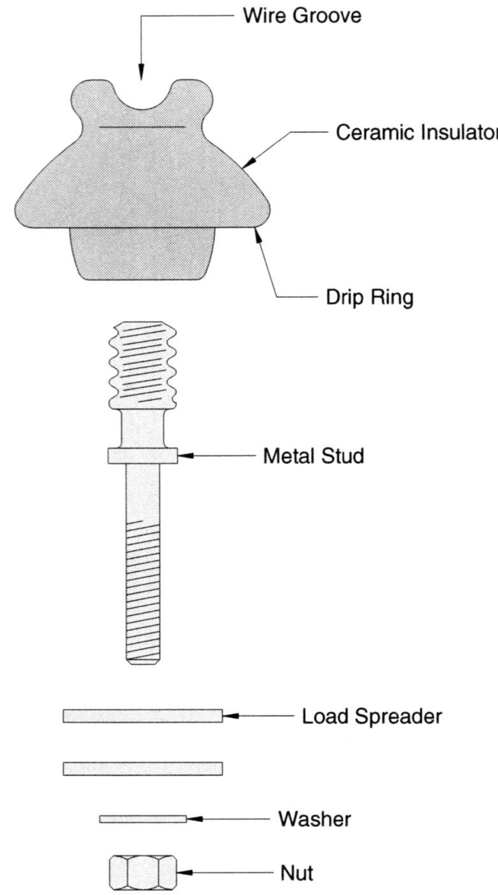

Figure 10-52 Modern Ceramic Pole Insulator

and are now sought after by antique collectors. Figure 10-51 shows a typical glass insulator of the early twentieth century. The transmission line was aligned with the wire groove and a second piece of wire was wrapped onto the line, around the opposite side of the groove and back around the line. These insulators were typically mounted on a threaded wooden post that was either nailed onto or driven into the pole.

Modern pole insulators do not differ much from their glass predecessors. Figure 10-52 shows a typical ceramic pole insulator. In order to support higher cable weights, the wire groove is on the top of the insulator and is laced onto the unit with a length of solid wire. These insulators are designed to screw directly onto a steel post, as shown. The steel post is generally inserted through a through hole in the cross beam and a pair of load spreaders, washer, and nut are used to secure the assembly.

Figure 10-53 shows a typical high-voltage pole insulator. High-voltage, cross country transmission lines require a greater level of isolation than a standard pole insulator can provide. The high-voltage unit is essentially a stack of standard insulators used to provide a greater stand-off voltage rating.

For high-tension applications, insulators are stacked as shown in Figure 10-54. Each insulator assembly can stand off a given voltage. Hanging ten insulators together multiplies the stand-off voltage by a factor of 10. As an example, if four insulators, with a stand-off voltage of 7500 each, are stacked together then the assembly can stand off 30,000 volts.

Figure 10-53 Ridged Mount, High-Voltage Insulator

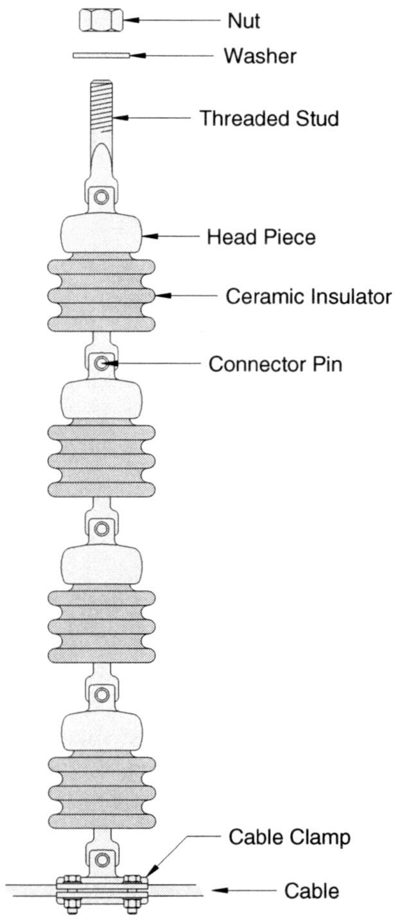

Figure 10-54 Stacked Hanging, High-Voltage Insulators

Antenna and guy wires must be isolated from ground and/or power lines. Antenna or *egg* insulators are commonly used for this application. Figure 10-55 shows two popular guy wire insulators. The upper illustration is designed for high-weight antennas and guy wires. The illustration at the bottom is for lighter weight loads such as amateur radio antennas.

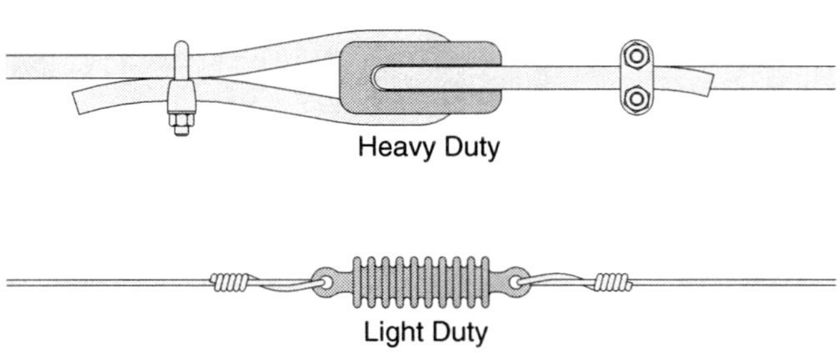

Figure 10-55 Antenna or Guy Wire Insulators

Electrical Feedthroughs

Various applications require that electrical signals pass through a hard barrier, such as a panel in a sealed cabinet or the wall of a pressure vessel. Bulk head fittings or feedthroughs are typically used for these applications.

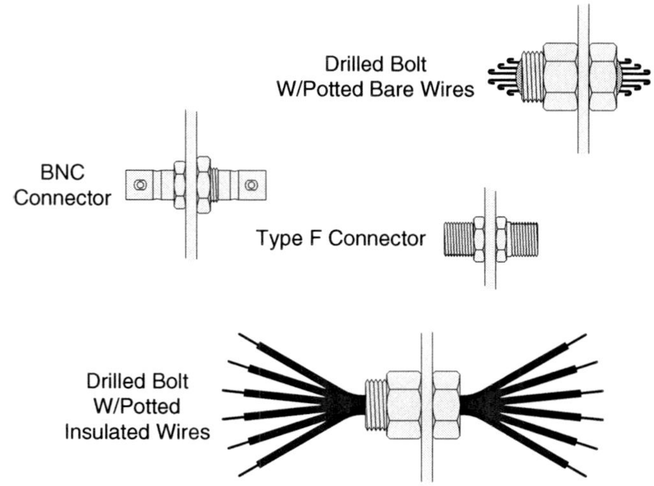

Figure 10-56 Electrical Bulk Head Fittings

Figure 10-56 shows a few commercial bulk head fittings. Many different types of connectors are available in bulk head configurations. These types of connectors are easy to install, provide an excellent seal, and are inexpensive. For soldered feedthroughs a bolt is drilled and wires are potted into the hole. Drilled bolt feedthrough may be constructed with insulated or uninsulated wires.

Low-differential pressure feedthroughs may use insulated wires, while high-pressure feedthroughs typically use bare, solid wires. Figure 10-57 shows a couple of standard high-pressure feedthroughs. The body is a through-drilled NPT (national pipe thread) bull plug. The wires are placed into the body and the through hole is flooded with a two-part, chemical set potting material.

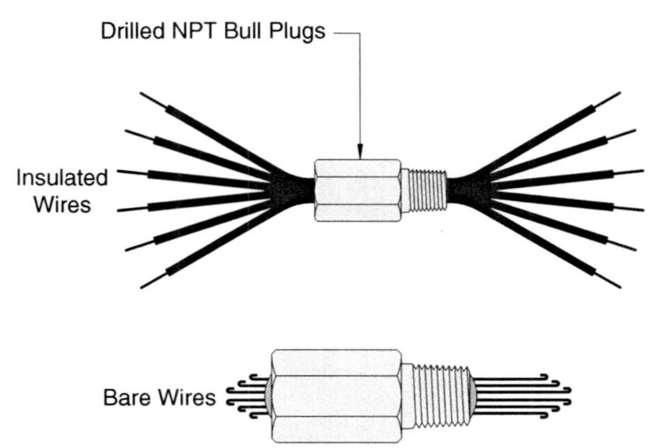

Figure 10-57 High-Pressure Feedthroughs

Building a high-pressure feedthrough is not a difficult proposition. Figure 10-58 can be used as a guide for constructing high-pressure feedthroughs. An NPT bull plug is drilled through on the center of its axis. A pattern of holes are drilled into a polytetrafluoroethylene (PTFE) plastic plate, which are spaced to align the necessary conductors. The conductors

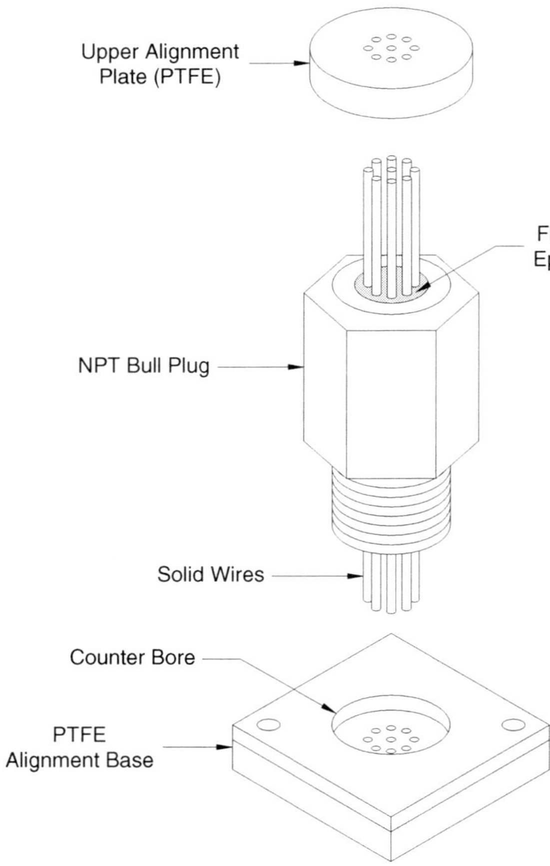

Figure 10-58 Bench Built High-Pressure Feedthrough

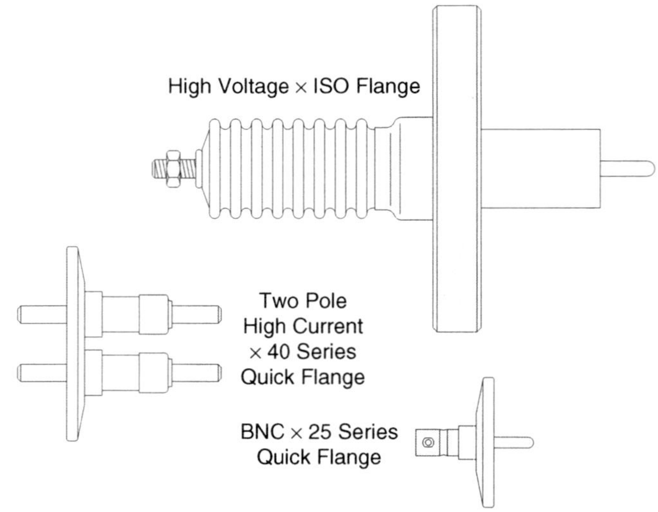

Figure 10-60 High Vacuum Feedthroughs

are placed into the holes and the plug is lowered into the counter bore. The counter bore is intended to center the plug around the conductors. The hole in the bull plug is flooded with a two-part chemical set glue or potting material. Before the glue sets, the upper alignment plate is set onto the conductors. After the glue sets, the plates are removed and the feedthrough is ready to be installed. It should be noted that PTFE plates are used because the glue will not adhere to this type of plastic and the fixtures can be easily removed and reused.

The same process can be used with a hose barb × NPT fitting, which allows an ordinary rubber hose to be used as the cable jacket. Figure 10-59 shows a hose barb fitting and hose assembly as a high-pressure feedthrough.

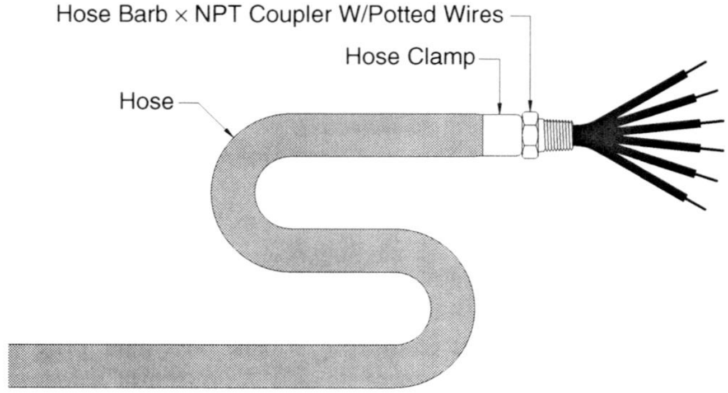

Figure 10-59 Water Tight Bench Built High-Pressure Feedthrough and Cable

High vacuum systems require a higher level of performance than glue or potting compounds can provide. In these cases, a combination of welding and soldering is generally used. Figure 10-60 shows a few high vacuum feedthroughs. The feedthrough is usually a connector or insulator assembly that is welded into a standard flange.

The ceramic-to-metal joint is made with an indium solder and provides an extremely clean and precise seal. Figure 10-61 shows a typical ceramic-to-metal soldered feedthrough. The stainless steel skirt is provided so that the assembly can be welded into a standard flange.

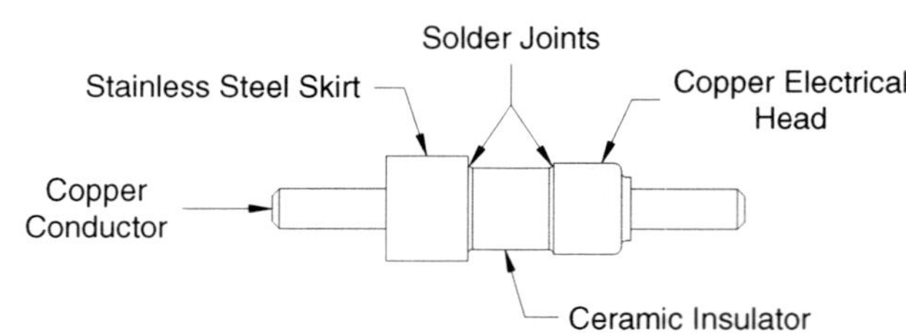

Figure 10-61 Ceramic-to-Metal Solder Joint

Conduit

Conduit is the pipe or tubing that electrical services are routed through. Not normally used in homes, it is required by electrical code in most commercial and industrial locations.

Conduit serves three basic functions, first it provides a guide through which conductors can be conveniently pulled. This makes installing large, convoluted distribution systems a much easier proposition. The second function is to protect the wires from being damaged from outside influences. Conduit is called on to provide mechanical, chemical, and weather protection. The third is to protect the outside world from the

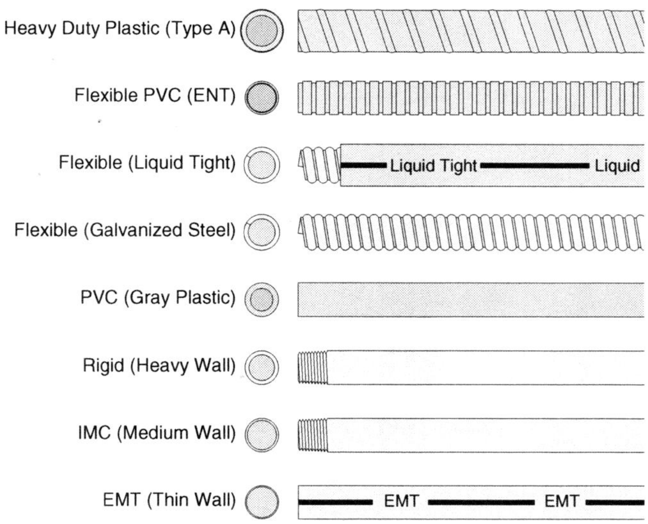

Figure 10-62 Common Commercial Conduit

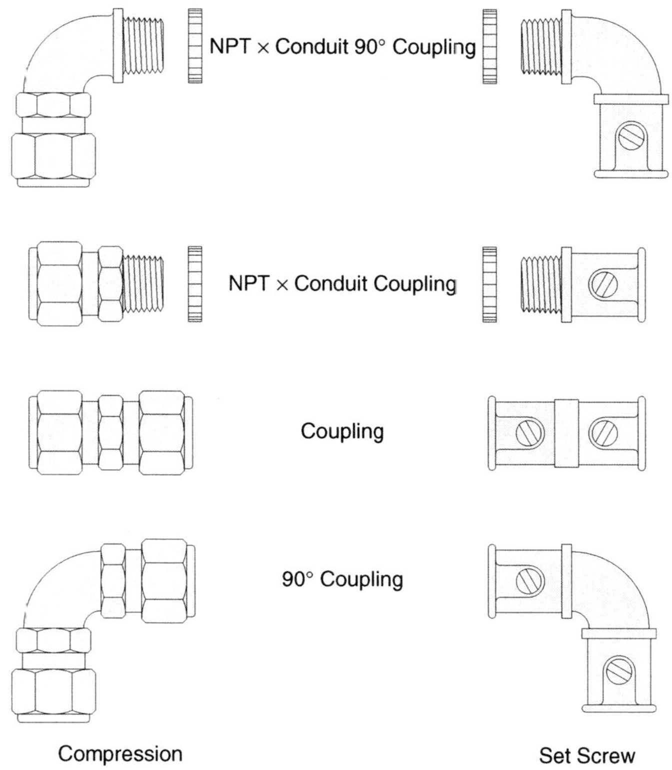

Figure 10-63 Commercial EMT Conduit Fittings

voltages the wires carry. Broken or frayed wires represent both an electrocution and fire hazard. Neither of these conditions can be tolerated in the residential, commercial, or industrial environments.

Figure 10-62 shows the most common conduit types. Electrical metallic tubing (EMT)conduit, polyvinyl chloride (PVC) conduit, and flexible galvanized conduit are the most common and are normally found in home, office, and commercial settings. Intermediate metal (IMC) conduit and rigid conduit are typically found in industrial applications. These conduits have considerably heavier wall than EMT and will provide much better protection in the harsh environments where they are installed. Intermediate metal conduit is the same material as schedule 10 pipe, while rigid is schedule 40 pipe. Liquid tight conduit is flexible metal conduit with an added plastic jacket. The jacket is water tight and provides moderate chemical protection. Flexible PVC conduit (electrical nonmetallic tubing, ENT) is generally reserved for wiring the internals of commercial equipment or for applications where protection requirements are minimal. Heavy duty plastic conduit (Type A) is generally used as a flexible power drop for commercial and industrial equipment. It is exceptionally rugged, liquid tight, chemical resistant, and provides vibration isolation.

Electrical metallic tubing conduit is generally connected with either a collet or set screw style fitting. Figure 10-63 shows a few commercial EMT fittings. The collet style fittings provide a stronger connection and are normally used in applications that are exposed; however they are difficult to connect. The set screw styles are generally used in protected settings and are very quick and easy to assemble.

Rigid and IMC fittings have NPT threads for assembly. This makes installing these types of conduit, essentially, a plumbing job. Figure 10-64 shows a selection of commercial rigid conduit fittings. These fittings are generally available in die cast aluminum and galvanized steel.

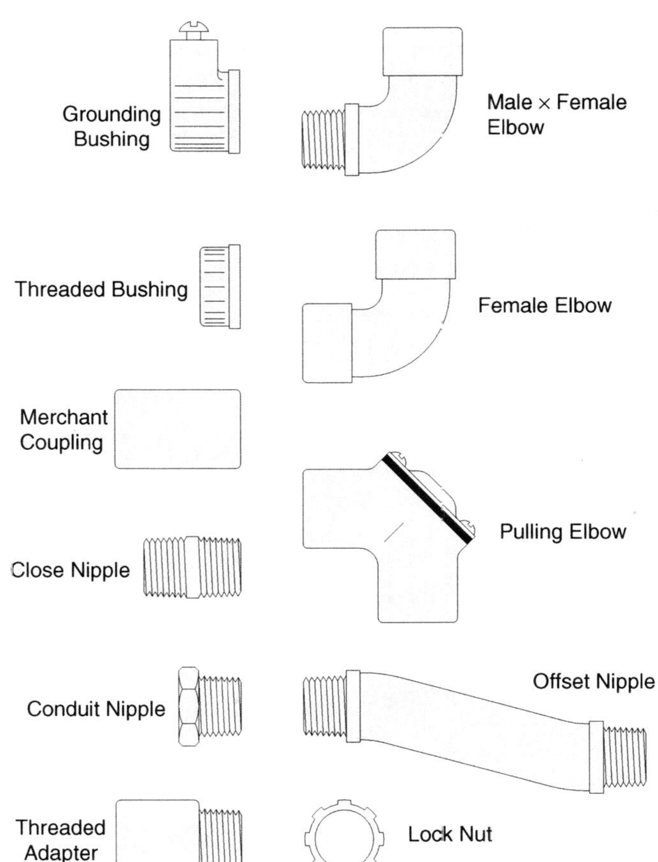

Figure 10-64 Commercial Rigid Conduit Fittings

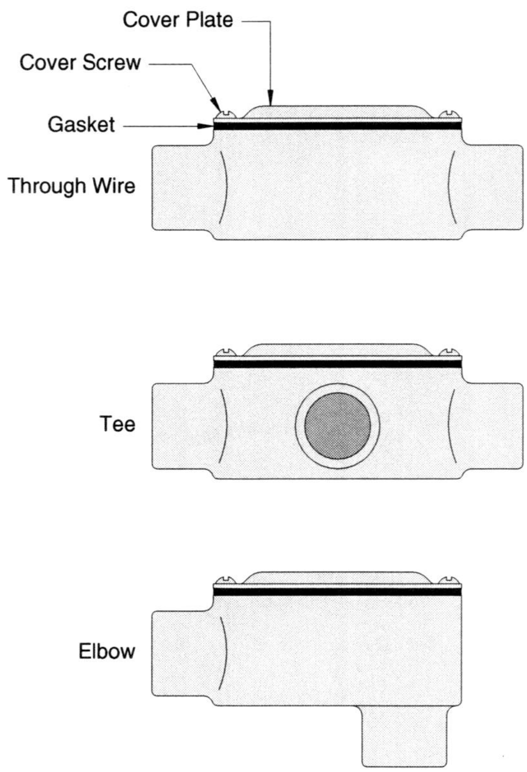

Figure 10-65 Conduit Access Ports

Figure 10-65 shows three basic conduit ports. These units provide ready access to the internal wires. They can be added to an existing system for expansion or they can be used to limit the length of wire pulls. These fittings are readily available for EMT, rigid, and PVC conduit.

Flexible metal conduit requires special fitting designs, as shown in Figure 10-66. These fittings are supplied in screw-on or screw clamp types and are commonly available through any electrical supply house or hardware store.

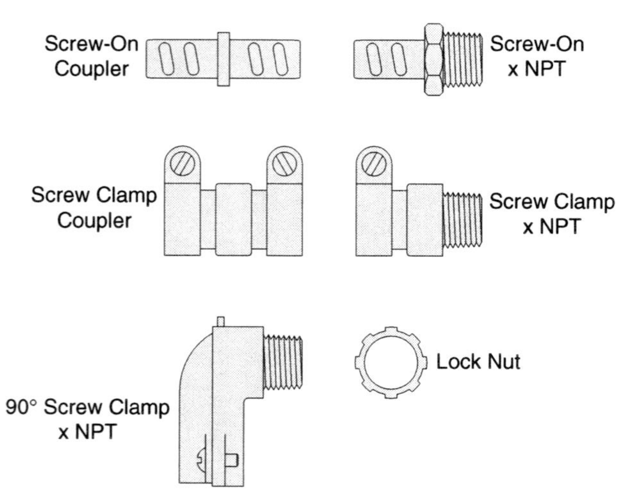

Figure 10-66 Flexible Metal Conduit Fittings

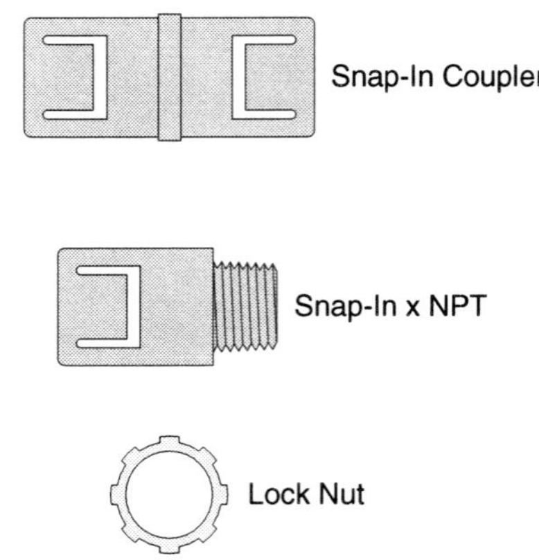

Figure 10-67 Flexible PVC Conduit Fittings

Flexible PVC conduit uses plastic snap-to-connect fittings, as shown in Figure 10-67. These fitting are exceptionally easy to connect. The conduit is cut to length and simply plugged into the fitting. To release the conduit, two tabs must be deflected out and away from the body of the fitting.

Figure 10-68 shows a few fittings that are used with liquid tight conduit. These fittings are collet type units with a rubber element to provide a liquid tight seal.

Heavy duty plastic conduit utilizes a series of special fittings, as shown in Figure 10-69. These fittings have a center piece to counter the crushing effect of the collet. They also have hand tight collet nuts and sealing collets.

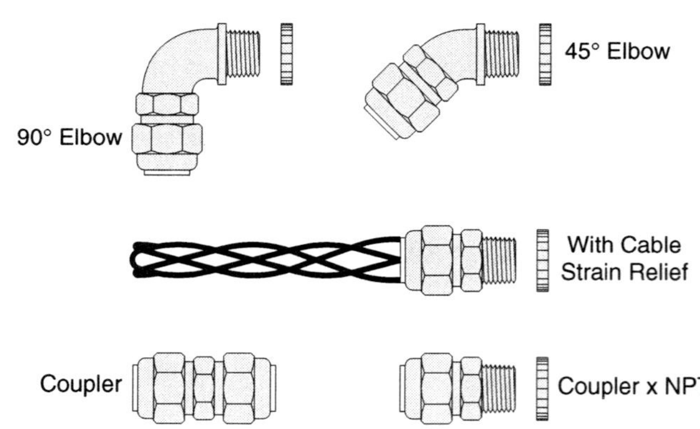

Figure 10-68 Liquid Tight Conduit Fittings

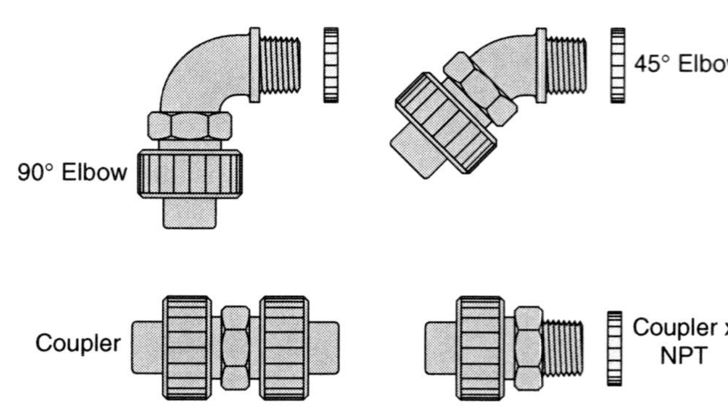

Figure 10-69 Heavy Duty Plastic Conduit Fittings

Service Heads

A service head is a specialized conduit fitting that is specifically designed to interface a building's electrical system to the power grid. Figure 10-70 shows a typical service head installation on a residential building. The head is typically mounted to the end of a piece of galvanized rigid conduit. The conduit penetrates the roof and is connected to the building's meter. The power cable is attached to an anchor insulator which is clamped to the conduit. The ground and power conductors are routed up and through the service head.

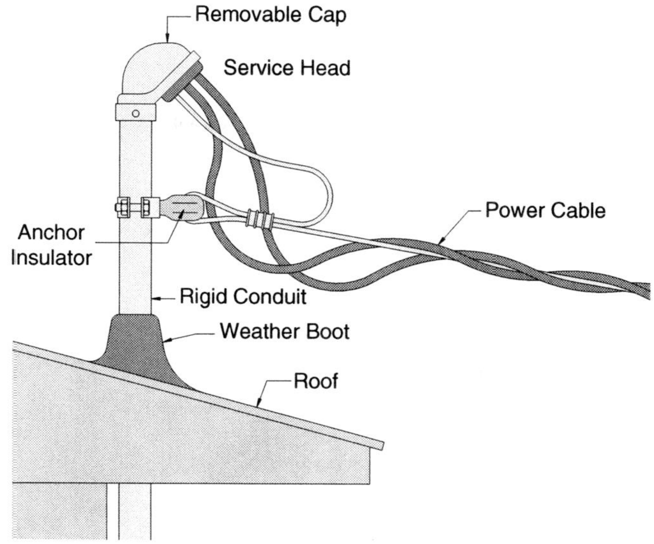

Figure 10-70 Service Head

Outlet and Switch Boxes

Figure 10-71 shows a selection of common switch and outlet boxes that are commonly available on the market today. Cover plates are as varied as the different devices to be mounted. Just a few of the more common cover plates are

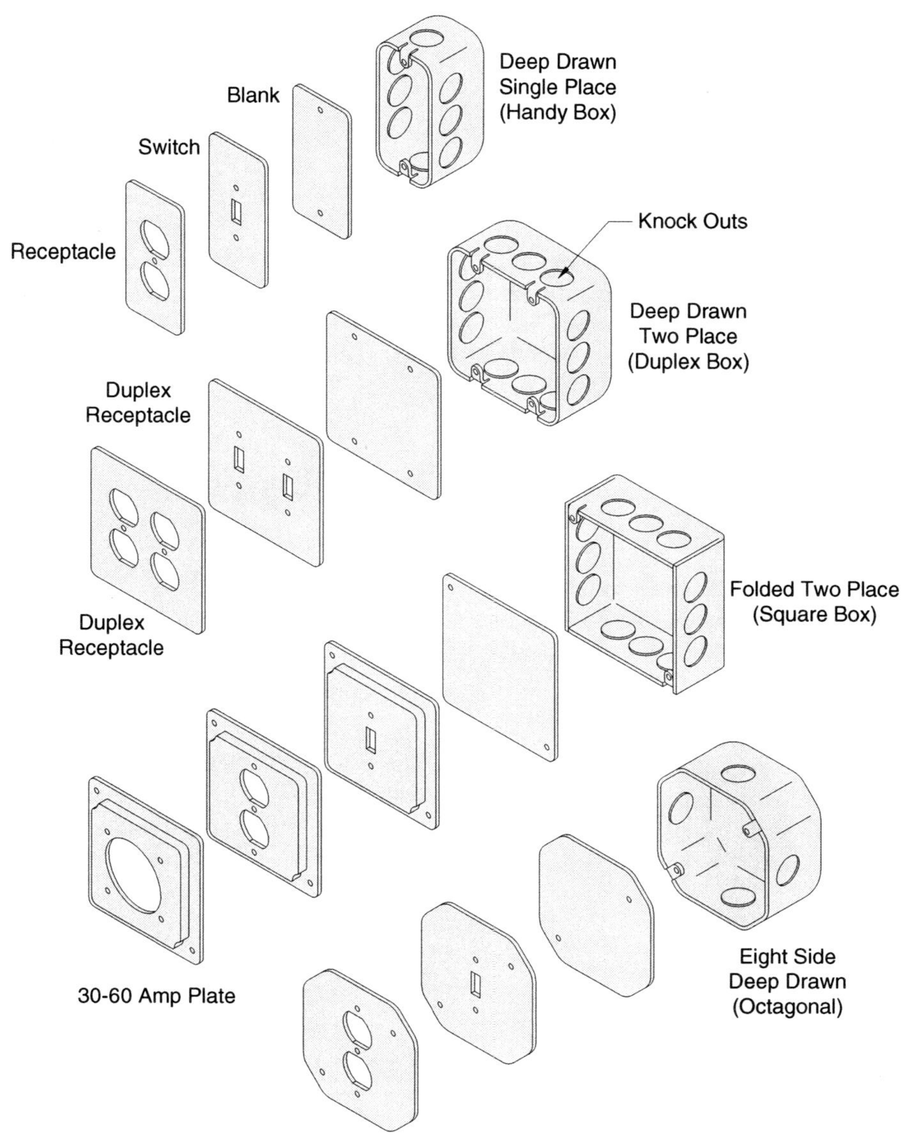

Figure 10-71 Outlet and Switch Boxes

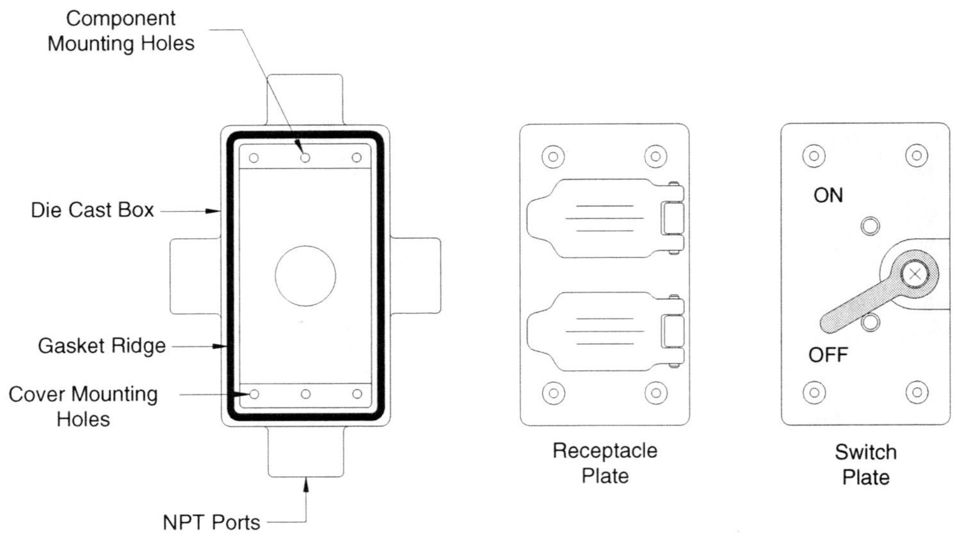

Figure 10-72 Outdoor Switch and Outlet Boxes

shown. These boxes are particularly handy for not only electrical wiring, but also as utility boxes for a number of bench projects. Most of these designs are also available in plastic and PVC versions.

Outdoor installations require that a certain amount of weather protection is provided. Outdoor outlet boxes are available that provide protection when they are not connected. These outlet boxes have a pair of spring loaded doors that close tight when the receptacle is not in use. Switch boxes use an ordinary switch and have a sealed plate with integral actuator that provides a weather seal. The box itself is generally equipped with NPT ports. Figure 10-72 shows a typical outdoor box, outlet plate and switch plate.

Standard NEMA Enclosures

The National Electric Manufacturers Association (NEMA) publishes standards for electrical enclosures. These standards cover construction requirements for various environmental conditions. Figure 10-73 shows examples of three of the most common NEMA enclosures. NEMA 1 is for indoor, relatively clean applications. NEMA 3R is designated as splash resistant and is good for protected outdoor locations. NEMA 4 and 4X are sealed cabinets and are appropriate for exposed outdoor environments and dirty industrial applications.

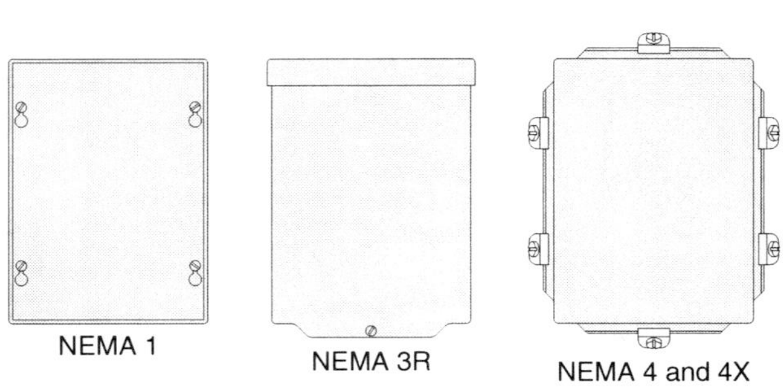

Figure 10-73 NEMA Standard Enclosures

Figure 10-74 provides a cross reference chart that matches standard NEMA enclosures with their environmental protection.

Hazardous locations, such as chemical plants, tank farms, shipboard, and grain silos, require explosion proof enclosures, as shown in Figure 10-75. These enclosures are generally a cast aluminum housing with a flange bolted top.

Installing Wire

After all the conduit is in place and all of the boxes and enclosures are installed, wire must be pulled through to inter-connect the system. To accomplish this a fish tape is used, as shown in Figure 10-76. The tape is a very stiff piece of flat steel that is coiled around a spool. The tape is rolled out and progressively pushed through the conduit until it protrudes out the far end. The wires are looped around the wire hook and taped smooth. The fish tape is then pulled back through the conduit, along with the wires.

For any given conduit size, a limited number of conductors can be installed. The chart given in Figure 10-77 shows the maximum number of thermoplastic high heat resistant nylon coated (THHN) conductors that any given size of conduit can typically support.

Raceway Systems

In some applications it is necessary or desirable to surface mount an electrical system. Raceway systems, as shown in

	NEMA Enclosure										
Environments	1	3R	4	4X	5	6	6P	7	9	12	13
Indoor	●				●					●	●
Indoor and Outdoor		●	●	●		●	●				
Rain and Light Splashing		●	●	●		●	●			●	
Dust, Lint and Fibers			●	●	●	●	●			●	●
Wash Down			●	●		●	●	●	●		
Light Oil and Coolant										●	●
Heavy Oil and Coolant											●
Corrosive				●			●				
Temporary Submersion						●	●				
Prolonged Submersion							●				
Hazardous Locations								●	●		
Class I, Div. 1, Group A,B,C and D								●			
Class II, Div. 1, Group E, F and G									●		

Figure 10-74 NEMA Standard Enclosure Environmental Guide Lines

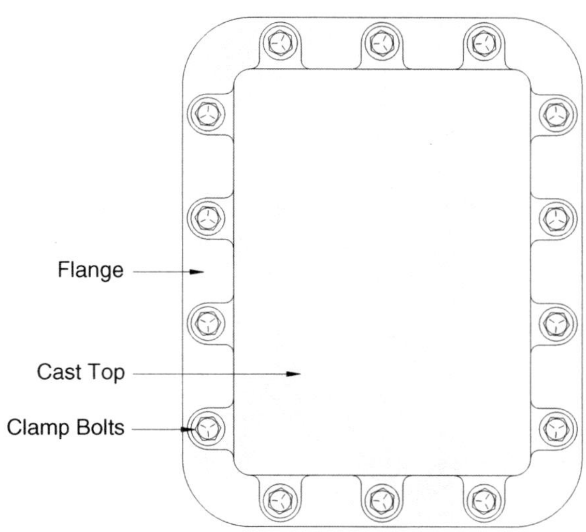

Figure 10-75 Explosion Proof Enclosure

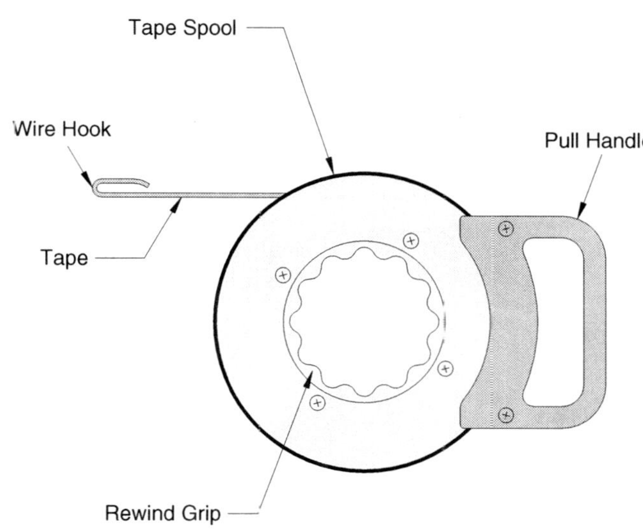

Figure 10-76 Fish Tape

	Conduit Size									
AWG	1/2 inchs	3/4 inchs	1 inchs	1-1/4 inchs	1-1/2 inchs	2 inchs	2-1/2 inchs	3 inchs	3-1/2 inchs	4 inchs
14	12	22	35	61	84	138	241	364	476	608
12	9	16	26	45	61	101	176	266	347	443
10	5	10	16	28	38	63	111	167	219	279
8	3	6	9	16	22	36	64	96	126	161
6	2	4	7	12	16	26	46	69	91	16
4	1	2	4	7	10	16	28	43	56	71
3	1	1	3	6	8	13	24	36	47	60
2	1	1	3	5	7	11	20	30	40	51
1	1	1	1	4	5	8	15	22	29	37

Figure 10-77 Maximum Number of Conductors in Conduit (THHN)

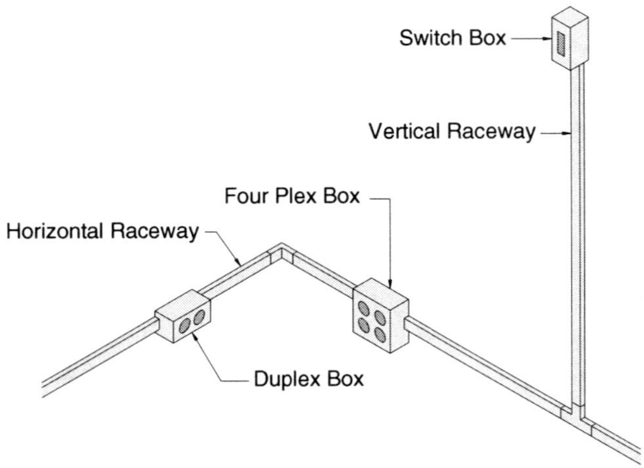

Figure 10-78 Raceway System

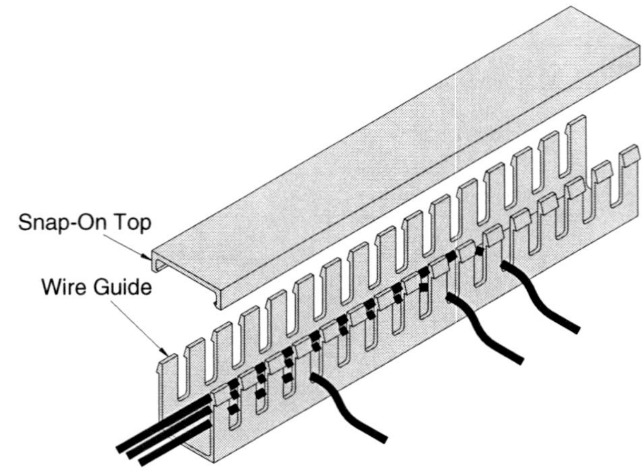

Figure 10-80 Wire guide

Figure 10-78, are designed for these applications. They are a complete system which consists of standard construction components that can easily be installed and wired.

Wire Duct

Wire duct is free standing ducting specifically designed to support wiring. These systems are commonly found on large equipment and in industrial environments. Figure 10-79 shows a few components of a typical wire duct system.

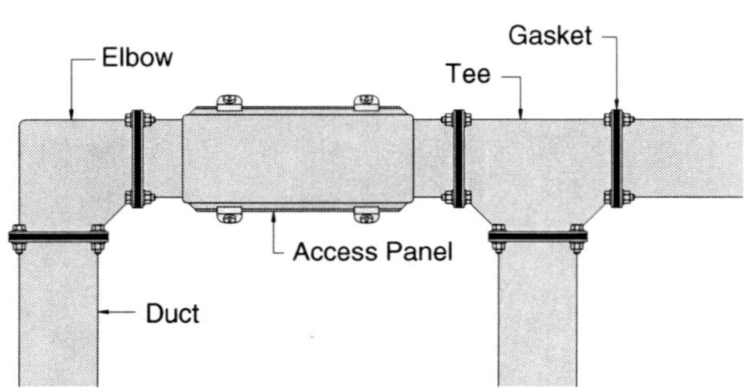

Figure 10-79 Wire Duct

Wire Guide

Wire guide is a convenient method to provide ducting for the inside of control cabinets. The guides are plastic trays with an array of vertical arms that make up the sides. Wires can be passed easily through the sides and down the guide. After wiring is complete, the guide has a snap-on top which provides a very clean appearance. Figure 10-80 shows a typical length of commercial wire guide.

Cable Protectors

Cable protectors are generally deployed anywhere a service cable must be routed across a heavy traffic area. Figure 10-81 shows two common cable protectors. Light duty protectors are made from molded or extruded plastic or rubber. They are appropriate for foot and light wheeled traffic. Heavy duty cable protectors are used to protect against vehicular traffic, such as light and heavy trucks and fork lifts.

For more permanent installations, a service trench is installed as shown in Figure 10-82. To prevent bending, the cover plate should be selected to support, at least, twice the highest anticipated load.

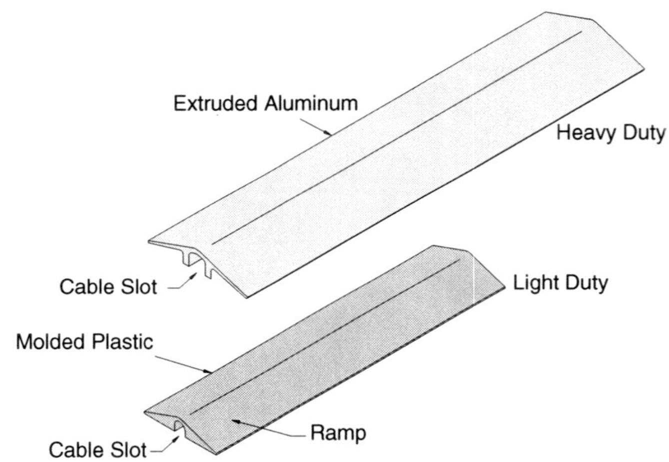

Figure 10-81 Cable Protector

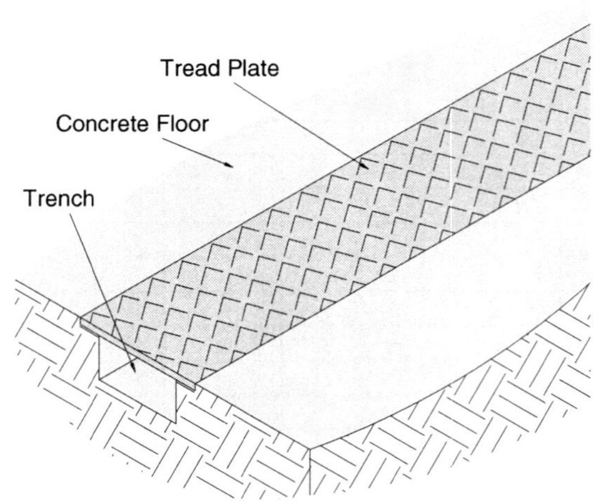

Figure 10-82 Service Trench

CHAPTER 11
ACOUSTIC DEVICES

One of the more common categories of electromechanical equipment is acoustic devices. An acoustic device is a piece of equipment that acts as an interface between electricity and sound waves. Therefore, any electromechanical device that is intended to produce or detect sound and/or vibration is considered an acoustic device. Things like bells, alarms, horns, loudspeakers, microphones, telephones, telegraphs, and vibrators are all acoustic in nature. These devices are all around us in our day-to-day lives. We hear the telephone when it rings, we pick it up and talk through the receiver. We listen to our radios and television sets, thanks to loudspeakers. When someone comes to our door, they press a button and we hear the door bell. Our microwave ovens alert us to the completion of the heating cycle with a beeping sound. And let's not forget that annoying alarm clock next to our bed.

Bells, Alarms, and Horns

There are a great many reasons to have audible signals that will alert us to various situations within our environments. As you stand talking to a coworker in the elevator, a gentle tone alerts you that you have arrived at your floor. The blaring siren of an emergency vehicle tells us to clear the way. The blast of a Klaxon tells the submarine crew that they are about to dive. Bells, alarms, and sirens are all around us and play an important role in our day-to-day lives.

Figure 11-1 shows a typical electric bell ringer. A bell is mounted to a base as shown. The base carries an electromagnet, point set, and clapper arm. When power is applied to the terminals, the electromagnet pulls the clapper arm and the clapper strikes the bell. At the same time, the point set is opened and the power to the magnet is disconnected. When the power is disconnected, the clapper arm is forced back into its original position by the leaf spring. When the arm moves back it closes the point set, which reconnects power to the magnet. A repeating cycle is generated, ringing the bell for as long as the power is connected to the terminals.

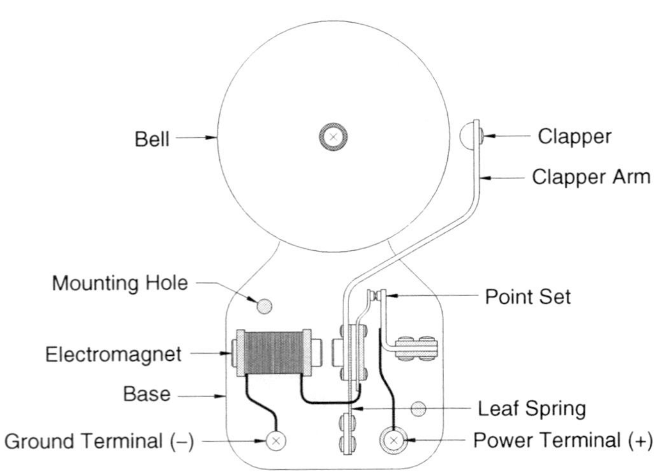

Figure 11-1 Bell Ringer

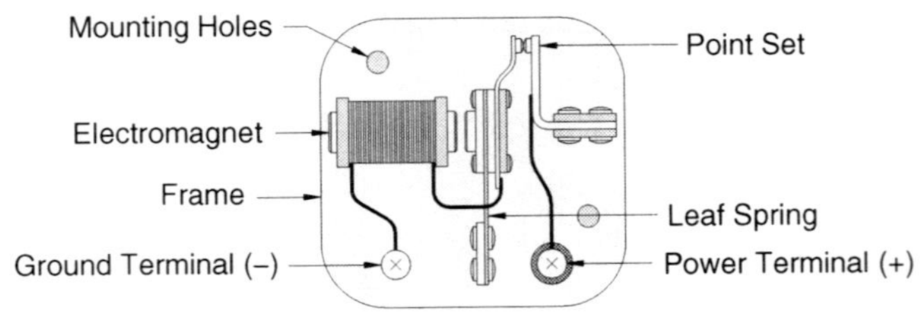

Figure 11-2 Buzzer

A buzzer, as shown in Figure 11-2, is essentially the same mechanism as the bell ringer. It operates on the same principle, except that the bell and clapper are deleted. A buzzer is generally specified in applications where background noise is low or in small, confined areas.

A two-tone door bell is an excellent example of the clever application of a solenoid coil. Figure 11-3 shows a typical two–tone door bell. When the door bell button is depressed, power is applied to the terminals and the plunger is pulled down, bouncing off the lower bar to generate the first tone. When the button is released, the recoil spring pulls the plunger up and it bounces off the upper bar to create the second tone.

Alarm horn mechanisms are used for applications that require a very loud audible signal. These units were commonly used to indicate diving and surfacing operations aboard submarines well into the 80s. Figure 11-4 shows a stylized schematic of an alarm horn mechanism. A metal diaphragm is attached to the base of a horn. A metal rivet is attached to the center of the diaphragm. A motor-driven wheel, which has a series of hammers around its circumference, is allowed to impact the rivet. While the motor is running, the hammering effect is amplified through the diaphragm and horn combination.

Figure 11-5 shows a typical marine alarm horn. These units are very robust and provide decades of service. Additionally, there are very few environments that will effectively drown out the noise produced by one of these devices.

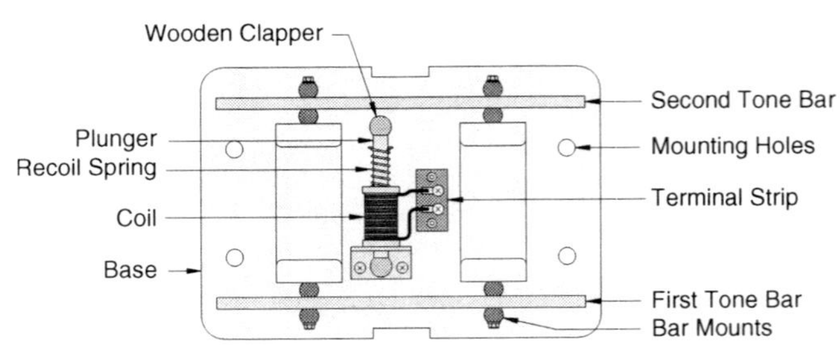

Figure 11-3 Two-Tone Door Bell

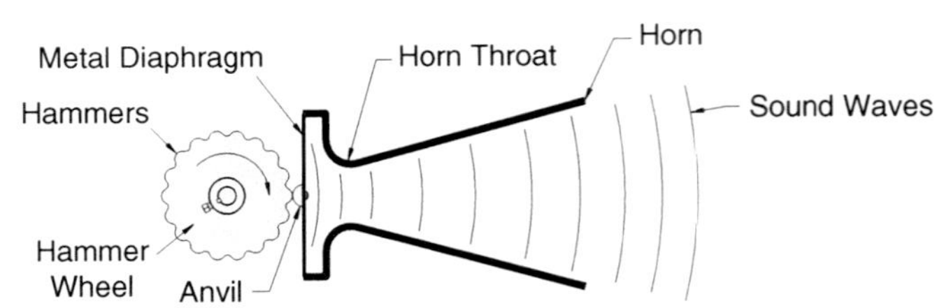

Figure 11-4 Alarm Horn Mechanism

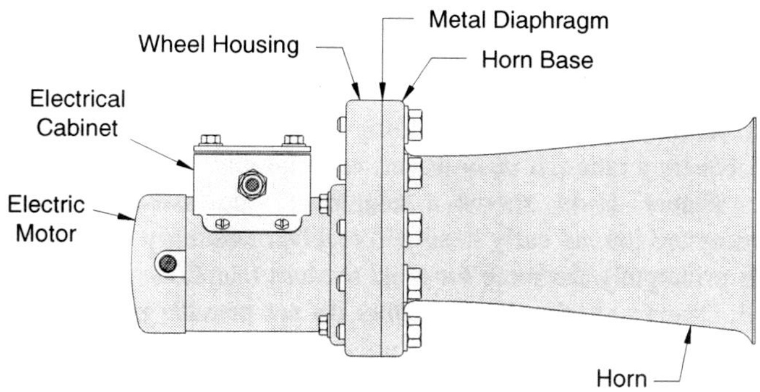

Figure 11-5 Marine Alarm Horn

Electric horns are particularly common in automotive applications. They produce a significant signal from a very compact, inexpensive, and reliable package. Figure 11-6 shows a stylized schematic representation of an electric horn. These units are similar to an alarm horn, except the rotary hammer is replaced with a solenoid mechanism that is similar in operation to the bell ringer or buzzer. A system like this allows much higher frequencies and smoother tones than an alarm horn or bell ringer. The tension of the return spring can be adjusted to change the tone of the horn.

Figure 11-7 shows a commercial electric horn such as might be found on small boats and trucks. These units are also an excellent replacement for the OEM (original equipment manufacturer) horn on most automobiles.

Loudspeakers

Most of us routinely receive audio information from loudspeakers. Audio is used in a myriad of applications, such as radios, televisions, stereo systems, telephones, public address systems, walkie-talkies, and even our personal computers.

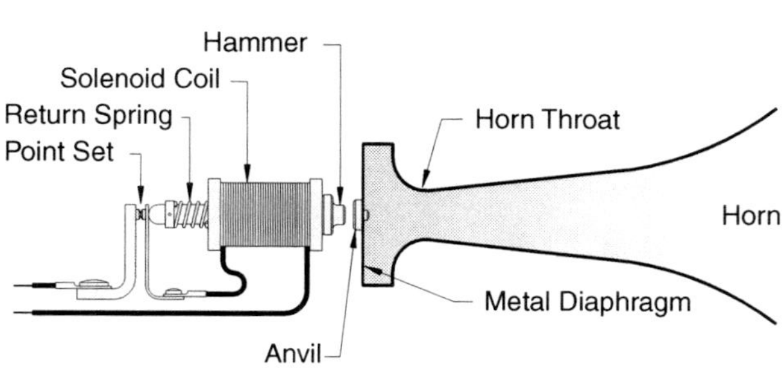

Figure 11-6 Electric Horn Mechanism

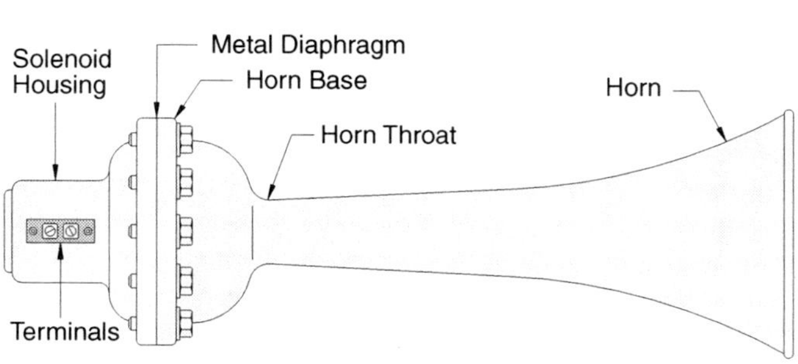

Figure 11-7 Electric Horn

In 1925, Chester Rice and Edward Kellogg of General Electric developed what is considered the modern, direct radiating, dynamic loudspeaker. This type of speaker has remained principally unchanged since its conception. Figure 11-8 shows a sectional view of a typical direct radiating, dynamic loudspeaker. A voice coil is positioned between the poles of a powerful permanent magnet. As a signal is applied to the coil, it is repelled or attracted to the magnet field in reference to the polarity and current of the signal. The coil form is fixed to the base of a conical diaphragm (cone). The cone movement is driven by the voice coil. As the cone moves, it pumps the air and creates sound pulses which mirror the electrical signal. The cone and coil assembly are suspended in a metal frame with two elements, the surround and the spider. The frame also carries a mounting flange and terminal strip.

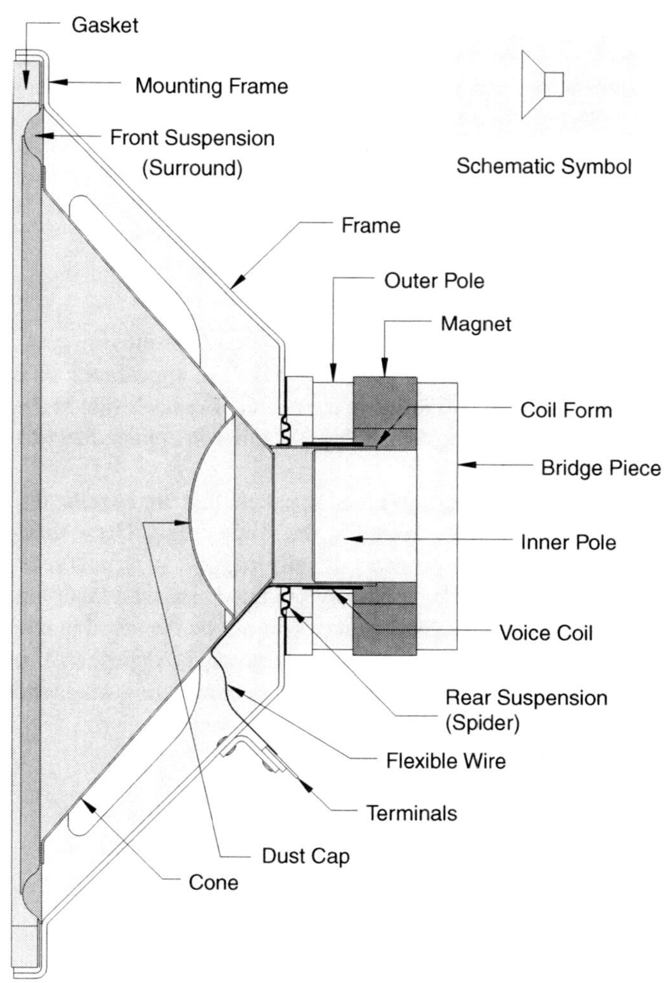

Figure 11-8 Dynamic Loudspeaker

The frequency range of any loudspeaker is limited by its diaphragm mass. To better reproduce sound in certain frequency ranges, designs are created to perform over limited ranges. Figure 11-9 shows a dynamic loudspeaker that is specifically designed to reproduce sound in the high frequency range. These units are generally referred to as tweeters. They are similar to the cone-type unit, except the cone is

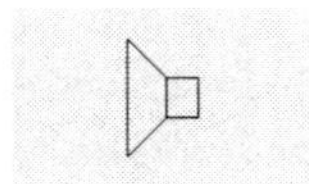

Schematic Symbol

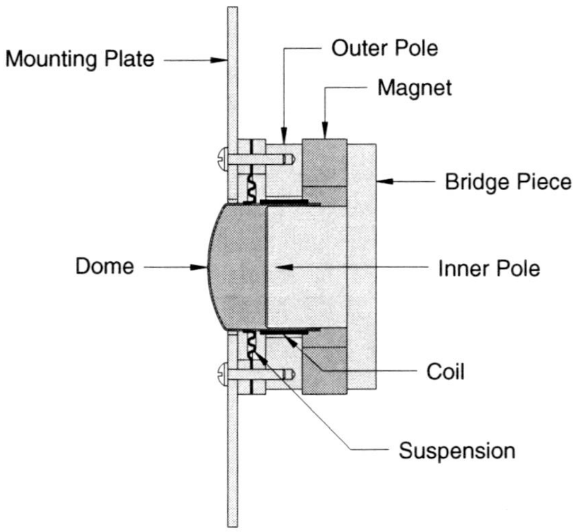

Figure 11-9 High Frequency Loudspeaker or "Tweeter"

eliminated so as to lower the overall mass of the moving components. These speakers generally have the appearance of a plate with a small soft dome in the center. The back side of the plate mounts to a magnet set that is similar in appearance to a cone unit.

Telephone receivers use loud speakers that are specifically designed to reproduce sound in the voice range. These units are also designed to be exceptionally rugged. A typical telephone receiver can be repeatedly pounded against a table top in frustration and the loudspeaker will not be damaged in any way. Figure 11-10 shows a typical telephone loudspeaker. The design is based on a pair of coils embedded into a magnetic

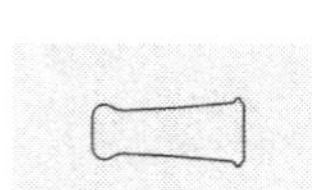

Schematic Symbol

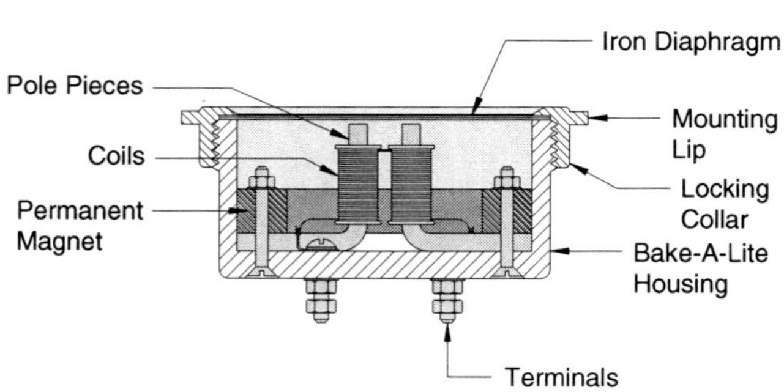

Figure 11-10 Telephone Receiver

field. The diaphragm is a tensioned piece of thin iron. When a signal is applied to the coils, the diaphragm deflects in direct reference to the varying field. Although this design is not exceptionally efficient, has fairly high distortion and limited frequency range, it is, however, very durable.

Figure 11-11 shows a telephone loudspeaker element mounted into an early handheld receiver assembly. Mounting is principally the same for most modern telephone handsets.

Direct radiating loudspeakers do not provide particularly good efficiency. In the home environment, this is usually not a problem. However, in the case of public address systems this can be a significant problem. To improve efficiency and, therefore, increase the volume of a loudspeaker, horns are matched to the driver as shown in Figure 11-12. The horn acts as an acoustic transformer and can significantly increase the output volume of the speaker.

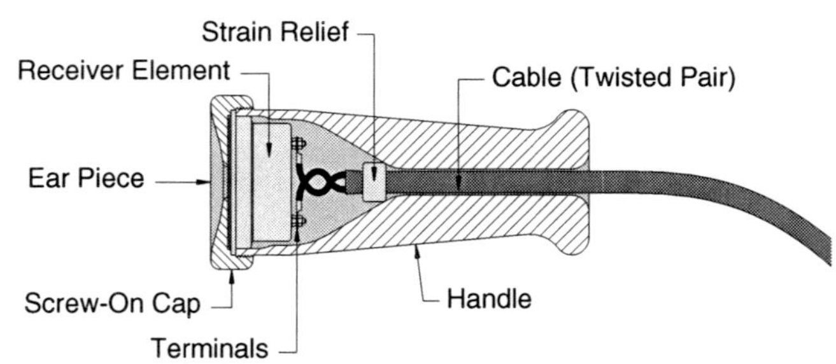

Figure 11-11 Handheld Telephone Receiver Assembly

Schematic Symbol

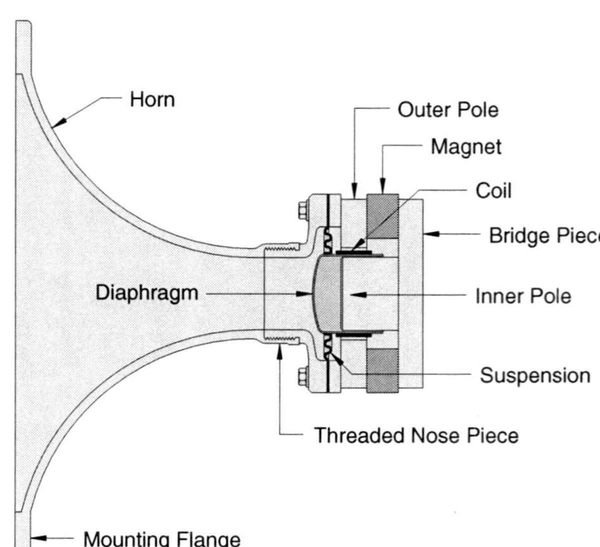

Figure 11-12 Horn Loaded Loudspeaker

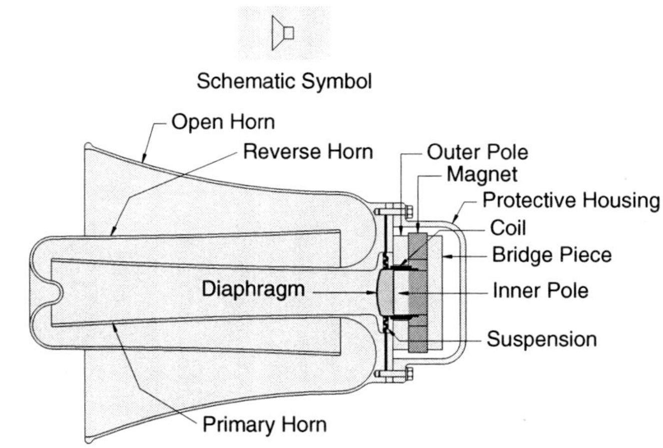

Figure 11-13 Folded Horn Loudspeaker

In some applications, the length and size of the horn can be prohibitive. In these cases, the horn is folded, as shown in Figure 11-13. The primary horn feeds a reverse horn which, in turn, feeds the open horn. This arrangement effectively shortens the horn length to one-third of its unfolded length. These types of speakers are very common in outdoor and industrial settings.

Another common use for the folded horn is in handheld public address systems, or megaphones. Figure 11-14 shows a typical commercial megaphone. The loudspeaker is mounted on the front of a housing that encloses an amplifier, battery, and microphone. A volume control and handle are mounted to the housing. Most units have a trigger switch to turn the amplifier on and off.

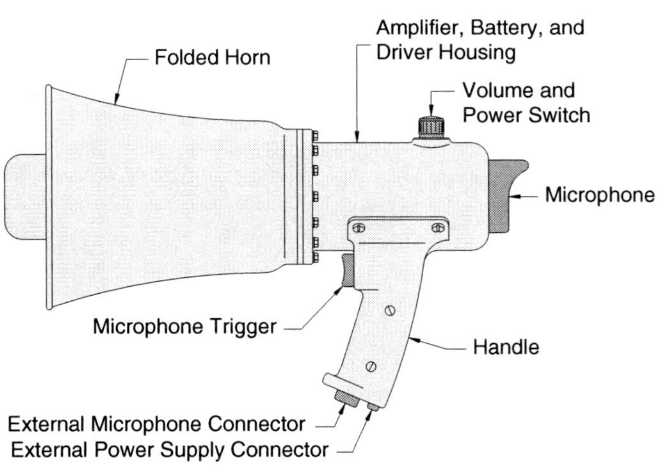

Figure 11-14 Handheld Public Address System or "Megaphone"

Ribbon element loudspeakers are generally used in high performance, high frequency, sound reproduction, such as home and studio applications. Figure 11-15 shows a stylized view of a typical ribbon tweeter. A corrugated, metalized ribbon is positioned between the poles of a strong permanent magnet. A signal is applied across the length of the ribbon and the foil deflects in reference to the polarity and current of the signal. As the ribbon moves, it pumps the air and creates sound pulses which mirror the electrical signal.

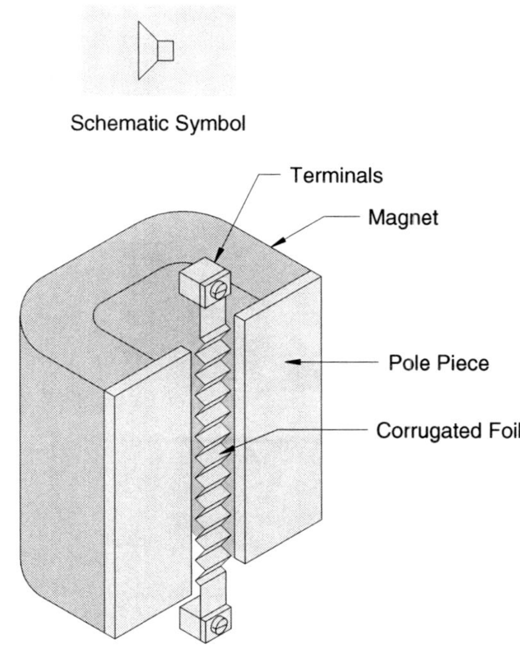

Figure 11-15 Ribbon Element Loudspeaker

Planar loudspeakers are made by stretching a large plastic diaphragm across a frame. Pancake coils are bonded to the diaphragm and a series of strip magnets are mounted in close proximity to the coils. As a signal is applied to the coils, the diaphragm deflects and produces sound. These types of speakers are reasonably efficient because of their large diaphragm. Figure 11-16 shows how a planar speaker is constructed.

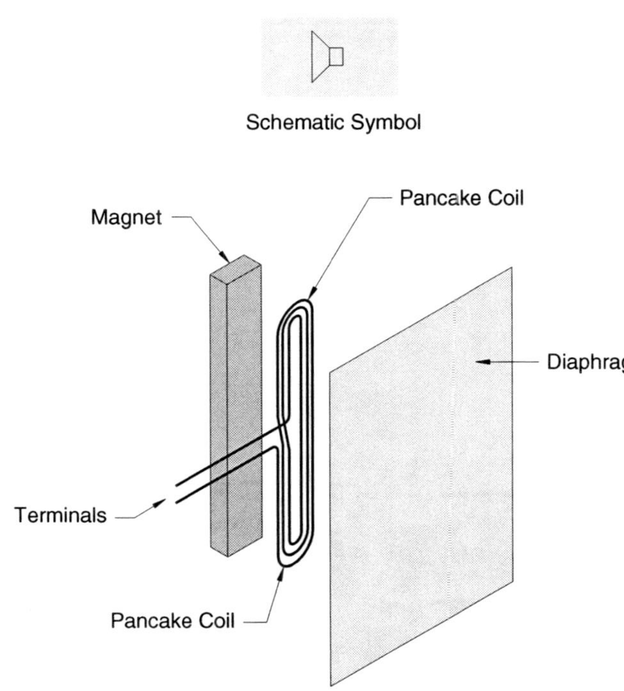

Figure 11-16 Planar Loudspeaker

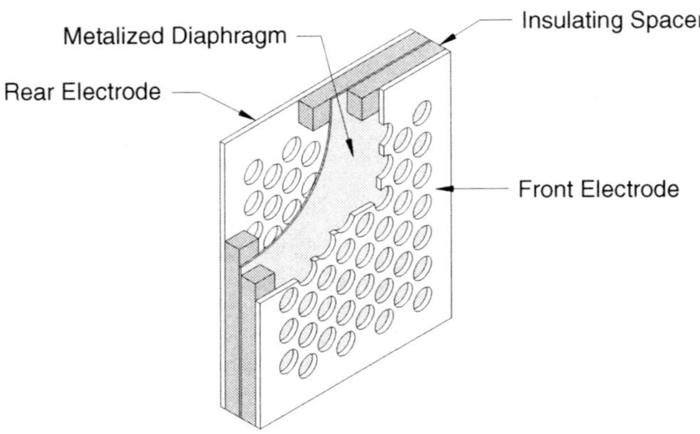

Figure 11-17 Electrostatic Loudspeaker Element

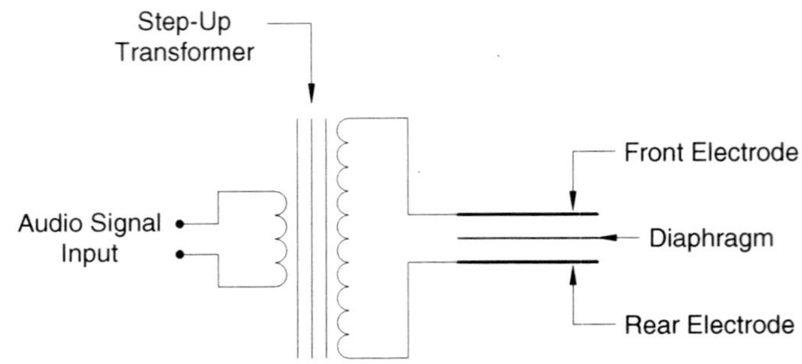

Figure 11-19 Electrit Loud Speaker Schematic

Electrostatic loudspeakers are a type of planar unit. In this case, a metalized diaphragm is spaced between two perforated electrodes, as shown in Figure 11-17. A signal is applied to the electrodes and the diaphragm deflects in reference to the polarity and voltage of the signal. As the diaphragm moves, it pumps the air and creates sound pulses which mirror the electrical signal.

Figure 11-18 shows a schematic representation of an electrostatic Loudspeaker system. The diaphragm requires a quality, high-voltage power supply, and the input signal is fed through a step-up transformer. These speakers offer good efficiency and excellent sound quality but because of their support equipment and internal voltages, they are typically rather expensive. They are generally only used in high-performance applications such as home or studio applications. One application where this technology performs exceptionally well, is high-performance headphones. The headphones are very light weight, can enclose the entire ear, and produce extremely high quality sound.

A variation of the electrostatic loudspeaker is the electric design. In this case the diaphragm is permanently charged, which eliminates the requirement for the high-voltage power supply. Electrit loudspeakers provide nearly as good sound quality as their electrostatic counterparts at a considerably lower cost. Figure 11-19 shows a schematic representation of an electric loudspeaker system.

Plasma loudspeakers are based on modulating an ionized plasma cloud. Figure 11-20 shows a schematic representation of a plasma speaker system. A power supply is used to create a plasma between two electrodes. A coupling transformer is placed in the output loop. The signal is applied to the input of the transformer which, in turn, modulates the plasma in reference to the polarity and current of the signal. As the plasma modulates, it couples with the air and creates sound pulses which mirror the input signal.

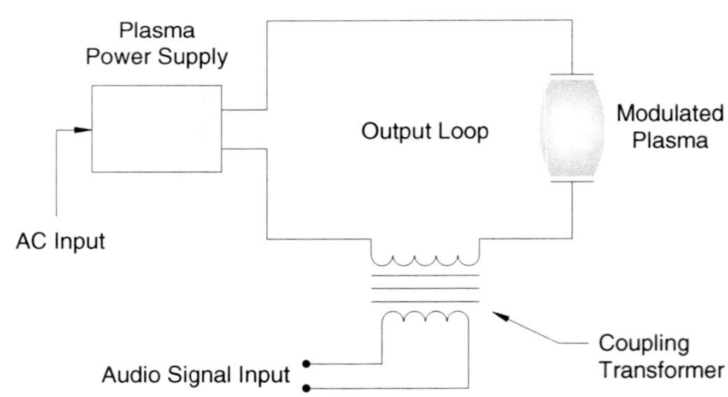

Figure 11-20 Plasma Loudspeaker Schematic

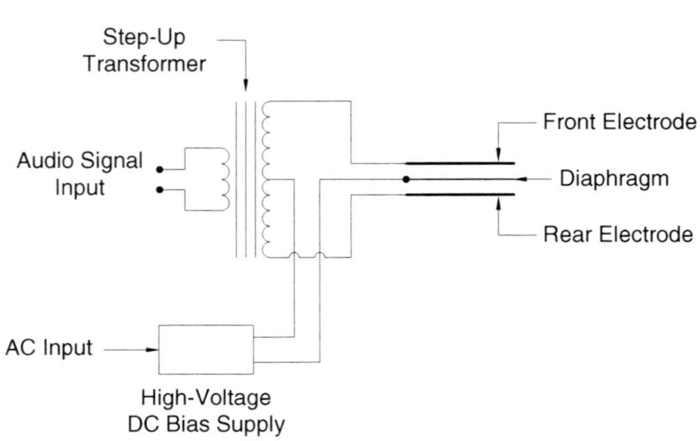

Figure 11-18 Electrostatic Loudspeaker Schematic

Full range, high-performance loud speaker systems, as used in the home and studio, are typically manufactured using two or more drivers. Each driver is selected to reproduce sound within a certain frequency range and to complement other drivers in the finished assembly. Figure 11-21 shows examples of both two- and three-way speaker cabinets.

To divide the frequencies of the electrical signal being fed to a multidriver system, a crossover network is deployed. Figure 11-22 shows a basic first-order crossover network. The network consists of a single conductor, which passes low-frequency power to the woofer, and a single capacitor, which passes high-frequency power to the tweeter.

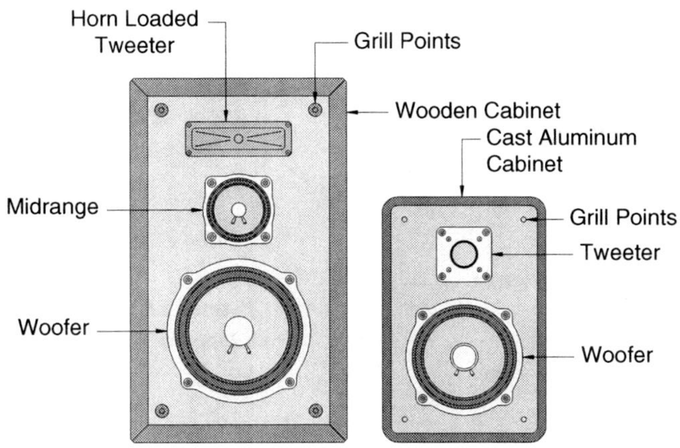

Figure 11-21 Two- and Three-Driver Loudspeaker Cabinets

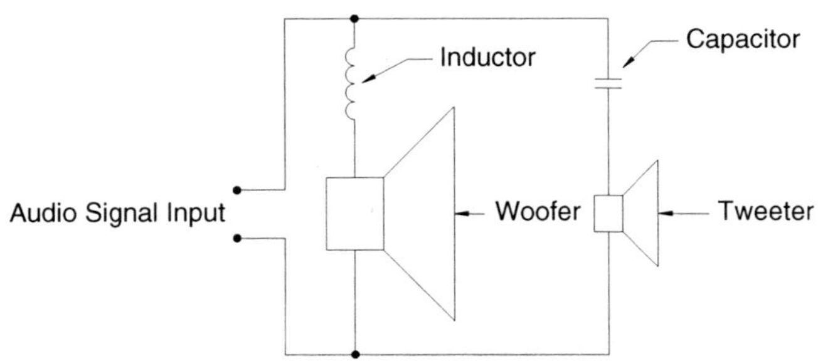

Figure 11-22 Two-Way Passive Crossover Schematic

Stereo sound reproduction systems are designed to provide a spatial sense or depth to the listing experience. This is accomplished by using a two-channel reproduction system, as shown in Figure 11-23. Careful attention must be paid to the placement of the speakers and the acoustics of the room to properly reproduce a stereo signal.

Another common application for loudspeakers is headphones. In this case, one or two small drivers are mounted on

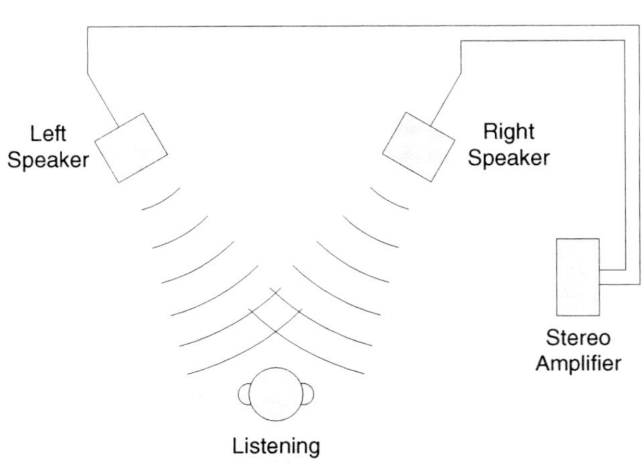

Figure 11-23 Stereo Sound Reproduction System

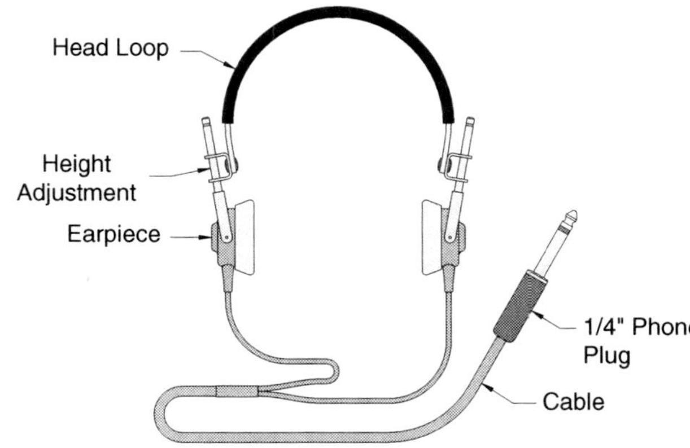

Figure 11-24 Headphones

an adjustable head loop, as shown in Figure 11-24. Headphones are particularly applicable in applications that have very low signal strength and in environments that have high ambient noise.

Microphones

Much like loudspeakers, but to a lesser extent, microphones also play an important part in our lives. The most noteworthy applications are our telephones. However, the sound reproduced by loudspeakers is almost entirely dependent on microphones. Without them, music couldn't be recorded, newscasters couldn't do their jobs, walkie-talkies would be of no value, and sporting events would be a lot more difficult to follow.

Figure 11-25 shows a sectional view of a basic carbon microphone. This is one of the oldest microphone designs still

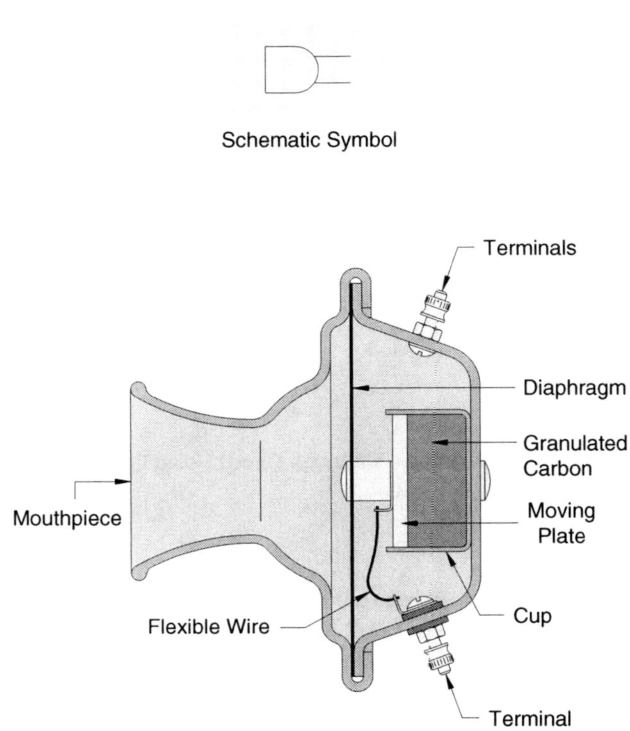

Figure 11-25 Carbon Microphone

in use today. A cup is packed with granulated carbon particles and capped with a moving plate. The moving plate is connected to a diaphragm mounted at the base of a mouthpiece. When a person speaks into the mouthpiece, the diaphragm vibrates and transfers those vibrations to the moving plate. As the plate moves, the carbon is packed tighter or allowed to relax based on the diaphragm vibrations. As the granules move, the resistance of the carbon charge changes in direct reference to the sound.

By placing a loudspeaker (receiver) and a pair of batteries in a loop with the microphone, the current of the loop can be controlled by speaking into the microphone. Figure 11-26 shows a carbon microphone circuit.

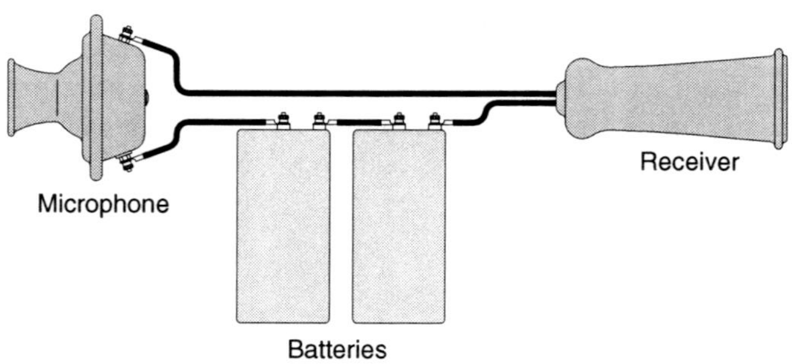

Figure 11-26 Carbon Microphone Circuit

For two-way communication, two sets consisting of a carbon microphone, battery, coupling transformer, and receiver can be arranged as shown in Figure 11-27. A simple system like this can provide a reasonably good communications link over several miles of cable. In some basic systems a "push-to-talk" button is added to disconnect the batteries when not in use.

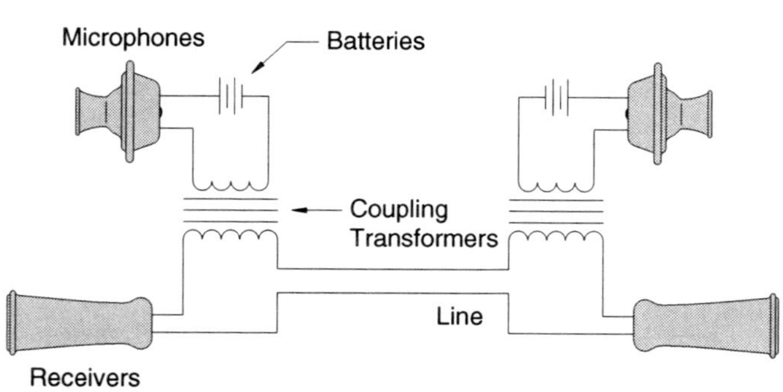

Figure 11-27 Two-Way Telephone Circuit Using Carbon Microphones

Dynamic microphones are very similar to a dynamic loudspeaker. In fact, many small loudspeakers are used as microphones in all sorts of commercial and industrial equipment. A prime example is walkie-talkies. The loudspeaker and microphone are usually the same component. Figure 11-28 shows a sectional view of a typical dynamic microphone.

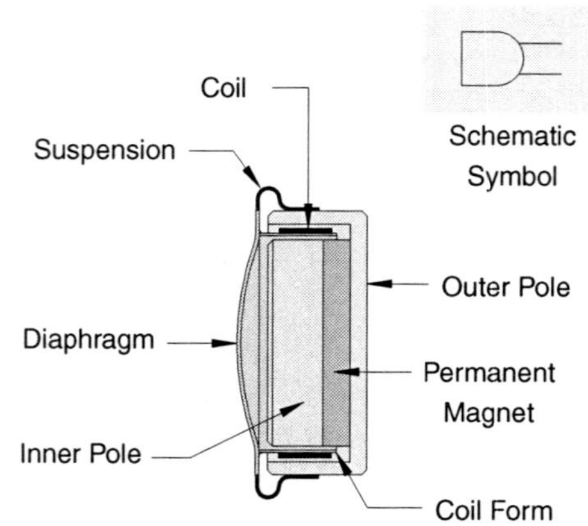

Figure 11-28 Dynamic Microphone Element

Dynamic microphones are normally found in use with public address systems. High-performance versions are used in recording studios and on stage. Figure 11-29 shows a high-performance, commercial dynamic microphone. Note the on/off switch and the use of an XLR connector.

Piezoelectric, or crystal, microphones have provided an inexpensive design for decades. These units take advantage of the piezoelectric effect. Some crystals, most notably Rochelle salt (potassium sodium tartrate), will produce an electrical signal if they are deflected. By connecting a diaphragm to the crystal, vibrations can be made to deflect the crystal and produce a signal in reference to sound. Figure 11-30 shows a schematic representation of a piezoelectric, or crystal, microphone.

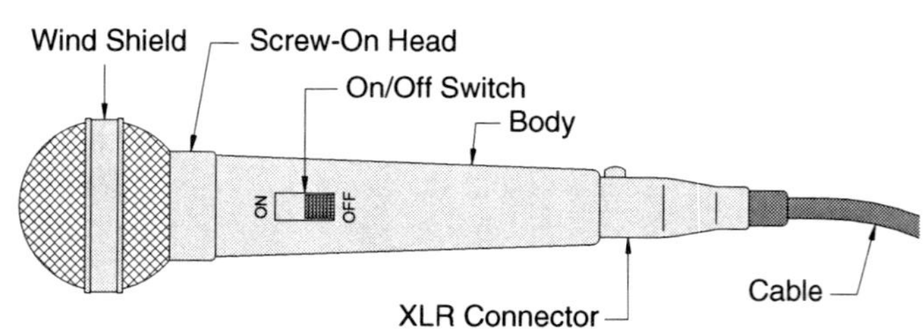

Figure 11-29 Typical Commercial Dynamic Microphone

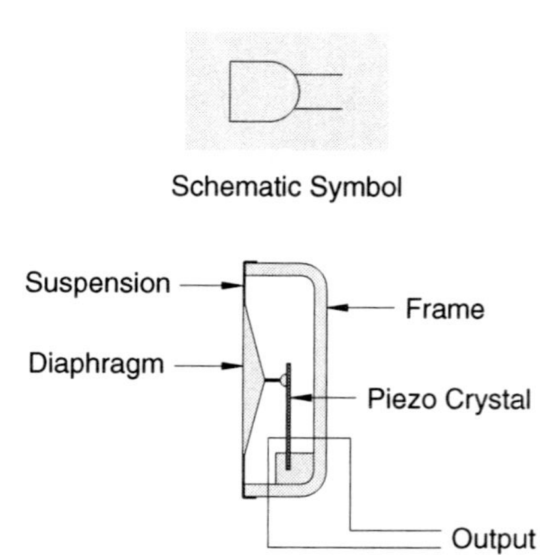

Figure 11-30 Piezoelectric or "Crystal" Microphone Element

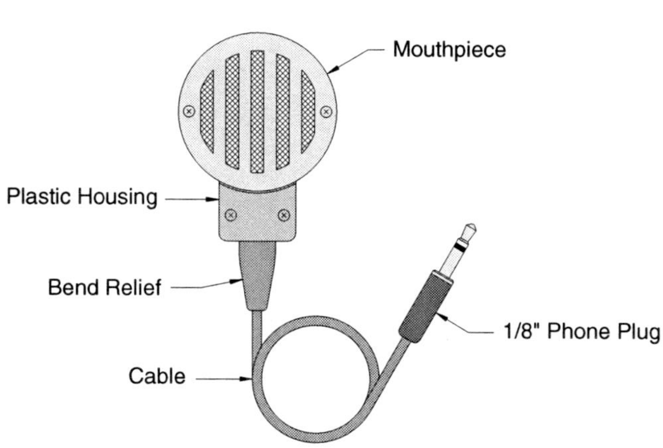

Figure 11-31 Piezo Crystal Microphone

Crystal microphones are most commonly found in inexpensive units, as shown in Figure 11-31, or as musical instrument pickups. These units are very inexpensive and generally produce a good quality output.

Condenser microphones operate on a variable capacitance principle. Figure 11-32 shows a schematic representation of a basic condenser microphone. A diaphragm is vibrated in reference to a fixed electrode. A small local battery provides power to the circuit. As the diaphragm vibrates, the capacitance of the circuit changes in reference to the sound. The output of the element is processed through a preamplifier and into the audio equipment.

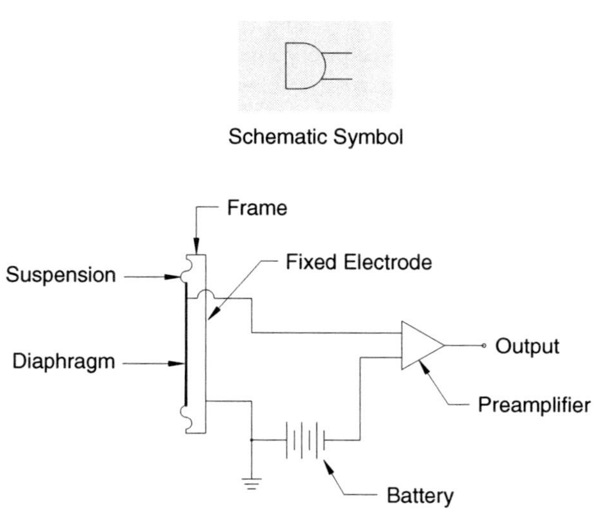

Figure 11-32 Condenser Microphone Schematic

Condenser microphones produce extremely good performance at a very low cost. The cartridge shown in Figure 11-33 cost only a few dollars to purchase and its performance rivals most professional recording units.

Figure 11-34 shows a typical commercial condenser microphone. These units are simply a cartridge installed into a housing with an on/off switch and battery compartment.

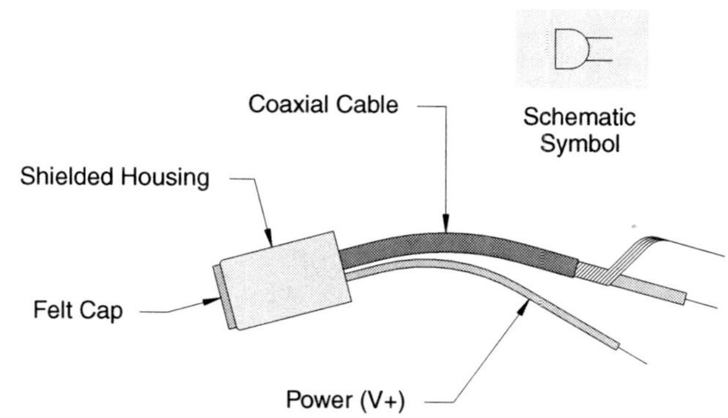

Figure 11-33 Condenser Microphone Cartridge

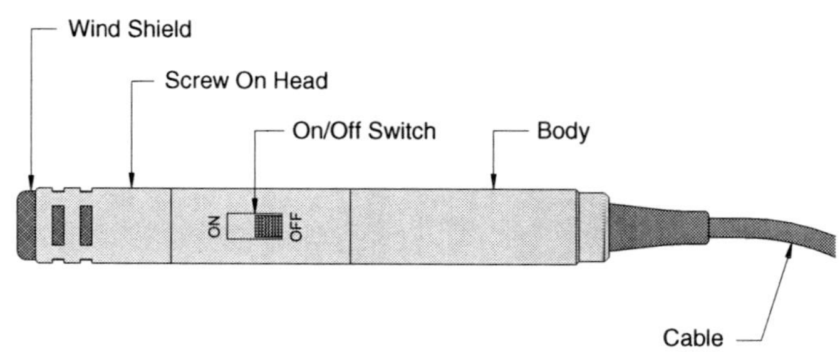

Figure 11-34 Commercial Condenser Microphone

Microphone sensitivity is important to understand. Sensitivity is generally measured radially from the microphone element. Figure 11-35 shows a typical sensitivity chart used to plot the performance of microphones. The curve shows the sensitivity of the unit at different locations surrounding the microphone.

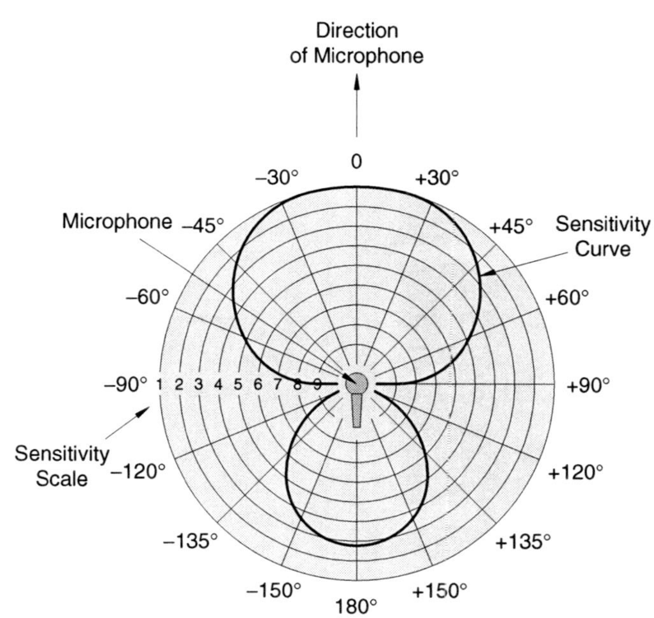

Figure 11-35 Microphone Sensitivity

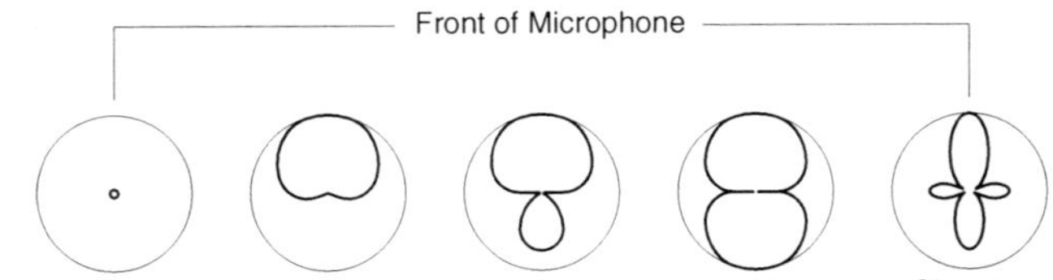

Figure 11-36 Microphone Patterns

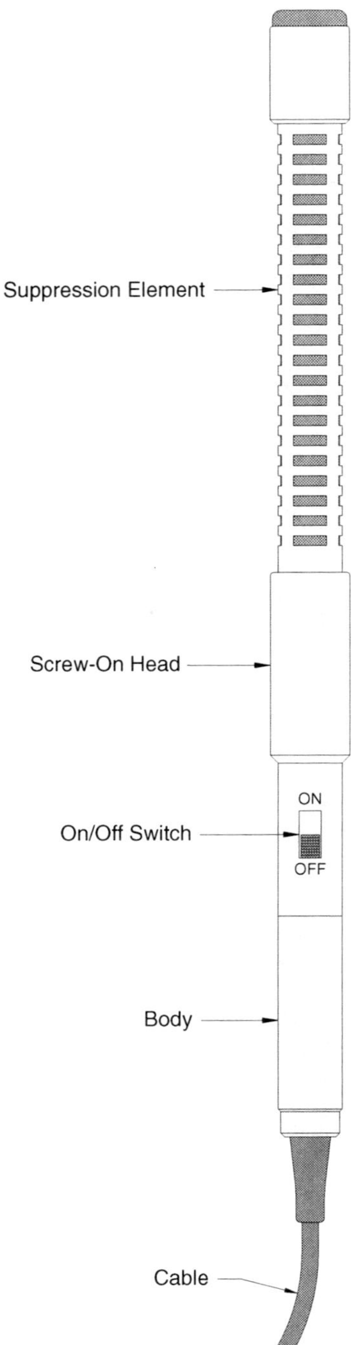

Figure 11-37 Commercial Shotgun Microphone

Microphones are generally designed to one of the five basic patterns shown in Figure 11-36. Cardioid and hypercardioid are the most common patterns and are generally found in most high-performance applications. Omnidirectional microphones are used in general area applications such as board and court rooms. Bidirectional patterns are typically produced by microphones that have their diaphragm exposed on two sides. Shotgun pattern is commonly used by the news and entertainment media. They have excellent side and back rejection which allows the microphone to be aimed at the sound source while having poor sensitivity to surrounding noise.

Figure 11-37 shows a condenser microphone equipped with a shotgun head. These types of microphones are often found on top of news cameras, at press conferences, and on sound stages.

To produce extreme direction ability, and to greatly improve the sensitivity, a cardioid or hypercardioid microphone element can be mounted at the focal point of a parabolic reflector, as shown in Figure 11-38. Side and rear rejection is very high and the sensitivity of the microphone can be increased as much as 100 fold. These units are often deployed to look for leaks in high or inaccessible piping. They are also used for studying wildlife, eavesdropping, and surveillance. One significant drawback to these units is that if they are inadvertently pointed at a loud sound source, the volume of the headphones can spike to unacceptable levels. In extreme cases, the headphones can be severely damaged. Higher cost units generally have limiting circuitry to prevent these mishaps.

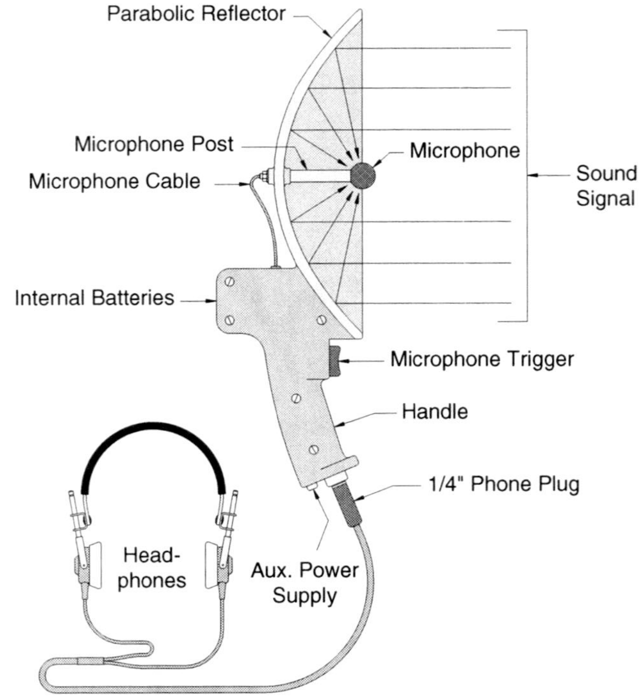

Figure 11-38 Parabolic Microphone Set

Geophones

Geophones are a type of dynamic microphone that is specifically designed to be sensitive to the low frequencies of the Earth's strata. The dirt that we walk on is a very poor transmitter of sound energy; therefore high-frequency sound is completely filtered out. The only sound that passes is very low frequency, that is, in the range of <1 to 20 Hz. Geophones consist of a large magnetic mass with a floating coil suspended and surrounding the core. The core vibrates in reference to the movement of the ground while the coil floats in a semifixed position. The differential movement of the magnet within the coil produces an electrical signal in reference to the movement of the ground. Figure 11-39 shows a section view of a typical geophone element.

Normally, a geophone is housed in a heavy duty plastic housing, as shown in Figure 11-40. These housings protect the relatively delicate element from the harsh environments where they are forced to operate.

Three-element geophones, as shown in Figure 11-41, are commonly used for seismic exploration. These units provide

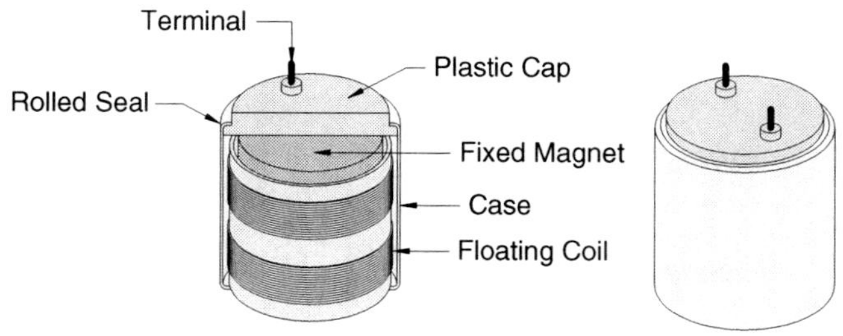

Figure 11-39 Geophone

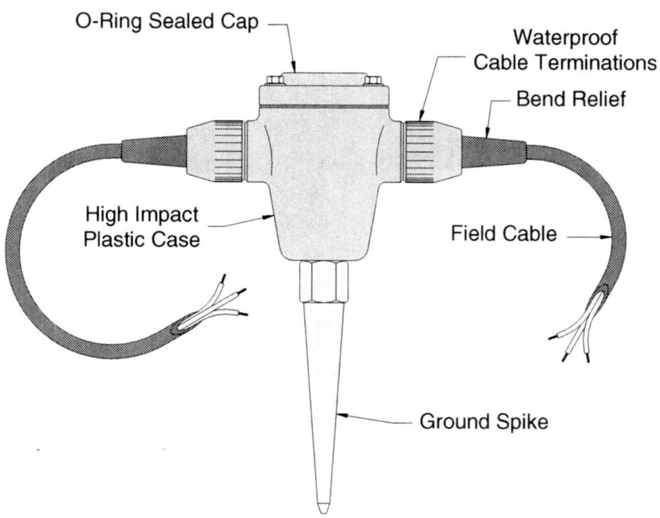

Figure 11-40 Geophone Field Assembly

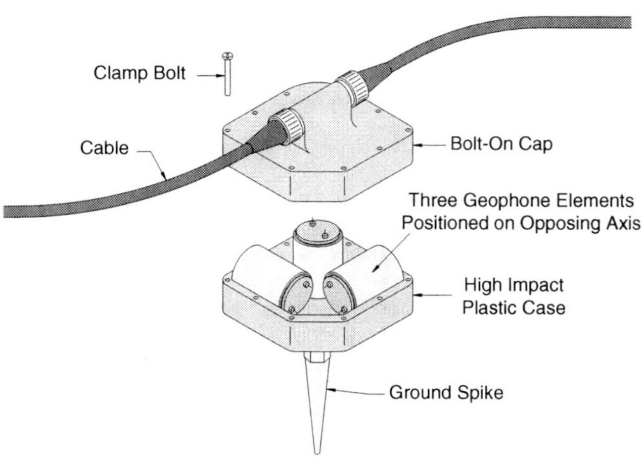

Figure 11-41 Three Element Geophone Assembly

higher accuracy data and are typically laid out in arrays consisting of several hundred to thousands of units over broad geographic areas. A single explosive pulse is detonated and the geophones pick up the reflected sound and transmit it to field recorders. The recorded data is later analyzed on a central computer system.

Hydrophones

Hydrophones are microphones that are specifically designed to operate under water. Typically, a hydrophone is a standard microphone that is housed in a waterproof housing. A simple hydrophone can be made by stretching a condom, or balloon, over a small standard microphone and sealing it around the cable. Care should be taken when using one of these homemade units, as they will not have very good resistance to depth. A hydrophone made in this fashion is probably good to about 1 atmosphere or 33 feet of depth.

Figure 11-42 shows an inexpensive hydrophone that is intended for general purpose underwater listening. It consists of a crystal microphone that is molded into a watertight housing. The unit is connected to an amplifier with a set of headphones and lowered into the water. These types of hydrophones are popular for listening to marine life and conducting acoustic inspections of underwater equipment.

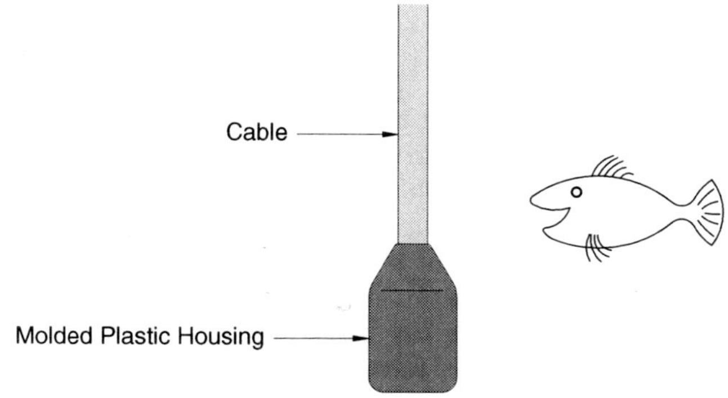

Figure 11-42 Hydrophone Assembly

For locating underwater targets, hydrophone arrays are deployed as shown in Figure 11-43. A ship will tow a string of hydrophones and then generate a sound pulse. The sound propagates through the water and bounces off the object to be located. As information from the hydrophones is fed into a shipboard computer, a bearing to the object can be instantly determined.

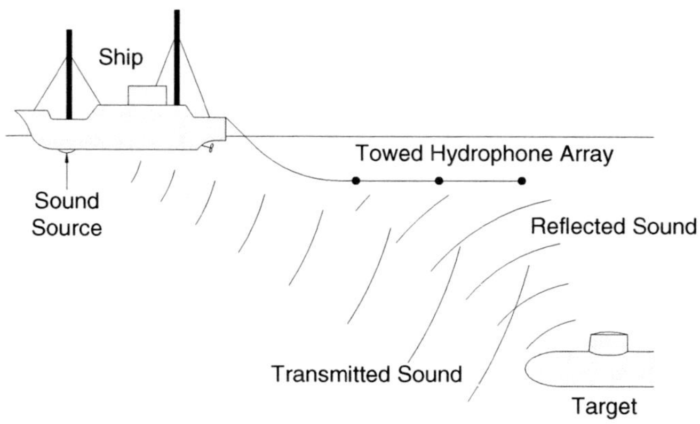

Figure 11-43 Towed Hydrophone Array

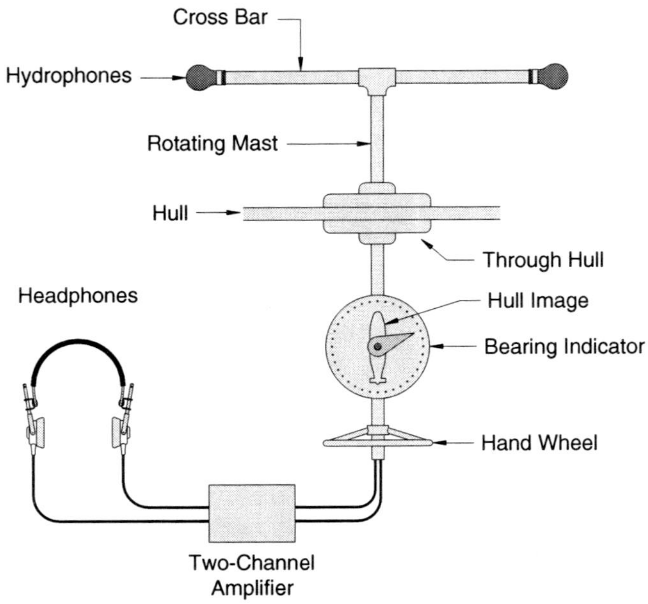

Figure 11-44 "T" Post Hydrophone Direction Finder

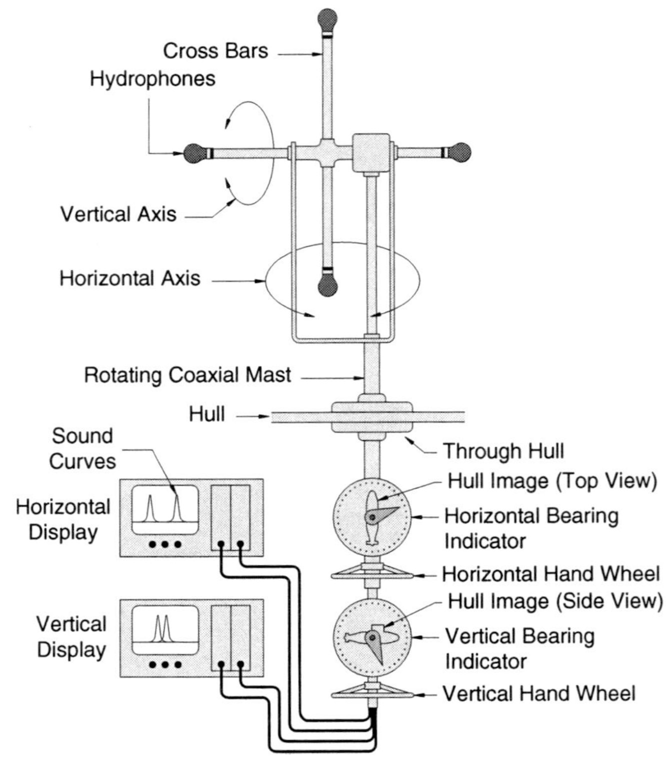

Figure 11-45 Cross Post Hydrophone Direction Finder

Submarines of the first and second World Wars typically used "T" post underwater direction finders, as shown in Figure 11-44. This simple system consisted of a pair of hydrophones mounted to the ends of a horizontal bar. The center of the bar was mounted to a rotating mast which extended into the interior of the boat. An operator could rotate the mast by turning a hand wheel. The mast also carried a bearing indicator which indicated the rotation position of the cross bar in reference to the hull. The hydrophones were connected to a two-channel amplifier, which fed the two different speakers of a headset. The operator listened to the sound from the headset and by rotating the mast and carefully matching the right and left signals, he could determine a bearing to the target.

A similar system can be configured using four hydrophones, which will add depth to the bearing. Figure 11-45 shows a cross post hydrophone direction finder. The two hydrophones on the horizontal bar sweep radially around the boat. The two hydrophones that are on the vertical bar are used to determine the depth of the target. The operator has two hand wheels with corresponding bearing indicators and the outputs of the hydrophones are connected to two-channel CRT (cathode ray tube) displays. One display shows horizontal information and the other vertical information. Each display shows a sound curve for each hydrophone. To determine a bearing, the operator turns the wheels until the two sound curves are aligned with one other. A system like this can be used in a passive or active role. In a passive role the system is used only to listen to sounds that the target emits. In an active role, a sound source is generated and the system listens to the sound that is reflected off of the target.

Modern submarines use spherical hydrophone arrays similar to the system shown in Figure 11-46. These are very sophisticated systems that rely on computer processing to determine target information. These systems can also operate in a passive or active role.

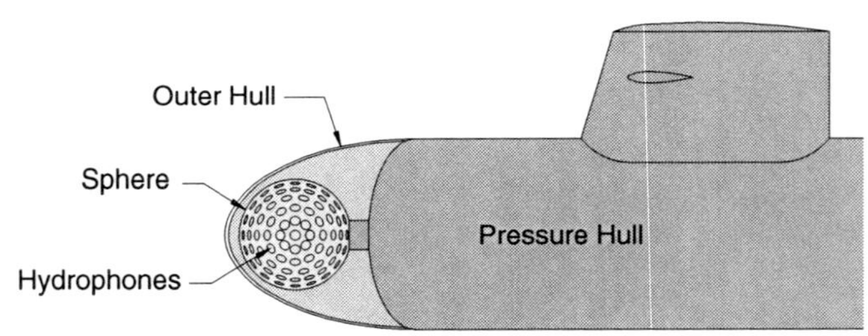

Figure 11-46 Spherical Hydrophone Array

Telegraph Systems

Until the advent of the telegraph, communications were restricted to mail or courier. The telegraph represented the first real time communication system. The telegraph relied on Morse code, a dot/dash system shown in Figure 11-47, to transmit information over great distances. An operator, who was trained in code, would take written information and transmit it to another station. The second operator would listen to the code and transcribe the message. The message was then sent by runner to the address specified in the message.

Figure 11-48 shows a basic telegraph system. Each station would have a key and sounder. One of the stations would be equipped with a battery set. The key has a send/receive switch. When the switch is opened, the station is in send mode. When the switch is closed, the station is in receive mode.

Alphabet

A	•—
B	—•••
C	—•—•
D	—••
E	•
F	••—•
G	——•
H	••••
I	••
J	•———
K	—•—
L	•—••
M	——
N	—•
O	———
P	•——•
Q	——•—
R	•—•
S	•••
T	—
U	••—
V	•••—
W	•——
X	—••—
Y	—•——
Z	——••

Numbers

0	—————
1	•————
2	••———
3	•••——
4	••••—
5	•••••
6	—••••
7	——•••
8	———••
9	————•

1. A dash is equal to three dots.
2. The space between parts of the same letter is equal to one dot.
3. The space between two letters is equal to three dots.
4. The space between two words is equal to five dots.

Figure 11-47 Morse Code

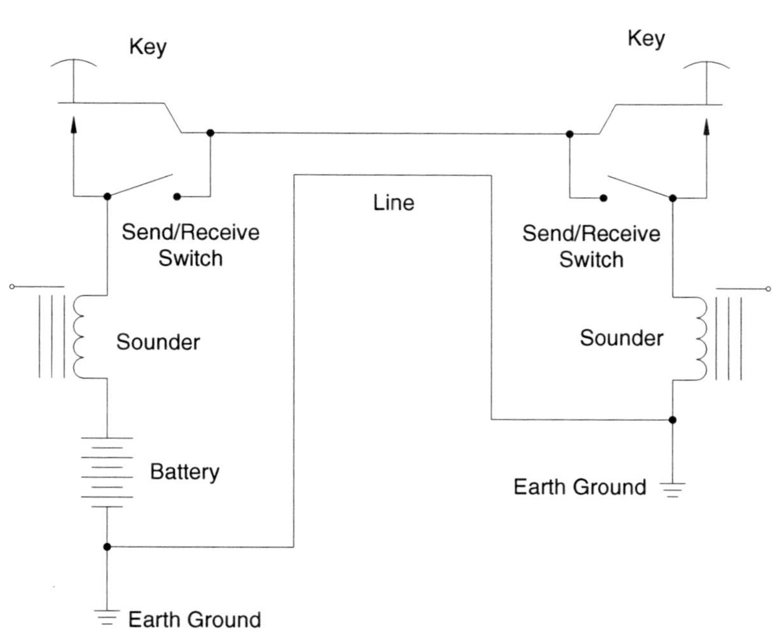

Figure 11-48 Basic Telegraph System

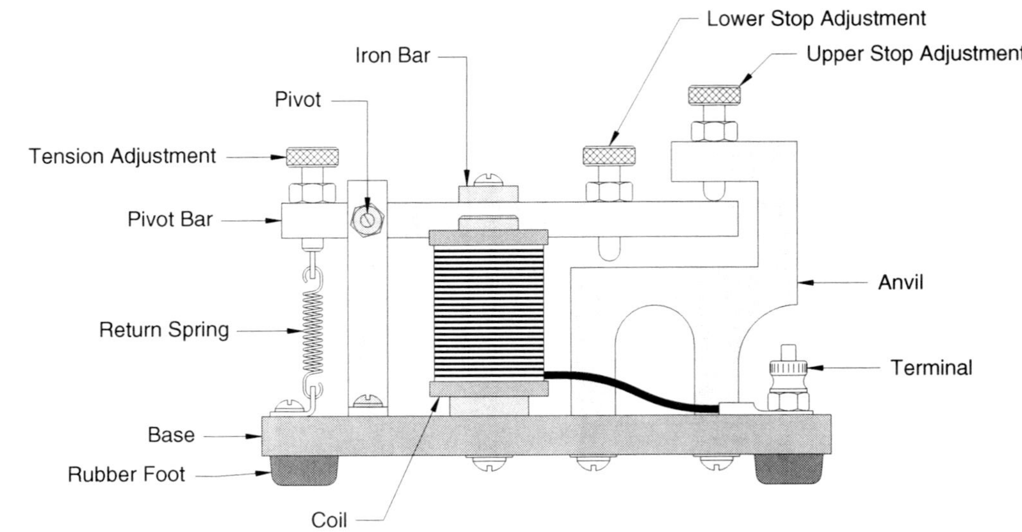

Figure 11-49 Telegraph Sounder

Figure 11-49 shows a typical sounder. These devices were a set of electromagnets that, when energized, would pull a spring loaded bar down against a stop and produce an audible click. When the magnet was de-energized the bar would return to an upper stop and produce a second click. Two clicks close to one another represented a dot and two clicks with a longer time between them represented a dash. In this manner, the operator could receive the message that he was transposing.

The key is a simple device that is designed to provide as much comfort to the operator as possible. The unit shown in Figure 11-50 is typical of a standard telegraph key. Note that the unit has a number of adjustments so that it can be maintained in top working order.

The batteries can only drive a system over a certain distance, after which the sounder will not receive enough power to operate because of line losses. To correct this problem, relay or repeater stations were established at various intervals over long transmission distances. Figure 11-51 shows a typical relay or repeater. These units are simply single-pole, single throw (SPST) relays that are designed to operate and switch the line voltage.

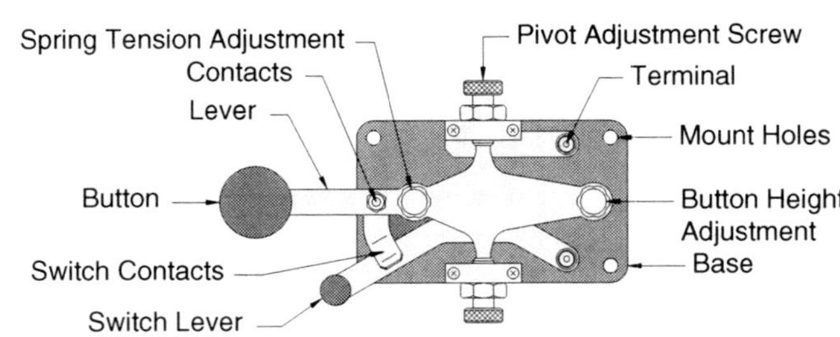

Figure 11-50 Telegraph Key

Figure 11-52 shows a schematic of a telegraph system with a single relay station. It should be noted that several relay stations may be deployed over hundreds of miles of transmission line.

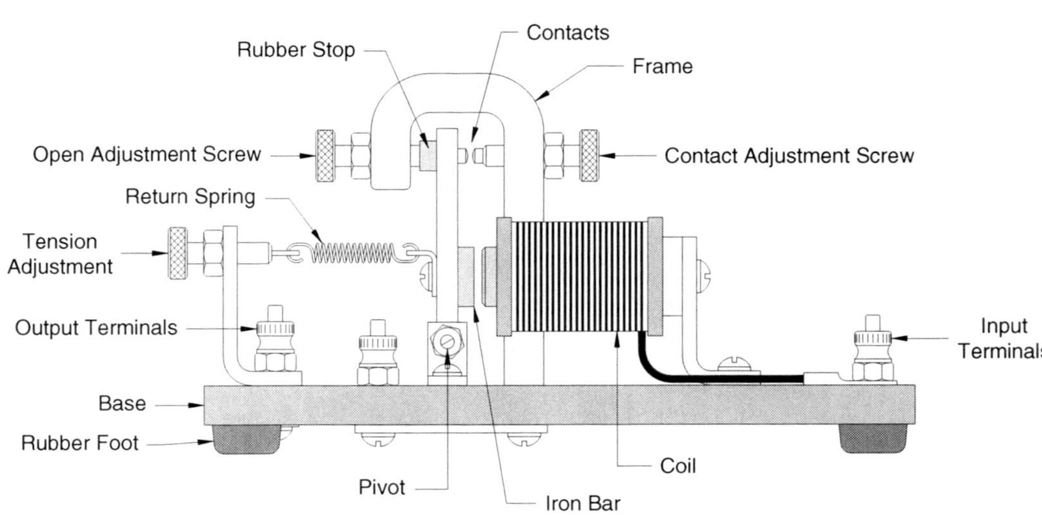

Figure 11-51 Telegraph Relay or Repeater

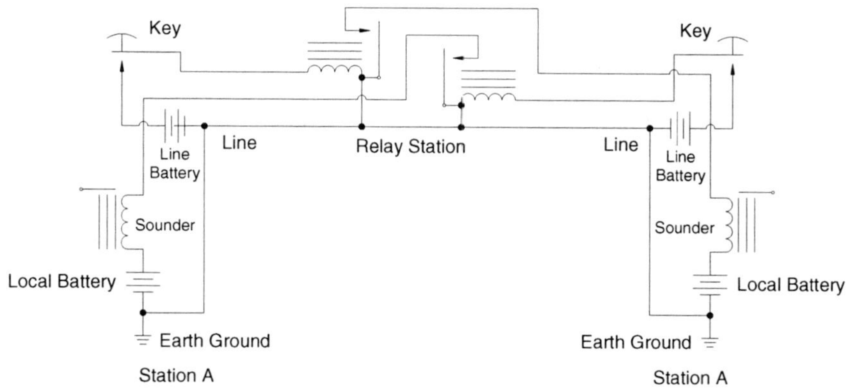

Figure 11-52 Two-Way Telegraph System with Relay Station

Telephones

Although we take them for granted, the telephone is probably the most important communication technology ever invented. In the early days of the telephone, coverage was limited to local areas and was in many ways, a novelty. The telegraph was still relied on to communicate over long distances. The advent of the carbon microphone and a reliable receiver made voice communications possible. Figure 11-53 shows a basic telephone circuit. When you speak into the transmitter (microphone), the sound is reproduced on both receivers (loudspeakers). This basic system made it possible to have a normal conversation over great distances.

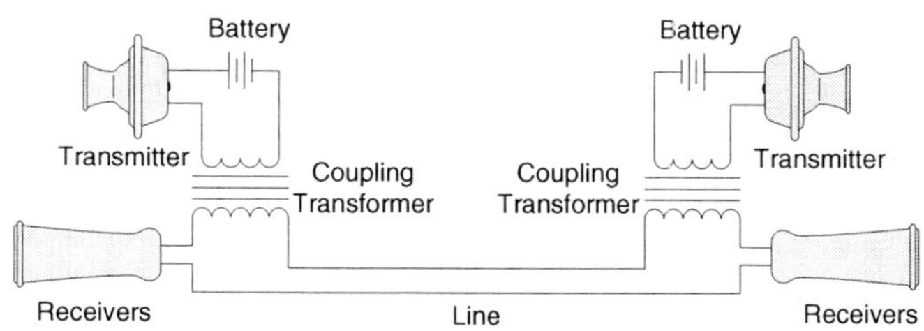

Figure 11-53 Two-Way Telephone Circuit

Figure 11-54 shows a schematic for a two-way telephone system with magneto ringers. Cranking the magneto (generator) on one set, will activate the ringer on the other set. This addition to the basic voice made it possible to produce a loud call at the other end of the line alerting the party that they are being called.

Figure 11-55 shows an illustration of an early wall hanging phone. These units were placed in thousands of homes and business in the early 1900s. To make a call, the subscriber cranked the magneto and the operator's ringer would sound. The operator would pick up and ask what number the caller would like to be connected to. The operator would ring that number and connect the two lines together.

Military phones of the World Wars were simply a basic telephone with magneto installed into a leather or canvas case, as shown in Figure 11-56. These were a complete, self-contained

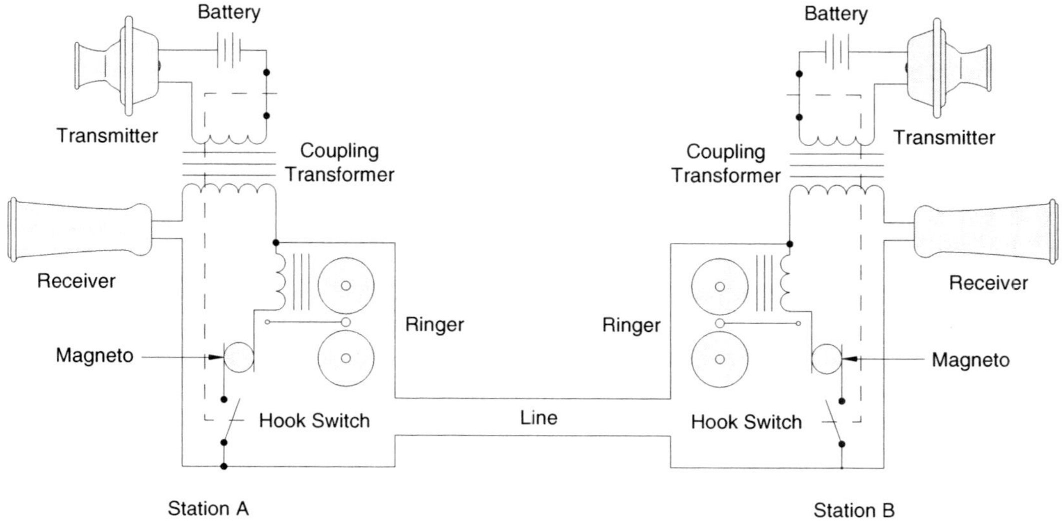

Figure 11-54 Two-Way Telephone Circuit with Magneto Ringers

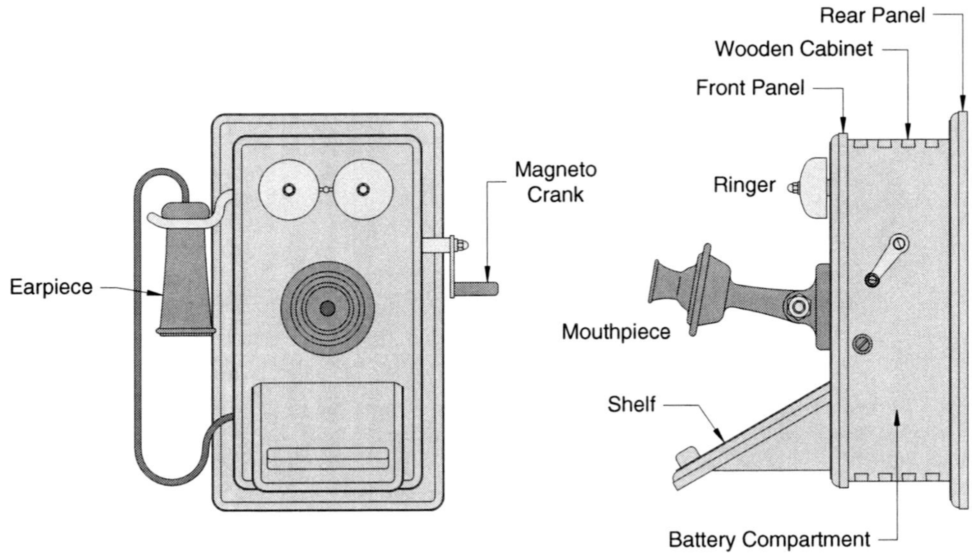

Figure 11-55 Early Wall Hanging Telephone

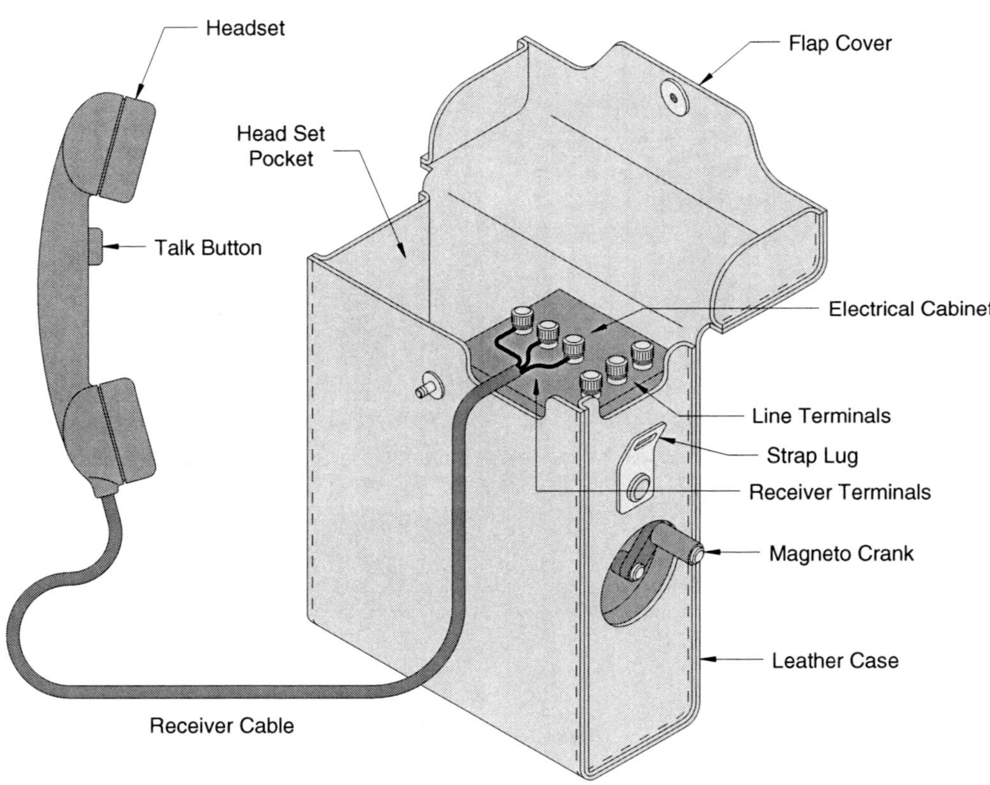

Figure 11-56 Military Field Telephone

system that could communicate with another phone set or a central switch board.

The system's operator controlled the connections between subscribers through the use of a switch board. The operator's phone was a headset with a magneto. The board had a single ringer which sounded whenever any subscriber was called in. Each subscriber's circuit had an indicator lamp which told the operator which line was ringing in. Figure 11-57 shows a schematic of a six-line switch board.

Figure 11-58 shows an illustration of a portable 10-line switch board. These units were used on the battle field, in mining operations, for trade shows or any other application where temporary communications were needed.

As telephone systems became more sophisticated, the operator could be called by simply patting the hook switch and the magneto was eliminated from the design. Figure 11-59 shows a schematic representation of a subscriber's telephone station without a magneto.

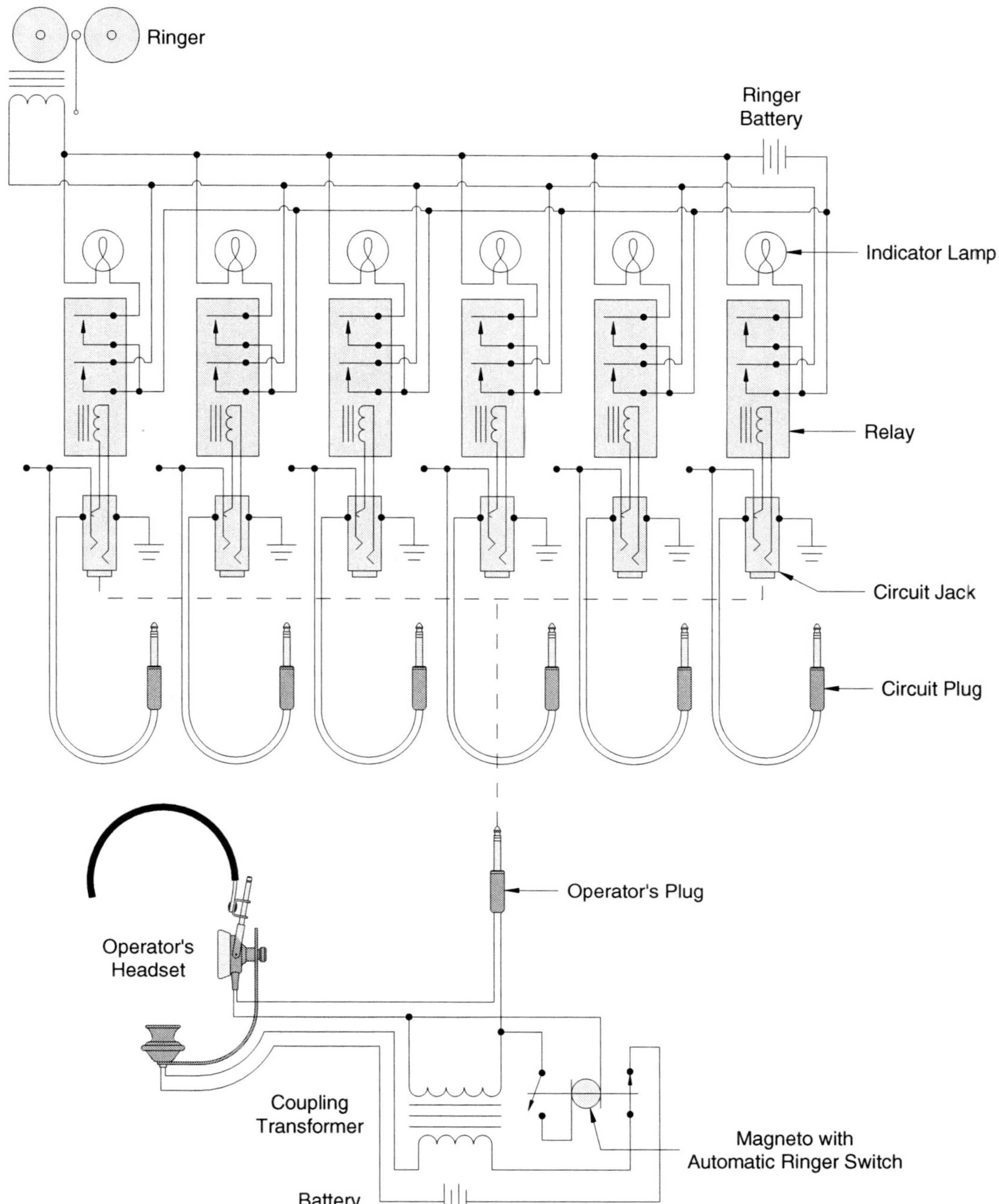

Figure 11-57 Simplified Operator's Switch Board Schematic

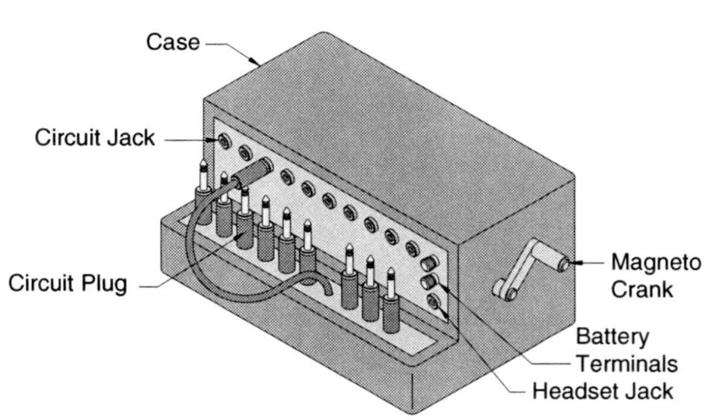

Figure 11-58 Ten-Channel Portable Switch Board

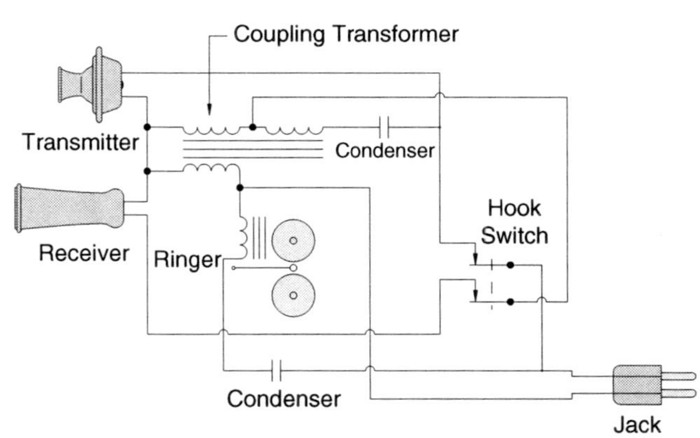

Figure 11-59 Subscriber's Telephone Station

The next significant advancement in telephone technology was the rotary dial system. This system was designed to allow the subscriber to call any other subscriber directly without calling the operator. The system used a dial that produced pulses that corresponded to the number on the dial. The pulses were translated into certain connections with the use of large arrays of sector relays. Figure 11-60 shows an early dial telephone.

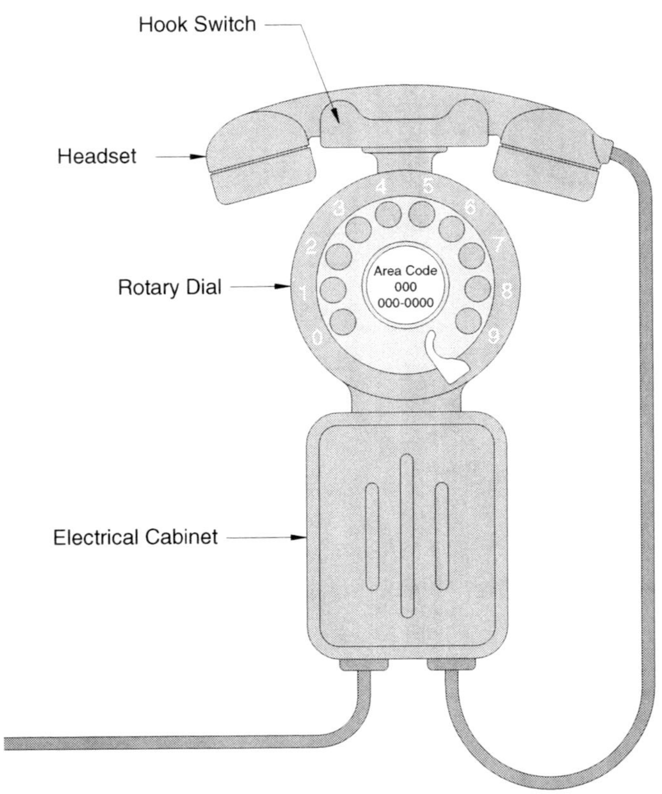

Figure 11-60 Early Rotary Dial Telephone

Vibrators

Vibrators are used in all manner of manufacturing. There are two principal types of vibrators—piston and rotary. Piston vibrators have a permanent magnet in the center of a solenoid coil, as shown in Figure 11-61. When an AC signal is applied to the coil, the magnet oscillates back and forth to create a vibration. Generally, piston vibrators are used in applications that require high frequency and low amplitude.

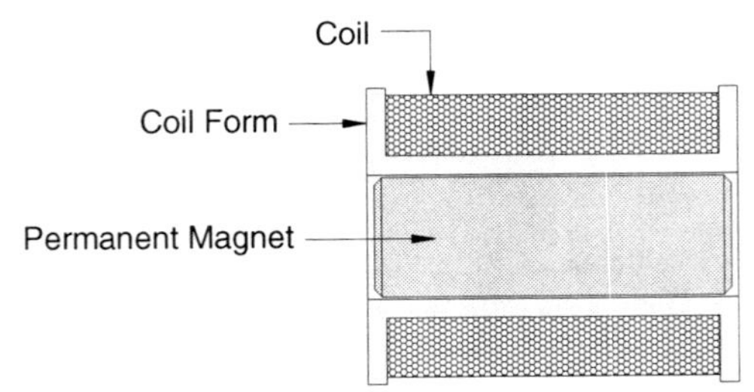

Figure 11-61 AC Vibrator

A rotary vibrator, as shown in Figure 11-62, is simply an electric motor with an off-center weight attached to its shaft. By varying the speed of the motor, the frequency of the vibration can be adjusted. By changing and/or adjusting the radius of the off-center weight, the amplitude can be adjusted. Generally, rotary vibrators are used in applications that require low frequency and high amplitude.

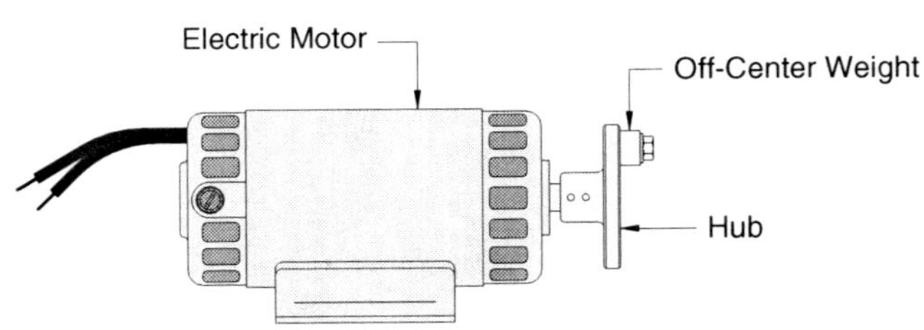

Figure 11-62 Rotary Vibrator

CHAPTER 12
LIGHTING

Most of us would agree that the electric light ranks as one of the most important technological developments that man has ever achieved. It was in the early 1800s that electric lighting began to appear in cities around the world. These early lamps used an electric arc to produce a brilliant light that was suitable for outdoor applications. However, because these lamps could only operate at high power levels, they were not suitable for indoor use. Early movie projectors relied on arc lamps to produce the high intensity light required to project motion pictures. During World War II, arc lamps were used extensively to spotlight enemy aircraft flying over England at night.

In the late 1800s, Thomas Edison invented the first practical incandescent light bulb. The difficulty in developing a light bulb came in the filament. Metals simply couldn't operate at the temperatures required without melting and then evaporating. After trying thousands of configurations with all types of plant matter, Edison finally developed a filament made up of a cotton thread that was coated and embedded with carbon granules. When placed in a low pressure, inert gas environment, the filament was able to glow brightly for long periods of time.

Incandescent Lights

Figure 12-1 shows an early incandescent light bulb and base. These units were supplied with a clear glass bulb which enclosed a long filament. The inside of the bulb is filled with a low pressure, inert gas. The filament is connected to two heavy wire terminals that are sealed into the base of the bulb. A third wire is used to support the rather fragile filament.

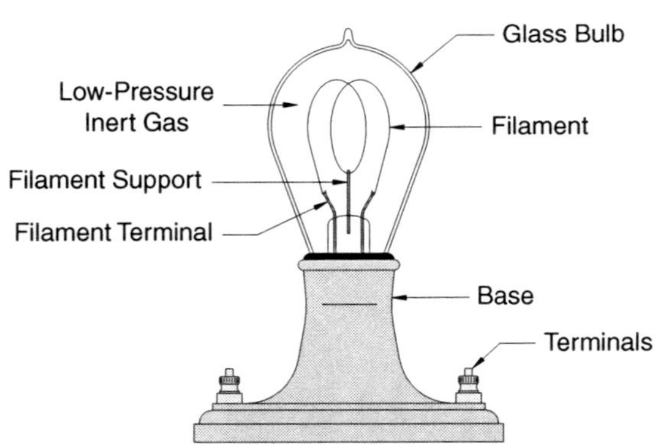

Figure 12-1 Early Incandescent Light Bulb

Building a simple incandescent light bulb, as shown in Figure 12-2, is a relatively simple matter. A fine tungsten wire is crimped to a pair of heavy terminal wires. The filament is inserted into a test tube and the inside of the tube is purged with Argon. A high temperature cork is lightly pushed into the end of the tube. Power from a variable power supply is then slowly increased until the filament glows brightly. After the bulb comes to full operating temperature, the cork is tightly seated to seal the gas.

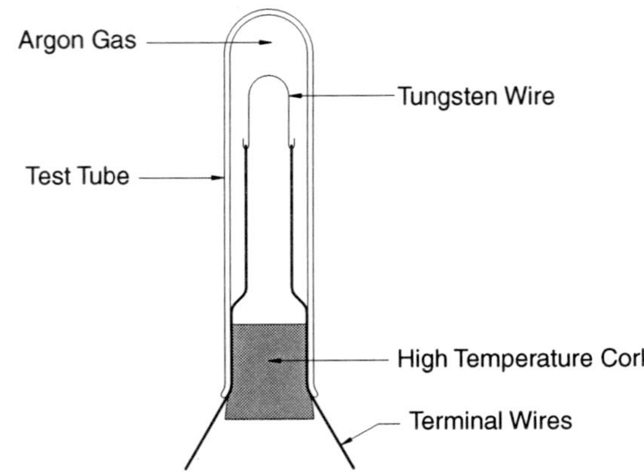

Figure 12-2 Simple, Bench Built Incandescent Light Bulb

Figure 12-3 shows a modern incandescent light bulb with a screw base. These bulbs differ very little from early units. They have a coiled tungsten filament, which operates in an inert gas environment. The gas is usually argon at approximately 80% of atmospheric pressure. The lower pressure is intended to bring the internal pressure back up to atmospheric pressure when the bulb is at full operating temperature. Modern bulbs generally have a white diffuser coating in the inside to better distribute and soften the light.

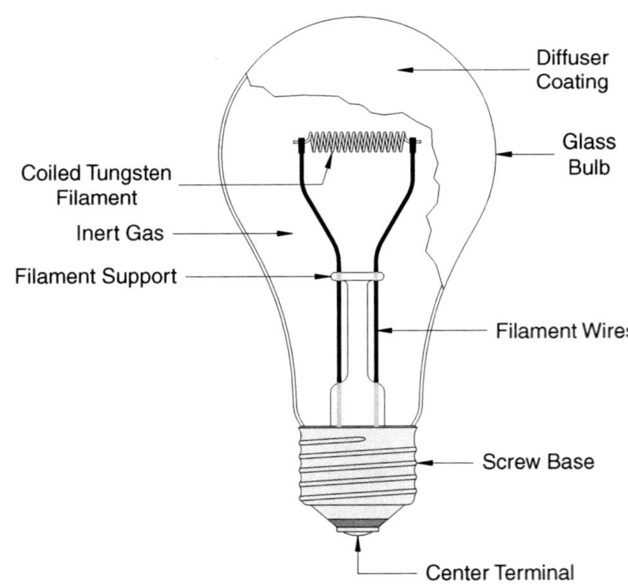

Figure 12-3 Commercial Incandescent Light Bulb

Fluorescent Lights

The second most common lamp is the fluorescent light bulb. These units are nearly as common as the incandescent lamp. They produce higher light emission per watt and are preferred in most offices and commercial applications. Figure 12-4 shows a typical fluorescent light bulb.

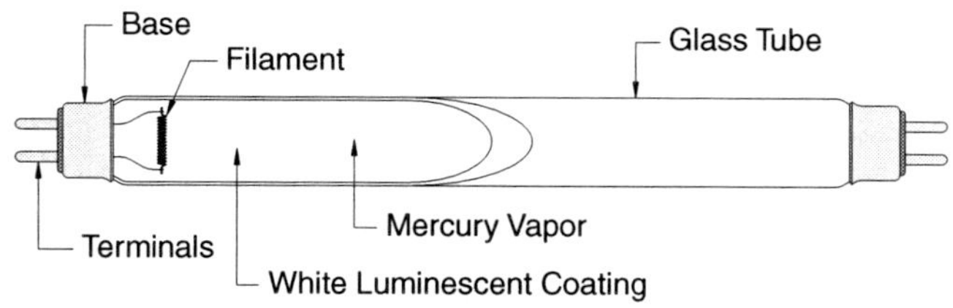

Figure 12-4 Fluorescent Light Bulb

These lamps consist of a long tube with a filament at both ends. The tube is filled with an argon/mercury atmosphere. The inside surface of the tube is coated with a white fluorescent material. To start the tube, power is fed to the filaments, which produce intense electron emission and heat. After the tube is heated up, a voltage is applied across the two different filaments and over the length of the tube. The gas within the tube becomes excited and produces ultraviolet light. The ultraviolet light excites the fluorescent coating which, in turn, produces visible light.

Figure 12-5 shows a simple starting circuit for a fluorescent tube. Pressing the start switch makes the filaments glow. After the tube heats up, the switch is released and power is redirected between the filaments and over the length of the tube, which, in turn, forces the gas charge to glow. To turn the lamp off, the power is disconnected.

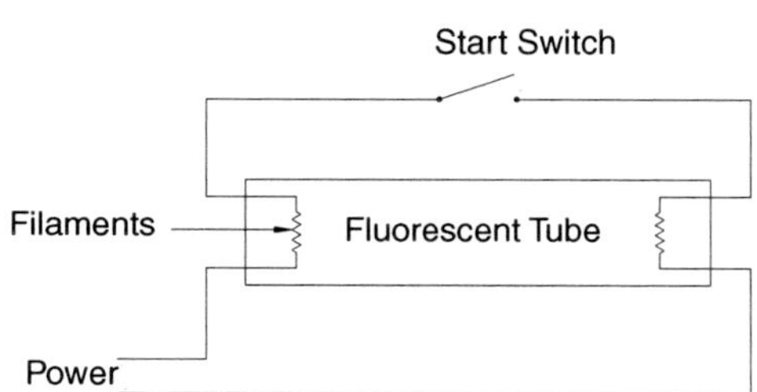

Figure 12-5 Fluorescent Light Bulb Starting Circuit

To start a fluorescent tube automatically, a glow switch starter, as shown in Figure 12-6, is generally utilized. The glow starter is a glass tube with a neon gas atmosphere. There are two contacts within the tube; one is fixed while the other is made from a bimetal strip.

Figure 12-7 shows a fluorescent tube circuit with a glow starter. When power is connected to the circuit, the starter glows and heats the bimetal strip. As the strip heats, it deforms, closes the contacts, and supplies power to the filaments. When the contacts close, the glow stops and the bimetal strip starts to cool. When the strip cools enough, the contacts open, power is disconnected from the filaments and the tube lights. The current drain on the circuit from the tube is enough to prevent the starter from glowing again. One of the most significant advantages of a starter circuit like this is that the lamp will automatically restart in the event of a momentary power outage.

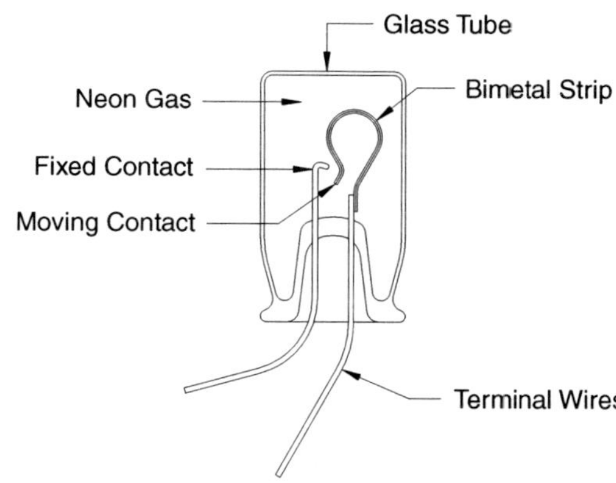

Figure 12-6 Glow Switch Starter

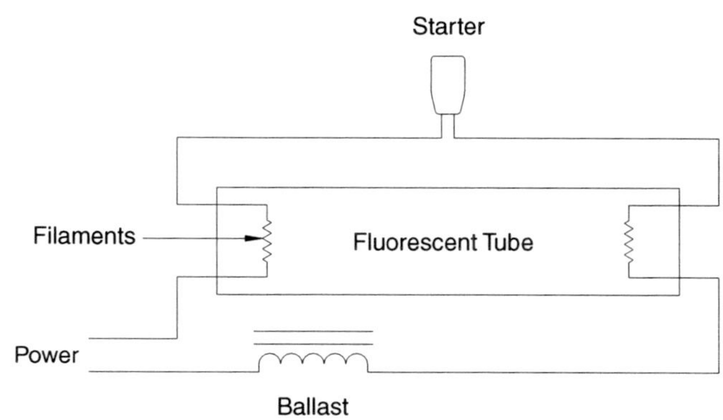

Figure 12-7 Fluorescent Light Bulb Starting Circuit with Ballast

Since a fluorescent tube has very little resistance when operating, it is necessary to use a ballast in the power circuit, as shown in Figure 12-7. The principal function of the ballast is to provide a high-voltage spike when the starter contacts open, and to limit the current once the lamp is operating. Figure 12-8 shows a typical commercial lamp ballast.

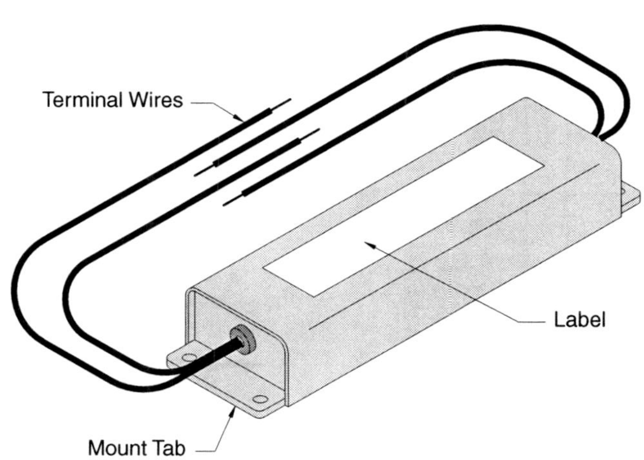

Figure 12-8 Fluorescent Lamp Ballast

Neon Lights

The neon lighting cycle is created when a voltage is applied across a pair of electrodes in a neon gas atmosphere. As electrons flow from one electrode to the other, the gas becomes excited and produces visible light. Most of us have seen neon lights used in advertising. The flashing open light at the end of a long dark road is an icon of the American cinema.

Figure 12-9 shows the most common type of neon lamp. These types of lamps are commonly used as night-lights and indicator lamps. It consists of a small glass bulb, which is purged with neon gas, and two electrodes. When power is applied to the electrodes, the lamp glows with a soft orange light.

Figure 12-10 shows a screw base neon lamp with shaped electrodes. The electrodes can take any shape that will fit into the bulb. When the lamp is turned on, it appears to have a flame and produces a subtle environment.

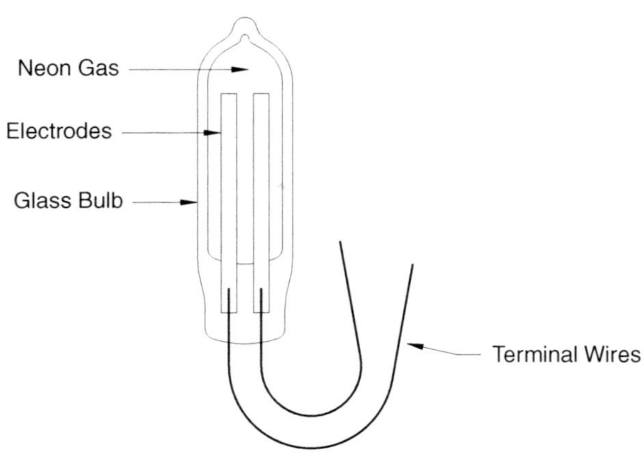

Figure 12-9 Neon Lamp

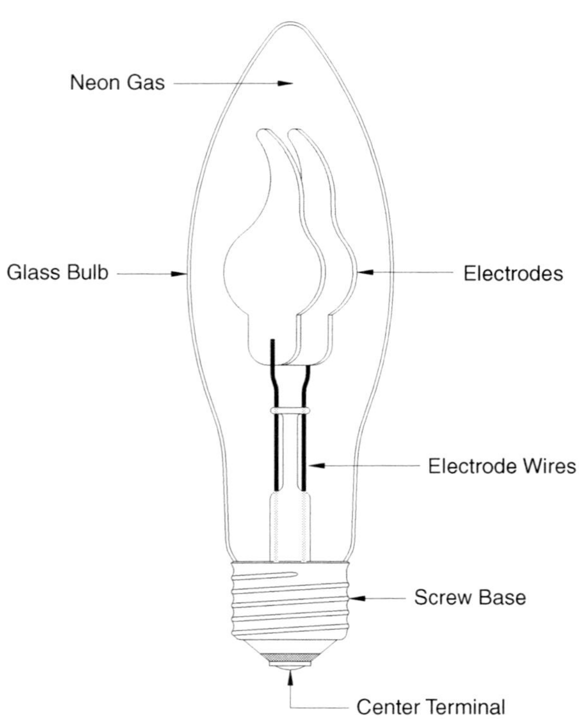

Figure 12-10 Shaped Electrode Neon Lamp

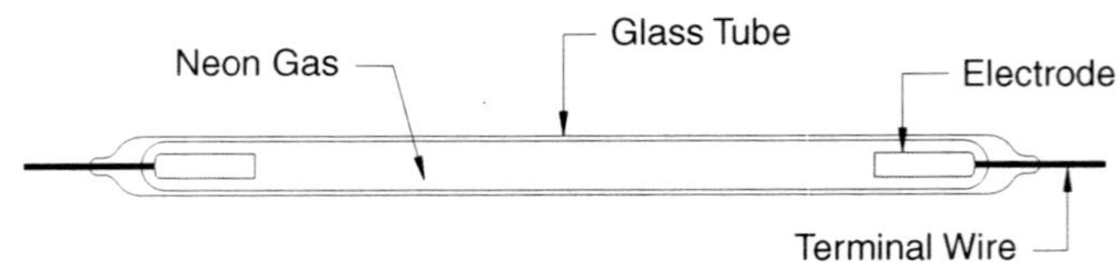

Figure 12-11 Neon Tube Lamp

The glow, or plasma, of a neon lamp can extend over rather long distances, if the tube is properly constructed. Figure 12-11 shows a straight neon tube. The plasma will extend over the entire length of the tube between the electrodes.

Another attribute of the plasma is that it will form around curves and bends in the tube. An open sign, as shown in Figure 12-12, is actually a single tube bent into the shape of the word. The connecting parts are blacked out and when the tube is energized the letters glow brightly.

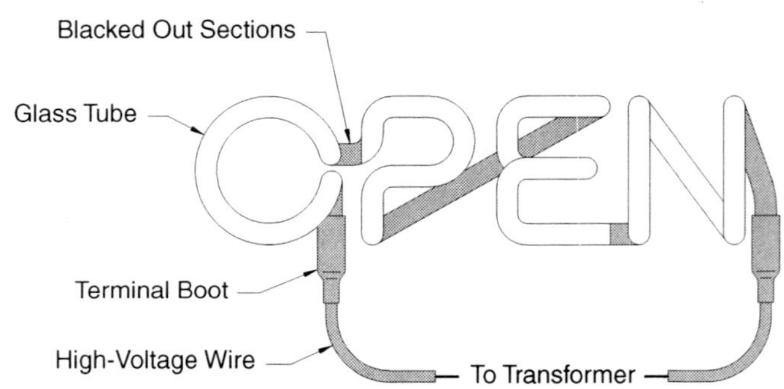

Figure 12-12 Commercial Neon Lamp

High voltage is required to start a neon tube because the electrodes are so far apart. A current-limited transformer, as shown in Figure 12-13, is typically used for these applications. The open voltage of the transformer is usually in the 20,000- to 45,000-volt range. When the voltage is applied to the electrodes, electrons flow from one electrode to the other.

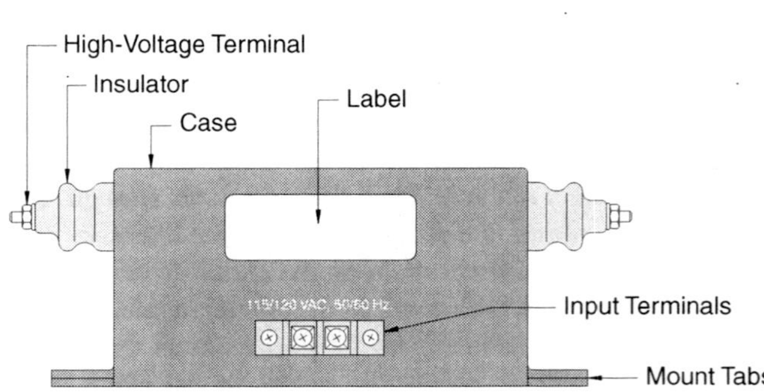

Figure 12-13 Commercial Neon Sign Transformer

This, in turn, ionizes the gas and it starts to glow. When the gas becomes ionized, its resistance lowers and the voltage of the transformer is pulled down to the operation voltage, usually around 400 volts.

Neon lights used in advertising are manufactured by shaping a glass tube into the desired letter, word, or image and fusing electrodes onto both ends. One of the electrodes will have a fusible port connected to a vacuum pump. The atmosphere is pumped out of the tube and the starting voltage is applied to the electrodes. The vacuum is valved off and neon gas is slowly bled into the tube cavity. When enough gas is present, the lamp will light. Additional gas is fed in to adjust the brightness of the tube. After the gas charge is adjusted, the fusible port is melted closed, creating a hermetically sealed neon tube. Figure 12-14 shows a schematic representation of a neon tube manufacturing system.

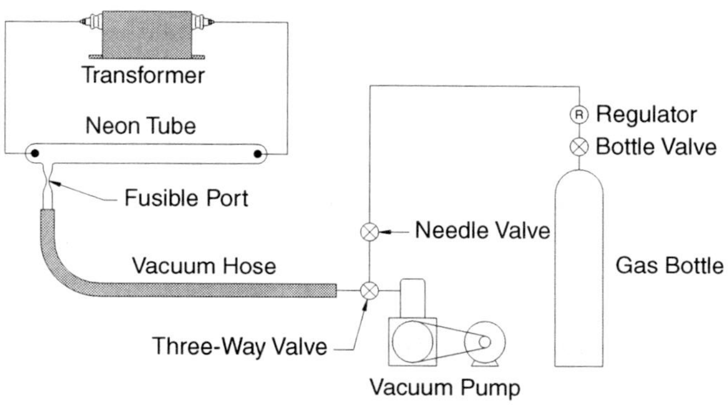

Figure 12-14 Neon Tube Manufacturing System

Halogen Lamps

The halogen lamp, as shown in Figure 12-15, is an improved incandescent light bulb. The halogen cycle continuously redeposits evaporated tungsten back onto the filament. This, in turn, produces a very brilliant lamp with an exceptional filament life. When the filament is at full temperature, about

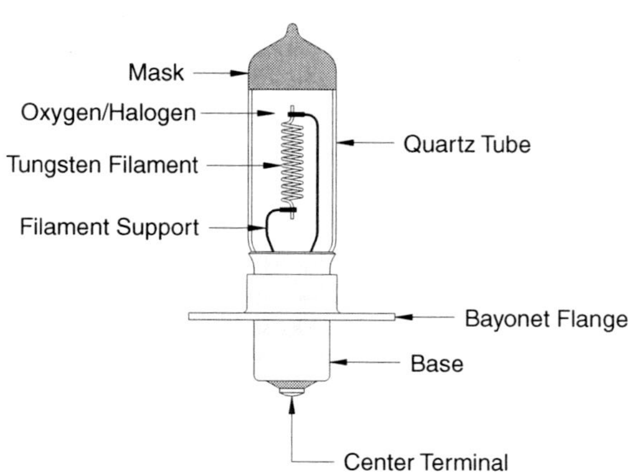

Figure 12-15 Halogen Lamp

5500°F (3040°C), it slowly evaporates and releases tungsten atoms. The tungsten atoms migrate towards the bulb, which is at approximately 1340°F (730°C). Near or at the bulb, the tungsten atoms combine with the oxygen and halogen atoms, forming tungsten oxyhalides. Convection currents within the bulb carry the tungsten oxyhalide back toward the filament. The heat from the filament breaks down the tungsten oxyhalide, and the oxygen and halogen atoms move back toward the bulb. The tungsten atoms recombine on the filament and the cycle starts over. In this manner the filament is continuously replenished.

Mercury Vapor Lamps

The first mercury vapor lamp was patented by Peter Hewitt, in 1901, and went into production the following year. Early mercury vapor lamps, as shown in Figure 12-16, were fairly simple devices. They consisted of a tube with a small reservoir at one end containing a pool of mercury and the lower electrode. At the opposite end of the tube was the upper electrode. When power was connected to the electrodes, the mercury vapor was excited and the tube glowed with brilliant bluish-green light. To start the lamp, it was simply rotated until the liquid mercury flowed along the length of the tube and created an electrical connection. The tube was then rotated back into its operation position.

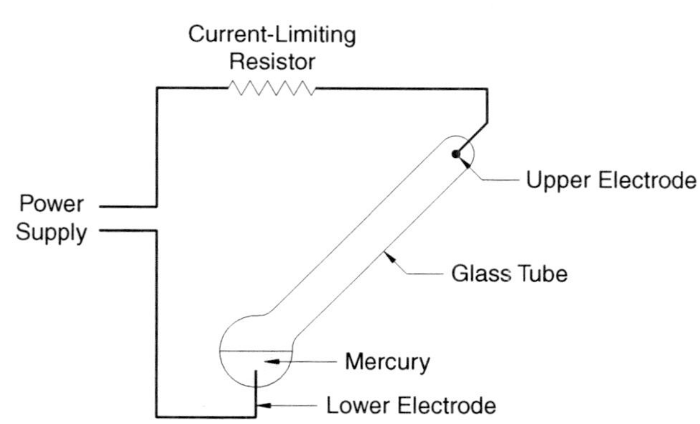

Figure 12-16 Basic Mercury Vapor Lamp

Mercury vapor lamps proved ideal for outdoor and industrial lighting applications and quickly became the standard for factories, roadway lighting, stadiums, parking lots, and the like. These lamps are still commonly in use today, an example of which is the street lights in your neighborhood.

Figure 12-17 shows a commercial mercury vapor lamp. The tube is located in the center of a glass bulb. The power terminal wires are used to support the assembly and a current-limiting resistor is usually included within the bulb.

Most modern mercury vapor lamps operate through a current-limited, step-up autotransformer, as shown in Figure 12-18. During start-up the transformer supplies the high voltage that the tube requires to initiate the plasma. After the plasma is

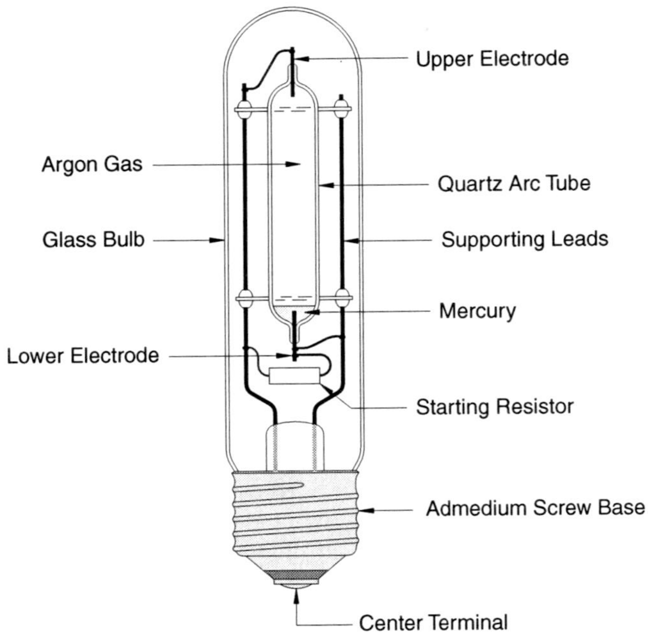

Figure 12-17 Commercial Mercury Vapor Lamp

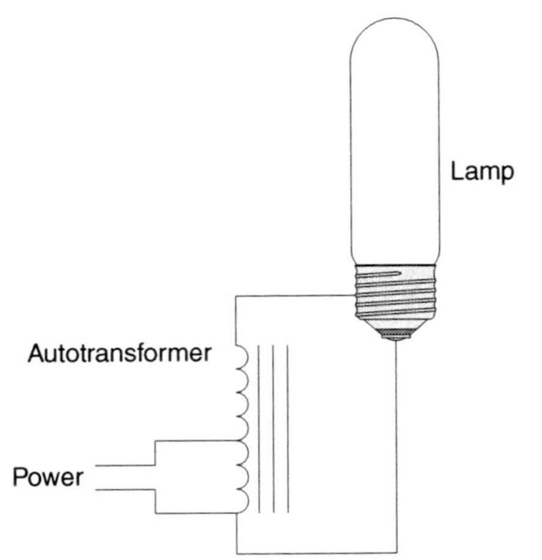

Figure 12-18 Mercury Vapor Lamp Power Supply

established, the low resistance pulls the transformer's voltage down to the operational voltage.

High-Pressure Sodium Vapor Lamps

Sodium vapor lamps have become the lamp of choice for highway lighting. These are the lights we see on our freeways that produce the golden-yellow light. This spectrum of light is more comfortable to the eye, producing softer shadows and less glare.

Figure 12-19 shows a typical high-pressure sodium vapor lamp. The lamp consists of a quartz tube mounted in the center of a vacuum jacketed bulb. The vacuum jacket bulb is

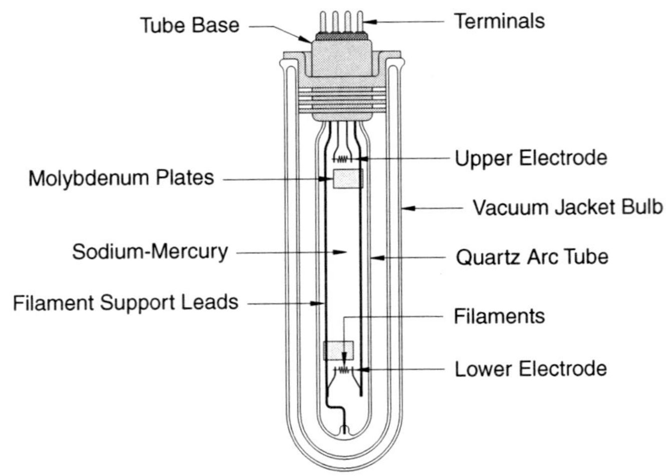

Figure 12-19 High-Pressure Sodium Vapor Lamp

necessary to provide insulation for the high temperatures at which these bulbs must operate. The quartz tube contains a small amount of sodium and neon gas. There are two filaments mounted on either end of the tube. Starting the lamp is similar to starting a fluorescent unit. The two filaments are heated, generating electron flow and heat. The heat vaporizes the sodium and, after a predetermined start period, the filaments are turned off, high voltage is applied to the filaments and the plasma initiate. The small molybdenum plates that are backing the filaments are intended to protect the filaments by carrying the bulk of the extreme heat that the lamp generates during operation. It takes about 30 minutes for a sodium vapor lamp to come up to full operating temperature, so it is important that the lamps are used in applications where a warm-up period is acceptable. One other note, these lamps are referred to as high pressure because the internal pressure of the quartz tube is at several atmospheres during operation.

Standard Lamp Bases

There are a variety of standard lamp bases that are commonly available when selecting a light bulb. Common sense will dictate most of these choices, as an example, most household lighting utilizes a medium screw base lamp while automobiles typically use a three lug pattern. It doesn't make much sense to buck the norm and place a bulb into a service for which it isn't designed.

Figure 12-20 shows the standard screw bases that are commonly used for incandescent lamps. The medium is the most common size, used for light bulbs in the 25 through 150 watt range. Intermediate and candelabra sizes are commonly used for decorative lamps, Christmas lights, indicator lamps, and nightlights. The miniature is found in flashlights, indicator applications, and model building. The medium skirt is typically used on lamps that are used in outdoor fixtures, such as flood lights. Admedium size is mostly used for higher-wattage lamps and mercury vapor lamps. The mogul base is used in industrial and high-wattage applications.

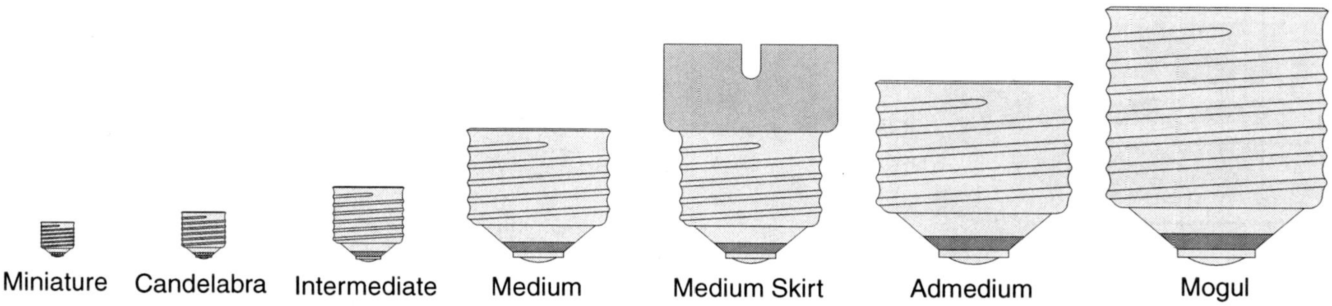

Figure 12-20 Standard Screw Bases

Bayonet bases are common in automotive and instrumentation applications. Figure 12-21 shows both double contact and single contact bases. The double contacts are generally reserved for dual-filament lamps, such as automotive tail lamps. One filament is for running lights while the second, brighter, filament is for brakes.

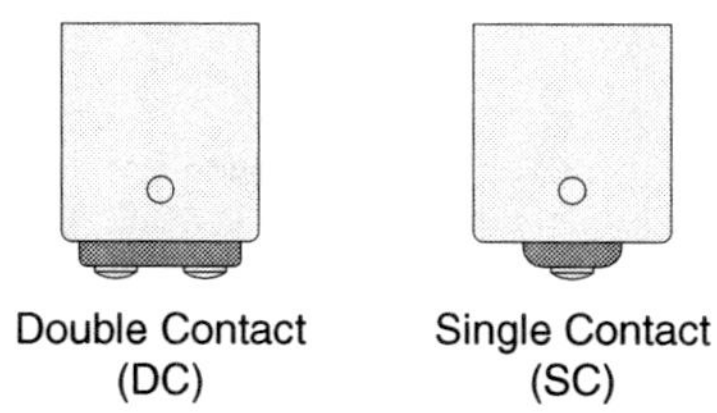

Figure 12-21 Standard Bayonet Bases

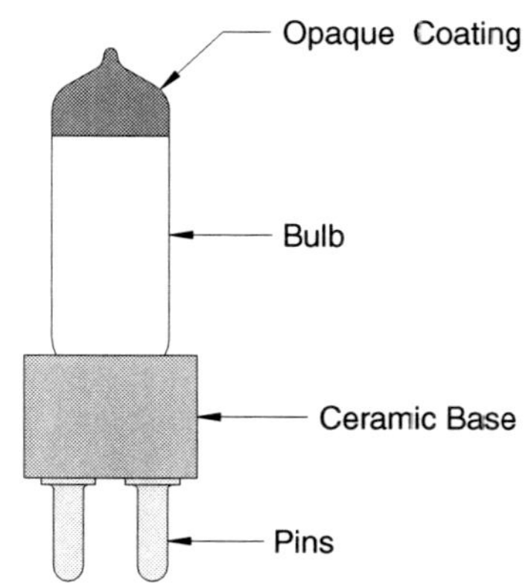

Figure 12-23 Two-Pin Base

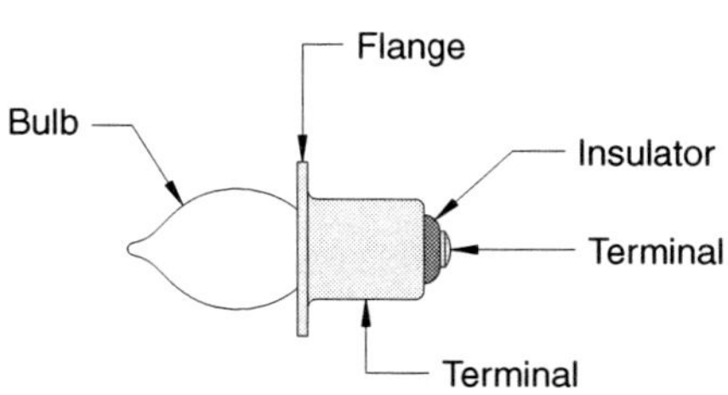

Figure 12-22 Flanged Base

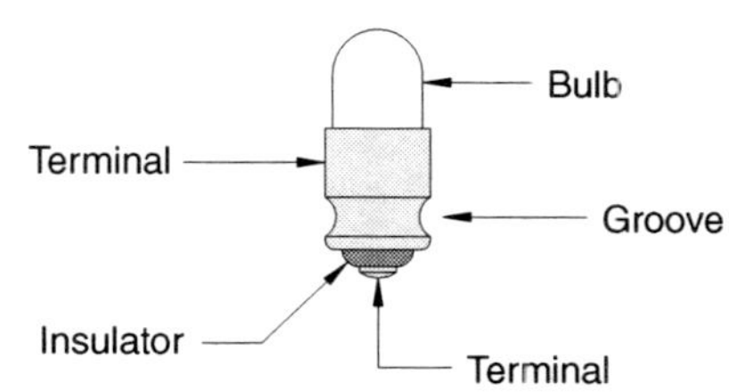

Figure 12-24 Grooved Base

The flanged base bulb, shown in Figure 12-22, is typically used in flash lights and indicator applications. The bulb is dropped into the socket and a screw-on cap is used to secure it in place.

Two-pin bases, as shown in Figure 12-23, are typically found on high intensity lamps, such as halogen units. These bases are commonly found in projection and audio visual equipment.

Grooved base bulbs are used in applications that require a very small incandescent bulb. These lamps snap into place and are retained with a spring loaded detent. Figure 12-24 shows a typical grooved base bulb.

Sealed beams, most commonly used in automotive, construction equipment, and marine applications, use one of four different base configurations. Figure 12-25 shows the typical base patterns for sealed beams. The two- and three-lug types are usually used with a standard connector. The space lugs are used with standard crimp-on connectors and the screw terminals are for connecting to either stripped wire or screw lugs.

Fluorescent tubes are most commonly supplied with either medium bi-pin or single pin bases, as shown in Figure 12-26. The recessed double contact is typically found in industrial

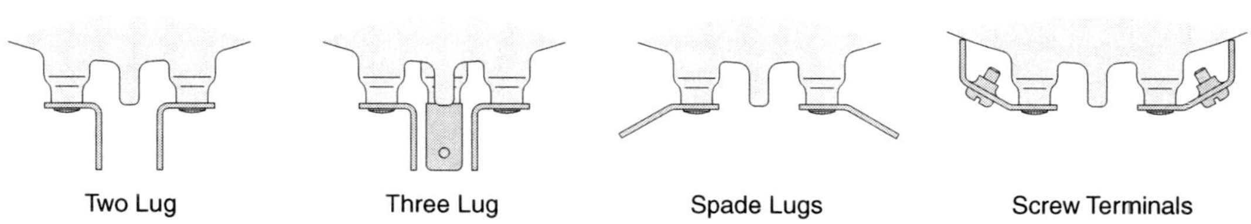

Figure 12-25 Sealed Beam Bases

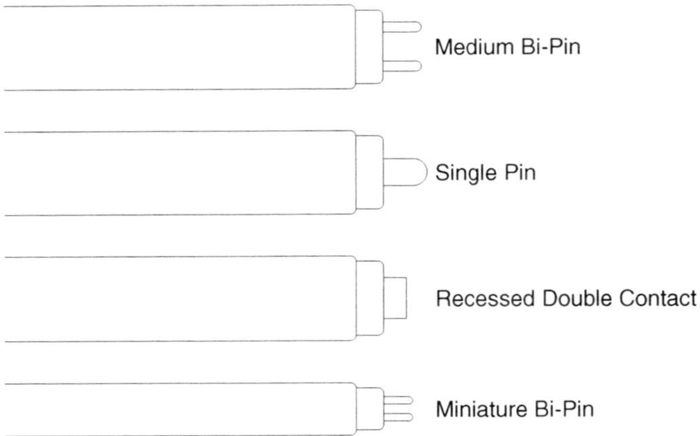

Figure 12-26 Fluorescent Tube Bases

applications, while the miniature bi-pin is used in small appliances and instrumentation.

Lamp Sockets

Figure 12-27 shows a few commercial lamp sockets. Sockets are commonly available with a variety of wattages, materials, switches, and mounting options. Also shown is a screw socket adaptor for mounting a candelabra base into a medium socket.

Bayonet bases are generally available with solder lug, angle mount, or flanged plastic bases. The solder lug socket is intended to support the lamp with the wires that electrically connect it. A better choice for these applications is the angle mount. The base can be secured with a nut and bolt or pop rivet. Figure 12-28 shows three typical bayonet bases.

Panel mount sockets, as shown in Figure 12-29, are commonly used in industrial equipment and instrumentation. Shown are two examples of panel mount lamp sockets, the left for bayonet and screw base lamps and the right for flange base lamps.

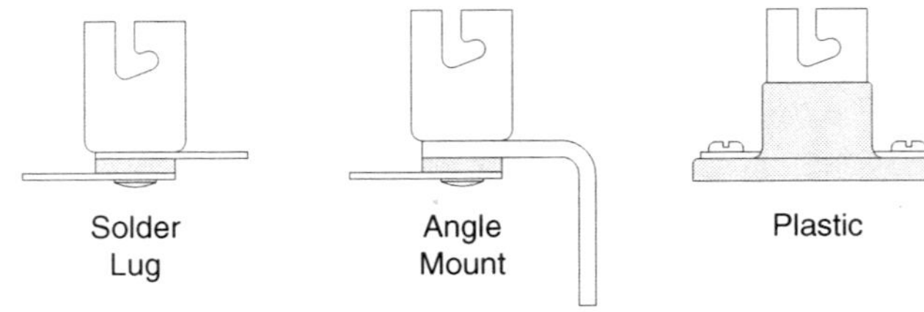

Figure 12-28 Bayonet Base Lamp Sockets

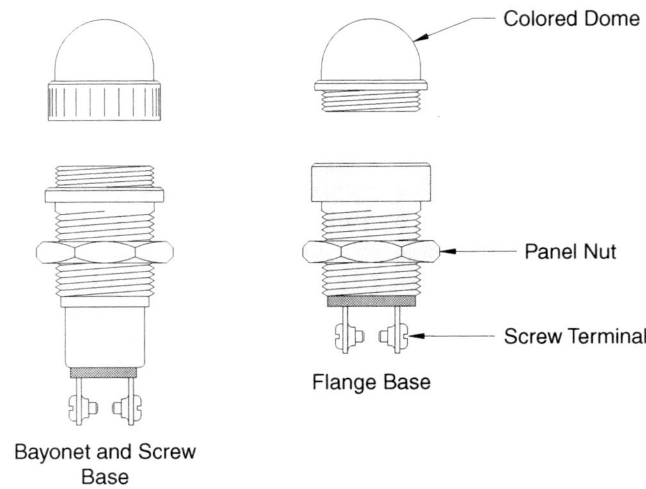

Figure 12-29 Bayonet Base Lamp Sockets

Bulb Shapes

Light bulb shapes are available in a wide variety of configurations designed for nearly every conceivable application. Figure 12-30 shows a few standard shapes that are commonly available for incandescent lamps. The type letter specifies the general shape. The type letter is normally followed by a number, which indicates the diameter of the bulb in $^1/_8$-inch increments. As an example, a G25 is a round globe that is 25 $\times$ 0.125

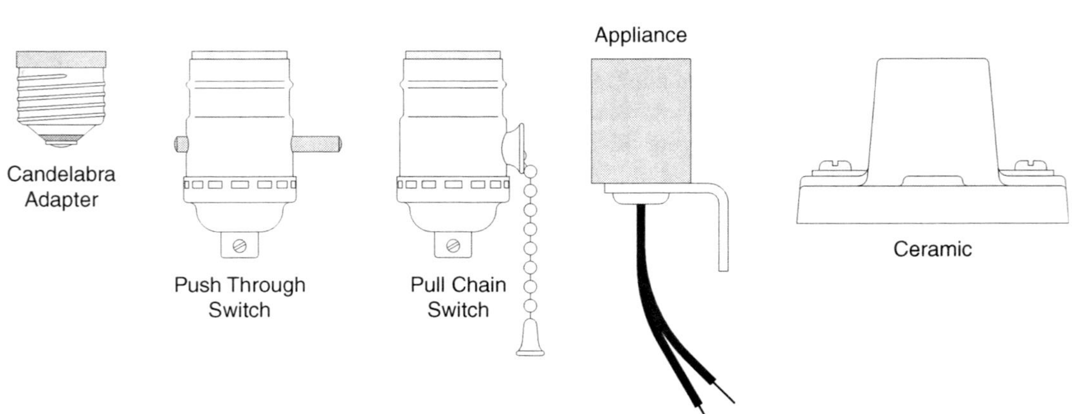

Figure 12-27 Medium Screw Base Lamp Sockets

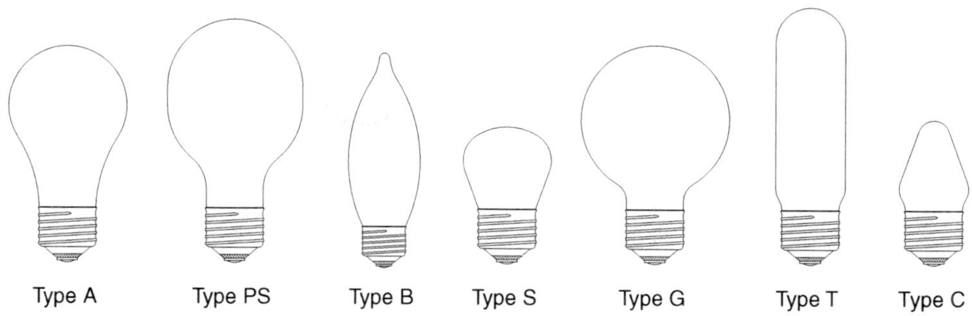

Figure 12-30 Standard Light Bulb Shapes

or 3-$^1/_8$ inch in diameter. A T10 is a cylindrical globe that is 1-$^1/_4$ inch in diameter.

Like standard incandescent bulbs, flood and spotlights have their own designators. Figure 12-31 shows a few of the more common flood and spotlights that are available in the market today.

Figure 12-32 shows designators and bulb shapes for mercury vapor and high-pressure sodium lamps. It should be noted that these bulbs are generally delivered with either admedium or mogul bases.

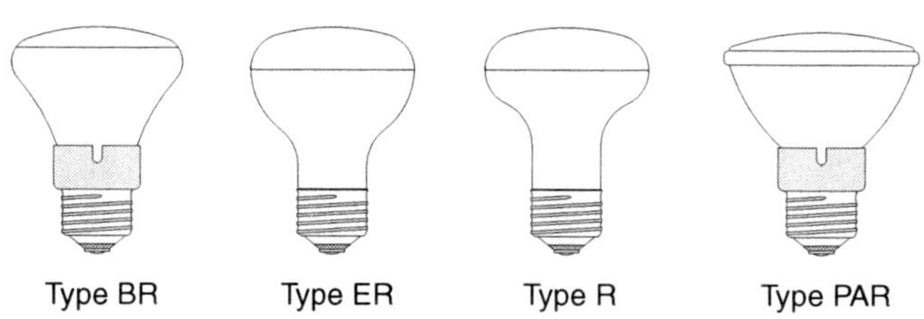

Figure 12-31 Standard Flood and Spot Light Shapes

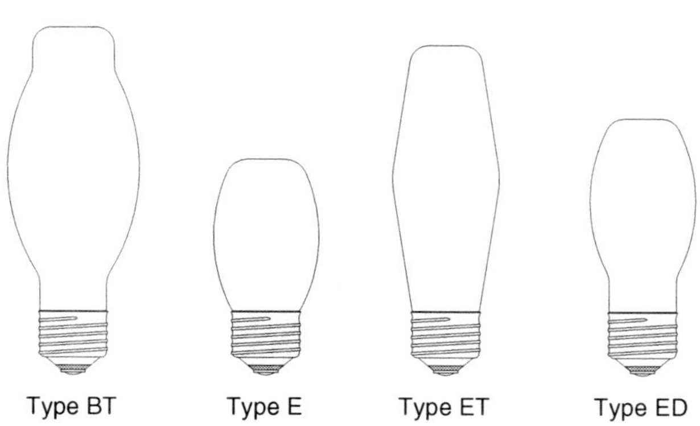

Figure 12-32 Standard High-Intensity Discharge Lamp Shapes

Compact fluorescent tubes are now commonly available in screw base packages, as shown in Figure 12-33. These lamps are gaining popularity as high efficiency replacements for their incandescent counterparts and can be purchased in any hardware or home improvement store.

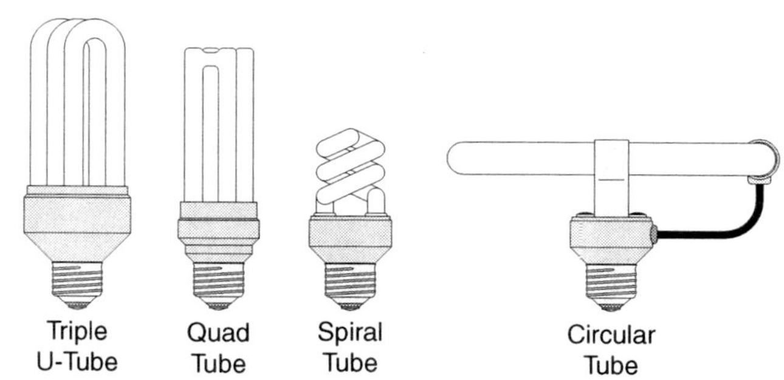

Figure 12-33 Screw Base Fluorescent Light Bulbs

Lamps that have an integral reflector fall into two different categories, flood and spot. Flood lights generally have a reflective surface behind the filament that is intended to reflect the light that the filament generates at the back towards the front of the lamp. The lens of the lamp acts a diffuser and is either frosted or carries a series of diffuser lenses, as shown in Figure 12-34.

Spot lights have a parabolic reflector that is intended to focus a point source of light into a powerful beam. These types of lights are available with an integral light bulb; however, they are

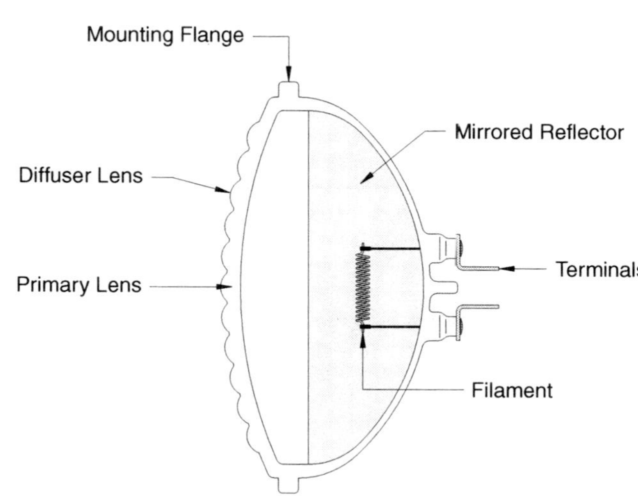

Figure 12-34 Multilens Diffuser

more commonly found as an assembly, as shown in Figure 12-35. In this case, the reflector is designed to mount a high intensity halogen lamp at the focal point of the reflector. The assembly has a stray light shield mounted to the center of the flat lens. The lens is intended to keep dust and dirt off of the reflector.

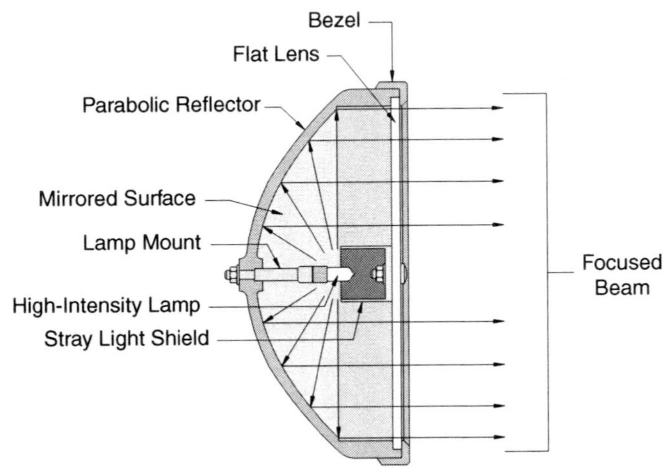

Figure 12-35 Parabolic Reflector Spot Light

Color Temperature

The specific wavelengths that any light bulb produces are generally described in K (Kevnin) or color temperature. The temperature rating does not refer to an actual temperature that the bulb may generate during operation. It refers to the temperature to which a black body must be heated to emit a certain wavelength of light. Red, on one end of the spectrum, is represented as roughly 1800 K and blue, on the opposite end of the spectrum, is represented as roughly 16000 K. In addition to specifying a lamp's output in K, industry has adopted terms for color temperature that are a little more intuitive to the typical buyer. Figure 12-36 shows the colors and industry terms that correspond with color temperatures.

Xenon Lamps

Most of us have experienced a xenon flash unit in our cameras. A xenon flash tube is an integral part of almost every camera manufactured today.

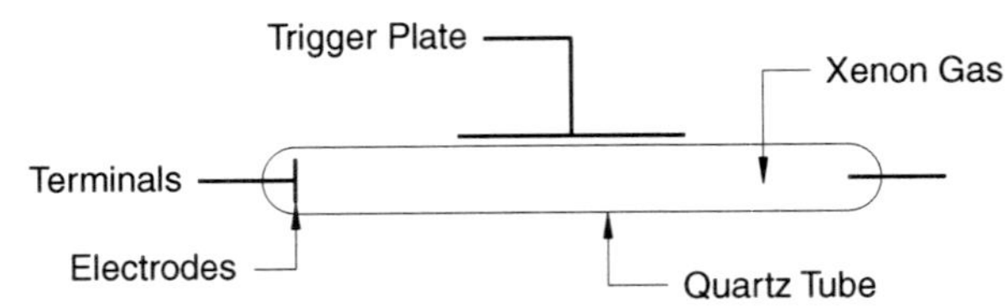

Figure 12-37 Xenon Flash Lamp

Figure 12-37 shows a schematic representation of a xenon flash tube. The glass tube is purged with xenon gas and has an electrode mounted on both ends. A trigger plate is affixed to the outside of the tube. The internal resistance is too high to initiate a plasma when a high voltage is applied to the terminals. The trigger plate is pulsed with a short duration signal which, in turn, ionizes the xenon gas in the tube and lowers its resistance. Once the resistance is lowered, the high voltage across the terminals can flow and a brilliant plasma is formed for a short duration.

Xenon flash tubes are most commonly supplied in either a straight or U-shaped tube, as shown in Figure 12-38. Notice that both tubes have a trigger plate affixed to the outside of the tube.

Figure 12-39 shows a basic schematic for a xenon flash tube. When voltage is applied to the circuit, both C_1 and C_2 are allowed to come up to full charge. When the trigger is closed, C_2 discharges, creating a pulse in the primary of T_1 and consequently a high-voltage pulse is generated in the secondary. The xenon gas is ionized, allowing the charge in C_1 to discharge and creating a brilliant flash. R_1 is used to prevent C_1 from discharging into C_2 when the trigger is closed.

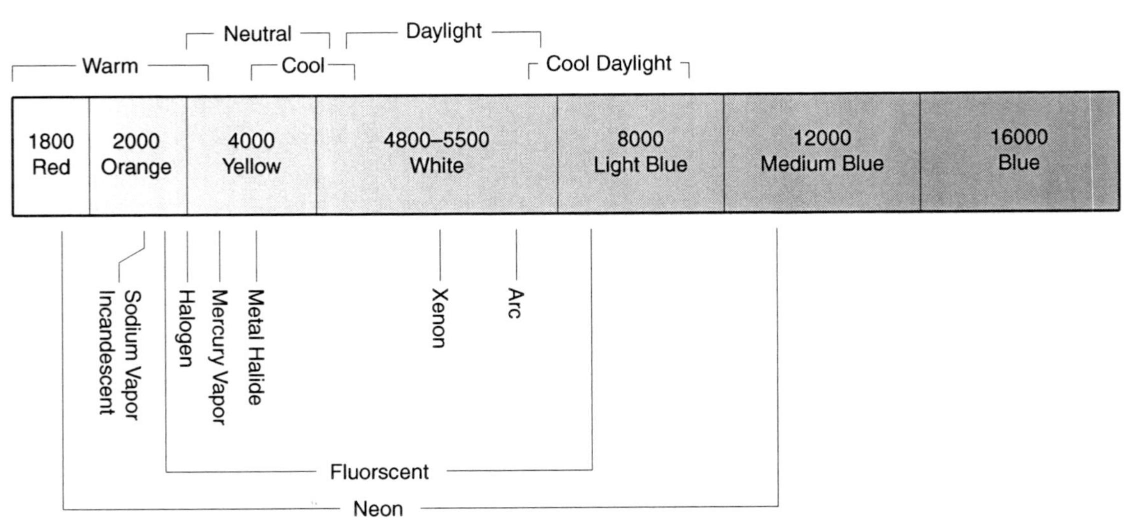

Figure 12-36 Color Temperature

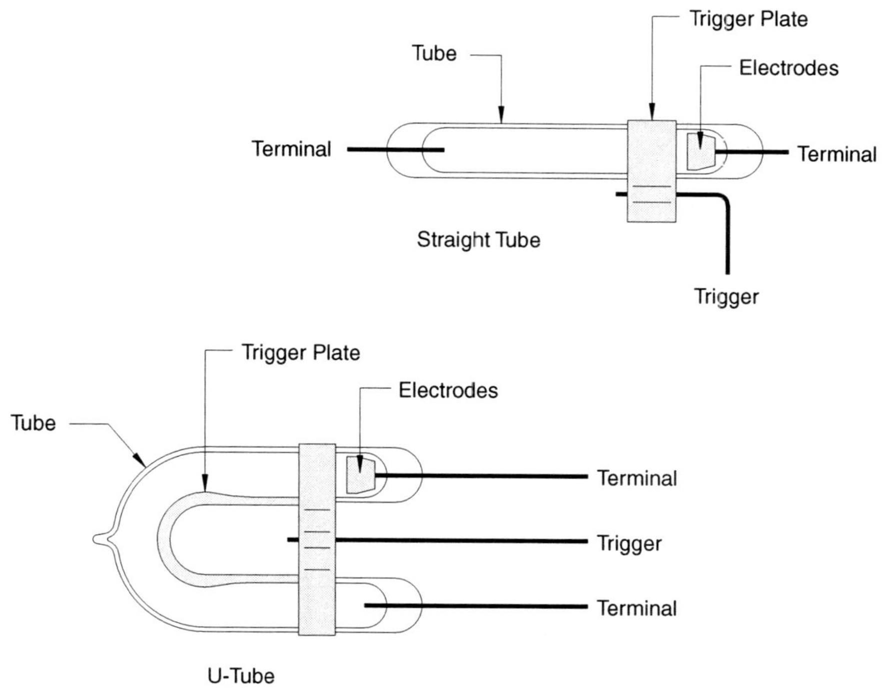

Figure 12-38 Commercial Xenon Flash Tubes

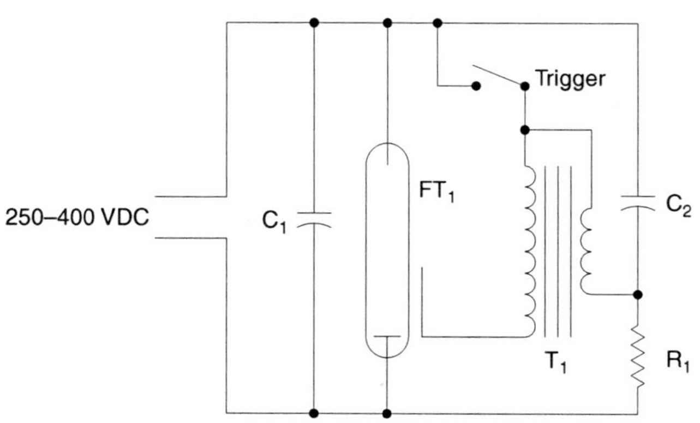

Figure 12-39 Xenon Flash Tube Circuit

Short arc xenon lamps are intended to operate in a steady state fashion and are deployed for applications that require an extremely high-intensity output and daylight color balance. The most noteworthy application of short arc xenon lamps is in motion picture projectors found at local movie theaters.

Figure 12-40 shows a typical short arc xenon lamp. A high-voltage pulse is applied to start the bulb and then a lower operation voltage is maintained during operation. Because of the extreme heat that these bulbs generate, most units are water cooled.

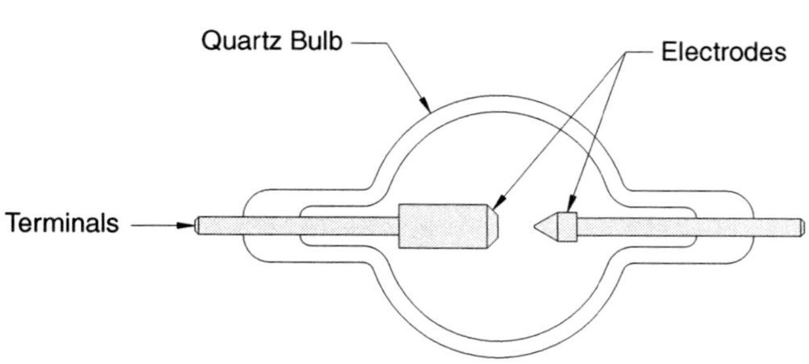

Figure 12-40 Short Arc Xenon Lamp

Carbon Arc Lighting

Carbon arc lighting has all but disappeared in modern applications. It has been principally replaced by short arc xenon lamps. In the early 1800s, the first street lighting was carbon arc. Probably the most noteworthy application of carbon arc lighting was the spotlights used during World War II to spot enemy planes flying over England at night. Figure 12-41 shows a schematic representation of a carbon arc spotlight with a high-voltage ignition starter.

Light Emitting Diodes (LED)

Light emitting diodes (LED's) have permeated our lives over the years. The little indicator lamp that shows your hard drive is working, the infrared source in your TV remote controller, and the red power indicator on your stereo, are all LED's.

Light emitting diodes are diodes that generate light emissions when they are energized. They are commonly supplied in two basic sizes—5 millimeter and 3 millimeter —as shown in Figure 12-42. Light emitting diodes are available in red, yellow, green, and white. There are super bright versions that are appropriate for lower level lighting applications. These super bright LED's are commonly found arrayed for traffic signals and automotive tail lights. They are also used for small inspection lights, being no bigger then a pen or can be hung on a key chain.

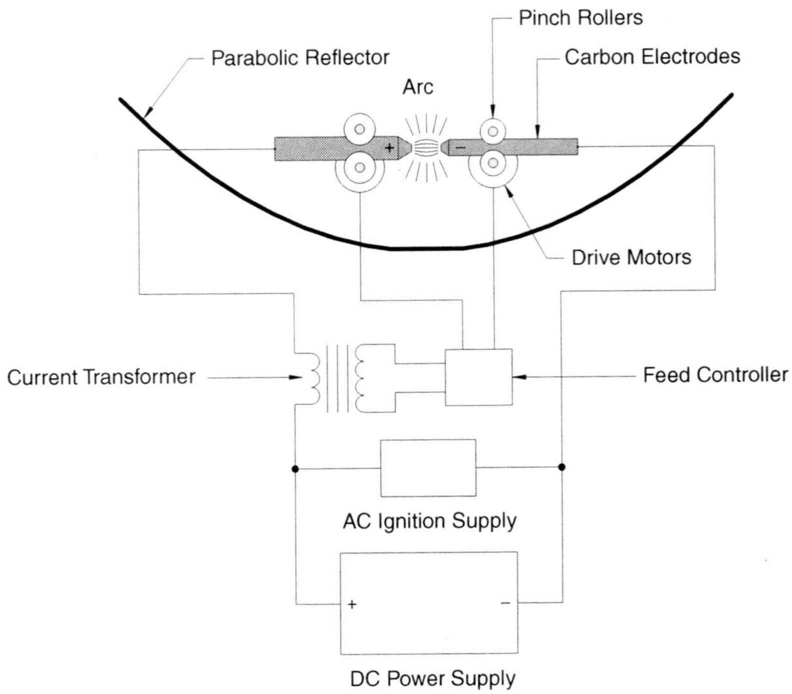

Figure 12-41 Carbon Arc Lamp Schematic

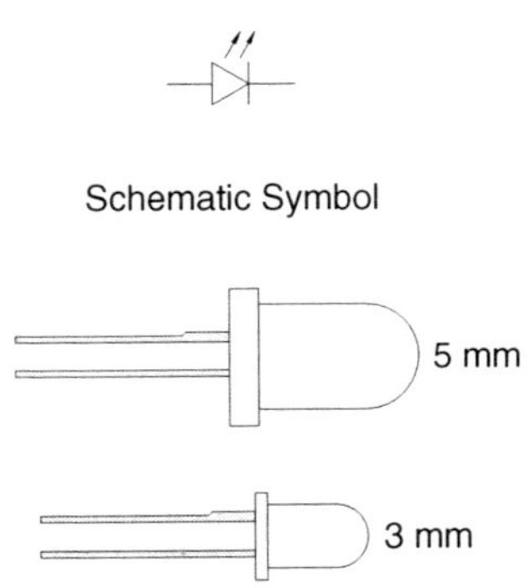

Figure 12-42 Discrete Light Emitting Diodes

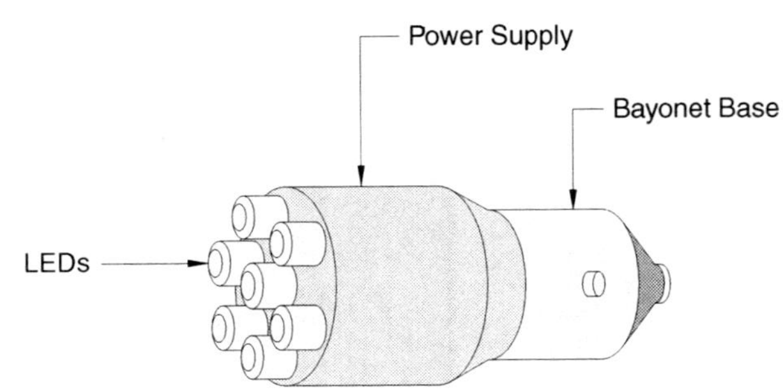

Figure 12-43 LED Cluster Lamp

Figure 12-43 shows an array of super bright LED's on a standard bayonet base. These units are intended to provide a long life, high efficiency replacement for standard incandescent bulbs.

Another common use for LED's is in seven segment displays, as shown in Figure 12-44. The most noteworthy application is in ordinary digital alarm clocks where the bright red numbers are as easily read during the night as in the daytime.

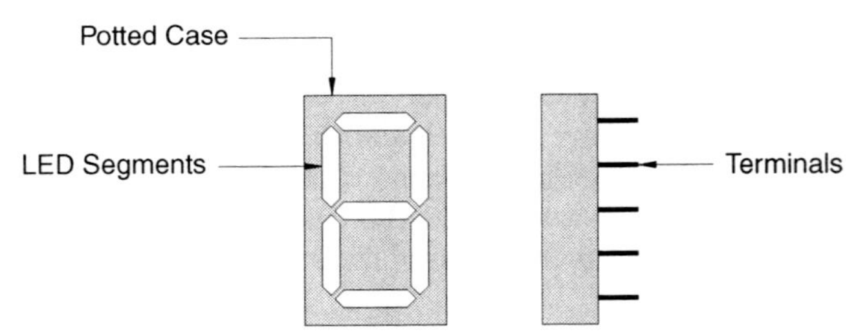

Figure 12-44 Seven Segment LED Display

CHAPTER 13
METERS

When working with electromechnical devices, it is imperative that the technician be able to gauge certain aspects of the electrical power and the signals being utilized by the equipment. This gauging may be as simple as connecting a light bulb to verify power or it may require a highly sensitive vacuum tube voltmeter to monitor extremely low voltages. Similarly, a signal may be verified with a simple loudspeaker or it may require the sophisticated display of an oscilloscope. In any case, a preliminary understanding of meters and their uses is extremely valuable information to have under your belt.

Compass

Let's start by examining one of the most basic electromagnetic instruments, the compass. Figure 13-1 shows a typical commercial compass such as may be found in any sports and outdoor store. This instrument is made by mounting a magnetized needle onto a precision pivot. The pivot allows the needle to freely align with the magnetic field of the earth. In doing so, the North Pole of the needle will always point toward the magnetic North Pole of the earth. The pivot is mounted in the center of a graduated face, which, in turn, is placed in the bottom a nonferrous case. The case is typically sealed with a glass window that prevents the needle from coming off the pivot when the instrument is transported in a pack or pocket.

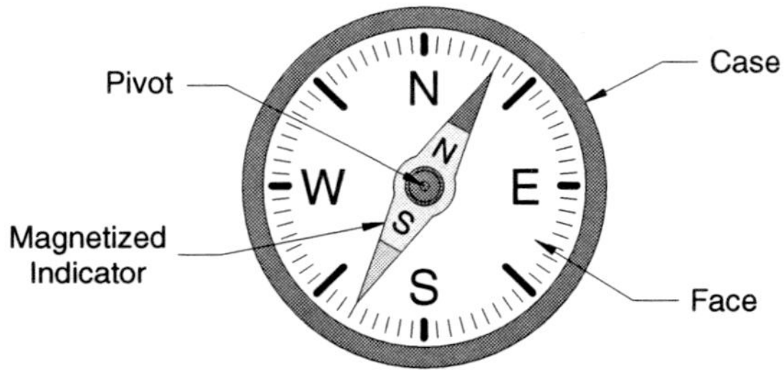

Figure 13-1 Magnetic Compass

Building a compass couldn't be easier. An upholstery needle is magnetized by stroking it with a permanent magnet. Once the needle is magnetized, it is forced through the center of a cork. The cork and needle assembly is then floated in a bowl of water. The needle will rotate until its North Pole is pointing towards the magnetic North Pole of the earth. Figure 13-2 shows a simple bench built compass.

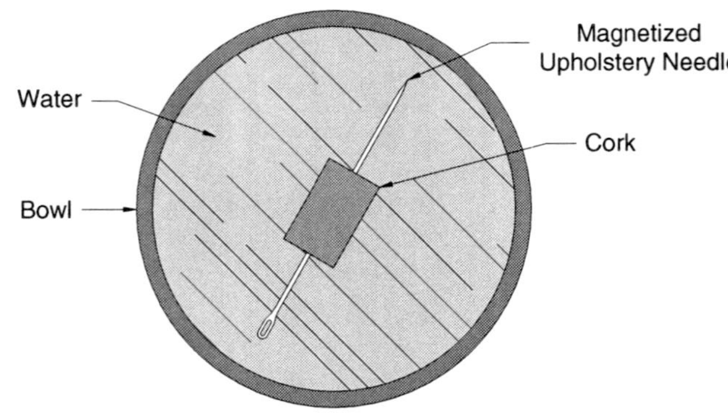

Figure 13-2 Bench Built Compass

Galvanometers

The earliest type of electrical meter was the fixed coil galvanometer. These instruments used a simple compass and a coil of wire to detect and measure electrical signals. Early fixed coil galvanometers consisted of a compass mounted on a pedestal or support post and then surrounded with a large coil of wire. The instrument was set up so that the needle pointed north and the coil position was adjusted to be parallel to the needle. When an electrical signal was applied to the coil, the compass needle deflected. These instruments could be made to detect extremely low signal levels. Figure 13-3 shows an early laboratory galvanometer.

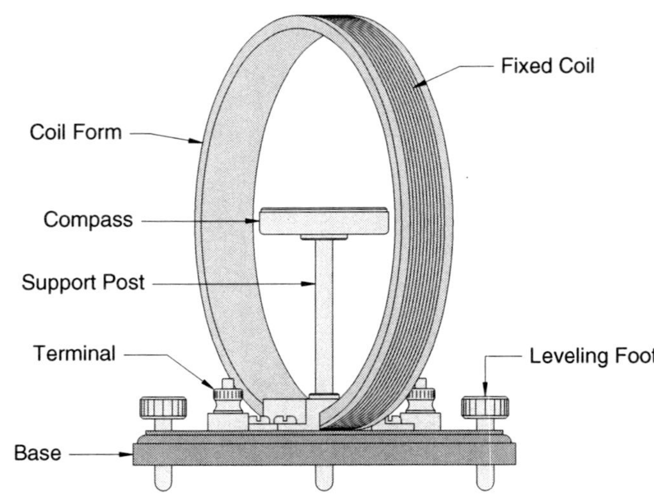

Figure 13-3 Fixed Coil Galvanometer

Building a fixed coil galvanometer can be accomplished by winding a coil around a piece of 6-inch polyvinyl chloride (PVC) pipe, as shown in Figure 13-4. The coil is mounted to

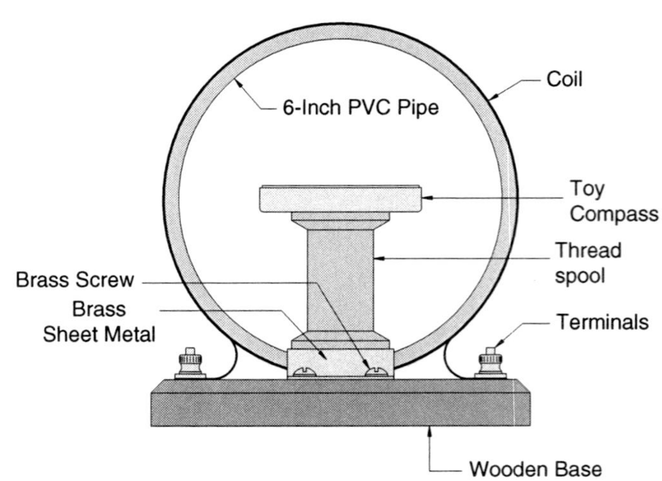

Figure 13-4 Bench Built Fixed Coil Galvanometer

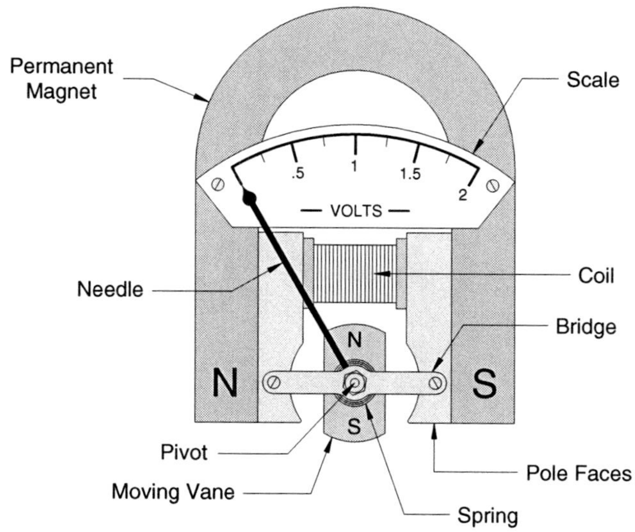

Figure 13-5 Permanent Magnet Galvanometer

a baseboard and a toy compass is placed on top of a thread spool in the middle of the coil.

The permanent magnet galvanometer is designed to operate independent of the earth's magnetic field. A magnet is added to counter the effects of stray magnetic fields, as shown in Figure 13-5. When the coil is energized, the instrument's field is altered and the needle deflects in direct proportion to the signal.

Figure 13-6 shows how to build a permanent magnet galvanometer. A toy compass is glued to the base of a plastic box. A curved, magnetized strip is placed around the magnet, as shown. The coil is then wrapped around the box, compass, and poles of the magnet. When a signal is applied to the terminals, the compass needle will deflect.

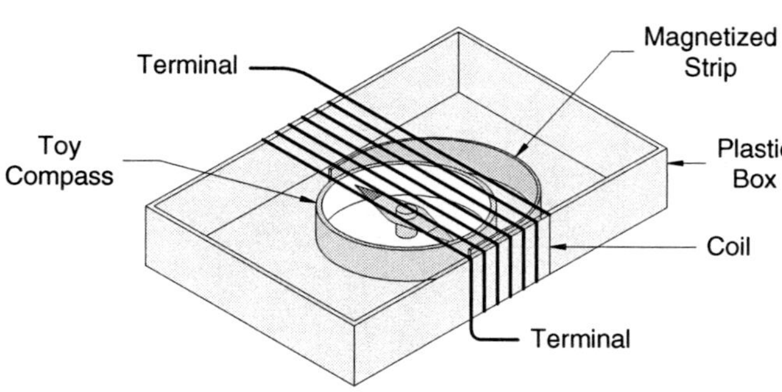

Figure 13-6 Bench Built Permanent Magnet Galvanometer

Moving coil galvanometers are the most common configuration for this class of instruments. Figure 13-7 shows an early moving coil galvanometer. A coil, with an iron core, is suspended from a fine wire so that it is located between the poles of a horseshoe magnet. Tension is maintained with a preload spring at the bottom of the coil. A needle, which points to a volts scale, is mounted to the top of the coil assembly.

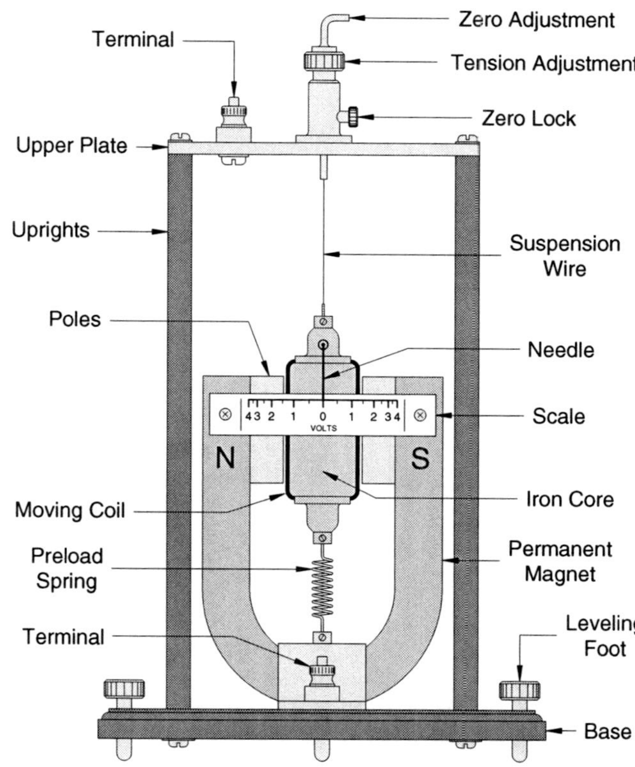

Figure 13-7 Moving Coil Galvanometer

When a signal is applied, the coil deflects and the needle indicates the applied voltage. To improve the sensitivity and resolution of these instruments, the needle is often replaced with a mirror. A focused light source is reflected off the mirror and onto a scale located at a distance from the instrument. The distance of the scale from the mirror amplifies any movement of the coil.

Moving Coil Voltmeters

The most common type of voltmeter is the moving coil design. This type of meter operates in the same fashion as a moving coil galvanometer. The principal difference between the two instruments is that the voltmeters are generally less sensitive and considerably more rugged. Their lower sensitivity is generally due to the higher resistance of the coil. These instruments are also more compact than a galvanometer because they are usually mounted into a panel or stand-alone equipment.

Figure 13-8 shows a stylized view of a typical moving coil voltmeter. A coil and an iron core are positioned between the poles of a permanent magnet. The coil/core assembly is allowed to rotate on two pivot points. A needle, or pointer, is affixed to the core and a small clock spring is used to return the mechanism back to a zero reading. The needle points to a scale mounted onto the magnet. When a signal is applied to the terminals, the coil generates a magnetic field and the coil/core assembly rotates to align with the field of the permanent magnet. The stronger the signal, the more the coil/core assembly rotates, which, in turn, generates a higher reading.

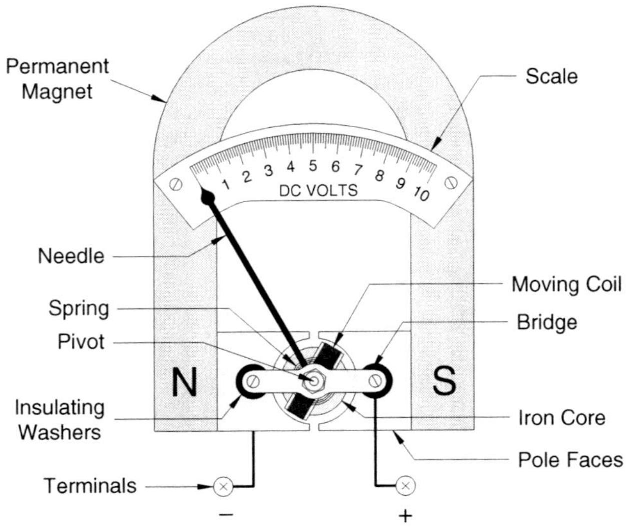

Figure 13-8 Moving Coil Voltmeter

By adjusting the position of the coil/core assembly and needle, as shown in Figure 13-9, it is possible to set up a voltmeter to indicate the polarity of the incoming signal. If the signal matches the polarity of the meter then the needle will deflect to the right. If the signal has a reverse polarity, then the needle will deflect to the left.

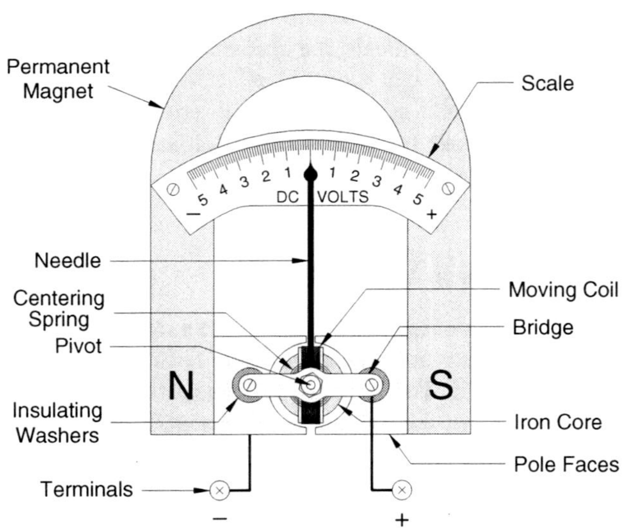

Figure 13-9 +/– Indicating Moving Coil Voltmeter

The range of any voltmeter can be adjusted to read higher voltages by adding a compensation resistor, as shown in Figure 13-10. In this example, the internal resistance of a 0 to 10 volt meter is 10,000 ohm. By adding a 900,000-ohm resistor, the effective resistance of the instrument is 10 times higher and, therefore, will read one-tenth of the input signal. To get a full reading at 10 volts, the input signal must be 100 volts. By simply adding the resistor, the 10-volt meter has been converted to a 100-volt meter. In this way, virtually any voltage can be measured on a relatively low voltmeter.

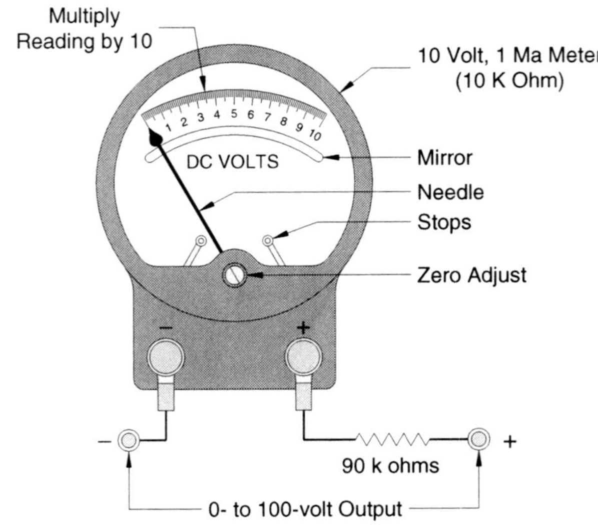

Figure 13-10 Voltmeter with Single Voltage Compensation resistor

Multirange voltmeters can be configured by setting up an array of resistors, as shown in Figure 13-11. In this case there is a common terminal and four voltage terminals. Each terminal is arranged with a resistor in series with the meter. The "Times 0" terminal doesn't require a resistor. The voltage reading is based on the multiplier associated with each terminal. As an example, if an 8-volt reading is shown while a voltage is connected across the "Times 100" terminal and the common, then the actual indication would be multiplied by 100. ($8 \times 100 = 800$ volts)

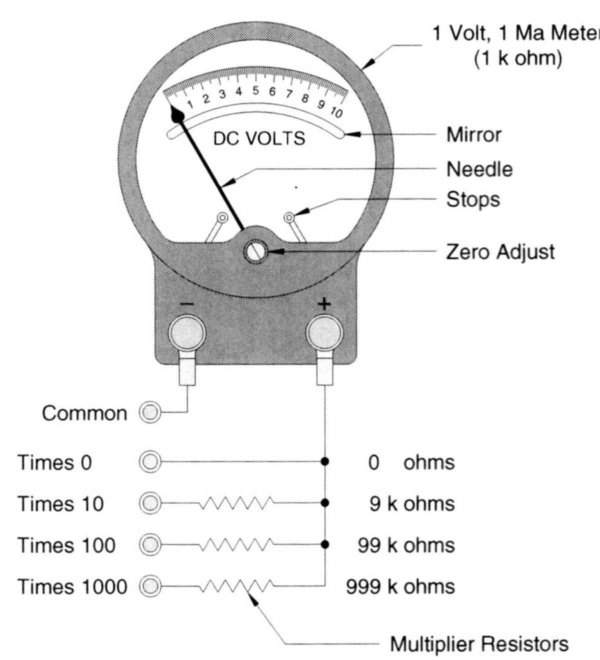

Figure 13-11 Voltmeter with Multi Range Multiplying Resistors

Another method to read higher voltages with a low voltmeter is to incorporate a voltage divider, as discussed in Chapter 4. Figure 13-12 shows a 10-volt meter configured to accept a 0- to 100-volt input signal. This method is normally not used on analog meters because the current loss over the circuit can be fairly high. This, in turn, affects the sensitivity of the meter. As an example, the circuit shown would require a 10-mA drive current to read full scale.

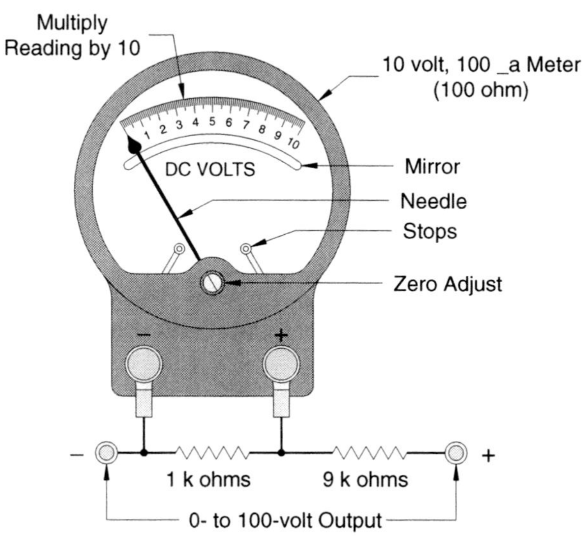

Figure 13-12 Voltmeter with Voltage Divider Compensation Resistors

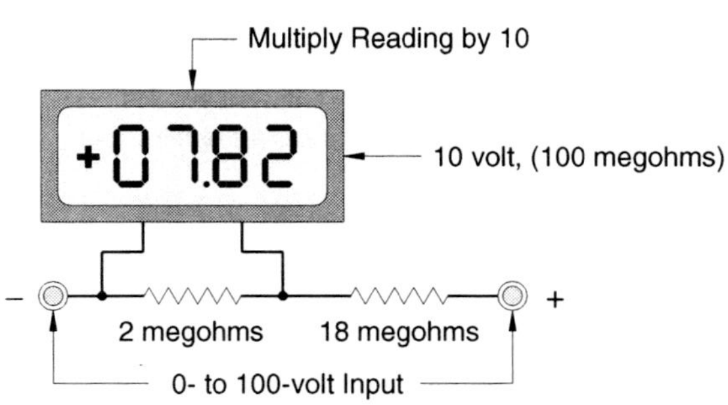

Figure 13-13 Digital Voltmeter with Voltage Divider Resistors

Voltage dividers are more commonly used on digital meters, as shown in Figure 13-13. Because a digital meter has an extremely high input impedance, the resistors that are used for the voltage divider can be in the megohm range and, therefore, require very low driving currents. As an example, the circuit shown would only require a 0.5-μA drive current to read full scale.

A multirange digital voltmeter can be set up using a voltage divider network coupled with a selector switch, as shown in Figure 13-14. In this case the input impedance is 100 megohms, which translates to an extremely low drive current.

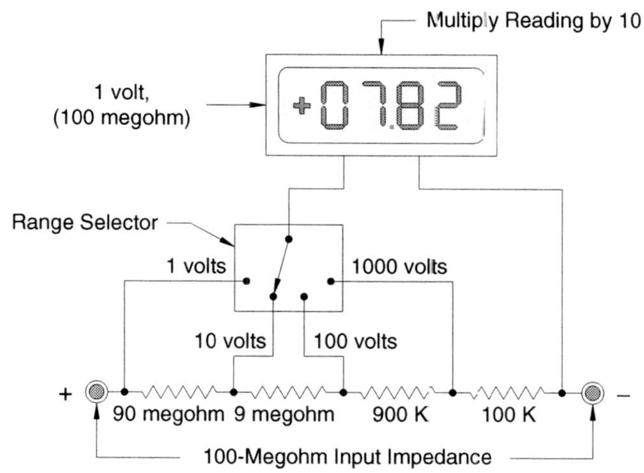

Figure 13-14 Digital Voltmeter with Four Range Voltage Divider

If an extremely high input impedance is required while using a moving coil voltmeter, then an amplifier must be incorporated, as shown in Figure 13-15. The voltage divider network is the same as with a digital voltmeter. The output of the selector switch is fed through a calibration potentiometer and then into an amplifier, which, in turn, drives the meter. The calibration adjustment is intended to tune the input signal to the amplifier so that the meter can be referenced against a standard voltage.

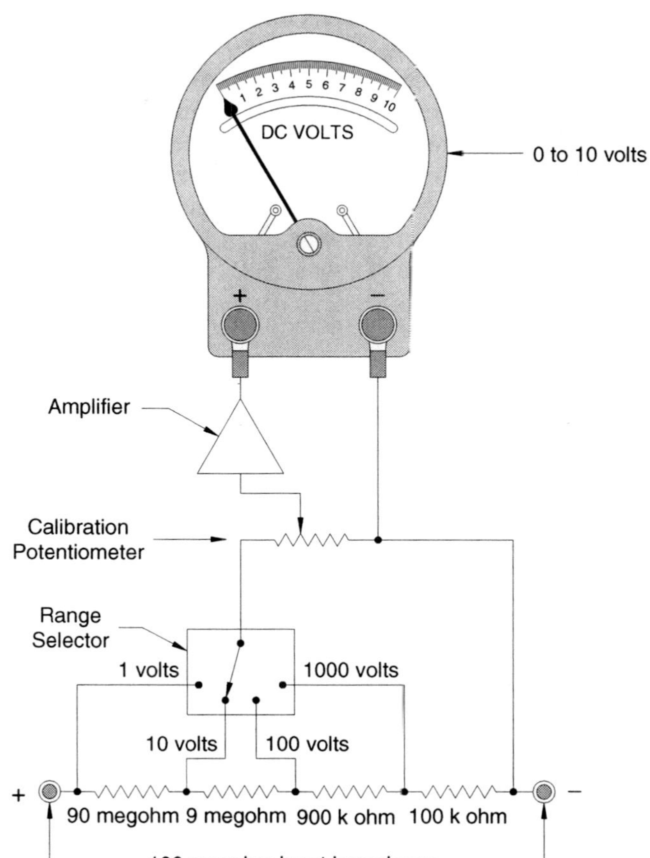

Figure 13-15 Amplified Analog Voltmeter with Four Range Voltage Divider

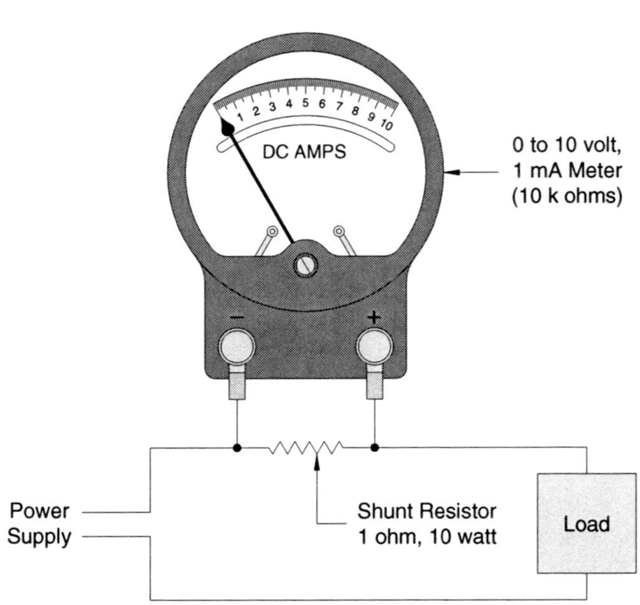

Figure 13-16 Voltmeter Configured to Indicate Amperes

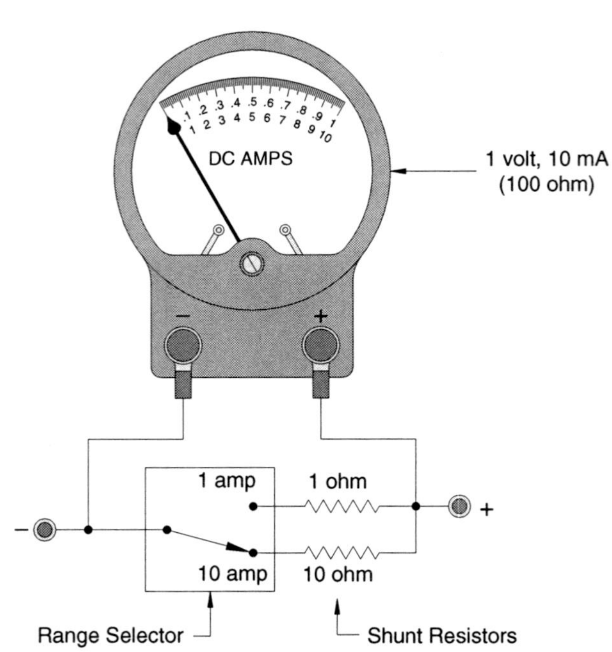

Figure 13-17 Voltmeter Configured to Indicate Three Current Ranges

Measuring current with a standard voltmeter is a simple proposition. A shunt resistor is added across the terminals of the meter, as shown in Figure 13-16. The resistance is matched to the meter so that 1 volt is equal to 1 amp. In this case a 1-ohm resistor is placed in parallel with the meter to act as a shunt. If the load pulls 8 amps, then the voltage drop across the resistor will be 8 volts and the meter will read 8 amps. It should also be noted that the resistor selected for this application must be able to carry a significant percentage of the current that is being tested. In this case, if the circuit requires 800 watts under normal operation and 1000 watts at peak meter deflection, then the shunt the resistor selected should have a minimum power rating of 10 watts.

To calculate the valve of a shunt resistor, use the following formula:

$$R_s = R_m \div [(D_s \div O_s) - 1]$$

where: R_s is the resistance of the shunt
R_m is the internal resistance of the meter
D_s is the desired current scale in amps
O_s is the original or meter current scale in amps

As an example, our circuit is calculated as follows:

$$[10\ (D_s) \div 0.001\ (O_s) - 1] \div 10{,}000\ (R_m) = 1\ (R_s)$$

Figure 13-17 shows a one voltmeter set up to indicate two different current ranges. The 1-amp range uses a 1-ohm shunt and the 10 range uses a 10-ohm shunt. The two different ranges are selected with a simple toggle switch.

A voltmeter can also be set up to measure resistance. A battery is placed in series with the resistance to be determined, and by knowing the voltage of the battery and the internal resistance of the meter; the unknown resistance can be determined. Figure 13-18 shows a schematic of a voltmeter set up for multirange resistance readings. The 0 ohms adjust is used to calibrate the meter before use. The range that is being used is connected directly to the common terminal. The meter will deflect to zero, but may not be exactly on zero. The zero adjust can then be used to tune the needle precisely to zero, calibrating the meter. The X100 range is actually the direct reading range. The X10 and X1 ranges are achieved by switching a shunt resistor across the meter terminals and

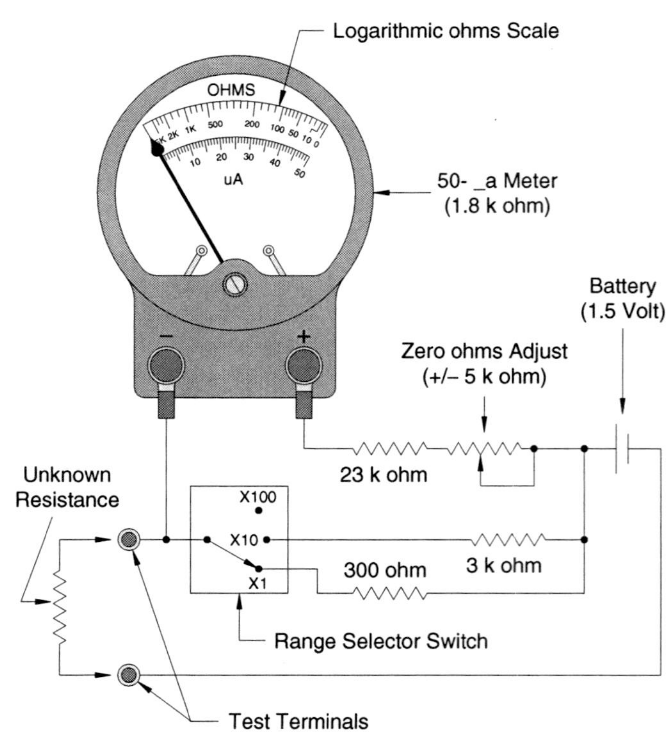

Figure 13-18 Microammeter Configured to Measure Ohms

scaling the current reading. When measuring ohms in this fashion the ohms scale is logarithmic so a conversion from the current reading can be calculated with Ohm's law. Use the following formula to convert the current reading of this circuit to ohms:

$$[1.5 \text{ (battery volts)} \div \text{(indicated current)}] \div \text{range} = \text{ohms}$$

Doing the math every time you measure a resistor is a little inconvenient, so a special meter face can be printed and glued over the existing face. The special face should have both current and Ohms scales as shown in the illustration. This will make the meter movement direct reading in the X1 range. The ohms indication is simply multiplied by the range for higher resistance values.

Plunger Type Voltmeters

Figure 13-19 shows a plunger type voltmeter mechanism. The movement is a needle that is affixed to an iron core piece. The bottom of the core carries an axle, which is mounted into a pivot set. A clock spring is utilized to return the movement back to zero. The iron core piece has a circular vane protruding from the right side. Just below the far end of the vane, a solenoid coil is positioned so that its magnetic field will act on the iron vane. When a signal is applied to the coil, the plunger is pulled into the coil in direct proportion to the strength of the magnetic field produced.

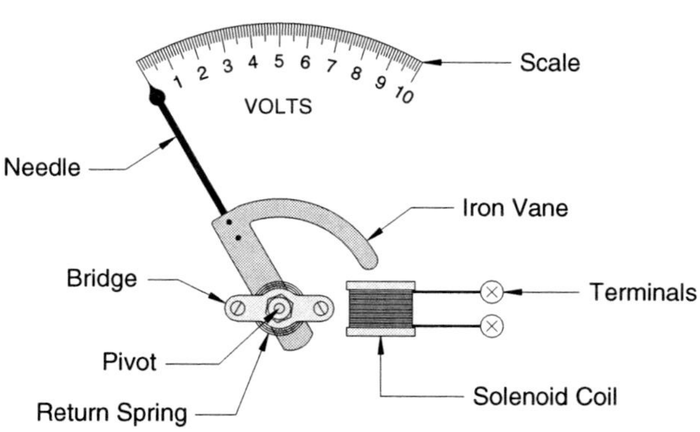

Figure 13-19 Plunger Type Voltmeter

Repulsion Vane Voltmeters

The repulsion vane mechanism consists of a coil of wire with two iron cores. One core is in a fixed position while the other is allowed to rotate about the axis of the coil. When a signal is applied to the coil, a magnetic field is generated causing the moving core to attempt to adopt a position in the field that will bring about a balance. The moving core rotates against the clock spring with the needle reading in direct proportion to the input signal. Figure 13-20 shows a repulsion vane voltmeter mechanism.

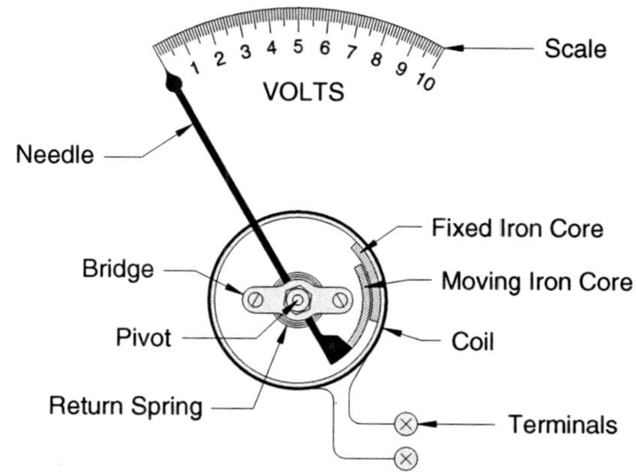

Figure 13-20 Repulsion Vane Voltmeter

Dynamometer Voltmeters

This type of voltmeter does not rely on permanent magnets or iron cores. In this arrangement, the signal itself generates the opposing magnetic fields to provide the requisite deflection. Three coils are used in the design, two are fixed and the third is a moving coil mounted in a pivot set with a clock spring. The two fixed coils are aligned so as to provide a uniform magnetic field. The moving coil is placed off-axis and in opposition to the fixed coils. When a signal is applied to the coils, the moving coil deflects in direct proportion to strength of the applied voltage. Figure 13-21 shows a dynamometer voltmeter arrangement.

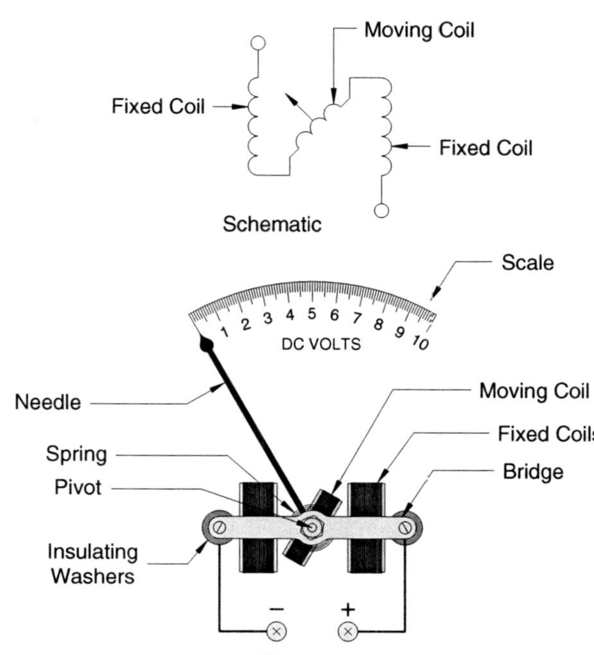

Figure 13-21 Dynamometer Voltmeter

Watt Meters

A watt meter is a dynamometer with the center coil driven independently from the two fixed coils, as shown in Figure 13-22. The moving coil is connected to the power feed, while the fixed coils are connected in series with the power source and the load. In this manner, the moving coil's deflection is based on the amount of current that the load uses while the fixed coils are based on the line voltage. Therefore, the meter indicates wattage.

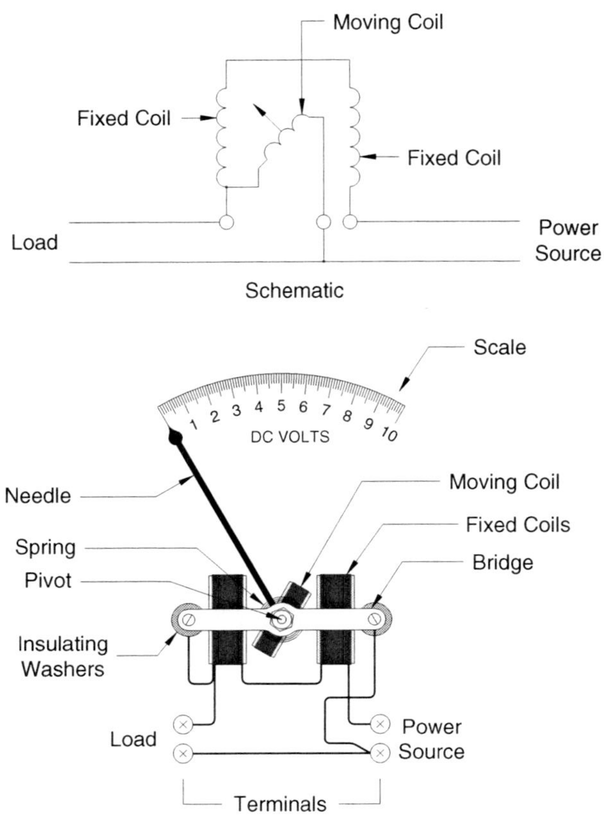

Figure 13-22 Watt Meter

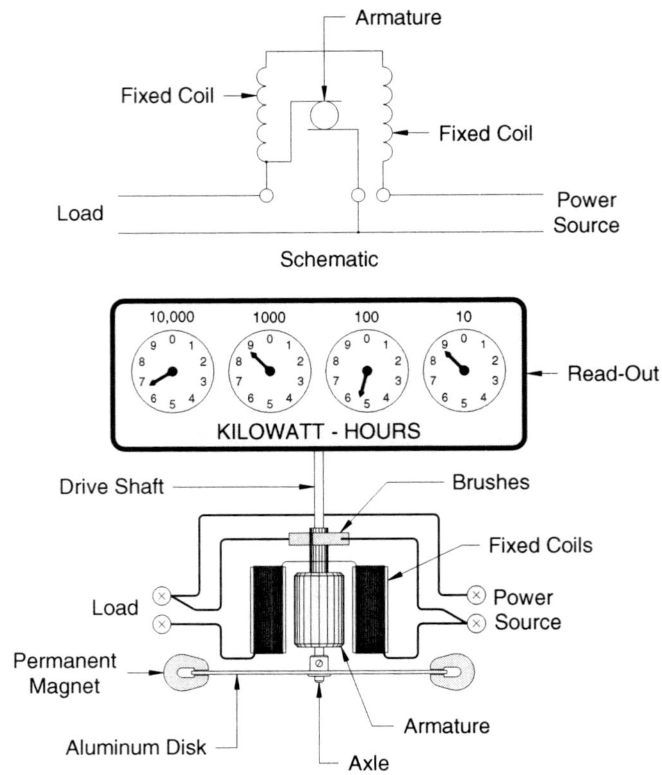

Figure 13-23 Watt-Hour Meter

Watt-Hour Meters

A watt-hour meter is effectively a watt meter with the moving coil replaced with a motor armature. The higher the load placed on the motor, the higher the speed at which the motor turns. The motor drives a totalizer mechanism that records the total wattage used during any given period. To prevent inaccurate transients, most watt-hour meters use an eddy current damper in the form of an aluminum disk with two permanent magnets. Figure 13-23 shows a stylized view of a watt-hour mechanism.

Hot Wire Meters

These meters rely on the expansion and contraction of a wire element in reference to its temperature. As a current is applied to the hot wire element, as shown in Figure 13-24, it expands. The traction wire pulls the hot wire down via the preload spring and the needle deflects in reference to the expansion of the hot wire. A hot wire meter is an excellent choice for applications where a slow or averaging response to signal changes is desired.

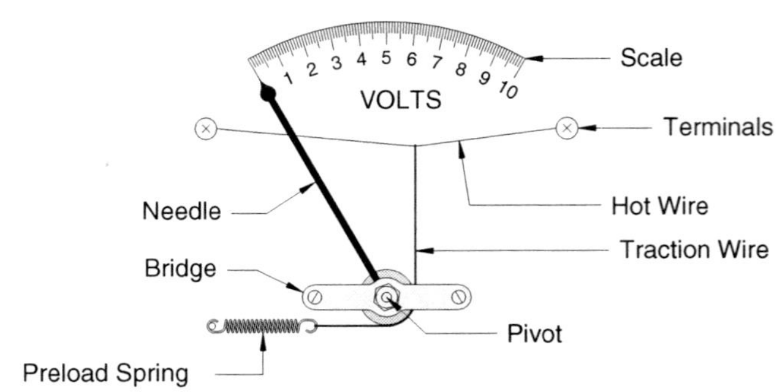

Figure 13-24 Hot Wire Meter

Multimeters

The single most important piece of electrical and electronic test equipment is the multimeter. These meters are designed to test for a broad range of voltages and values. Figure 13-25 shows a typical analog multimeter. These units have an analog meter movement, a network of matching resistors, and a selector switch housed in a compact, high-impact plastic

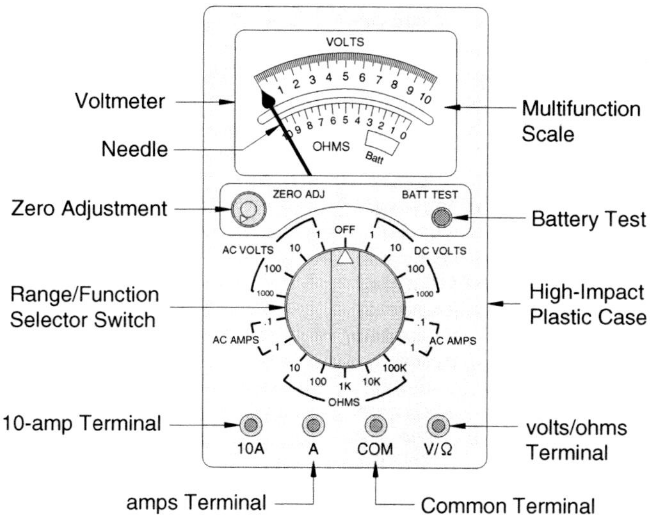

Figure 13-25 Portable Analog Multimeter

housing. They are easily placed in a toolbox or the technician's pocket.

Most multimeters will read AC volts, DC volts, AC amps, DC amps, and ohms. Many units also provide a beeper for continuity testing and a battery check function. As previously discussed, analog units will have a fairly low impedance, so sensitivity to very low voltages will be poor.

Vacuum Tube Voltmeters

Vacuum tube voltmeters, or VTVM's as they are sometimes referred to, are a type of multimeter that is designed to provide high sensitivity. Figure 13-26 shows the front panel of a typical VTVM. These meters generally have the same functions as a typical multimeter, but with boarder ranges. They also have a high input impedance and are accurate over a broad frequency range. To accomplish the broader parameters, the high input impedance and broad frequency range, these meters are equipped with an internal amplifier similar to the circuit shown in Figure 13-15. Another feature that is common to VTVM's is a 0- to 10-volt output. Regardless of the input range selected, the output mirrors a 0- to 10-volt signal. This allows easy interfacing to other instruments.

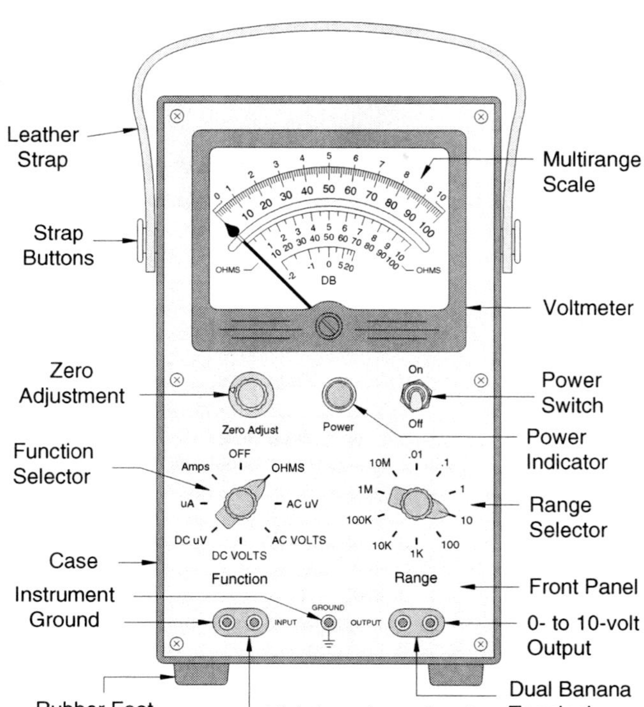

Figure 13-26 Vacuum Tube Voltmeter

Digital Multimeters

Most modern multimeters are equipped with a digital readout, as shown in Figure 13-27. A digital multimeter combines the compactness of an analog unit with the sensitivity of a VTVM. These instruments have become very affordable and are an excellent addition to any toolbox or workbench.

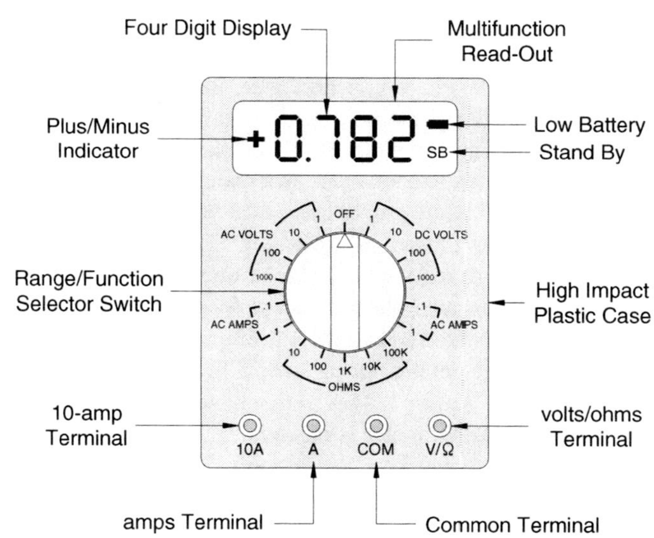

Figure 13-27 Portable Digital Multimeter

Bench Built Multimeter

Building a simple multimeter is an excellent way to gain some experience with electromechanical devices. This is a very inexpensive project that can be built in just a few evenings. In addition, when you are finished you will have a useful instrument that will provide support for future efforts.

Figure 13-28 shows a schematic and list of components for a basic analog multimeter. The meter is a 50 μA, panel mount movement with an internal resistance of 1800 ohms. The selector switch is simply a banana jumper set. The volt and ohm resistors are common 2% carbon units. The current resistors are high wattage 5% units. The battery is an ordinary 1.5-volt AA cell.

Item	Description
B1	1.5-volt Dry Cell Battery
D1	1500 PIV Diode
J1	Banana Jumper
M1	50- _a, 1.8-k ohm, Meter
P1	10-k ohm Potentiometer
R1	23 k ohm
R2	300 ohm
R3	3 k ohm
R4	20 k ohm
R5	200 k ohm
R6	2 megohm
R7	20 megohm
R8	9 ohm
R9	0.09 ohm
R10	0.009 ohm
R11	– ohm
S1	SPDT Toggle Switch

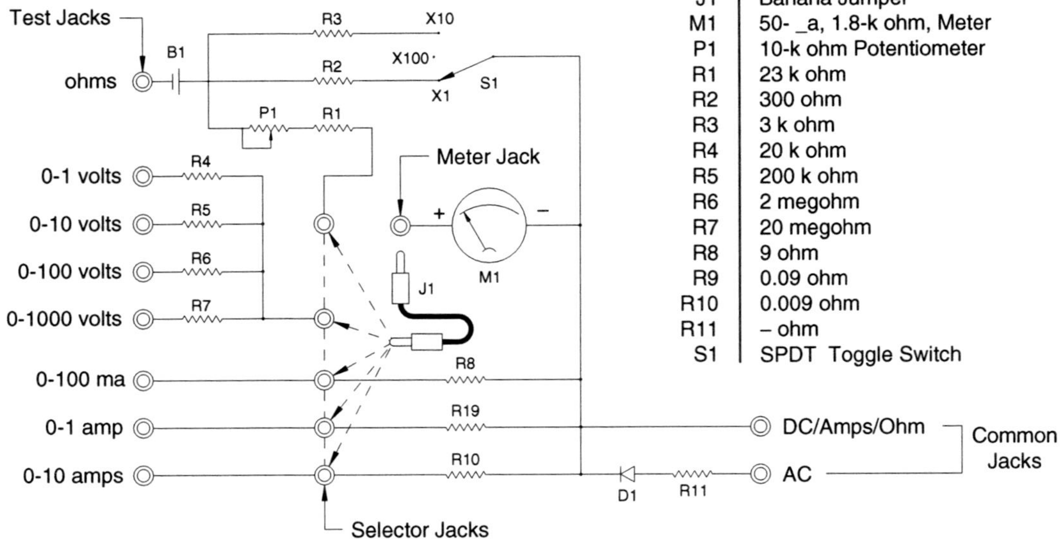

Figure 13-28 Bench Built Multimeter Schematic

Figure 13-29 shows a suggested layout for the front panel. If compactness is not a concern, then increase the size to accommodate your personal desires. The banana selector is made by mounting six panel jacks, as shown. The distance from the center jack to each outer jack should be ³/₄ of an inch. This spacing will allow you to use a dual banana plug as your jumper and prevent misconnection.

Figure 13-30 shows the back side of the finished panel. All of the connections are soldered except for the meter. R_{13} and R_{14} should be mounted so that there is ample clearance around them, as they may get hot during normal operation. Be certain to select a sturdy battery holder so that the cell is not knocked out when the finished meter is moved.

Figure 13-31 shows the case assembly. The top and bottom panels can be plastic or laminated Masonite. The box itself is a simple frame made from 1 inch × 2 inch #1 pine boards, held together with drywall screws.

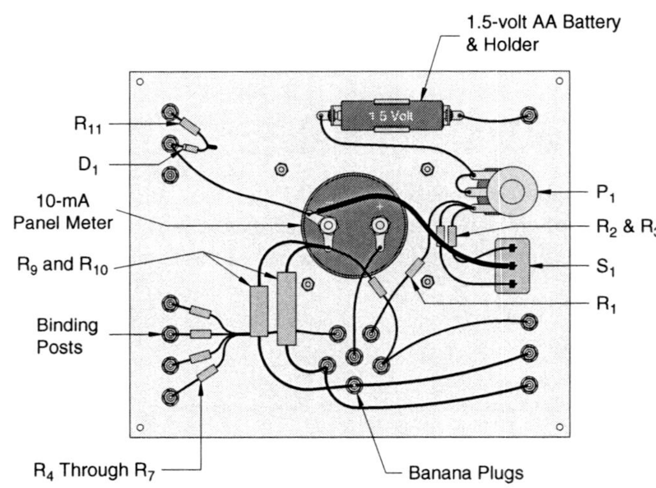

Figure 13-30 Bench Built Multimeter Panel Wiring

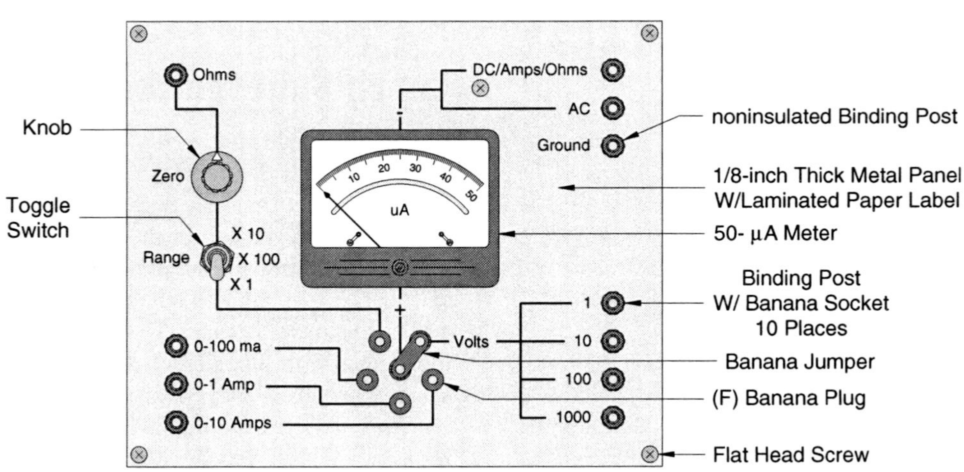

Figure 13-29 Bench Built Multimeter Panel

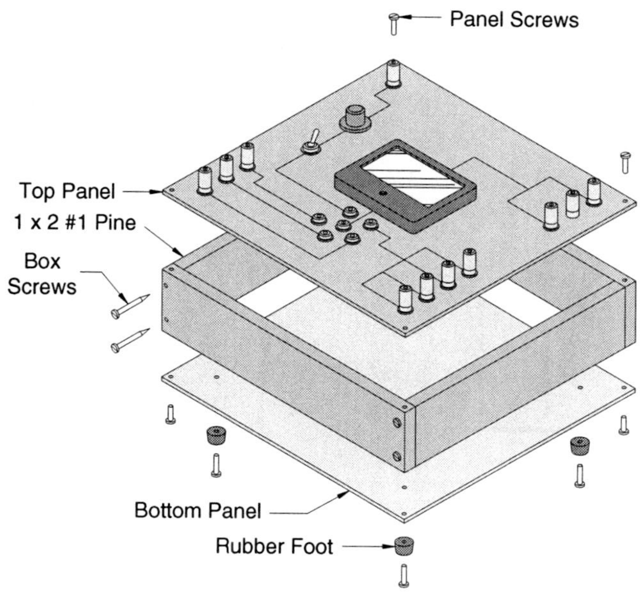

Figure 13-31 Bench Built Multimeter Cabinet Assembly

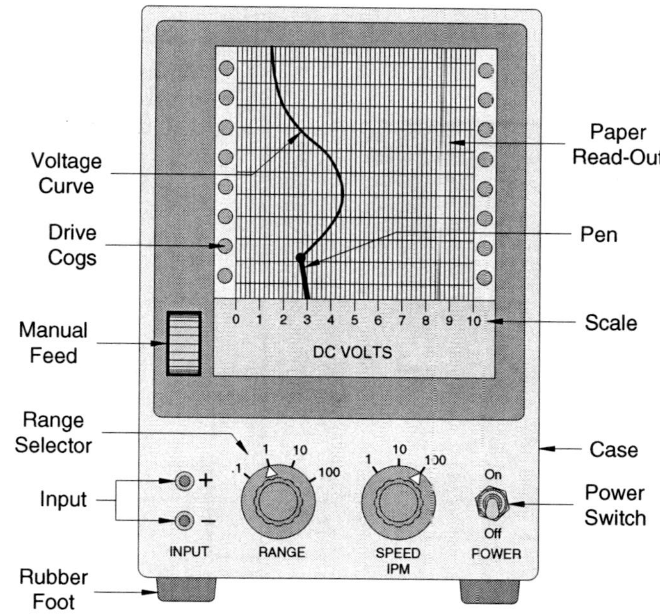

Figure 13-33 Strip Chart Recorder

The probe set shown in Figure 13-32 is made by pressing a brass rod, with a wire soldered to one end, into a heavy-wall plastic tube. After the rod is in place use a file to sharpen the ends. The opposite ends of the leads are equipped with standard banana plugs.

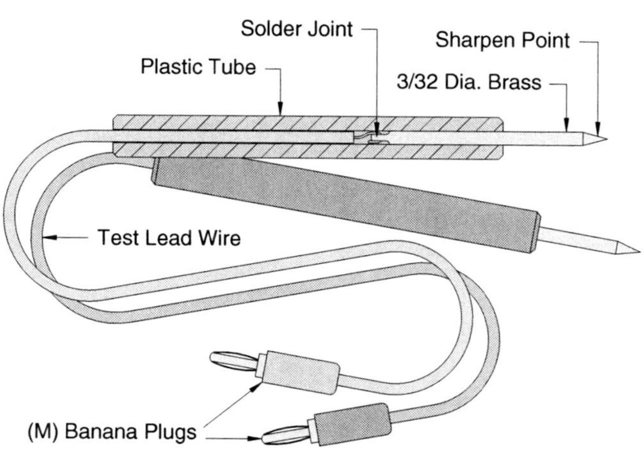

Figure 13-32 Bench Built Multimeter Test Probes

Strip Chart Recorders

For applications that require monitoring, a strip chart recorder is often used. These instruments are simply a voltmeter with an ink pen replacing the needle. A roll of paper is continuously moved under the pen and a continuous record is maintained. Strip chart recorders are generally multirange voltmeters with a speed range selector to control the paper feed. Figure 13-33 shows a typical strip chart recorder.

Circular Chart Recorders

Another type of recording instrument is the circular chart recorder. These recorders provide the same function as a strip chart unit, except that they are generally used for long-term monitoring. Speeds on the units are usually in hours, days, and/or weeks. Figure 13-34 shows a typical circular chart recorder.

Because of the internal mechanism, chart recorders are not particularly sensitive. To monitor lower level signals some sort of amplifier is required. Figure 13-35 shows a strip chart recorder being driven from the 0-to 10-volt output of an ordinary VTVM.

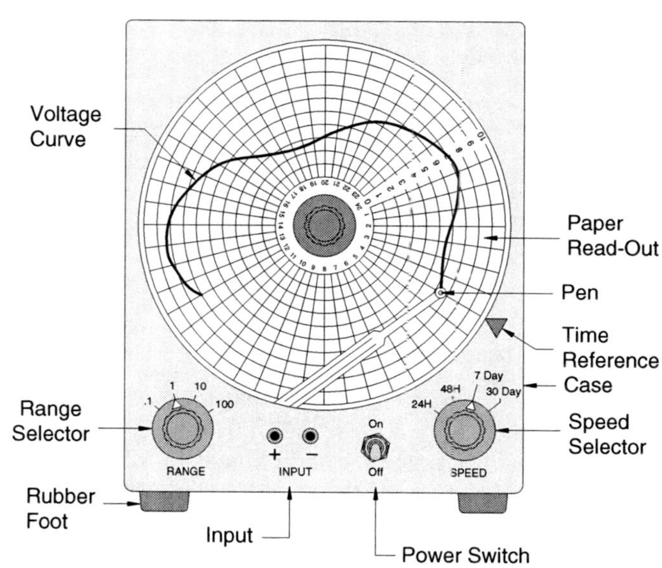

Figure 13-34 Circular Chart Recorder

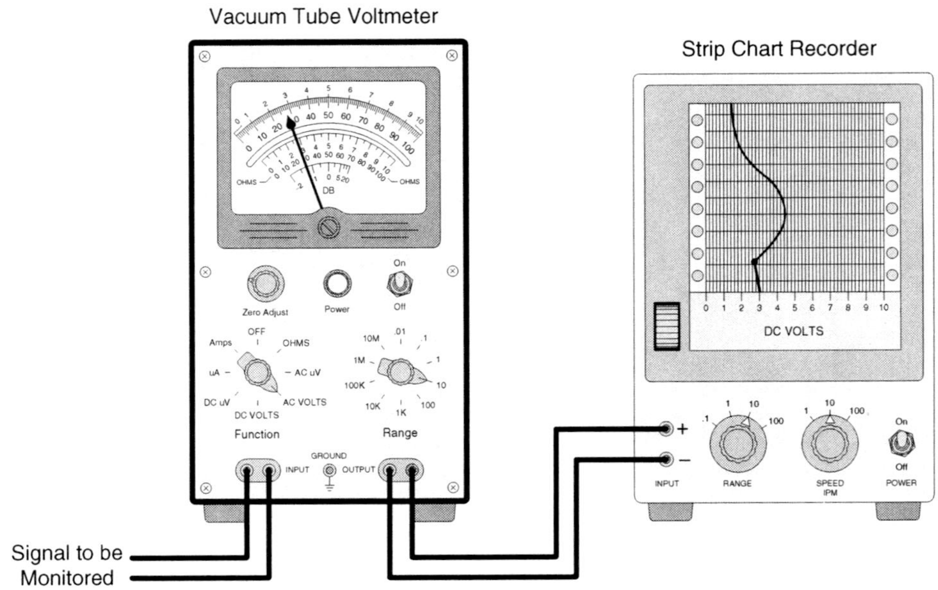

Figure 13-35 Using a Strip Chart Recorder with a Vacuum Tube Voltmeter

Meter Accessories

A basic probe set is by far the most important accessory that any meter can have. An ordinary probe set, as shown in Figure 13-32, can be built or purchased and will provide suitable performance for most situations. A better choice is a commercial set with interchangeable tips, as shown in Figure 13-36. The variety of tips provides better access to intricate circuits and hands-off use.

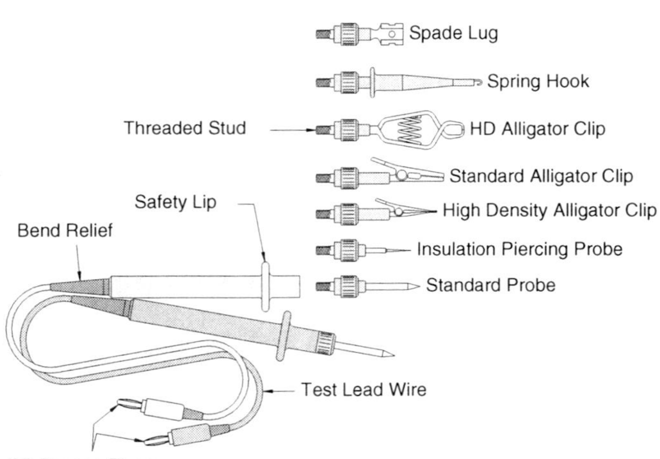

Figure 13-36 Multi Purpose Test Probe Set

The high-voltage probe shown in Figure 13-37 is used to gauge voltages in excess of the meter's range. These probes use a dropping resistor in the head of the body that generally drops the output voltage to 1 volt per 1000. That is to say that if the probe is connected to an 8000-volt source the meter will read 8 volts. Before using a high-voltage probe, it should be carefully inspected for any damage and should be clean. If the probe is damaged in any way, it should be immediately discarded. A high-voltage probe with even a small crack in the housing can be lethal. Always follow the manufacturer's recommendations when using high-voltage probes.

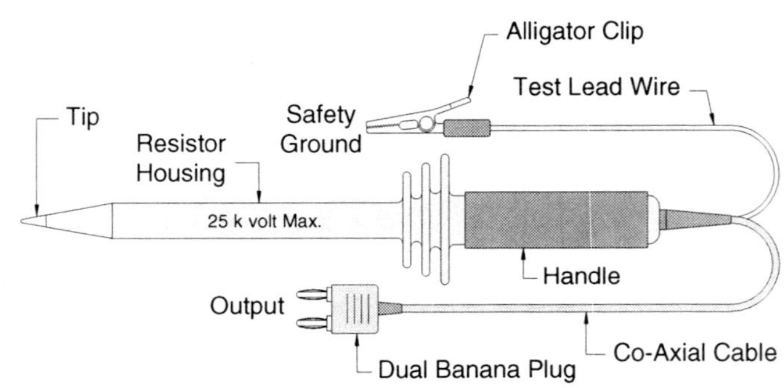

Figure 13-37 High-Voltage Probe

A clamp-on AC current probe, as shown in Figure 13-38, is an excellent accessory for any multimeter. These units make reading current very easy. Simply clamp the jaws around the wire to be surveyed and read the voltage on the meter. Generally, these units output 1 volt per amp. It should be noted that at maximum current, the voltage at the banana plugs can be dangerously high and great care should be taken not to disconnect the probe from the meter while the head is clamped onto a cable.

Figure 13-39 shows a schematic representation of a current probe. The jaw set is actually the core of a transformer. The primary is the cable being measured and the secondary provides the output signal. It should be noted that if the cable is looped twice around the core, the output voltage will double.

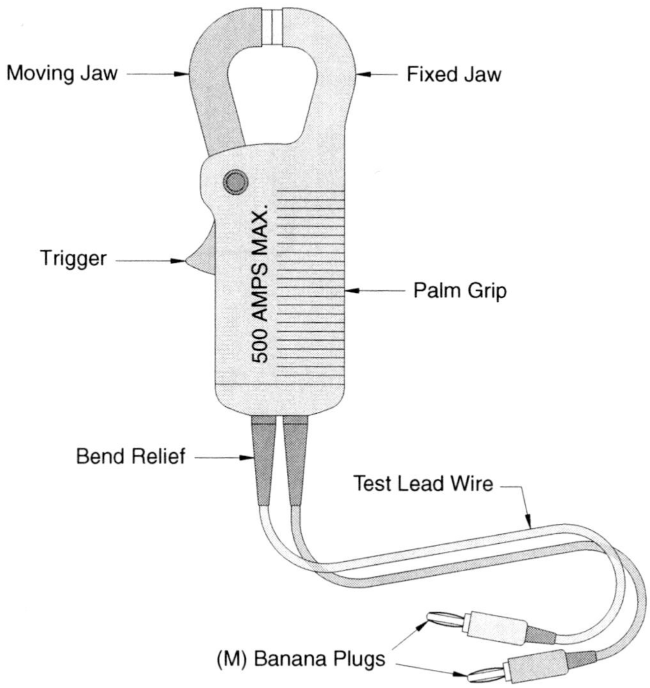

Figure 13-38 Inductive Pickup

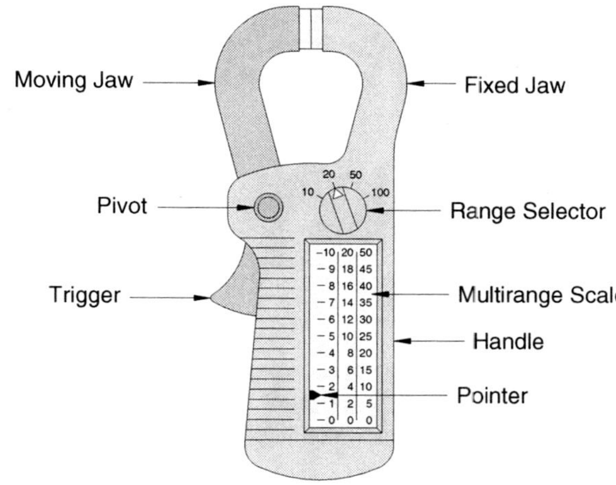

Figure 13-40 Hand Held Inductive Current Meter

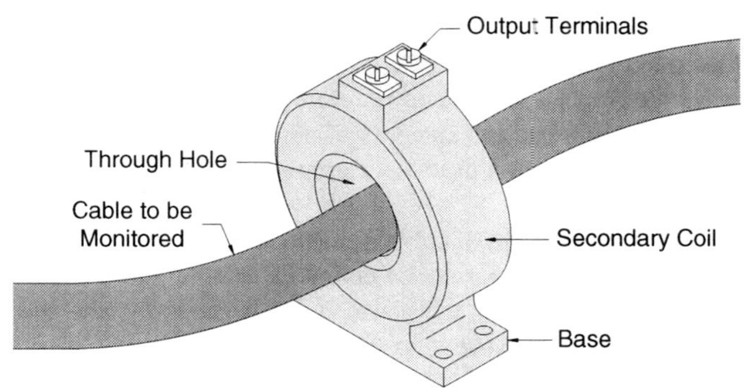

Figure 13-41 Commercial Current Transformer

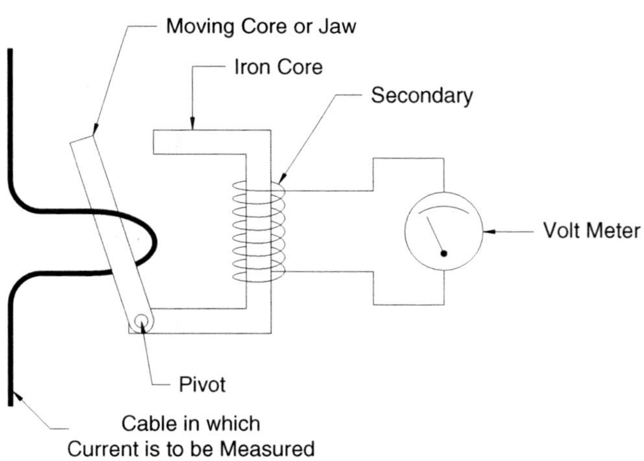

Figure 13-39 Inductive Pickup Schematic

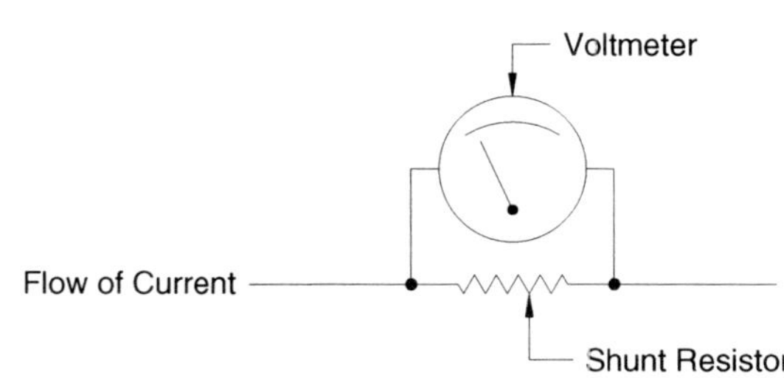

Figure 13-42 Current Shunt Schematic

Current probes that are complete, self-contained instruments are available, as shown in Figure 13-40. These instruments are very popular with technicians in most industries and are used to gauge the performance of all types of equipment.

For fixed applications, component current transformers are available, as shown in Figure 13-41. In this case, the current transformer is mounted in a location appropriate to conveniently route the high-current cable in the through hole. The output of the transformer is wired to a remote voltmeter. Using these devices throughout a plant, and routing their outputs to a central location, allows one technician to monitor a rather substantial facility.

Using a voltmeter to read amps is as simple as adding a shunt resistor, as shown in Figure 13-42. Oftentimes the real problem is finding a resistor with a low enough resistance and a high enough current capacity to do the job.

Figure 13-43 shows a typical commercial current shunt. These shunts are delivered with meter terminals that are properly spaced on the resistor. The shunt should also specify the volts per amp it is designed to output. For high current shunts this is usually 0.1 volts per amp. Therefore, a 600-amp shunt would output 0 to 60 volts.

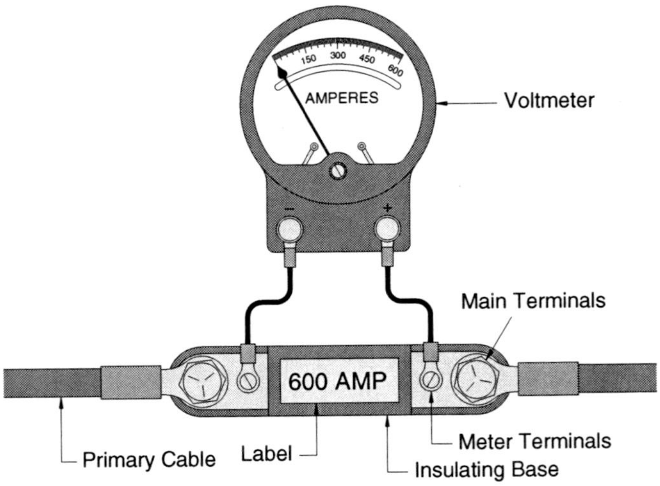

Figure 13-43 Commercial Current Shunt

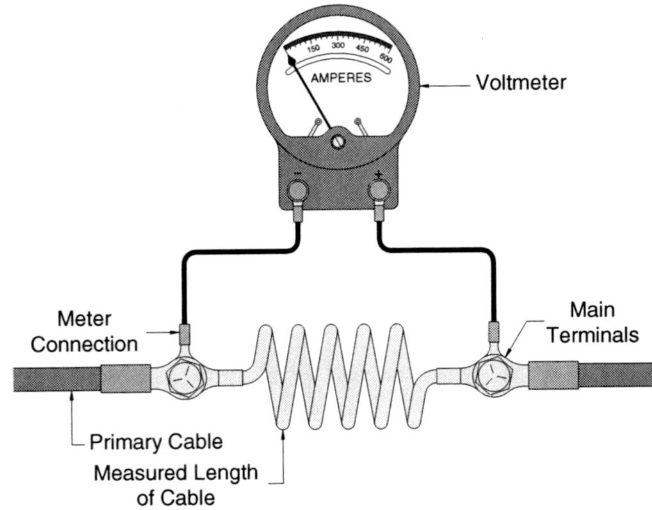

Figure 13-45 Cable Shunt

A shunt may be constructed using a copper buss bar, as shown in Figure 13-44. A voltage drop over the spacing of the meter terminals is calculated in reference to the resistance of the copper buss bar and an appropriate meter is selected. This type of shunt is often used for extremely high-current applications.

A measured length of cable can also act as a current shunt. Figure 13-45 shows a piece of coiled cable acting as a shunt. Like the buss bar shunt, this arrangement is generally reserved for extremely high-current applications.

Continuity Testers

Continuity testing is probably the most common test conducted on electrical and electronic equipment. It is an indispensable technique, which verifies the electrical soundness of a conductor. Continuity testing is also used extensively to trace and diagnose circuits.

Figure 13-46 shows two common commercial continuity testers. The combination flashlight and continuity tester is a popular tool among industrial service technicians. The basic tester is a very handy device when working in cramped environments. In both of these units, a battery is connected to a light bulb and a pair of test leads. When the test leads are touched together, the lamp turns on.

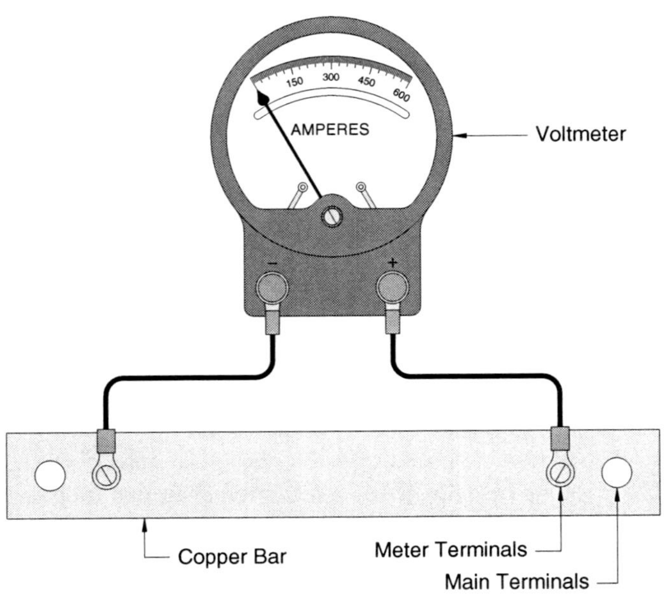

Figure 13-44 Buss Bar Shunt

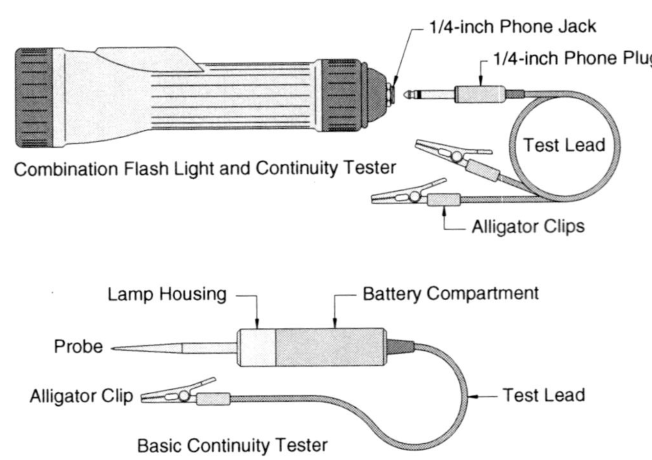

Figure 13-46 Continuity Testers

A basic continuity tester can be easily built by mounting a lamp base and battery holder onto a baseboard, as shown in Figure 13-47. The test leads are connected with banana plugs and combination binding posts. This provides for the use of the finished instrument in semipermanent applications on the bench.

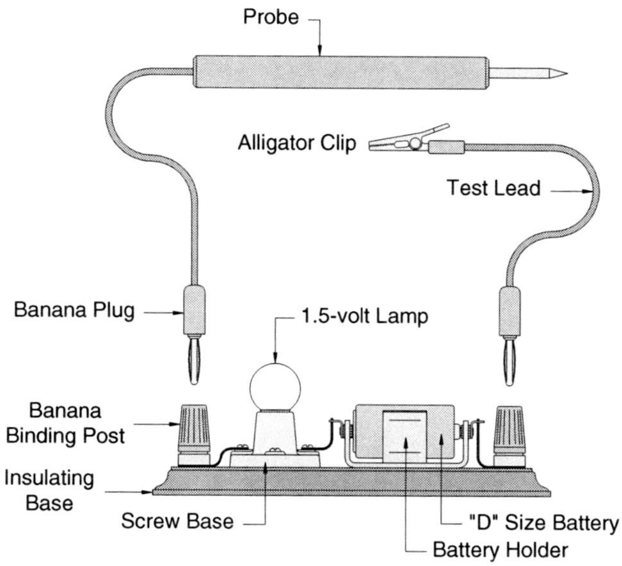

Figure 13-47 Bench Built Continuity Tester

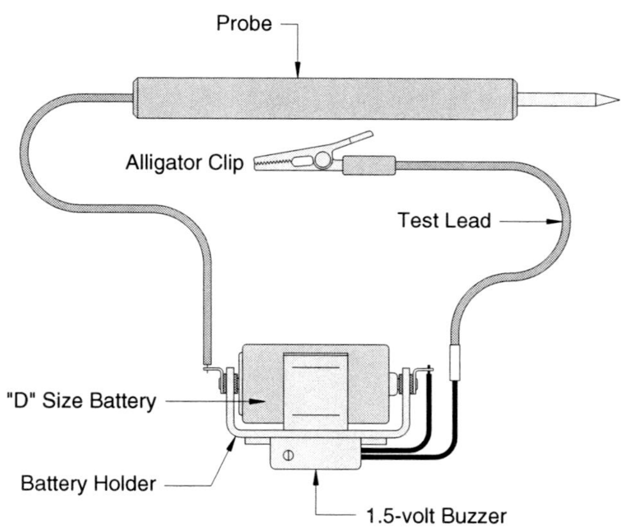

Figure 13-48 Bench Built Continuity Tester with Buzzer

Figure 13-48 shows a continuity tester made from a battery holder and audible buzzer. Instead of relying on a visual indicator, the buzzer alerts the technician of continuity.

Power Indicators

Power indicators are very inexpensive and quite handy to have in your pocket or toolbox. These devices allow a technician to quickly determine whether or not a circuit has power. Figure 13-49 shows three common commercial power indicators, a single voltage unit, a dual voltage unit, and dual voltage unit with a ground fault indicator.

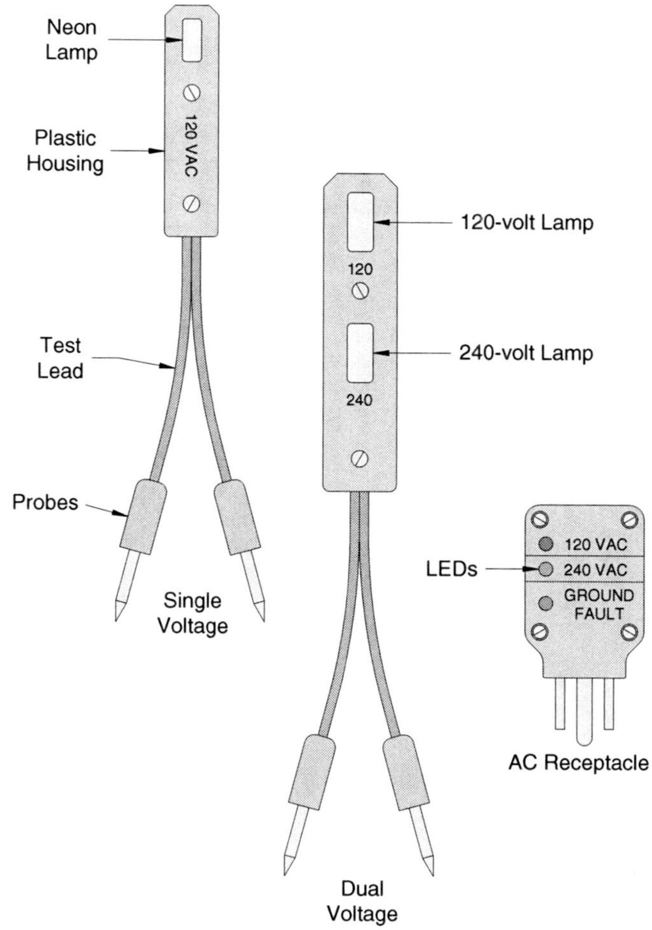

Figure 13-49 Power Indicators

In days gone by, technicians often used what they referred to as a service light. This device is simple a rubberized screw base with a 40-watt incandescent light bulb. The socket is equipped with a cage to protect the bulb and the base is wired to two probes. Figure 13-50 shows a typical, bench built, service light.

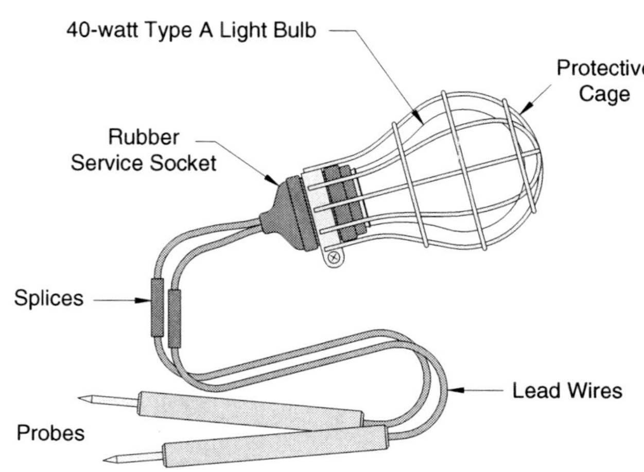

Figure 13-50 Bench Built Power Indicator or Service Light

Capacitor Function Test

An analog multimeter can be used to perform a basic test on a capacitor, as shown in Figure 13-51. The meter is set to ohms and one probe is clipped to one lead on the capacitor to be tested. When the other probe is touched to the opposite lead, the needle on the meter will jump up and then settle back down towards zero. This indicates that the internal resistance of the capacitor is initially nearly zero and, as it charges, the resistance climbs to a higher value. This is only a relative test and is not suitable for determining the actual value of the capacitor.

Measuring Resistance

Using a multimeter to measure resistance is a simple matter. Figure 13-52 shows a basic resistance test. The meter is set to an appropriate Ohms scale and the probes are connected to the leads of the resistor. The resistance reads out in ohms on the meter.

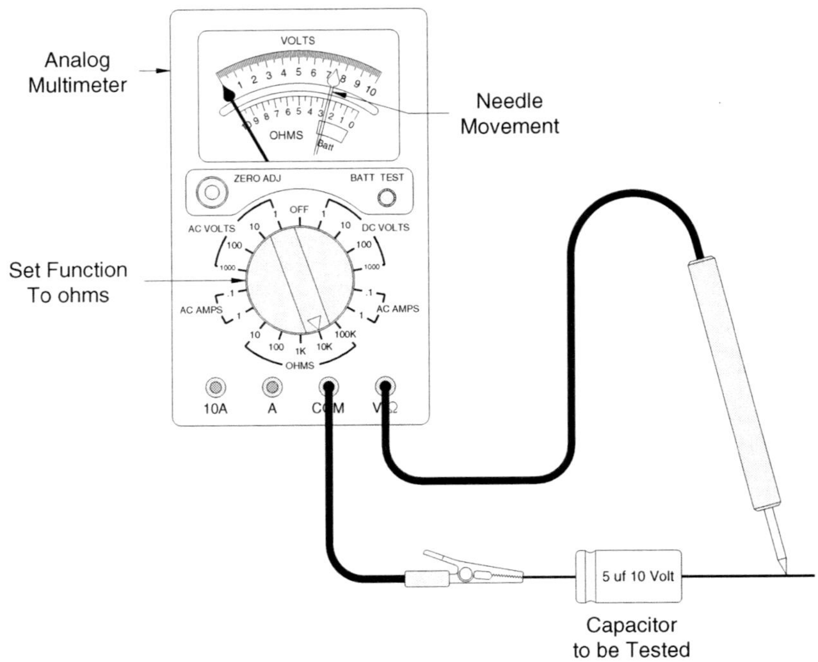

Figure 13-51 Basic Capacitor Function Test

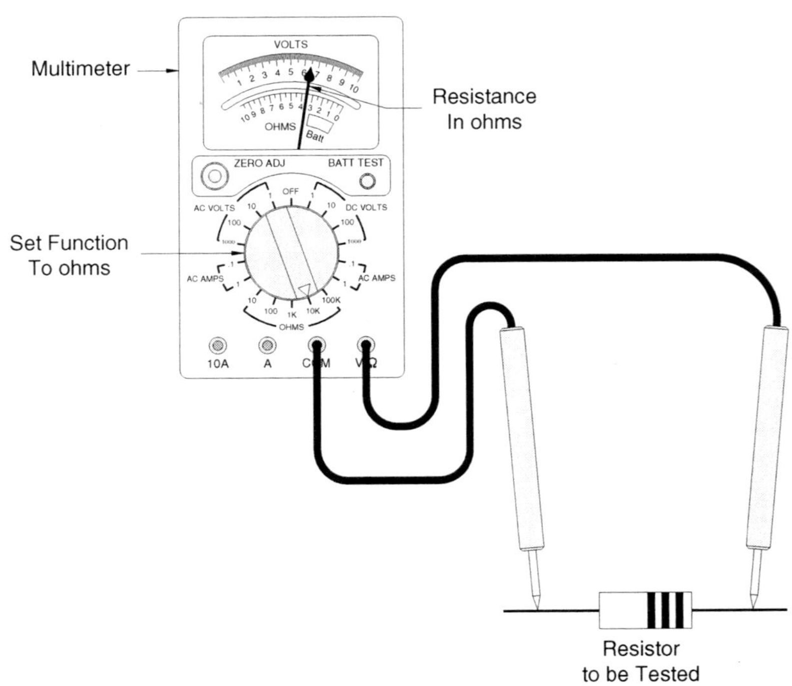

Figure 13-52 Measuring a Resistor

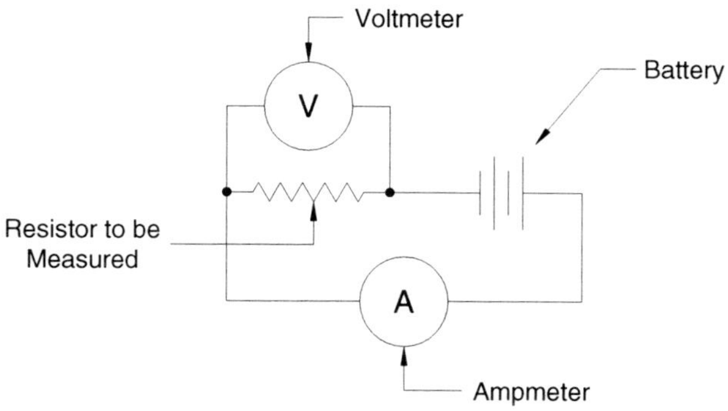

Figure 13-53 Measuring Resistance with an Amp and Voltmeter

Another method to measure resistance is by measuring the current and voltage drop across a resistor and calculating the resistance using Ohm's law, as outlined in Chapter 1. Figure 13-53 shows a schematic for a current/voltage resistance measurement.

The Wheatstone Bridge

For more accurate resistance measurements, a Wheatstone bridge can be utilized, as shown in the schematic of Figure 13-54. The Wheatstone bridge is comprised of four resistors arranged in a closed pattern. Two of the resistors are of known value, one resistor is variable, and the fourth resistor is the device to be tested. A voltmeter is set up to bridge the junction between the known resistors and the junction between the variable and unknown resistors. When a voltage is applied across the junction between the known and variable resistors, and between the known and unknown resistors, current flows through the bridge. The voltmeter will deflect in direct proportion to the imbalance in resistance between the known resistors and the variable/unknown resistors. By adjusting the resistance of the variable resistor until the voltmeter reads zero, it can be matched to the unknown resistor. The resistance reading of the variable unit is then equal to the resistance of the unknown unit.

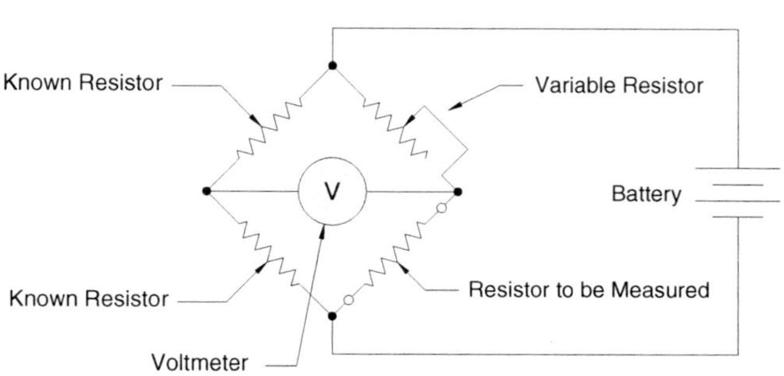

Figure 13-54 Measuring Resistance with a Wheatstone Bridge

Normally a decade resistance box, as discussed in Chapter 4, is set up as one of the known resistors. In this manner, the range of the bridge can easily be adjusted. The variable resistor is generally a calibrated test unit. Using a digital multimeter and progressively selecting lower voltage ranges can make an extremely accurate measurement.

The SlideWire Bridge

A slide wire bridge is a high accuracy version of the Wheatstone bridge. Figure 13-55 shows a schematic representation of a slide wire bridge. Like the Wheatstone bridge, the known resistor is usually a decade resistance box. The unknown resistor is placed opposite the known resistor. The upper known and variable resistors are replaced with a slide wire. A slide wire usually consists of a 36-inch length of resistance wire, a sliding contact, and a scale. The resistance of the slide wire, and thus the balance of the bridge, is adjusted by moving the contact along the scale. This arrangement provides an extremely accurate method for measuring resistance.

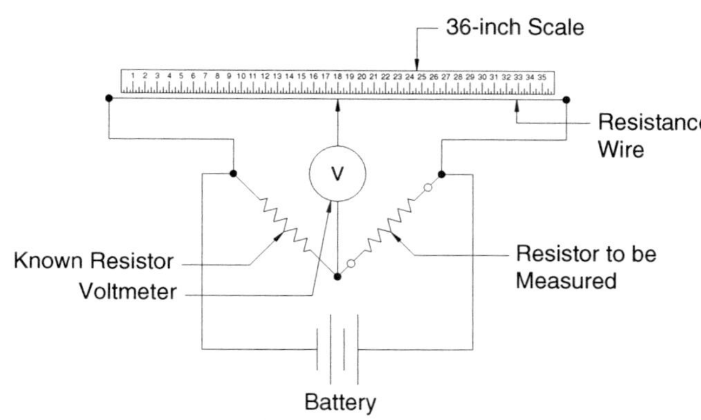

Figure 13-55 Measuring Resistance with a Slide-Wire Bridge

Other Useful Test Equipment

Although most electromechnical equipment can be gauged with a continuity tester, power indicator, or multimeter, there are times that more sophisticated test equipment must be utilized. The following briefly reviews some of the more common instruments and how they may be applied.

Circuit Tracers

Circuit tracers are instruments that are used to follow and map a signal through a live circuit. The most fundamental circuit tracer is a set of headphones that is equipped with a capacitor, as shown in Figure 13-56. The alligator clip is connected to the common and the lead from the capacitor is used to probe

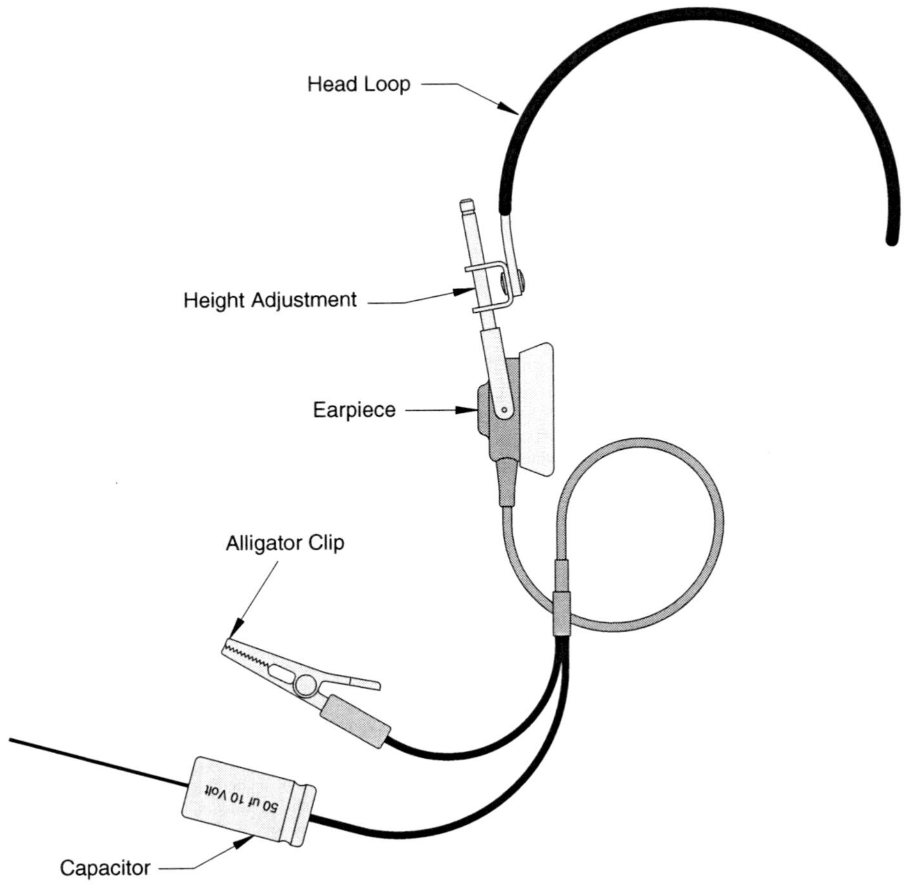

Figure 13-56 Head Set Circuit Tracer

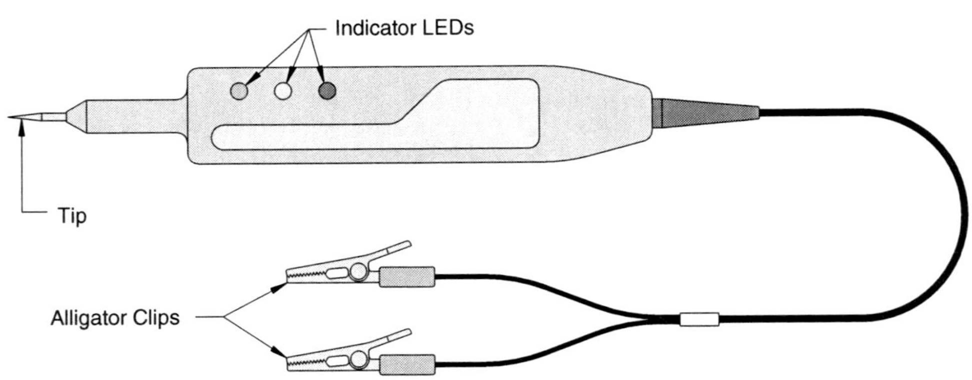

Figure 13-57 Logic Probe

the circuit. The capacitor's function is to protect the headphones from being damaged by high-powered signals.

Logic Probes

Logic probes are circuit tracers that are specifically designed to operate with digital circuitry. Figure 13-57 shows a typical logic probe.

Oscilloscopes

An oscilloscope is literally a television that allows the technician to view the particulars of an electrical signal. These instruments are invaluable tools in the electronics industry, allowing detailed analysis of complex signal and waveforms. Figure 13-58 shows a typical commercial oscilloscope.

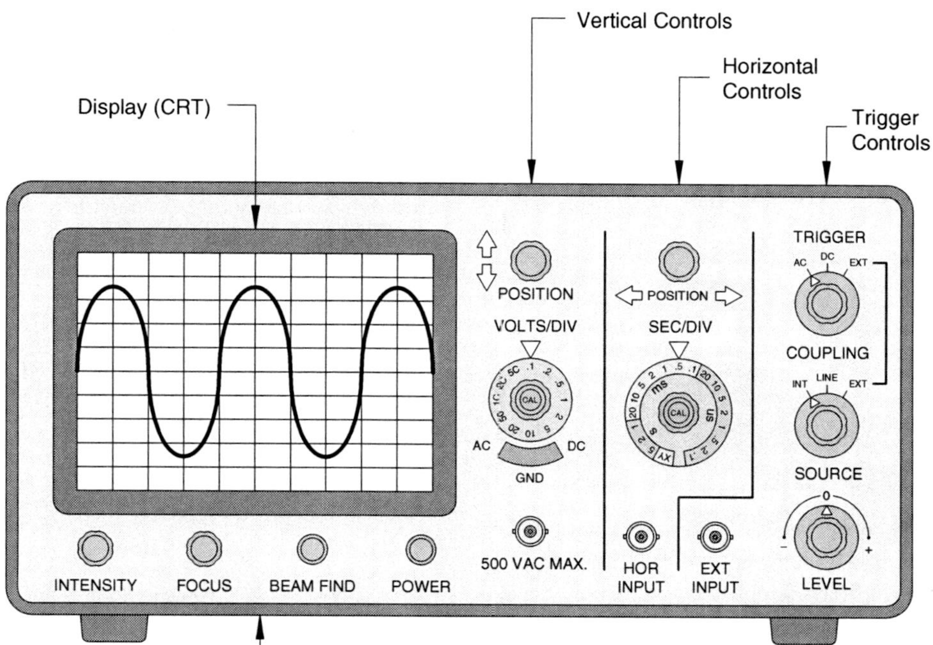

Figure 13-58 Oscilloscope

Sine Wave Generators

These instruments are used to generate a standard sine wave at any frequency that the technician may desire. The sine wave is the base waveform for power generation, audio equipment, and motor controllers. Figure 13-59 shows a typical bench type sine wave generator

Function Generators

The function generator takes the sine wave generator a step further. These instruments will produce several waveforms and any frequency that the technician may desire. Standard waveforms that a typical function generator will produce are sine, triangle, sawtooth, and square. Figure 13-60 shows a typical commercial function generator.

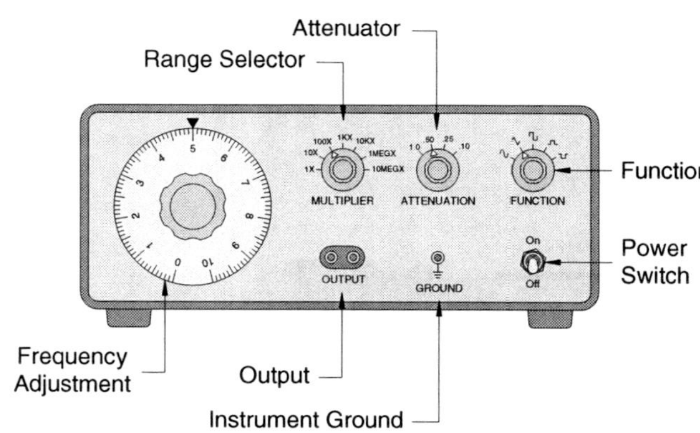

Figure 13-60 Function Generator

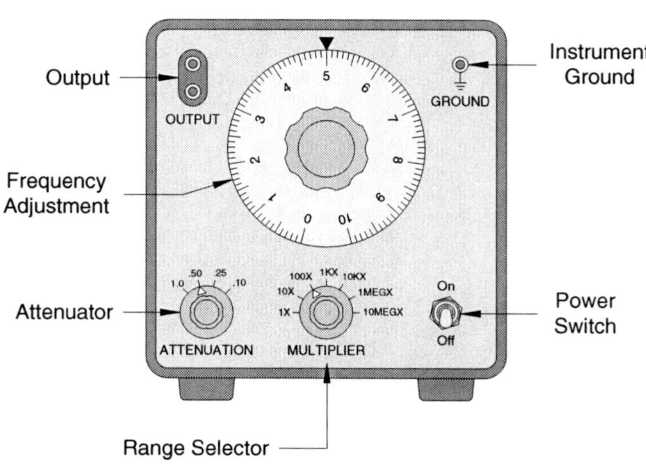

Figure 13-59 Sine Wave Generator

Frequency Counters

In some applications, principally radio frequency (RF) and digital, it is necessary to determine the frequency at which a circuit is operating. In these cases a frequency counter, as shown in Figure 13-61, is used. These instruments generally have an LED (light emitting diode) display that provides a direct frequency reading.

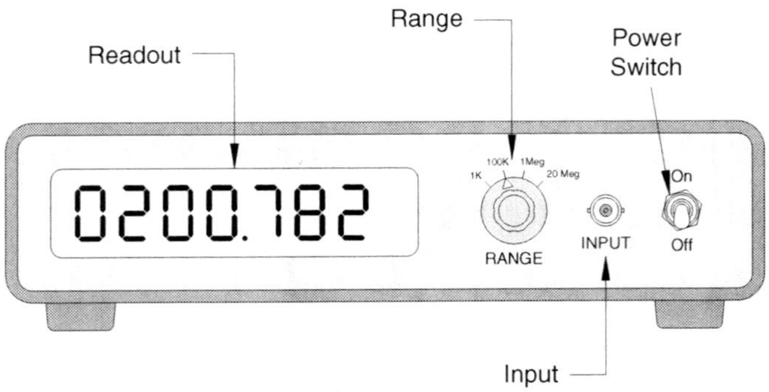

Figure 13-61 Frequency Counter

Insulation Testers (Meggers)

Insulation testers, or meggers as they are sometimes referred to, are used to test and verify the effectiveness of electrical insulation and isolation. These units have a high-voltage generator and a meter that displays the leakage of the voltage across the insulation being tested. The leakage can be directly converted into ohms and the gauge will normally read in megohms. Figure 13-62 shows a typical insulation tester. It should be noted that while crank sets are quite common; these instruments are also available with battery powered high-voltage supplies.

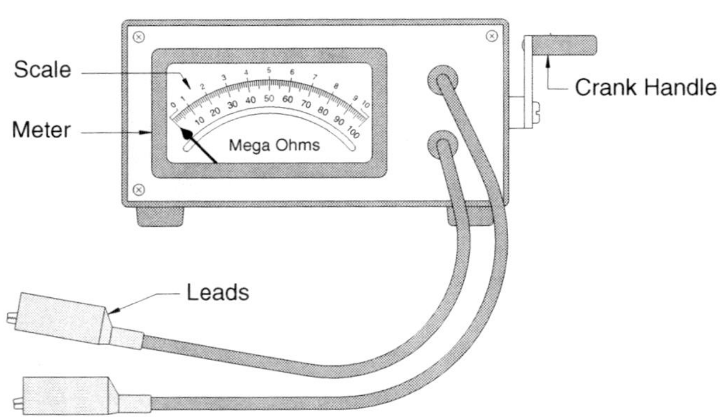

Figure 13-62 Insulation Tester (Megger)

Sound Level Meters

Various industrial, public address, theater, and home audio applications can benefit from the ability to gauge the output level of sound producing equipment. To accomplish this, a sound level meter, as shown in Figure 13-63, is commonly used. These instruments consist of a microphone, amplifier, and readout packaged in a single unit. The readout is in decibels (dB). Most sound level meters are also equipped with a fast/slow response switch and A and B weighting selector. The range is selected until the needle is as near to zero as possible. The value off zero is either added or subtracted from the selected range value to provide the sound level.

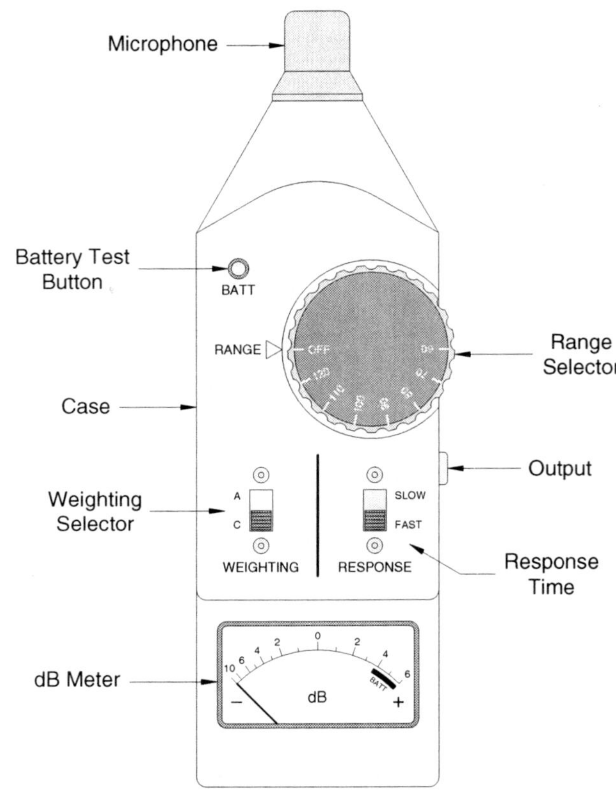

Figure 13-63 Sound Level Meter

CHAPTER 14
VACUUM TUBES

Most people consider vacuum tubes to be electronic devices; however, they are very much electromechanical devices. A vacuum tube takes advantage of the fact that a glowing filament, when properly biased, will emit electrons. Figure 14-1 shows a demonstration of this phenomenon. The glowing filament is a piece of coiled nichrome wire. When the switch is closed, the filament starts to glow and a voltage can be observed on the voltmeter. When the switch is opened, the filament cools off and the voltage returns to zero. When the coil and plate are in air, most of the electrons emitted will instantaneously reattach to the gases that make up the air. To eliminate reattachment and improve the flow of electrons, the inside of a vacuum tube is evacuated of all gases. In the absence of any gases, the electrons are free to travel across the components of the vacuum tube.

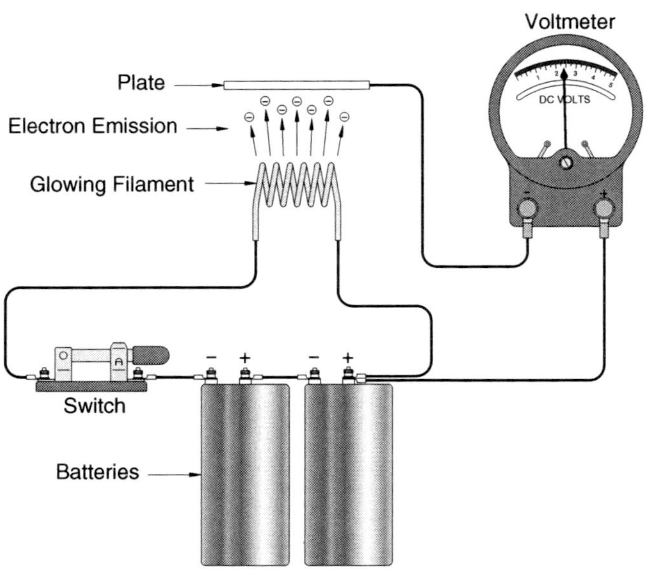

Figure 14-1 Electron Emission

Thomas Edison first observed this phenomenon when experimenting with his incandescent bulbs. He did not, however, carry the investigations very far and the effect, which came to be known as the Edison effect, remained a mystery. It wasn't until 1905 that John Fleming of England was able to develop the first diode vacuum tube and, in turn, launched an entire industry that would take advantage of the effect.

Diodes

The vacuum tube diode is the simplest form of the vacuum tube. These devices serve the same purpose and have been principally replaced by the solid-state diode discussed in Chapter 4. Figure 14-2 shows a vacuum tube circuit schematic. When the filament is heated, it emits electrons that flow across to the anode (+). In effect, this closes the meter circuit and a deflection will be observed. When the filament supply is disconnected, the filament cools and electron emission stops, in effect, opening the meter circuit. In this arrangement the filament forms the cathode (–) of the tube.

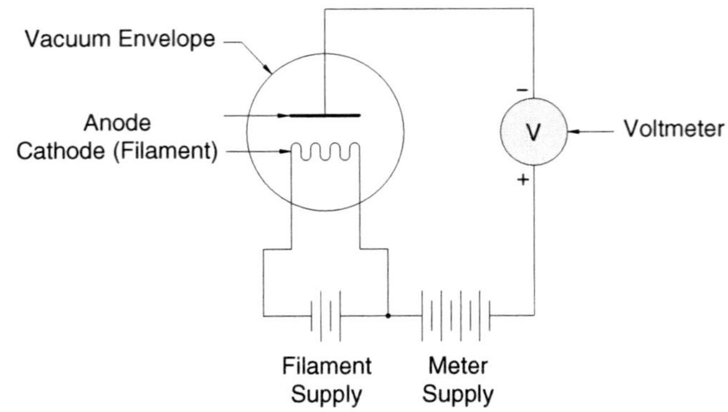

Figure 14-2 Vacuum Tube Diode Circuit

Figure 14-3 shows a simple vacuum tube diode. The base has two pins to connect the filament with the anode terminal located on top of the glass tube or envelope.

To provide isolation from the filament supply, some diode tubes use a separate cathode, as shown in Figure 14-4. In this

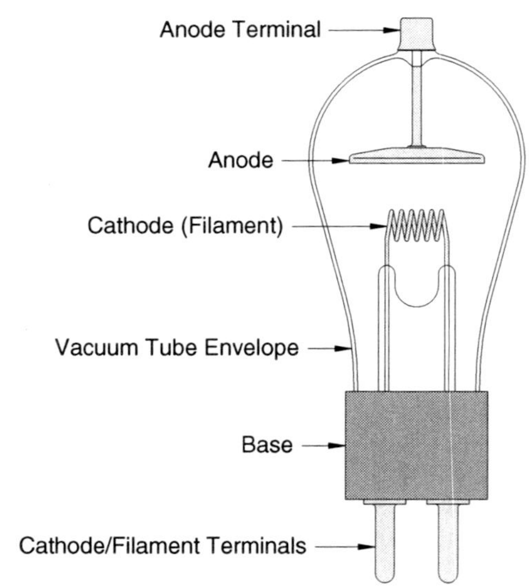

Figure 14-3 Vacuum Tube Diode

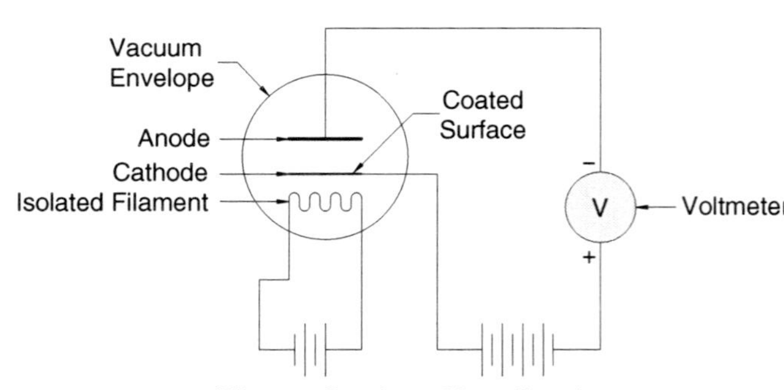

Figure 14-4 Vacuum Tube Diode Circuit with an Isolated Filament

arrangement the cathode is indirectly heated by the filament. The emission surface of the cathode plate is generally coated with a material that will improve electron emission. Notice that the filament and its supply are electrically isolated from the meter circuit.

Figure 14-5 shows a half-wave DC power supply using one vacuum tube diode. The transformer has a dual secondary, one dedicated to the filament supply and one for the output. The output of the tube will be pulsed DC and therefore a filter network, made from two capacitors and a choke, is usually required.

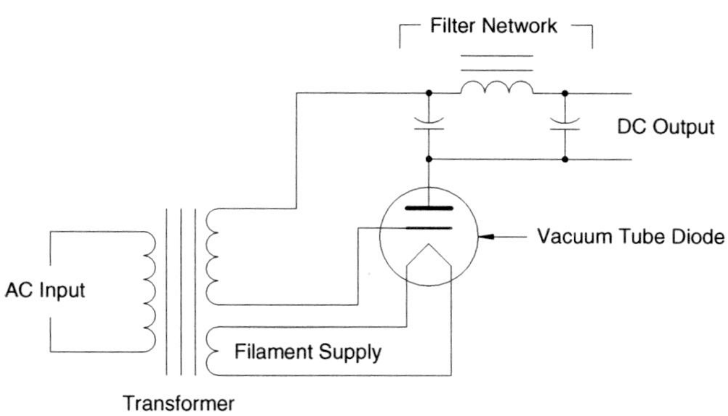

Figure 14-5 Half-Wave DC Power Supply Schematic

Figure 14-6 shows a full-wave DC power supply using two vacuum tube diodes. In this case, the transformer also has a dual secondary, except the output side has a center tap. The center tap makes up the positive side of the DC output, with the outer terminals connected to the anodes of the tubes. The cathodes are connected and make up the negative terminal of the DC output. The output of a full-wave power supply is smoother than a half-wave unit; however, a filter network is usually used on these designs as well.

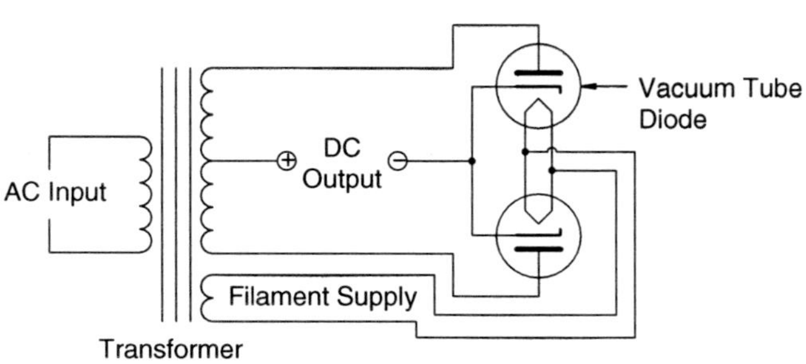

Figure 14-6 Full-Wave DC Power Supply Schematic

The octal base dominated the vacuum tube industry in the 40s and 50s and was rather common well into the 60s. Figure 14-7 shows an octal base vacuum tube with base.

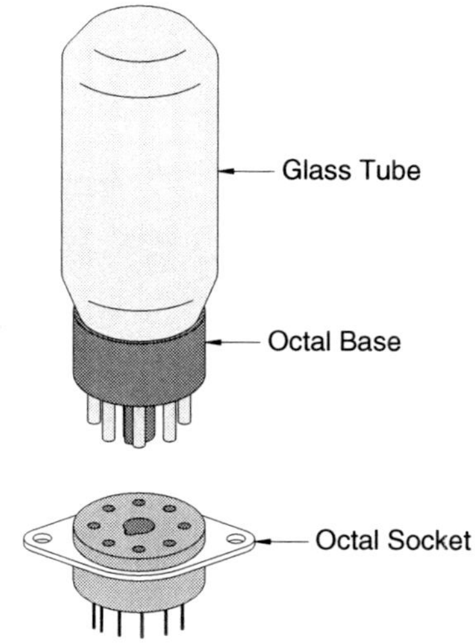

Figure 14-7 Vacuum Tube with Octal Base

Grids

A perforated plate or screen placed between the cathode and anode of a vacuum tube is referred to as a grid. Figure 14-8 shows a vacuum tube with a grid plate. These units are referred to as triodes because of their three elements, cathode, grid, and anode. The flow of electrons from the cathode to the anode can be controlled by adjusting a bias voltage applied to the grid. In this manner, the voltage across the tube can be variably controlled.

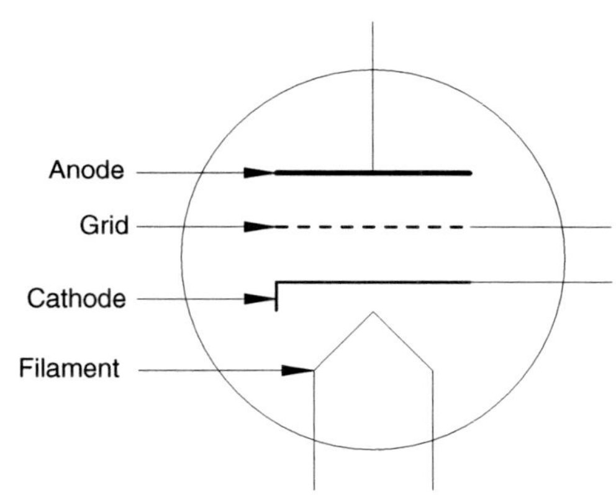

Figure 14-8 Vacuum Tube with Grid (Triode)

The neutral illustration of Figure 14-9 shows that if the bias voltage is the same as the cathode, then medium voltage will flow. The suppression shows that if a negative voltage is applied to the grid, then the electron flow is deflected and low voltage flows. The acceleration illustration shows that if a positive bias voltage is applied to the grid, then full electron flow is realized and high voltage flows. By adjusting the voltage and polarity of the grid supply, the current flow through of the tube can be controlled.

Figure 14-10 shows a vacuum tube amplifier. The input is connected to the primary winding of T_1. As the input signal is varied, the transformer's secondary mirrors these changes at a

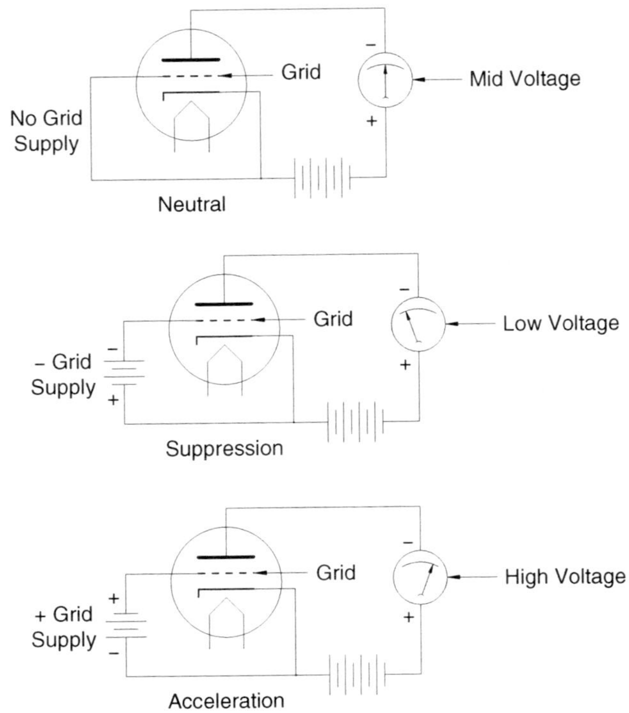

Figure 14-9 Grid Function and Effect

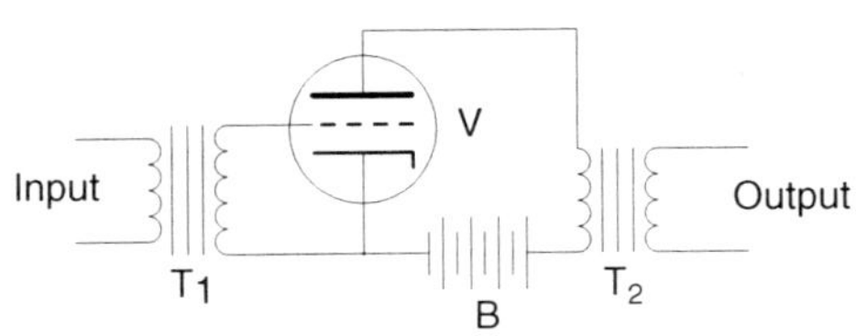

Figure 14-10 Basic Single Tube Amplifier Schematic

higher voltage, which is, in turn, used to bias the grid. As the bias voltage on the grid varies, the electron flow from the cathode to the anode is controlled and the output of the tube mirrors the input at a much higher power level. T_2 is an impedance matching transformer and is generally required on the output of any vacuum tube amplifier.

Tubes with grids are generally classified as triodes (one grid), tetrodes (two grids), and pentodes (three grids). Figure 14-11 shows schematic representations of the three types of vacuum tubes with grids.

The 50s and 60s brought about the systematic miniaturization of vacuum tubes. Figure 14-12 shows the three standard

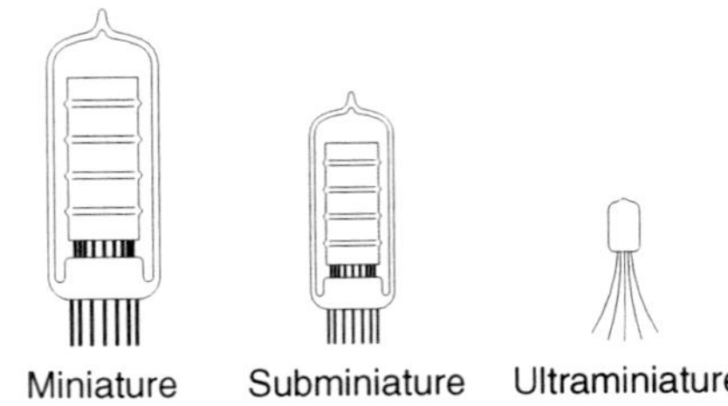

Figure 14-12 Miniature Vacuum Tubes

packages that were commonly supplied during these times. The ultraminiature tubes were principally used in high-cost test instruments, avionics, and military equipment. Of course, the advent of solid-state electronics spelled the certain death of vacuum tubes in everyday appliances.

Mercury Vapor Rectifiers

The early part of the 1900s saw a variety of mercury vapor rectifiers. By using mercury vapor in the tube, much higher currents could be used and these types of diodes were commonly used for industrial power supplies. However, the only remaining mercury vapor rectifier still in common use is the ignitron, and even these have been principally displaced by the advent of high-power silicon controlled rectifiers (SCRs).

The ignitron is simply an envelope with a pool of mercury as its cathode. The top of the envelope is equipped with an anode. A small needle electrode, or igniter, is located on the side of the envelope and touching the surface of the mercury. When a high-voltage signal is applied to the igniter, a small amount of mercury is vaporized. The vaporized mercury is enough to short the cathode and anode. As long as current is flowing across the cathode and anode, the mercury maintains a vapor state and a low resistance junction is formed. Figure 14-13 shows a schematic representation of an ignitron rectifier.

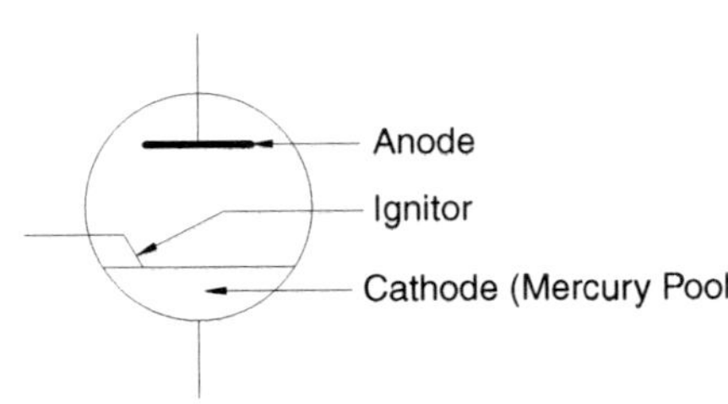

Figure 14-13 Ignitron Rectifier Symbol

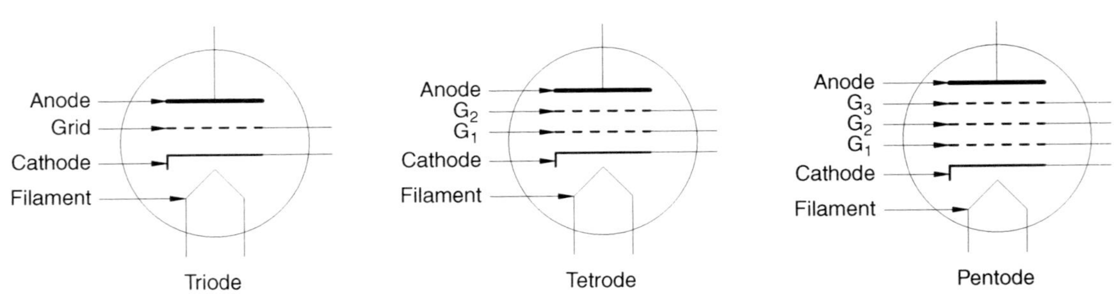

Figure 14-11 Standard Vacuum Tube Types

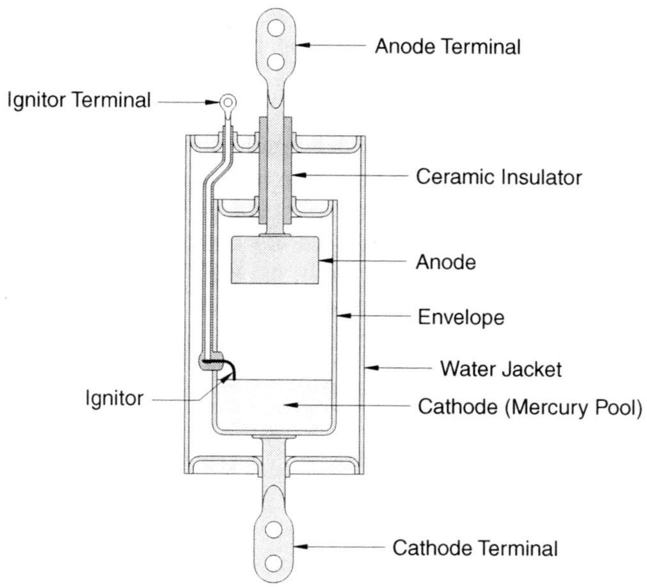

Figure 14-14 Sectional View of an Ignitron Tube

Figure 14-14 shows a sectional view of a medium current ignitron tube. Notice the water jacket surrounding the envelope. Cooling is imperative because of the high currents that ignitrons are designed to switch. These units are often found in equipment that must switch extremely high currents, such as industrial spot welders.

Cathode Ray Tubes (CRT)

We have all watched TV or sat staring at the screens of our computers. The displays for these devices are actually large vacuum tubes, referred to as a cathode ray tube or CRT.

The CRT, as shown in Figure 14-15, has a cathode grid similar to an ordinary vacuum tube, except that the geometries are designed to produce an electron beam. The beam is directed through a set of focusing plates and finally through an acceleration plate. The result is a high-energy, focused electron that impinges on a coated screen. The coating fluoresces at any point where the beam hits. By sweeping the beam both vertically and horizontally and turning it on and off at precisely timed intervals, an image can be generated on the screen. Figure 14-16 shows a commercial CRT of the type that might be found in an oscilloscope.

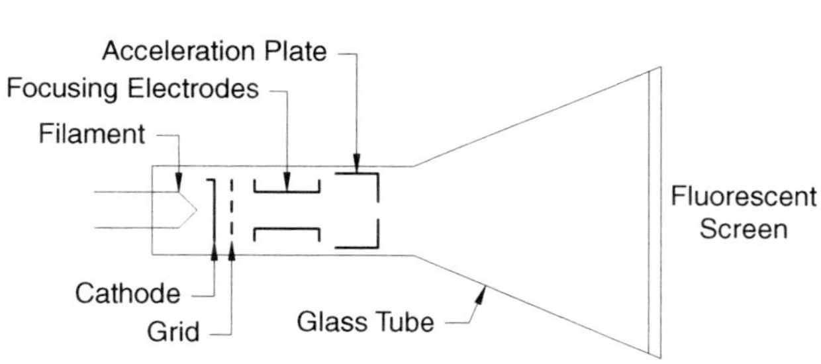

Figure 14-15 Cathode Ray Tube (CRT)

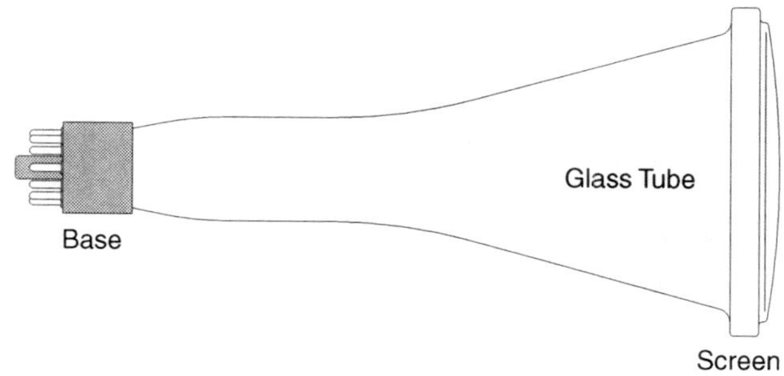

Figure 14-16 Commercial CRT

Photosensitive Tubes

Certain materials exhibit the characteristic of emitting electrons when exposed to light. In the case of a photosensitive vacuum tube, as shown in Figure 4-17, electrons are ejected as light impacts the cathode. If a bias voltage is applied across the cathode and anode, then current flows when the tube is exposed to light and doesn't flow when it is in the dark. Similarly, the rate of electron flow can be controlled by the amount of light to which the tube is exposed.

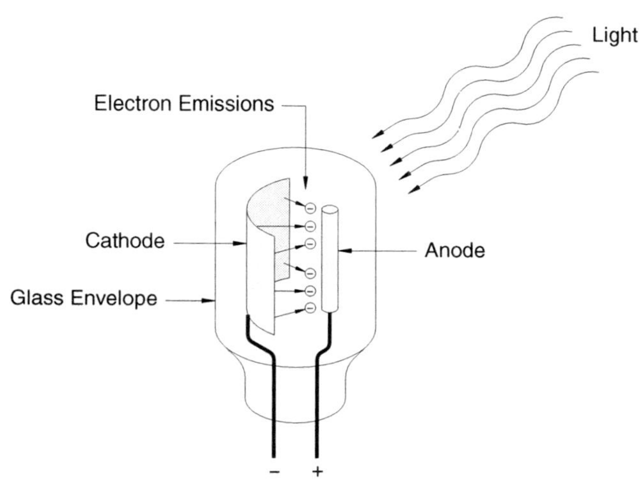

Figure 14-17 Photosensitive Vacuum Tube

Magnetrons

To generate the microwaves used in your microwave oven, a magnetron vacuum tube is utilized. These are special tubes that are designed to emit high-frequency power from a very compact and inexpensive package. Figure 14-18 shows a typical commercial magnetron tube such as may be found in a home microwave oven. These units are a self-contained system that requires only a high-voltage power source.

Figure 14-19 shows the internal geometry of a typical magnetron tube. An electrode beam forms around the central cathode. The beam resonates in the resonator cavities and generates a microwave signal.

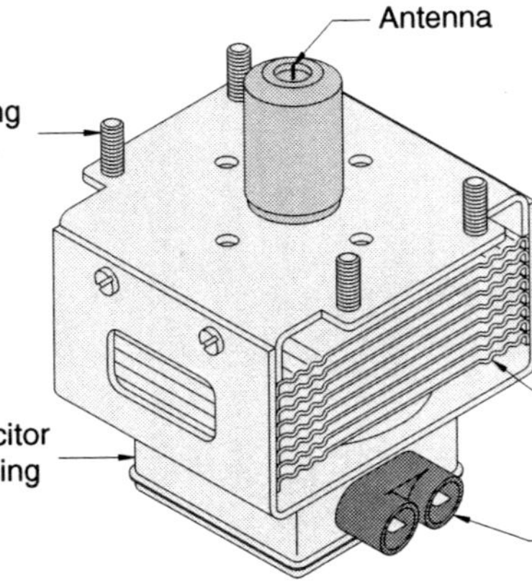

Figure 14-18 Commercial Magnetron Tube

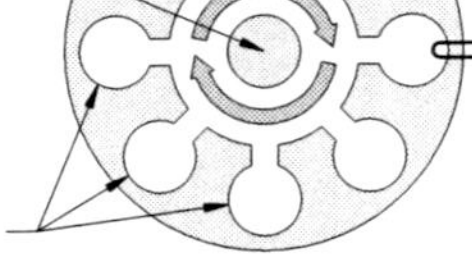

Figure 14-19 Internal Magnetron Geometry

Klystrons

Klystrons are designed to produce high-power microwaves from a relatively low-power input. These tubes consist of a beam tube with two toroidal resonator cavities. The tube has a cathode and anode that are set up to produce a high-energy electron beam. An RF signal is introduced into the left hand cavity and, in turn, modulates the electron beam. The modulated electron beam resonates in the right hand cavity and generates a modulated microwave output. These tubes are commonly used on military radar sets. Figure 14-20 shows a schematic representation of a Klystron tube.

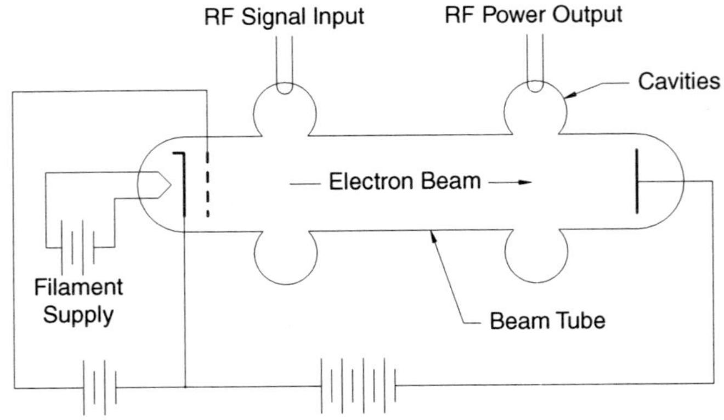

Figure 14-20 Klystron Tube Schematic

CHAPTER 15
SENSORS

The electromechanical marriage plays an important role in detecting the real world. Most electrical sensors are employed to detect or monitor a physical attribute and, therefore, must be electromechanical in nature. The simplest form of these sensors is the limit switch, such as the button that detects when the refrigerator's door is open. Your automobile has a wide variety of sophisticated transducers that monitor all aspects of engine operation. Home heating and air-conditioning systems rely on various sensors to keep them operating at peak efficiency. Sensors provide a necessary interface between the mechanical and electrical worlds.

Proximity Sensors

The simplest type of sensor is the limit switch, as discussed in Chapter 4. These units simply throw a switch element when they come in contact with a mechanical component. A variation of the limit switch is the magnetic proximity sensor, as shown in Figure 15-1. These sensors are simply a read switch with a permanent magnet affixed to the contact set. When the switch is in close proximity to a ferrous material, the magnet pulls the contact into the switched position, providing an indication. Most magnetic position sensors are supplied with both normally open and normally closed contacts.

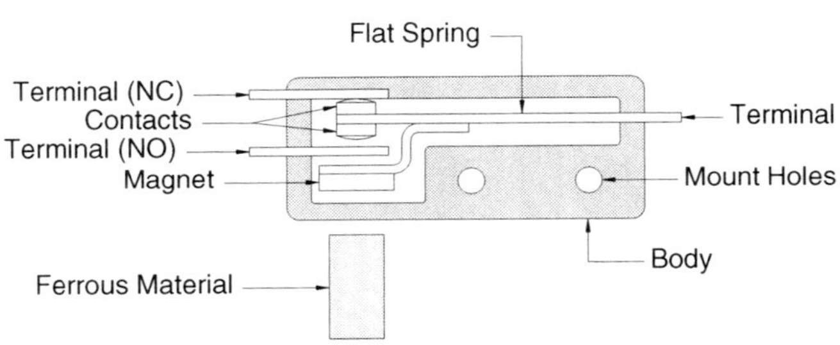

Figure 15-1 Magnetic Proximity Sensor

Inductive sensors are generally coils of wire that detect motion by the change in their inductance. Figure 15-2 shows a basic inductive proximity sensor. When a ferrous component is placed near the pole faces, the magnetic circuit is altered and the resistance of the inductor changes.

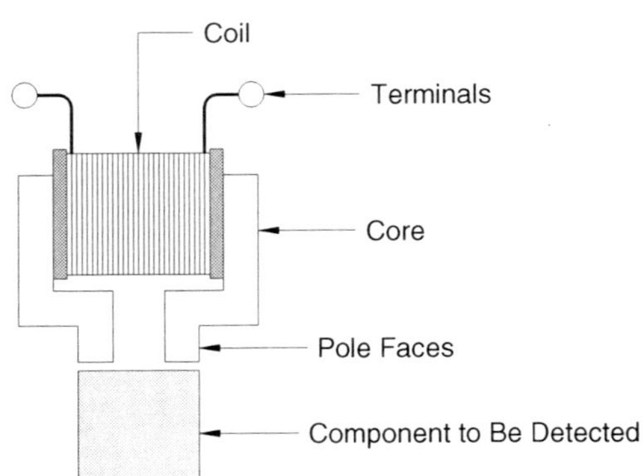

Figure 15-2 Inductive Proximity Sensor

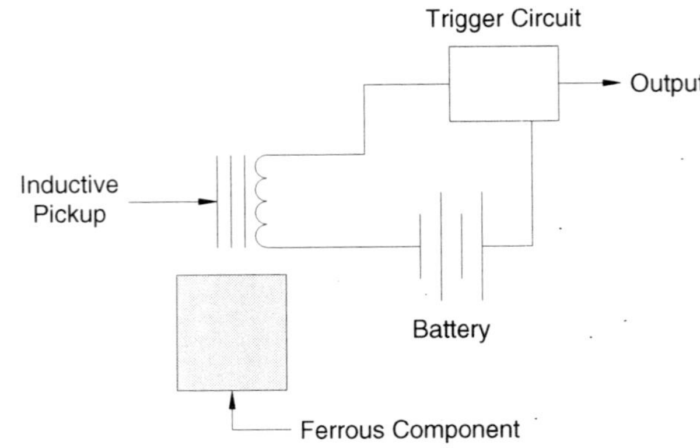

Figure 15-3 Inductive Proximity Sensor Schematic

Figure 15-3 shows a basic schematic of an inductive sensor. These devices will detect motion as well as proximity. When properly configured, the sensors can detect the relative position and speed of the ferrous component.

Another method of using an inductive sensor is to generate an output voltage by sweeping it with a magnet, as shown in Figure 15-4. The output can be rectified and a voltage displayed on a voltmeter to provide an average reading or the pulses can be counted to provide a precise indication. One area where pulse-type arrangements are commonly used is in modern automobile distributors.

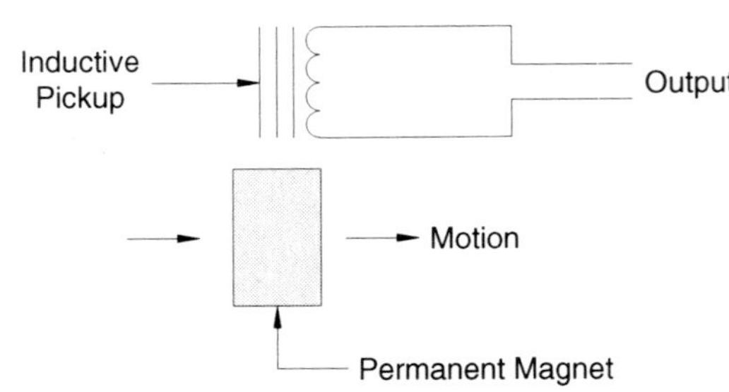

Figure 15-4 Magnetic Induction Proximity Sensor Schematic

Capacitive proximity sensors are used where a nonferrous material must be detected. The capacitance of these sensors changes when any conductive material is placed in close proximity to the plates. This change in capacitance is monitored with a trigger circuit that, in turn, outputs a signal. Figure 15-5 shows a basic capacitive proximity sensor schematic.

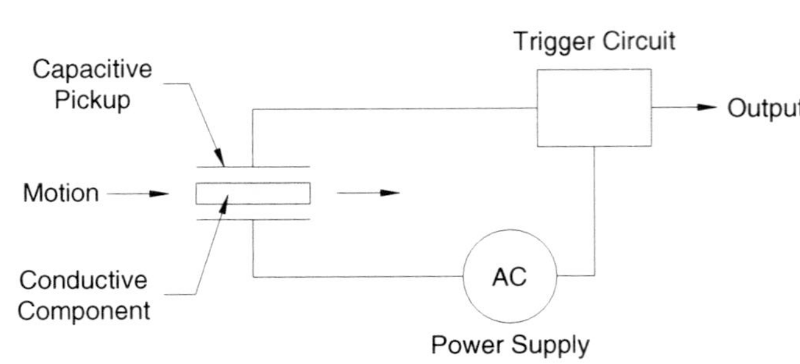

Figure 15-5 Capacitive Proximity Sensor Schematic

A Hall effect sensor takes advantage of the characteristic of some materials to change resistance when placed in a magnet field. These types of sensors are very inexpensive and, consequently, very common. Figure 15-6 shows a Hall effect sensor schematic. Like the inductive sensor, these sensors are also commonly used in automobile distributors.

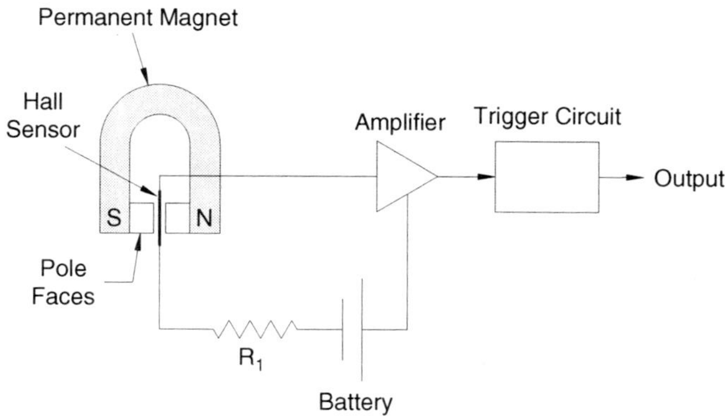

Figure 15-6 Hall Effect Proximity Sensor Schematic

Optical sensors are typically used to detect nonconductive materials. These units normally consist of a light emitting diode (LED) and photosensitive transistor packaged into a small plastic housing. The LED is powered to provide constant light. The transistor becomes conductive when exposed to the light. Placing an opaque object between the LED and transistor opens the circuit, while removing the object closes the circuit. Figure 15-7 shows a basic opto-coupled sensor schematic.

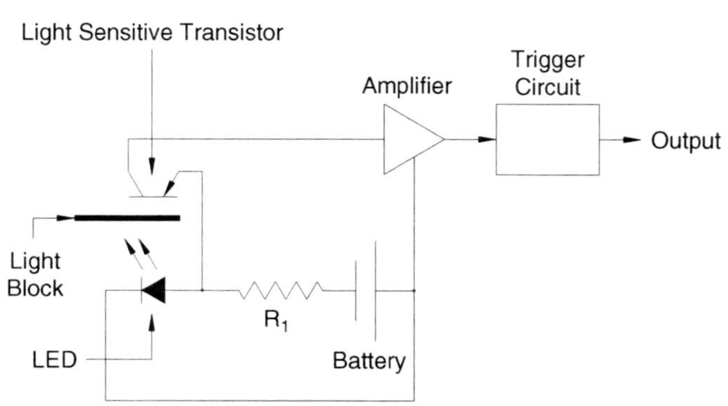

Figure 15-7 Opto-Coupled Sensor Schematic

Opto-sensors are generally supplied in three different forms, as shown in Figure 15-8. The top illustration shows a reflective unit while the middle illustration shows an interruption unit. The bottom illustration shows an opto-isolator. These units are used to electrically isolate two circuits that must communicate with one another. Opto-isolators are commonly found providing interfacing functions between industrial equipment and computers.

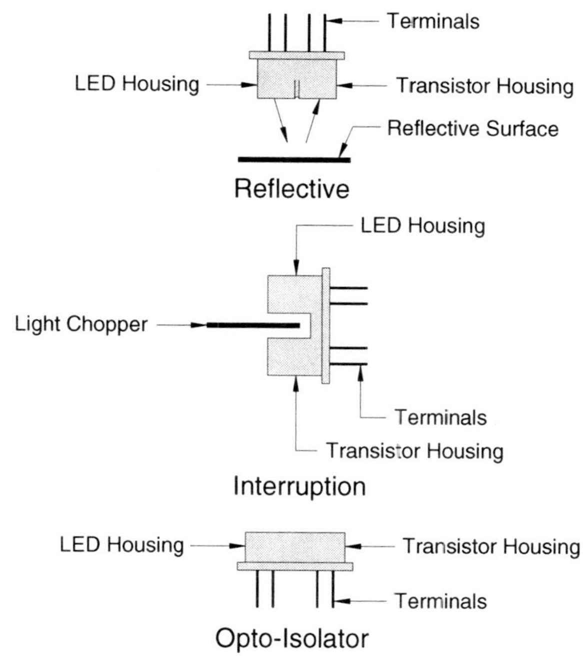

Figure 15-8 Opto-Coupled Sensors

Rotational Sensors

The most common rotational sensor is the shaft resolver, as shown in Figure 15-9. These devices are simply an opaque circular disk mounted to a hub. The outer portion of the disk has a series of perforations extending around the full diameter. A single opto set is used to communicate the rotational position of the disk. To increase resolution the resolver may be used through a ratio reduction transmission. The output of a shaft resolver is a pulse signal, usually +5 volts, and feeds a counter circuit that, in turn, outputs a higher level signal.

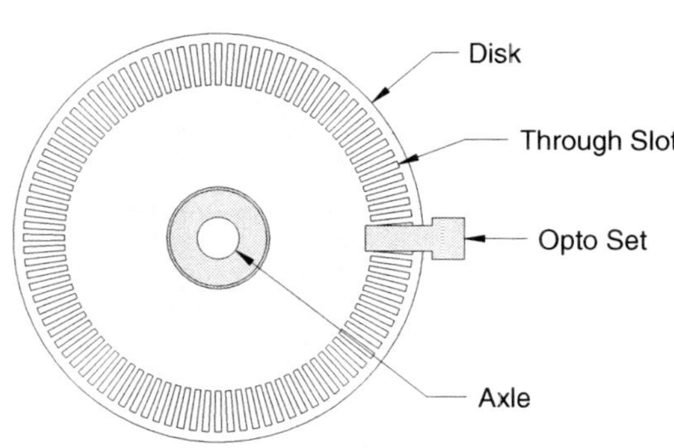

Figure 15-9 Shaft Resolver

As computer technologies encroach on common electronics, digital potentiometers are becoming increasingly common. These units have the same outward appearance as an analog potentiometer, except they carry a multi-pin output connector. The unit shown in Figure 15-10 is actually an eight-bit rotary shaft encoder. The internal element has eight

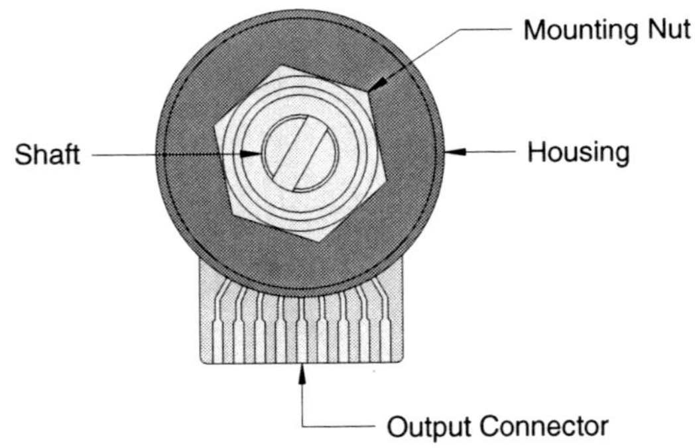

Figure 15-10 Digital Potentiometer

tracts and outputs a base two signal corresponding to the rotational position of the shaft. Digital potentiometers have become very popular in consumer and industrial electronics. They are also quite useful as a shaft resolver and can be applied to a variety of motion control applications.

A simple purpose built shaft position indicator can be built by drilling small holes in an opaque disk, as shown in Figure 15-11. An array of opto-sensors is positioned at the surface of the disk and a light source is mounted on the opposite side. The number of channels and location of the holes can be determined for the specific project at hand.

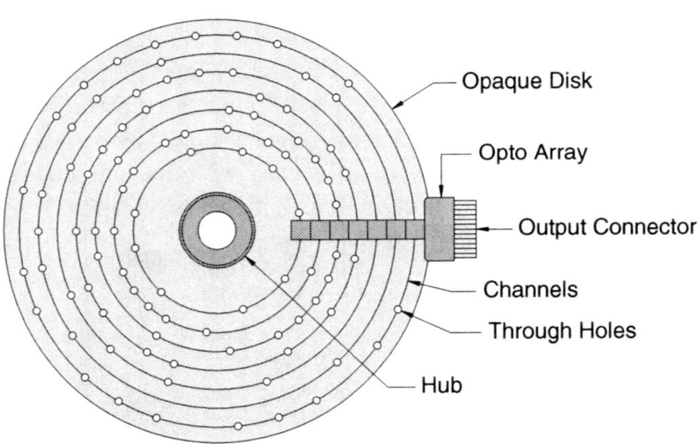

Figure 15-11 Perforated Disk with Opto Limit Sets

A similar position indicator can be built using reflective opto sets, as shown in Figure 15-12. In this case the disk is painted with a nonreflective coating. Reflective dots or lines are added as necessary.

In applications that cannot support a disk-type resolver, a barrel or drum-type arrangement can be configured. Figure 15-13 shows a drum-configured resolver. The outside diameter of the drum can be painted as described in the previous Chapter 14, or can be applied as a tape. The latter being a very convenient method if programming changes must be made from time-to-time.

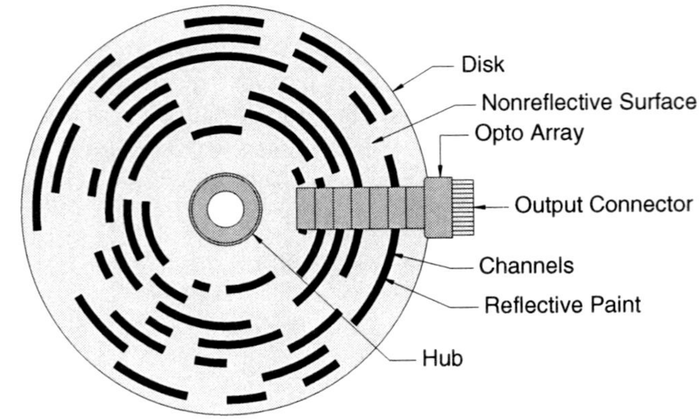

Figure 15-12 Painted Disk with Reflective Opto Limit Array

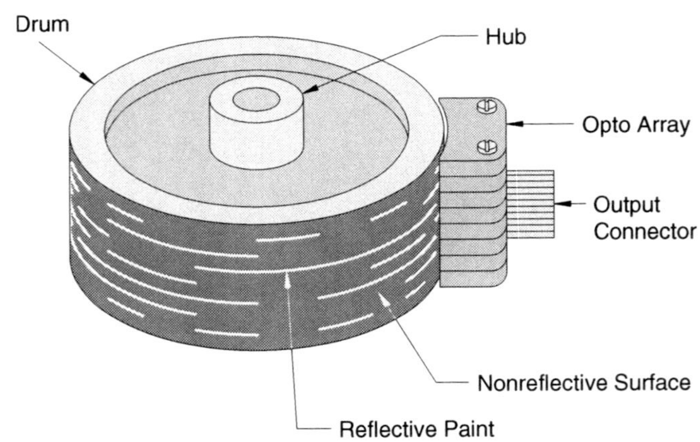

Figure 15-13 Painted Drum with Reflective Opto Limit Array

Drum resolvers with brush sets can be easily constructed, as shown in Figure 15-14. This resolver is similar in construction to the drum timer shown in Chapter 4. For low-speed, low-performance applications, such as industrial equipment, these types of encoders can provide a simple and reliable solution to position indication.

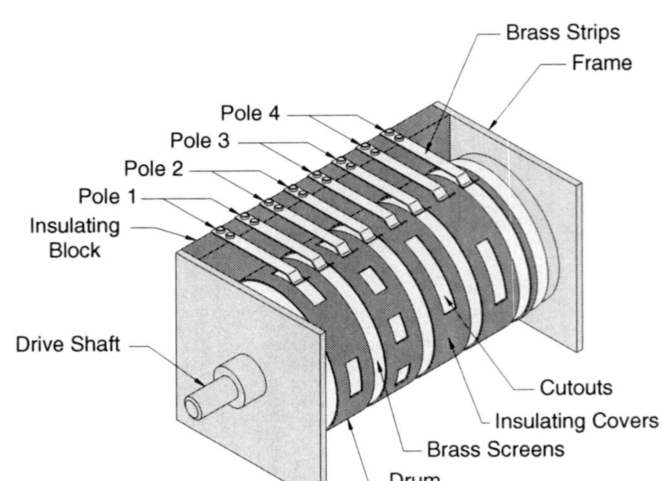

Figure 15-14 Drum with Brush Sets

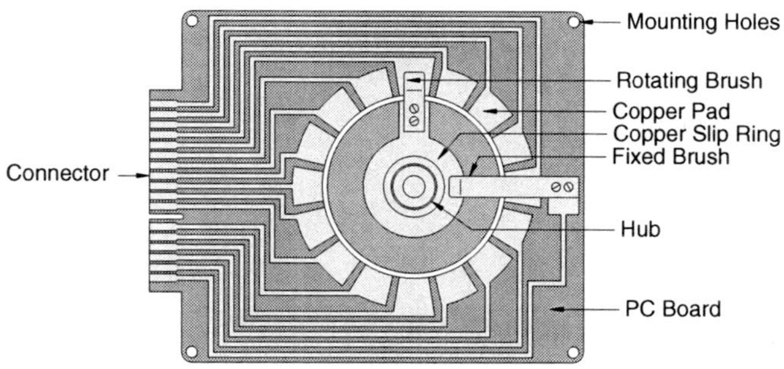

Figure 15-15 PC Board Rotary Position Indicator

Printed Circuit (PC) board rotary position indicators provide a simple method to construct low-cost resolvers. Figure 15-15 shows a typical PC board rotary position indicator. This particular unit is a single-pole, 15-position unit. It should be noted that these types of positioners can be configured to have several tracts, with specific positions and/or base two outputs.

Figure 15-16 shows a typical application for a PC board position indicator. The hub of the sensor is mounted to the shaft of a wind vane. The outputs are wired to a circular array of panel lights, which indicate the points of the compass. The panel lights continuously indicate the position of the wind vane.

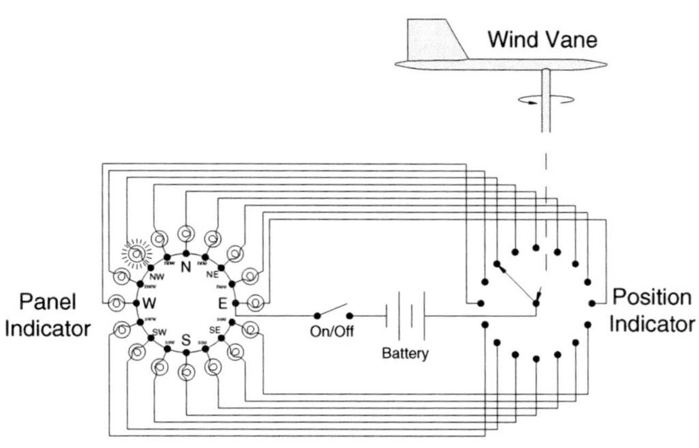

Figure 15-16 Wind Direction Indicator

A wind position indicator can be configured using a digital potentiometer, as shown in Figure 15-17. In this case the output of the potentiometer is connected to a controller or computer. The controller drives a panel display that indicates wind direction.

Detecting motion in three dimensions is generally the duty of a gyroscope. These detectors are most commonly found in aircraft, where they supply base line attitude information to the navigation system and pilot. Figure 15-18 shows a three-axis gyroscope equipped with digital potentiometers. The outputs of the potentiometers are connected to a digital controller and altitude information is generated for the navigation system and displayed for the pilot.

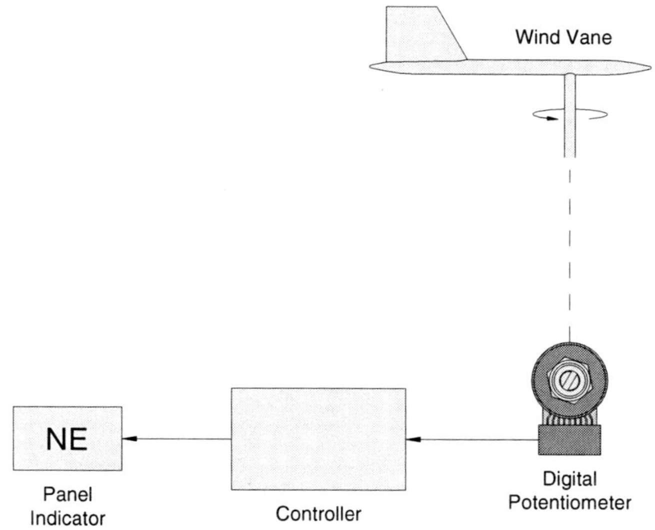

Figure 15-17 Digital Wind Direction Indicator

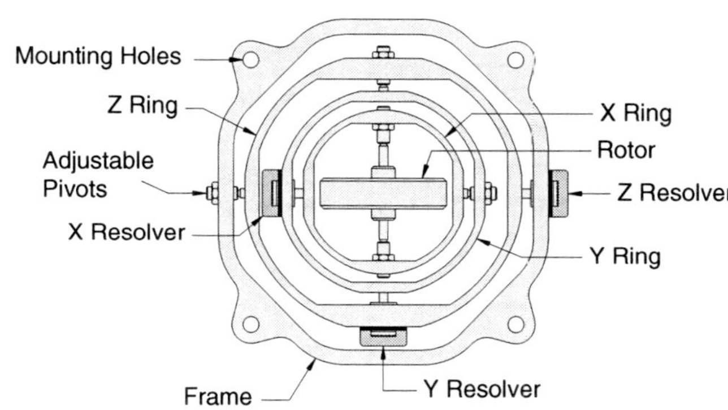

Figure 15-18 Digital Gyroscope Position Indication

One of the most common rotational parameters to monitor is revolutions per minute (RPM). This can easily be accomplished by using a generator and voltmeter, as shown in Figure 15-19. The faster the generator is turned, the higher the voltage reading. The displayed volts correspond directly with the input RPM of the generator. Standard generators rarely

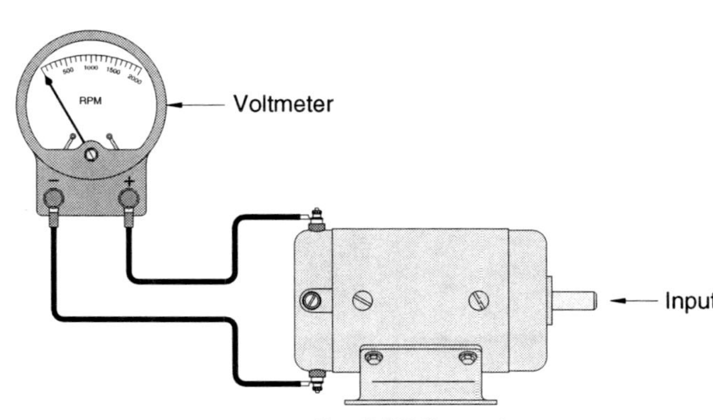

Figure 15-19 Generator RPM Indicator

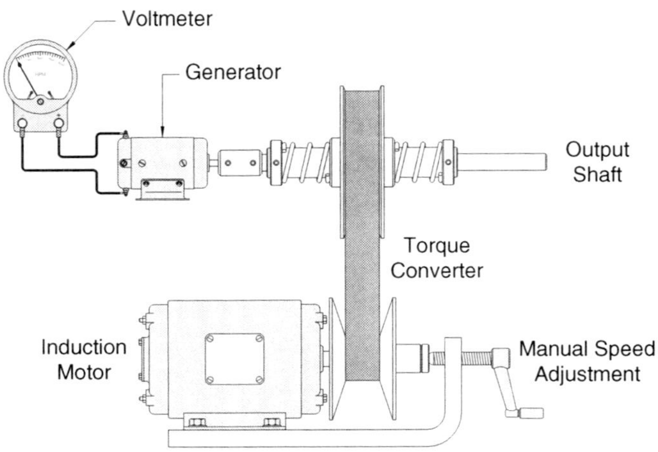

Figure 15-20 RPM Indicator for Torque Converter Output

have a linear or consistent voltage/RPM output. To combat this tendency, special tachometer generators are manufactured that will produce an accurate and predictable voltage output per input RPM.

Figure 15-20 shows a typical use for a generator tachometer. The generator is driven off of the output shaft of the torque converter. The output shaft RPM is directly displayed on the voltmeter. The output of the tachometer generator may also be fed into an analog-to-digital converter, providing information for a control computer.

Figure 15-21 shows a small tachometer generator used in a wind speed indicator. The propeller and generator are matched to produce 1 volt per 10 miles per hour (MPH) of wind speed. The generator feeds a 0 to 10 volt meter, which directly displays wind speed in MPH.

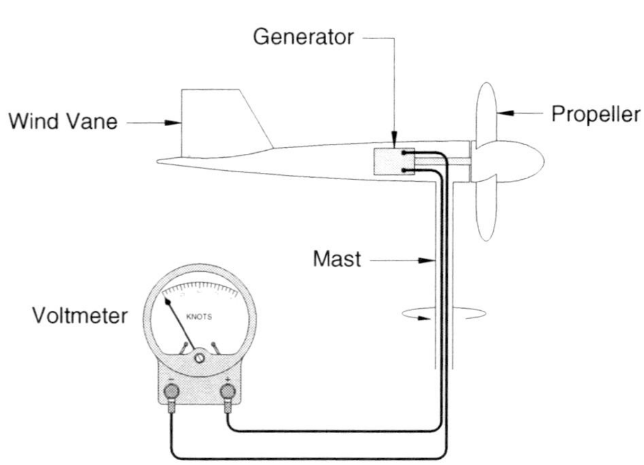

Figure 15-21 Wind Speed Indicator

Linear Position Sensors

Detecting the position of a linear motion is an important function in the electromechanical world. There are a variety of methods deployed to accomplish this, each one having its own merits.

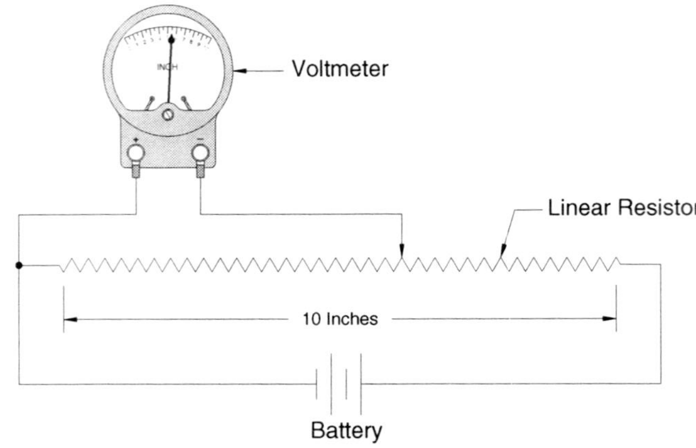

Figure 15-22 Voltage Divider Position Indicator Schematic

Figure 15-22 shows a basic voltage divider position indicator schematic. A high precision linear resistor is fixed and the wiper is allowed to traverse the length of the resistor, based on the motion of the equipment to be monitored. The output drives a voltmeter calibrated in inches. The voltmeter directly reads the position based on the location of the wiper.

Figure 15-23 shows a +/– indicating position indicator based on a slide wire bridge. The slide wire bridge is further discussed in Chapter 13. The linear resistor is fixed with the wiper centered. If the wiper is moved off-center, the voltmeter displays its +/– position in inches. The calibration resistor is used to adjust zero on the voltmeter.

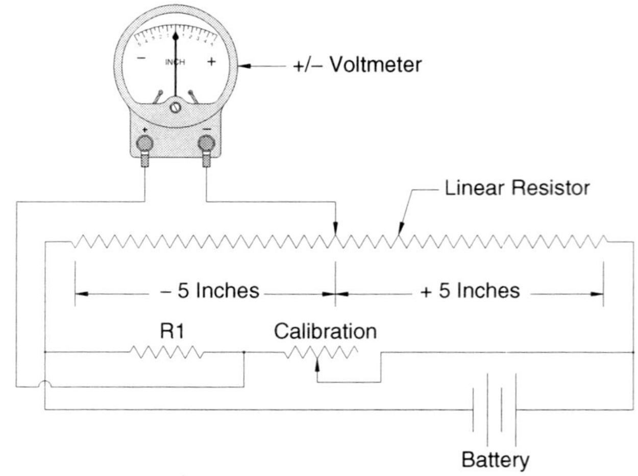

Figure 15-23 +/– Slide Wire Position Indicator Schematic

By using a potentiometer and belt arrangement, as shown in Figure 15-24, a simple linear position indicator can be assembled that is suitable for most applications. The potentiometer can be resistive or digital, or it can be substituted with a rotational shaft resolver. For higher accuracies and improved repeatability, a toothed belt should be used.

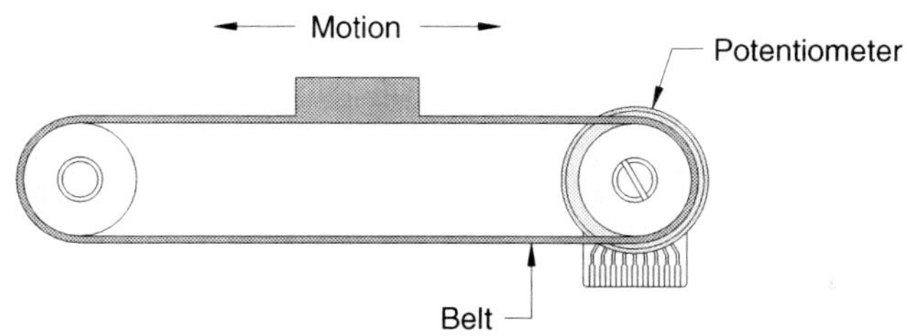

Figure 15-24 Belt Linear Position Indicator

For longer distances a cable spool driving resolver can be configured as shown in Figure 15-25. A lightweight cable is set up on a spring return spool. The axle of the spool drives a 10-turn potentiometer, digital potentiometer or shaft resolver. It should be noted that over long distances the cable may sag, creating inaccurate readings. For longer distances the cable should be supported to minimize the effects of sagging.

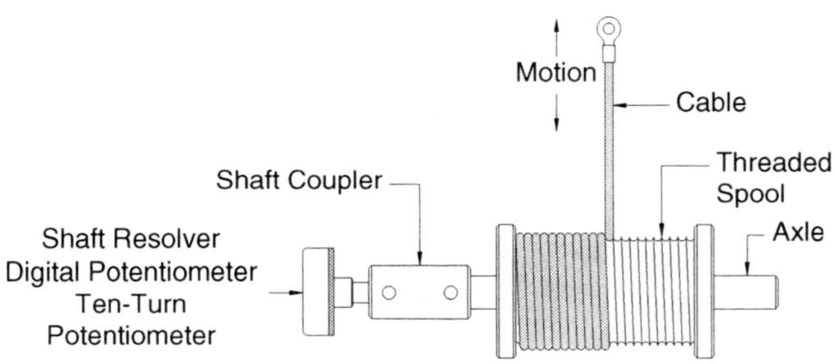

Figure 15-25 Cable Spool Linear Position Indicator

The most common linear position indicator is the magnetic scale, as shown in Figure 15-26. These units are generally a nonferrous bar with a magnetic strip affixed to the centerline of one side. The magnetic strip is recorded with a uniform signal. The carriage is equipped with a head that reads and counts the signal pulses. The count is displayed on an operator panel, as shown, or it may be fed to a control computer.

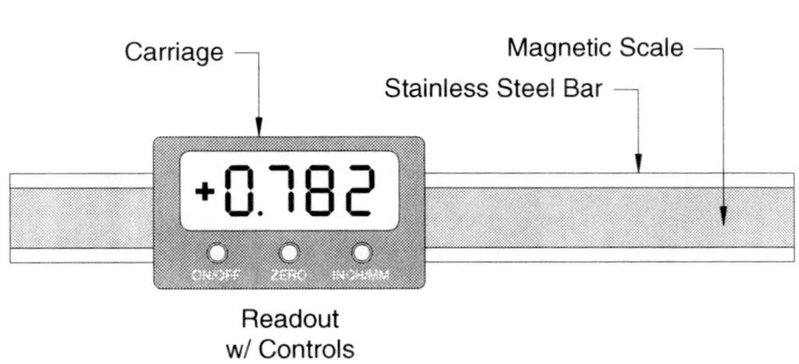

Figure 15-26 Magnetic Scale

Linear variable differential transformers (LVDT) are a type of transducer that relies on the coupling effect of a transformer to produce positioning information. These units are extremely versatile and generally provide exceptional accuracy. Figure 15-27 shows a basic schematic of an LVDT. The unit has a single primary coil and two secondary coils. The secondary coils can either act independently or be wired in series, as shown. The core is the moving element of these devices. As the core moves, the coupling between one of the secondary coils diminishes and the output voltage drops in proportion to the position of the core. By manipulating the size, length, and shape of the core, a broad range of applications can be served with these transducers.

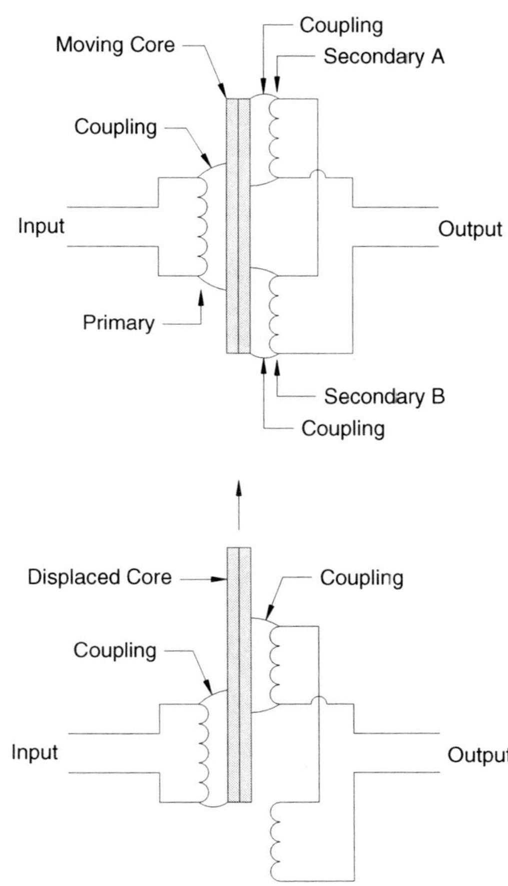

Figure 15-27 Linear Variable Differential Transformer Schematic

Figure 15-28 shows a typical LVDT. It consists of three coils wound on a common form with a moving core through the center. Applying an AC input voltage will produce an output voltage that corresponds to the position of the core. Applying a DC voltage to the input will allow the LVDT to be used as a vibration sensor. As the core moves, a magnetic field is induced by the primary coil and, in turn, generates a voltage in the secondary coils.

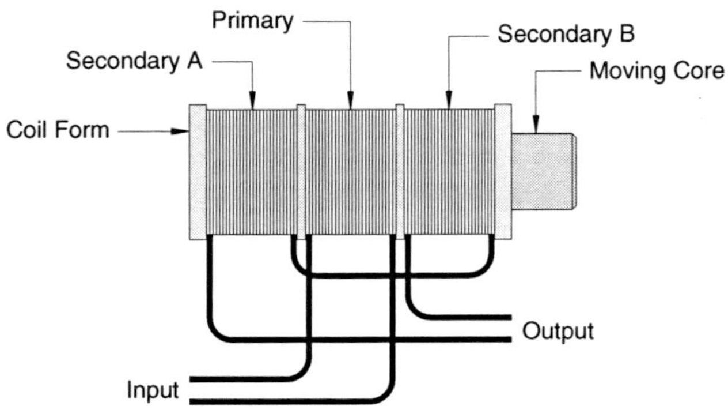

Figure 15-28 Linear variable Differential Transformer

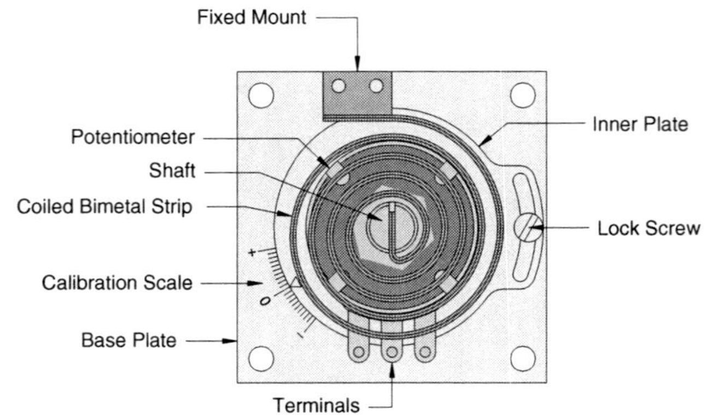

Figure 15-30 Bimetal Strip Temperature Transducer

Temperature Sensors

Temperature sensors generally fall into three different categories, bimetal strips, bulb-type, and thermocouples. Bimetal strip sensors are the most common, being used in most adjustable thermostats and temperature controllers. The most common example of a bimetal strip unit is found in an ordinary home thermostat.

Figure 15-29 shows a bimetal strip thermostat with a double throw mercury switch. The trip temperature is set by adjusting the pointer to the desired setting on the reference scale. If the temperature rises, the mercury switch closes in one position and if the temperature drops, the switch closes in the opposite position.

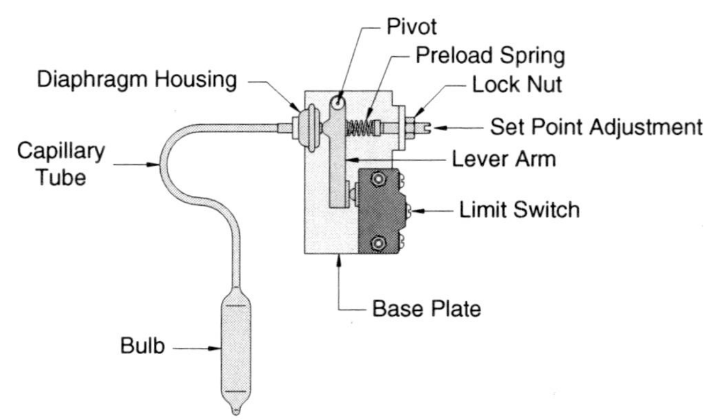

Figure 15-31 Bulb-Type Thermostat

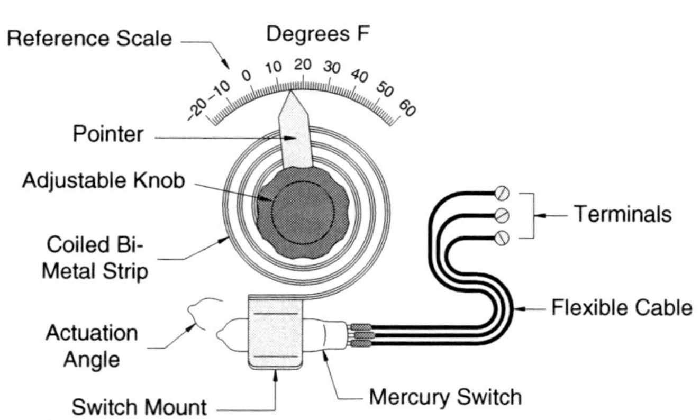

Figure 15-29 Adjustable Bimetal Strip Thermostat

A bimetal strip temperature transducer, as shown in Figure 15-30, is configured by fixing the outer end of a coiled element to a base plate and allowing the inner end to rotate the shaft of a potentiometer. The potentiometer is mounted to the center of a secondary base plate whose position can be adjusted in reference to the outer base plate. To calibrate the transducer, the inner plates are rotated against one another until the output voltage matches the current temperature condition. The output of the potentiometer can drive either a voltage divider or a Wheatstone bridge.

Bulb-type temperature thermostats, as shown in Figure 15-31, operate in reference to the expansion of a captured fluid. The bulb, capillary tube, and diaphragm housing are filled with a fluid with a high thermal expansion rate. As the temperature of the bulb increases, the fluid expands and forces the diaphragm to extend. The set point is adjusted by tensioning the preload spring, which counters the movement of the diaphragm,

Figure 15-32 shows a bulb-type temperature transducer. The bulb and capillary tube feed a bellows set. As the fluid

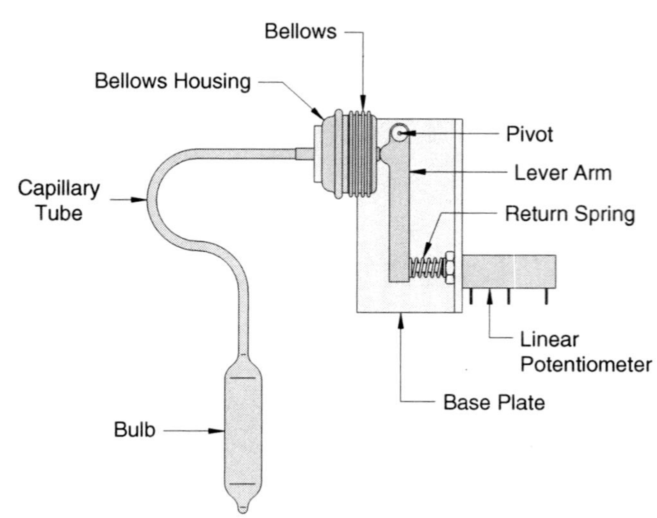

Figure 15-32 Bulb-Type Temperature Transducer

expands, the bellows extends and the lever arm actuates the potentiometer. The output of the potentiometer can drive either a voltage divider or a Wheatstone bridge.

When two dissimilar metals, such as iron and copper, are placed in contact with one another and heated they will produce a voltage. This junction is referred to as a thermocouple. The characteristics of thermocouple junctions are well understood and predictable, making them ideal as temperature transducers. A simple temperature transducer can be constructed by simply twisting two wires of dissimilar metals together and connecting them to a voltmeter, as shown in Figure 15-33. The output of a thermocouple is very low, so an amplifier must be deployed to drive controls of any consequence.

Figure 15-34 shows a digital thermocouple readout. These units generally consist of a high-impedance digital voltmeter that has been calibrated to match the characteristics of the specific thermocouple used.

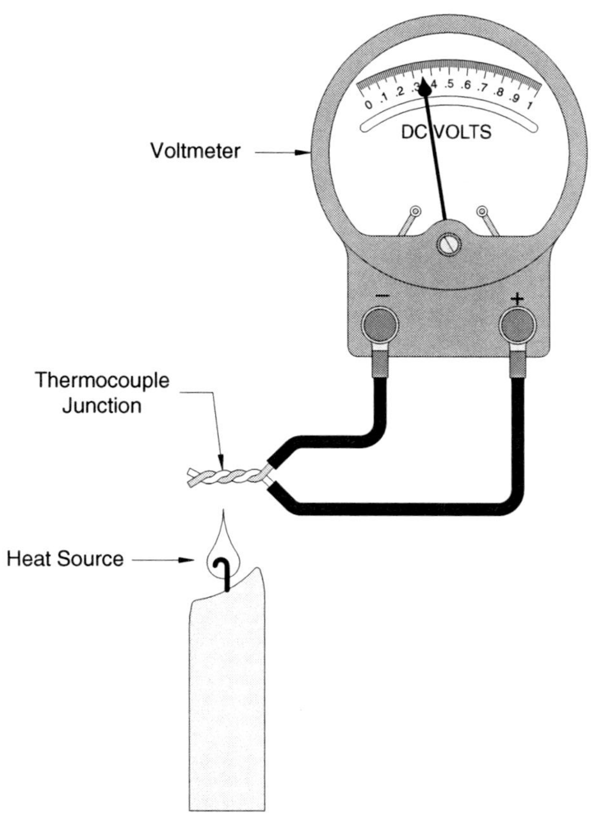

Figure 15-33 Thermocouple Temperature Sensor

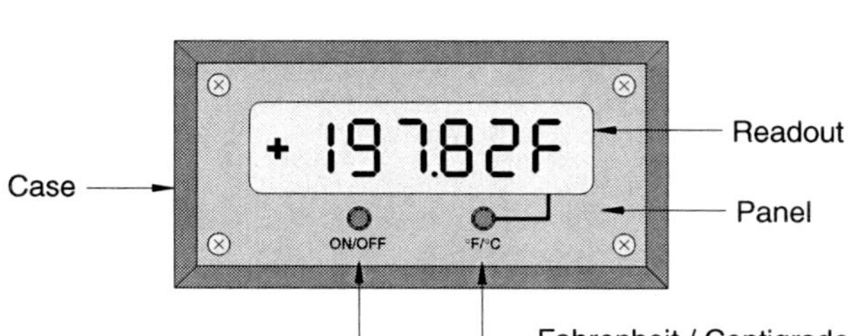

Figure 15-34 Thermocouple Temperature Readout

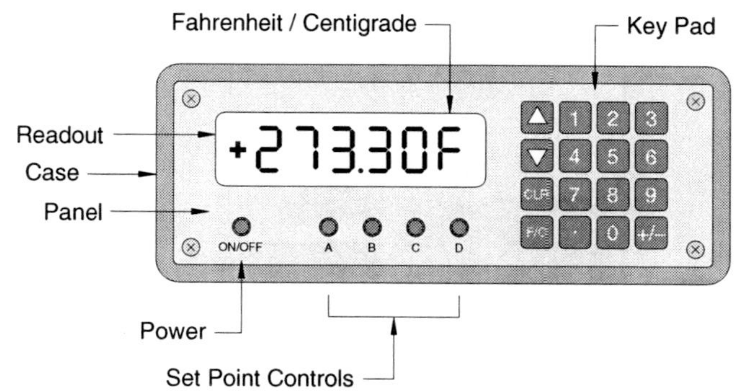

Figure 15-35 Thermocouple Temperature Controller

Figure 15-35 shows a commercial thermocouple temperature controller. These units normally provide a readout, serial port, and one or more adjustable set points. Many units also provide programming functions that allow time/function/temperature curves to be programmed.

Level Sensors

The simplest types of fluid level indicators are float switches. Principally, there two different types, as shown in Figure 15-36. Top mount switches have a cylindrical float that carries a magnet. The float is fitted around a central core containing a magnetic reed switch. The normal condition of the switch can be changed by removing the float, flipping it, and reinstalling it.

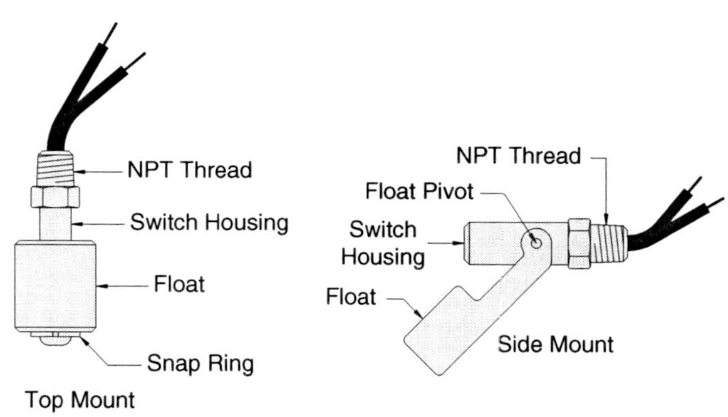

Figure 15-36 Fluid Level Switches

Side mount switches have a hinged float, which carries a magnet. When the float is aligned with the end of the body, the switch activates. Orienting the float either up or down can change the normal condition of the switch.

Level indication can be provided by installing a float switch array, as shown in Figure 15-37. In this case the fluid level controls a simple string of panel lamps. It should be

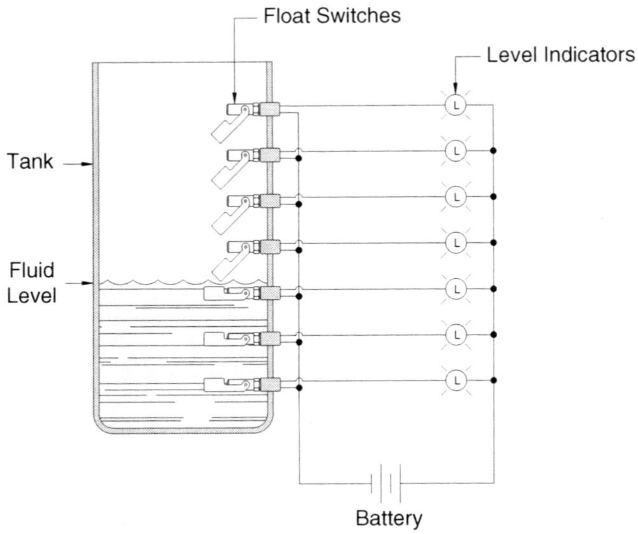

Figure 15-37 Float Switch Level Indication

noted that the float switches can be used to control any number of electrical controls.

A potentiometer fluid level indicator, as shown in Figure 15-38, is simply a resistive element whose wiper is controlled by a float. The elements are generally wire wound and configured in a semicircular pattern. The float rod is bent so as to allow the mechanism to remain dry. The most common use of these types of lever indicators is in automobile fuel tanks. In automotive cases, the resistive element is designed to operate while immersed in the fuel.

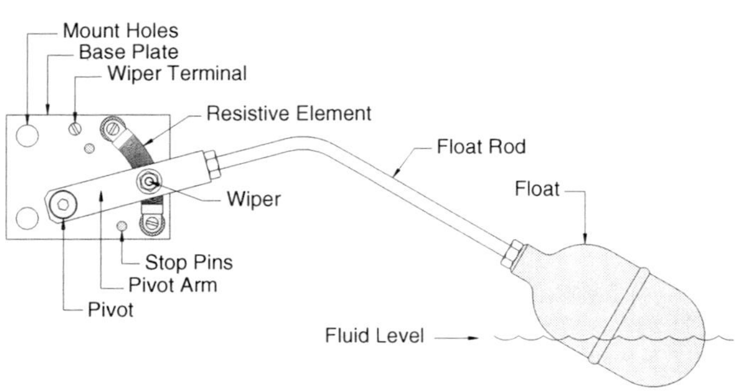

Figure 15-38 Potentiometer Fluid Level Indicator

Figure 15-39 shows a typical voltage divider schematic, commonly used with potentiometer fluid level indicators.

Capacitance fluid level transducers operate using the variation that a changing capacitor has on the throughput of a circuit. Figure 15-40 shows a capacitance fluid level schematic. The moving element is a conductor that is allowed to move between the plates of a capacitor. As the fluid level changes, the capacitance changes providing a signal variation to the voltmeter.

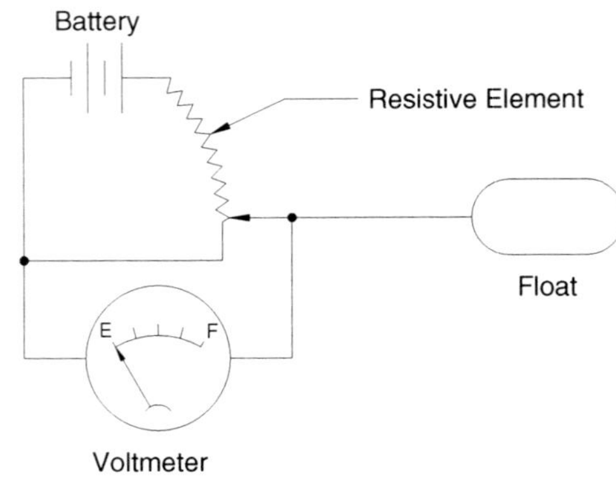

Figure 15-39 Potentiometer Fluid Level Schematic

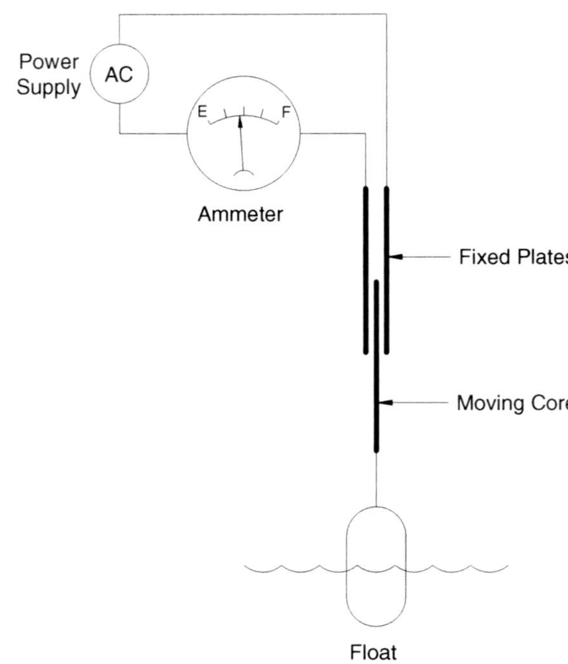

Figure 15-40 Capacitance Fluid Level Schematic

If the fluid column is sufficient, then the level can be monitored using only a pressure gauge, as shown in Figure 15-41. This method of level indication is not particularly accurate and is only used in applications where critical readings are not necessary. This type of fluid level monitoring is principally used in municipal water towers. The water utility measures the system pressure and can calculate the level in the tower at any given time.

Despite all the buzz that high-temperature superconductors originally created, they have not delivered on the speculations that the technology originally inspired. However, there is one application where this material has provided an excellent solution, as level monitors for cryogenic fluids.

A cryogenic fluid is a gas that becomes liquid when cooled to very low temperatures. Among the most common cryogenic

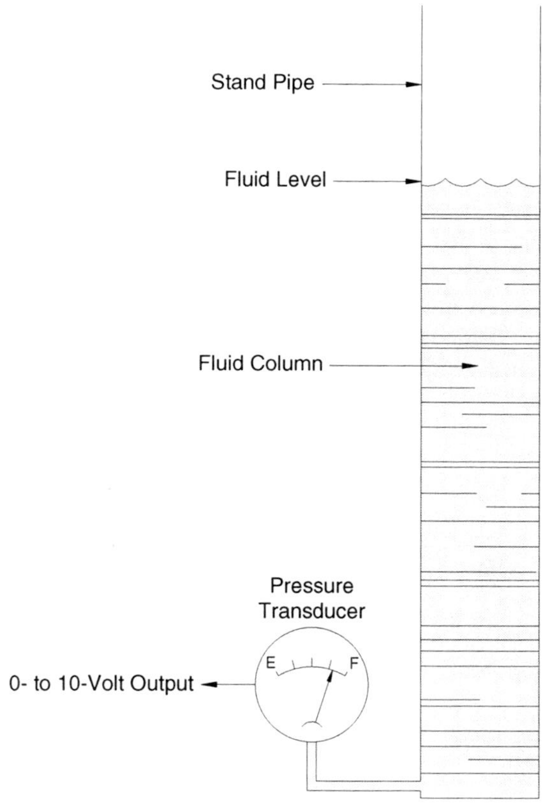

Figure 15-41 Pressure Fluid Level Schematic

fluids are liquefied oxygen, nitrogen, helium, and argon. The natural temperatures at which these gases exist in a liquid state, is well below the zero resistance level of the super conductor. Therefore, if a loop of high-temperature super conducting wire is immersed into a quantity of cryogenic fluid, as shown in Figure 15-42, the section below the fluid level will have zero electrical resistance. The section above the fluid level retains its noncooled resistance. By measuring the overall resistance of the loop, the amount of cooled versus noncooled wire can be calculated and the level of the fluid can be determined.

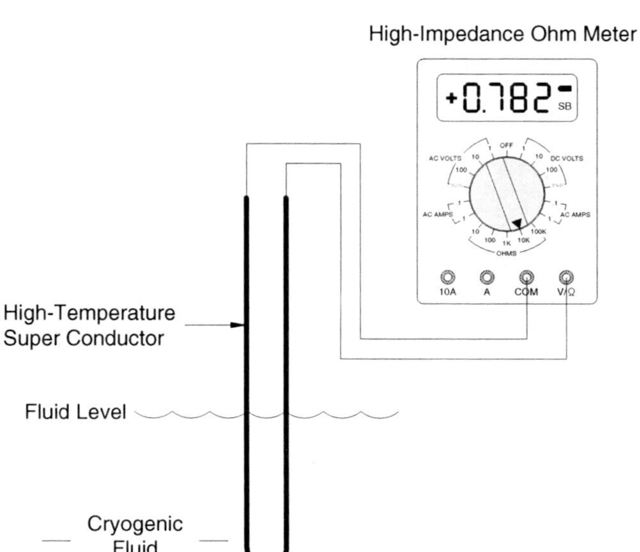

Figure 15-42 Low-Temperature Super Conductor Fluid Level Schematic

Pressure Sensors

Detecting pressure is an important function in all manner of consumer, commercial, and industrial equipment. The basic pressure switch, as shown in Figure 15-43, consists of a pressure housing with a diaphragm. A plunger style limit switch is arranged, so the diaphragm button will activate the unit when a preset pressure is reached. The switch is equipped with a through actuator, which is countered with a preload spring and adjustment screw. The pressure actuation, or set point, can be adjusted by turning the screw until the switch actuates.

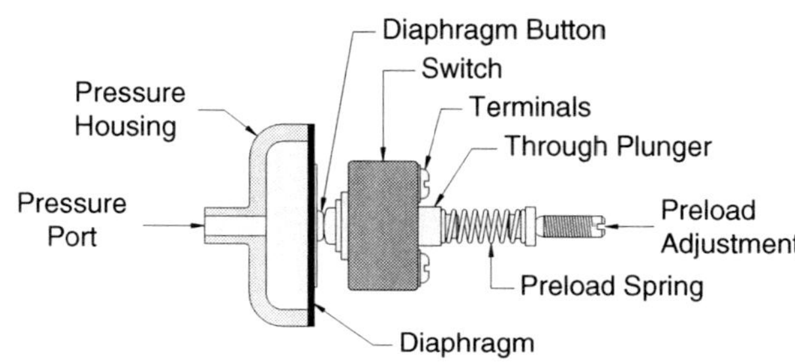

Figure 15-43 Pressure Switch with Adjustable Set Point

Figure 15-44 shows a commercial pressure switch with adjustable set point. These units are generally equipped with an NPT thread, two terminals, and a top mounted adjustment screw.

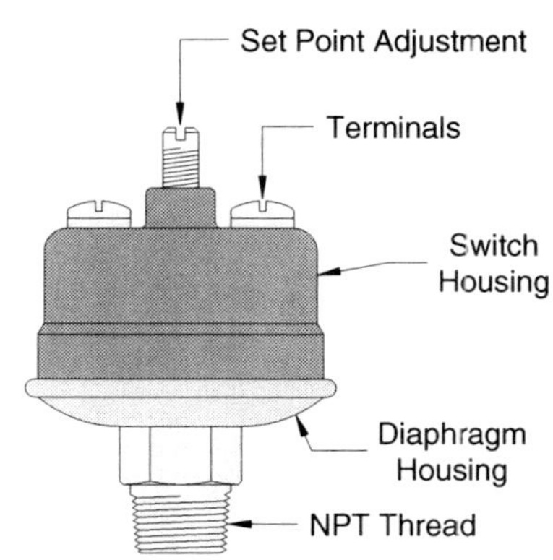

Figure 15-44 Commercial Pressure Switch with Adjustable Set Point

Diaphragm pressure transducers are similar in construction to their switch counterparts. The switch assembly is replaced with a potentiometer assembly, which will produce a continuously variable output. Figure 15-45 shows a diaphragm pressure transducer layout.

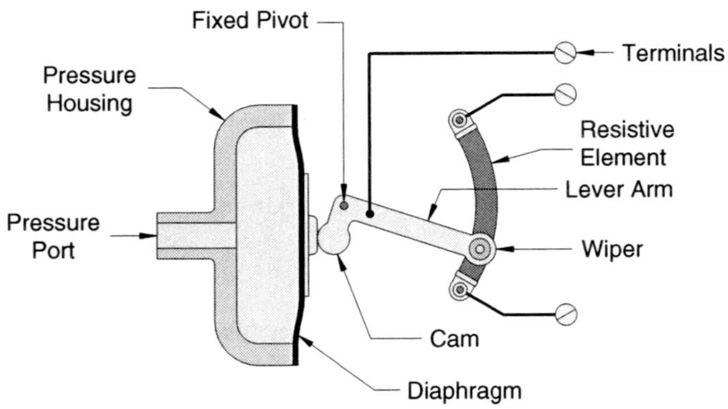

Figure 15-45 Diaphragm Pressure Transducer

which also drives the gauge needle. As pressure is applied to the unit, the Borden tube tries to straighten out and activates the gear mechanism that turns the needle and potentiometer. The output of the potentiometer can drive either a voltage divider or a Wheatstone bridge.

For more sensitive pressure readings, a bellows pressure transducer is generally specified, as shown in Figure 15-48. Pressure is fed to the input and the bellows extends. The opposite end of the bellows is connected to a linear potentiometer. As the pressure varies, so does the bellow's length and the position of the potentiometer. The output of the potentiometer can drive either a voltage divider or a Wheatstone bridge.

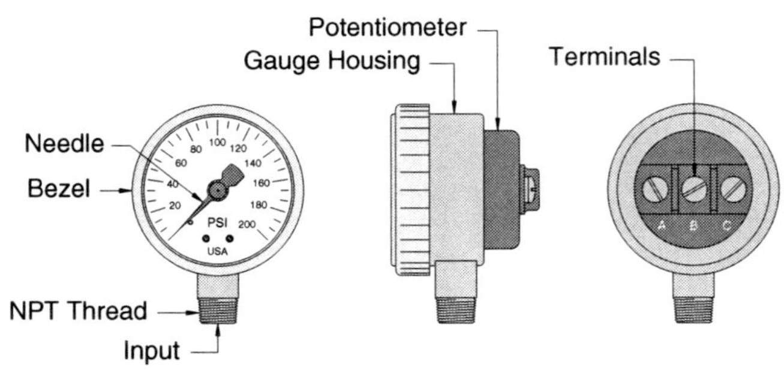

Figure 15-46 Commercial Pressure Gauge with 0- to 10-Volt Output

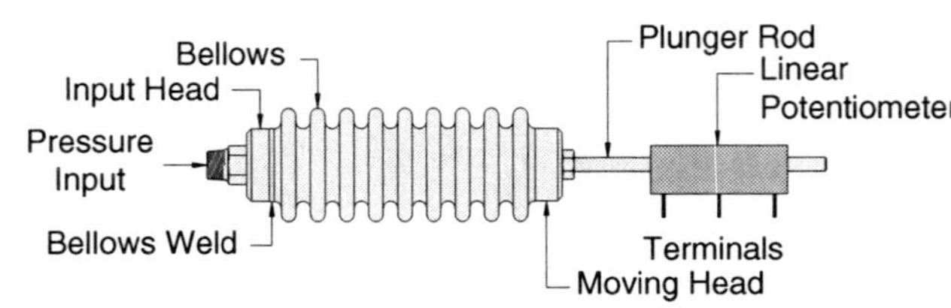

Figure 15-48 Bellows Pressure Transducer

Pressure transducers are also available in a pressure gauge package, as shown in Figure 15-46. These units have principally the same appearance as a standard pressure gauge, except that they have an extended rear housing and three terminals. The rear housing accommodates a potentiometer element. Information supplied by the manufacturer should be used for wiring the transducer, although the center terminal is generally the wiper.

Figure 15-47 shows the internals of a pressure gauge transducer. The mechanism is a standard Borden tube arrangement. A potentiometer is added to the rear of the pinion gear,

A bellows-type pressure transducer can be constructed using a spring return cylinder, as shown in Figure 15-49. The output shaft of the cylinder is connected to the shaft of a linear potentiometer with a connecting nut. A cylinder arrangement will not be as sensitive to minute changes as a bellows because the piston and cylinder will exhibit a certain amount of stiction.

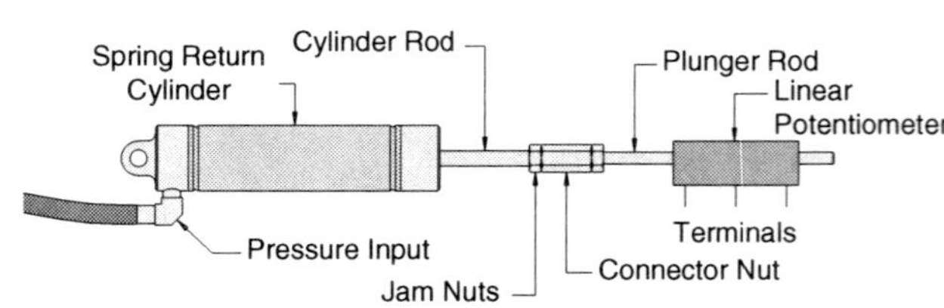

Figure 15-49 Cylinder Pressure Transducer

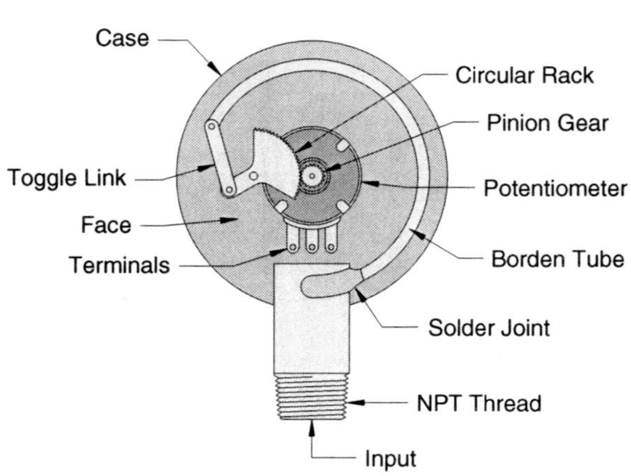

Figure 15-47 Pressure Gauge Transducer Internals

Differential Pressure Sensors

Differential pressure can be measured by arranging two bellows sets and a linear potentiometer, as shown in Figure 15-50. The output of the potentiometer drives a slide wire bridge. The zero adjustment is used to zero the meter after the unit is placed in service.

A differential pressure transducer can also be constructed using a pair of spring return cylinders, as shown in Figure 15-51. The output shaft of the cylinders is connected to the shaft of a linear potentiometer with connecting nuts. As with the

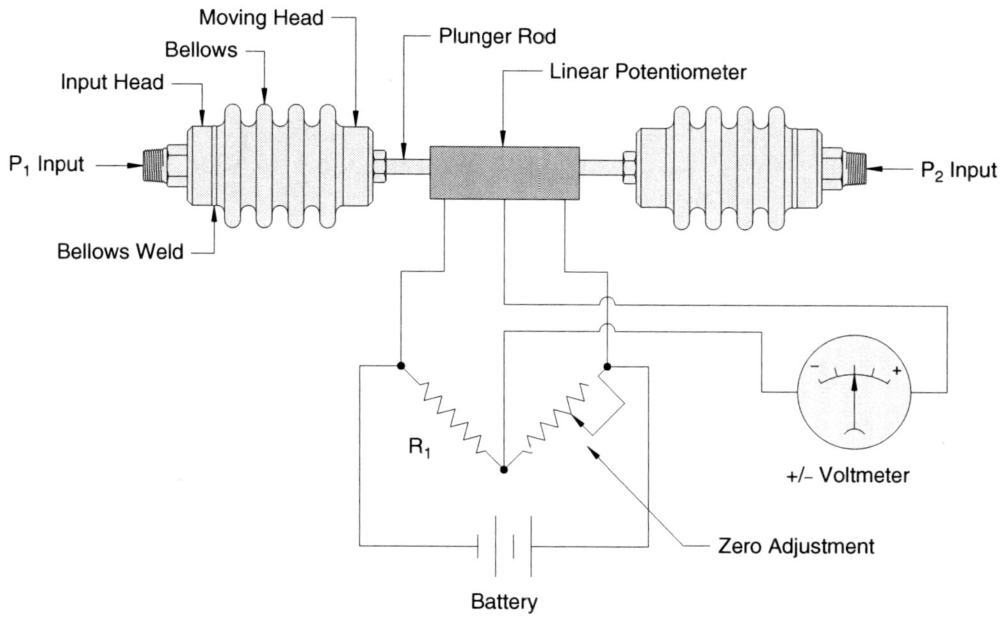

Figure 15-50 Differential Bellows Transducer with Slide Wire Bridge Output

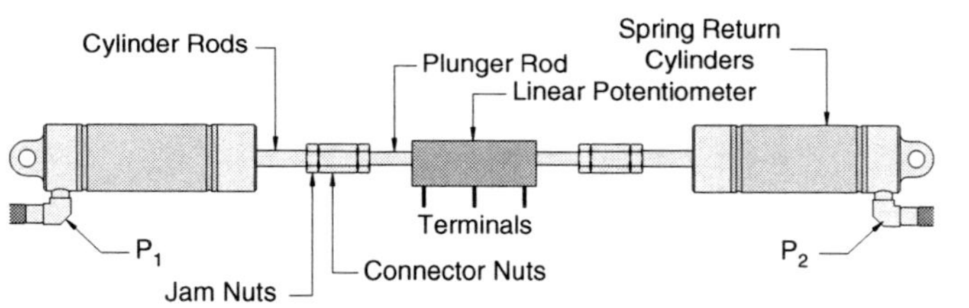

Figure 15-51 Cylinder Differential Pressure Transducer

cylinder pressure transducer, the sensitivity of this cylinder arrangement will not be as sensitive to minute changes as a bellows unit because the pistons and cylinders will exhibit a certain amount of stiction.

For the most sensitive differential pressure applications, a capacitance manometer is generally specified. These transducers are extremely sensitive to minute changes in pressure and are often found on vacuum systems to monitor and control the blending of partial pressure gases. Figure 15-52 shows a schematic representation of a capacitance manometer.

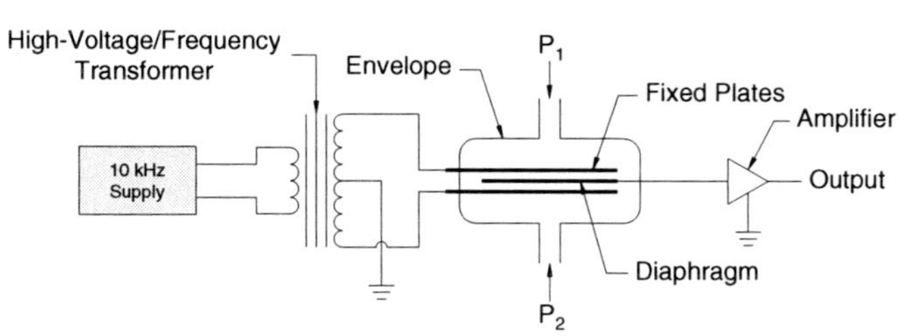

Figure 15-52 Capacitance Manometer Differential Pressure Gauge Schematic

Vacuum Sensors

Vacuum Sensors generally fall into two different categories—negative pressure, or vacuum, and random molecular motion, or high vacuum. The negative pressure regime is usually considered to be between 14.7 pounds per square inch absolute (PSIA), or atmosphere, and 0 PSIA (0 to 30 inch Hg.). In this pressure range the molecules are in constant contact with one another and when gases are added or removed there is a cause and effect that is mechanically linked. Because the gas molecules are in contact with one another, the pressure can be gauged using mechanical methods.

The most common vacuum gauge is the Borden tube gauge. These units use the same basic mechanism as their pressure counterparts, except that they are set up to read vacuum. Figure 15-53 shows a typical Borden tube vacuum gauge with a potentiometer output.

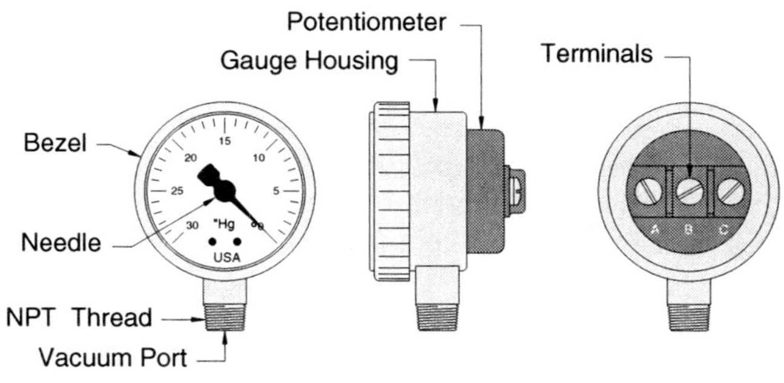

Figure 15-53 Commercial Vacuum Gauge with 0- to 10-Volt Output

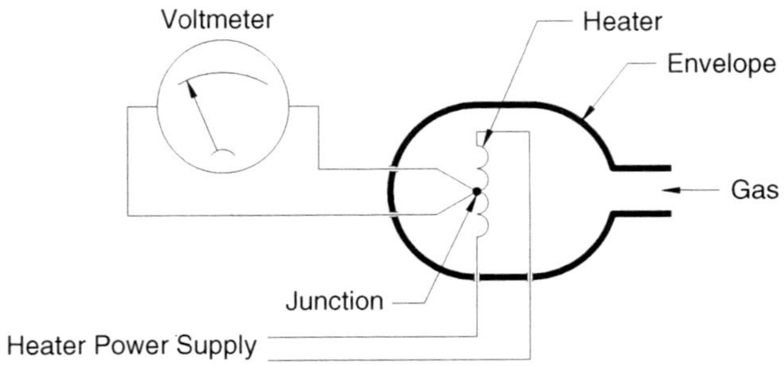

Figure 15-54 Thermocouple Vacuum Gauge Schematic

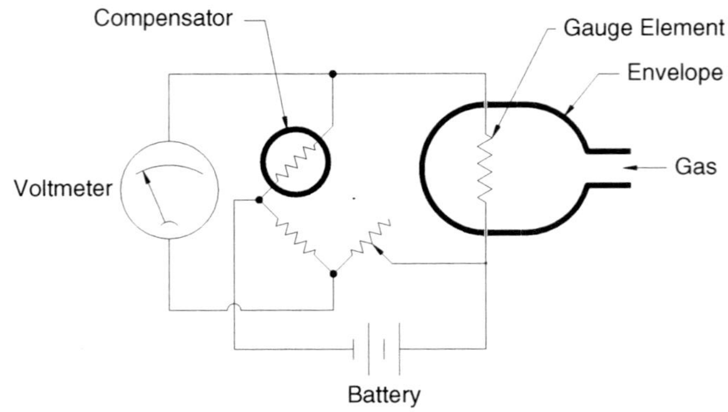

Figure 15-56 Pirani Vacuum Gauge Schematic

As pressure decreases, the effectiveness of mechanical gauges diminishes. To improve accuracy, thermal elements are employed. Figure 15-54 shows a schematic representation of a thermocouple vacuum gauge. In this case a heating element is fed a constant power. At the center of the element a thermocouple is attached. When the element is exposed to gas, a certain amount of heat is drawn off and the output of the thermocouple junction changes in direct proportion to the gas pressure.

Figure 15-55 shows a typical commercial thermocouple gauge. These units are generally supplied with a 1/8-inch NPT (national pipe thread) and octal base. The transducer will also require a power supply and readout calibrated in pressure units.

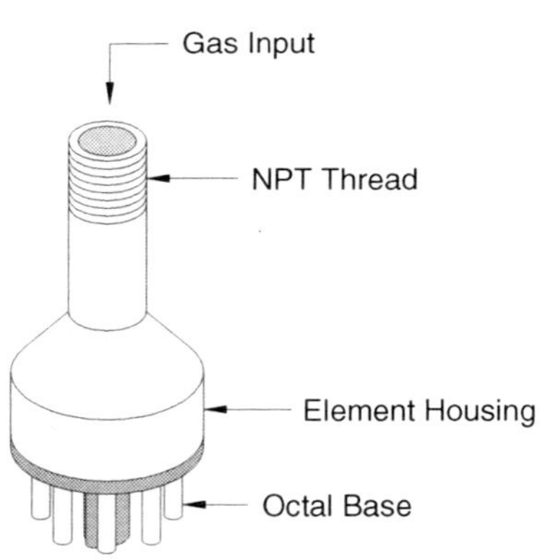

Figure 15-55 Commercial Thermocouple Vacuum Sensor

To provide even lower pressure accuracy, a Pirani gauge is typically specified. This particular gauge element will normally bridge the vacuum and high vacuum regimes. Figure 15-56 shows a schematic representation of a Pirani gauge. The gauge element makes up one resistor in a Wheatstone bridge and is exposed to the pressure to be monitored. A compensator resistor is placed into a vacuum envelope opposite the element. The element and compensator are matched at high vacuum pressures. A battery is used to heat the gauge element. Any gas coming in contact with the element will cool it and, therefore, change the resistance. This change is reflected by the voltmeter and, thereby, the pressure of the system can be gauged.

Figure 15-57 shows a typical commercial Pirani vacuum transducer. These units are generally supplied with a quick disconnect vacuum flange, cable, and DIN connector. The calibration controls are usually placed directly on the gauge head. Like the thermocouple gauge, these transducers also require a power supply and readout calibrated in pressure units.

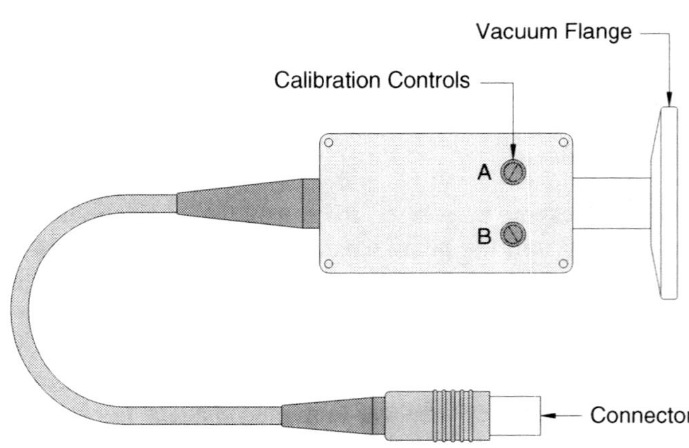

Figure 15-57 Commercial Pirani Vacuum Gauge

High Vacuum Sensors

The second vacuum regime is high vacuum, or random molecular motion. In this regime the gas molecules are not in constant contact with one another and are moving at random in space. When gases are added or removed there is a cause and effect that is statistically linked. Because the gas molecules are not in constant contact with one another, mechanical measuring methods are ineffective; therefore, the gas content must be measured statically.

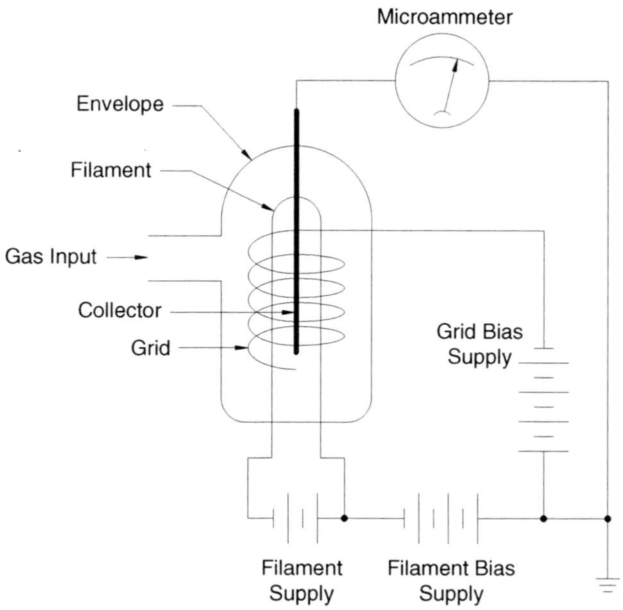

Figure 15-58 Ionization Vacuum Gauge Schematic

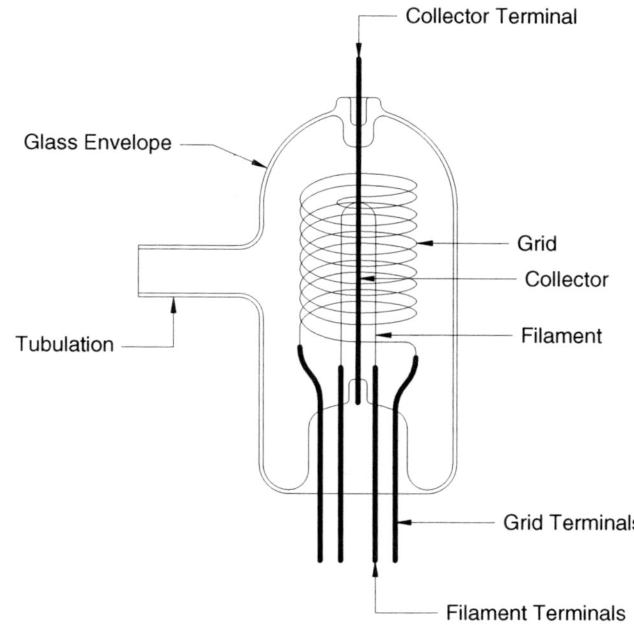

Figure 15-59 Bayard-Alpert Ionization Vacuum Gauge

Figure 15-58 shows a schematic representation of an ionization vacuum gauge. An ionization gauge is a vacuum tube that is specifically designed to collect and concentrate gas atoms and molecules (particles). In this case, a filament is heated and a bias voltage is placed across the filament, grid, and collector. A higher bias voltage is placed across the filament and the grid, creating a steady stream of electrons across the two elements. When a gas particle drifts into the electron field it is ionized, that is to say that the particle is charged. The charged state of the particle propels it to the collector and the microammeter reads the additional current flow that is created during this action. The number of particles impacting the collector can be displayed on the microammeter as pressure units.

Figure 15-59 shows the most popular ionization transducer, the Bayard-Alpert gauge. These units are fairly inexpensive and provide exceptional accuracy at high vacuum pressures.

The other type of common high vacuum gauge is the cold cathode gauge. Figure 15-60 shows a schematic representation of a cold cathode gauge. In this case, a cylindrical emitter with a central collector is biased with a high voltage, which creates an electron flow. As a particle drifts into the electron field, it becomes charged and is drawn to the collector. The microammeter reads the additional current generated by this action and provides a readout in pressure units. Most cold cathode gauges have a large permanent magnet surrounding the envelope in order to improve focusing of the charged particles onto the collector.

Figure 15-61 shows a typical commercial cold cathode vacuum gauge head. These are normally supplied with a metal seal vacuum flange. It should also be noted that these units and their controllers are generally rather expensive, being 5 to 10 times as costly as their ionization counterparts.

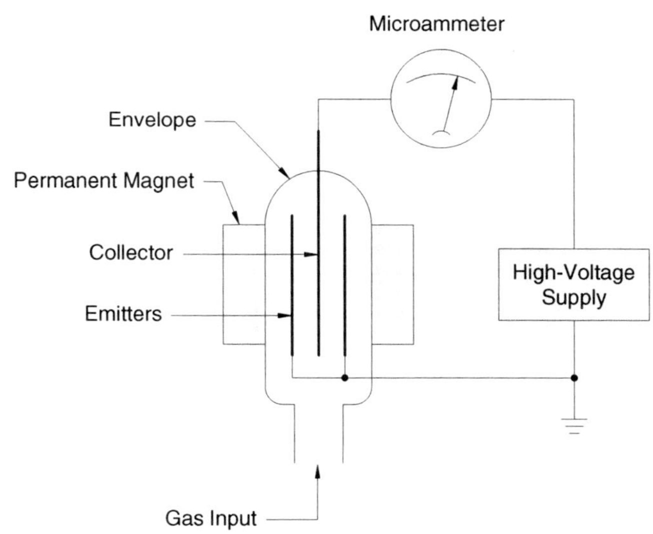

Figure 15-60 Cold Cathode Ionization Vacuum Gauge Schematic

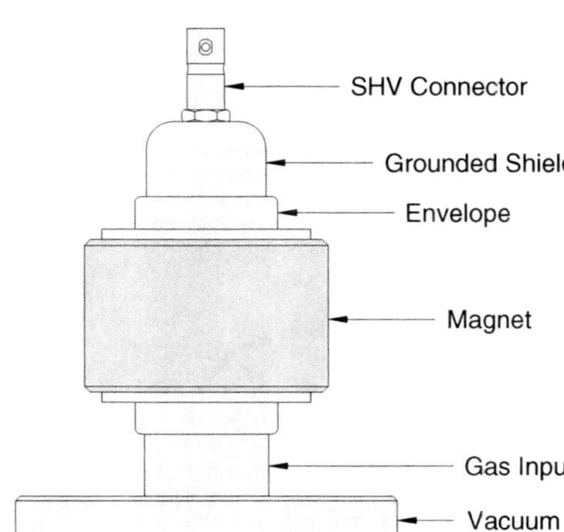

Figure 15-61 Commercial Cold Cathode Vacuum Gauge

Flow Sensors

Sensing the flow of gases and fluids is critical for many commercial and industrial processes. Sensing flow can be as simple as viewing a pressure gauge. If pressure is present at one end of an open conduit, then flow must be occurring. If the pressure changes, then there will be a corresponding change in the flow rate. This method is commonly used for applications where low accuracy is acceptable, such as pumping irrigation water.

For applications that require a high degree of accuracy, there are a number of methods that are commonly employed. Figure 15-62 shows a basic paddle wheel flow meter. A paddle wheel is placed in the flow of a liquid and allowed to rotate. The output of the paddle wheel drives a tachometer generator that, in turn, outputs a proportional signal to a voltmeter. These types of flow meters are very inexpensive and provide good performance. Because of the generator drive, they do create a certain amount of backpressure and, therefore, restrict the flow to a certain degree.

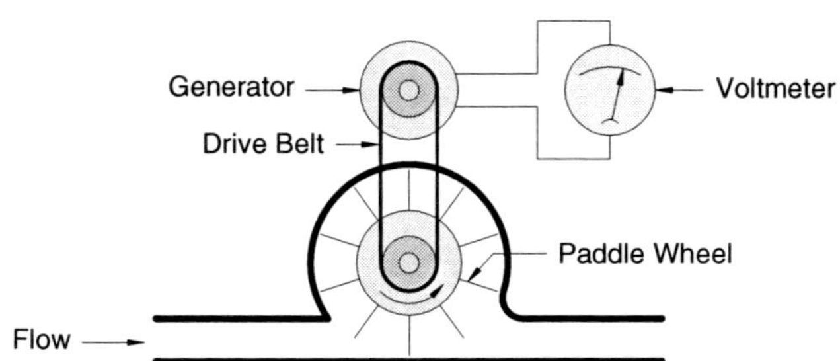

Figure 15-62 Paddle Wheel Flow Meter

To limit the backpressure of a paddle wheel flow meter, the generator can be replaced with a proximity sensor, as shown in Figure 15-63. The sensor detects the passing paddle and outputs a pulse each time. The sensor is interfaced to a controller that drives a voltmeter calibrated in flow units. Another attribute to flow meters with proximity sensors is that the moving element is sealed within the housing and there are no rotational seals to create leaks.

A variation of the paddle wheel flow meter is the turbine unit. These transducers operate in the same manner as the paddle wheel, except that the wheel is replaced with a turbine rotor. The number and pitch of the blades allows the turbine rotor to be configured for a broad range of flow rates and, therefore, is a much more popular design. Figure 15-64 shows a stylized sectional view of a pulse generating turbine flow meter. Figure 15-65 shows a typical commercial pulse generating flow transducer.

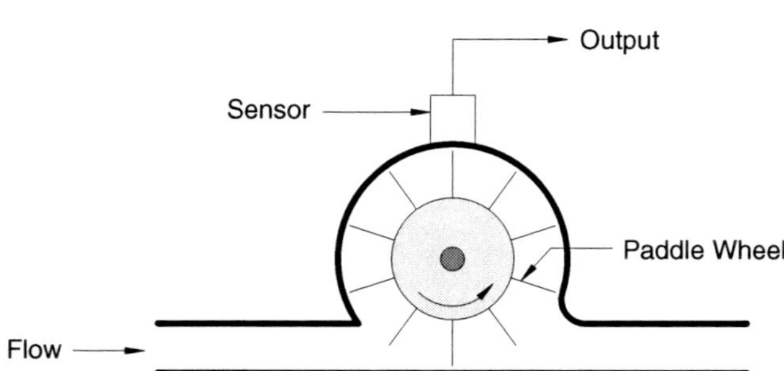

Figure 15-63 Pulse Generating Paddle Wheel Flow Meter

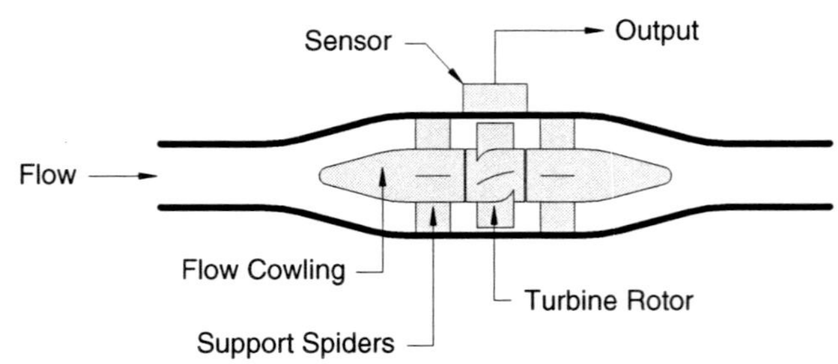

Figure 15-64 Pulse Generating Turbine Flow Meter

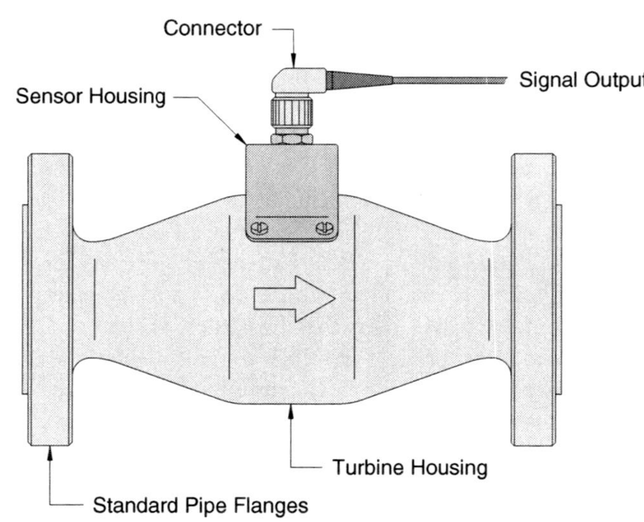

Figure 15-65 Commercial Turbine Flow Meter

One of the most common methods for measuring flow is through pressure drop. By measuring the pressure drop through a restriction, the flow rate of a given fluid can be determined with a high degree of accuracy.

Figure 15-66 shows a stylized sectional view of a differential pressure flow meter. The fluid is forced to flow through a restrictor orifice, which is bridged with two resistive pressure transducers. The pressure transducers make up two elements of a Wheatstone bridge and the meter reads only the voltage generated by differential resistance of the two transducers.

Figure 15-67 shows a typical commercial differential pressure flow meter and readout. These units are often constructed by sandwiching an orifice plate between two standard flanges, as shown. To change the flow characteristics of the transducer, one only needs to swap out the orifice plate.

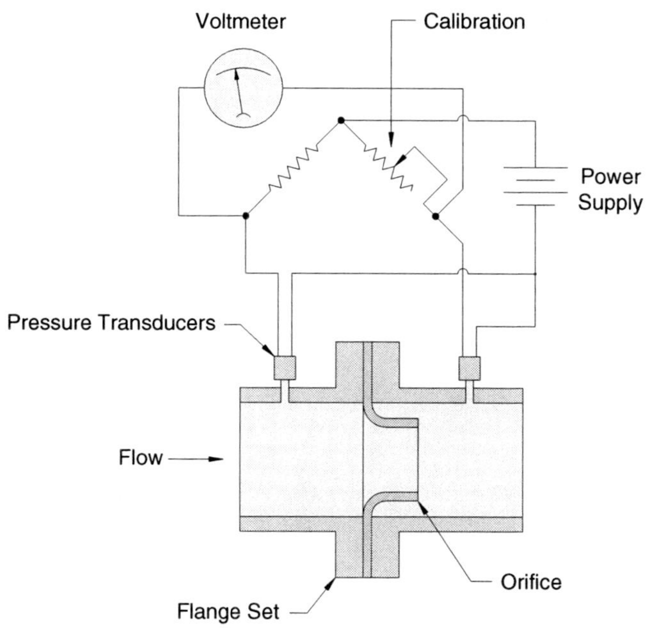

Figure 15-66 Differential Pressure Flow Meter

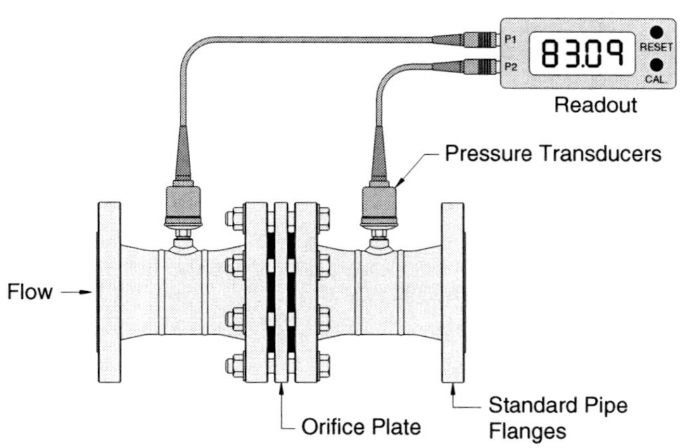

Figure 15-67 Commercial Differential Pressure Flow Meter

Gauging the flow of gas is little more difficult, especially at low flow rates. To achieve accurate gas flow rates a hot wire element is generally used, as shown in Figure 15-68. An element is placed in the flow of the gas to be measured and is heated to a fixed level. As the rate of the gas flowing over the element increases, it has a cooling effect on the element with its resistance lowering in proportion to the flow rate. The hot wire element makes up one element of a Wheatstone bridge with the out voltage directly proportional to the flow rate of the gas.

A second method of gauging gas flow rate is with a paddle flow meter, as shown in Figure 15-69. In this method, a transducer paddle interrupts the flow of the gas. As the flow increases, the paddle deflects actuating the wiper on a potentiometer. The unit can drive a voltage divider circuit or a Wheatstone bridge.

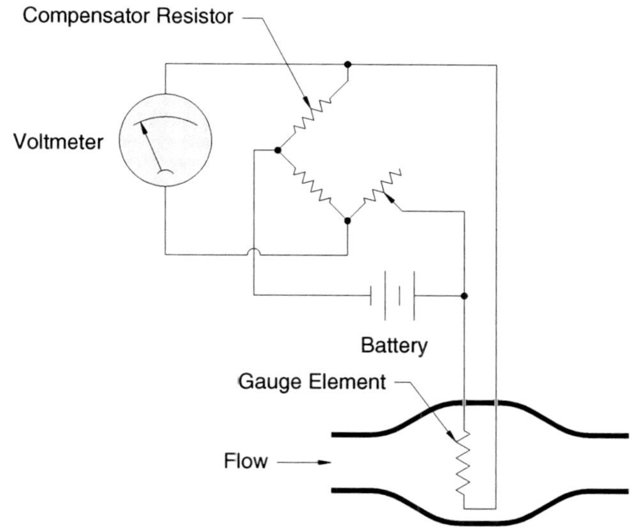

Figure 15-68 Hot Wire Flow Meter

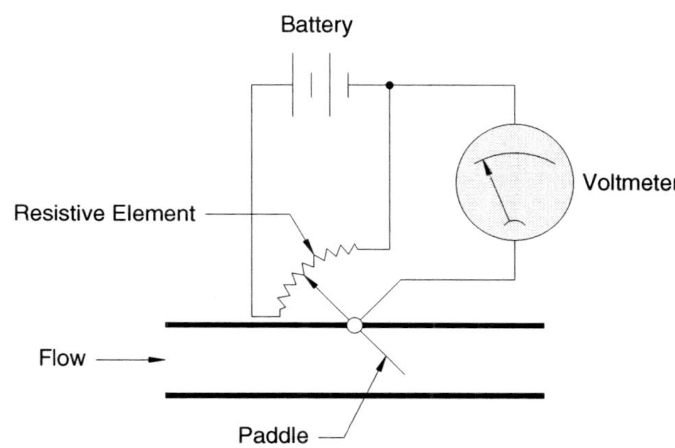

Figure 15-69 Paddle Flow Meter

Paddle flow sensors are most commonly found in automotive applications where they are referred to as mass flow sensors. Figure 15-70 shows a typical automotive style mass flow sensor. These units gauge the air flow that the engine is drawing and matches the fuel delivery rate to achieve maximum efficiency.

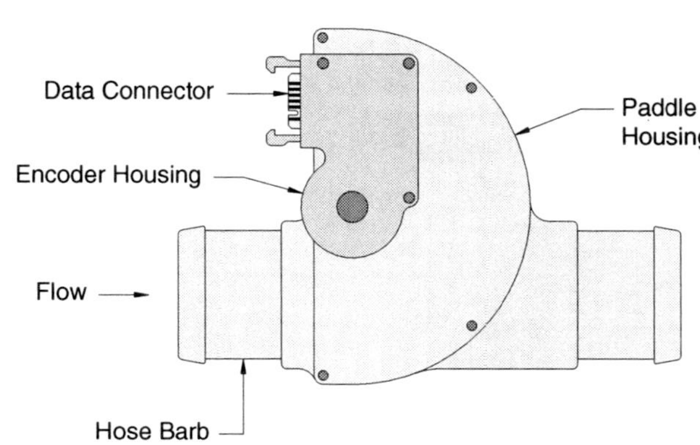

Figure 15-70 Automotive Mass Flow Sensor

Light Sensors

There are a variety of applications that call for the detection of light. Most notably is in the area of opto-sensor sets. The photosensitive vacuum tube, as shown in Figure 15-71, is most commonly used in movie projectors that use an optical sound strip. A bias voltage is applied across the cathode and anode. When the cathode is exposed to light it emits electrons in proportion to the intensity of the light.

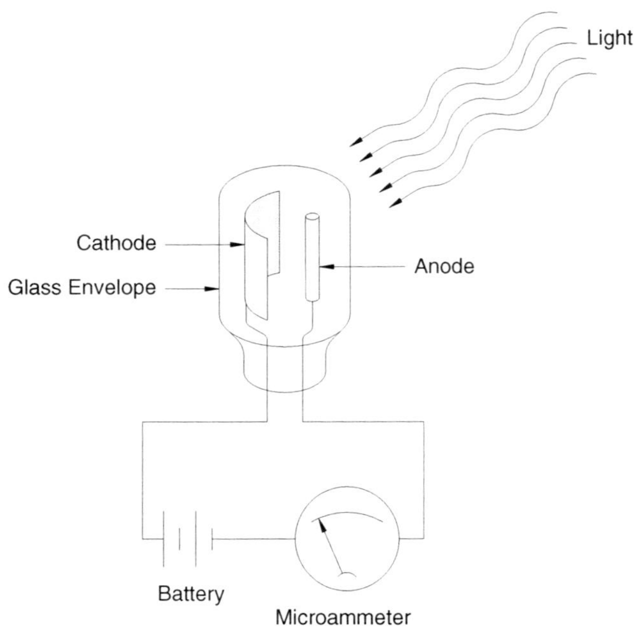

Figure 15-71 Photosensitive Vacuum Tube Schematic

Figure 15-72 shows a photosensitive diode. These units operate in the same fashion as a diode, except that they can be turned on and off by exposure to light. These devices are used in most opto sets and are very inexpensive.

Most transistors are photosensitive and can be modified to operate as photo sensors by simply drilling a hole in the top of the can that packages the device. The base wire is removed

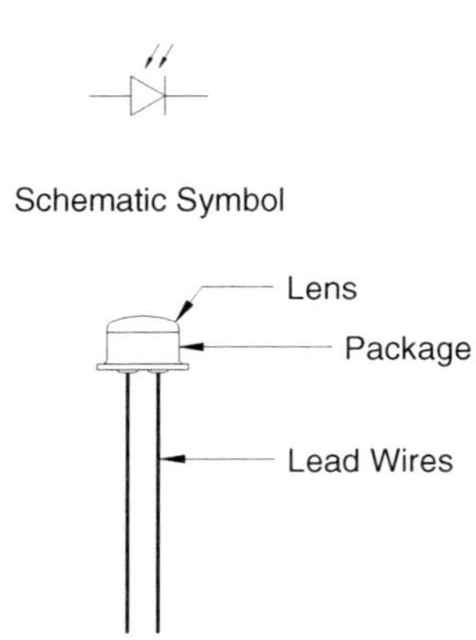

Figure 15-72 Photosensitive Diode

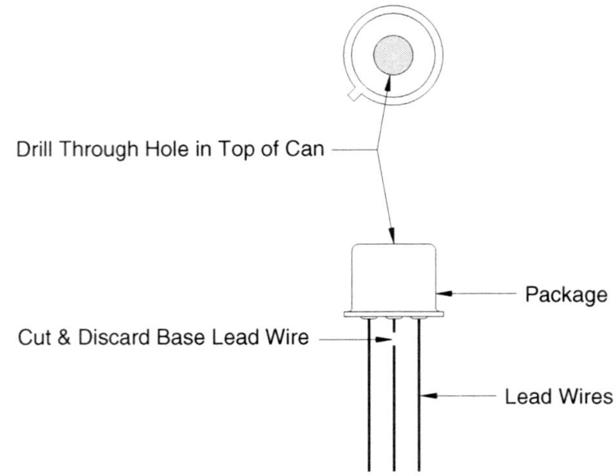

Figure 15-73 Photo Sensor Made from an Ordinary Transistor

and the light operates as the base signal. Figure 15-73 shows an ordinary transistor that has been modified to operate as a photo sensor.

Vibration Sensors

Detecting vibration can provide important information as to the internal condition of many kinds of industrial equipment. A vibration sensor generally consists of a floating coil suspended within a magnetic field, as shown in Figure 15-74. The base of the sensor is fixed to the equipment to be monitored. As the equipment vibrates, so does the base and permanent magnet assembly. The coil is suspended between the poles of the magnet, at the end of a leaf spring, which is fixed to the base. The inertial mass of the coil has a tendency to resist any movement and, therefore a voltage is generated by the differential movement of the magnet.

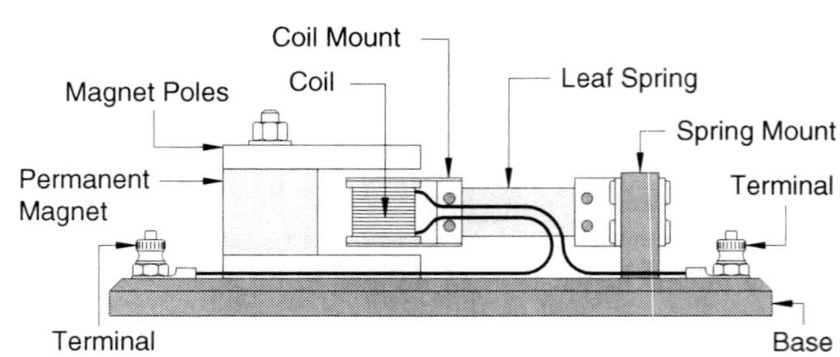

Figure 15-74 Moving Coil Vibration Sensor

The output of the vibration sensor can be rectified and the signal is fed to a voltmeter. The meter will display the relative amplitude of the sensor. A rising voltage indicates a deteriorating condition within the equipment being monitored. Figure 15-75 shows a schematic of a vibration sensor circuit. The filter capacitor is intended to smooth out the pulses and provide a more steady voltage reading.

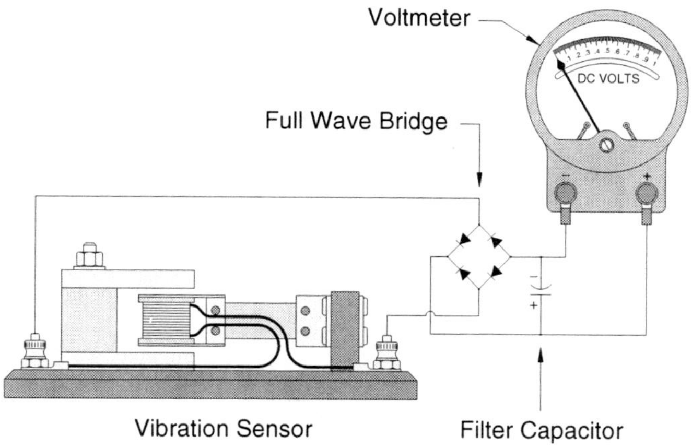

Figure 15-75 Averaged Vibration Reading

If a higher order analysis is desired, then the vibration sensor can be set up with an oscilloscope and frequency counter, as shown in Figure 5-76. The oscilloscope provides a clear indication of the waveform that the equipment is generating while the frequency counter indicates the resonant frequency. Either of these two instruments can be used independently with a vibration sensor.

Accelerometers

When an object is moving in a circular arc, a certain amount of centrifugal force is generated. In the case of an automobile traversing a corner, this force has a tendency to through the driver toward the outside of the curve. This force is generally described in G, for gravity. One G equals the equivalent force of gravity. Therefore, an object weighing 10 pounds and accelerated at 20 G will have the equivalent weight of 200 pounds.

Figure 15-77 shows a pendulum accelerometer. A dead weight is placed at the end of a lever arm and allowed to hang down. If a force is applied that makes the pendulum swing 22.5°, then a $^1/_2$-G acceleration is indicated. If the pendulum swings 45°, then a 1-G acceleration is indicated.

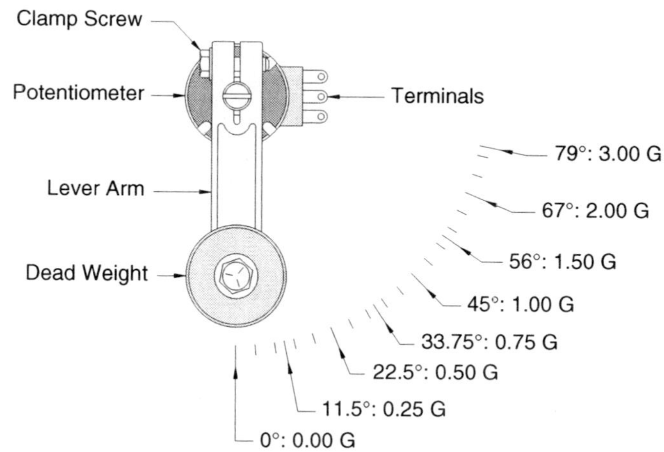

Figure 15-77 Pendulum Accelerometer

An accelerometer can be configured with a dead weight, two centering springs, and a linear potentiometer, as shown in Figure 15-78. The tension on the springs can be tuned for sensitivity and the position of the springs is adjusted to center the potentiometer. The sensitivity of this type of sensor can also be increased by increasing the size of the dead weight.

The output of the potentiometer in an accelerometer is generally used as the moving element of a slide wire bridge, as shown in Figure 15-79. Sensitivity adjustment is achieved by configuring a potentiometer in a voltage divider arrangement. This circuit is appropriate for either a pendulum or spring accelerometer.

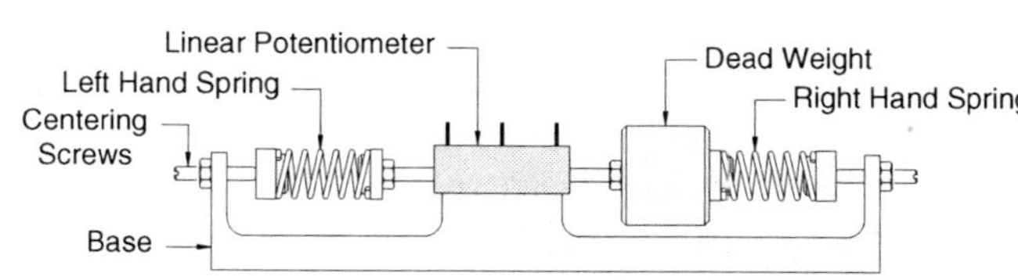

Figure 15-78 Spring Centered Accelerometer

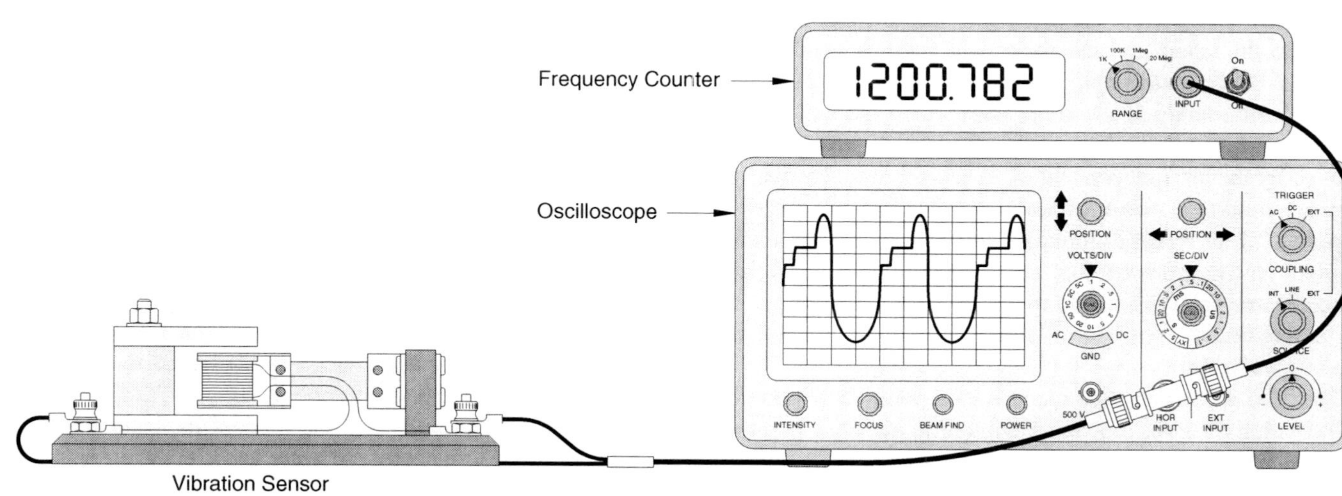

Figure 15-76 Vibration Analysis Setup

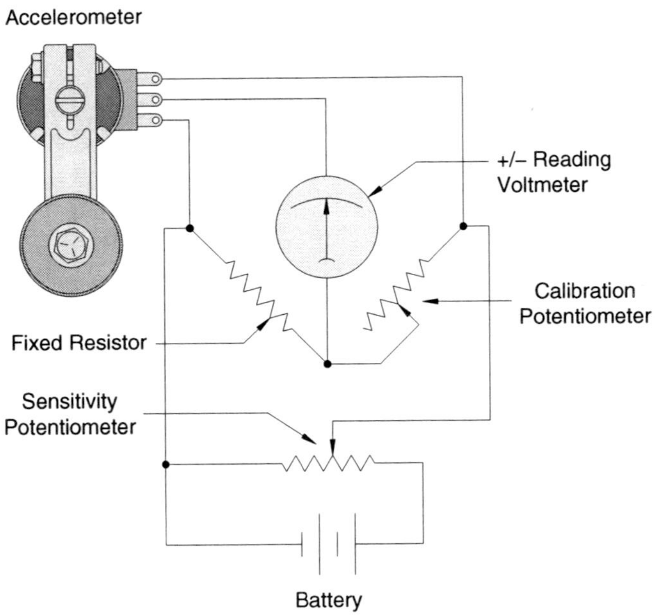

Figure 15-79 Accelerometer Schematic

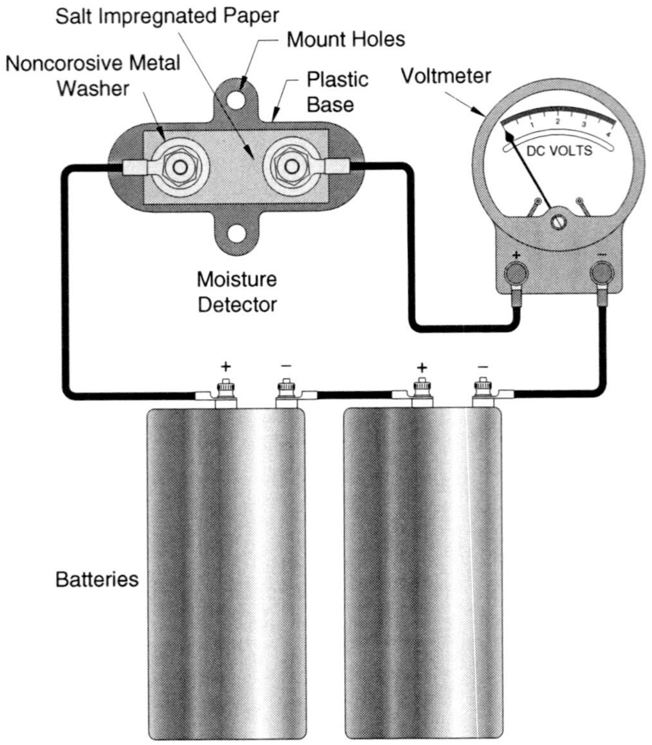

Figure 15-80 Salt Paper Moisture Detector

Moisture Detectors

There are three fundamental scenarios that call for the detection and monitoring of moisture, the presence of water, relative humidity, and dew point. Each of these arenas has specific applications in commercial and industrial equipment.

Detecting the presence of water is primarily used in leak detection. Normally, a detector is placed in the bottom of a drip pan beneath the equipment to be monitored. If a leak develops, the water is caught in the drip pan and the detector signals an alarm panel. These types of detectors are difficult to find and, oftentimes, more difficult to implement.

A simple moisture detector can be constructed by soaking a piece of paper in a salt-water solution. After the paper is thoroughly saturated, it is removed and dried under a low heat. The paper is then mounted to an insulating base with a terminal at either end of the paper. As long as the paper remains dry it will have an extremely high resistance, in effect, will be an open circuit. However, if the paper gets wet, the salt will dissolve into the water and the paper will become conductive. Figure 15-80 shows a simple salt paper moisture detector.

Relative humidity is the arena with which most of us are familiar. Relative humidity is the amount of water vapor retained by the air, independent of its temperature. Normally, relative humidity is measured with a hair hygrometer and displayed in "% of relative humidity." A humidity transducer is simply a hair hygrometer that has a potentiometer replacing its needle and scale, as shown in Figure 15-81. The output of the potentiometer can drive either a voltage divider or Wheatstone bridge.

The third water detection arena is dew point. Dew point is an indication of the actual water content in air. Dew point is the temperature at which the water content in air will start to condense. This parameter is measured with a dew point analyzer and is typically displayed in °F or °C.

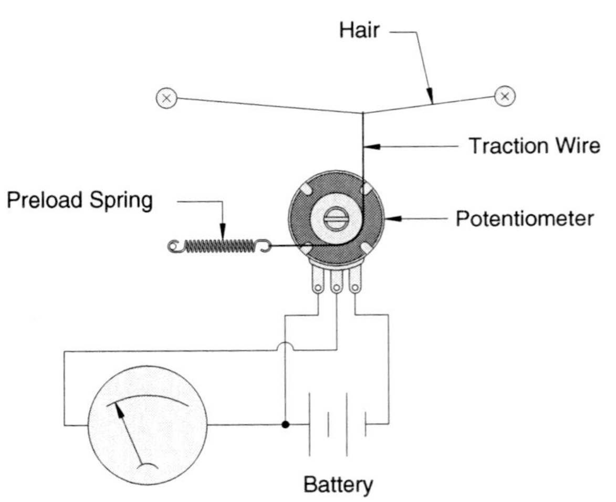

Figure 15-81 Hair Hygrometer Detector

Figure 15-82 shows a basic dew point schematic. A light source is projected through a glass plate at an angle. As the beam passes through the plate, it diffracts and projects through an optical slit onto an opto-sensor. The glass is equipped with a readout that monitors the temperature of the glass at all times. The glass is also equipped with a peltea device (solid-state heater), which is connected to a variable power supply. As the power supply is adjusted, the glass plate gets progressively cooler and the readout indicates the plate temperature. At some point the water vapor in the air will start to condense onto the glass plate, causing the refractive index to change. The angle of the refracted light beam changes and

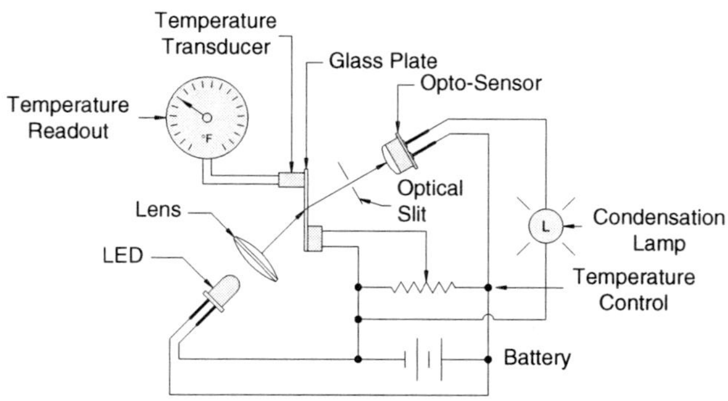

Figure 15-82 Dew Point Detector

moves off the opto-sensor. When this happens the condensation lamp turns off. The temperature indicated at the time that the lamp turns off is the dew point.

Viscometers

Viscosity is the term used to describe the thickness of a liquid. As an example, honey is a high viscosity liquid and alcohol is a low viscosity liquid. One of the accepted methods of measuring viscosity is to rotate a cylinder which is completely submerged in a liquid. By comparing the torque required to rotate the cylinder in the fluid being tested versus the torque required to rotate the same cylinder in water, a viscosity can be determined. Figure 15-83 shows a typical rotating viscometer setup. The torque is measured by gauging the current draw on the motor. The thicker the fluid, the higher the current reading.

A second method of measuring viscosity is the falling weight instrument, as shown in Figure 15-84. These instruments

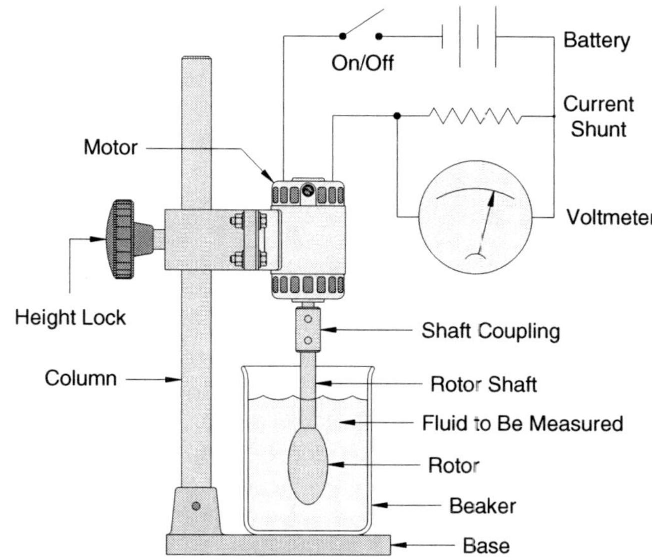

Figure 15-83 Rotating Viscometer

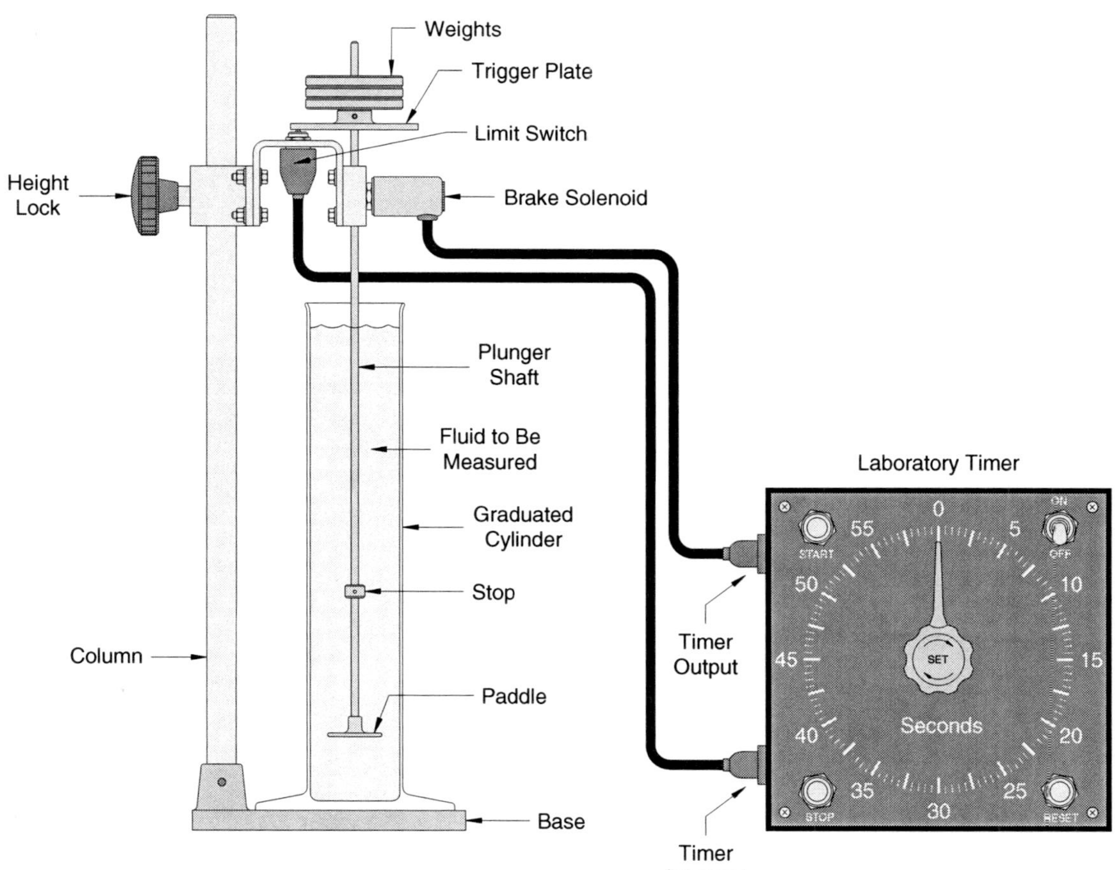

Figure 15-84 Falling Weight Viscometer

measure the time it takes for a known weight to fall through a column of liquid. The set up shown uses a graduated cylinder placed on an ordinary laboratory stand. A weighted rod with a paddle on one end is suspended in the center of the cylinder. The rod is pulled up and held in place by a solenoid-operated brake assembly. The solenoid and a limit switch communicate with an ordinary laboratory timer. When the start button on the timer is pressed, the solenoid releases the brake and the rod starts to fall at the same instant the timer starts. When the rod reaches the bottom, it trips the limit switch, which turns off the timer. By gauging the displayed time, the viscosity of the fluid can be determined.

Load Cells

Measuring force has far reaching applications, from gauging the load on a crane, to accurately applying a torque to a fastener. Until the advent of load cells, mechanical scales were the only real method for measuring force. Mechanical scales have been all but replaced by the used of load cells.

Figure 15-85 shows a typical load cell schematic. The load cell is some sort of frame that mounts a strain gauge element in a fashion that allows a micro amount of flexure if a load is placed on the device. In the case of the illustration, as a pulling force is applied to the length of the frame, it has a tendency to stretch. The element is placed at a bridging point of an asymmetric cutout, which is designed to introduce a shear load to the gauge. As the element changes shape, its resistance changes in direct proportion to the amount of force being applied. The gauge is set up as one element of a Wheatstone bridge and the voltmeter is calibrated in pounds of force.

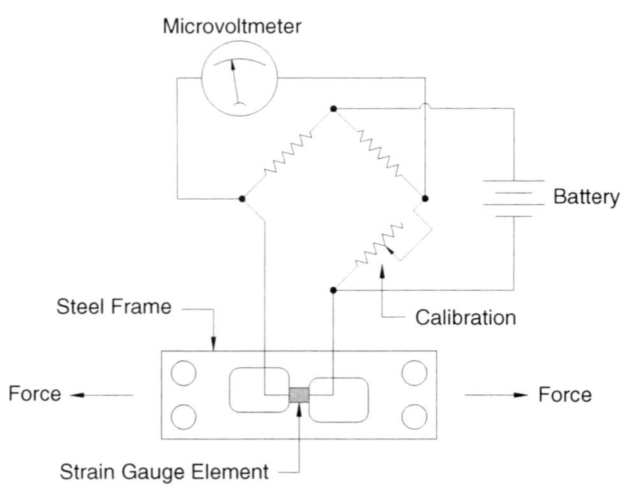

Figure 15-85 Tensile Load Cell

Reading forces in a bidirectional application is done with an S-type load cell, as shown in Figure 15-86. These units are generally inexpensive and provide exceptional performance.

Washer-type load cells, as shown in Figure 15-87, are used to accurately gauge forces that are placed on rods and fasteners. In applications that require critical clamping forces, gauging

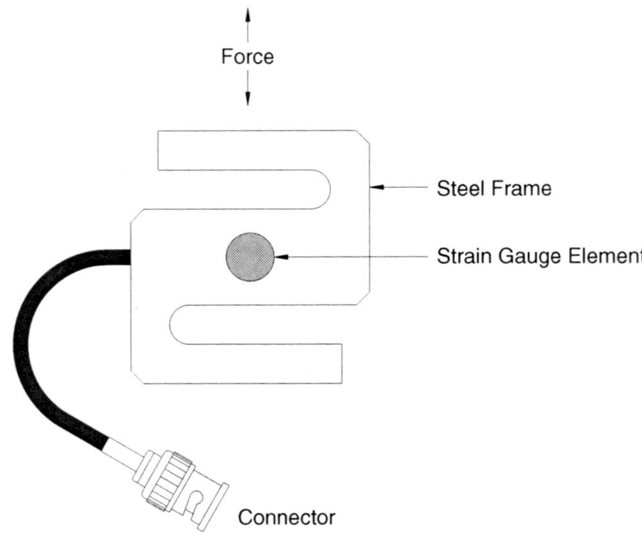

Figure 15-86 S-Type Load Cell

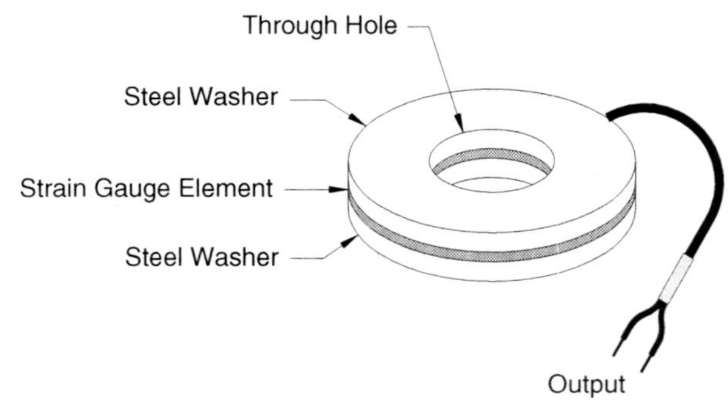

Figure 15-87 Washer-Type Load Cell

the torque of a fastener may not be adequate. In these cases, a washer load cell can be deployed to gauge the actual clamping force, independent of torque.

Chip Detector

In engines and power transmission equipment, the accumulation of small microchips is a good indicator that regular service intervals should be observed. The chip detector, as shown in Figure 15-88 is a permanent magnet that is straddled with two

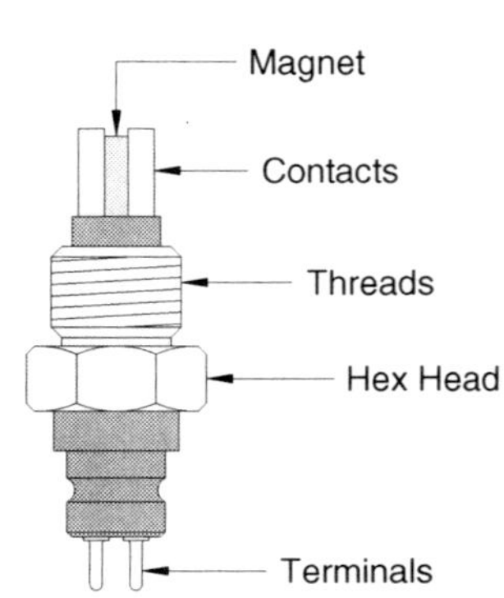

Figure 15-88 Chip Detector

contacts. The detector is located in the bottom of the equipment's oil sump where it slowly collects ferrous particles. When enough particles have been collected, the contacts become conductive and alert the operator that a critical condition exists.

Light Spectrometer

A simple light spectrometer can be configured by using an array of opto-sensors and an ordinary prism, as shown in Figure 15-89. A senor like this can be used to gauge the performance of all sorts of light sources. As an example, the light emitted by an ordinary spark plug can indicate surface condition, gap, contamination, voltage, and the like. Oil can be gauged by creating a spark through a film of the oil and studying the light spectrum that it produces.

Filter Clog Indicator Switch

One method to detect a clogged filter is shown in Figure 15-90. Most oil filter systems operate on suction with the oil drawn up and through the filter element. In the event the filter becomes clogged, a bypass loop is generally incorporated into the design. The idea being that it is better to get unfiltered oil than no oil at all. The bypass loop is usually a spring-loaded disk valve that will open if the suction becomes too great. By extending the disk valve shaft out of the bypass loop, a limit switch can be activated. If the disk valve opens, then the limit switch closes and the indicator lamp turns on.

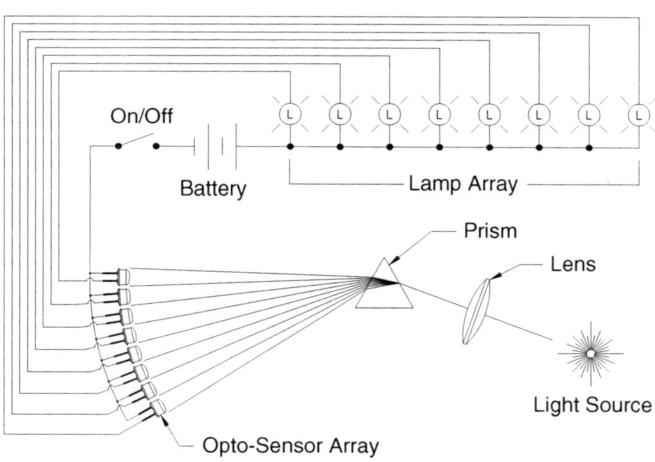

Figure 15-89 Light Spectrometer

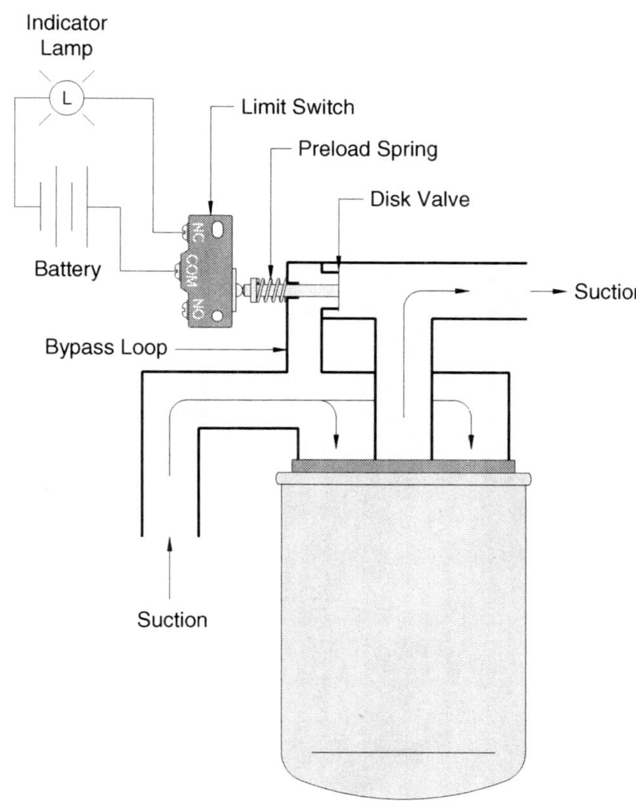

Figure 15-90 Filter Clog Indicator Switch

CHAPTER 16
ELECTROSTATICS

Simply stated, electrostatics is the study of electrical energy at rest. Any time a conductor is at a potential, it exhibits a natural tendency to equalize the charge state with an opposite charge. Similarly, the conductor exhibits a natural tendency to repel like charge states.

Electrostatic Voltmeters

As an example, let's review the operation of the electroscope shown in Figure 16-1. This device is intended to detect either positive or negative charge states. The instrument is simply a glass jar with a piece of folded foil hanging from a metal frame, which is connected to a terminal. When a charge is applied to the terminal, the entire metal frame and both tabs of the foil adopt the same charge potential. The hanging tabs, having the same charge potential, repel each other and the tab's visible deflection is in direct proportion to the voltage applied. Electroscopes are very simple devices to build and provide an excellent introduction to the study of electrostatics.

As an example of the attractive tendencies of unlike charge states, we can review the electrostatic voltmeter shown in Figure 16-2. In this case, a high-voltage potential is applied to a fixed plate, while a moving plate, which is suspended above, is connected to ground. If the potential applied to the fixed plate is positive, it attracts the moving plate and draws it down in direct proportion to the voltage applied. Similarly, if a negative charge is applied to the fixed plate, the moving plate will be repelled up in direct proportion to the voltage applied. In this manner the meter can be configured to be a +/− indicating instrument.

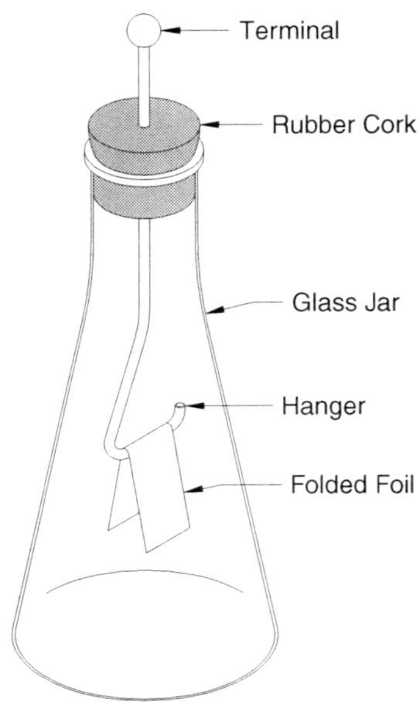

Figure 16-1 Electroscope

Electrostatic Air Filtration

Probably the most mainstream use of electrostatics is in air filtration. Figure 16-3 shows a schematic representation of a typical electrostatic air filter. These appliances are available in sizes ranging from small personal units through large units intended to filter air in expansive arenas and auditoriums.

These units operate by creating a flow of electrons across a flow of air. As the dust particles pass through the electron cloud, they build a charge state. The charged dust particles

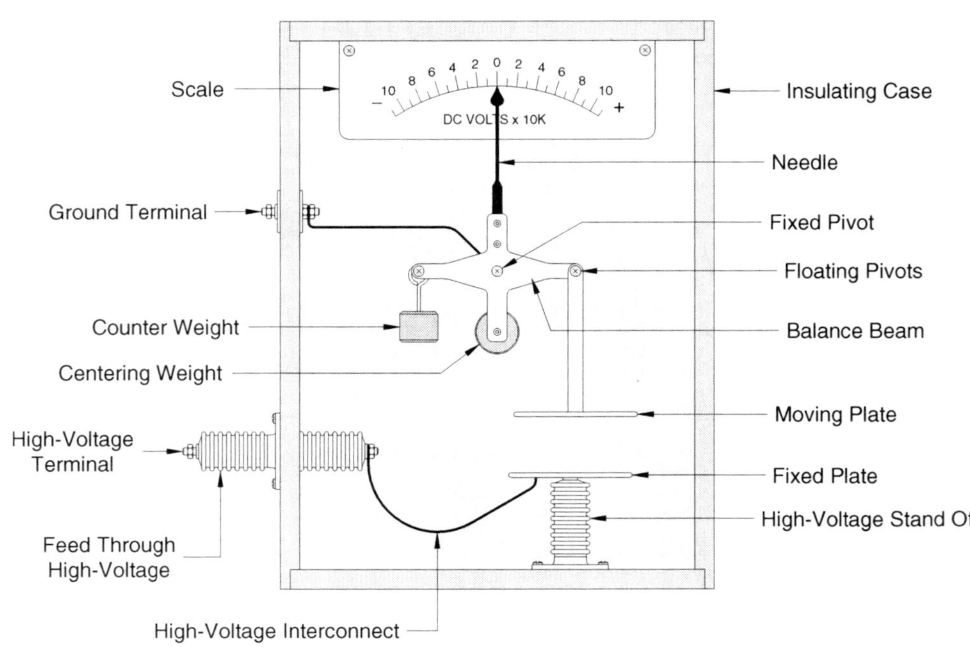

Figure 16-2 100,000-Volt Electrostatic Voltmeter

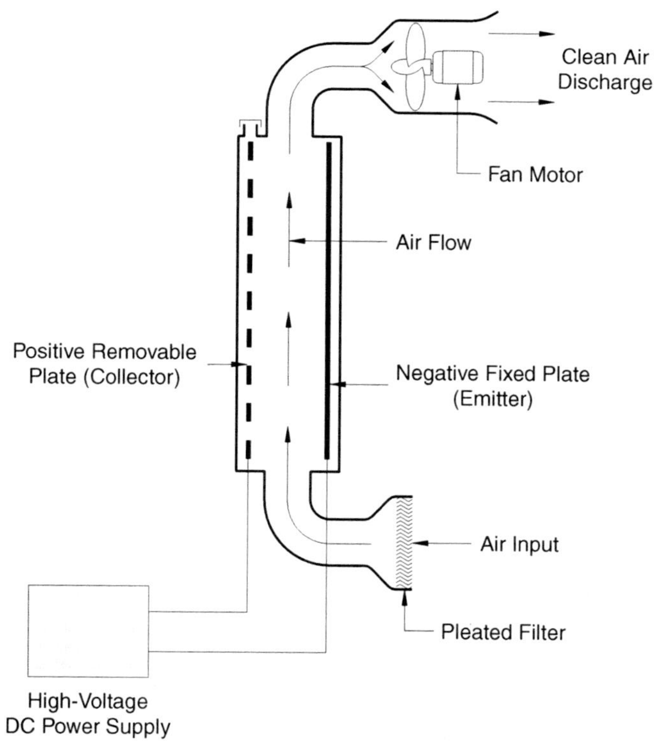

Figure 16-3 Electrostatic Ionizing Air Filter Schematic

are then electrostatically attracted to the collector plate. The collector plate is removable so that it can be periodically cleaned.

Electrostatic Loudspeakers

Figure 16-4 shows a schematic representation of an electrostatic loudspeaker system. The diaphragm requires a quality, high-voltage power supply and the input signal is fed through a step-up transformer. As a signal is applied to the input of the transformer, the attractive and repulsive forces allow the diaphragm to vibrate in direct relation to the input signal. This, in turn, generates sound.

These speakers offer good efficiency and excellent sound quality but because of their support equipment and internal voltages, they are typically rather expensive. They are generally used only in high-performance applications such as home or studio applications. One application where this technology performs exceptionally well is high-performance headphones. The headphones are very lightweight and can enclose the entire ear, produceing extremely high quality sound.

High-Voltage Isolation

Generally, when dealing with electrostatics, you are dealing with high voltages. Low voltages simply do not provide enough potential to be useful. It is important to respect the voltages involved and to provide proper isolation between the energized equipment and its operators. Generally speaking, the cat doesn't want to be shocked every time it brushes up against your electrostatic air filter. In a more serious light, most electrostatic equipment carries extremely high voltages and if isolation is not designed and maintained properly the result can be serious injury or death.

In one instance that I am aware of, a technician was issued a high-voltage probe so that he could troubleshoot the power supplies of a large vacuum deposition system. It turned out that the probe had an internal crack through the insulator, which protects the operator from high voltage. When the technician touched the first power supply output with the probe, he was immediately electrocuted. It seems that on the previous day another technician had dropped the probe, and rather than discarding it or turning it on for inspection and requalification,

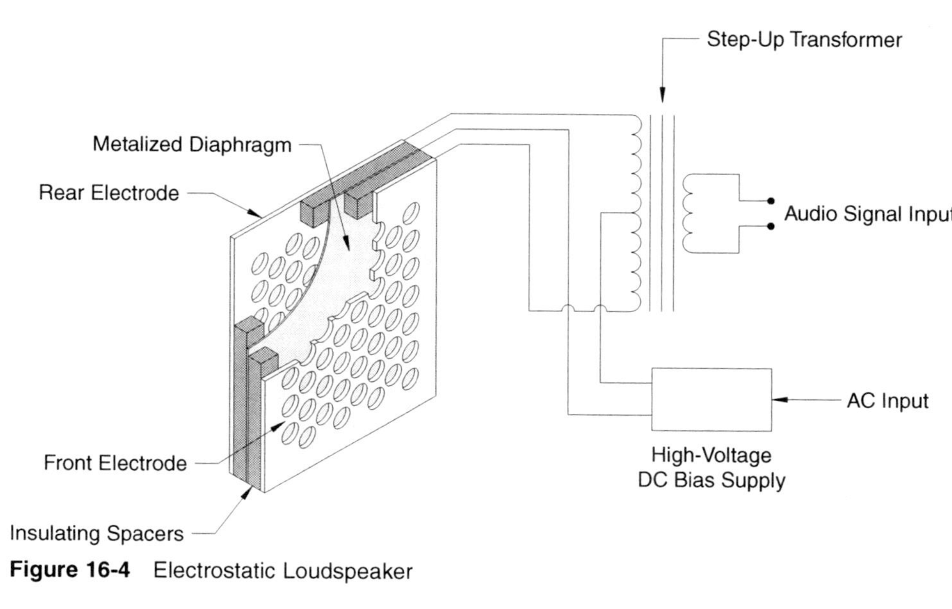

Figure 16-4 Electrostatic Loudspeaker

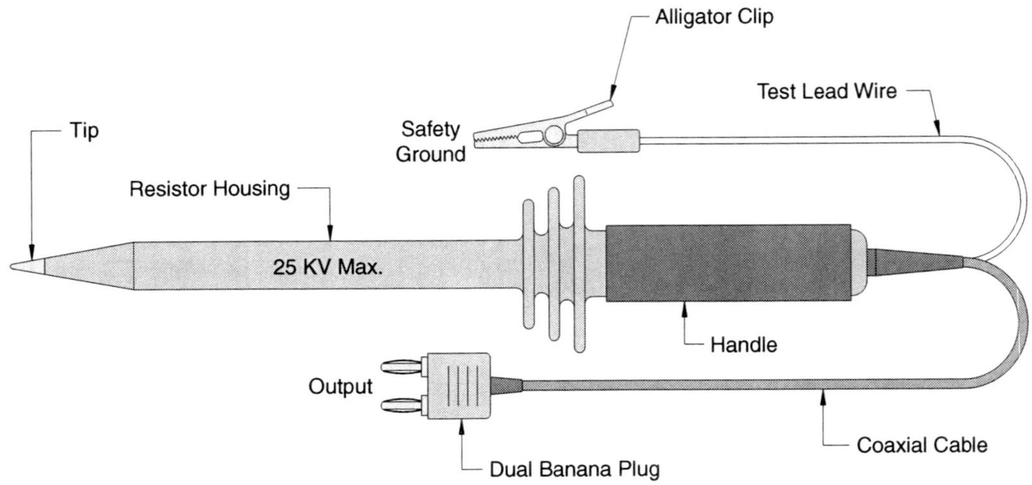

Figure 16-5 High-Voltage Probe

he simply returned it to the instrument room. To make matters worse, the technician who was killed was working alone. Had he followed proper safety procedures, he would have had a safety technician present who could have applied cardiopulmonary resuscitation (CPR) and, most likely, saved his life. Figure 16-5 shows a typical high-voltage probe. These devices should be cleaned, inspected, and qualified on a regularly scheduled basis. If the probe receives any damage whatsoever, it should be immediately taken out of service. Never use a high-voltage probe without connecting its safety ground clip to a known ground.

Isolating high voltages is principally a function of distance. Any material or media has a distance or thickness per volt that it will stand off. As an example, the general rule of thumb is that air will stand off 20,000 volts per inch. Many factors, such as humidity, airborne dust, magnetic fields, and the like, can affect this figure.

Controls are typically isolated by using plastic drive or push shafts, as shown in Figure 16-6. The shaft should be sufficiently long to prevent any chance of arcing along its length. The control panel should be constructed from a grounded piece of sheet metal with metal shaft collars. In this manner,

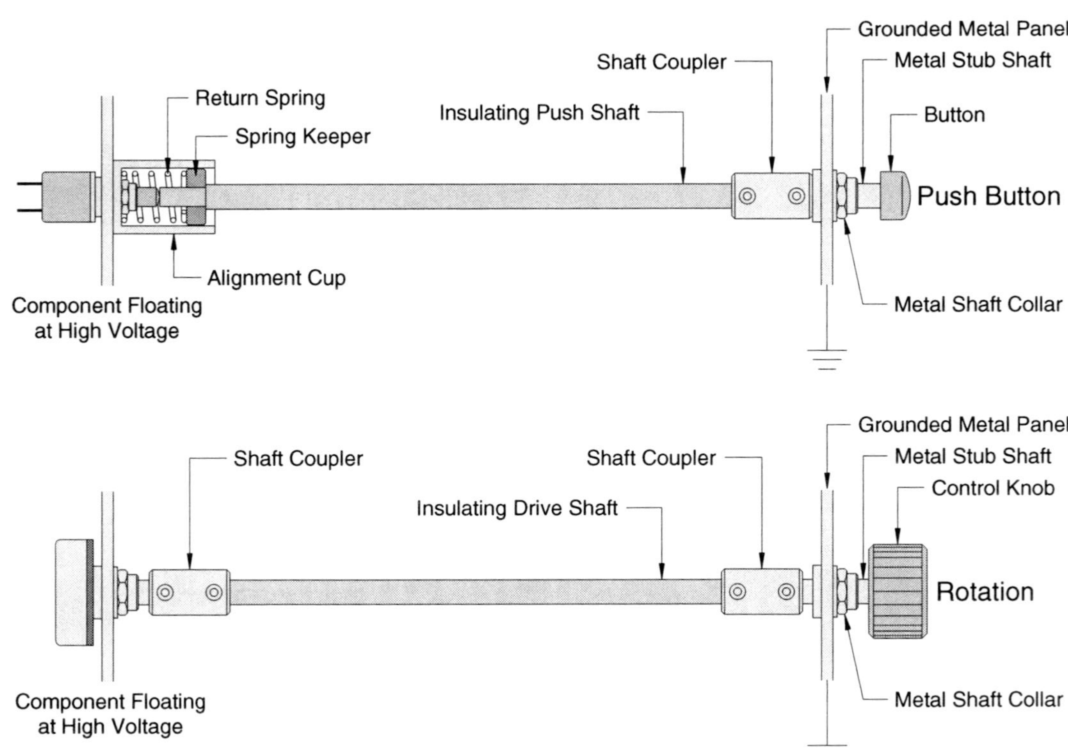

Figure 16-6 High-Voltage Isolation Methods

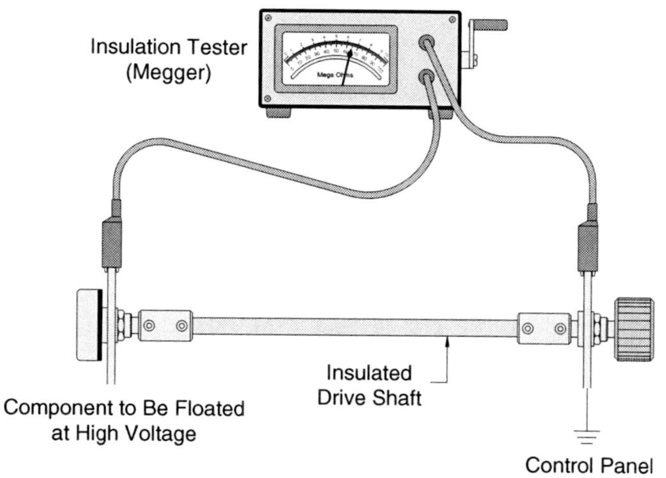

Figure 16-7 Verifying High-Voltage Isolation with a Megger

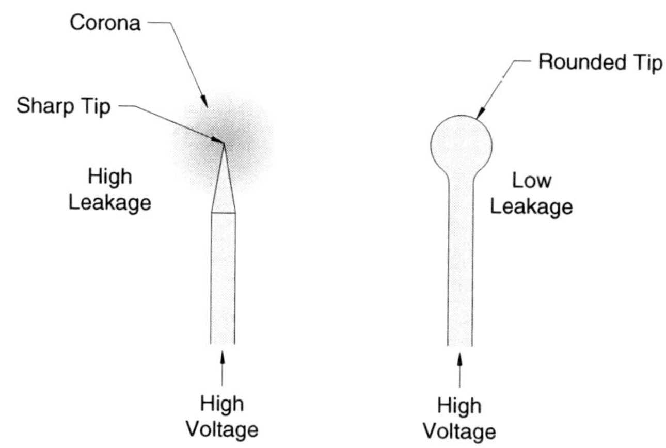

Figure 16-8 Corona Discharge

any high voltage that may leak to the panel will be immediately grounded and, therefore, the operator will be protected.

Testing the stand-off voltage capability of a design can be accomplished by using an insulation tester, or megger, as shown in Figure 16-7. These tests should be carried out only while the equipment is de-energized. Additionally, any high-voltage isolation system should undergo regularly scheduled inspections.

Corona Discharge and High-Voltage Leaks

Corona discharge happens when a conductor is forced to carry more electrons than it can accommodate. In these instances the electrons jump into the air where they flow to any neutralizing charge that they may find. Near the conductor the electrons have sufficient energies to ionize the air and force it to glow, or in other terms produce a corona.

The most common reason for electron concentration is sharp points on the conductor. The electrons concentrate at the diminishing point and eventually are forced off the conductor. As an example, the sharp tip of the conductor on the left hand side of Figure 16-8 will have a great deal of leakage at high voltage and will, most likely, glow with a low purple light. On the other hand, the conductor on the right, with its rounded tip, will have very little leakage and will stand off much higher voltages.

Solder joints are always a trouble spot in high-voltage circuits. In one instance I designed and constructed a 10,000-volt DC power supply for an industrial flash unit. When I energized the circuit, it was outputting only 2000 volts. As I probed the circuit I noticed that the voltage got progressively lower as I worked through from the transformer secondary to the output of the supply. This was puzzling. Without any other ideas, I turned off the lights and noticed that the entire circuit was glowing brightly from numerous corona discharge sites. I turned on the lights, de-energized the circuit and carefully applied high-voltage silicone rubber to all of the soldered and bolted joints. The next morning I re-energized the circuit and it produced its full 10,000-volt output. The solder and bolted joints were leaking 80% of the power supply's voltage to air! Figure 16-9 shows an example of poor high-voltage solder joint and the use of high-voltage silicon rubber to suppress leakage.

The slow build up of dirt can also produce high-voltage leaks. This is particularly problematic with equipment that is forced to operate in dirty and dusty environments. High-performance cabinet filtration will help mitigate the effects of dirt built-up. Figure 16-10 shows how an exposed stand off insulator can be protected from dirt build-up by using a protective boot. In this case, the leakage would be from the terminals to the grounded case. If the case is floating, the leakage would be between the terminals forming a bypass of the component.

Carbon paths, as shown in Figure 16-11, are an insidious problem that will build over time. There are two principal ways that carbon paths can develop. The first being during

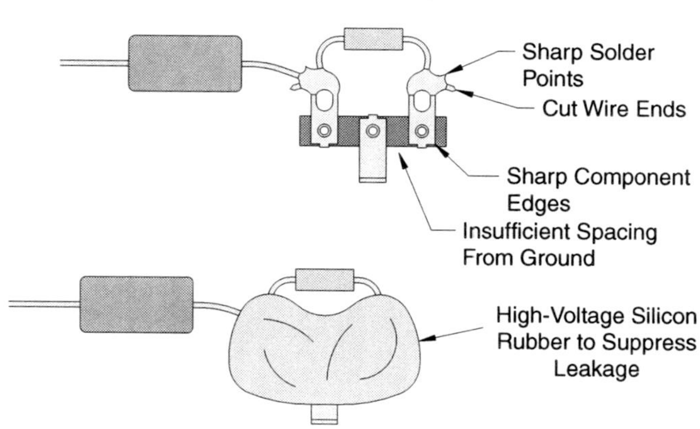

Figure 16-9 Suppressing Corona Discharge

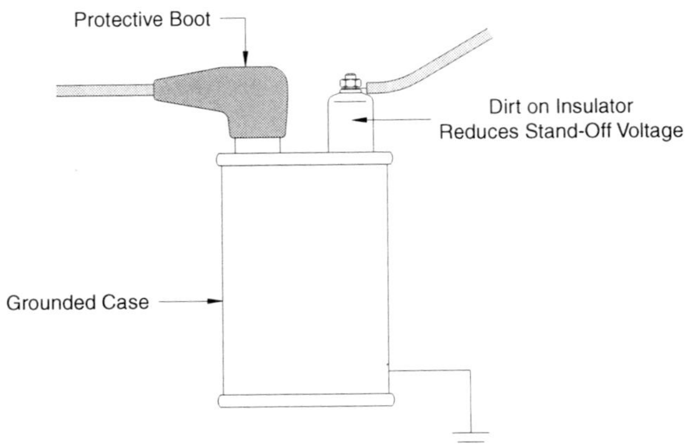

Figure 16-10 Leakage Due to Dirt

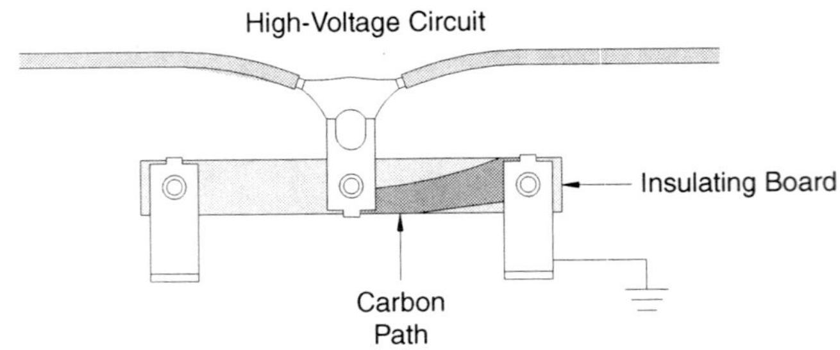

Figure 16-11 Leakage Due to Carbon Paths

uncontrolled arcing. The arc produces a great deal of heat and burns the surface of the insulator, forming carbon. If the arc has enough energy and the insulator is sufficiently burned, a permanent short can be produced. Glass or ceramic insulators with a glazed surface resist this type of damage, while phonolic and plastics are particularly susceptible.

The second process that forms carbon paths is the slow carbonizing of dirt deposits. As dirt builds on the insulator, small micro arcs accrue and produce localized carbon sites. As time progresses, the carbon builds until its leak rate is detrimental to the circuit's continued operation.

CHAPTER 17
ELECTROMECHANICAL MECHANISMS

Electromechanical mechanisms can be extremely complex assemblies. Consider an automobile, a clothes washer, your computer printer, or the air conditioner, all are just big electromechanical components.

This chapter of the book is intended to expose the reader to a few miscellaneous electromechanical components and assemblies that haven't been reviewed in the previous chapters.

Solenoid Door Latch

Figure 17-1 shows a simple solenoid-activated door latch. The bolt is spring-loaded and interfaces with a striker, so the system will automatically latch when the door is closed. To unlock the mechanism, the solenoid is energized and the plunger toggles the link, which, in turn, pulls the bolt back.

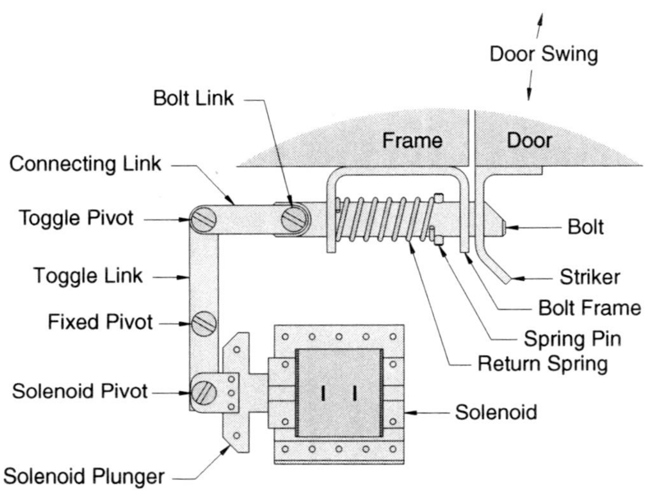

Figure 17-1 Solenoid Latch

Hinge Cable

Electrically bridging a hinged assembly is a simple matter that seems to give a lot of people trouble. Simply anchor a cable loop, as shown in Figure 17-2 between two screw-on blocks. It is important to allow enough wire in the loop to accommodate the throw of the door.

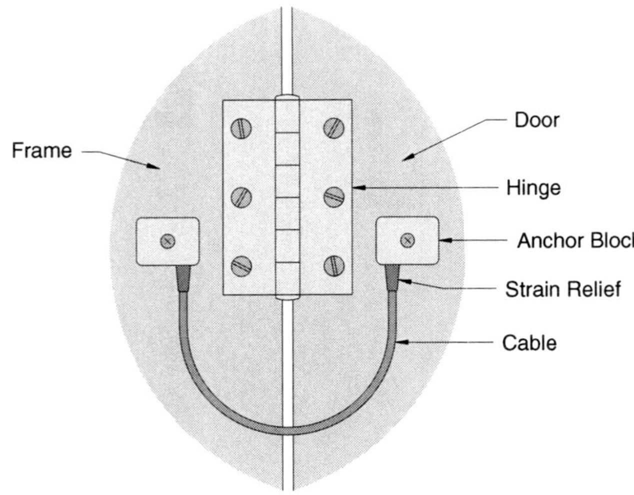

Figure 17-2 Hinge Cable

Explosive Bolts

Explosive bolts are used in any application where an emergency or rapid release of a bolted component is necessary. Military aircraft use explosive bolts to release the canopy as part of a controlled sequence just prior to pilot ejection. Remote piloted deep submersibles use explosive bolts to attach their ballast. If the control umbilical fails or is severed, the bolts fire and drop the ballast. The vehicle floats back to the surface where it can be recovered and repaired.

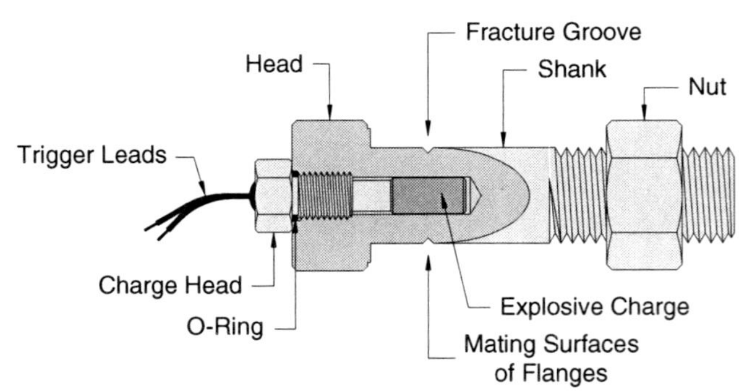

Figure 17-3 Explosive Bolt

Figure 17-3 shows a typical explosive bolt design. A cap screw is drilled and tapped to accept an electrically activated charge. A fracture groove is cut into the shank of the bolt. The location of the groove corresponds to the mating surfaces of the components to be bolted. This assures a proper release and minimizes the chance of jamming. The charge is held in place with a sealed bolt. The charge must be serviceable, otherwise the entire fastener will have to be changed out at the end of its service life. It should also be noted that the ends of an explosive bolt should be retained during detonation. The end pieces can blow out with a great deal of force and can seriously damage any equipment that they may impact.

Traction Elevator

Most of us have ridden in an elevator from time-to-time. Almost all passenger elevators are of the traction design, as shown in Figure 17-4. In these cases, a traction is powered with a simple transmission system. The car cable is looped over the top of the spool and carries a counter weight on the opposite end. The automatic controls are mounted in the car and provide an intuitive interface that can be operated even by

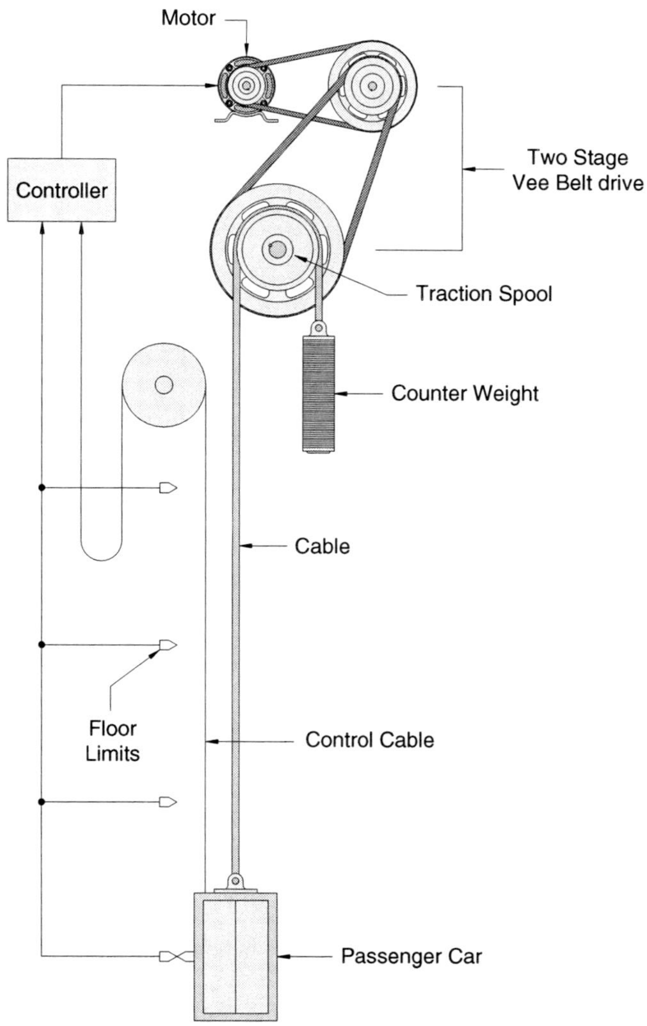

Figure 17-4 Basic Traction Elevator

someone who has never been in an elevator before. The floor locations are detected with ordinary limit switches or opto sets. The controller is generally mounted adjacent to the lifting machinery and a control cable is routed to the car.

Dash Pots

Dash pots are mechanical shock absorbers that are intended to smooth out the actions of a sensor or drive. Figure 17-5 shows two common dash pots—a hydraulic unit and a pneumatic unit.

Pneumatic units are generally used for low-load applications, such as damping the motion of a turn table tone arm or filtering out high-frequency signals on a vibration sensor. These units typically consist of a small cylinder with a loosely fitting piston and rod, as shown in the upper illustration (A). Air is allowed to leak between the gap formed around the outside of the piston and the inside diameter of the cylinder. At low speeds the flow rate through this gap is sufficient to allow

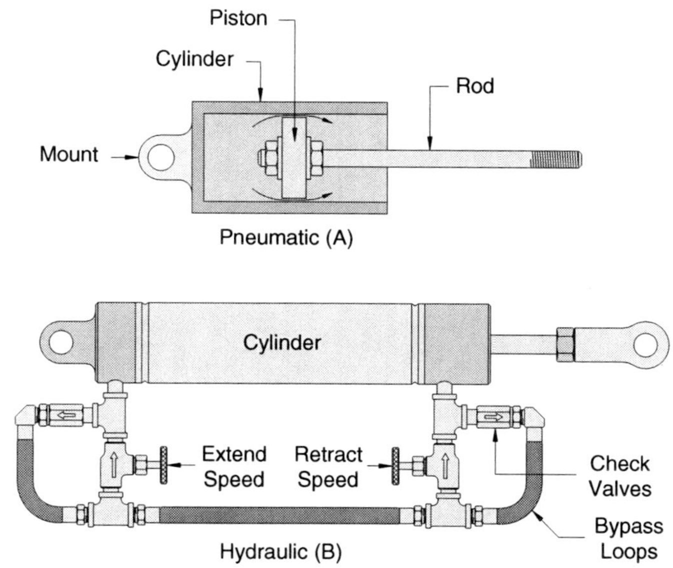

Figure 17-5 Dash Pot Shock Absorbers

the piston and rod assembly to move unimpeded. At higher speeds the gap restricts the flow and, in turn, places a load on the motion of the piston and rod.

Hydraulic dash pots operate in much the same manner as their pneumatic counterparts, except the flow is controlled through a bypass loop, as shown in the lower illustration (B). The bypass loop can be set up with a pair of needle valves and check valves, which allows the damping characteristics of both the extend and the retract to be tuned independently.

Figure 17-6 shows a pneumatic dash pot used to dampen the motion of a pendulum accelerometer. The dash pot will limit sudden impulse loads, while allowing long duration loads to be monitored.

Figure 17-7 shows a hydraulic dash pot used to limit the speed at which a solenoid-activated knife switch throws.

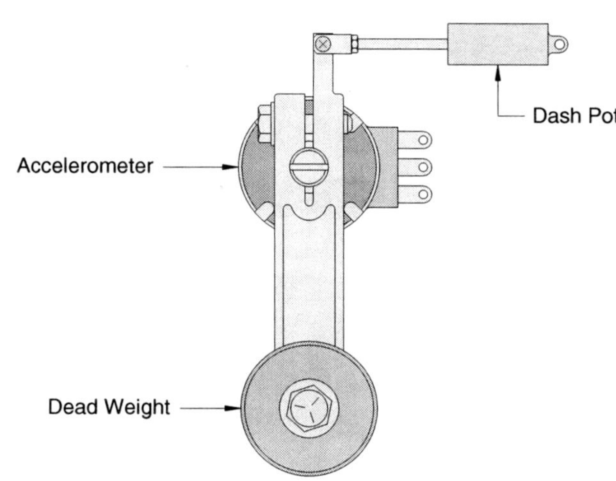

Figure 17-6 Accelerometer Equipped with a Dash Pot

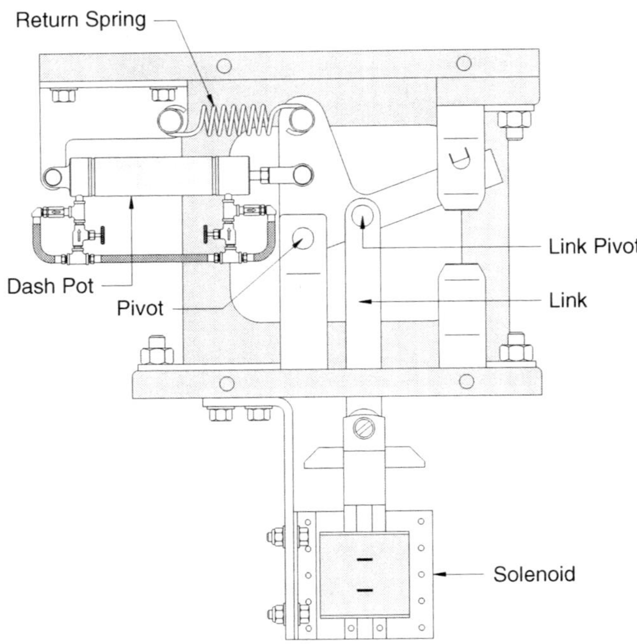

Figure 17-7 Powered Knife Switch with Hydraulic Dash Pot

Spark Plugs

Spark plugs are simply a pressure feedthrough that is configured for a special purpose. These devices are excellent high-pressure, high-voltage feedthroughs that can be used in all sorts of equipment. The electrical terminal is simple, reliable, and can comfortably handle voltages as high as 40,000 volts. When using a spark plug as a feedthrough, it is important to select a plug without an internal resistor, as shown in Figure 17-8. These units generally have an "R" in the code printed on the insulator.

Dynamic Braking

A permanent magnet or shunt wound DC motor can be used as a brake in certain applications. The idea is that the motor is

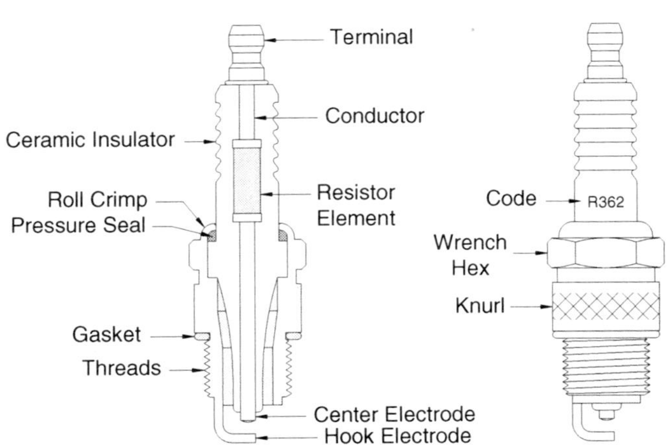

Figure 17-8 Spark Plug

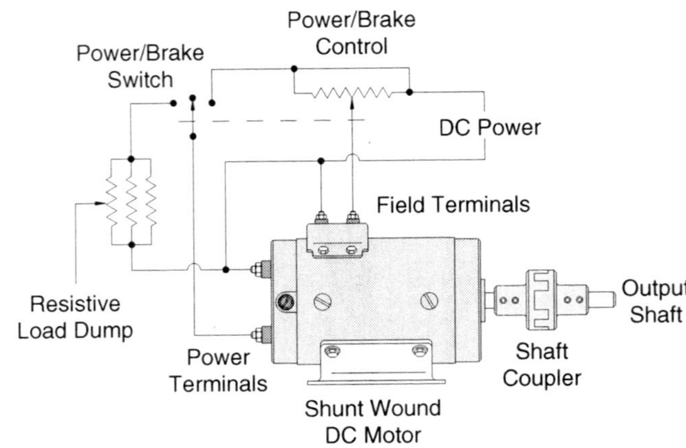

Figure 17-9 Dynamic Braking Schematic

allowed to act as a generator and the power is dumped into a set of high-capacity resistors.

Figure 17-9 shows a dynamic braking system with a shunt wound DC motor. The field current is controlled by the Power/Brake control rheostat. The operation of the motor (run or brake) is controlled by the Power/Brake switch. In the power mode, the motor is fed DC power and operated as a normal electric motor. The speed of the motor is controlled by adjusting the field current. In the brake mode, the motor is disconnected from the DC power and is connected to a resistive load dump. During this time, the spinning motor acts as a generator and the rotational energy that is being introduced into the output shaft is removed in the form of heat. The braking effect can be controlled by adjusting the field current. By integrating the switch and rheostat into a common assembly, a single lever throttle/brake control can be configured.

Three Door Bell System

Figure 17-10 shows how to wire a three door bell system using a bell, buzzer, two single-pole buttons, and a double-pole button. The bell is used for the primary door (front) and the buzzer is used for the secondary door (back). The double-pole button is mounted on the third door (side) and is wired to operate both the bell and the buzzer simultaneously. The power supply is a transformer with a 120-volt primary and an 18-volt secondary.

Utility Transformer

In many situations it is advantageous to have a 120-VAC receptacle adjacent to a major equipment installation. A 120-volt utility receptacle allows maintenance equipment and powered hand tools to be used without the hassle of running several hundred feet of extension cord. However, capital equipment is generally wired with a service that does not offer this utility voltage (240 volt, delta three phase or 480-volt

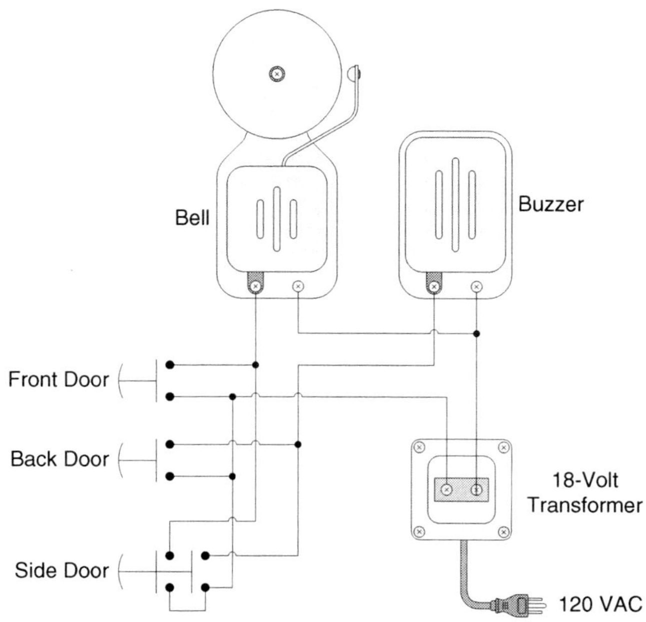

Figure 17-10 Three Door Bell System

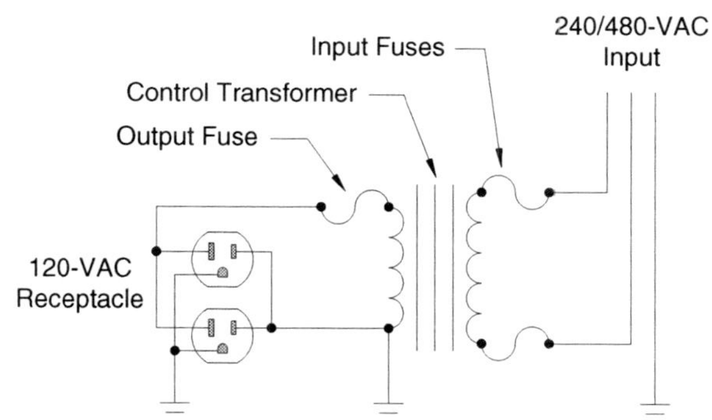

Figure 17-12 120-VAC Utility Transformer Schematic

three phase). In these cases, a simple utility transformer can be configured, as shown in Figure 17-11. A suitable control transformer is selected and mounted in a NEMA (National Electric Manufacturers Association) cabinet along with a 120-VAC receptacle. Control transformers are readily available with dual-voltage inputs and integral fuse sets. The cabinet can be mounted directly onto or adjacent to the power disconnect that services the equipment.

Figure 17-12 shows a schematic representation of the utility control transformer. It is important to use both input and output fuses , as shown.

String Drives

Figure 17-13 shows a typical arrangement used in radio receivers to adjust the frequency with a variable capacitor. A string is wrapped around a small capstan mounted on the back of a knob. The string is routed around an idler and the large tuner pulley. Over the length of the string, a pointer is mounted to indicate the relative frequency on the scale. Although these types of drives are most commonly found on radios, they are applicable to a variety of other applications.

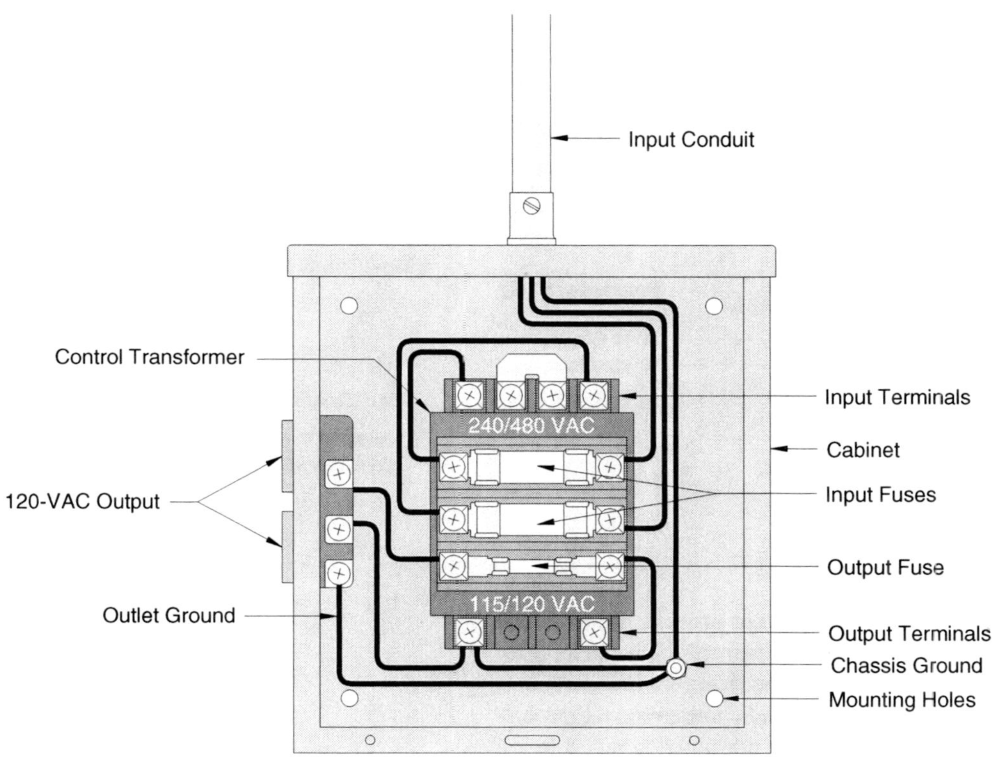

Figure 17-11 120-VAC Utility Transformer

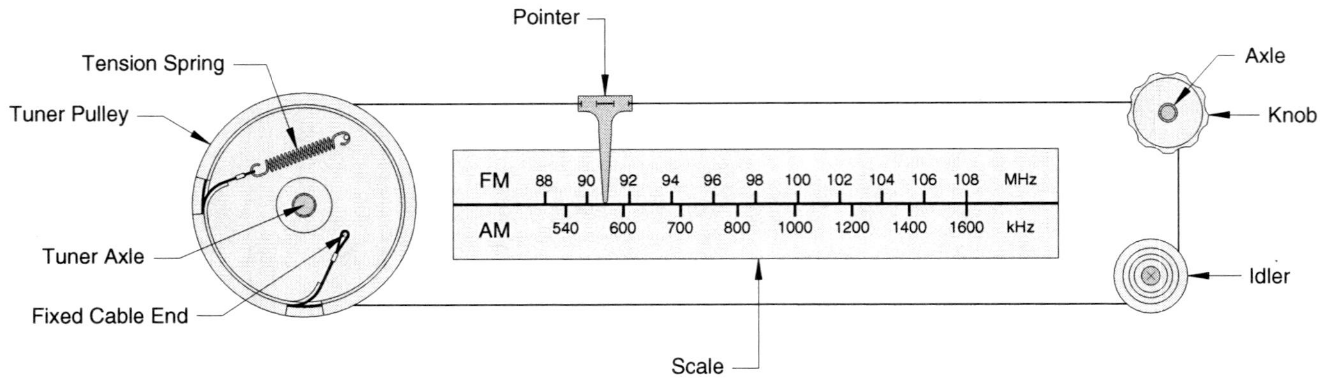

Figure 17-13 String Tuner Drive

When the variable capacitor is replaced with a potentiometer, the scale can indicate voltage, resistance, volume, balance, and the like.

Motorized Locking Systems

For high security systems, large pins or bolts are commonly used to lock a heavy door in place. Figure 17-14 shows a worm drive locking system with four bolts. When the door is closed, the motor is activated and the driven gear forces the bolts out into a corresponding frame. When the motor polarity is reversed, the bolts are retracted back into the door. In this manner a relatively small gear motor can be used to lock a rather substantial door.

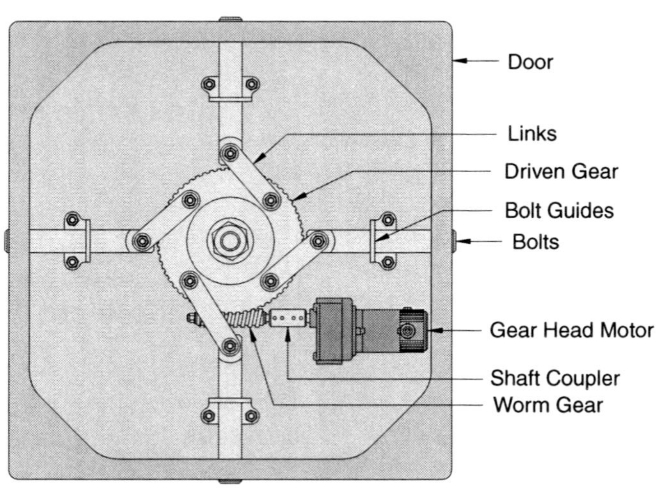

Figure 17-14 Motorized Locking Pins

Air Compressor Control

A typical reciprocating air compressor provides an excellent example of how simple it is to control high-horsepower motors with relatively low power, and therefore low cost components. Figure 17-15 shows a typical commercial reciprocating air compressor. These units are normally supplied in the 7.5- through 30-horsepower range. They turn on when the air

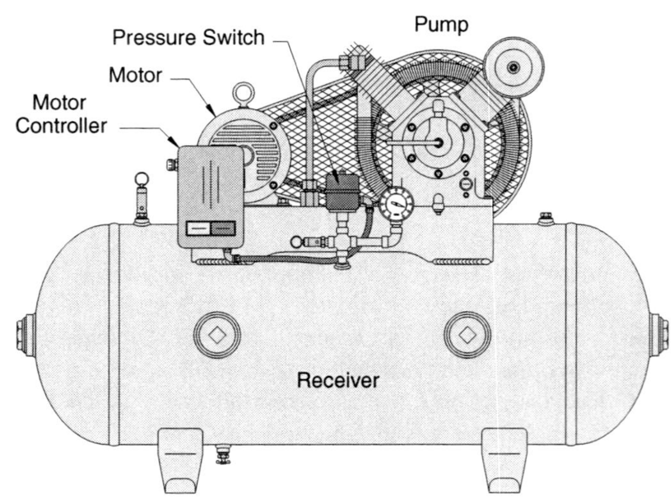

Figure 17-15 Packaged Air Compressor

pressure in the receiver is below a preset lower limit and turn off when the receiver pressure reaches a preset upper limit.

Figure 17-16 shows the electrical schematic for the compressor. The motor is connected to the power source through a motor controller with a set of overload heaters. The coil is controlled with an upper/lower limit pressure switch. The control circuit is normally operated from a 120-VAC control transformer, as shown.

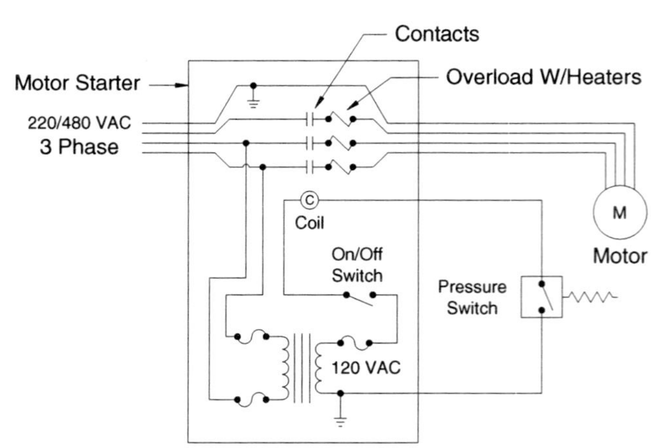

Figure 17-16 Packaged Air Compressor Schematic

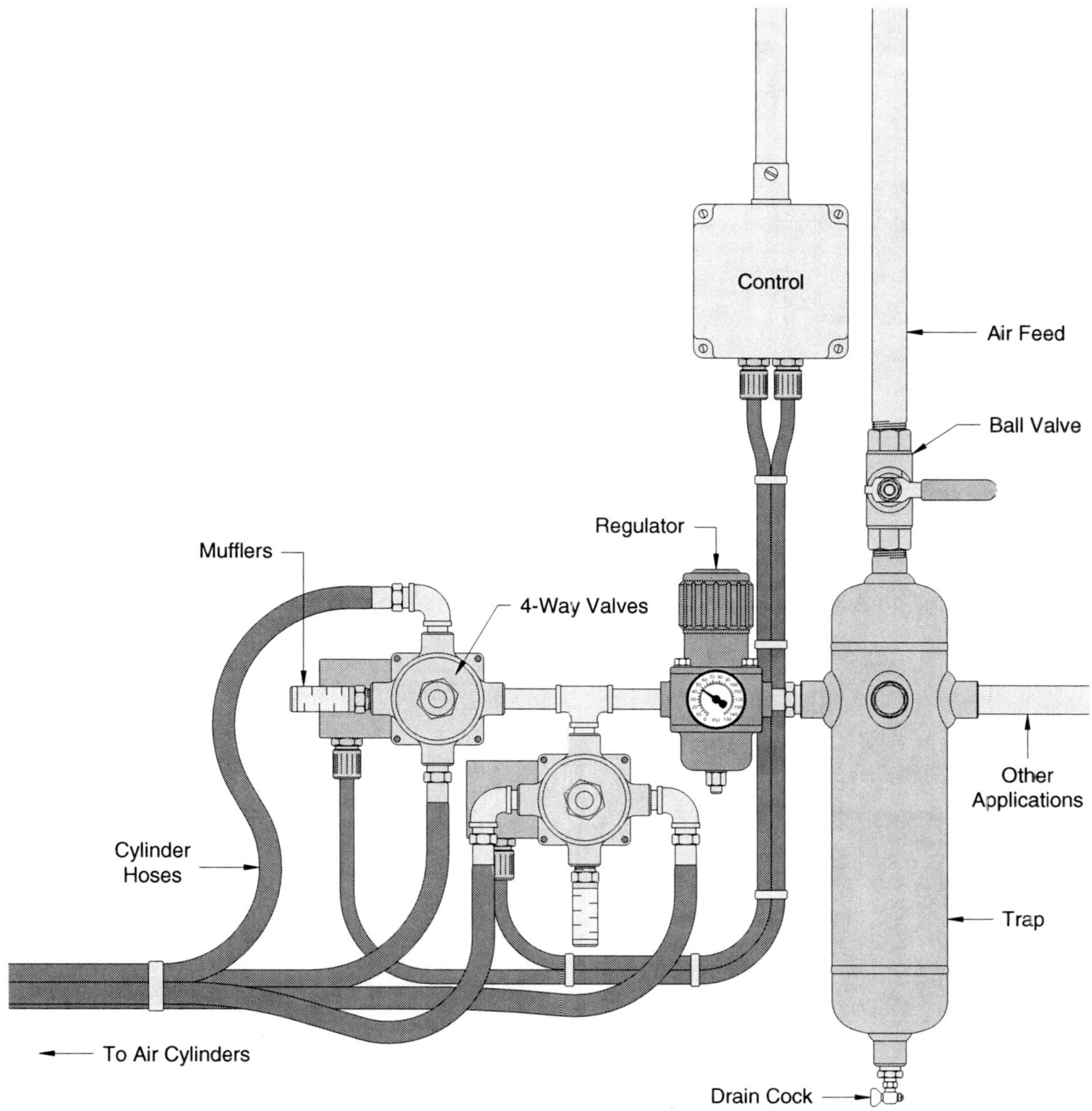

Figure 17-17 Pneumatic Control Station

Pneumatic Control Stations

Figure 17-17 shows a pneumatic control station configured to control the positions of two air cylinders on a piece of nearby equipment. A pair of four-way, venting solenoid valves is mounted to the output of a pressure regulator. The solenoids receive their signals from a plant-wide control loop.

Fuel Injector Nozzles

Virtually all modern automobiles use electronically timed fuel injection. Any other fuel induction method simply won't meet the stringent pollution standards that are called for by our government. The modern fuel injection system centers around a set of valved injector nozzles, as shown in Figure 17-18. A nozzle is mounted into each intake port on an engine. The valves are opened and closed via a signal provided from a central computerized controller.

The nozzle itself consists of a poppet valve that is controlled by an electrical pulse. The fuel flows through the center of the poppet and is stopped at the valve seat. When the coil receives a pulse, the poppet raises and the fuel is allowed to spray into the port. The amount of fuel that flows is controlled by the duration of time that the valve is energized.

Spot Welders

Spot welders join metals by introducing a high-energy electrical pulse into a confined area. The amount of energy is high enough to melt and fuse the base metals, forming a single piece. Figure 17-19 shows a typical spot welding circuit. To accomplish a weld, two pieces of sheet metal are pinched

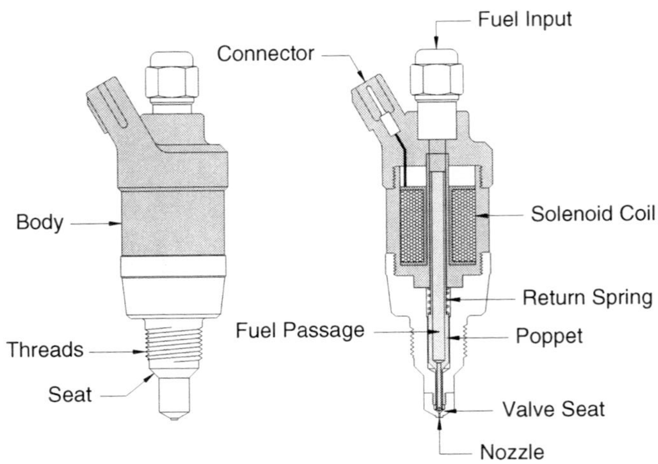

Figure 17-18 Electronic Fuel Injection Nozzle

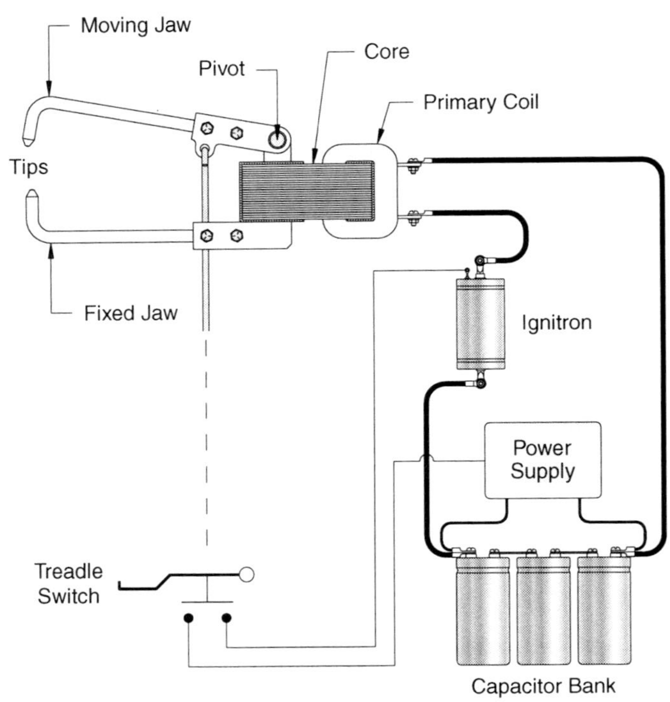

Figure 17-19 Spot Welder Circuit

between a pair of tips. When the tips are closed they form the secondary winding of a transformer. The primary winding is connected to a bank of storage capacitors. The capacitors are slowly charged with a small power supply. When the capacitors reach full charge, they are switched into the primary coil circuit via an ignitron and they dump their entire power into the transformer and, consequently, into the weld site. For more information on transformers, see Chapter 5. For more information on ignitrons, see Chapter 14.

Toasters

One electromechnical device that we have all experienced is the ordinary bread toaster. These are clever devices that will perfectly toast a slice of bread every time. Figure 17-20 shows a schematic representation of a typical bread toaster. The bread is placed into the slot and rests on a bread tray. When the tray is lowered, it closes a limit switch and is latched into place. As the heaters cook the bread, the coiled bimetal strip heats up and eventually pulls the latch open, allowing the bread tray to pop up. By adjusting the preload on the coiled bimetal strip, the down time can be adjusted and the brownness of the toast can be controlled.

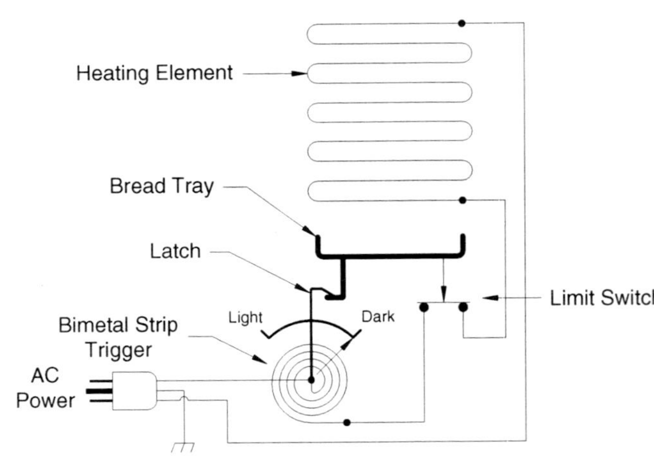

Figure 17-20 Bread Toaster

CHAPTER 18
ELECTRICAL SCHEMATICS

In the fields of electronics and electrics, the written language is the schematic. A schematic is a graphical representation of an electronic or electrical assembly or installation. These drawings may be as simple as a small printed label glued onto the inside of a toaster case or as complex as hundreds of engineering drawings representing the complex power distribution and control systems for a petrochemical plant. Opening the case on an ordinary stereo will usually put you face-to-face with a single-sheet schematic which describes the circuitry with sufficient details to aid a repair technician in his task of troubleshooting. On the other hand, a modern airliner will have something on the order of 10 to 20, 3-inch thick binders that are nothing but electrical schematics and diagrams describing every aspect of the electronics and electrical systems onboard. In addition to the circuit diagrams and schematics, these binders will also contain test, calibration, and inspection procedures. Military ships are so complex that their printed electrical schematics and diagrams are usually stored in a special room designed specifically for the application.

Drawing electronic and electrical schematics is not unlike other engineering disciplines in that standard methods exist. Also like other engineering disciplines, most designers apply a certain amount of leeway when drawing a schematic. There are specific standards that are published by many nations and international organizations. However, more often then not, rigidly adhering to one of these standards produces a schematic that is actually more difficult to interpret. To further confuse matters, many designers use a combination of standards ranging from international, national, industrial, military, and even obsolete.

The thing to remember when reading or drawing a schematic is for whom it is intended. If you're a military technician looking at a commercial power distribution diagram, you're probably going to have a lot of questions. Similarly, if you're a designer tasked with drawing a schematic for a consumer audio power amplifier kit, remember that the customer is probably an amateur with no formal training in electronics. An electrical designer that produces industrial control panels can easily produce a cryptic drawing that is 100% electrically accurate. However, do you really want the shop technicians deciding where to place the components, what gauge wire to use, and how to cable the assembly? A little quality time spent developing a consistent drawing that is easily understandable will provide substantial returns in the future.

Figures 18-1 through 18-6 provide a list of standard symbols that are commonly used in electronic and electrical schematics. Notice that there are some duplications, such as the symbol for a galvanometer, Figure 18-2, and the symbol for a generator, Figure 18-6, which are the same. Also note that there are devices that may have several different symbols, such as an incandescent lamp, Figure 18-3, which is commonly shown in all three different versions.

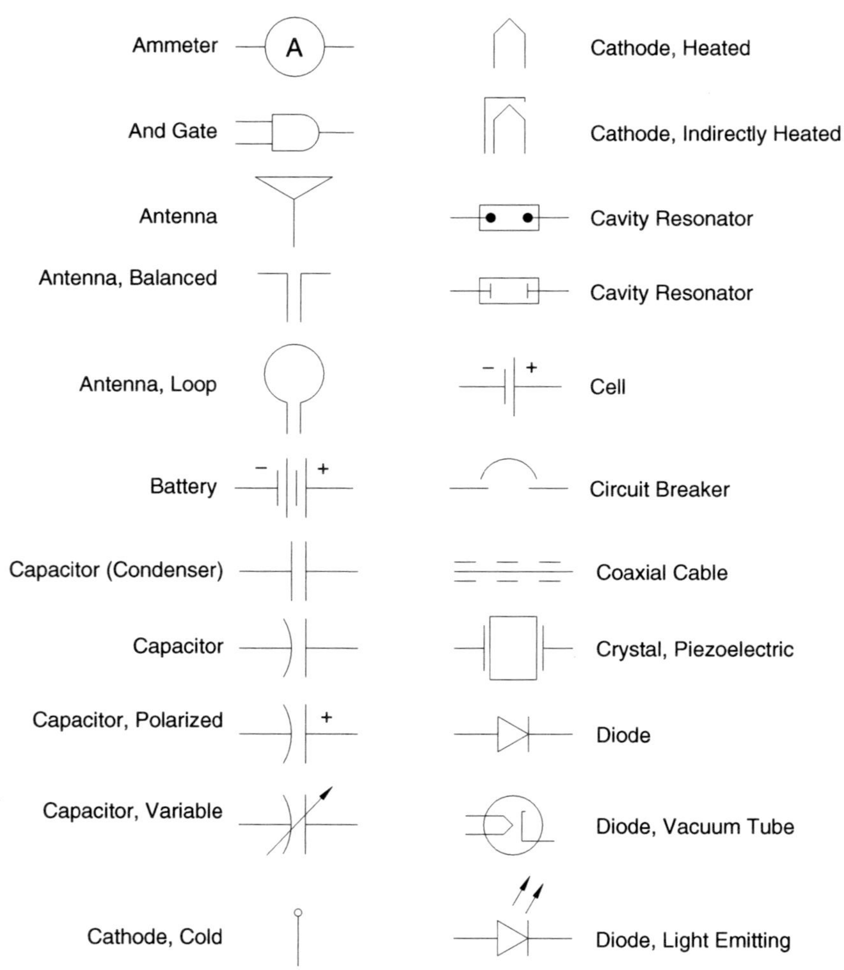

Figure 18-1 Standard Schematic Symbols

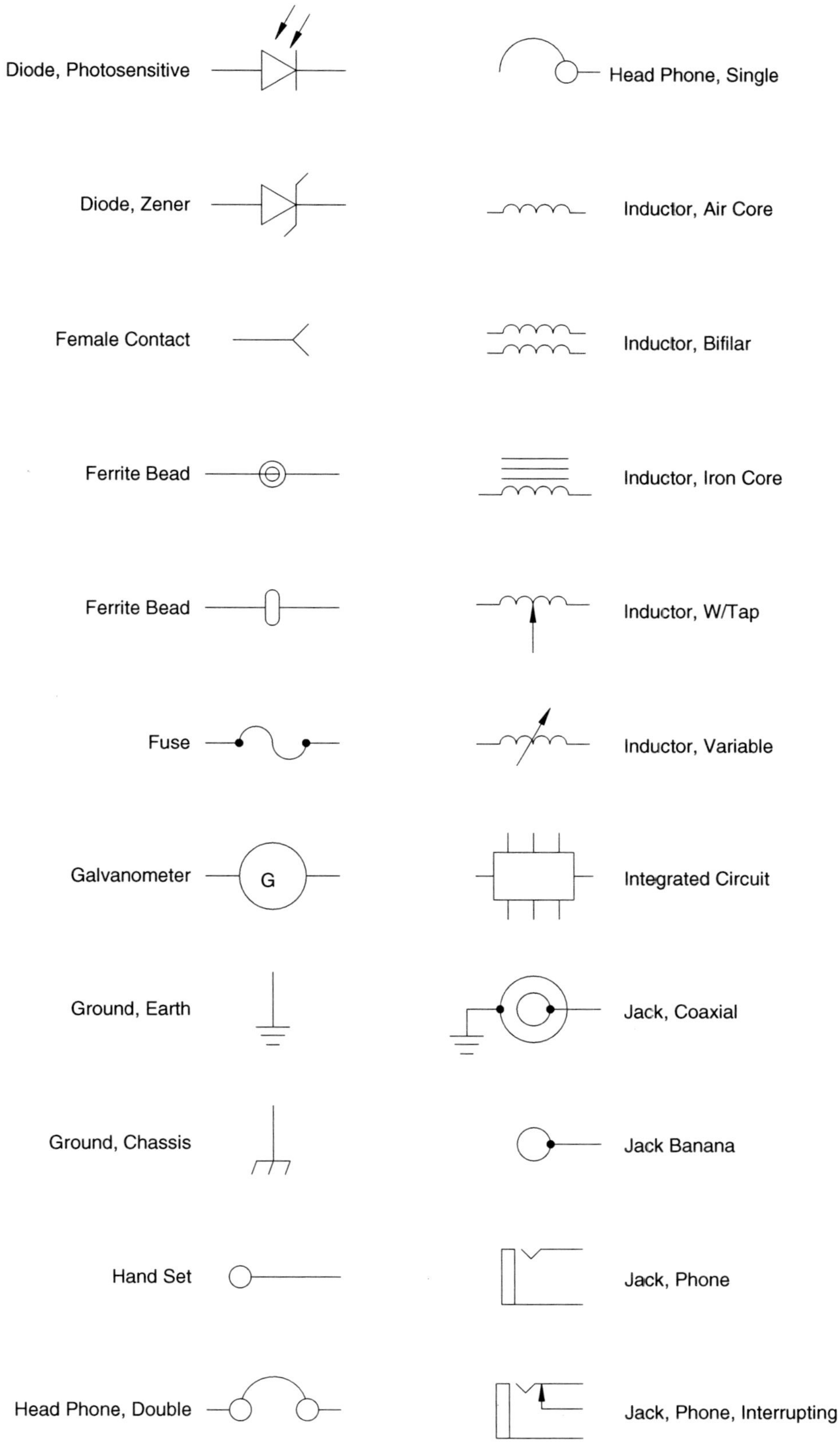

Figure 18-2 Standard Schematic Symbols

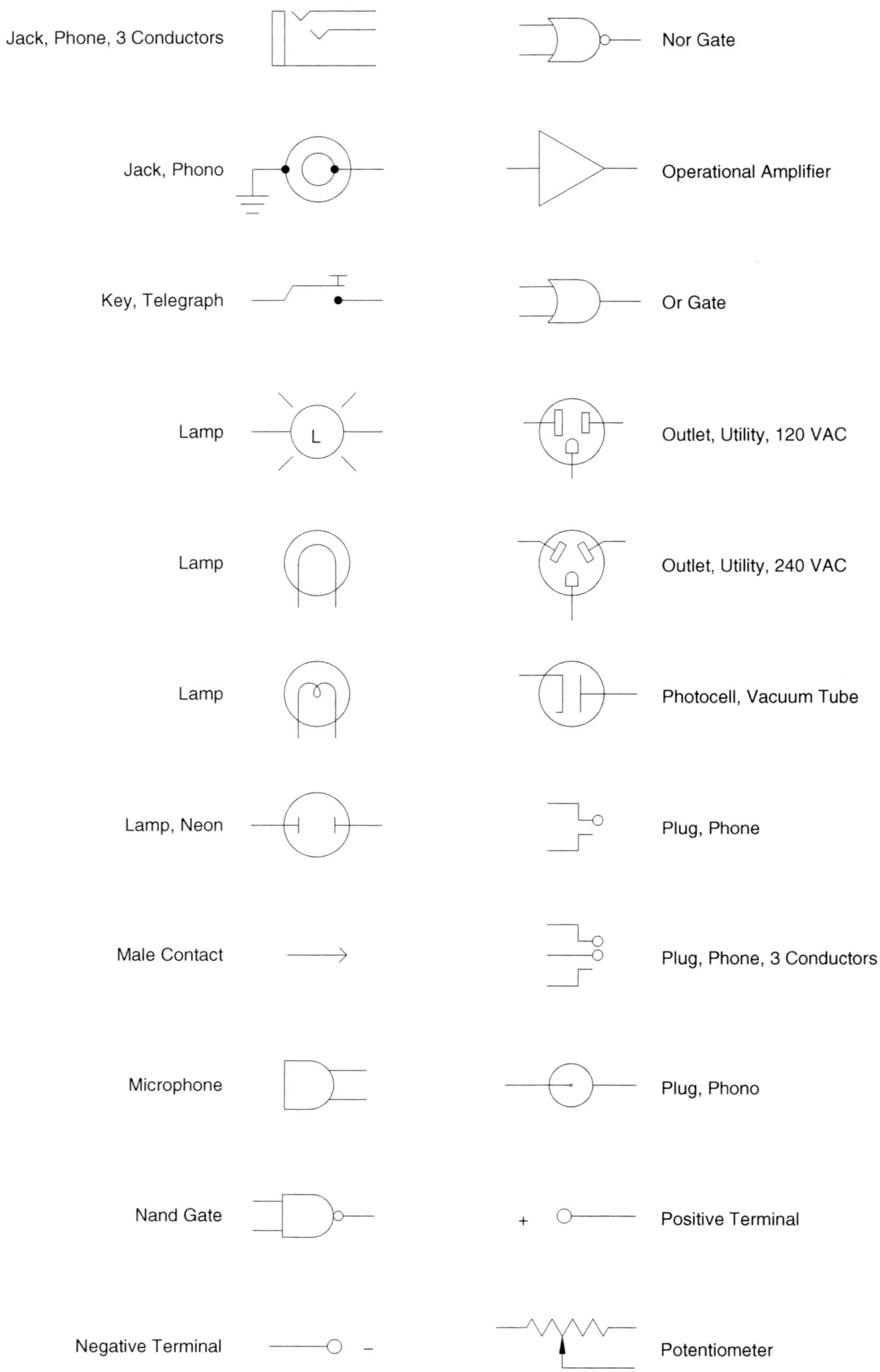

Figure 18-3 Standard Schematic Symbols

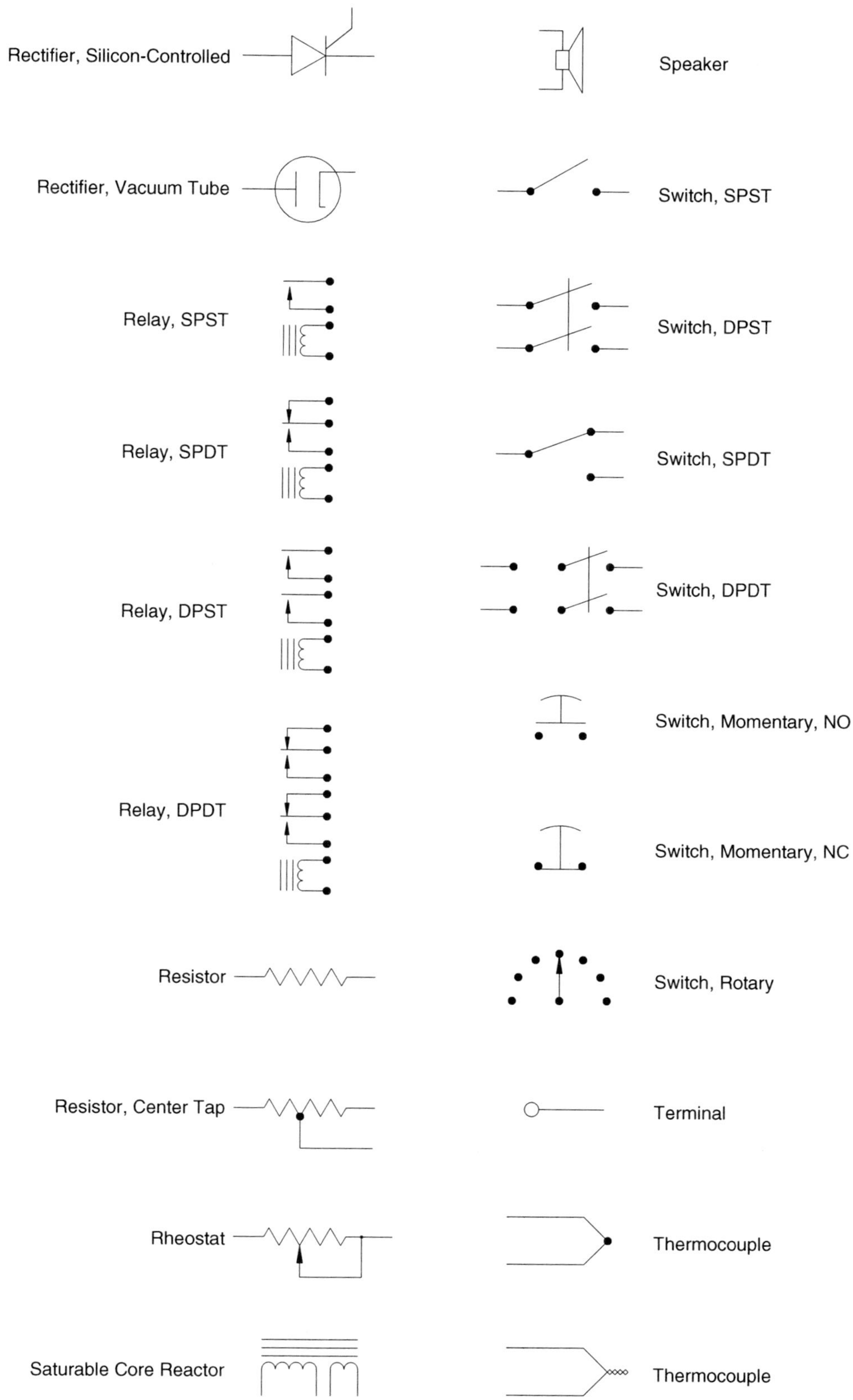

Figure 18-4 Standard Schematic Symbols

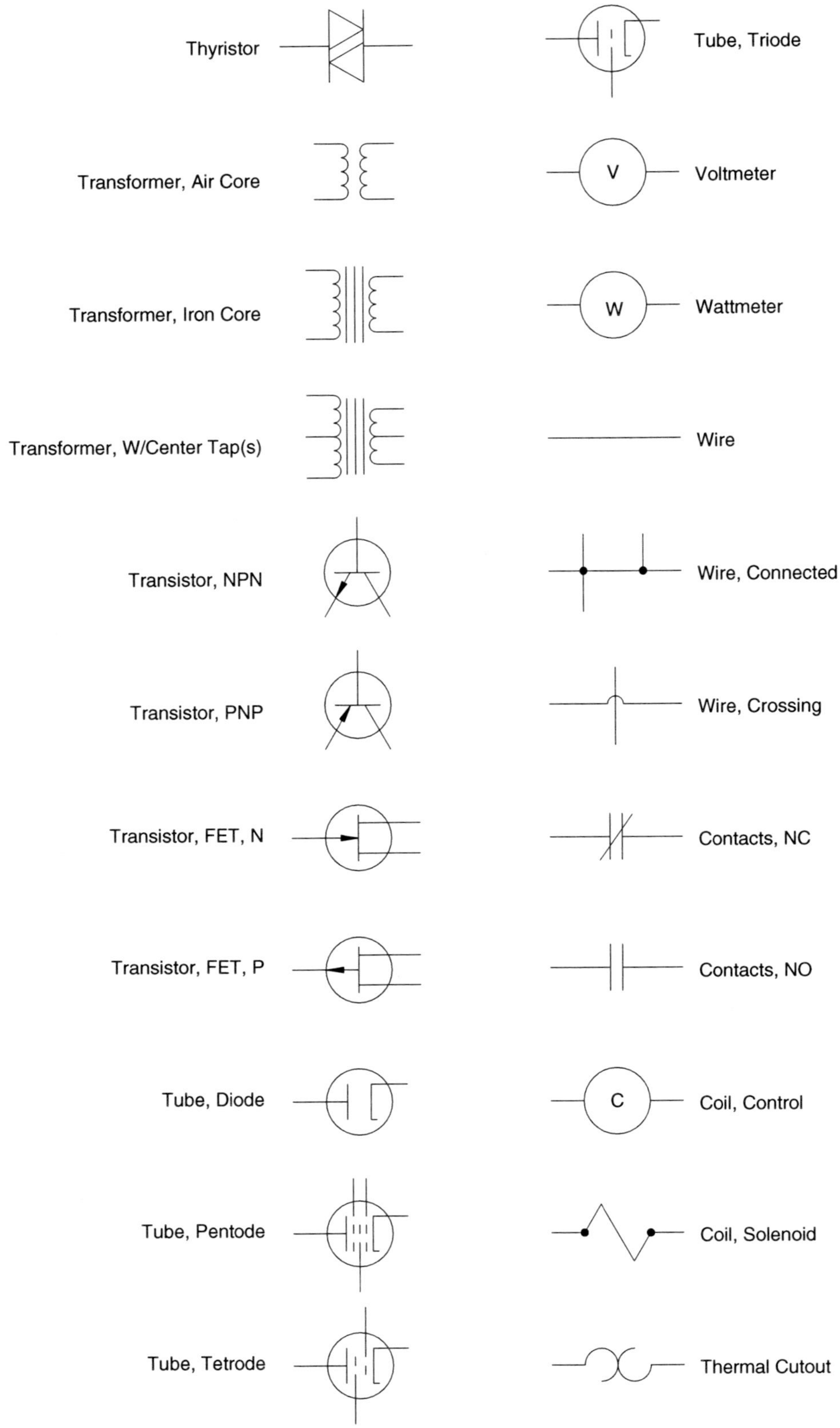

Figure 18-5 Standard Schematic Symbols

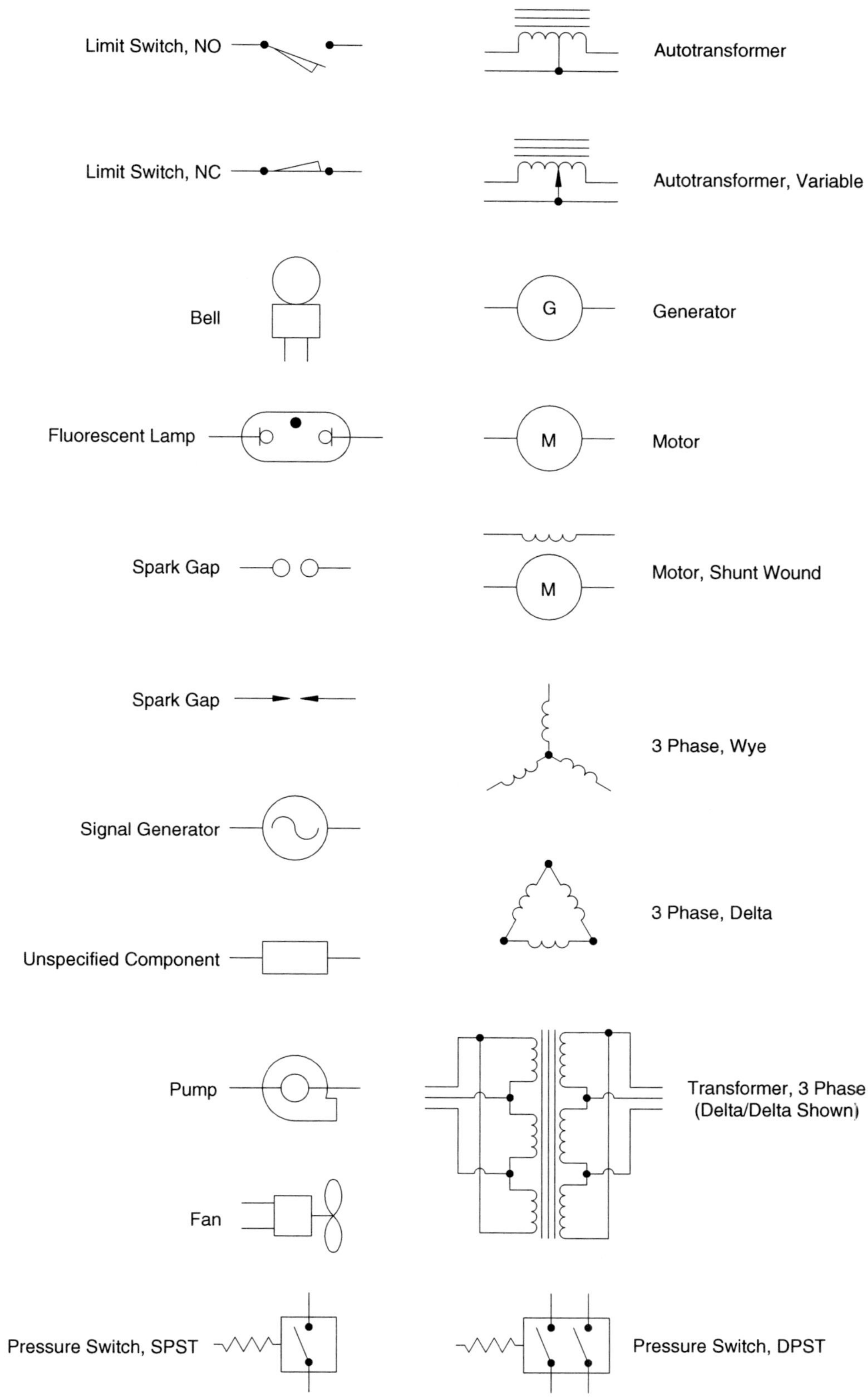

Figure 18-6 Standard Schematic Symbols

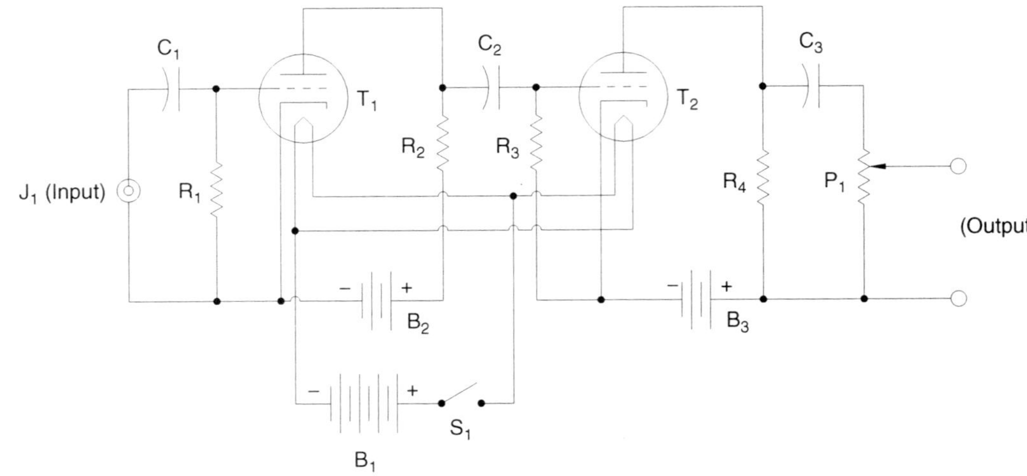

Figure 18-7 Typical Tube Amplifier Schematic

A simple electronic schematic for a two-stage vacuum tube amplifier is shown in Figure 18-7. Notice that all of the components are clear and easily readable, the wires are spaced at a suitable distance from adjacent wires and components. Also note that all of the components have an identifying code next to them, which corresponds to a list that is attached to the schematic. For small schematics, the component list may be printed on the same piece of paper. The component list should provide all of the necessary information for any given part, including electrical data, part number, vendor, and the like.

Figure 18-8 shows a typical unregulated bench power supply schematic. Once again, notice that all of the components are clear and easily readable, the wires are spaced at a suitable distance from adjacent wires, and all of the components have an identifying code next to them. The component list on this schematic is printed above the schematic. A small schematic like this can be easily folded up and tucked into the chassis of the finished assembly.

It is common practice to shrink the schematic down to a size that allows it to be glued to the inside of a panel. Although this does provide a schematic that is difficult to misplace, in many instances the reduction required is so great that the schematic is virtually unreadable. Before proceeding with this method, reduce the schematic and print it out. Get it dirty and then try to read it in poor lighting. If it's not legible, then it probably won't be much use to a technician when the equipment breaks down in 20 years.

Representational Schematics

Figure 18-9 shows an air compressor control schematic. In this case all of the components are shown and the circuit is electrically accurate. However, all of the components are not labeled and those that are, have a poor description. If you compare the schematic with the actual chassis, shown in Figure 18-10, it becomes apparent that even this very simple circuit would be difficult for a technician to follow. To add to the confusion, the finished chassis would have bundled wires instead of the neatly arranged wires as shown.

C_1	100-MFD, 100-volt Capacitor
$D_{1\text{-}4}$	400-PIV Diode
F_1	5-amp Type MDL
F_2	8-amp Type MDL
P_1	120-VAC Plug
S_1	SPST Toggle Switch
T_1	0- to 120-VAC Variable Autotransformer
T_2	Transformer, 120-volt Primary, 50-volt Secondary

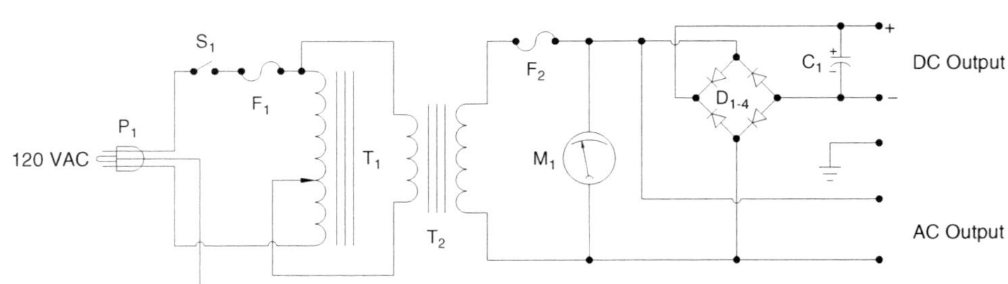

Figure 18-8 Typical Bench Power Supply Schematic

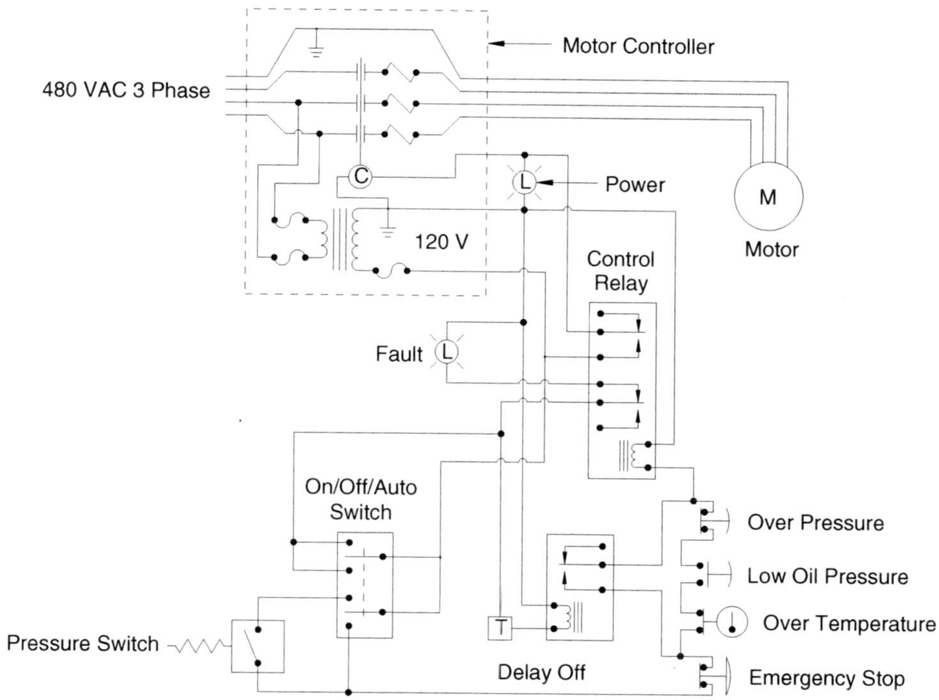

Figure 18-9 Air Compressor Schematic

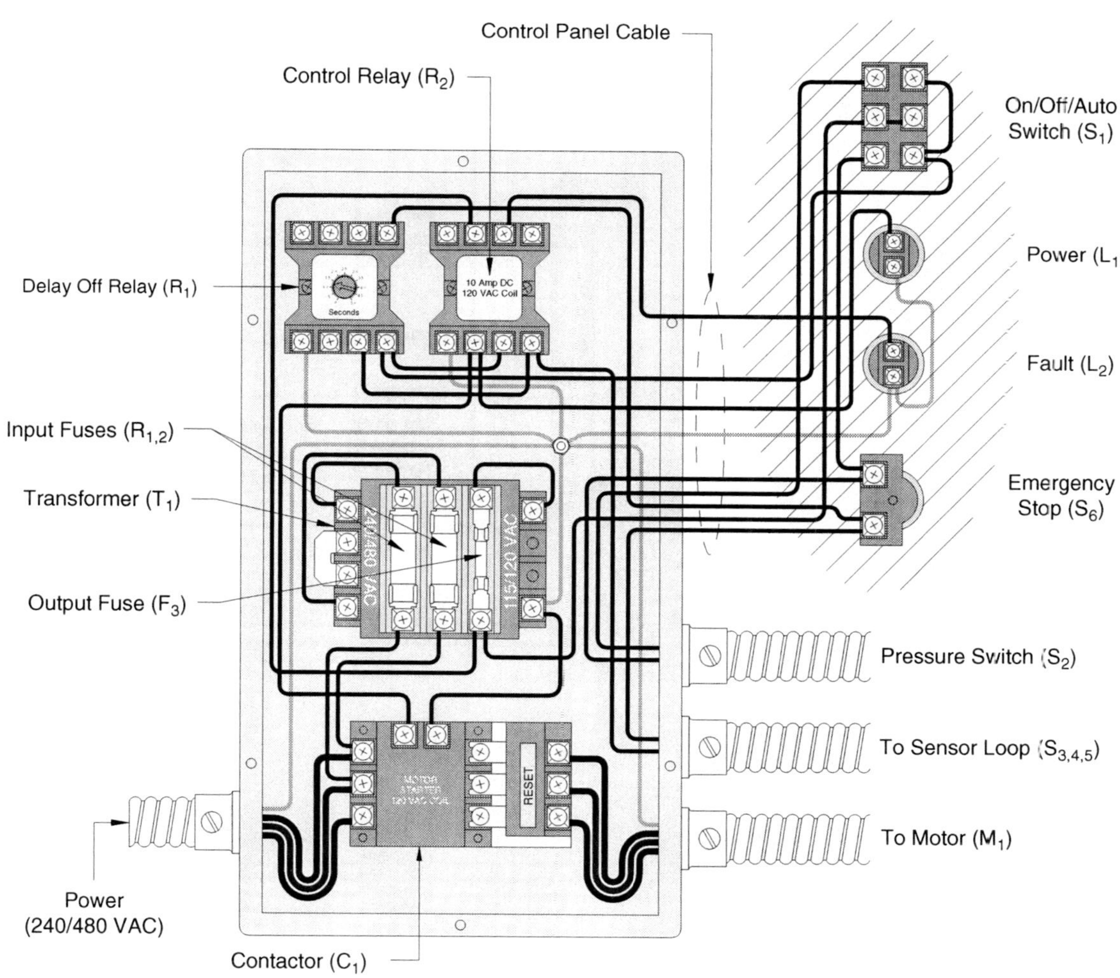

Figure 18-10 Air Compressor Control Chassis

Figure 18-11 shows how the circuit is drawn using a representational schematic. Notice that all of the components are arranged in principally the same locations as in the finished chassis. The wires are routed using the same paths and locations as the finished chassis. Also note that all of the components are clearly labeled and a component list is present. The wire colors and gauges are clearly labeled. Components that require greater detail, such as R_1 and R_2 are detailed in a secondary drawing, as shown in the upper right hand corner. Even though there are a number of deviations from standard symbols and methods, a technician presented with this information would have no difficulty deciphering the actual circuit.

C_1	Motor starter, 30 HP, 120 VAC Coil
F_1	2 Amp Type FNW
F_2	2 Amp Type FNW
F_3	3 Amp Type MDL
L_1	120 Volt Incandescent
L_2	120 Volt Incandescent
M_1	30 HP Motor, 480 VAC, 3 Phase
R_1	Delay On Relay, SPST, NO
R_2	Control Relay, DPDT
S_1	Switch, DPDT
S_2	Pressure Switch, SPST
S_3	Over Pressure Switch
S_4	Low Pressure Switch
S_5	Over temperature Switch
S_6	Emergency Stop Switch
T_1	Control Transformer

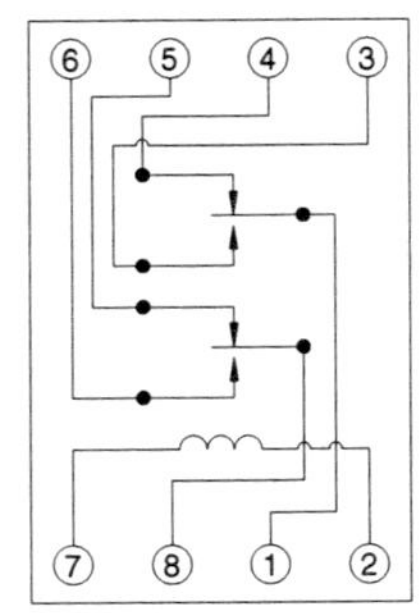

R_1 & R_2

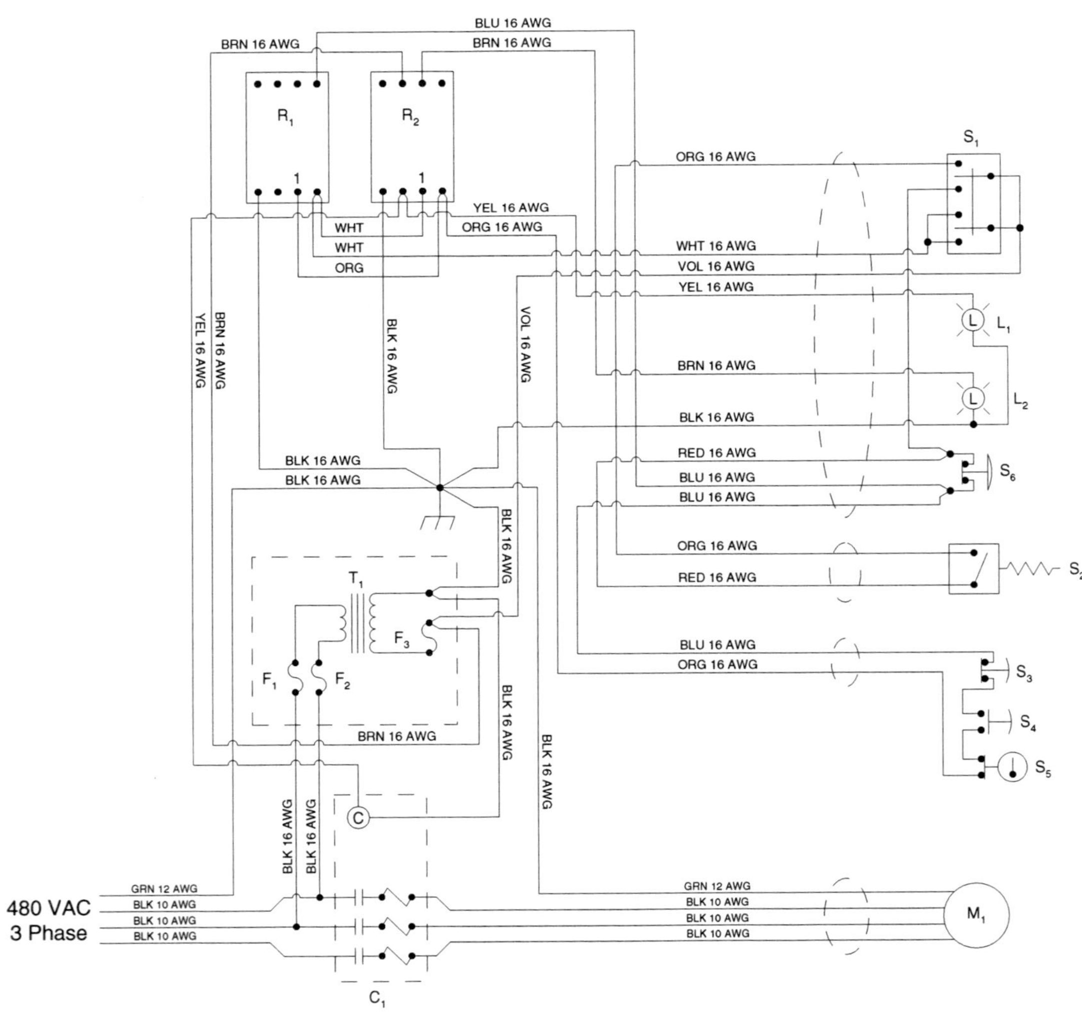

Figure 18-11 Air Compressor Representational Schematic

Figure 18-12 shows how a component is drawn in a representational schematic. The basic component is shown in the middle, in this case an octal base relay. The standard method of drawing is shown on the left and the representational method is shown on the right. The basic schematic and electrical information is the same for both methods, except in the representational method, the socket is shown with its terminals arranged as they are on the actual component. This provides a great deal of clarity for a technician reading the finished schematic.

Figure 18-13 shows a triode vacuum tube drawn in both standard and representational methods. Once again, the electrical information is the same except the representational methods also provides the pin lay out of the socket.

To better illustrate representational schematics, compare the schematic in Figure 18-14 to Figure 18-7. At first glance the representational schematic seems more complicated; however, it is only because it provides substantially more information. If you compare the two schematics with the finished chassis, shown in Figure 18-15, it becomes clear that the representational method is much easier to follow.

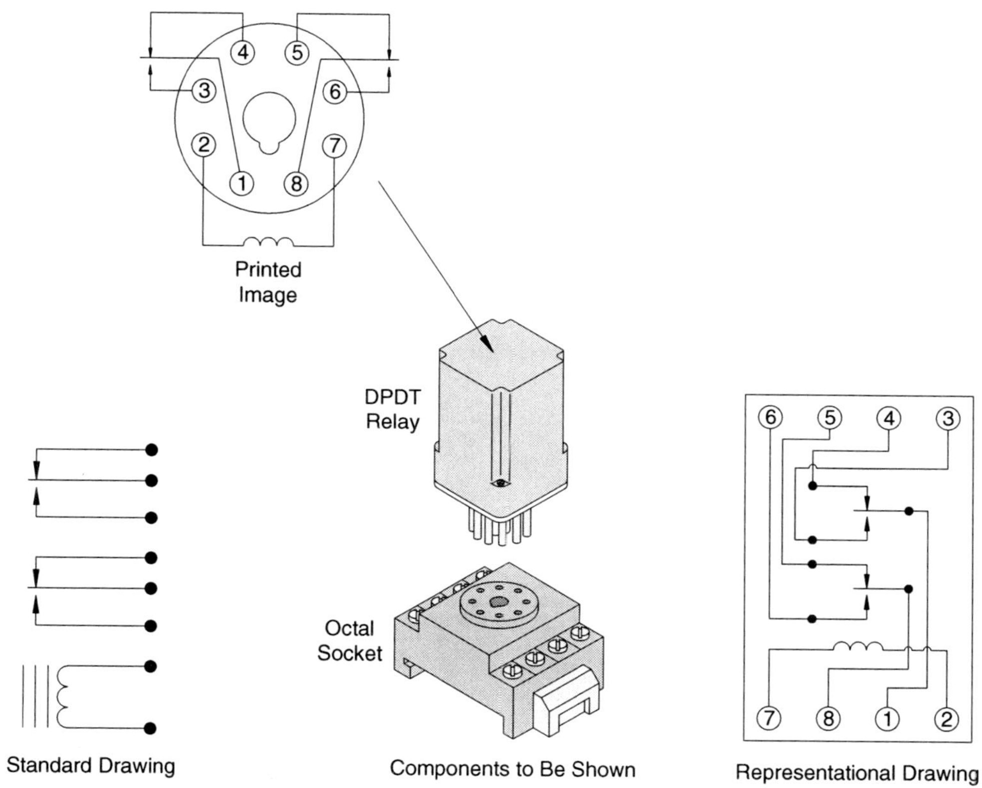

Figure 18-12 Standard versus Representational Relay Illustrations

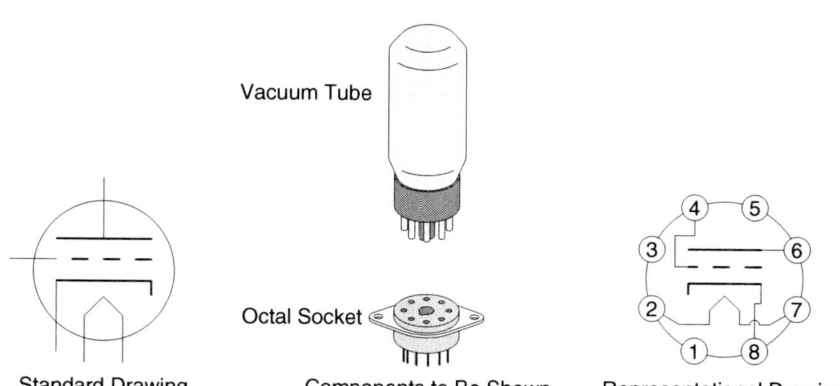

Figure 18-13 Standard versus Representational Vacuum Tube Illustrations

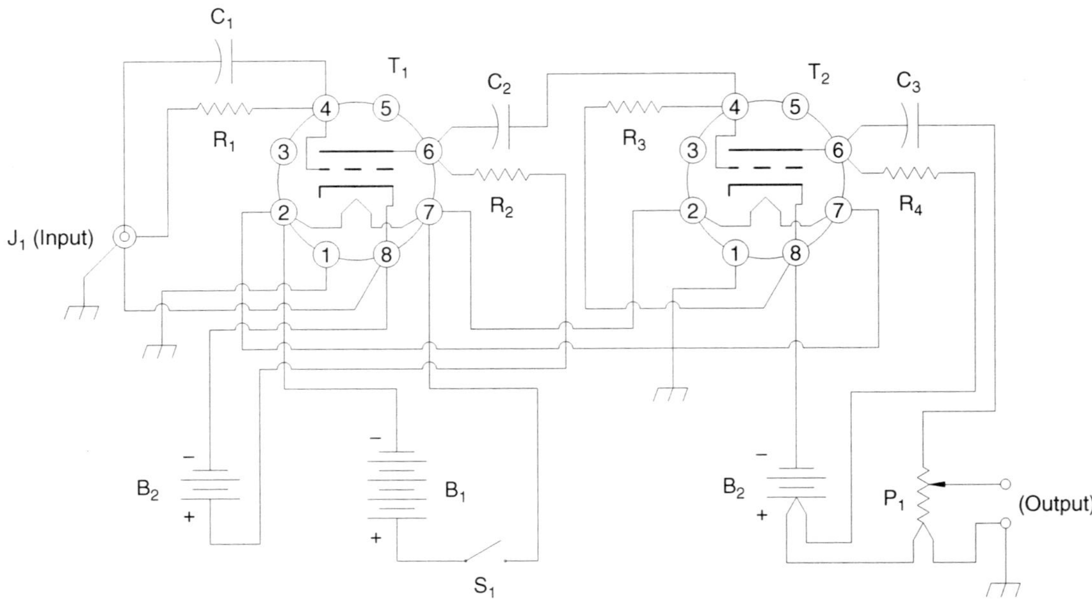

Figure 18-14 Tube Amplifier Representational Schematic

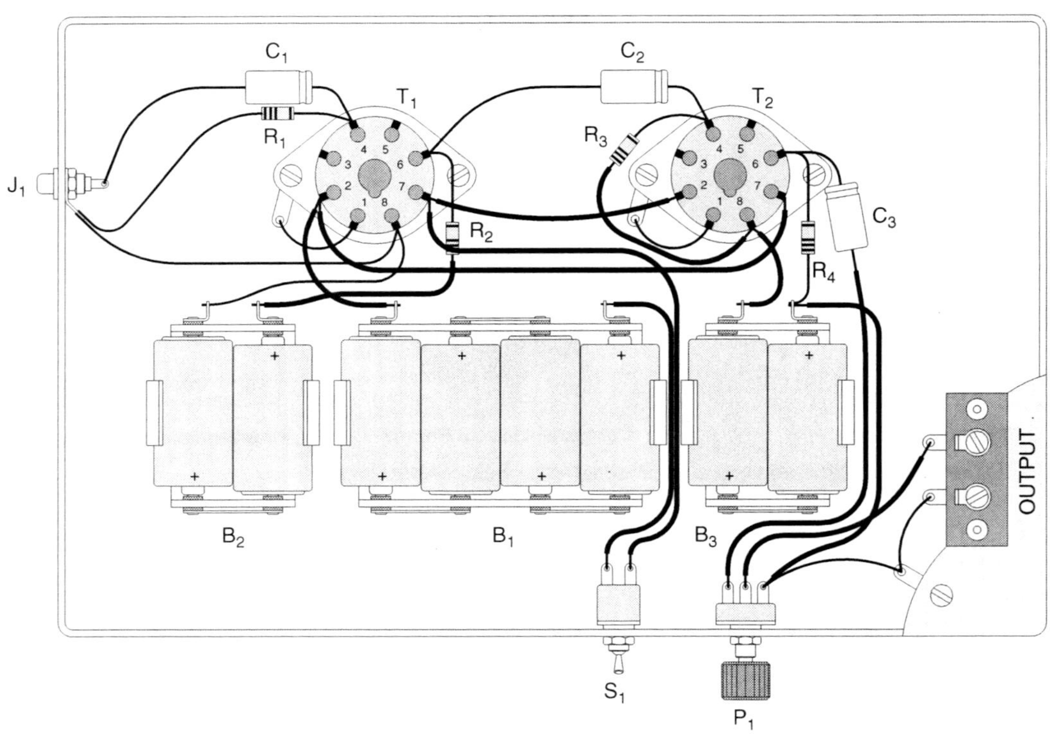

Figure 18-15 Tube Amplifier Chassis Layout

GLOSSARY

A

AC Abbreviation for alternating current
accelerometer An instrument to measure acceleration
acorn tube An acorn shaped vacuum tube
acoustic Pertaining to sound
actuator A device that moves a load
adaptor A plug or jack designed to change the configuration of a connector
alarm A device that generates an audible signal for the express purpose of alerting an operator to a specific condition
alligator clip A spring loaded clip used for testing
am Abbreviation for amplitude modulation
ammeter A meter that is intended to display current
amp An abbreviation for ampere
amperage The number of amps flowing through a circuit
ampere A measure of electrical current
amp-hour Amps per hour
amplifier A device for proportionately increasing an electrical signal
analog Continuously variable signal
anode Positive electrode
antenna A wire or frame intended to pick up or transmit electrostatic signals
arc A high-energy discharge between two conductors
armature The moving element in an electromechnical device
armored cable Two or more wires within a protective jacket
ASCII Abbreviation for American Standard Code for Information Interchange
attenuator A device for reducing an electrical signal
AWG Abbreviation for American wire gauge

B

batt Abbreviation for battery
battery A device for storing DC electrical energy
bell An alarm which used a resonating metal element
bias A electrical or magnetic reference signal
bimetal strip Two different types of metal fixed together for the sole purpose of distorting in reference to temperature change
binding post A nut an bolt arrangement for connecting wires
black box Used to describe a complex component that is intended to be replaced rather than repaired
break down The voltage at which the insulation fails
breaker points A set of mechanical contacts
bridge rectifier A series of diodes arranged so that AC is converted into DC
brush A device that is intended to transmit an electrical signal from a moving element to a nonmoving element
buss bar A bar generally used to provide multiple power connections
button A switch that is actuated by pushing
BX cable Flexible metal jacketed wires

C

cable A general term for multiconductor bundles or heavy wire
CAD Abbreviation for computer aided design
capacitor A device that is intended to provide short-term storage of an electrical signal
cathode Negative electrode
CRT Abbreviation for cathode ray tube
cell Single unit battery
center tap A connection to the center of a coil or resistive element
ceramic insulator A nonconductive component made from a high-fired material
channel One path within a multipath circuit
charge The electrical state of a conductor
choke term for an inductor
circuit Term for an electrical assembly
circuit breaker A reusable device that is intended to automatically protect a circuit form over-current conditions
CB Abbreviation for Citizens Band radio
coaxial A configuration where one conductor surrounds another
coil A continuous length of wire wound in a circular pattern
coil form A piece that is intended to aid in winding and maintaining a coil
cold cathode An electron emitter that does not require heat
collector A conductor that is specifically designed to collect electrons
component Any single or grouped piece of an assembly
condenser Term for capacitor
conductivity The efficiency of a conductor to transmit an electrical signal without resistive loss
contact Two conductors forced together for the propose of an electrical connection
continuity To describe a continuous electrical path
control transformer A transformer that is intended to provide a lower control voltage, usually 120 volts
collector A components that receives electron flow
corona A visible leakage of electrons into the surrounding gas
crimp To make an electrical connection by the defamation of a metal
cross talk The leakage of an electrical signal into an adjacent circuit
current Amperage

current limiter A device that is intended to limit the amount of current that is intended to pass through a circuit
Cps Abbreviation for cycle per Seconds, frequency

D

damping The action of reducing voltage transients
dashpot A pneumatic or hydraulic shock absorber generally used on instrumentation
db Abbreviation for decibels (sound level)
DC Abbreviation of direct current
decade box A test device that allows the internal value to be adjusted in tens
Delta Greek letter, a three phase circuit that is arranged in a loop
detector An electronic device that is intended to detect a physical condition
dielectric The insulating barrier between two conductors
digital A circuit that operated at two states, off/on, +5 volts/0 volts
diode A semiconductor device that acts as one way valve for electricity
DIP Abbreviation for Duel In-line Package
distribution system A system intended to distribute power from a central location
DPDT Abbreveation for double-pole, double throw
DPST Abbreviation for double-pole, single throw
drum switch A switch activated from a rotating drum
dry cell A single cell battery with a past-type electrolyte
dry battery A multicell battery with a past-type electrolyte
duty cycle The amount of on time a device can operate during a predetermined period, i.e. 2 minutes of run time within a 10-minute period is a 20% duty cycle
dynamic In motion

E

E-core An iron core shaped in the form of an E
eddy currents Currents that develop within a conductor due to changing magnetic fields
electrode The termination of a circuit
electrolyte A chemical that relies on electrochemical action for conductivity
electromagnetic Magnetism that is generated by electricity
electron The negatively charged particle
electron beam A beam comprised of electrons
electron field An area with elevated electron activity
electron emission The act of emitting electrons
element A single component of an assembly
emitter A conductor which emits electrons
envelope The vacuum or atmosphere container surrounding a device

F

ferrite A powered metal used for high-frequency transformers and inductors
FET Abbreviation for field effect transistor
filament The active element in a light bulb
flash tube A gas discharge tube for producing short-duration, high-intensity light
FM Abbreviation for frequency modulation
fuse A sacrificial device that is intended to protect a circuit form over-current situations
fuse holder A reusable mount for standard fuses

G

galvanometer An instrument for measuring low-level electrical signals
generator A device for producing electrical energy from rotation motion
giga A prefix meaning of billion
glow discharge Light emitted through the electrical excitation of a gas
grid A biased plate that is used to control the flow of electrons in a vacuum tube
ground An electrical connection to the earth
ground buss A grounded bar providing a series of connections
ground lug A solder lug attached to a grounded chassis
guy wire A wire or cable used to steady a tower

H

half-wave One-half of a full AC cycle
handset The speaker/microphone assembly of a telephone
headphones Two small speakers mounted onto a loop intended to be worn on the head
headset Headphones
henery Unit of measure of inductance
hermetic Permanent sealing
hertz Frequency, abbreviated Hz
high frequency AC signal with a frequency above 3 MHz and below 30 MHz, abbreviated hf
high tension High voltage
high voltage A relative term, generally a voltage that is higher then normal
Horseshoe magnet A permanent magnet in the shape of a U
hydrophone A microphone designed to operate under water

I

ignition coil A transformer designed specifically to fire spark plugs
ignitron A type of mercury pool rectifier
impedance References the internal resistance of a circuit
incandescent lamp A lamp with a hot wire filament
induced voltage A voltage that is generated from a dynamic magnetic field
inductor A coil of wire intended to provide short-term storage of delay of current
interface A circuit that is used to connect two incompatible circuits
interference An induced voltage that is picked up from stray signals
ionization The act of introducing a charge state on particles
isolation Operating a circuit with no connections to the ground or external power sources

isolation transformer A transformer whose primary and secondary coils are electrically isolated

J

J Abbreviation for joule
jack Female connector
jones plug A class of multi pin connectors
joule A unit of measure describing energy
junction A connection between conductor or semiconductors

K

k Abbreviation for 1000
kelvin An absolute temperature scale using the same unit division as centigrade
kHz The abbreviation for kilo hertz
klystron A vacuum tube used to amplify or generate high frequencies
knife switch A switch constructed from a pivoting bar
knot Nautical mile (6,080.2 feet)
kV Abbreviation for kilovolt
kW Abbreviation for kilowatt

L

laminated core A iron core made by stacking thin pieces of sheet metal
lamp An electrical device intended to produce visible light
laser Abbreviation for Light Amplification by Stimulated Emission of Radiation
lead-acid cell A liquid electrolyte battery
leak A condition that creates a high resistance short to ground
lightning arrestor A device that will shunt a lighting strike to earth ground
loop antenna A wire arranged in a loop for receiving electrostatic transmissions
loudspeaker A device designed to reproduce sound in reference to an electrical signal
low frequency AC signals with a frequency between 30 Hz to 300 kHz

M

magnetic amplifier A device that takes advantage of the saturable core characteristics of a transformer to provide amplification
magnetic circuit The path of magnetic flux
magnetic field The magnetic circuit that surround a magnet
magnetron A vacuum tube used to generate high-frequency signals
matching transformer A transformer specifically designed to match impendence of two incompatible circuits
meg Abbreviation for mega-ohm
mega Abbreviation for 1 million
mercury vapor lamp A high intensity lamp which uses gaseous mercury
mercury contact A contact set that is wetted with mercury
metalized A layer of metal deposited onto an insulator
μfd Abbreviation for microfarad
mH Abbreviation for millihenery
micro Abbreviation for 1/1,000,000
microwave AC signal from 1 gigahertz and up
milli Abbreviation for 1/1000
milliamp 1/1000 of an ampere
morse code A communications protocol which uses dots and dashes
motor-generator set A motor driven generator
moving coil A coil which generated a signal by moving in reference to a fixed magnet

N

nano Abbreviation for 1 billionth
negative The – side of a circuit
NEMA Abbreviation for National Electric Manufactures Association
neon bulb A light bulb which generates a glow within a neon gas atmosphere
NC Abbreviation for normally closed
NO Abbreviation for normally open

O

octal Base eight
ohm The unit of measure for electrical resistance
ohmmeter A voltmeter specifically configured to read ohms
open circuit A circuit that does not have a return path
open voltage The voltage present in the absence of a load
opto-isolator A device that is intended to electrically isolate a communications line using light transmission
oscillator AC signal source
oscilloscope An instrument designed to provide a visual display of an electrical signal

P

pad A solder point on a printed circuit board
parallel connection Two or more components arranged in a parallel configuration
patch cord A multiconductor cable with plugs on both ends
peak voltage The highest voltage in a signal
permanent magnet A permanently magnetized material
phone jack/plug A type of plug that is common in audio circuits
phono jack/plug A type of plug that is common in low-level audio circuits
photocell A device that produces electricity from light
piezoelectric The action of producing electrical energy from bending a crystals
plasma A glow discharge in a gas
plate A term for the anode in a vacuum tube
potentiometer A variable resistor
primary winding The input coil of a transformer
printed circuit board An insulating board with foil conductors laid out on the surface
probe A test terminal for hand operation
pyrometer An instrument for measuring temperature

Q

Q Abbreviation for electrical charge

R

reactor A magnetic device designed to introduce resistance
reed switch A switch constructed from two flexible metal strips
regulator A device designed to produce a fixed voltage output
relay A switch that is activated with a solenoid
resistance The opposition to electric flow
resistance bridge A device that is intended to adjust the resistance in a circuit
RF Abbreviation for radio frequency

S

saturable core An iron core that can be fully charged with a magnetic signal
SCR Abbreviation for silicon controlled rectifier
secondary winding The output winding of a transformer
shield A grounded conductor protecting a circuit from stray signals
shunt A device that is intended to carry high-current loads
shunt wound motor A DC motor with an electromagnet field
sine wave An AC signal that is expressed as a sine of a linear function of time
solar cell photo cell
solenoid A coil with a moving iron core
spark gap Two points intended to support an electrical spark
speaker Loudspeaker
spot welding The joining of two metal plates via an electrical pulse
standard cell A special battery that serves as a reference voltage
standard resistor A special resistor that serves as a reference resistance
strain gauge A device whose electrical resistance changes proportionally as force is applied
switch A device intended to interrupt a circuit

T

tachometer An electrical device used to measure revolutions
telegraph An early communication system that utilized dots and dashes
telephone A hard wired, voice communication system
terminal A point at which a wire can be connected to a circuit
terminal block A series of terminals arranged at a single location
terminal strip A series of terminals arranged in a straight line
test point A terminal in a circuit for the express purpose of testing
thermostat An adjustable switch that activates in reference to temperature
toggle switch A switch with a cylindrical actuator which is flipped from one position to the other
transformer A device that is intended to change the voltage of an AC signal
trickle charger A charger that is intended to provide a low-charge rate

U

uhf Abbreviation for ultra high frequency

V

V Abbreviation for volts
vacuum The absents of any gas
vacuum envelope The vessel used to contain a vacuum
vacuum tube An electronic device which utilizes and manipulates the free flow of electrons
vhf Abbreviation for very high frequency
volts The unit of measure for electromotive force
voltage divider A resistance circuit that is used to extract lower voltages from a high-voltage source
voltage regulator A device that is intended to produce a fixed voltage output
Vtvm Abbreviation for vacuum tube voltmeter
VU Abbreviation for volume unit

W

W Abbreviation for watt
watt Unit of measure for electrical power
Western Union Joint A procedure of splicing solid wire
Wheatstone bridge An arrangement of resistive elements that produce a null reading
wiper The moving element on a potentiometer
wire wound resistor A resistor made from a coil of wire

X

Xenon A rare earth gas
Xenon flash tube An arc discharge tube with xenon atmosphere

Y

Yagi antenna A type of directional antenna
yoke A coil used to deflect an electron beam

Z

Zener diode A type of diode that is used as a voltage regulator
zero crossing The point where there is zero voltage during an AC cycle

INDEX

C

D

I

J

K

L

M

N

O

P

T

U

V

W

X

CPSIA information can be obtained at www.ICGtesting.com
Printed in the USA
LVOW091454010513

331840LV00010B/190/P

9 780071 477529